Frontiers of Cellular Bioenergetics

Molecular Biology, Biochemistry, and Physiopathology

Frontiers of Cellular Bioenergetics

Molecular Biology, Biochemistry, and Physiopathology

Edited by

Sergio Papa and
Ferruccio Guerrieri
University of Bari
Bari, Italy

and

Joseph M. Tager
University of Amsterdam
Amsterdam, The Netherlands

KLUWER ACADEMIC / PLENUM PUBLISHERS
NEW YORK, BOSTON, DORDRECHT, LONDON, MOSCOW

Library of Congress Cataloging in Publication Data

Frontiers of cellular bioenergetics: molecular biology, biochemistry, and physiopathology /
   edited by Sergio Papa, Ferruccio Guerrieri, and Joseph M. Tager.
      p.   cm.
   Includes bibliographical references and index.
   ISBN 0-306-45851-9
      1. Energy metabolism. 2. Cell metabolism. 3. Bioenergetics. 4. Mitochondrial pathology.
   I. Papa, S. (Sergio) II. Guerrieri, Ferruccio. III. Tager, J. M.
   QP176.F76   1999   98-49078
   572′.43—dc21   CIP

ISBN 0-306-45851-9

© 1999 Kluwer Academic / Plenum Publishers
233 Spring Street, New York, N.Y. 10013

10 9 8 7 6 5 4 3 2 1

A C.I.P. record for this book is available from the Library of Congress.

Printed in the United States of America

Contributors

L. Mario Amzel   Department of Biophysics and Biophysical Chemistry, and Department of Biological Chemistry, Johns Hopkins School of Medicine, Baltimore, Maryland 21205

Susanne Arnold   Fachbereich Chemie, Philipps-Universität, D-35032 Marburg, Germany

Gerald T. Babcock   Department of Biochemistry, Michigan State University, East Lansing, Michigan 48824

J. A. Berden   E. C. Slater Institute, BioCentrum, University of Amsterdam, Plantage Muidergracht 12, 1018 TV Amsterdam, The Netherlands

Paolo Bernardi   CNR Unit for the Study of Biomembranes, Department of Biomedical Sciences, University of Padova Medical School, I-35121 Padova, Italy

Mario A. Bianchet   Department of Biophysics and Biophysical Chemistry, and Department of Biological Chemistry, Johns Hopkins School of Medicine, Baltimore, Maryland 21205

Domenico Boffoli   Institute of Medical Biochemistry and Chemistry, University of Bari, I-70124 Bari, Italy

Maurizio Brunori   Department of Biochemical Sciences and CNR Center of Molecular Biology, University of Roma "La Sapienza," 00185 Rome, Italy

Gerard Buse   Institut für Biochemie, Rheinisch-Westfälische Technische Hochschule, 52057 Aachen, Germany

Palmiro Cantatore   Department of Biochemistry and Molecular Biology, University of Bari, 70125 Bari, Italy

Nazzareno Capitanio   Institute of Medical Biochemistry and Chemistry, University of Bari, I-70124 Bari, Italy

H. Cock Department of Clinical Neurosciences, Royal Free Hospital School of Medicine, London NW3 2PF, England

Johann Deisenhofer Howard Hughes Medical Institute and Department of Biochemistry, University of Texas, Southwestern Medical Center, Dallas, Texas 75235

Kai-Ping Deng Department of Biochemistry and Molecular Biology, Oklahoma State University, Stillwater, Oklahoma 74078

Emilio D'Itri Department of Biochemical Sciences and CNR Center of Molecular Biology, University of Roma "La Sapienza," 00185 Rome, Italy

Stefan Exner Fachbereich Chemie, Philipps-Universität, D-35032 Marburg, Germany

Shelagh Ferguson-Miller Department of Chemistry, Michigan State University, East Lansing, Michigan 48824

Adrian Flierl Wissenschaftliche Nachwuchsgruppe, Theodor-Boveri-Institut, 97074 Würzburg, Germany

Francesca Forti Division of Biochemistry and Genetics, National Neurological Institute "C. Besta," Milan 20133, Italy

Viola Frank Fachbereich Chemie, Philipps-Universität, D-35032 Marburg, Germany

Masamitsu Futai Department of Biological Sciences, Institute of Scientific and Industrial Research, Osaka University, Ibaraki, Osaka 567, Japan

Antonio Gaballo Institute of Medical Biochemistry and Chemistry, University of Bari, I-70124 Bari, Italy

Maria N. Gadaleta Department of Biochemistry and Molecular Biology, University of Bari, 70125 Bari, Italy

Alessandro Giuffrè Department of Biochemical Sciences and CNR Center of Molecular Biology, University of Roma "La Sapienza," 00185 Rome, Italy

Ferruccio Guerrieri Institute of Medical Biochemistry and Chemistry, and Centre for the Study of Mitochondria and Energy Metabolism, University of Bari, 70124 Bari, Italy

A. F. Hartog E. C. Slater Institute, BioCentrum, University of Amsterdam, Plantage Muidergracht 12, 1018 TV Amsterdam, The Netherlands

Youssef Hatefi Division of Biochemistry, Department of Molecular and Experimental Medicine, The Scripps Research Institute, La Jolla, California 92037

Curtis W. Hoganson Department of Biochemistry, Michigan State University, East Lansing, Michigan 48824

Gerhard Hummer Theoretical Biology and Biophysics Group, Los Alamos National Laboratory, Los Alamos, New Mexico 87545

Maik Hüttemann Fachbereich Chemie, Philipps-Universität, D-35032 Marburg, Germany

So Iwata Max-Planck-Institut für Biophysik, D-60528 Frankfurt am Main, Germany

Anatoly M. Kachurin Department of Biochemistry and Molecular Biology, Oklahoma State University, Stillwater, Oklahoma 74078

Bernhard Kadenbach Fachbereich Chemie, Philipps-Universität, D-35032 Marburg, Germany

Hoeon Kim Howard Hughes Medical Institute and Department of Biochemistry, University of Texas, Southwestern Medical Center, Dallas, Texas 75235

Angela M. S. Lezza Department of Biochemistry and Molecular Biology, University of Bari, 70125 Bari, Italy

Dietmar Linder Biochemisches Institut am Klinikum, Justus Liebig Universität, D-35385 Giessen, Germany

Thomas A. Link Universitätsklinikum Frankfurt, ZBC, Institut für Biochemie I, Molekulare Bioenergetik, D-60590 Frankfurt/Main, Germany

Rebecca Lucas Department of Chemistry, Michigan State University, East Lansing, Michigan 48824

Hartmut Michel Max-Planck-Institut für Biophysik, D-60528 Frankfurt am Main, Germany

Denise A. Mills Department of Chemistry, Michigan State University, East Lansing, Michigan 48824

Joel E. Morgan Helsinki Bioenergetics Group, Department of Medical Chemistry, Institute of Biomedical Sciences and Biocentrum Helsinki, University of Helsinki, FI-00014 Helsinki, Finland

Monica Munaro Division of Biochemistry and Genetics, National Neurological Institute "C. Besta," Milan 20133, Italy

Peter Nicholls Department of Biological Sciences, Brock University, St. Catharines, Ontario L2S 3A1, Canada; *present address*: Department of Biological and Chemical Sciences, University of Essex, Wivenhoe Park, Colchester CO4 3SQ, England

Ben van Ommen Department of Pharmaco-Biology, Laboratory of Biochemistry and Molecular Biology, University of Bari, I-70125 Bari, Italy

Hiroshi Omote Department of Biological Sciences, Institute of Scientific and Industrial Research, Osaka University, Ibaraki, Osaka 567, Japan

Christian Ostermeier Max-Planck-Institut für Biophysik, D-60528 Frankfurt am Main, Germany

Ferdinando Palmieri Department of Pharmaco-Biology, Laboratory of Biochemistry and Molecular Biology, University of Bari, I-70125 Bari, Italy

Sergio Papa Institute of Medical Biochemistry and Chemistry, University of Bari, I-70124 Bari, Italy

Peter L. Pedersen Department of Biophysics and Biophysical Chemistry, and Department of Biological Chemistry, Johns Hopkins School of Medicine, Baltimore, Maryland 21205

Harvey Penefsky   State University of New York, Syracuse, New York 13210; and Public Health Research Institute, New York, New York 10028

Vittoria Petruzzella   Unit of Molecular Medicine, Children's Hospital "Bambino Gesù," Rome 00165, Italy

Jean-Paul di Rago   Centre de Génétique Moléculaire, CNRS, 91190 Gif-sur-Yvette, France

Annette Reith   Fachbereich Chemie, Philipps-Universität, D-35032 Marburg, Germany

Peter R. Rich   Department of Biology, University College, London WC1E 6BT, England

Cecilia Saccone   Department of Biochemistry and Molecular Biology, C.S.M.M.E. CNR, University of Bari, I-70125 Bari, Italy

Paolo Sarti   Department of Biochemical Sciences and CNR Center of Molecular Biology, University of Roma "La Sapienza," 00185 Rome, Italy

Richard C. Scarpulla   Department of Cell and Molecular Biology, Northwestern Medical School, Chicago, Illinois 60611

A. H. V. Schapira   Department of Clinical Neurosciences, Royal Free Hospital School of Medicine, and Department of Clinical Neurology, Institute of Neurology, London NW3 2PF, England

Martina Schliebs   Wissenschaftliche Nachwuchsgruppe, Theodor-Boveri-Institut, 97074 Würzburg, Germany

Ulrich Schulte   Institut für Biochemie, Heinrich-Heine-Universität, D 40225 Düsseldorf, Germany

Peter Seibel   Wissenschaftliche Nachwuchsgruppe, Theodor-Boveri-Institut, 97074 Würzburg, Germany

Sudha K. Shenoy   Department of Biochemistry and Molecular Biology, Oklahoma State University, Stillwater, Oklahoma 74078

Wenjun Shi   Department of Biochemistry, Michigan State University, East Lansing, Michigan 48824

Kyoko Shinzawa-Itoh   Department of Life Science, Himeji Institute of Technology, Kamigohri, Akoh Hyogo 678-12, Japan

Vladimir P. Skulachev   Department of Bioenergetics, A. N. Belozersky Institute of Physico-Chemical Biology, Moscow State University, Moscow 119899, Russia

Piotr P. Slonimski   Centre de Génétique Moléculaire, CNRS, 91190 Gif-sur-Yvette, France

Abdul-Kader Souid   State University of New York, Syracuse, New York 13210; and Public Health Research Institute, New York, New York 10028

Tewfik Soulimane   Institut für Biochemie, Rheinisch-Westfälische Technische Hochschule, 52057 Aachen, Germany

J.-W. Taanman   Department of Clinical Neurosciences, Royal Free Hospital School of Medicine, London NW3 2PF, England

Bernard L. Trumpower Department of Biochemistry, Dartmouth Medical School, Hanover, New Hampshire 03755

Tomitake Tsukihara Institute for Protein Science, Osaka University, Suita, Osaka 565, Japan

Michael I. Verkhovsky Helsinki Bioenergetics Group, Department of Medical Chemistry, Institute of Biomedical Sciences and Biocentrum Helsinki, University of Helsinki, FI-00014 Helsinki, Finland

Gaetano Villani Institute of Medical Biochemistry and Chemistry, University of Bari, I-70124 Bari, Italy

Douglas C. Wallace Center for Molecular Medicine, Emory University School of Medicine, Atlanta, Georgia 30322

Hanns Weiss Institut für Biochemie, Heinrich-Heine-Universität, D 40225 Düsseldorf, Germany

Mårten Wikström Helsinki Bioenergetics Group, Department of Medical Chemistry, Institute of Biomedical Sciences and Biocentrum Helsinki, University of Helsinki, FI-00014 Helsinki, Finland

William H. Woodruff Chemical Science and Technology Division, Los Alamos National Laboratory, Los Alamos, New Mexico 87545

Di Xia Howard Hughes Medical Institute and Department of Biochemistry, University of Texas, Southwestern Medical Center, Dallas, Texas 75235

Ting Xu Institute of Medical Biochemistry and Chemistry, University of Bari, I-70124 Bari, Italy. *Permanent address*: Institute of Biophysics, Academia Sinica, Beijing, 100101, China

Shinya Yoshikawa Department of Life Science, Himeji Institute of Technology, Kamigohri, Akoh Hyogo 678-12, Japan

Chang-An Yu Department of Biochemistry and Molecular Biology, Oklahoma State University, Stillwater, Oklahoma 74078

Linda Yu Department of Biochemistry and Molecular Biology, Oklahoma State University, Stillwater, Oklahoma 74078

Franco Zanotti Institute of Medical Biochemistry and Chemistry, University of Bari, I-70124 Bari, Italy

Massimo Zeviani Unit of Molecular Medicine, Children's Hospital "Bambino Gesù," Rome 00165, Italy

Li Zhang Department of Biochemistry and Molecular Biology, Oklahoma State University, Stillwater, Oklahoma 74078

Yuejun Zhen Department of Chemistry, Michigan State University, East Lansing, Michigan 48824

Foreword

One of the most obvious and fundamental properties of living things is their ability to utilize the potential energy in their environment for their own ends. It is not surprising therefore that the mechanism of this biotransformation of energy—bioenergetics—has long exercised the minds of the leading scientists of their day—from Lavoisier's recognition that respiration is fundamentally the same as combustion, Priestley's discovery of photosynthesis, and Spallanzani's thesis that respiration is an intracellular process.

Until about 30 years ago, the field of bioenergetics was the center of biochemistry, attracting many of the great biochemists of the century, such as Warburg, Keilin, Meyerhof, Engelhardt, Krebs, Lipmann, and Calvin. After the main pathways of intermediary metabolism and electron transfer were established, however, further progress in elucidating the mechanistic details was slow. This slowdown was due to the absence of sufficient knowledge of the structure of the catalysts involved, even after the innovative proposal by Mitchell on the fundamental role of the translocation of protons across membranes, which revolutionized the way in which the mechanism of energy transduction was envisaged and for which he was awarded the Nobel Prize for chemistry in 1978. Competition from other fields of biochemistry in which progress was more spectacular has caused funding difficulties for those working in bioenergetics.

Recently, however, significant advances in our knowledge of the structure of most of the large proteins in energy-transducing membranes, even the very large mammalian cytochrome c oxidase (Warburg's *atmungsferment*), has opened new perspectives for solving in fine detail the mechanism of electron transfer and oxidative phosphorylation. This has been recognized in the award of the Nobel prizes in chemistry in 1988 for the determination of the three-dimensional structure of a photosynthetic reaction center and in 1997 for the structure and mechanism of action of ATPases.

The detailed study of the properties of mitochondria and chloroplasts, the energy-transducing organelles of respiration and photosynthesis, respectively, led bioenergeticists in the 1960s to the discovery of extranuclear DNAs, functioning as a cytoplasmic genetic system, which was foreshadowed by Ephrussi. In particular, the mitochondrial genome was found to code for a number of hydrophobic membrane-associated subunits of the major proteins involved in respiration and the oxidative phosphorylation, a finding that has greatly extended the frontiers of bioenergetics. The discovery that certain rare inherited human diseases are due to mutations in the mitochondrial DNA has opened up the study of the physiopathology of mitochondria as a new frontier in bioenergetics. The hypothesis that random mutations during the course of a human life is a component of the normal aging process has added a new dimension to these studies. In addition, the development of extraordinarily sensitive methods of comparing the nucleotide sequences of DNA molecules has allowed this maternally inherited mitochondrial genome to be used as in the fields of metazoan evolution, archeology, history, and anthropology.

E. C. Slater
University of Southampton
United Kingdom

Preface

During an advanced course of the Federation of the European Biochemical Societies on Oxidative Phosphorylation held in Bari, Italy, in 1996, the experts convened on that occasion discussed the desirability of publishing a monograph on this topic. As a result of that exchange of ideas, and in light of the major advances made in the field of cellular bioenergetics and encouraged by the interest of Kluwer Academic/Plenum Publishers in publishing this type of volume, F. Guerrieri, J. M. Tager, and I decided to act as editors for the present monograph. We promptly registered the availability of authors representing major groups working in the field to contribute chapters on selected topics.

Cellular bioenergetics stands today as one of the most advanced sectors of biological and medical research. Life, growth, and specialized endoergonic functions of cells depend on membrane energy transfer processes. Energy transfer systems are made up of complex oligomeric proteins with masses of hundred-thousands of daltons, which catalyze intricate electrochemical reactions involving numerous protein subunits, multiple redox centers, and cooperative and proton-motive processes. Advanced spectroscopy, molecular genetics, and, more recently, X-ray crystallographic analysis now provide the necessary information to examine, at an atomic level, the mechanism of energy transfer in cellular systems. Remarkable examples of these achievements are the functional studies, mutational analysis, and X-ray crystallographic analysis of prokaryotic and mitochondrial cytochrome c oxidase, mitochondrial ATP synthase, and ubiquinol cytochrome c oxidase. This volume presents chapters on topics from the authors who contributed directly to these important advances. Other chapters deal with mitochondrial metabolite carriers and genetics and biogenesis of mitochondrial oxidative phosphorylation systems covering, in particular, the evolution of the mitochondrial genome and regulation of the expression of respiratory enzymes. In the last decade, numerous genetic and phenotypic defects in mitochondrial energy transfer enzymes have been identified that are associated with a variety of inborn and/or degenerative human diseases, in particular, in postmitotic tissues. Furthermore, a general decline of oxidative phosphorylation systems has been found to occur in human aging and increase in mitochondrial DNA mutations. Chapters in this volume deal with these physiopathological aspects of mitochondria as well as with the role of the mitochondrial permeability transition pore in apoptosis.

It is felt that the coverage of these aspects of cellular bioenergetics, contributed by leading experts in the various areas extending from the latest advancements in understanding, at an atomic level, the structure–function relationship in energy transfer membrane proteins, to the more recent perspectives on the implication of their genetic and phenotypic defects in human pathologies, makes this compendium an important and informative source for all those, from undergraduate to graduate students, researchers and clinicians, interested in obtaining a comprehensive appraisal of the latest achievements in the field.

We would like to point out that the collection of the chapters for this book started in the autumn of 1996, but the last contributors were invited and manuscripts received in the autumn of 1997. At the same time, the opportunity was given to the contributors to make some revision, where appropriate, to their chapters. We adopted this policy so as to compile, as far as possible, an up-to-date book, in consideration of the rapid and important progress very recently made in cellular bioenergetics.

Nevertheless, the reader will realize that there are still some aspects of cellular bioenergetics that are not covered in the book. It was, however, necessary to limit its size and production time. We hope that the reader, while appreciating what has been covered, will forgive the editors for any omissions.

We would like to take this opportunity to thank all the authors for their outstanding contributions, Dr. Helena Kirk for her secretarial assistance, and Kluwer Academic/Plenum Publishers, in particular Miss Joanna Lawrence, for helping us publish this volume.

Sergio Papa, Ferruccio Guerrieri, and Joseph M. Tager
University of Bari
Bari, Italy

Contents

Chapter 3

Proton Pumps of Respiratory Chain Enzymes 49

Sergio Papa, Nazzareno Capitanio, and Gaetano Villani

Chapter 4

Uncoupling of Respiration and Phosphorylation 89

Vladimir P. Skulachev

Chapter 5

**Crystallization, Structure, and Possible Mechanism of Action of
Cytochrome *c* Oxidase from the Soil Bacterium *Paracoccus
denitrificans* 119**

Hartmut Michel, So Iwata, and Christian Ostermeier

Chapter 6

**The Structure of Crystalline Bovine Heart Cytochrome *c*
Oxidase 131**

Shinya Yoshikawa, Kyoko Shinzawa-Itoh, and Tomitake Tsukihara

Chapter 7

Electron and Proton Transfer in Heme–Copper Oxidases 157

Yuejun Zhen, Denise A. Mills, Curtis W. Hoganson, Rebecca Lucas, Wenjun Shi, Gerald T. Babcock, and Shelagh Ferguson-Miller

Chapter 8

Mechanism of Proton-Motive Activity of Heme–Copper Oxidases 179

Peter R. Rich

Chapter 9

**Oxygen Reduction and Proton Translocation by the Heme–
Copper Oxidases** 193

Mårten Wikström, Joel E. Morgan, Gerhard Hummer,
William H. Woodruff, and Michael I. Verkhovsky

Chapter 10

**Transient Spectroscopy of the Reaction between Cytochrome *c*
Oxidase and Nitric Oxide: A New Hypothesis on the Mechanism
of Inhibition and Oxygen Competition** 219

Alessandro Giuffrè, Paolo Sarti, Emilio D'Itri, Gerhard Buse,
Tewfik Soulimane, and Maurizio Brunori

Chapter 11

**Energy Transduction in Mitochondrial Respiration
by the Proton-Motive Q-Cycle Mechanism of the Cytochrome
bc_1 Complex**

Bernard L. Trumpower

Chapter 12

**The Crystal Structure of Mitochondrial Cytochrome bc_1
Complex**

Chang-An Yu, Li Zhang, Anatoly M. Kachurin, Sudha K. Shenoy,
Kai-Ping Deng, Linda Yu, Di Xia, Hoeon Kim,
and Johann Deisenhofer

Chapter 13

Structural Aspects of the Cytochrome bc_1 Complex 291

Thomas A. Link

Chapter 17

**Mutational Analysis of ATP Synthase: An Approach to Catalysis
and Energy Coupling** 399

Masamitsu Futai and Hiroshi Omote

Chapter 18

**Analysis of the Nucleotide Binding Sites of ATP Synthase and
Consequences for the Catalytic Mechanism** 423

J. A. Berden and A. F. Hartog

Chapter 19

**Coupling Structures and Mechanisms in the Stalk of the Bovine
Mitochondrial F_0F_1-ATP Synthase** 459

Sergio Papa, Ting Xu, Antonio Gaballo, and Franco Zanotti

Chapter 20

The Mitochondrial Carrier Protein Family 489

Ferdinando Palmieri and Ben van Ommen

Chapter 21

Structure and Evolution of the Metazoan Mitochondrial Genome 521

Cecilia Saccone

Chapter 22

**Nuclear Transcription Factors in Cytochrome *c* and Cytochrome
Oxidase Expression** 553

Richard C. Scarpulla

Chapter 23

**Suppressor Genetics of the Mitochondrial Energy Transducing
System: The Cytochrome bc_1 Complex** 593

Jean-Paul di Rago and Piotr P. Slonimski

Chapter 24

**Tissue-Specific Expression of Cytochrome c Oxidase Isoforms
and Role in Nonshivering Thermogenesis** 621

Bernhard Kadenbach, Viola Frank, Dietmar Linder, Susanne Arnold,
Stefan Exner, and Maik Hüttemann

Chapter 25

Mitochondrial DNA Mutations and Nuclear Mitochondrial Interactions in Human Disease

H. Cock, J.-W. Taanman, and A. H. V. Schapira

Chapter 26

Strategy toward Gene Therapy of Mitochondrial DNA Disorders

Peter Seibel, Martina Schliebs, and Adrian Flierl

Chapter 30

Aging and Degenerative Diseases: A Mitochondrial Paradigm ... 751

Douglas C. Wallace

Chapter 31

**Perspectives on the Permeability Transition Pore,
a Mitochondrial Channel Involved in Cell Death** 773

Paolo Bernardi

CHAPTER 1

The Mitochondrial and Bacterial Respiratory Chains

From MacMunn and Keilin to Current Concepts

Peter Nicholls

Untwisting all the chains that ty
The hidden soul of harmony
John Milton, L'Allegro

I break all chains and set all captives free
Florence Earle Coates, The Christ of the Andes

1. CELLULAR RESPIRATION BEFORE THE CHAIN

Science is often technology driven. The experimenter needs to await the equipment. For at least four centuries, respiratory technology has been entwined with emergent ideas of respiration. The air pump was a driving force in the development of an understanding of the atmosphere we breathe and our need to breathe it. Robert Boyle and Thomas Hobbes engaged in a prolonged debate as to its usefulness; a debate in which Boyle, for all his conventional piety and alchemical dabblings, seems a contemporary, while Hobbes, for all his agnosticism and bleakly modern view of mankind, seems the traditionalist (Shapin and Schaffer, 1985). Joseph Wright's 1768 Tate Gallery picture "An

Peter Nicholls • Department of Biological Sciences, Brock University, St. Catharines, Ontario L2S 3A1, Canada.   *Present address*: Department of Biological and Chemical Sciences, University of Essex, Wivenhoe Park, Colchester CO4 3SQ, England.

Frontiers of Cellular Bioenergetics, edited by Papa *et al.* Kluwer Academic/Plenum Publishers, New York, 1999.

experiment with an air pump" remains a compelling image of the remorseless impact of the new science, even before Priestley and Cavendish had completed their revolution. Unmistakably English but with a technique borrowed from Caravaggio, Wright portrayed the ambiguity inherent in the scientific endeavor with a clarity that remains unsurpassed by later and more sophisticated painters. Just as the air pump was the technology of the 18th century, the spectroscope was the new respiratory technology of the mid-19th century.

The microspectroscope, developed by Browning (1882) and others, was one of a group of instruments that permitted study of the structure of atoms, the composition of stars (Lockyer, 1873), and the nature of biological pigments. This modest family of technical tricks followed Newton with a delay of almost two centuries and provided the intellectual liberation that led to much of today's science.

New pieces of technology commonly give rise to information overload, as illustrated by the large collection of spectra published by Charles MacMunn (1880). MacMunn, an Irish physician working in Wolverhampton, England, was one of a number of "amateurs" who moved into the field of spectroscopy in the latter part of the century, and so created the possibility of conflict with "professionals." In MacMunn's case, the professional was Felix Hoppe-Seyler, discoverer of the chemical nature of blood and tissue hemoglobins. Hoppe-Seyler (1865) had been among the first to look at the spectral behavior of hemoglobin with the microspectroscope and partially to identify the several species involved, all derivatives of the iron-containing prosthetic group, heme.

MacMunn proceeded to broaden our understanding of the range of heme iron-containing pigments by describing a range of hemoglobin-related compounds, among them the *histohematins*. Bold enough to publish some of his results in Hoppe-Seyler's own journal (MacMunn, 1890), he provoked the now notorious controversy between himself and the editor on the origin of the pigments that ended with Hoppe-Seyler's dismissive footnote. This exchange effectively closed the correspondence for the next 40 years, and left histohematins as an obscure minor pigment family rather than the key to cell respiration that they later proved to be.

2. THE COMING OF THE RESPIRATORY CHAIN

Otto Warburg (cf. Krebs, 1981), son of Emil Warburg the photochemist, believed that he could mimic the phenomenon of cell respiration using the test bed of his era, blood charcoal. Blood charcoal was obtained by carbonizing blood to a charcoal residue that retained the hemoglobin iron as a surface catalyst. This charcoal catalyzed the autoxidation of various organic and biochemical substances in a narcotic and cyanide-sensitive fashion (Warburg,

1924, 1948). Warburg's blood charcoal studies were interrupted by a session of blood and iron in cavalry duty on the eastern front in World War I. He was supposedly only persuaded to come home by a letter from Albert Einstein to his father in which Einstein (not yet the pacific humanist of later years) pointed out that others could fight in the cavalry but only Otto could do useful science at home (Krebs, 1981).

Warburg proceeded to develop the concept of the *Atmungsferment* (see Warburg, 1948) as the oxygen-activating mechanism responsible for cell respiration, raising the chemical reactivity of molecular oxygen so that it was capable of directly oxidizing otherwise stable biological compounds such as amino acids. The alternative viewpoint was that of Heinrich Wieland (1922), whose model for cell respiration was based on the idea of "hydrogen" activation. According to Wieland and fellow theorists such as Torsten Thunberg, it was not the oxygen that required activation but the hydrogen-bearing substrate. They showed that systems in cell extracts or brei existed that could activate substrate hydrogen and reduce dyes such as methylene blue. To follow these processes, another piece of technology was needed that derived almost directly from Boyle's air pump: the Thunberg tube (Thunberg, 1930).

3. THE DISCOVERY OF THE CYTOCHROME SYSTEM

David Keilin was a student of Maurice Caullery at the Sorbonne. Legend has it that this association was brought about by Keilin's medical instructions not to expose himself unnecessarily to the chance of catching cold (a lifelong asthmatic, his disease finally overcame him in concert with the stress of an academic meeting). Rain interrupted his walk to a Henri Bergson class and he ducked into the nearest lecture hall. There was Caullery, one of France's most distinguished parasitologists and a student himself of Alfred Giard, who had introduced Darwinian ideas to the French biological and intellectual worlds. With some melioration of the law of natural selection, Giard's evolution ("transformism") was said to be "Darwin matiné de Renan" (Le Dantee, 1909). When German guns were on the outskirts of Paris in 1915, Keilin left Caullery to continue parasitology studies at the Quick Laboratory in Cambridge, England. Commercial pressures had led to the establishment of a number of parasitology centers whose indirect role was to facilitate European and especially English imperial penetration into the heart of Africa. James Quick, the founder–donor of the Quick Laboratory, was a coffee planter. His laboratory was transformed into the Molteno Institute after World War I with the aid of funds from South Africa itself. The Moltenos were wealthy members of an old, established Cape family. Sir John Molteno had been the first premier of Cape Province. But their involvement supposedly came as the result of a semi-chance meeting on a train with the Quick Professor, George Henry Falkiner Nuttall, one of whose personal rules was "Always travel first class—You are sure to meet useful people."

At the opening of the new institute, Lord Buxton called for studies to eliminate parasites that rendered parts of the Empire "uninhabitable by the white man and his animals" and precluded a "supply of healthy native labour" (Buxton, 1921). Sydney Buxton's politics today might be thought to be a trifle right wing, but he was a Liberal friend of Gladstone who had upset local opinion in South Africa during his tenure as Governor General by meeting with black African leaders on a social basis.

Put to work on one of the white man's animals—the horse—Keilin sought a better understanding of the physiology of the horse botfly and of the hemoglobin present in its larval form. In the course of this study he found that the hemoglobin spectrum was continually obscured by a set of other absorption bands of an unknown kind. Keilin (1925, 1966) thus rediscovered the histohematins, the existence of which he was unaware, and renamed them *cytochromes*. They were soon found to be a much more important family of heme compounds than the more narrowly distributed hemoglobins. In 1925, biochemistry had advanced to a point where this discovery could be appreciated. But Keilin immediately became embroiled in a controversy with Warburg, who claimed that Keilin's original cytochromes, which do not react with oxygen, can only be degraded forms of the true respiratory enzyme—an echo of Hoppe-Seyler's claim that MacMunn's histohematins were degraded forms of hemoglobin. Using bacterial preparations, Warburg himself soon found that oxygen-reactive pigments of a cytochromelike character could be seen. He identified these as respiratory enzymes, while Keilin appropriately enough treated them as degraded cytochromes (because the criterion of a "native" cytochrome was that, unlike artificial hemochromogens, it did not react with oxygen). This debate was only effectively ended much later with the discovery of cytochrome a_3 by Keilin and Hartree (1939). This cytochrome is a member of the set that is nevertheless oxygen-reactive and that also forms complexes with inhibitors such as carbon monoxide and cyanide, which Warburg had employed as definitive probes of the respiratory enzyme.

Keilin's resolution of the Wieland–Warburg debate was a classic of dialectical thought. Both the Warburg and Wieland models were seen to be correct but incomplete. Keilin's respiratory chain, shown in Fig. 1, linked the two activating enzymes, for hydrogen and for oxygen, with a cytochrome-containing "bridge" that accepted oxidant from the one and reductant from the other. As a result of this resolution of a scientific conflict by someone outside the field of the controversy, Keilin developed a model of the outsider as innovator in science generally, a model that was certainly applicable to the second great step in understanding cell respiration taken by Peter Mitchell in developing the chemiosmotic theory.

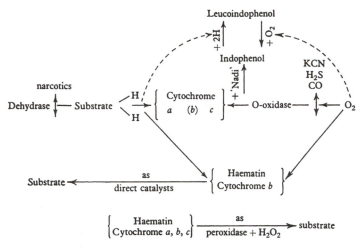

FIGURE 1. Keilin's respiratory chain. From Keilin (1966).

At essentially the same time that Keilin developed his respiratory system model, Warburg (1948) also had proposed a respiratory chain by which oxidizing equivalents are passed from oxygen to substrates. At the same time, he had discovered the carriers that precede the cytochromes: flavoproteins and coenzymes I and II. Warburg's initial respiratory chain is shown in Fig. 2. In both this and Keilin's models there was a transfer of "oxygen" in some form "up" the chain as well as reducing equivalents "down" that chain. But slowly the former idea—that oxygen itself was somehow transferred—was replaced by the idea of electron transfer, down the chain.

The final steps in determining the relationship between the cytochrome system and oxygen were taken both by Warburg and by Keilin. The two groups decided to examine the effects of classic inhibitors: cyanide and carbon monoxide. Cyanide was the tool first used by Warburg against Keilin. Warburg pointed out that the original cytochromes, unlike *Atmungsferment*, did not react with cyanide. He thought that the *Atmungsferment* must therefore be different from the cytochromes, which can at best have a subsidiary role. He followed

$$O_2 \rightarrow Fe^{2+} \rightleftarrows Fe^{3+} \rightarrow Fe^{2+} \rightleftarrows Fe^{3+} \rightarrow Fe^{2+} \rightleftarrows Fe^{3+} \rightarrow Fe^{2+} \rightleftarrows Fe^{3+} \rightarrow Fe^{2+} \rightleftarrows Fe^{3+} \rightarrow$$
$$\underbrace{\phantom{Fe^{2+} \rightleftarrows Fe^{3+}}}_{1} \quad \underbrace{\phantom{Fe^{2+} \rightleftarrows Fe^{3+}}}_{2} \quad \underbrace{\phantom{Fe^{2+} \rightleftarrows Fe^{3+}}}_{3} \quad \underbrace{\phantom{Fe^{2+} \rightleftarrows Fe^{3+}}}_{4} \quad \underbrace{\phantom{Fe^{2+} \rightleftarrows Fe^{3+}}}_{5}$$

1: oxidase 2: intermediary iron enzyme 3, 4, 5: cytochrome components

FIGURE 2. Warburg's respiratory chain. Adapted from Warburg (1948) and Warburg *et al.* (1933) (as modified in Nicholls, 1963).

this up with work using carbon monoxide. The respiratory inhibitor CO normally exerts its effect by binding to reduced hemoglobin. But it is also an inhibitor of respiration that acts at higher concentrations when effects on hemoglobin have been neutralized. The earlier observation of photosensitivity in the complex between hemoglobin and CO led Warburg (1948) to carry out one of his most remarkable experiments. By letting yeast and tissue slices respire in the presence of CO, he was able to measure the release of the CO-inhibited respiration by light of varying wavelength. He could thus build up indirectly a spectrum of the photosensitive species involved, which he recognized as involving a heme compound like the oxygen carrier of *Spirographis* (the sabellid tube worm): chlorocruorin. This then was the image of the *Atmungsferment*, now reified, and looking rather different from any of Keilin's cytochromes.

Keilin and his new co-worker Edward Hartree then turned the tables on Warburg by using the same two inhibitors to deconvolute part of the cytochrome spectrum (Keilin and Hartree, 1939). Cytochrome a proved to be a mixture of two species: the classic cytochrome a and a cytochrome a_3 that binds cyanide in both its ferric and ferrous forms and provides the site of cyanide sensitivity. Renaming Warburg's terminal oxidase as a cytochrome of course brought it back into the cytochrome-dominated "chain."

At the same time others had begun to study these systems. Among these were the Japanese groups led by Tamiya and Yakushiji, who, with their co-workers Ogura and Okunuki, examined a number of respiratory systems and attempted to focus more clearly on the nature of the terminal oxidase. They eventually decided that it must be a multiple species, and that the carbon monoxide- and cyanide-reacting components were not the same. This gave rise to modified respiratory chains, such as that in Fig. 3, by Yakushiji and Okunuki (1940). Their conclusion that the oxidase was a complex moiety was an important insight, even if their actual proposals for the chain were still inadequate.

All these discussions had been taking place in the darkening shadows of the late 1930s. Part of this story is told by Helmut Beinert in the 10th Keilin Memorial Lecture (Beinert, 1985). This review reproduces the remarkable

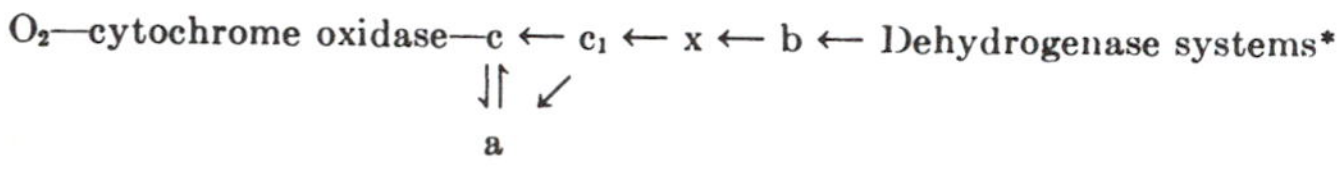

$$O_2—\text{cytochrome oxidase}—c \leftarrow c_1 \leftarrow x \leftarrow b \leftarrow \text{Dehydrogenase systems*}$$

* Succinic dehydrogenase, or the diaphorase–coenzyme–dehydrogenase–substrate system.

FIGURE 3. Okunuki's respiratory chain. From Yakushiji and Okunuki (1940).

guest signature page from the Kaiser-Wilhelm Institute in Heidelberg on the occasion of the meeting on biological oxidations held from March 21 to 23, 1932. Nearly all the signatories can be recognized as contributors to the field, and they include Keilin, Willstätter, Neuberg, Franck, Haber, Haldane (J. B. S.), Warburg, Krebs, Gaffron, and Polanyi. While the participants were discussing respiratory enzymes, Hitler was planning his campaign in the second leg of the elections for Reichs Chancellor. This led to the famous aerial criss-crossing of Germany ("Hitler über Deutschland"), which eventually left him with the greatest number of votes ever achieved by the Nazi party in more-or-less free elections, although still slightly behind Hindenburg's tally. Within months of the meeting many of the signatories were refugees.

The coming of World War II brought to a close the discussions between the English, German, and Japanese schools. Its ending reopened some of the questions, now with significant participation from the New World.

4. HIGH NOON FOR THE RESPIRATORY CHAIN

The hiatus of World War II was followed by a period of Kuhnian "normal science," which lasted from 1945 until the mid 1960s, in which details of the respiratory chain were elaborated schematically and kinetically by a number of workers, but especially by the groups of Chance (Chance and Williams, 1956) and Slater (1958). Until the mid-1960s, the carriers between NADH or succinate and oxygen were seen as a set of relatively mobile species associated with mitochondrial (and later bacterial) membranes that interact by electron transfer from one carrier to the next. Some individual carriers were "transformers" that could accept two electrons and donate them one at a time ("step-down," e.g. flavoproteins) or accept electrons one at a time and donate them in groups of four ("step-up," e.g. cytochrome a_3 or cytochrome oxidase). But the semi-mobile carriers were not thought of as arranged in any particular way, either on any particular "side" of the membrane or in any special stoichiometry.

Slater's model of the respiratory chain (Slater, 1958) derived from that of Keilin with some elaboration of the region between cytochromes *b* and *c*—the location of the putative "Slater factor" (Fig. 4). Both he (1953) and Chance (1961), whose respiratory chain is shown in Fig. 5, were also interested in the linkage between respiration and phosphorylation, something that had not concerned the earlier workers at all. That link was only shown after the groundwork had been done, and then by Belitser and Tsibakova (1939), who were outside the contemporary community of electron transfer workers.

Cytochromes are not the only species in the respiratory chain. They are merely those most easily detectable with the visual microspectroscope. The development of other methods of detecting electron carriers led to the discov-

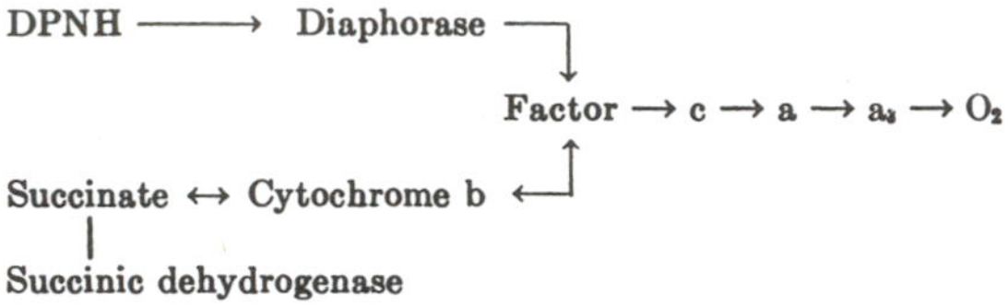

DPNH $\longrightarrow$ Diaphorase $\neg$

Factor $\rightarrow$ c $\rightarrow$ a $\rightarrow$ a$_3$ $\rightarrow$ O$_2$

Succinate $\leftrightarrow$ Cytochrome b $\leftarrow\!\rfloor$

Succinic dehydrogenase

FIGURE 4. Slater's respiratory chain. From Slater (1958).

ery of the iron–sulfur proteins, with roles in respiration, photosynthesis, bacterial metabolism, and nitrogen fixation. Helmut Beinert's story of the world of the 1930s was written as part of an introduction to a review of his own work on these proteins. Beinert's move into the iron–sulfur (non-heme) area also represented something of an "academic" escape. At the same time it required a substantial increase in the number of carriers to be considered, as summarized in Fig. 6. Such complexity sounded a death knell for all models with single, simple, and independent carriers acting as "links" in a chain of molecules and reacting with each other by bimolecular collisions.

5. MEMBRANE LOCATION OF THE RESPIRATORY CHAIN

The next important idea, whose clearer formulation took place entirely after 1945, was that the chain had to be membrane linked. Although some hemoproteins and flavoproteins are water soluble, the ones associated with respiration are structure dependent. In some cases, such as cytochrome P450 and prostaglandin synthase, the membrane relationship is comparatively modest, with one or two strands of α-helix anchoring an otherwise soluble protein in the bilayer. With the cytochromes and the other members of the full respiratory chain, the relationship is more intimate. Warburg (1948) had early defined the *Atmungsferment* as "iron in combination similar to haematin ...," but in association with a surface like blood charcoal. And Keilin (cf. Keilin and Hartree, 1940) always emphasized the role of "mutual accessibility" of the components of the respiratory chain, including cytochrome c oxidase, but specifying neither the membranous nature of the accessibility-ensuring structure nor the cellular organelles involved.

Others were implicated in linking the structured respiratory chains of eukaryotes with mitochondria. Claude (1946) described the fractionation of liver cells into their soluble and particulate components shortly after World War II. These discoveries, made by cell biologists rather than biochemists, were then picked up and integrated into the bioenergetic picture in ways that had not been envisaged by the founders. Oxidative phosphorylation was a

FIGURE 5. Chance's respiratory chain. From Chance and Williams (1956).

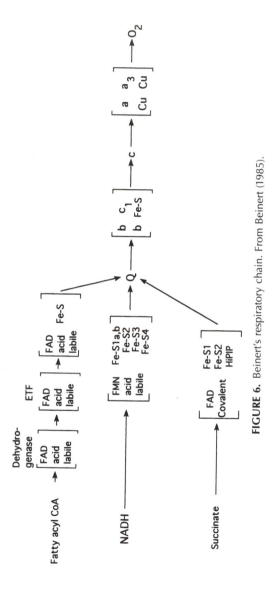

FIGURE 6. Beinert's respiratory chain. From Beinert (1985).

function of the mitochondria (cf., e. g., Lehninger, 1951), which could not be duplicated in simpler systems such as the submitochondrial fragments isolated as "Keilin–Hartree" particles (Keilin and Hartree, 1940). The two links—(1) between cell biology and chemistry and, (2) between electron transfer and ATP synthesis—were essential for the paradigm change created almost single-handedly by Peter Mitchell, student of James Danielli and admirer of David Keilin. Mitchell's chemiosmotic theory was the first to link successfully membranes and respiration in a functionally significant way (Mitchell, 1961) and led to the award of the 1978 Nobel prize (Mitchell, 1979).

6. BREAKING THE RESPIRATORY CHAIN INTO COMPLEXES

But for these conceptual changes to become fixed in the biochemical psyche, a set of parallel changes in ideas was needed (cf. the changes between the reviews by Nicholls, 1963, 1987). First, the discovery of coenzyme Q showed that the "billiard table" model of membrane–enzyme interactions could not be completely correct. This was almost immediately followed by the development of the idea of respiratory complexes in the laboratory of David Green, a progressive discovery in which Youssef Hatefi himself (see Chapter 2, this volume) played a seminal role (see also Green and Hatefi, 1961). David Green had come to England to work in Cambridge during the tense period when others (including Hans Krebs) were coming there to escape the darkening situation in Germany. He then returned to the United States, eventually to build an exceptionally successful but academically somewhat isolated team of colleagues in Madison, Wisconsin. By that time, Green was both in the heartland of the new wave of respiratory chain ideas, the United States, but he also remained an "outsider," for reasons that probably need more detailed historical analysis. Keilin, following his own success in reconciling the Warburg–Wieland controversy by discovery of the cytochromes, always thought of himself as an outsider in respiratory biochemistry who was for that reason able to break new ground. Green, who admired Keilin (see dedication to Green, 1940), fitted the Keilin model.

In the early 1950s, the Madison group began developing an idea of the respiratory chain as a functional insoluble multienzyme complex (Green *et al.*, 1954a) in which soluble components such as cytochrome c might not play a necessary role (Mackler *et al.*, 1954). The functioning of the respiratory chain was closely linked to that of the Krebs cycle enzymes in Green's thinking, and he tried to develop the idea of a "cyclophorase complex" in which all the bioenergetic roles of the mitochondrion might be united (Green *et al.*, 1954b).

For the first time there was a serious attempt to relate mitochondrial membrane structure with biochemical function, a question that had been addressed but in rather cavalier fashion by earlier workers. The possibility of looking at the structure of the mitochondrion with the electron microscope reified what had previously been the subject of generalized speculations.

These ideas from the late 1950s were summarized in a review article (Green, 1959) and in the plenary address to the Fifth International Biochemistry Congress in Moscow (Green, 1961), where Green achieved a degree of recognition often denied him at home (and perhaps for that reason). There was then a period of reassessment that depended on the new ability to fractionate the respiratory chain system into functional complexes. Hatefi *et al.* (1962a) were able to dissect the family of complexes into the "right" sets by bile salt and detergent fractionation that gave active fractions, not only at the oxidase level, already achieved by Smith and Stotz (1954), but also at the reductase level (Hatefi *et al.*, 1962b) where the roles and relations of the flavoproteins involved had previously been poorly understood.

The contemporary improvements in electron microscopy enabled Fernandez-Moran and the Wisconsin group (Fernandez-Moran *et al.*, 1964) to show what we now know to be the F_1 components of the mitochondrial ATPase *in situ*. These were initially identified however, as the so-called "elementary particles" of respiration in the mitochondrial membrane and provided the basis for Green's (1964) remarkable *Scientific American* review, later immortalized by Tab in his cartoon of the history of the system in David Nicholls' bioenergetics textbook (Nicholls, 1982). This review managed to develop an innovative picture of the respiratory chain with important new concepts, while getting almost all the details wrong. The existence of the four discrete complexes, the intermediate quinone and cytochrome *c* carriers and the transmembrane movement of electrons were all introduced in this review to a wider audience. At the same time, the proposed localization of the complexes and of the Krebs cycle enzymes and the postulated relationship of the latter to the respiratory system were all guessed quite wrongly. And the existence of a separate ATPase was nowhere suspected. Indeed, Green remained determinedly "classical" in his views on ATP formation, seen at this time as directly driven by the electron transfer complexes themselves. This was in direct conflict with the known existence of at least one form of ATPase in the mitochondrial membrane, the subject of intensive study of Kagawa and Racker (1966) and their colleagues (Racker *et al.*, 1965). The identity of the inner membrane "knobs" or headpieces (respiratory chain or ATPase components?) was the subject of the debate between Green and Racker at the 1964 Albany meeting (cf. the discussion section of Racker *et al.*, 1965).

Green's ideas on both energy conservation, as he moved toward a "conformational" model of this process and the "elementary particle" were summarized and extended in Chapters 9 and 10 of his book (Green and Goldberger, 1967). At the same time, the Chance school, while rejecting Green's identification of the ATPase knobs as containing all the respiratory components, came up with their own model of the respiratory chain in which all the processes were still linked together in a molecular device termed the *oxysome* (Chance *et al.*, 1963, 1965). This mechanism was still determinedly one involving electron transfer only parallel to the membrane surface, although in one version parallel chains on the two surfaces were considered (Chance *et al.*, 1963), and energy transfer was entirely chemical rather than conformational or chemiosmotic in nature. Increasingly strong evidence for the individual complexes (McConnell *et al.*, 1966) as well as the theoretical need for transmembrane electron transfer (Mitchell, 1961) were either ignored or discounted. Using his skills as a kinetic analyst, Chance was able to retain his classic linear model through the end of the 1960s (Chance, 1967).

7. RECONSTITUTING THE RESPIRATORY CHAIN FROM ITS PARTS

The accepted respiratory chain model, however, was slowly changing from the sequential set of reactions in Figs. 1–6 to the series of black boxes in Fig. 7, in which only cytochrome *c* remained as a classic mobile carrier. This change of paradigm was an essential prerequisite for (1) the isolation and reconstitution and (2) the ultimate crystallization of such complexes to determine their molecular structure.

Isolation of the components of the chain and their reconstitution into functional systems had originally been achieved for cytochrome *c* by Keilin (1930). Subsequent efforts at isolation and reconstitution failed and were sometimes seen as almost impossible enterprises (Keilin and Hartree, 1947). But Tsoo King, who had been influenced in his thinking by discussions with David Green, was able to prepare a soluble form of succinate dehydrogenase in Keilin's laboratory and to demonstrate its functional reconstitution to the latter's critical satisfaction (Keilin and King, 1958). The long debate between King (1966) and Singer (cf. Nicholls, 1987), in which King (1963a,b) may have won the discussion over succinate dehydrogenase but Singer won the one over NADH dehydrogenase (Watari *et al.*, 1963), may have been productive in so far

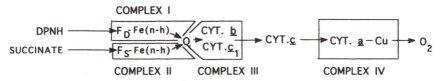

FIGURE 7. Green's respiratory chain. From McConnell *et al.* (1966).

as it emphasized the possibilities of reconstituting complexes technically awkward to handle and the need to make sure of the purity and state of the isolated material before trying reconstitution.

Development of an appreciation of the uniqueness of cytochrome c oxidase or complex IV is not easy to chronicle with precision. Certainly, Keilin and Hartree (1939) understood that both cytochromes a and a_3 are involved in the oxidation of cytochrome c, but the relationship and stoichiometry of these three cytochromes remained uncertain even after the isolation of solubilized enzyme (Smith and Stotz, 1954; Yonetani, 1960). The Wisconsin group (Mackler *et al.*, 1954) doubted the essential substrate role of cytochrome c, and Smith and Conrad (1956) initially interpreted complex formation between the soluble c and the insoluble a and a_3 cytochromes as entirely inhibitory. The action of the oxidase as an enzyme toward cytochrome c as a specific substrate was placed on a firm kinetic footing by Minnaert (1961). Nicholls (1976) also later made an effort to describe the cytochrome chain at the oxidase level in terms of a permanently associated cytochrome c–cytochrome oxidase complex as the site of electron acceptance from the preceding complex, ubiquinol–cytochrome c reductase or complex III. The complex pattern of interactions between complexes I, II, and III, interpreted by the Wisconsin group as specific "supercomplex" formation, was better elucidated by Ragan and Heron (1978) in terms of both specific complex–complex interactions and a pool function of membrane-dissolved coenzyme Q.

The idea that cytochromes a and a_3 might themselves be separable and then reconstitutable (Horie and Morrison, 1963) never proved experimentally fruitful. Now we know that this project was a will-o'-the-wisp, as the two heme components are bound in a single polypeptide subunit and share a common transmembranous α-helix. However, Trumpower and Edwards (1979) succeeded in dissociating and reconstituting complex III using the isolated FeS subunit and an iron–sulfur protein-depleted complex, in an analogous fashion to King's reconstitution of the succinate dehydrogenase system.

8. TRANSFORMING THE RESPIRATORY CHAIN INTO A TRANSMEMBRANOUS ELECTRON AND PROTON TRANSFER SYSTEM

The pattern of electron transfer was then redescribed by Mitchell (1961). Referring to one of Green's papers as "farsighted," Mitchell adopted Green's emphasis on "supramolecular" structural features but added his own characteristic biophysical treatment to the respiratory machinery and the ATPase, both of which are now regarded as transmembranous rather than lateral in character. Anisotropy of the respiratory chain was adapted from the earlier proposal of Lundegårdh (1940), but a corresponding transmembranous arrangement of the ATPase was Mitchell's own idea. Also for the first time, Mitchell (1962) used the bacterial membrane as a serious model of the mitochondrial system. The bacterial respiratory systems had first been examined by Keilin (1925, 1966) in his review of the occurrence of cytochromes throughout the living world. At the time of their discovery the interest was to show that such systems are far more widely distributed than the hemoglobins originally suggested as their precursors. Despite their early reexamination by many skilled workers, including Lucile Smith (1954) and Martin Kamen (Kamen and Vernon, 1955), the relationships between the bacterial systems and the eukaryotic respiratory chains had remained obscure. Three developments were needed to bring the prokaryotes and eukaryotes under the same conceptual roof:

1. The demonstration by Mitchell and Moyle (1956) that the bacterial systems in the plasma membrane were behaving topochemically in an analogous way to the systems in the mitochondrial inner membrane.
2. The establishment of the idea that eubacteria and mitochondria have a common evolutionary origin, the latter being former bacteria that associated symbiotically with the precursor of the eukaryotic cell (Margulis, 1981).
3. The development of the molecular tools of sequencing and mutation analysis that showed the relationships between cytochromes previously regarded as distinct, such as the cytochrome bo complex of *Escherichia coli*, now known to be closely similar to eukaryotic and prokaryotic cytochrome aa_3 (Garcia-Horsman *et al.*, 1994).

The evolving concept that bacteria are "wild" mitochondria (or that mitochondria are descendants of symbiotic bacteria) was also gaining ground, facilitating the success of analogies between the two types of respiratory system. The apparently exposed character of the bacterial plasma membrane, as well as observations on common gram-negative forms such as *E. coli*, led the present author to predict (Nicholls *et al.*, 1971) that despite the general analogies, no cytochrome c of the eukaryotic type would be found in bacteria.

This prediction was soon dramatically falsified (Dickerson and Timkovich, 1975). Although *E. coli* has no c cytochromes, and despite its metabolic versatility cannot link a protoheme to the cysteine residues on an apocytochrome expressed transgenically, other bacteria have many cytochromes c, including water-soluble forms evolutionarily and functionally related to those of eukaryotes (Bolgiano *et al.*, 1988). *Paracoccus denitrificans* became the paradigm of a wild mitochondrion, with a terminal oxidase closely similar to that of eukaryotes and both soluble and membrane-bound cytochrome c species. The only remnant of the "no soluble bacterial c cytochrome" hypothesis has been the observation that in a number of cases the bacterial aa_3 oxidase also contains a c heme resulting from the fusion of a c-type cytochrome with one of the membranous oxidase peptides (Fee *et al.*, 1980; Sone and Yanagita, 1982; Saraste *et al.*, 1991). Even in *Paracoccus*, the need to restrict the electron transfer function to the membrane and at the same time carry out numerous types of such electron transfer has led to the evolution of several membrane-linked c cytochromes, which can be isolated both alone and as parts of a supercomplex of the type earlier postulated by Green and co-workers. Berry and Trumpower (1985) found that the coenzyme Q (CoQ) oxidase activity of *Paracoccus* membranes behaved as if it was the property of such a super-complex. This heralded a return of a multienzyme complex idea (Joliot *et al.*, 1993)—another Green doctrine—that essentially had been abandoned because of the success of the Madison group's other idea: that of the four respiratory complexes linked by membrane-soluble (CoQ) or water-soluble (cyt. c) carriers.

Hydrogen and electron transfers were envisaged by Mitchell (1961, 1979) as alternating processes in the membrane, creating a respiratory chain that was successively electroneutral and electrogenic in character (Fig. 8). Proton trans-fer was an accompaniment to electron transfer because hydrogen transfer requires both protons and electrons. Mitchell's Nobel lecture (1979) paid

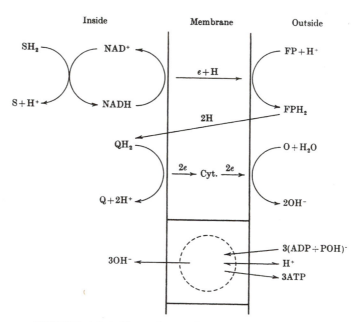

FIGURE 8. Mitchell's respiratory chain. From Mitchell (1962).

homage to Keilin in its title, "David Keilin's respiratory chain concept and its chemiosmotic consequences." But his detailed mechanism of chain action was flawed in its narrow requirements for alternating electron and hydrogen transfer and the consequent limitations on the "pumping" stoichiometry that this involved. Models of the respiratory chain soon began to appear (Papa *et al.*, 1974) in which the Green–Hatefi complexes acted as "proton pumps." Do respiratory chains behave as sequences of Mitchellian loops or do the individual complexes act as proton pumps?

The most striking discrepancy between the needs of the loop model and the known composition of the chain was at the complex III (CoQ–cyt. *c* reductase) level. Here the only hydrogen carrier seemed to be coenzyme Q located at the substrate side of the mechanism. All the known redox components in the complex were electron carriers. Yet, there were kinetic anomalies in this region of the respiratory chain, especially the oxidant-induced reduction of cytochrome *b* (Chance, 1972), that were difficult to reconcile with the idea of a linear sequence of carriers. To sustain his classic model at the complex III level, Mitchell introduced the concept of the proton–motive Q loop (Mitchell, 1975, 1976), as reviewed by Trumpower (Chapter 11, this volume). This theoretical tour de force, which was introduced at a conference by Mitchell armed with a wooden model of his mechanism, solved both his problem (the need for a hydrogen carrier at both oxidizing and reducing "sides" of the *b* cytochromes) and Chance's and Slater's problem (the anomalous kinetic behavior of those cytochromes). It since has been elaborated in various versions to become the standard explanatory model of bc_1 and $b_6 f$ complexes in mitochondria, prokaryotes, and chloroplasts.

The second problematic area was that of complex IV, the terminal oxidase. Despite the large redox drop between cytochrome *c* and oxygen, the original loops terminated at the oxidase with only a single electron transfer process moving two negative charges per oxygen atom from the P to the N side of the membrane. Proton pumping by the oxidase, which actually moves extra protons from the N side to the P side during electron transfer, remained undetected by Peter Hinkle, despite his being the first to dissect and reconstitute the oxidase into proteoliposomes showing a high degree of respiratory control (Hinkle *et al.*, 1972). Wikström, an early advocate of proton pump modeling to account for the high stoichiometries of H^+ translocated to oxygen taken up, was able to demonstrate proton release from the P side of the membrane in mitochondrial and proteoliposomal systems, containing cytochrome oxidase (Krab and Wikström, 1978). These observations brought him into conflict with Mitchell, who opposed the postulation of any pumps (Mitchell and Moyle, 1983), as well as others, including some who accepted pumps at the complex III level but denied them to the oxidase (Papa *et al.*, 1983). The reconciliation came after a long session on the topic at Glynn House, as Mitchell was persuaded by several colleagues, not least among them Jennifer Moyle, that to accept proton extrusion by the oxidase did not mean abandoning a chemically comprehensible analysis of proton movement (West *et al.*, 1986). Indeed, Mitchell used his defeat on this technical issue to begin developing new ideas in the chemical pathways of proton movement, especially the copper zoop models (Mitchell *et al.*, 1985; Mitchell, 1988), on which he was still working when he died in 1992.

Does transmembrane electron transfer, in its Mitchellian form, occur in any of the respiratory chain complexes? Complex I is transmembranous (Walker, 1992) and pumps protons (De Jonge and Westerhoff, 1982), but its mechanism of action remains uncertain. Charge movement may represent electron transfer between its various iron–sulfur centers or proton transfer via proton channels. Complex II is transmembranous but apparently does not transfer protons under normal conditions. A combination of proton transfer to the quinones as they are reduced to quinols and electron transfer between the two b hemes accounts for charge separation by complex III. Cytochrome c oxidase, once the paradigmatic transmembrane electron transferring complex (Mitchell and Moyle, 1983), is now thought to act as a pure proton transfer system (Iwata *et al.*, 1995). The two heme groups, unlike those of the bc_1 complex, are arranged at the same depth in the membrane. Indeed, the iron of cytochrome a may be slightly closer to the N side of the membrane than that of cytochrome a_3 to which it transfers electrons. Electron transfer in cytochrome oxidase, after a brief sojourn at the Cu_A level and partially transmembranous movement to cytochrome a (Hill, 1994), proceeds through the Fea, Fea_3, and Cu_B centers in a direction almost parallel to the plane of the membrane (*pace* Chance). Only the protons, whose existence was for so long debated, traverse the membrane: both those involved in oxygen reduction (the scalar protons that are not really scalar) and those that are pumped.

A current textbook chain is given in Fig. 9, which may be compared with Fig. 1, to summarize changes that have occurred over the last 75 years.

9. OPENING THE BLACK BOXES THAT HAVE REPLACED THE CHAIN

What are the conceptual shifts between the two pictures (Figs. 1 and 9)?

1. The respiratory chain became a sequence of separate unique complexes.
2. The initially scalar processes became membrane linked, first parallel to the membrane surface and then perpendicular to it.
3. The hydrogen or electron transfer became coupled H^+ and e^- transfer.

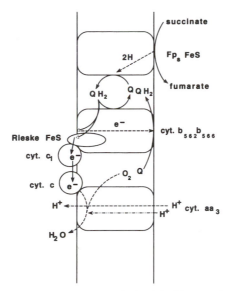

FIGURE 9. The contemporary respiratory chain. A leaf from an imaginary textbook.

These conceptual shifts were essential prerequisites for the functional dissection and eventual crystallization of the complexes. We now have X-ray crystal structural information, not only for the simplest molecule, cytochrome c (Takano and Dickerson, 1980), but also for its complex enzyme partners, the reductase (Berry *et al.*, 1992; Yu *et al.*, 1996) and the terminal oxidase (Iwata *et al.*, 1995; Tsukihara *et al.*, 1995, 1996). The latter success with both bacterial and mitochondrial (Chapters 5 and 6, respectively, this volume) terminal oxidases completes the theoretical change begun by Keilin and elaborated by so many workers in the period 1965–1995. The determination of the crystal structure of cytochrome c oxidase is the first for a truly membrane-linked enzyme. The mechanistic analysis of events within this and the other respiratory complex black boxes—an interpretation in functional terms of the pathways of electron and proton transfer—is the outstanding remaining task for bioenergeticists and students of electron transfer. It will be achieved during the 21st century. I anticipate a new century, and a new millennium, more creative and less destructive than the present ones. The 20th century's achievements in pure and applied sciences were so remarkable, but its wars and conflicts—dependent as they were on that technology and science—proved among the cruelest in all history.

ACKNOWLEDGMENTS. My researches into the mechanism of action of the *Atmungsferment*, complex IV, cytochrome aa_3, or cytochrome c oxidase, and into the background for our ideas about this enzyme, have been supported for the past 20 years by the Canadian Natural Sciences and Engineering Research Council. My doctoral supervision by David Keilin and postdoctoral and subsequent research periods with Tsoo E. King, E. C. Slater, and Britton Chance gave me a privileged access to the thinking of some of the key participants in the development of ideas about the respiratory chain, for which I am grateful. I regret never having met Otto Warburg, but even more so my failure to talk more extensively with David Green, who was always exceptionally friendly to me as Keilin's "Benjamin," despite my subsequent association with the Chance school in the debate over the respiratory chain. Perhaps the strongest recent influence on my thinking was that of Peter Mitchell, with whom I met regularly when visiting my mother in Cornwall, only a brief drive from the Glynn Research Institute. I am sorry that both the research at Glynn and this intellectually stimulating era have now finally come to a close.

10. REFERENCES

Beinert, H., 1985, *Biochem. Soc. Trans.* **14:**527–533.
Belitser, V. A., and Tsibakova, E. T., 1939, *Biokhimiya* **4:**516–535.
Berry, E. A., and Trumpower, B. L., 1985, *J. Biol. Chem.* **260:**2458–2467.
Berry, E. A., Huang, L., Earnest, T. N., and Jap, B. K., 1992, *J. Mol. Biol.* **224:**1161–1166.
Bolgiano, B., Smith, L., and Davies, H. C., 1988, *Biochim. Biophys. Acta* **933:**341–350.
Browning, J., 1882, *How to Work with the Spectroscope*, John Browning, London.
Buxton, S., 1921, quoted in *The African World*, Dec. 3.
Chance, B., 1961, *J. Biol. Chem.* **236:**1569–1576.
Chance, B., 1967, *Biochem. J.* **103:**1–18.
Chance, B., 1972, *FEBS Lett.* **23:**1–20.
Chance, B., and Williams, G. R., 1956, *Adv. Enzymol.* **17:**65–134.
Chance, B., Estabrook, R. W., and Lee, C.-P., 1963, *Science* **140:**379–380.
Chance, B., Erecinska, M., and Lee, C.-P., 1965, *Proc. Natl. Acad. Sci.* **66:**928–935.
Claude, A., 1946, *J. Exp. Med.* **84:**51–89.
Criddle, R. S., Bock, R. M., Green, D. E., and Tisdale, H., 1962, *Biochemistry* **1:**827–842.
De Jonge, P. C., and Westerhoff, H. V., 1982, *Biochem. J.* **204:**515–523.
Dickerson, R. E., and Timkovich, R., 1975, in: *The Enzymes*, XI (P. D. Boyer, ed.), pp. 397–547, Academic Press, New York.
Fee, J. A., Choc, M. G., Findling, K. L., Lorence, R., and Yoshida, T., 1980, *Proc. Natl. Acad. Sci. USA* **77:**147–151.
Fernandez-Moran, H., Oda, T., Blair, P. V., and Green, D. E., 1964, *J. Cell. Biol.* **22:**63–100.
Garcia-Horsman, J., Barquera, B., Rumbley, J., Ma, J., and Gennis, R. B., 1994, *J. Bacteriol.* **176:**5587–5600.
Green, D. E., 1940, *Mechanisms of Biological Oxidations*, Cambridge University Press, Cambridge, England.
Green, D. E., 1959, *Adv. Enzymol* **21:**73–129.

Green, D. E., 1961, in: *Proceedings of the 5th International Congress of Biochemistry*, Vol. 9 (V. A. Engelhardt *et al.*, eds.), pp. 9–33, Pergamon Press, New York.

Green, D. E., 1964, *Sci. Am.* (January), 3–11.

Green, D. E., and Goldberger, R. F., 1967, *Molecular Insights into the Living Process*, Academic Press, New York.

Green, D. E., and Hatefi, Y., 1961, *Science* **133**:13–19.

Green, D. E., Mackler, B., Repaske, R., and Mahler, H. R., 1954a, *Biochim. Biophys. Acta* **15**:435–437.

Green, D. E., Loomis, W. F., and Auerbach, V. H., 1954b, *J. Biol. Chem.* **172**:389–403.

Hatefi, Y., Haavik, A. G., Fowler, L. R., and Griffiths, D. E., 1962a, *J. Biol. Chem.* **237**:2661–2669.

Hatefi, Y., Haavik, A. G., and Griffiths, D. E., 1962b, *J. Biol. Chem.* **237**:1676–1680.

Hill, B. C., 1994, *J. Biol. Chem.* **269**:2419–2425.

Hinkle, P. C., Kim, J. J., and Racker, E., 1972, *J. Biol. Chem.* **247**:1338–1339.

Hoppe-Seyler, F., 1865, *Handbuch der Physiologisch und Pathologisch Chemischen Analyse*, August Hirschwald, Berlin.

Horie, S., and Morrison, M., 1963, *J. Biol. Chem.* **238**:2859–2865.

Iwata, S., Ostermeier, C., Ludwig, B., and Michel, H., 1995, *Nature* **376**:660–669.

Joliot, P., Verméglio, A., and Joliot, A., 1993, *Biochim. Biophys. Acta* **1141**:151–174.

Kagawa, Y., and Racker, E., 1966, *J. Biol. Chem.* **241**:2475–2482.

Kamen, M., and Vernon, L. P., 1955, *Biochim. Biophys. Acta* **17**:10–22.

Keilin, D., 1925, *Proc. Roy. Soc. B* **98**:312–339.

Keilin, D., 1930, *Proc. Roy. Soc. B* **104**:418–444.

Keilin, D., 1966, *The History of Cell Respiration and Cytochrome* (J. Keilin, ed.), Cambridge University Press, Cambridge, England.

Keilin, D., and Hartree, E. F., 1939, *Proc. Roy. Soc. B* **127**:167–191.

Keilin, D., and Hartree, E. F., 1940, *Proc. Roy. Soc. B* **129**:277–306.

Keilin, D., and Hartree, E. F., 1947, *Biochem. J.* **41**:500–502.

Keilin, D., and King, T. E., 1958, *Nature* **181**:1520–1522.

King, T. E., 1963a, *J. Biol. Chem.* **238**:4032–4036.

King, T. E., 1963b, *J. Biol. Chem.* **238**:4037–4051.

King, T. E., 1966, *Adv. Enzymol.* **28**:155–236.

Krab, K., and Wikström, M., 1978, *Biochim. Biophys. Acta* **504**:200–214.

Krebs, H. A., 1981, *Otto Warburg, Cell Physiologist, Biochemist and Eccentric*, Clarendon Press, Oxford.

Le Dantee, F., 1909, *Bull. Scient. de la France et Belgique* **42**:III–XIII.

Lehninger, A. L., 1951, *J. Biol. Chem.* **190**:345–359.

Lockyer, J. N., 1873, *The Spectroscope and Its Applications*, Macmillan, London.

Lundegårdh, H., 1940, *Lantbr. Hogsk. Ann.* **8**:233–404.

Mackler, B., Kohout, P. M., and Green, D. E., 1954, *Biochim. Biophys. Acta* **15**:437–438.

MacMunn, C. A., 1880, *The Spectroscope in Medicine*, Lindsay and Blakiston, Philadelphia.

MacMunn, C. A., 1890, *Hoppe-Seyler's Zeitschrift* **14**:328–329.

Margulis, L., 1981, *Symbiosis in Cell Evolution*, W. H. Freeman, San Francisco.

McConnell, D., MacLennan, D., Tzagoloff, A., and Green, D. E., 1966, *J. Biol. Chem.* **241**:2373–2382.

Minnaert, K., 1961, *Biochim. Biophys. Acta* **50**:23–34.

Mitchell, P., 1961, *Nature* **191**:144–148.

Mitchell, P., 1962, *Biochem. Soc. Symp.* **22**:142–168.

Mitchell, P., 1975, *FEBS Lett.* **56**:1–6.

Mitchell, P., 1976, *J. Theor. Biol.* **62**:327–367.

Mitchell, P., 1979, *Nobel Lecture, Reimpression de les Prix Nobel en 1978*, pp. 135–172, Nobel Foundation, Stockholm.

Mitchell, P., 1988, *Ann. NY Acad. Sci.* **550**:185–198.

Mitchell, P., and Moyle, J., 1956, *Discuss. Farad. Soc.* **21**:258–265.

Mitchell, P., and Moyle, J., 1983, *FEBS Lett.* **151**:167–178.

Mitchell, P., Mitchell, R., Moody, A. J., West, I. C., Baum, H., and Wrigglesworth, J. M., 1985, *FEBS Lett.* **188**:1–7.

Nicholls, D. G., 1982, *Bioenergetics*, Academic Press, London.

Nicholls, P., 1963, in: *The Enzymes*, Vol. **VIII**, 2nd ed., (P. Boyer, H. Lardy, and K. Myrbäck, eds.), pp. 3–40, Academic Press, New York.

Nicholls, P., 1976, *Biochim. Biophys. Acta* **430**:30–45.

Nicholls, P., 1987, in: *Advances in Membrane Biochemistry and Bioenergetics* (C. H. Kim, H. Tedeschi, J. J. Diwan, and J. C. Salerno, eds.), pp. 13–20, Plenum Press, New York.

Nicholls, P., Kimelberg, H. K., Mochan, E., Mochan, B. S., and Elliott, W. B., 1971, in: *Probes of Structure and Function of Macromolecules and Membranes*, Vol. I (B. Chance *et al.*, eds.), pp. 431–436, Academic Press, New York.

Papa, S., Guerreri, F., and Lorusso, M., 1974, *BBA Library* **13**:417–432.

Papa, S., Lorusso, M., Capitanio, N., and De Nitto, E., 1983, *FEBS Lett.* **157**:7–13.

Racker, E., Tyler, D. D., Estabrook, R. W., Conover, T. E., Parsons, D. F., and Chance, B., 1965, in: *Oxidases and Related Redox Systems* (T. E. King, H. S. Mason, and M. Morrison, eds.), pp. 1077–1101, J. Wiley & Sons, New York.

Ragan, C. I., and Heron, C., 1978, *Biochem. J.* **174**:783–790.

Saraste, M., Metso, T., Nakari, T., Jalli, T., Lauraeus, M., and Van der Oost, J., 1991, *Eur. J. Biochem.* **195**:517–525.

Shapin, S., and Schaffer, S., 1985, *Leviathan and the Air-Pump: Hobbes, Boyle and the Experimental Life*, Princeton University Press, Princeton, NJ.

Slater, E. C., 1953, *Nature* **20**:147–199.

Slater, E. C., 1958, *Adv. Enzymol.* **20**:147–199.

Smith, L., 1954, *Bacteriol. Rev.* **18**:106–130.

Smith, L., and Conrad, H. E., 1956, *Arch. Biochem. Biophys.* **63**:403–413.

Smith, L., and Stotz, E., 1954, *J. Biol. Chem.* **209**:819–828.

Sone, N., and Yanagita, Y., 1982, *Biochim. Biophys. Acta* **682**:216–226.

Takano, T., and Dickerson, R. E., 1980, *Proc. Natl. Acad. Sci. USA* **77**:6371–6375.

Thunberg, T., 1930, *Qt. Rev. Biol.* **5**:318–347.

Trumpower, B. L., and Edwards, C. A., 1979, *J. Biol. Chem.* **254**:8697–8706.

Tsukihara, T., Aoyama, H., Yamashita, E., Tomizaki, T., Yamaguchi, H., Shinzawa-Itoh, K., Nakashima, R., Yaono, R., and Yoshikawa, S., 1995, *Science* **269**:1069–1074.

Tsukihara, T., Aoyama, H., Yamashita, E., Tomizaki, T., Yamaguchi, H., Shinzawa-Itoh, K., Nakashima, R., Yaono, R., and Yoshikawa, S., 1996, *Science* **272**:1136–1144.

Walker, J. E., 1992, *Qt. Rev. Biophys.* **25**:253–324.

Warburg, O., 1924, *Biochem. Z.* **152**:479–494.

Warburg, O., 1948, *Heavy Metal Prosthetic Groups and Enzyme Action*, Oxford University Press, Oxford, England.

Warburg, O., Negelein, E., and Maas, E., 1933, *Biochem. Z.* **266**:1–8.

Watari, H., Kearney, E. B., and Singer, T. P., 1963, *J. Biol. Chem.* **238**:4063–4073.

West, I. C., Mitchell, R., Moody, A. J., and Mitchell, P., 1986, *Biochem. J.* **236**:15–21.

Wieland, H., 1922, *Erg. Physiol.* **20**:477–518.

Yakushiji, E., and Okunuki, K., 1940, *Proc. Imp. Acad. Japan* **16**:299–302.

Yonetani, T., 1960, *J. Biol. Chem.* **235**:845–852.

Yu, C.-A., Xia, J.-Z., Kachurin, A. M., Yu, L., Xia, D., Kim, H., and Deisenhofer, J., 1996, *Biochim. Biophys. Acta* **1275**:47–53.

CHAPTER 2

The Mitochondrial Enzymes of Oxidative Phosphorylation

Youssef Hatefi

1. INTRODUCTION

Today, the amino acid sequences of proteins of unknown function are deduced from the DNA before they are isolated, the receptors of unknown hormones are identified, and antibodies that catalyze nonbiological reactions are generated. Forty years ago, when this author began work on the mitochondrial oxidative phosphorylation system, it was necessary to have a well-defined and assayable biological reaction before one could attempt to isolate the enzyme(s) that catalyzed it. The fortunate circumstance that led to the discovery of the respiratory chain enzyme complexes occurred in 1957, when F. L. Crane, R. L. Lester, C. Widmer, and the author isolated coenzyme Q (ubiquinone) from bovine heart mitochondria and obtained evidence for its role as a respiratory chain electron carrier (Hatefi, 1963; Crane *et al.*, 1957). Consequently, one could think of the mitochondrial electron transport system in terms of the following four assayable reactions, and one could attempt to isolate the four enzymes or enzyme systems that catalyzed them:

Youssef Hatefi • Division of Biochemistry, Department of Molecular and Experimental Medicine, The Scripps Research Institute, La Jolla, California 92037.

Frontiers of Cellular Bioenergetics, edited by Papa *et al.* Kluwer Academic/Plenum Publishers, New York, 1999.

$$NADH + Q + H^+ \rightleftharpoons NAD^+ + QH_2 \tag{1}$$

$$Succinate + Q \rightleftharpoons fumarate + QH_2 \tag{2}$$

$$QH_2 + 2\ ferricyt.\ c \rightleftharpoons Q + 2\ ferrocyt.\ c + 2H^+ \tag{3}$$

$$2\ ferrocyt.\ c + 2\ H^+ + \tfrac{1}{2}O_2 \rightleftharpoons 2\ ferricyt.\ c + H_2O \tag{4}$$

The fact that cytochrome c oxidase, which catalyzes reaction (4), had already been isolated (Hatefi, 1958; Okunuki et al., 1958; Smith and Stotz, 1954) suggested the feasibility of this endeavor. Soon, it became possible to isolate from the same batch of bovine heart mitochondria the enzymes that catalyzed reactions (1–3) as well (Hatefi et al., 1985). Furthermore, it was shown that these four preparations could be combined in the presence of cytochrome c to reconstitute the entire respiratory chain, with a turnover capability for the oxidation of NADH and succinate by molecular oxygen comparable to that of intact submitochondrial particles (SMP) (Hatefi et al., 1962a,b,c, 1985). More importantly, the fractionation pattern and the chemical properties of the four preparations convinced this author that each is a discrete multiprotein–lipid complex of unvarying composition (Hatefi et al., 1961, 1962). The idea that the preparations we had isolated were discrete structural units of function was new in 1961, but later it was recognized as a universal design for many integral membrane enzymes, especially when it was found subsequently that the genes that encoded the subunits of some of these enzyme complexes constituted an operon in bacterial genomes.

Mitchell's (1961) chemiosmotic hypothesis was also introduced in 1961. This novel concept predicted, among other things, the existence of a membrane-bound reversible ATPase, of which the catalytic sector, F_1, had been isolated in the previous year by Racker and co-workers (Pullman et al., 1960). In 1968, Tzagoloff et al. (1968) isolated the bovine mitochondrial ATPase complex, whose hydrolytic activity could be inhibited, as in SMP, by the antibiotic oligomycin. However, this preparation did not catalyze ATP–$^{32}P_i$ exchange, an activity that was a characteristic of well-coupled SMP. The intact ATP synthase complex, with oligomycin-sensitive ATPase and uncoupler-sensitive ATP–$^{32}P_i$ exchange activities, was first isolated in our laboratory in 1974, again from the same batch of mitochondria that also yielded the four respiratory chain enzyme complexes (Galante et al., 1979; Hatefi et al., 1974).

This volume contains authoritative chapters on the structure, function, and various molecular biological aspects of the above enzyme complexes from mammalian and bacterial sources. Therefore, to avoid repetition, the following discourse will be confined in the limited space allowed to certain features that may complement what is found in greater detail elsewhere in this compendium. For the same reason, cytochrome oxidase will not be discussed at all, because the reader will find in this volume a thorough and awe-inspiring account of the recent progress on the structure and function of this important enzyme complex.

2. COMPLEX I (NADH–QUINONE OXIDOREDUCTASE)

2.1. Composition

The bovine mitochondrial NADH–ubiquinone oxidoreductase is composed of >40 unlike polypeptides (Fearnley and Walker, 1992; Walker, 1992), of which seven are encoded by the mitochondrial DNA and synthesized within the mitochondria (Chomyn et al., 1985, 1986), and the remainder are cyto-ribosomal products and are imported. Complex I from Neurospora crassa has ≥32 subunits (Schulte et al., 1994), from Paracoccus denitrificans 14 subunits (Yagi et al., 1993), from Escherichia coli 13 subunits (Leif et al., 1995; Blatter et al., 1997), and from Thermos thermophilus 14 subunits (Yano et al., 1997). The P. denitrificans genes are located in a cluster together with six unidentified reading frames (URFs). The complex I genes of E. coli and T. thermophilus are also arranged in clusters, but without URFs, which suggests that each cluster is an operon. There is considerable sequence homology among the bacterial complex I subunits and their bovine and Neurospora counterparts (for details, see Fearnley and Walker, 1992; Yano et al., 1997).

All proton translocating NADH–quinone oxidoreductases contain as prosthetic groups one flavin mononucleotide (FMN) and several iron–sulfur clusters. In bovine complex I, there may be as many as eight iron–sulfur clusters, of which five are electron paramagnetic resonance (EPR) visible, and are designated N1a, N1b, N2, N3, and N4 (Ohnishi, 1979). The estimated reduction potentials of these redox centers are listed in Table I, and data for the bacterial iron–sulfur clusters are compiled in Sled' et al. (1993). The 51-kDa subunit of bovine complex I binds NAD(H) (Deng et al., 1990), and studies on P. denitrificans and E. coli complex I have shown that the bacterial analogue of this subunit also binds NAD(H) and is the site of FMN binding (Fecke et al., 1994; Yano et al., 1996). The amino acid sequences of bovine complex I

Table I
The EPR-Visible Iron–Sulfur Clusters
of Bovine Heart Complex I

Iron–sulfur cluster	E_m (mV)[a]	Likely location
N1a[2Fe-2S]	−370	24 kDa, FP
N1b[2Fe-2S]	−250	75 kDa, IP
N2[4Fe-4S]	−150	HP
N3[4Fe-4S]	−250	51 kDa, FP
N4[4Fe-4S]	−250	75 kDa, IP

[a]The E_m data are from Sled' et al. (1993). The E_m of N2 is for pH 8.0.

subunits indicate that five of these subunits contain one or more cysteine motifs required for housing iron–sulfur clusters (Fearnley and Walker, 1992). Early studies on the fragmentation of bovine complex I (Ohnishi *et al.*, 1985) and more definitive recent studies on the expression in *E. coli* of *P. denitrificans* complex I subunits (Yano *et al.*, 1994, 1995, 1996) have identified three subunits that house iron–sulfur clusters (Table I).

2.2. Structure

In 1967, it was shown that bovine complex I is made up of three distinct structural domains, later designated FP (flavoprotein), IP (iron–sulfur protein), and HP (hydrophobic protein) (Hatefi and Stempel, 1967). FP contains three subunits of molecular mass 51, 24, and 9 kDa. As mentioned earlier, the 51-kDa subunit binds NAD(H) and contains FMN and a tetranuclear iron–sulfur cluster (N3). The 24-kDa subunit contains one binuclear iron–sulfur cluster, presumably N1a, and the 9-kDa subunit, which does not have a bacterial counterpart, is devoid of redox carriers. Isolated FP catalyzes the oxidation of NADH (and slow oxidation of NADPH) in the presence of quinoid compounds (including ubiquinone) and ferric ion complexes (including cytochromes *c*) as electron acceptors (Galante and Hatefi, 1979).

IP contains seven major polypeptides with molecular masses of 75, 49, 30, 18, 15, 13, and 11 kDa, of which the latter two polypeptides comigrate upon sodium dodecyl sulfate (SDS) gel electrophoresis, and the 75-kDa polypeptide contains a tetranuclear, a binuclear, and possibly another tetranuclear iron–sulfur cluster (Yano *et al.*, 1995). FP and IP are water soluble, whereas HP is highly water insoluble and contains the seven mtDNA encoded subunits of complex I. Among the >30 subunits of HP are a 23-kDa polypeptide with two cysteine motifs for housing two tetranuclear iron–sulfur clusters and a 20-kDa polypeptide with the possible potential (a partial motif of three cysteines) for another tetranuclear cluster. Other investigators include the 20- and the 23-kDa subunits in the IP domain (Walker, 1992). It is possible that they are located in the interface between HP and IP, and depending on the procedure used they fractionate more into one or the other fraction (Fig. 1) (see, for example, Finel, 1993).

The tripartite FP–IP–HP organization of the subunits of bovine complex I appears to be the structural blueprint for complex I from *N. crassa* and *E. coli* (see Chapter 14 this volume), possibly also for the enzyme from *P. denitrificans* (Takano *et al.*, 1996). In *N. crassa*, inhibition of mitochondrial protein synthesis results in the accumulation of a water-soluble NADH dehydrogenase corresponding in composition to the bovine mitochondrial FP + IP. Electron microscopic analysis of membrane crystals of *N. crassa* complex I and its peripheral (FP + IP) and integral (HP) segments have indicated that complex I

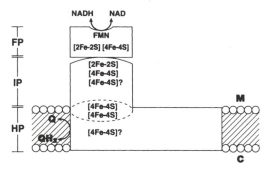

FIGURE 1. The FP–IP–HP tridomain structure of bovine heart complex I, showing the locations of FMN and the eight possible iron–sulfur clusters. The IP–HP interface region delineated by dashed lines may contain the 20- and the 23-kDa subunits. M and C are, respectively, the matrix and the cytosolic sides of the mitochondrial inner membrane. For other details, see text and Table I.

is L-shaped, with one arm (corresponding to FP + IP) located outside of the membrane, and the other arm (corresponding to HP) located within the membrane (Hofhaus *et al.*, 1991).

2.3. Enzymatic Properties

Complex I prepared from bovine heart mitochondria oxidizes NADH at a rate of 350–400 sec^{-1} in the presence of ubiquinone-1 as electron acceptor, and at a rate of ~10,000 sec^{-1} in the presence of potassium ferricyanide as electron acceptor (Hatefi and Stiggall, 1976; Hatefi *et al.*, 1962b). The former but not the latter activity is inhibited by rotenone, piericidin A, capsaicin, barbiturates, demerol, N,N'-dicyclohexylcarbodiimide (DCCD), myxothiazol and related structures, and annonaceous acetogenins. The latter are highly potent inhibitors of complex I, with K_i values ≤1.0 nM (Degli Esposti *et al.*, 1994). They have in common a substituted unsaturated γ-lactone attached to a long saturated hydrocarbon chain (24 carbons), which is interrupted near the center of the chain by two vicinal or nonadjacent tetrahydrofurans. It has been shown that rotenone and DCCD bind to the mitochondrial DNA (mtDNA)-encoded subunit, ND1, at nonoverlapping sites (Yagi and Hatefi, 1988; Earley *et al.*, 1987). The binding sites of the other inhibitors are not known, but they all appear to inhibit the last electron transfer step in complex I, apparently between iron–sulfur cluster N2 and ubiquinone.

Little is known about the mechanism of electron transfer by complex I, and even less about proton translocation. It is a safe assumption that NADH delivers a hydride ion to FMN; then, in single electron transfer steps through the multiple iron–sulfur clusters, electrons are conveyed to ubiquinone, which binds to HP. Using protein cross-linking reagents, it was shown that only the 51-kDa subunit of FP cross-links to only the 75-kDa subunits of IP (Yamaguchi and Hatefi, 1993). Recent ligand-blotting results agree with this finding. These ligand-blotting experiments also showed that IP (or certain subunits thereof) binds to the 42-, 39-, 23-, 20-, and 16-kDa subunits of HP (Belogrudov and Hatefi, 1996). Should these 23- and 20-kDa subunits of HP be the same as the putative iron–sulfur proteins discussed above, then one may speculate that the path of electrons in complex I might be NADH → 51- and 24-kDa subunits of FP → 75-kDa subunit of IP → 23 (and 20?) kDa subunit of HP → ubiquinone (Fig. 1).

In general, the mechanism of proton translocation by the enzymes of oxidative phosphorylation is unknown. In the case of complex I, the proton stoichiometry of the system is also unclear. However, the available values ($H^+/e \geq 2$) preclude a mechanism whereby a single electron transfer step would be coupled to the translocation of one proton. Therefore, coupling would have to take place in more than a single one-electron/one-proton step or there would have to be a Q-cycle-type mechanism in which the transfer of a single electron is coupled to the translocation of more than one proton, such as single electron oxidation of ubiquinol to ubisemiquinone being coupled to the translocation outward of the two protons of ubiquinol (see Section 4). Accordingly, several Q-cycle type mechanisms involving FMN and/or ubiquinone as proton carriers have been proposed (Vinogradov, 1993; Weiss and Friedrich, 1991; Krishnamoorthy and Hinkle, 1988; Ragan, 1987), for which there is no experimental evidence, however.

Another way to rationalize a proton stoichiometry other than unity would be to consider systems like the ATP synthase complex (Hatefi, 1993) and the nicotinamide nucleotide transhydrogenase (Hatefi and Yamaguchi, 1996). Neither contains a proton carrier like FMN or ubiquinone, which upon reduction would be able to accept protons from one side of the membrane and upon oxidation would deliver them to the opposite side. In the ATP synthase and the transhydrogenase, substrate-binding energy alters the conformation of the enzyme, thereby resulting in proton translocation, presumably by pK_a changes of appropriate amino acid residues on opposite side of the membrane. This mechanism has the advantage that in principle it can be applied also to Na^+ translocating enzymes (Pfenninger-Li *et al.*, 1996), whereas mechanisms involving FMN and ubiquinone as proton carriers cannot. Considering the possibility of energy transduction via protein conformation change, we have shown that in energy-promoted reverse electron transfer from succinate to NAD, increase of proton-motive force in SMP increases the affinity of complex I (V_{max}/K_m) for NAD by 20-fold (Hatefi *et al.*, 1982). We have also shown that reduction of bovine complex I by NADH or NADPH results in changes in the extent (increase or decrease) of cross-linking between the 51-kDa subunit of FP and the 75-kDa subunit of IP, among the IP subunits, and between the IP and HP subunits (Belogrudov and Hatefi, 1994). Whether the energy-promoted increase in affinity of complex I for NAD and the cross-linking changes that result from the reduction of complex I by NAD(P)H and involve the subunits of FP, IP, and HP are indicative of confirmational coupling remains to be seen.

3. COMPLEX II (SUCCINATE–QUINONE OXIDOREDUCTASE)

Succinate–quinone oxidoreductases are membrane-bound enzymes involved in both the citric acid cycle and the terminal electron transport system of aerobic organisms. They are located in the mitochondrial inner membrane of eukaryotes and in the plasma membrane of prokaryotes, but they do not translocate protons. The counterparts of succinate–quinone reductases in anaerobic organisms and facultative aerobes are the quinol–fumarate reductases, which in composition and overall mechanism of action are very similar to succinate–quinone oxidoreductases. Recent, excellent reviews on quinol–fumarate reductases, which cannot be discussed here for lack of space, are available (Ackrell *et al.*, 1992; Hederstedt and Ohnishi, 1992).

3.1. Composition and Structure of Complex II

The mitochondrial succinate–ubiquinone oxidoreductase is composed of four subunits of molecular mass 70, 27, 15 and 13 kDa (Fig. 2) (Hatefi and Galante, 1980). These subunits are referred to as FP (flavoprotein) or A, IP (iron–sulfur protein) or B, C, and D, respectively. In certain species, complex II

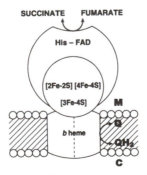

FIGURE 2. Topography, subunits, and the redox elements of complex II. M and C are the same as in Fig. 1. For details, see text.

is composed of only three subunits: A, B, and C (Hederstedt and Ohnishi, 1992). Complementary DNA sequence data are available for FP (Morris *et al.*, 1994; Birch-Machin *et al.*, 1992), IP (Au *et al.*, 1995; Kita *et al.*, 1990), and C (Cochran *et al.*, 1994) from mammalian sources. Also available are DNA sequences for the entire *sdh* operon from *E. coli* (four genes, *ABCD*) (Ackrell *et al.*, 1992) and *Bacillus subtilis* (three genes, *ABC*) (Hederstedt and Ohnishi, 1992).

The largest subunit, FP, is the site of dicarboxylic acid binding and contains a molecule of FAD, which is bound covalently to a histidyl residue of the protein as 8α-(*N*-3-histidyl)FAD. The IP subunit is considered to house three iron–sulfur clusters, a [2Fe-2S] cluster, a [4Fe-4S] cluster, and a [3Fe-4S] cluster, designated as centers S1, S2, and S3, respectively (Ohnishi, 1987). The midpoint potentials of these centers in bovine heart complex II have been estimated as zero mV for S1, -260 mV for S2, and $+60$ to $+120$ mV for S3 (Hederstedt and Ohnishi, 1992). Another redox component of complex II is cytochrome *b*. The cytochrome *b* of bovine heart complex II has been purified (Hatefi and Galante, 1980). The preparation is composed of the two small subunits C and D of complex II. The cytochrome *b* of *E. coli* complex II is also composed of subunits C and D (Ackrell *et al.*, 1992). In this case, it has been shown that the *b* heme is ligated by two axial histidyl residues, one contributed by subunit C and the other by subunit D. Complex II from *B. subtilis* contains two *b* hemes in subunit C, both hemes being ligated by axial histidyl residues (Hägerhäll and Hederstedt, 1996).

Structurally, complex II is made up of two domains: a water-soluble peripheral domain composed of FP and IP and a water-insoluble integral domain composed of subunits C and D (Fig. 2). FP and IP are easily separated from C and D with the use of chaotropic salts (Davis and Hatefi, 1971), which destructure water and as a consequence destabilize hydrophobic interactions (Hatefi and Hanstein, 1974). Together, FP and IP make up what is known as succinate dehydrogenase, which has been the subject of intensive past investigation (Singer *et al.*, 1973; Singer and Kearney, 1963). Models for the transmembrane topology of the *b* cytochromes of complex II of different organisms and the arrangements of the mono- and *bis*-hemes have been published (Hägerhäll and Hederstedt, 1996).

In bovine heart complex II, the interaction between the peripheral domain (succinate dehydrogenase) and the integral domain (cytochrome *b*) is easily altered by changing the medium water structure. As mentioned above, water structure-breaking chaotropic salts resolve complex II into the two domains. The extent of this resolution depends on the concentration and the potency of the chaotrope used; the removal of the chaotrope or addition to the medium of a water structure-forming ion, such as sulfate or phosphate, reverses the resolution. Similarly, suspension of complex II in ^2H_2O, which is more structured than H_2O, lends stability to complex II and makes it more resistant to resolution by chaotropes (Hatefi and Stiggall, 1976; Hatefi and Hanstein, 1974).

3.2. Enzymatic Properties

Long after the discovery of complex II, the peripheral domain of this enzyme, i.e., FP + IP or succinate dehydrogenase, was still considered a component enzyme of the citric acid cycle. It is now clear that succinate dehydrogenase, like the three-subunit FP segment of complex I, is not a discrete enzyme and does not catalyze a physiological reaction. Succinate dehydrogenase catalyzes the oxidation of succinate to fumarate in the presence of an artificial electron acceptor, such as ferricyanide or 5-N-methyl phenazonium sulfate (PMS). Nevertheless, this water-soluble preparation was very useful in early studies on the kinetics of oxidation of succinate to fumarate and the mode of action of various inhibitors (Singer and Kearney, 1963).

In the presence of ubiquinone-2 and 2,6-dichlorophenol indophenol as electron acceptors, bovine complex II catalyzes the oxidation of succinate at a rate of 180–200 sec^{-1} (Hatefi and Stiggall, 1976). This reaction is inhibited at the level of FP by oxaloacetate, malonate, other structural analogues of succinate, and by thiol modifiers. It is also inhibited between the iron–sulfur centers and ubiquinone by 2-thenoyltrifluoroacetone (TTFA), carboxins, and 2-alkyl-4,6-dinitrophenols (Yankovskaya et al., 1996). Among these, alkylating agents and oxaloacetate are the most potent. The former bind to a nonessential cysteine at the active site of FP. Oxaloacetate binds in a 1:1 stoichiometry to FP with a K_d of about 0.02 μM for the oxidized enzyme (Ackrell et al., 1992). This tight binding was thought to be due to interaction of oxaloacetate with the active site thiol to form a weak thiohemiacetal. However, this view was abandoned when it was found that the B. subtilis enzyme, which has an alanine in the position corresponding to the bovine cysteine, also binds oxaloacetate very tightly (Hederstedt and Henden, 1989).

Whereas the highly reactive cysteine of bovine succinate dehydrogenase is nonessential, there are two amino acid residues in the same region of FP that are essential. One is an arginine on the caboxyl side of the reactive cysteine (Arg255 in bovine FP), which has been suggested to form an ion pair with a substrate caboxyl group (Kotlyar and Vinogradov, 1984). In E. coli fumarate reductase, the absence of this arginyl residue (Arg248) nearly completely abolishes both fumarate reduction and succinate oxidation (Ackrell et al., 1992; Schröder et al., 1991). The other residue is a histidine (His239 in bovine FP) in a conserved His-Pro-Thr triad. The bovine enzyme was shown to be inhibited by treatment with the histidine-modifying reagents Rose Bengal and ethoxyformic anhydride (Hederstedt and Hatefi, 1986; Vik and Hatefi, 1981). The latter inhibition was partially reversed when the inhibited enzyme was treated with hydroxylamine, thus indicating that the inhibition involved modification of histidine. The effect of pH on succinate oxidation and fumarate reduction by the bovine enzyme showed mirror image profiles with pK_a near pH 7.0, which agreed with the possible involvement of a histidyl residue in catalysis. Because enzyme activity was highest in succinate oxidation at pH $\geqslant 7$ and in fumarate reduction at pH $\leqslant 7$, it was proposed that the unprotonated imidazole moiety of a histidyl residue may form a hydrogen bond with a methylene hydrogen of succinate, thus allowing electron rearrangement and labilization of a H: on the second methylene group and its transfer to FAD (Fig. 3). In the reverse direction, the protonated histidyl residue would donate a proton for the reduction of fumarate to succinate. These authors also predicted that in such a scheme the histidyl residue may play a less significant role in fumarate reduction as compared to succinate oxidation (Vik and Hatefi, 1981). This proposed mechanism was strongly supported by mutation of His232 to Ser in the His-Pro-Thr triad of *E. coli* fumarate reductase FP. It was shown that this H232S mutation abolished succinate oxidation, but had only a small effect on the fumarate reductase activity of the enzyme (Schröder *et al.*, 1991).

As mentioned above, the soluble bovine succinate dehydrogenase (FP + IP) reacts with PMS and ferricyanide as electron acceptors. The latter exhibits two types of binding as suggested by a high K_m and a low K_m (~250 μM). The low K_m ferricyanide reductase activity is unmasked when the soluble FP + IP are separated from the membrane-anchoring subunits C and D, and requires the intactness of iron–sulfur center S3 (Ackrell *et al.*, 1992; Hederstedt and Ohnishi, 1992; Vinogradov *et al.*, 1975). In the absence of succinate, center S3 in soluble succinate dehydrogenase is destroyed by oxygen, resulting in the loss of its EPR signal and in parallel losses of the low K_m ferricyanide reductase activity and the ability of the soluble enzyme to reconstitute succinate–ubiquinone reductase activity when combined with subunits C and D. It is therefore thought that center S3 is the immediate electron donor to ubiquinone.

EPR data have indicated the presence in complex II of two ubisemi-

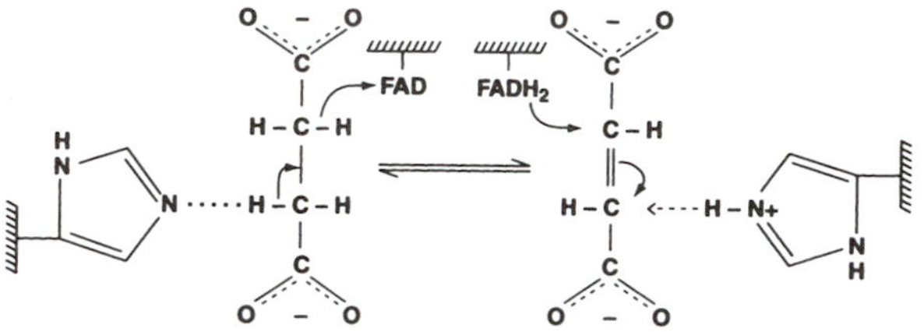

FIGURE 3. Proposed role of His239 in the conserved active-site triad His–Pro–Thr of succinate dehydrogenase in succinate oxidation and fumarate reduction. For details, see text and Vik and Hatefi (1981).

quinones (Salerno and Ohnishi, 1980; Ruzicka *et al.*, 1975), one of which (Q_A) is considered to be close to center S3, and the others (Q_B) about 8 Å away from Q_A in an edge to edge arrangement (Ruzicka *et al.*, 1975). By analogy to the mechanism of Q reduction in the reaction centers of plants and photosynthetic bacteria, it is thought that Q_A is the primary acceptor of electrons from the iron–sulfur clusters. These electrons are then passed singly to Q_B, which when fully reduced exchanges with oxidized ubiquinone in the Q-pool (Ackrell *et al.*, 1992).

3.3. Cytochrome *b* of Complex II

It was shown in 1972 that the cytochrome *b* of bovine complex II is distinct from the *b* cytochromes of complex III on the basis of its spectral properties (Davis *et al.*, 1973). Subsequently, it was demonstrated in *N. crassa* that the cytochrome *b* of complex II was a cytoribosomal product (Weiss and Kolb, 1979), whereas those of complex III (two hemes bound to a single polypeptide) in *Neurospora* and in *Saccharomyces cerevisiae* were known to be encoded by the mitochondrial DNA. The *b* cytochrome of bovine complex II was purified in 1980, and was shown to consist of subunits C and D (Hatefi and Galante, 1980). This preparation recombined with pure succinate dehydrogenase in a 1:1 molar ratio to reconstitute a highly active succinate–ubiquinone oxidoreductase. Another preparation of bovine subunits C + D was also reported in the same year (Ackrell *et al.*, 1980). This preparation was essentially devoid of heme, but nevertheless recombined with soluble succinate dehydrogenase to reconstitute appreciable quinone reductase activity.

The E_m of the cytochrome *b* of bovine complex II has been reported to be −185 mV (Yu *et al.*, 1987). Because of this low E_m, succinate reduces the *b* heme to a very small extent (Hatefi and Galante, 1980). However, when chemically reduced with $Na_2S_2O_4$, the *b* heme of bovine complex II can be oxidized rapidly by fumarate or ubiquinone-2. The reoxidation of the reduced *b* was shown to be inhibited by the mercurial reagent, mersalyl, when the electron acceptor was fumarate, but not when it was ubiquinone. The latter reaction was also not inhibited by TTFA (Hatefi and Galante, 1980). These data suggested that (1) the *b* heme is not required for electron transfer from succinate to ubiquinone, and (2) the *b* heme is in electronic communication with FP as well as with ubiquinone, the latter via a path not inhibitible by TTFA. Point (1) is in agreement with the more recent finding that the *E. coli* quinol–fumarate reductase is devoid of a *b* heme (Cecchini *et al.*, 1986). By contrast, the three-subunit *B. subtilis* succinate–quinone (menaquinone) oxidoreductase (Hederstedt and Ohnishi, 1992) and the three-subunit *Wolinella succinogenes* menaquinol–fumarate oxidoreductase (Körtner *et al.*, 1990) contain two *b* hemes each in their respective subunit C. The E_m values of the hemes in the *B.*

subtilis enzyme are $+65$ and -95 mV, and in the *W. succinogenes* enzyme -20 and -200 mV (Hederstedt and Ohnishi, 1992). In these and other enzymes also, where the E_m values of the substrate couple (whether succinate–fumarate or quinol–quinone) are not too unfavorable with respect to that of one or both *b* hemes, the heme is partially or completely reducible by the substrate. Therefore, it might be considered that the *b* hemes have (or have had at some stage in evolution) a role in these enzymes, possibly as entry points for electrons from an as yet unknown ancillary redox system.

4. COMPLEX III (QUINOL–CYTOCHROME *c* OXIDOREDUCTASE)

Quinol–cytochrome *c* (c_2, plastocyanine) oxidoreductases are present in the respiratory chains of all aerobic organisms as well as in the bacterial and plant photosynthetic electron transfer systems (Brandt and Trumpower, 1994; Gennis *et al.*, 1993). In mitochondria, complex III (bc_1 complex) is the recipient of electrons via the Q-pool from complexes I and II, the fatty acid oxidation system, and α-glycerophosphate dehydrogenase. These electrons are then passed on to cytochrome *c* in a process that is coupled to outward proton translocation with a stoichiometry of $H^+/e = 2$.

4.1. General Features

The bovine mitochondrial complex III is composed of 11 unlike polypeptides. It is dimeric with a monomer molecular mass of 240 kDa (Hatefi, 1985). The enzymes from *P. denitrificans* and *Rhodobacter capsulatus* have three subunits each (Gennis *et al.*, 1993; Yang and Trumpower, 1986) and that from *Rhodobacter sphaeroides* has four subunits (Yu and Yu, 1991; Andrews *et al.*, 1990). Common to these various forms of complex III are three redox proteins, a *bis*-heme cytochrome *b*, a [2Fe-2S] iron–sulfur protein, and cytochrome c_1 (or cytochrome *f* in chloroplasts). Cytochrome *b* is largely membrane intercalated, whereas the iron–sulfur protein and cytochrome c_1 are largely extramembranous on the cytosolic side of the inner membrane. In bovine complex III, the two *b* hemes have different midpoint potentials. The low-potential heme, b_L, has an E_m of ~ -30 mV and is located near the cytosolic surface of the mitochondrial inner membrane. The high-potential heme, b_H, has an E_m of $\sim +90$ mV, and is located 20 Å away from b_L near the center of the membrane (Ohnishi *et al.*, 1989). The iron–sulfur protein, abbreviated hereafter as ISP, has an E_m of $+280$ mV and an EPR signal centered at a g of 1.90. One of its two iron atoms is ligated to two cysteine sulfurs, and the other to two histidines via their δ imidazole nitrogens (Britt *et al.*, 1991; Gurbiel *et al.*, 1989, 1991). Cytochrome c_1 has an $E_{m,7}$ of $+230$ mV. X-ray diffraction data of crystallized bovine complex III dimer to a resolution of 2.9 Å have been published, showing among other things the positions of hemes b_H, b_L, and c_1, as well as the FeS cluster of ISP and the binding sites of antimycin and myxothiazol (Xia et al., 1997). Heme b_L is 21 Å apart from b_H, and 27 Å away from ISP. The latter is 31 Å away from c_1 iron. The crystals contain 0.75–0.85 mole Q_{10}/mole c_1 or b, but the location of Q has not been shown.

Complex III was originally purified from bovine heart mitochondria, using bile salts as detergents (Hatefi et al., 1962c). In the presence of ubiquinol-2 as electron donor, this preparation reduced cytochrome c at a rate of 2500–3000 sec^{-1} at 38°C (Rieske, 1967, 1976). Several different procedures were subsequently devised, which employed Triton X-100. These preparations exhibited very low activities (see Table 1 in Hauska et al., 1983), whereas a more recent procedure, which used dodecyl maltoside as detergent, resulted in a more active enzyme (Ljungdahl et al., 1987). Among other possible deleterious effects, Triton X-100 labilizes the ISP of complex III (Engel et al., 1980, 1983; von Jagow et al., 1978). It is therefore important, especially with complex enzyme systems, that structural and mechanistic studies are done only with preparations that have best preserved the turnover capability of the enzyme in the unfractionated membrane.

4.2. Mechanisms of Electron Transfer and Proton Translocation

Among several mechanisms that have been proposed for electron transfer and proton translocation by complex III, the Q-cycle hypothesis, proposed by Mitchell (1975, 1976), is now accepted by most workers in the field. This mechanism, which will be discussed in greater detail elsewhere in this volume, has four principal steps (see Fig. 4):

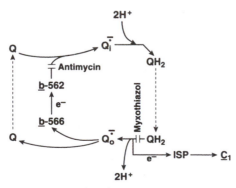

FIGURE 4. The proton-motive Q-cycle (after Trumpower, 1990). Hemes b_{562} and b_{566} are b_H and b_L, respectively. For details, see text.

1. Single electron transfer from QH_2 (lower right in Fig. 4) to ISP and cytochrome c_1, and translocation outward of the two protons of QH_2.
2. Transfer of the second electron from the ubisemiquinone product of step 1 to heme b_L (b_{566}).
3. Transfer of electron from b_L to b_H (b_{562}).
4. Transfer of electron from b_H to Q (upper center in Fig. 4) to form ubisemiquinone.

A second cycle of single-electron transfer through the b hemes to Q and acquisition of two protons from the inside reforms QH_2 (upper right in Fig. 4). The net result in two cycles is the transfer of two electrons through complex III to cytochrome c and cytochrome oxidase and the translocation of $4H^+$ from the inside of the membrane (e.g., mitochondrial matrix) to the outside (e.g., mitochondrial intermembrane space). Two sets of inhibitors are known, which appear from the study of bacterial-resistant mutants to bind near the inside and the outside surfaces of cytochrome b (Brandt and Trumpower, 1994; Gennis *et al.*, 1993; Link *et al.*, 1993; von Jagow and Link, 1986). Those binding on the inside, as represented by antimycin, are considered to inhibit the single-electron reduction of Q by reduced b_H. Those binding on the outside, as represented by myxothiazol, are considered to block the oxidation of QH_2 by ISP, as shown in Fig. 4.

As will be discussed elsewhere in this volume, the Q-cycle hypothesis, with its assigned sites of inhibition by antimycin and myxothiazol, agrees with many experimental observations. However, it is important to note that there is as yet no direct and unequivocal proof for any one of the four principal steps of the Q-cycle enumerated above, and there is as yet no evidence, direct or indirect, for the postulate that Q is the vehicle for proton translocation from one side of the membrane to the other. Furthermore, as seen in Fig. 4, the role of cytochrome b in the Q-cycle is to recycle an electron back to the Q-pool. One may well wonder whether this is not too insignificant a task for the only transmembranous redox protein of complex III with its elaborate system of two b hemes of widely different E_m values.

These and other concerns plus an observation made long ago that could not be rationalized in terms of the Q-cycle led us to the following recent findings. It is well known that antimycin and myxothiazol are stoichiometric inhibitors of complex III, each exerting its highest degree of inhibition at 1 mole/mole of complex III monomer. Phenomenologically, however, antimycin and myxothiazol (as well as other inhibitors that bind near the inside or the outside surface of cytochrome b) inhibit three steps each in the redox reactions of the *bis*-heme cytochrome b, and all three inhibitions are incomplete, allowing electrons to leak through at various slow rates (Matsuno-Yagi and Hatefi, 1996):

1. In submitochondrial particles treated with NADH or succinate, antimycin or myxothiazol inhibits cytochrome b oxidation.
2. When both inhibitors are added, the reduction of b is also inhibited. However, because both inhibitions leak when both antimycin and myxothiazol are added, cytochrome b becomes slowly reduced, indicating that these reagents inhibit the oxidation of b more strongly than its reduction.
3. The slow reoxidation of b through the leak allowed by either antimycin or myxothiazol is biphasic, with b_L undergoing reoxidation at least 10 times faster than b_H.

These results are difficult to rationalize in terms of the Q-cycle, which proposes that (1) antimycin and myxothiazol inhibit a single step each in the electron transfer system of complex III, and (2) the direction of electron transfer is from b_L to b_H. According to our results, antimycin or myxothiazol inhibits electron transfer most from b_H to b_L, next from b_L to ISP/c_1, and least from QH_2 to the b hemes. One could rationalize this pleiotropic effect of these stoichiometric inhibitors by considering the possibility that each inhibitor alters the conformation of the cytochrome b molecule, thus restricting the reactivities of the b hemes in different ways. This possibility of inhibitor-induced conformation change is supported by experimental evidence (McCurley et al., 1990; Rieske et al., 1967). In addition, it was shown in Q-extracted bovine SMP, in which the molar ratio of Q to complex III monomer was reduced from 12.5 to ~0.06, as well as in SMP from a Q-deficient yeast mutant, that cytochrome b could be reduced by reverse electron transfer via ISP/c_1 (Matsuno-Yagi and Hatefi, 1996). This finding also could not be rationalized in terms of the Q-cycle, which places Q as an obligatory electron carrier between b and ISP/c_1. These recent results are summarized in the scheme shown in Fig. 5, where I_1 and I_2 are the two classes of inhibitors and the transmembranous cytochrome b molecule is proposed to act as a proton pump.

Among the peculiar electron transfer features of complex III is the phenomenon known as the oxidant-induced transient extra reduction of cytochrome b (Chance, 1974; Rieske, 1971). In early studies of complex III, it was shown that in antimycin-treated preparations containing prereduced ISP/c_1, cytochrome b was slowly and incompletely reduced by QH_2. When ferricyanide was added to reoxidize ISP/c_1, there followed a transient rapid further reduction of cytochrome b. This phenomenon is explained by the Q-cycle, which requires an oxidized ISP to accept an electron from QH_2, thereby resulting in the formation of ubisemiquinone, which is the electron donor to b_L (see Fig. 4). We have recently shown in antimycin-treated SMP that prereduction of ISP/c_1 does not inhibit the subsequent rapid reduction of b_H by substrates. Only b_L reduction under these conditions becomes slow and partial.

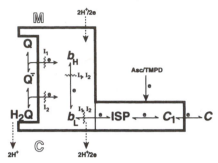

FIGURE 5. Scheme showing the electron transfer pathways in complex III that are inhibited by the stoichiometric inhibitors antimycin and myxothiazol. These inhibitors are designated in the figure as I_1 and I_2 without specifying which is antimycin and which myxothiazol. The scheme also shows that reduction of b via ISP/c_1 by ascorbate + TMPD (N,N,N',N'-tetramethyl-p-phenylenediamine) does not require ubiquinone. M and C are the same as in Fig. 1. From Matsuno-Yagi and Hatefi (1996), with permission.

In addition, we have shown that membrane energization by addition of ATP also results in incomplete reduction of cytochrome b by either NADH or succinate. At this point, deenergization of SMP by addition of an uncoupler or oligomycin to inhibit ATP hydrolysis results in further reduction of cytochrome b. Furthermore, this deenergization-induced extra reduction of cytochrome b takes place regardless of whether ISP/c_1 have been prereduced (Matsuno-Yagi and Hatefi, 1997). The author will not expand further on these complicated data. Suffice it to say, however, that these recent results do not fit the Q-cycle either, and emphasize the need for further study before a scheme for electron transfer and proton translocation by complex III can be devised that would rationalize all the experimental observations.

5. COMPLEX V (ATP SYNTHASE COMPLEX)

The ATP synthase complexes of mitochondria, chloroplasts, and bacterial plasma membranes are mushroomlike tridomain structures, composed of a globular headpiece (F_1), which is the catalytic sector; a membrane-bound base piece (F_O), which is concerned with transmembrane proton translocation and transduction of the proton-motive force into protein conformational energy; and a stalk that connects F_1 to F_O and is concerned with conformational energy transfer between them. This conformational energy is used by the catalytic β subunits of F_1 to alter the binding energies of ATP and ADP + P_i on different β subunits (Hatefi *et al.*, 1982), resulting in the tight binding and dehydration of ADP + P_i, and in the release of the ATP so formed (Boyer, 1993; Hatefi, 1993; Boyer *et al.*, 1973).

5.1. Composition and Structure

Complex V from bovine heart mitochondria is composed of 16 unlike subunits (Belogrudov *et al.*, 1996; Collinson *et al.*, 1994), twice as many as the subunits of the ATP synthase from *E. coli* (Table II). Isolated F_1 from various species contains five unlike subunits with the stoichiometry $\alpha_3\beta_3\gamma\delta\varepsilon$ (Penefsky and Cross, 1991). F_1 from rat liver (Amzel *et al.*, 1992) and bovine heart have been crystallized, and the structure of a major portion of the latter ($\alpha_3\beta_3\gamma$) has been resolved to 2.8 Å (Abrahams *et al.*, 1994). The crystals contained one molecule of 5-adenylyl-imidodiphosphate (AMP-PNP) bound to each α and to one of the three β subunits, a second β contained ADP, and the third β was empty. Thus, the three β subunits differed in conformation and faced different aspects of γ, which runs through the length of the F_1 core and into the stalk as two loosely entwined α-helices. The authors suggest that the different conformations of the three β subunits represent an instant in the catalytic cycle where one β is empty and ready to receive substrate, the second β contains ADP + P_i,

Table II
Bovine Heart Mitochondrial ATP Sythase Subunits

Subunit[a]	*E. coli* equivalent	Molecular mass (kDa)	Mole/mole enzyme
α	α	55.2	3
β	β	51.6	3
γ	γ	30.1	1
δ	ε	15.1	1
ε		5.7	1
IF_1		9.6	1
b	b	24.7	1 or 2
OSCP	δ	21.0	1
d		18.6	1
F_6		9.0	1 or 2
A6L		8.0	1
a	a	24.8	1
c	c	7.4	Several
e		8.2	?
f		10.2	?
g		11.3	?

[a]IF_1, ATPase inhibitor protein. Subunits a and c are mtDNA encoded.

and the third β contains *de novo* synthesized ATP (Abrahams *et al.*, 1994; see also Cross, 1992).

The structures of the remaining 13 subunits of the bovine ATP synthase are not known. However, nuclear magnetic resonance data regarding the solution structures of subunits ε and *c* of the *E. coli* enzyme are available. The *E. coli* ε subunit, which is considered to be analogous to the bovine δ subunit, is made up of two domains. The NH_2-terminal 84 amino acids form a flattened, 10-stranded β-barrel, and the COOH-terminal 48 residues form two antiparallel, hairpinlike α-helices. The latter interacts with the αβ subunits in the vicinity of the so-called DELSEED (single-letter designations of amino acids) region of β and the corresponding region of α. The open end of the β-barrel interacts with F_O near the extramembranous loop of subunits *c*, and a lateral face of the β-barrel interacts with γ (Wilkens *et al.*, 1995).

There are in the *E. coli* ATP synthase 9–12 copies of subunit *c*. This highly hydrophobic protein has a hairpinlike structure, with the two arms of the hairpin forming antiparallel α-helices that traverse the plasma membrane and extend slightly out on the periplasmic side (Fillingame, 1996; Girvin and Fillingame, 1993, 1994, 1995). The bend of the hairpin is hydrophilic and protrudes from the F_1 side of the membrane, where it interacts with γ and ε. There is a DCCD-reactive carboxyl group (Asp61 in *E. coli*, Glu58 in bovine subunit *c*) near the center of the membrane in helix 2, which is located in a hydrophobic pocket, has a high pK_a of 7.1 (hence the DCCD reactivity), and is essential for proton translocation (Assadi-Porter and Fillingame, 1995).

In the *E. coli* ATP synthase, F_O contains, in addition to the multiple copies of subunit *c*, subunit *a*, which traverses the membrane several times, and portions of two copies of subunit *b* (Fillingame, 1996). This subunit appears to have a short hydrophilic NH_2-terminus, which is exposed on the F_1 side. It then traverses the membrane twice, emerging once again on the F_1 side. The two *b* subunits are considered to act as an F_1 stator by binding at one end to subunit *a* in the membrane and at the other to δ, which appears to be bound to the top outside surface of one of the α subunits (Ogilvie *et al.*, 1997; Zhou *et al.*, 1997).

In the bovine ATP synthase, the stator may contain oligomycin sensitivity-conferring protein (OSCP), *d*, F_6, the COOH-terminal half of A6L, and the COOH-terminal 130 amino acid residues of *b*. The remainder of *b* appears to be anchored to the membrane via two hydrophobic α-helices as in the *E. coli* subunit *b*, with the NH_2-terminal 30 residues exiting the membrane on the F_1 side. The near-neighbor relationships of the aforementioned subunits of bovine ATP synthase have been studied in SMP by cross-linking. Cross-linked products identified by immunoblotting with subunit-specific antibodies have suggested the following. Subunits *b*, OSCP, *d*, and F_6 interact with the α/β subunits of F_1, but not with γ or δ. OSCP, *d*, and F_6 interact with *b*, but not with one another. The extramembranous part of A6L interacts with subunit *d*. In F_1-

depleted SMP, d cross-links to F_6, and a product suggestive of a $b-b$ dimer also appears (Belogrudov et al., 1995).

Bovine F_O contains subunit a, multiple copies of c, the NH_2-terminal half of A6L, the hydrophobic α-helices of b, and possibly the short hydrophobic NH_2-terminus of d. In addition, F_O contains portions of subunits e, f, and g. The COOH-termini of these subunits are extramembranous on the cytosolic side of the inner membrane, and the NH_2-termini of f and g protrude from the matrix side (Belogrudov et al., 1996). Cross-linking studies in SMP have shown the following products involving these small subunits: A6L-f, f-g, and g-e (Belogrudov et al., 1996).

5.2. Enzymatic Properties

In the process of ATP synthesis or hydrolysis, the three β subunits of F_1 display negative cooperativity in their substrate affinity and positive catalytic cooperativity in the sense that substrate occupation of the second and third catalytic sites greatly increases the turnover rate (Hatefi, 1993). In ATP hydrolysis, when ATP is added (in the presence of Mg^{2+}) at a concentration substoichiometric with respect to isolated bovine F_1, the ATP binds very tightly to a single catalytic site ($K_a = 10^{12}M^{-1}$), is hydrolyzed to enzyme-bound ADP and P_i with an equilibrium constant $K = 0.5$, and the products are released at the reported slow rate of 4×10^{-3} sec^{-1} (Penefsky and Cross, 1991). When the MgATP concentration is raised to millimolar, this slow unisite rate is converted to the multisite rate of 600 sec^{-1}, with an apparent K_m^{ATP} in the range of 10^{-4} to 10^{-3} M. The unisite and multisite rates of ATP hydrolysis by the ATP synthase of bovine SMP are, respectively, 0.12 and 400–500 sec^{-1} (Hatefi, 1993; Matsuno-Yagi and Hatefi, 1993a).

Data for ADP binding (in the presence of saturating concentrations of P_i and Mg^{2+}) and ATP synthesis by bovine heart SMP are summarized in Fig. 6. It is seen that ADP binds to the highest and the intermediate affinity sites with K_d values, respectively, of $<10^{-8}$ and $\sim10^{-7}$ M, with no or an extremely slow rate of ATP production. Only after the ADP concentration is raised to fill the

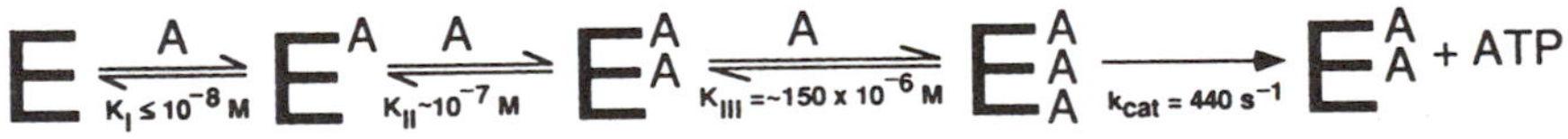

FIGURE 6. Scheme showing the dissociation constants K_I, K_{II}, and K_{III} for ADP (A) binding to the ATP synthase (E) of bovine heart SMP in the process of ATP synthesis. K_{III} is equivalent to K_m for ADP. The last step shows k_{cat}, which is the turnover number of the ATP synthase of bovine heart SMP for ATP synthesis at saturating $[P_i]$, $[Mg^{2+}]$ and proton-motive force. For details, see Matsuno-Yagi and Hatefi (1988, 1990) and Hatefi (1993).

lowest affinity site (see K_m in Fig. 6) does rapid ATP production ensue (Hatefi, 1993). The values of K_m and k_{cat} depend on the magnitude of proton-motive force (PMF), which appears to behave as a Michaelis substrate for the ATP synthase (Hatefi, 1993). At saturating PMF and P_i, K_m for ADP (equivalent to K_{III}) is 120–160 μM, and k_{cat} is 440 sec^{-1} for bovine heart SMP. At very low PMF, K_m^{ADP} reaches a low limit of 2–4 μM, and k_{cat} is ~8 sec^{-1}.

At high PMF, addition of venturicidin or organotin compounds to SMP converts the high K_m^{ADP} and high k_{cat} in ATP synthesis to low K_m^{ADP} and low k_{cat} values (Matsuno-Yagi and Hatefi, 1993b). This is because the inhibition of F_0 by these reagents is incomplete, allowing a slow rate of proton flow through F_0. In multisite ATP hydrolysis, this incomplete inhibition also converts the kinetics of the reaction to a slow process with a k_{cat} of ~1.5 sec^{-1} and apparent $K_m^{ATP} = 0.25$ μM (Matsuno-Yagi and Hatefi, 1993a). In ATP synthesis, it has been possible to show that the PMF-induced changes in the kinetics of oxidative phosphorylation occur at constant V_{max}/K_m^{ADP}, thus suggesting that the K_m^{ADP} changes are mainly a function of PMF-induced changes in k_{cat} (Hatefi, 1993).

5.3. Conformational Energy Transfer in the ATP Synthase

In bovine SMP, it has been shown that modification of F_0 by oligomycin or DCCD greatly decreases the affinity of F_1 for ATP at a distance of about 100 Å (Matsuno-Yagi et al., 1985; Penefsky, 1985). These results illustrated in principle how conformation changes induced in F_0 by protonation could affect the release of de novo synthesized and tightly bound ATP from F_1.

The path of conformational energy transfer between F_1 and F_0 has been investigated in the E. coli ATP synthase in a series of elegant experiments, mainly in the laboratories of Capaldi, Cross, and Fillingame. It was initially shown that Ser108 in the loop between the two α-helices of the ε subunit could be cross-linked to Glu391 in the DELSEED region of β by the water-soluble carbodiimide, 1-ethyl-3-[3(dimethylamino)propyl]carbodiimide (Mendel-Hartvig and Capaldi, 1991). Subsequently, Cys residues were introduced by mutation in appropriate positions on subunits α, β, ε, and c to achieve intersubunit proximity of the Cys thiols. Addition of an oxidizing agent resulted in the formation of a disulfide cross-link between the extramembranous hydrophilic loop of subunit c and the bottom of the barrel of ε (Zhang and Fillingame, 1995), between a Cys at position 108 of ε and a Cys at position 381 of β in the DELSEED region or at position 411 of α in a corresponding region (Haughton and Capaldi, 1995; Wilkens et al., 1995), as well as between a Cys at position 380 of β in the DELSEED region and the wild-type Cys87 of the γ subunit (Duncan et al., 1995). These results indicated that the bottom of the 10-stranded β-barrel of the ε subunit contacts the extramembranous loop of subunit c, and the loop between the two helices at the other end of ε contacts the DELSEED region of β or the corresponding region of α. Furthermore, the DELSEED region of β also comes into contact with the γ subunit in the core of F_1. The facts that these cross-links were often substrate dependent and inhibitory and that the inhibition was reversed upon cleavage of the disulfide cross-link by reduction to dithiol indicated that ATP hydrolysis required freedom of motion or conformational change of the regions of the enzyme molecule that were immobilized by intersubunit cross-linking. In agreement with these findings, it was shown that γ and ε undergo substrate-dependent positional changes relative to the α and β subunits (Aggeler and Capaldi, 1996; Capaldi *et al.*, 1994; Gogol *et al.*, 1990). Indeed, the cross-linking experiments involving γ and β suggested that γ may rotate relative to the three β subunits during ATP hydrolysis, and that during catalysis γ interacts equally with all three β subunits (Duncan *et al.*, 1995).

This conclusion agrees with our results for ATP synthesis (Hatefi, 1993) and those of Senior and co-workers for ATP hydrolysis (Weber and Senior, 1996; Weber *et al.*, 1993) that rapid enzyme turnover involves the participation of all three β subunits in catalysis. More recently, the ingenious experiments of Noji *et al.* (1997) and Sabbert *et al.* (1997) have established that indeed subunit γ rotates within F_1 during ATP hydrolysis, in complete agreement with the mechanism earlier proposed by Boyer (1997).

ACKNOWLEDGMENTS. The work of the author's laboratory was supported by the United States Public Health Service Grants DK08126 and GM24887. This is publication 10364-MEM from The Scripps Research Institute, La Jolla, California.

6. REFERENCES

Abrahams, J. P., Leslie, A. G. W., Lutter, R., and Walker, J. E., 1994, *Nature* **370**:621–628.

Ackrell, B. A. C., Ball, M. B., and Kearney, E. B., 1980, *J. Biol. Chem.* **255**:2761–2769.

Ackrell, B. A. C., Johnson, M. K., Gunsalus, R. P., and Cecchini, G., 1992, in: *Chemistry and Biochemistry of Flavoenzymes*, vol. 3 (F. Müller, ed.), pp. 229–297, CRC Press, London.

Aggeler, R., and Capaldi, R. A., 1996, *J. Biol. Chem.* **271**:13888–13891.

Amzel, L. M., Bianchet, M. A., and Pedersen, P. L., 1992, *J. Bioenerg. Biomembr.* **24**:429–433.

Andrews, K. M., Crofts, A. R., and Gennis, R. B., 1990, *Biochemistry* **29**:2645–2651.

Assadi-Porter, F. M., and Fillingame, R. H., 1995, *Biochemistry* **34**:16186–16193.

Au, H. C., Ream-Robinson, D., Bellew, L. A., Broomfield, P. L. E., Saghbini, M., and Scheffler, I. E., 1995, *Gene* **159**:249–253.

Belogrudov, G., and Hatefi, Y., 1994, *Biochemistry* **33**:4571–4576.

Belogrudov, G. I., and Hatefi, Y., 1996, *Biochem. Biophys. Res. Commun.* **227**:135–139.

Belogrudov, G. I., Tomich, J. M., and Hatefi, Y., 1995, *J. Biol. Chem.* **270**:2053–2060.

Belogrudov, G. I., Tomich, J. M., and Hatefi, Y., 1996, *J. Biol. Chem.* **271**:20340–20345.

Birch-Machin, M. A., Farnsworth, L., Ackrell, B. A. C., Cochran, B., Jackson, S., Bindoff, L. A., Aitken, A., Diamond, A. G., and Turnbull, D. M., 1992, *J. Biol. Chem.* **267**:11553–11558.

Blattner, F. R., Plunkett, G., III, Bloch, C. A., Perna, N. T., Burland, V., Riley, M., Collado-Vides, J., Glasner, J. D., Rode, C. K., Mayhew, G. F., Gregor, J., Davis, N. W., Kirkpatrick, H. A., Goeden, M. A., Rose, D. J., Mau, B., and Shao, Y., 1997, *Science* **277**:1453–1462.

Boyer, P. D., 1993, *Biochim. Biophys. Acta* **1140**:215–250.

Boyer, P. D., 1997, *Annu. Rev. Biochem.* **66**:717–749.

Boyer, P. D., Cross, R. L., and Momsen, W., 1973, *Proc. Natl. Acad. Sci. USA* **70**:2837–2839.

Brandt, U., and Trumpower, B., 1994, *Crit. Rev. Biochem. Mol. Biol.* **29**:165–197.

Britt, R. D., Sauer, K., Klein, M. P., Knaff, D. B., Kriauciunas, A., Yu, C.-A., Yu, L., and Malkin, R., 1991, *Biochemistry* **30**:1892–1901.

Capaldi, R. A., Aggeler, R., Turina, P., and Wilkens, S., 1994, *Trends Biochem. Sci.* **19**:284–289.

Cecchini, G., Ackrell, B. A. C., Deshler, J. D., and Gunsalus, R., 1986, *J. Biol. Chem.* **261**:1808–1814.

Chance, B., 1974, in: *Dynamics of Energy-Transducing Membranes* (L. Ernster, R. W. Estabrook, and E. C. Slater, eds.), pp. 553–578, Elsevier Science, Amsterdam.

Chomyn, A., Mariottini, P., Cleeter, M. W. J., Ragan, C. I., Matsuno-Yagi, A., Hatefi, Y., Doolittle, R. F., and Attardi, G., 1985, *Nature* **314**:592–597.

Chomyn, A., Cleeter, M. W. J., Ragan, C. I., Riley, M., Doolittle, R. F., and Attardi, G., 1986, *Science* **234**:614–618.

Cochran, B., Capaldi, R. A., and Ackrell, B. A. C., 1994, *Biochim. Biophys. Acta* **1188**:162–166.

Collinson, I. R., Runswick, M. J., Buchanan, S. K., Fearnley, I. M., Skehel, J. M., Van Raaij, M. J., Griffiths, D. E., and Walker, J. E., 1994, *Biochemistry* **33**:7971–7978.

Crane, F. L., Hatefi, Y., Lester, R. L., and Widmer, C., 1957, *Biochim. Biophys. Acta* **25**:220–221.

Cross, R. L., 1992, in: *Molecular Mechanisms in Bioenergetics* (L. Ernster, ed.), pp. 317–330, Elsevier, Amsterdam.

Davis, K. A., and Hatefi, Y., 1971, *Biochemistry* **10**:2509–2516.

Davis, K. A., and Hatefi, Y., Poff, K. L., and Butler, W. L., 1973, *Biochim. Biophys. Acta* **325**:341–356.

Degli Esposti, M., Ghelli, A., Ratta, M., Cortes, D., and Estornell, E., 1994, *Biochem. J.* **301**:161–167.

Deng, P. S. K., Hatefi, Y., and Chen, S., 1990, *Biochemistry* **29**:1094–1098.

Duncan, T. M., Bulygin, V. V., Zhou, Y., Hutcheon, M. L., and Cross, R. L., 1995, *Proc. Natl. Acad. Sci. USA* **92**:10964–10968.

Earley, F. G. P., Patel, S. D., Ragan, C. I., and Attardi, G., 1987, *FEBS Lett.* **219**:108–113.

Engel, W. D., Schägger, H., and von Jagow, G., 1980, *Biochim. Biophys. Acta* **592**:211–222.

Engel, W. D., Michalski, C., and von Jagow, G., 1983, *Eur. J. Biochem.* **132**:395–402.

Fearnley, I. M., and Walker, J. E., 1992, *Biochim. Biophys. Acta* **1140**:105–134.

Fecke, W., Sled', V. D., Ohnishi, T., and Weiss, H., 1994, *Eur. J. Biochem.* **220**:551–558.

Fillingame, R. H., 1996, *Curr. Opin. Struct. Biol.* **6**:491–498.

Finel, M., 1993, *J. Bioenerg. Biomembr.* **25**:357–366.

Galante, Y. M., and Hatefi, Y., 1979, *Arch. Biochem. Biophys.* **192**:559–568.

Galante, Y. M., Wong, S.-Y., and Hatefi, Y., 1979, *J. Biol. Chem.* **254**:12372–12378.

Gennis, R. B., Barquera, B., Hacker, B., Van Doren, S. R., Arnaud, S., Crofts, A. R., Davidson, E., Gray, K. A., and Daldal, F., 1993, *J. Bioenerg. Biomembr.* **25**:195–210.

Girvin, M. E., and Fillingame, R. H., 1993, *Biochemistry* **32**:12167–12177.

Girvin, M. E., and Fillingame, R. H., 1994, *Biochemistry* **33**:665–674.

Girvin, M. E., and Fillingame, R. H., 1995, *Biochemistry* **34**:1635–1645.

Gogol, E. P., Johnston, E., Aggeler, K., and Capaldi, R. A., 1990, *Proc. Natl. Acad. Sci. USA* **87**:9585–9589.

Gurbiel, R. J., Batie, C. J., Sivaraj, M., True, A. E., Fee, J. A., Hoffman, B. M., and Ballou, D., 1989, *Biochemistry* **28**:4861–4871.

Gurbiel, R. J., Ohnishi, T., Robertson, D. E., Daldal, F., and Hoffman, B. M., 1991, *Biochemistry* **30**:11579–11584.

Hägerhäll, C., and Hederstedt, L., 1996, *FEBS Lett.* **389**:25–31.

Hatefi, Y., 1958, *Biochim. Biophys. Acta* **30**:648–650.

Hatefi, Y., 1963, *Adv. Enzymol.* **25**:275–328.

Hatefi, Y., 1985, *Annu. Rev. Biochem.* **54**:1015–1069.

Hatefi, Y., 1993, *Eur. J. Biochem.* **218**:759–767.

Hatefi, Y., and Galante, Y. M., 1980, *J. Biol. Chem.* **255**:5530–5537.

Hatefi, Y., and Hanstein, W. G., 1974, *Methods Enzymol.* **31**:770–790.

Hatefi, Y., and Stempel, K. E., 1967, *Biochem. Biophys. Res. Commun.* **26**:301–308.

Hatefi, Y., and Stiggall, D. L., 1976, in: *The Enzymes*, vol. 13 (P. D. Boyer, ed.), pp. 175–297, Academic Press, New York.

Hatefi, Y., and Yamaguchi, M., 1996, *FASEB J.* **10**:444–452.

Hatefi, Y., Haavik, A. G., and Griffiths, D. E., 1961, *Biochem. Biophys. Res. Commun.* **4**:441–446, 447–453.

Hatefi, Y., Haavik, A. F., Fowler, L. R., and Griffiths, D. E., 1962a, *J. Biol. Chem.* **237**:2661–2669.

Hatefi, Y., Haavik, A. G., and Griffiths, D. E., 1962b, *J. Biol. Chem.* **237**:1676–1680.

Hatefi, Y., Haavik, A. G., and Griffiths, D. E., 1962c, *J. Biol. Chem.* **237**:1681–1685.

Hatefi, Y., Stiggall, D. L., Galante, Y. M., and Hanstein, W. G., 1974, *Biochem. Biophys. Res. Commun.* **61**:313–321.

Hatefi, Y., Yagi, T., Phelps, D. C., Wong, S.-Y., Vik, S. B., and Galante, Y. M., 1982, *Proc. Natl. Acad. Sci. USA* **79**:1756–1760.

Hatefi, Y., Ragan, C. I., and Galante, Y. M., 1985, in: *The Enzymes of Biological Membranes*, vol. 4 (A. N. Martonosi, ed.), pp. 1–70, Plenum Press, New York.

Haughton, M. A., and Capaldi, R. A., 1995, *J. Biol. Chem.* **270**:20568–20574.

Hauska, G., Hurt, E., Gabellini, N., and Lockau, W., 1983, *Biochim. Biophys. Acta* **726**:97–133.

Hederstedt, L., and Hatefi, Y., 1986, *Arch. Biochem. Biophys.* **247**:346–354.

Hederstedt, L., and Henden, L-O., 1989, *Biochem. J.* **260**:491–497.

Hederstedt, L., and Ohnishi, T., 1992, in: *Molecular Mechanism in Bioenergetics* (L. Ernster, ed.), pp. 163–198, Elsevier, Amsterdam.

Hofhaus, G., Weiss, H., and Leonard, K., 1991, *J. Mol. Biol.* **221**:1027–1043.

Kita, S. T., Oya, H., Gennis, R. B., Ackrell, B. A. C., and Kasahara, M., 1990, *Biochem. Biophys. Res. Commun.* **166**:101–108.

Kotlyar, A. B., and Vinogradov, A. D., 1984, *Biochem. Int.* **8**:545–552.

Körtner, C., Lauterbach, F., Tripier, D., Unden, G., and Kröger, A., 1990, *Mol. Microbiol.* **4:** 855–860.

Krishnamoorthy, G., and Hinkle, P. C., 1988, *J. Biol. Chem.* **263**:17566–17575.

Leif, H., Sled', V. D., Ohnishi, T., Weiss, H., and Friedrich, T., 1995, *Eur. J. Biochem.* **230:** 538–548.

Link, T. A., Haase, U., Brandt, U., and von Jagow, G., 1993, *J. Bioenerg. Biomembr.* **25**:221–232.

Ljungdahl, P. O., Pennoyer, J. D., Robertson, D. E., and Trumpower, B. L., 1987, *Biochim. Biophys. Acta* **891**:227–241.

Matsuno-Yagi, A., and Hatefi, Y., 1988, *Biochemistry* **27**:335–340.

Matsuno-Yagi, A., and Hatefi, Y., 1990, *J. Biol. Chem.* **265**:82–88.

Matsuno-Yagi, A., and Hatefi, Y., 1993a, *J. Biol. Chem.* **268**:1539–1545.

Matsuno-Yagi, A., and Hatefi, Y., 1993b, *J. Biol. Chem.* **268**:6168–6173.
Matsuno-Yagi, A., and Hatefi, Y., 1996, *J. Biol. Chem.* **271**:6164–6171.
Matsuno-Yagi, A., and Hatefi, Y., 1997, *J. Biol. Chem.* **272**:16928–16933.
Matsuno-Yagi, A., Yagi, T., and Hatefi, Y., 1985, *Proc. Natl. Acad. Sci. USA* **82**:7550–7554.
McCurley, J. P., Miki, T., Yu, L., and Yu, C.-A., 1990, *Biochim. Biophys. Acta* **1020**:176–186.
Mendel-Hartvig, J., and Capaldi, R. A., 1991, *Biochemistry* **30**:1278–1284.
Mitchell, P., 1961, *Nature* **191**:144–148.
Mitchell, P., 1975, *FEBS Lett.* **56**:1–6.
Mitchell, P., 1976, *J. Theor. Biol.* **62**:327–367.
Morris, A. A. M., Farnsworth, L., Ackrell, B. A. C., Turnbull, D. M., and Birch-Machin, M. A., 1994, *Biochim. Biophys. Acta* **1185**:125–128.
Noji, H., Yasuda, R., Yoshida, M., and Kinosita, K., Jr., 1997, *Nature* **386**:299–302.
Ogilvie, I., Aggeler, R., and Capaldi, R. A., 1997, *J. Biol. Chem.* **272**:16652–16656.
Ohnishi, T., 1979, in: *Membrane Proteins in Energy Transduction*, (R. A. Capaldi, ed.), pp. 1–87, Marcel Dekker, New York.
Ohnishi, T., 1987, *Curr. Topics Bioenerg.* **15**:37–65.
Ohnishi, T., Ragan, C. I., and Hatefi, Y., 1985, *J. Biol. Chem.* **260**:2782–2788.
Ohnishi, T., Schägger, H., Meinhardt, S. W., LoBrutto, R., Link, T. A., and von Jagow, G., 1989, *J. Biol. Chem.* **264**:735–744.
Okunuki, K., Sekuzu, I., Yonetani, T., and Takemori, S., 1958, *J. Biochem.* **45**:847–854.
Penefsky, H. S., 1985, *Proc. Natl. Acad. Sci. USA* **82**:1589–1593.
Penefsky, H. S., and Cross, R. L., 1991, *Adv. Enzymol.* **64**:173–214.
Pfenninger-Li, X. D., Albracht, S. P. J., Van Belzen, R., and Dimroth, P., 1996, *Biochemistry* **35**:6233–6242.
Pullman, M. E., Penefsky, H. S., Datta, A., and Racker, E., 1960, *J. Biol. Chem.* **235**:3322–3329.
Ragan, C. I., 1987, *Curr. Topics Bioenerg.* **15**:1–36.
Rieske, J. S., 1967, *Methods Enzymol.* **10**:239–245.
Rieske, J. S., 1971, *Arch. Biochem. Biophys.* **145**:179–193.
Rieske, J. S., 1976, *Biochim. Biophys. Acta* **456**:195–247.
Rieske, J. S., Baum, H., Stoner, C. D., and Lipton, S. H., 1967, *J. Biol. Chem.* **242**:4854–4866.
Ruzicka, F. J., Beinert, H., Scheplev, K. L., Dunham, W. R., and Sands, R. H., 1975, *Proc. Natl. Acad. Sci. USA* **72**:2886–2890.
Sabbert, D., Engelbrecht, S., and Junge, W., 1997, *Proc. Natl. Acad. Sci. USA* **94**:4401–4405.
Salerno, J. C., and Ohnishi, T., 1980, *Biochem. J.* **192**:769–781.
Schröder, I., Gunsalus, R. P., Ackrell, B. A. C., Cochran, B., and Cecchini, G., 1991, *J. Biol. Chem.* **266**:13572–13579.
Schulte, U., Fecke, W., Krüll, C., Nehls, U., Schmiede, A., Schneider, R., Ohnishi, T., and Weiss, H., 1994, *Biochim. Biophys. Acta* **1187**:121–124.
Singer, T. P., and Kearney, E. B., 1963, in: *The Enzymes*, vol. 7 (P. D. Boyer, H. Lardy, and K. Myrbäck, eds.), pp. 383–445, Academic Press, New York.
Singer, T. P., Kearney, E. B., and Kenney, W. C., 1973, *Adv. Enzymol.* **37**:189–272.
Sled', V. D., Friedrich, T., Leif, H., Weiss, H., Meinhardt, S. W., Fukumori, Y., Calhoun, M. W., Gennis, R. B., and Ohnishi, T., 1993, *J. Bioenerg. Biomembr.* **25**:347–356.
Smith, L., and Stotz, E., 1954, *J. Biol. Chem.* **209**:819–829.
Takano, S., Yano, T., and Yagi, T., 1996, *Biochemistry* **35**:9120–9127.
Trumpower, B. L., 1990, *J. Biol. Chem.* **265**:11409–11412.
Tzagoloff, A., Byington, K. H., and MacLennan, D. H., 1968, *J. Biol. Chem.* **243**:2405–2412.
Vik, S. B., and Hatefi, Y., 1981, *Proc. Natl. Acad. Sci. USA* **78**:6749–6753.
Vinogradov, A. D., 1993, *J. Bioenerg. Biomembr.* **25**:367–375.

Vinogradov, A. D., Gavrikova, E. V., and Goloveshkina, V. G., 1975, *Biochem. Biophys. Res. Commun.* **65**:1264–1269.

von Jagow, G., and Link, T. A., 1986, *Methods Enzymol.* **126**:253–271.

von Jagow, G., Schägger, H., Engel, W. D., Riccio, P., Kolb, H. J., and Klingenberg, M., 1978, *Methods Enzymol.* **53**:92–98.

Walker, J. E., 1992, *Q. Rev. Biophys.* **25**:253–324.

Weber, J., and Senior, A. E., 1996, *Biochim. Biophys. Acta* **1275**:101–104.

Weber, J., Wilke-Mounts, S., Lee, R. S.-F., Grell, E., and Senior, A. E., 1993, *J. Biol. Chem.* **268**:20126–20133.

Weiss, H., and Friedrich, T., 1991, *J. Bioenerg. Biomembr.* **23**:743–754.

Weiss, H., and Kolb, H. J., 1979, *Eur. J. Biochem.* **99**:139–149.

Wilkens, S., Dahlquist, F. W., McIntosh, L. P., Donaldson, L. W., and Capaldi, R. A., 1995, *Nature Struct. Biol.* **2**:961–967.

Xia, D., Yu, C. A., Kim, H., Xia, J. Z., Kachurin, A. M., Yu, L., and Diesenhofer, J., 1997, *Science* **277**:60–66.

Yagi, T., and Hatefi, Y., 1988, *J. Biol. Chem.* **263**:16150–16155.

Yagi, T., Yano, T., and Matsuno-Yagi, A., 1993, *J. Bioenerg. Biomembr.* **25**:339–345.

Yamaguchi, M., and Hatefi, Y., 1993, *Biochemistry* **32**:1935–1939.

Yang, X., and Trumpower, B. L., 1986, *J. Biol. Chem.* **261**:12282–12289.

Yankovskaya, V., Sablin, S. O., Ramsay, R. R., Singer, T. P., Ackrell, B. A. C., Cecchini, G., and Miyoshi, H., 1996, *J. Biol. Chem.* **271**:21020–21024.

Yano, T., Sled', V. D., Ohnishi, T., and Yagi, T., 1994, *FEBS Lett.* **354**:160–164.

Yano, T., Yagi, T., Sled', V. D., and Ohnishi, T., 1995, *J. Biol. Chem.* **270**:18264–18270.

Yano, T., Sled', V. D., Ohnishi, T., and Yagi, T., 1996, *J. Biol. Chem.* **271**:5907–5913.

Yano, T., Chu, S. S., Sled', V. D., Ohnishi, T., 1997, *J. Biol. Chem.* **272**:4201–4211.

Yu, L., and Yu, C.-A., 1991, *Biochemistry* **30**:4934–4939.

Yu, L., Xu, J.-X., Haley, P. E., and Yu, C.-A., 1987, *J. Biol. Chem.* **262**:1137–1143.

Zhang, Y., and Fillingame, R. H., 1995, *J. Biol. Chem.* **270**:24609–24614.

Zhou, Y., Duncan, T. M., and Cross, R. L., 1997, *Proc. Natl. Acad. Sci. USA* **94**:10583–10587.

CHAPTER 3

Proton Pumps of Respiratory Chain Enzymes

Sergio Papa, Nazzareno Capitanio, and Gaetano Villani

1. INTRODUCTION

Redox and photoredox enzymes, plugged through the osmotic barrier of the coupling membranes, the mitochondrial inner membrane of eukaryotic cells, the plasma membrane of bacteria, and the thylakoid membrane of chloroplasts, convert chemical and light energy, respectively, into proticity, thus creating a transmembrane protonic potential difference, Δp (Fig. 1). In turn, Δp is drawn up by proticity utilizers and plugged through the same coupling membranes (Mitchell, 1966, 1987a; Papa, 1976; Boyer *et al.*, 1977). Proticity is essentially used to drive phosphorylation of ADP to ATP by the $F_O F_1$-ATP synthase of coupling membranes (Fig. 1). It also can drive solute transport by proton-coupled transporters (Papa *et al.*, 1970) and mechanical work as in bacterial flagella (Macnab, 1992), and can result in heat production in brown adipose tissue mitochondria (Nicholls and Locke, 1992).

Advanced spectroscopy, molecular genetics, mutational inspection, and more recently crystallographic analysis of proticity producers and proticity utilizers have begun to provide the necessary information to examine at a

Sergio Papa, Nazzareno Capitanio, and Gaetano Villani • Institute of Medical Biochemistry and Chemistry, University of Bari, I-70124 Bari, Italy.

Frontiers of Cellular Bioenergetics, edited by Papa *et al.* Kluwer Academic/Plenum Publishers, New York, 1999.

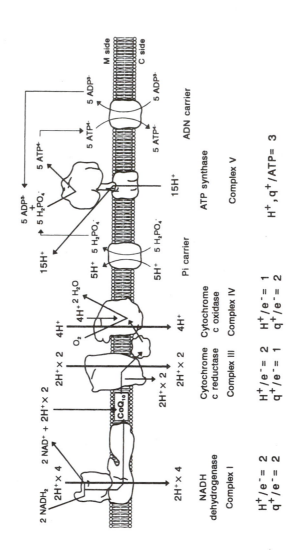

molecular level the mechanism of energy transfer by proticity. Remarkable examples of these achievements are the mutational analysis of light-driven (Khorana, 1988) and redox proticity producers (Chapter 7, this volume) and of the F_OF_1 H^+-ATP synthase (Chapter 17, this volume); the X-ray crystallographic analysis of the photoreaction center of purple bacteria (Deisenhofer and Michel, 1989); and the F_1 moiety of the mitochondrial H^+-ATP-synthase (Abrahams *et al.*, 1994; Chapter 15, this volume), the prokaryotic (Chapter 5, this volume) and mitochondrial (Chapter 6, this volume) cytochrome *c* oxidase, and the mitochondrial ubiquinol cytochrome *c* oxidoreductase (Chapters 12 and 13, this volume).

One aim of this chapter is to describe how the concepts on energy coupling in proto-motive respiratory enzymes have developed to the present state of knowledge. Models for redox proton pumps have been classified as "direct" and "indirect." In the direct mode, the pump is conceived as a direct consequence of the arrangement and mobility in the membrane of the same redox groups involved in the catalysis, i.e., electron transfer metals, heme–iron, non-heme–iron, Cu and organic hydrogen carriers, NADP, FMN, and quinones (Mitchell, 1966, 1987a, 1988). In the indirect mode, proton pumping will result from thermodynamic (Papa, 1976; Papa *et al.*, 1973) and/or kinetic (Malmström, 1989; Blair *et al.*, 1986) linkage between primary redox catalysis and proton transfer by other chemical groups in the enzyme. It has been argued that if the "other" chemical groups involved are the same ligands of the redox metals, i.e., histidine ligands of the heme–iron and Cu, cysteine ligands of Cu and non-heme–iron, these coupling models should be considered direct, while

←——————————————————————————————————————

FIGURE 1. Protonic coupling of oxidative phosphorylation enzymes in the inner membrane of mitochondria. The shapes of complex I and III are drawn from the three-dimensional electron microscopy image reconstruction of Hofhaus *et al.* (1991) and the X-ray crystallographic analysis of Yu *et al.* (Chapter 12, this volume), respectively. The shape of complex IV is from the X-ray crystallographic analysis of the whole bovine heart cytochrome *c* oxidase (Tsukihara *et al.*, 1996). The shape of complex V is derived from X-ray crystallographic analysis of the F_1 catalytic moiety (Abrahams *et al.*, 1994) and information on the F_O protein composition (Chapter 19, this volume). Complex I consists of around 41 subunits, of which seven are encoded by the mtDNA and the remaining by nuclear genes (Walker, 1992). Complex III consists of 11 subunits, of which only cytochrome *b* is coded by the mtDNA (Trumpower and Gennis, 1993). Complex IV consists of 13 subunits, of which three are encoded by the mtDNA (Trumpower and Gennis, 1993; Fergusson-Miller, 1993). The F_1 moiety of complex V is composed of five nuclear encoded subunits; the F_O moiety of 10 subunits, of which two are encoded by the mtDNA (Collinson *et al.*, 1994). The maximal H^+/e^- and q^+/e^- ratios attainable for the three redox complexes and the H^+/ATP ratio for the ATP synthase are given at the bottom of the scheme. Proton and charge translocation for the import of $H_2PO_4^-$ and ATP^{3-} with export of ATP^{4-} is also shown. The overall balance refers to a P/O_2 ratio of 5 for the oxidation of two molecules of $NADH_2$ by one molecule of dioxygen.

indirect models would be those involving protolytic groups in the protein distant from the redox centers (Wikström, 1994). It should be kept in mind, however, that direct and indirect coupling events are not necessarily exclusive to one another, but they could be combined to various extent.

Mitchell (1966) originally proposed the proton-motive activity of the respiratory chain to derive directly from primary catalysis at the redox centers organized in consecutive redox loops (alternating hydrogen and electron carriers). Aerobic oxidation of quinol by terminal cytochrome in particular was conceived to result in the effective translocation of one H^+/e^- from the N to the P aqueous phase. It was found, however, that in mitochondria two protons are released per electron flowing from ubiquinol to cytochrome c (Papa *et al.*, 1975), and that further electron flow from cytochrome c to oxygen results in the additional release of one H^+/e^- (Wikström, 1977). To explain these H^+/e^- stoicheiometries and related features of electron transfer in terms of direct model, the proton-motive quinone cycle for the bc_1 complex (Mitchell, 1976) and the Cu loop for cytochrome c oxidase (Mitchell, 1987b) were developed (Fig. 2).

Based on the principle of cooperative thermodynamic linkage of solute

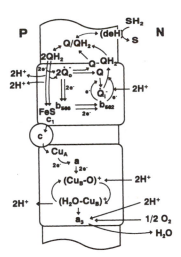

FIGURE 2. Direct models for redox and proton-motive activity of the cytochrome system of mitochondria–ubiquinone cycle for the bc_1 complex and Cu_B loop for the cytochrome c oxidase complex. The ubiquinone cycle is based on cycling of one of the two electrons donated by ubiquinol of the pool to the bc_1 complex and might involve exchange of protein-bound quinone with the quinone of the pool (Mitchell, 1976, 1987a). In the Cu_B loop proton translocation from the N to the P side is mediated by reorientation around Cu_B of OH^- (or O^-) and H_2O (Mitchell, 1987b). Reproduced with permission from Papa *et al.* (1994).

binding at separate sites in allosteric proteins (Wyman, 1968) and the observation of linkage between electron transfer at the metals and protolytic events (pK shifts) in cytochromes (Clark, 1960), in 1973, Papa and co-workers proposed a cooperative model for proton pumping in respiratory enzymes (Papa *et al.*, 1973; Papa, 1976). By analogy to the cooperative linkage phenomena in hemoglobin, known as the Bohr effect (Wyman, 1968; Kilmartin and Rossi-Bernardi, 1973), the redox linkage in cytochromes and the derived cooperative model for proton pumping were denominated redox-Bohr effects and vectorial Bohr mechanism, respectively. The H^+/e^- linkage in cytochromes is likely to arise from modification of the coordination bonds of metal centers associated with a change in their valence state. The linkage could be confined locally and could involve pK shifts and exchange of axial ligands (Wikström, 1994; Iwata *et al.*, 1995). The H^+/e^- linkage could also involve porphyrin substituents (Babcock and Callahan, 1983; Woodruff *et al.*, 1991) and conformational propagation of primary effects over long distances in the protein, as in the oxygen Bohr effect of hemoglobin. In the formulation of the vectorial Bohr mechanism, the cooperative events were conceived to be extended over the transmembrane span of the protein to promote proton uptake from the N aqueous phase, translocation to and from the redox catalytic domain, and release to the P aqueous phase (Fig. 3). The vectorial Bohr mechanism thus encompasses by definition cooperative pK shifts of direct ligands of the redox metals as well as pK shifts of distant residues in the proton input and proton output pathways (see Section 5). [See the case of the light-driven proton pump of bacteriorhodopsin (Henderson *et al.*, 1990).]

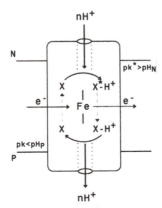

FIGURE 3. Scheme describing the minimal steps for proton pumping by the redox carrier based on linkage between oxidoreduction of the metal and protolytic events in the apoprotein (redox Bohr effect, vectorial Bohr mechanism) (Papa *et al.*, 1973; Papa, 1976). Reproduced with permission from Papa *et al.* (1981).

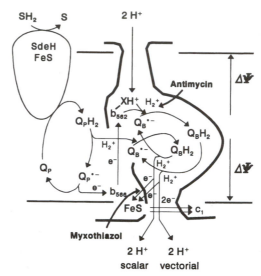

FIGURE 4. Q-gated proton pump for the bc_1 complex. The model envisages a linear split pathway for electron transfer mediated by nonexchangeable protein-stabilized quinol–semiquinone, which provides pumping of 2 H^+/e^- (for more details see Section 6, and Papa *et al.*, 1983, 1989).

The cooperative vectorial Bohr mechanism was first applied by Papa and co-workers to the low potential b cytochrome of the bc_1 complex of the respiratory chain (Papa, 1976; Papa *et al.*, 1973; for further versions, see von Jagow and Sebald, 1980; Wikström *et al.*, 1981). Later, Papa *et al.* (1989), developed a combined model in which redox Bohr effects in cytochrome b and the Rieske Fe-S protein are conceived to operate in a series with proton-motive redox catalysis by a protein-bound quinol–semiquinone couple (Q-gated proton pump) (see Section 6) (Papa *et al.*, 1983, 1989) (Fig. 4). A cooperative proton pump model for heme–copper oxidases as well as possible mechanisms of the proton pump in complex I and III will be presented in sections 5 and 6, respectively.

2. THE OUTPUT EFFICIENCY OF REDOX PROTON PUMPS

Direct models of redox proton pumps predict, in principle, integral and fixed H^+/e^- stoichiometries when electrons flow through a coupling site (Mitchell, 1987a). The stoichiometry could decrease below the predicted mechanistic value, however, if a fraction of the electrons flow through a parallel decoupled redox pathway. Indirect models can produce variable stoichiometries due to electron and/or proton slips (decoupling of the pump) (Blair et al., 1986; Azzone and Luvisetto, 1988; Malmström, 1989). The observed nonlinearity of the steady-state relationship between respiratory rate and Δp is taken by some authors as evidence of a slip in proton pumps (Pietrobon et al., 1983; Murphy and Brand, 1988; Luvisetto et al., 1991) and by others as due to a nonohmic increase of membrane proton conductance at high Δp (leak) (Brown and Brandt, 1986; Brown, 1989).

The H^+/e^- stoichiometry of redox pumps is measured directly by two methods: the pulse method and the rate method. With the pulse method, based on determination of the extent of H^+ release associated with consumption of a known amount of oxygen or a reductant (Mitchell et al., 1979), it is difficult to adjust the rate off electron flow and the ensuing Δp. On the other hand, these parameters can be easily adjusted in the rate method. Using this approach, Papa and co-workers have carried an extensive study of the influence of kinetic and thermodynamic factors on the H^+/e^- stoichiometry of the bc_1 complex and cytochrome c oxidase in mitochondria and in the isolated–reconstituted state (Papa et al., 1991; Capitanio et al., 1991, 1996; Cocco et al., 1992, 1997). Measurements in mitochondria (Fig. 5A) showed that under level flow conditions (valinomycin plus K^+ was present to prevent the build up of $\Delta\psi$ and to leave Δp as ΔpH), the H^+/e^- ratio for succinate respiration varies from a minimum of two at extremely high and low respiratory rates to about three at intermediate rates, with a rate dependence attributable to that of cytochrome c oxidase for which the H^+/e^- varies from around zero at extreme high and low rates to around one at intermediate rates (Papa et al., 1991). It should be noted that the range of electron transfer rates in which the H^+/e^- stoichiometry changes corresponds to the physiological range of respiratory rates in tissues (Fitzgerald, 1976) and isolated rat cardiac myocytes (Kennedy and Jones, 1986). These rates amount to around 10% of the maximal turnover of purified cytochrome c oxidase from bovine heart (Papa et al., 1987). Furthermore, in vivo measurements of cytochrome oxidase activity in human cells have shown that the oxidase capacity is slightly higher than that required to support the endogenous respiration rate, thus pointing to a tight control of respiration by the mitochondrial enzyme (Villani and Attardi, 1997).

Measurements carried out on the isolated redox complexes reconstituted in liposomes (Capitanio et al., 1991, 1996) showed that the H^+/e^- ratio for the reconstituted oxidase (COV) exhibits a rate dependence (Fig. 5B) similar to that observed in mitochondria. The H^+/e^- stoichiometry for the bc_1 complex at level flow is, on the contrary, both in mitochondria (Fig. 5A) and in the reconstituted state (Fig. 7) constantly two, independent of the rate of electron flow (Cocco et al., 1992).

A large number of measurements of the H^+/c^- stoichiometry in COV

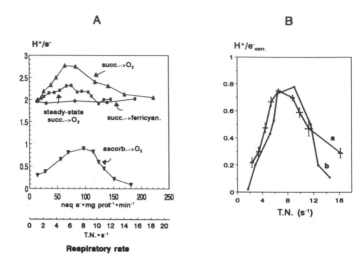

FIGURE 5. Dependence of the H$^+$/e$^-$ ratio on the rate of electron flow in cytochrome *c* reductase and cytochrome *c* oxidase in intact mitochondria and in cytochrome *c* oxidase vesicles. (A) Rat liver mitochondria; the electron flow rate from succinate to oxygen, at level flow or respiring steady-state, or to ferricyanide, in the presence of KCN, was adjusted with malonate. The respiratory rate with ascorbate was adjusted varying the concentration of TMPD. For level flow measurements, the H$^+$/e$^-$ ratios were obtained from the initial rates elicited by the addition of the substrates. The H$^+$/e$^-$ ratios in the respiring state were obtained from the rates of respiration and proton translocation measured at the aerobic steady state. For details, see Papa *et al.* (1991). (B) (a) Cytochrome *c* oxidase liposomes (COV) were supplemented with cytochrome *c* and ascorbate *plus* TMPD. The respiratory rate was varied changing the concentration of ferricytochrome *c*. (b) The oxidase vesicles were supplemented with cytochrome *c*, duroquinol, and a trace of soluble cytochrome *c* reductase whose concentration was changed to vary the overall respiratory rate. Valinomycin (*plus* K$^+$) was present to collapse aerobic $\Delta\Psi$. The points of all the curves represent the mean of six or more experiments (Capitanio *et al.*, 1996).

have been carried out in different laboratories using the reductant pulse method. In this technique, in which aerobic COV are pulsed with a molar excess of ferrocytochrome *c*, "optimal experimental conditions" are generally used that give H$^+$/e$^-$ ratios approaching 1 (Wikström and Krab, 1979; Nicholls and Wrigglesworth, 1982; Proteau *et al.*, 1983; Papa *et al.*, 1987). With the pulse technique, however, the H$^+$/e$^-$ ratio varies with the pH, ionic composition of the medium, and modalities of activation of electron flow (Proteau *et al.*, 1983; Papa *et al.*, 1987). In any event, the prevailing conditions used in the pulse method are such that this technique does not provide information on the aspects dealt with systematically in the studies with the rate method.

It seems to be agreed that, in order to be coupled to proton pumping in the oxidase, electron flow has to follow the sequence cytochrome *c* $\rightarrow$ Cu$_A$ $\rightarrow$

heme $a \rightarrow$ heme a_3–Cu_B (Babcock and Wikström, 1992). The a_3–Cu_B binuclear center is the site where the dioxygen reduction chemistry takes place and this process is considered to be directly coupled to H^+ pumping (Babcock and Wikström, 1992; Chapters 5 and 9, this volume). A breakthrough toward the understanding of the molecular mechanism of proton-motive catalysis in cytochrome c oxidase has been provided by the crystallographic resolution of the structures of cytochrome c oxidase from *Paracoccus denitrificans* (Iwata *et al.*, 1995) and bovine heart mitochondria (Tsukihara *et al.*, 1995, 1996), which gave remarkably similar structures for the conserved subunits I, II, and III. The crystallographic structures, which confirmed structural predictions based on mutational analysis of the prokaryotic oxidases (see Chapter 7, this volume), show that heme a and heme a_3–Cu_B are bound in subunit I to histidine residues of transmembrane helices II, X, VI, and VII, respectively, in regions extending toward the P surface (Fig. 6). The propionate groups of both hemes a and a_3, which are perpendicular to the plane of the membrane, point toward and communicate with the C-terminal domain of subunit II through conserved residues in subunit I, holding the two copper atoms of the Cu_A center. Residues have been located in subunit I and subunit II that can provide a hydrogen bond–

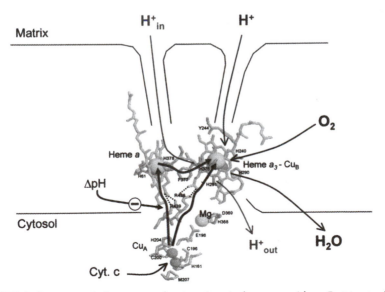

FIGURE 6. Sequence of electron transfer steps in cytochrome c oxidase. Proton-coupled electron transfer steps are shown by red arrows, decoupled electron transfer (slip) by blue arrows. The structure and arrangement of the metal centers and associated residues is drawn from Tsukihara *et al.* (1996). For details, see text and Chapter 6, this volume. For color representation of this figure, see color tip facing page 68.

ion pair network serving as electron transfer path between Cu_A and heme a. In this network, His 204, a Cu_A ligand in subunit II, is hydrogen bonded to Arg 438 and Arg 439 (bovine numbering) in the loop XI–XII of subunit I; the latter is bonded to the propionate groups of heme a. Tsukihara *et al.* (1996) have also identified two bond networks in the bovine heart oxidase involving the propionate groups of heme a_3, Arg 438 in the loop XI–XII, and His 368 of subunit I; the latter is coordinated by Mg to Glu 198 in subunit II, which is liganded to the lower Cu_A atom. These bond networks could provide direct electron transfer from Cu_A to the heme a_3–Cu_B center (Tsukihara *et al.* 1996; Chapter 6, this volume). The distance from the lower Cu_A to the heme a Fe is some 2 Å shorter than that to the heme a_3 Fe. With this difference, the rate of electron transfer from Cu_A to heme a would be much faster than to heme a_3, as in fact seems to be suggested by kinetic data (Hill, 1994). Tsukihara *et al.* (1996) believe that the structural and kinetic data, however, can yet be compatible with electron transfer from Cu_A to heme a as well as to heme a_3. It also should be kept in mind that the Cu_A center resides in a different subunit from that holding hemes a and a_3. The distances between Cu_A and the two hemes have been measured in the oxidized enzyme in the absence of cytochrome c, whose binding might affect the structure of the oxidase. It is conceivable that in the oxidase turning over in the steady state the contacts between subunits I and II, and hence the distance between Cu_A and heme a and a_3, can be changed. In particular, the steady-state $\Delta\mu_{H^+}$ might affect these intersubunit distances.

Combining the structural information with the finding of a variable H^+/e^- stoichiometry in cytochrome c oxidase, our group has proposed that electron transfer from Cu_A to the heme a_3–Cu_B center can take place along two pathways (Fig. 6) (Papa *et al.*, 1994; Capitanio *et al.*, 1996). The first via heme a will be associated through cooperative linkage (redox Bohr effect) to proton pumping. The second pathway will consist of direct electron transfer from Cu_A to the binuclear a_3–Cu_B center, bypassing heme a and resulting in decoupling of the proton pump.

The occurrence of these two electron transfer pathways is supported by the observation that in the various states of the respiratory proton pump in COV, the reduction level of Cu_A is found to be lower than that of heme a and is enhanced less than that of heme a by raising the rate of electron delivery to the oxidase (Capitanio *et al.*, 1996). The actual H^+/e^- in the oxidase will be determined by the relative contribution of the two electron transfer pathways and can vary from a maximum of 1 to 0 (Papa *et al.*, 1994). The relative contributions of the two pathways can be dictated by kinetic and thermodynamic factors. Under level flow conditions, the H^+/e^- ratio should only be influenced by kinetic factors. At initial rates below 4 electrons/sec per aa_3, the H^+/e^- ratio approaches 0. This might reflect the fact that in order to perform proton pumping, the oxidase has to experience a full turnover with the passage of 4 electrons/mole of aa_3 (Babcock and Wikström, 1992). With enhancement of the rate of electron flow, the H^+/e^- increases and, at intermediate rates, tends to the maximal obtainable value of 1. The kinetic situation could be adjusted here to have the electrons flowing almost exclusively through the coupled pathway via heme a. An the electron pressure and the rate of electron transfer in the oxidase are further increased, the uncoupled electron transfer pathway, directly from Cu_A to the binuclear center, can become more important, with a marked decrease in the H^+/e^- stoichiometry. The atomic resolution of the structure of cytochrome c oxidase provided by the X-ray crystallographic analyses now offers the basis for site-directed mutational analysis to verify the proposed occurrence of the two pathways for electron transfer from Cu_A to the binuclear a_3–Cu_B center.

Proton pumping, both in the bc_1 complex and in cytochrome c oxidase, is, at the steady state, partially decoupled by the ΔpH component of Δp (Fig. 7) (Lorusso et al., 1995; Capitanio et al., 1996). This decoupling should not be confused with the Δp-dependent membrane leak. Promotion of the steady-state proton leak by a protonophore, which decreases the ΔpH, has been shown to alleviate ΔpH-dependent slip in both bc_1 vesicles and COV (Fig. 7) (Cocco et al., 1992; Capitanio et al., 1996). At the steady state, alkalinization of the N phase can result in proton slip due to loss of protonation asymmetry of the critical protolytic group(s) in the pump in the input state. This protonation step might have a limited kinetic capacity (Hallen and Nilson, 1992; Hallen et al., 1994). An analogous situation seems to be met in the light-driven pump of bacteriorhodopsin, where the input proton pathway is the principal electrical barrier to the H^+ pump (Henderson et al., 1990). There are observations indicating that also this pump might "slip" at high Δp (Westerhoff and Dancshazy, 1984). Of particular interest in this respect is the recent observation that in bc_1 vesicles, weak acids like arachidonate and azide reverse the decoupling effect exerted on the proton pump by the steady-state Δp (Cocco et al., 1997). A similar recoupling effect was shown to be effected by azide in a bacteriorhodopsin mutant in which the light-driven proton pump was decoupled by the mutation Asp96→Asn (Tittor et al., 1989). It is possible that in the bc_1 complex, as in bacteriorhodopsin, there is a critical acidic residue at the N side of apocytochrome b; a good candidate could be the Asp 229 belonging to helix E (Brasseur, 1988), which would mediate the transfer of H^+ in the input channel of the pump.

In the case of cytochrome c oxidase, ΔpH can exert an additional decoupling effect by exerting an inhibitory back pressure on the proton-coupled electron transfer from Cu_A to the a_3–Cu_B center via heme a without obviously affecting the putative uncoupled electron transfer directly from Cu_A to the binuclear center. For other possible mechanisms of decoupling of the pump, see Section 5.

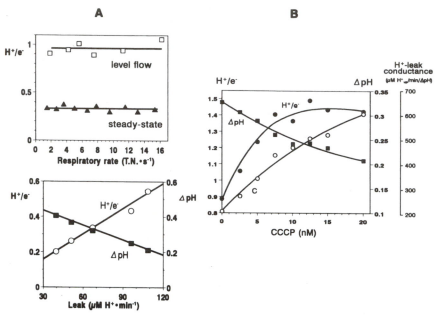

FIGURE 7. Effect of the proton leak on the steady-state H$^+$/e$^-$ ratios in cytochrome *c* reductase and cytochrome *c* oxidase vesicles. (A) The upper panel shows a comparison between the H$^+$/e$^-$ ratios measured in cytochrome *c* reductase vesicles under level flow and steady-state conditions at different rates of electron transfer rates; the lower panel shows the effect of increasing the proton leak with nanomolar concentrations of CCCP on the steady-state H$^+$/e$^-$ ratios and ΔpH in cytochrome *c* reductase vesicles. (B) The effect of CCCP on proton leak, ΔpH, and H$^+$/e$^-$ ratios measured under respiring steady-state conditions in cytochrome *c* oxidase vesicles. For experimental details, see Cocco *et al.* (1992, Experiment A) and Capitanio *et al.* (1996, Experiment B).

It should be noted that the loss of energy conversion effected in the proton pumps by ΔpH and also in the oxidase by high respiratory rates is alleviated in mitochondria by two processes: (1) direct Δp generation arising from the membrane anisotropy of reduction of O$_2$ to H$_2$O in the oxidase, and (2) proton-coupled uptake of phosphate and respiratory substrates and proton influx for ATP synthesis in phosphorylating mitochondria; similarly to what shown in Fig. 7 for CCCP in COV and *bc*$_1$ vesicles, these contribute to prevent establishment of a large transmembrane ΔpH so that proton pumping in the phosphorylating steady state can be preserved.

Intrinsic decoupling (slip) in the proton pump of cytochrome *c* oxidase might be of physiological relevance. A decrease of the ATP–O ratio in yeast mitochondria was observed when the electron flux was increased, which was ascribed essentially to a decrease in the H^+/e^- stoichiometry in cytochrome c oxidase (Fitton *et al.*, 1994). Decoupling of the oxidase at high electron pressure and high Δp and phosphate potential could contribute to prevent excessive electronegativity of redox carriers in complex I and III, which, above a threshold level, can lead to production of deleterious oxygen radicals. Enhanced oxidase activity will prevent accumulation of toxic O_2^- and other free radicals deriving from it, and also maintain low cellular oxygen tension (Skulachev, 1994) and possibly directly contribute to oxidation and/or dismutation of O_2^- (Markossian *et al.*, 1978).

3. REDOX BOHR EFFECTS AND COOPERATIVE COUPLING IN CYTOCHROME c OXIDASE

Redox Bohr affects in cytochromes, i.e., thermodynamic linkage between oxidoreduction of the metal centers and the pKs of metal ligands and/or distant residues in the apoproteins, result in pH dependence of the midpoint redox potentials (Wilson *et al.*, 1972) and in proton transfer associated with oxidoreduction of the metal centers (Papa *et al.*, 1986). Analysis of the pH dependence of H^+ transfer associated with redox transitions of the metal centers in the soluble cytochrome c oxidase purified from bovine heart, in the unliganded and CN- and CO-liganded state (Fig. 8), has identified four groups undergoing reversible redox-linked pK shifts (Table I) (Capitanio *et al.*, 1997a). Two protolytic groups with pK_{ox} and pK_{red} values around 7 and >12, respectively, result in being linked to redox transitions of heme a_3. One protolytic group with pK_{ox} and pK_{red} values around 6 and 7, respectively, is apparently linked to Cu_B (cf. Mitchell and Rich, 1994). The fourth group with pK_{ox} around 6 and pK_{red} around 9 is linked to oxidoreduction of heme a.

The assignment of the pK shifts of the four protolytic groups to individual metal centers in the oxidase does not exclude the possibility that the pK of a single group could also be influenced by cooperative interactions of different metals. The correspondence between the pK values of the four groups obtained in the free and the liganded oxidase would indicate, however, that, if occurring, these interactions do not alter significantly the pK shifts arising from the individual linkages.

Redox Bohr effects are likely to play a role in the exchange of protons between aqueous phases and protolytic reactions in the membrane environments, i.e., protonation of the intermediates of oxygen reduction to H_2O in the oxidase and protonation–deprotonation of protein-stabilized quino–semiquinones in the bc_1 complex. The cooperative model of proton pump, proposed by Papa *et al.* (1973), is based on vectorial organization of redox Bohr effects in membrane-associated respiratory enzymes. Recently, considerable attention has been paid

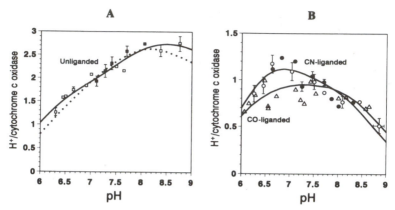

FIGURE 8. Measured pH dependence of redox-Bohr effects (H+/COX coupling number) in soluble cytochrome *c* oxidase and the corresponding best-fit analysis. (A) Unliganded enzyme; open and closed squares refer to the experiments carried out using succinate (in the presence of a "catalityc" amount of mitochondria *plus* cytochrome *c*) or ascorbate (in the presence of TMPD plus cytochrome *c*) as ultimate reductants, respectively. Solid line, best-fit analysis of the experimentally determined H+/COX ratios. Dotted line, curve constructed by using the mean of the pK values obtained as described in Table IC. (B) CN-liganded enzyme; open and closed circles refer to experiments with succinate or ascorbate as reductants; triangles, CO-liganded enzyme. The curves represent the best-fits obtained using the pK values reported in Table IA (see text for further details). Where indicated by vertical bars the points represent the mean ±SE of three or more H+/COX measurements at the given pHs. For experimental details, see Capitanio *et al.* (1997).

by different groups to protolytic events associated with oxidoreduction of metal centers in cytochrome oxidase (Hallen and Nilson, 1992; Mitchell and Rich, 1994), as well as to the possible involvement of redox-linked pK shifts of porphyrin substituents (De Paula *et al.*, 1990) and/or metal ligands in proton pumping (Papa *et al.*, 1994; Rich, 1995). Recently, a histidine cycle has been put forward to explain coupling in cytochrome *c* oxidase between oxidoreductions at the binuclear heme a_3–Cu_B center and proton pumping (Chapters 5 and 9, this volume). This model is based on redox-linked binding changes at Cu_B of an invariant histidine ligand that cycles between Cu_B-bound imidazolate (Im^-) and free imidazolium ($ImHH^+$) (see Fig. 6, Chapter 9, this volume). Furthermore, distinct putative proton pathways have been detected in the crystal structure of the oxidase for the transfer of the scalar protons (to be consumed in the protonation of the intermediates of oxygen reduction to H_2O) and the pumped protons, respectively (see Section 5) (Iwata *et al.*, 1995; Tsukihara *et al.*, 1995, 1996; Chapter 9, this volume).

Therefore, it appears of critical relevance to these issues to identify the

Table I

pK Values ± SD Obtained from "Best-Fit" Analysis of the pH Dependence of the Measured H$^+$/COX Ratios for Redox-Linked Scalar H$^+$ Transfer in Free and Liganded Soluble Cytochrome c Oxidase from Bovine Heart

Experimental conditions[a]	Active redox centers	pK$_{ox}$ ± SD	pK$_{red}$ ± SD	Possible linkage
(A)				
A. Unliganded	a Cu$_A$ a_3 Cu$_B$	6.0 ± 0.9	7.2 ± 0.6	Cu$_B$
		6.0 ± 0.9	9.4 ± 0.3	a
		7.2 ± 0.6	≥12.0	a_3
		7.7 ± 0.3	≥12.0	a_3
B. CN liganded	a Cu$_A$ Cu$_B$	6.2 ± 0.2	6.8 ± 0.1	Cu$_B$
		6.1 ± 0.2	8.7 ± 0.1	a
C. CO liganded	a Cu$_A$	5.8 ± 0.1	8.9 ± 0.1	a
(B)				
A-B	a_3	6.8 ± 0.2	≥12.0	
		7.4 ± 0.2	≥12.0	
A-C	Cu$_B$	6.3 ± 0.2	7.3 ± 0.7	
	a_3	7.4 ± 0.7	≥12.0	
		7.4 ± 0.5	≥12.0	
B-C	Cu$_B$	6.5 ± 0.2	7.0 ± 0.1	
(C)	a	6.0 ± 0.2	8.8 ± 0.1	
	a_3	7.1 ± 0.4	≥12.0	
		7.4 ± 0.4	≥12.0	
	Cu$_B$	6.4 ± 0.2	7.0 ± 0.3	

[a](A) refers to the pH dependence of the experimentally measured H$^+$/COX ratios presented in Fig. 8. (B) shows the pKs used for the "best-fit" analysis of the differences among the measured values under the three conditions shown in (A). (C) gives the means of the pK values for four groups calculated in B, C, A-B, A-C, and B-C. For further details, see Capitanio et al. (1997).

four protolytic groups whose pKs have been found to be linked to redox transitions of the metals in the oxidase as well as their membrane topology. One of the two protolytic groups linked to heme a_3 could be an H$_2$O molecule ligated to the heme Fe (Mitchell et al., 1992). The other, or both, may be protolytic residues in subunit I of the oxidase or in the porphyrin ring of the heme. The group linked to Cu$_B$ with pK$_{ox}$ and pK$_{red}$ values around 6 and 7, respectively, could be a histidine in the imidazole–imidazolium state or a H$_2$O/OH$^-$ molecule (Fann et al., 1995). The pK shift seems to be too small to represent an energetically relevant step in the pump. These observations seem, in principle, incompatible with the proposed binding to Cu$_B$ of a histidine in the imidazole–imidazolate state (pK around 14) as postulated in the histidine cycle. Another difficulty for the histidine cycle, at least in its original formulation, seems to be given by the fact that the crystal structure of the beef heart enzyme shows three imidazoles as the ligands to Cu_B both in the oxidized (Tsukihara *et al.*, 1996) and the reduced state (Yoshikawa, 1997; Ostermeier *et al.*, 1997).

The group linked to heme *a* would be represented by a protolytic residue (in subunit I) in direct or indirect cooperative communication with the heme. Mutational analysis of prokaryotic oxidases and the crystal structures of the eukaryotic and prokaryotic oxidases show that there are various conserved protolytic residues in subunit I in the proximity of heme *a* and the binuclear heme a_3–Cu_B center, which could contribute to pK shifts linked to the redox transitions. Amino acid residues distant from the redox centers but still connected to these by polarizable hydrogen bonds also could be involved in the redox Bohr effects (Papa, 1976; Papa *et al.*, 1994; Tsukihara *et al.*, 1996). In this respect, the data obtained by direct measurement of proton transfer linked to oxidoreduction of the metal centers could provide further clues when electrostatic calculations based on X-ray structures become available.

The membrane topology of the redox Bohr effects has been analyzed in the author's laboratory by direct measurement of proton transfer associated to oxidoreduction of the metal centers in the cytochrome *c* oxidase from bovine heart reconstituted in liposomes (COV) (Capitanio *et al.*, 1997b). Figure 9A and Table II show the results obtained with the CO-liganded reconstituted oxidase. The oxidase was fully reduced by succinate in the presence of a trace of broken mitochondria, cytochrome *c*, and CO (see Fig. 3 in Capitanio *et al.*, 1997a). Addition of a small amount of ferricyanide, substoichiometric with the redox metals in the oxidase, led to oxidation of heme *a*, Cu_A, and cytochrome *c*; oxidation of Cu_B and heme a_3 was blocked by CO. The oxidation of heme *a* and Cu_A was accompanied by a synchronous release of H^+, which continued, as expected, during the re-reduction of the metal centers by succinate. The overall reduction of added ferricyanide by succinate should have resulted in the 1-to-1 stoichiometric release of H^+ in the external medium. The observed acidification, however, was significantly larger than the amount of ferricyanide added. There is an additional source of H^+ that appears to be associated with the transient oxidation of heme *a*; the oxidation of cytochrome *c* and Cu_A is irrelevant in this respect as they are pH independent. It is thus evident that the Bohr protons associated with the oxidation of heme *a* are released in the external aqueous phase. If the redox Bohr protons were then taken up by the oxidase from the same external aqueous phase upon re-reduction of heme *a* by succinate, no net excess of H^+ release, with respect to the ferricyanide added, should have been left when heme *a* was fully re-reduced, as it was instead observed. The remaining extra-acidification, which was equal to the rapid H^+ release associated with heme *a* oxidation, shows that the Bohr protons associated with re-reduction of heme *a* are taken up from the inner aqueous phase. This was confirmed by the observation that in the presence of CCCP, which equilibrates the inner and outer pH changes, the same initial rapid acidification was observed upon oxidation heme of *a* by ferricyanide, as in the coupled

A

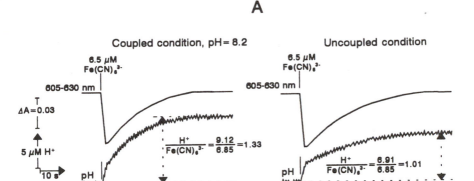

B

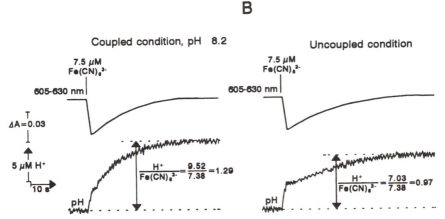

FIGURE 9. Measurement of the sidedness of H^+ transfer associated with redox transitions of metal centers in cytochrome c oxidase vesicles. (A) 2.5 μM COV were suspended in 0.15 M KCl and supplemented with 2.5 μM cytochrome c, 1 μg/ml valinomycin, and 0.1 mg/ml frozen–thawed beef heart mitocondria, pH 8.2. The suspension was gently bubbled with CO for 2 min and then covered with a layer of mineral oil to prevent further gas exchange. Addition of 3 mM succinate produced anaerobiosis and full reduction of cytochrome c oxidase occurred in 10–15 min. Where indicated, anaerobic ferricyanide [substoichiometric with respect to the active redox centers, cytochrome c, heme a, and Cu_A (heme a_3 and Cu_B were clamped in the CO-liganded reduced state)] was added and heme a absorbance and pH changes were monitored simultaneously (the latter followed by a fast responding electrode). Difference spectra collected after addition of succinate showed the characteristics of the fully reduced CO-liganded aa_3 and, upon addition of ferricyanide, those of the mixed valence CO-liganded aa_3. (B) 1.5 μM COV was suspended in the same medium as in A, supplemented with 0.5 μM cytochrome c and 0.1 mg/ml of mitochondria, pH 8.2. The suspension was bubbled with N_2 and treated as in A. The traces on the right refer to the experimental conditions of A and B but in the presence of 2 μg/ml of valinomycin plus 3 μM CCCP. See text and Table II for further details.

Table II
Analysis of the Sidedness of H[+] Transfer Linked
to Redox Transitions of the Metal Centers
in Reconstituted Cytochrome *c* Oxidase Vesicles

Experimental conditions[a]	H^+_T/Ferricyanide added[b]	ΔH^+/COX[c]	H^+_i/COX[d]
CO-liganded, pH 7.4			
Coupled	1.25 ± 0.02	0.86 ± 0.07	0.90 ± 0.10
Uncoupled	0.90 ± 0.05		0.84 ± 0.08
Unliganded, pH 8.2			
Coupled	1.21 ± 0.02	0.92 ± 0.08	2.17 ± 0.09
Uncoupled	1.02 ± 0.01		2.37 ± 0.20

[a]For experimental conditions relative to CO-liganded and unliganded cytochrome *c* oxidase, see Fig. 9.

[b]H^+_T/Ferricyanide: total amount of H^+ released following the oxidation–reduction cycle elicited by the addition of ferricyanide, divided by the amount of ferricyanide added. Internal measurements showed that the total amount of redox carriers oxidized was equivalent to the amount of ferrycianide added. Oxidized cytochrome *c* and heme *a* were directly estimated by absorbance changes, the equivalents of Cu_A undergoing oxidoreduction were assumed to be equal to those measured for heme *a* in the CO-liganded and unliganded conditions; the equivalents Cu_B undergoing oxidoreduction were assumed to be equal to those measured for heme *a* in the unliganded conditions.

[c]ΔH^+/COX refers, under coupled conditions, to the ratio between the extra H^+ release with respect to the amount of ferricyanide added and the amount of cytochrome *c* oxidase undergoing oxidoreduction.

[d]H^+_i/COX refers to the ratio between the initial H^+ release associated with the rapid oxidation of metal centers and the amount of cytochrome *c* oxidase.

system; however, at the completion of the re-reduction of heme *a*, no extra-acidification, with respect to the ferricyanide added, was observed.

The same measurements were also carried out with unliganded COV (Fig. 9B and Table II). The metal centers in the oxidase were reduced by succinate. Their oxidation by anaerobic ferricyanide resulted in the release of 2.2 H^+ per oxidase molecule at pH 8.2, which was quite close to the H^+ release corresponding to Bohr effects associated with oxidation of heme *a* and a_3. At this pH there is no contribution of Bohr effect of Cu_B (see Fig. 5 in Capitanio *et al.*, 1997a). Also, in the unliganded oxidase, when re-reduction by succinate of the metal centers of the oxidase was completed, an extra-acidification, with respect to the ferricyanide added, was left. This extra-acidification was equivalent to that observed with the CO-liganded oxidase. It can thus be concluded that both in the unliganded and CO-liganded oxidase the redox Bohr effect linked to heme *a* is vectorial, i.e., protons are taken up by the enzyme from the inner aqueous phase upon reduction of heme *a* and released in the external aqueous phase upon its oxidation, just as proposed in the vectorial Bohr mechanism (see Fig. 3). The redox Bohr protons linked to heme a_3 appear, on the other hand, to

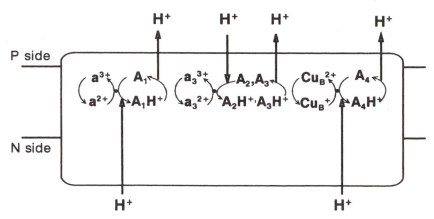

FIGURE 10. Membrane sidedness of Bohr protons linked to oxido-reduction of heme a, heme a_3, and Cu_B. For details, see Fig. 9, Table II, and text.

be taken up and released from the external aqueous phase (Fig. 10). It should be noted that at the pHs used for the experiments summarized in Table II, there is no contribution of the Bohr effect linked to Cu_B (Capitanio *et al.*, 1997a). Measurements are being carried out to examine the situation for the Bohr protons linked to oxidoreduction of this center; however, these never amount to more than 0.25 H^+/oxidase (Fig. 5 in Capitanio *et al.*, 1997a).

The membrane vectoriality of the redox Bohr effect linked to Cu_B was clarified by a statistical analysis of the redox Bohr effects in the unliganded COV at pH 6.7 where the redox Bohr effect for Cu_B reaches the peak of 0.3 H^+/COX (Capitanio *et al.*, 1997b). The data obtained at this pH indicated that the Bohr protons linked to oxidation of Cu_B are released, as those of heme a and heme a_3, in the external space; they are taken up, however, as those linked to heme a, from the inner space upon its re-reduction (Capitanio *et al.*, 1997a; Papa and Capitanio, 1998). It should be reminded, however, that at pHs around 7.4, no H^+ transfer is associated to redox transition of Cu_B (Capitanio *et al.*, 1997b).

The information resulting from these experiments is summarized in Fig. 10. The proton transfer resulting from the Bohr effects linked to heme a and Cu_B displays membrane vectorial asymmetry, that is, protons are taken up from the inner space upon reduction and released in the external space upon oxidation of the metals. This direction of the proton uptake and release is just what is expected from a vectorial Bohr mechanism in which the redox-linked cooperative events are conceived to be extended over the transmembrane span of the enzyme so as to result in proton uptake from the inner space and their release in the external space.

4. ROLE OF NONCATALYTIC SUBUNITS IN HEME–COPPER OXIDASES

Cytochrome *c* oxidase of *Paracoccus denitrificans* consists of four subunits. Subunits I, II, and III are conserved in the other members of the superfamily of heme–copper oxidases, whereas subunit IV does not show similarity to any subunit of the other oxidases. Mutational analysis of the *Paracoccus* oxidase indicates that subunits I and II alone are sufficient for both electron transfer and proton pumping in this enzyme (Haltia *et al.*, 1991). What is then the role of subunit III and IV as well as of the supernumerary subunits, which in higher eukaryotes reach the number of ten (Kadenbach *et al.*, 1983; Chapter 24, this volume)?

Haltia *et al.* (1991) presented evidence showing that subunit III may play a role in the assembly of the *P. denitrificans* cytochrome *c* oxidase. This has been confirmed by mutational analysis of the four subunit aa_3–quinol oxidase of *Bacillus subtilis*, which showed that deletion of the gene coding for subunit III (ΔqoxC) resulted in defective membrane assembly of the enzyme (Villani *et al.*, 1995) (Table III). The same study showed that deletion of the gene encoding subunit IV (ΔqoxD) of the *B. subtilis* quinol oxidase depressed respiration and

Table III
Phenotype Characterization of *Bacillus subtilis* Mutant Strains

Parameters	Value in strain[a]				
	Bs 168	Δqox	ΔqoxC	ΔqoxD	Y284F (qoxB)
Heme a^b	0.35	ND	0.13	0.35	0.44
Heme b^b	0.71	0.58	0.62	0.51	0.36
Heme c^b	NR	0.19	ND	0.18	0.20
O_2 cons.^{c}	182	125	299	88	82
I_{50}, KCN^d	35	177	395	76	68
I_{50}, HQNO^d	0.07	0.4	1.0	1.0	1.0
H^+/e^-	2.0	1.0	0.9	1.2	1.4
$t_{1/2}$ (sec)^e	15	18	18	18	16
Power output^f	365	125	269	106	115
Strept. res.^{g}	–	++	+	+++	+++

[a]Δqox, full deletion of qox operon; ΔqoxC, deletion of the *qoxC* gene (subunit III); ΔqoxD, deletion of the *qoxD* gene (subunit IV); Y284F, site-directed mutagenesis on qoxB (subunit I). For further details see Villani *et al.* (1995) and Papa *et al.* (1994).
[b]nmoles of heme–mg membrane protein.
[c]μM e$^-$/min per OD$_{600}$.
[d]I_{50}, μM concentrations.
[e]$t_{1/2}$ refers to the rate of anaerobic proton back flow.
[f]μM H$^+$ release/min per OD$_{600}$.
[g]Streptomycin resistence tested in a range of 10 to 100 mg/ml.
ND, not detectable.

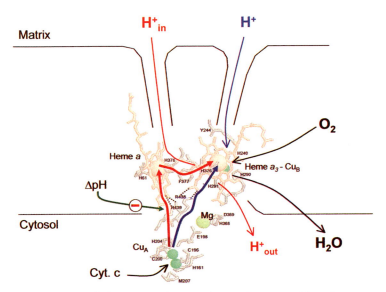

FIGURE 6. Sequence of electron transfer steps in cytochrome c oxidase. Proton-coupled electron transfer steps are shown by red arrows, decoupled electron transfer (slip) by blue arrows. The structure and arrangement of the metal centers and associated residues is drawn from Tsukihara *et al.* (1996). For details, see text and Chapter 6, this volume.

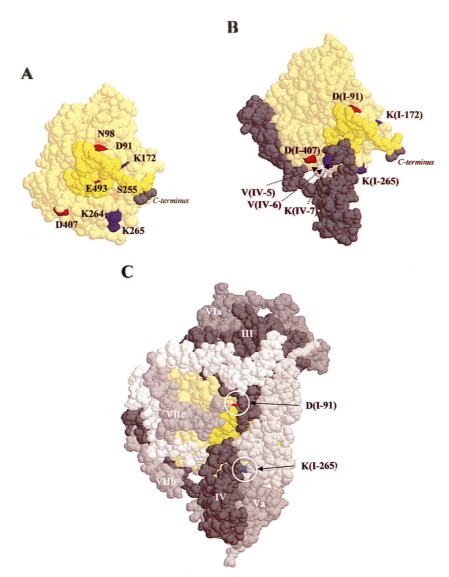

FIGURE 11. Space-fill view of the mass (A, B, and C) subunit I, (B) subunit I and IV, and (C) subunits I, III, IV, Va, Vb, VIa, VIIa, VIIb, VIIc, and VIII of bovine heart cytochrome *c* oxidase protruding at the N side of the membrane. Data from the PDB coordinates of the crystal structure (from Tsukihara *et al.*, 1996) are drawn using the RasMol 2.6 program. The conserved protolytic residues emerging at the N surface, in different colors and with their numbering, of subunits I (yellow) and IV (gray) are shown. The C-terminus of subunit I is also shown in more intensive yellow. For other details, see text.

proton pumping but did not affect the content of aa_3 cytochrome in the membrane. Subunit IV appears, then, to be essential for the activity of the proton pumping aa_3-quinol oxidase. This was confirmed by Saiki *et al.* (1996), who showed that the C-terminal two thirds of subunit IV is required for functional expression of the quinol oxidase and that this subunit assists the Cu_B binding to the binuclear center in subunit I during biosynthesis or assembly of the enzyme. Mutation Y284F in subunit I produced a depression in the catalytic and pumping activity, as well as modification of the CO-reduced minus reduced spectrum similar to that observed in the ΔqoxD mutant of *B. subtilis* (Table III) (Villani, 1995). Recently, it has been proposed (Ostermeier *et al.*, 1997) that in the corresponding Y280 *P. denitrificans* could be covalently bonded to H276, which is one of the Cu_B ligands (see also Fig. 13A).

A role of subunit III in the diffusion of oxygen to the heme a_3–Cu_B binuclear center in subunit I also has been proposed (Chapter 5, this volume). It should be recalled that subunit III but not IV shows a significant degree of sequence similarity in the prokaryotic and eukaryotic members of the superfamily of heme–copper oxidases.

A structural role of subunit IV in proton pumping in bovine heart cytochrome *c* oxidase is suggested by limited proteolysis of the enzyme (Capitanio *et al.*, 1994). Removal by trypsin digestion of the first seven residues from the N-terminus of subunit IV, which protrudes at the inner N-side of the membrane (Figs. 11 and 12), resulted in decoupling of the proton pump. Thermolysin, which cleaved off only the first four residues from the N-terminus of this subunit, did not affect proton pumping (Table IV). The sequence Val5-Val6-Lys7 of the peripheral N-terminus of subunit IV appears to be critical to maintaining the pumping activity of the oxidase. Inspection of the crystal structure of the bovine heart cytochrome *c* oxidase (Tsukihara *et al.*, 1996) shows that the N-terminus of subunit IV forms an extrinsic loop, which interacts with the C-terminal segment of subunit I protruding at the N side of the membrane by means of hydrogen bonds and hydrophobic forces (Figs. 11 and 12). On the other hand, a loop of the carboxy-terminal part of subunit I shapes one side of the access to the conserved D91 and N98 (Fig. 11), which are essential for the entrance of protons in the channel (channel D) that has been proposed to translocate pumped protons (Iwata *et al.*, 1995; Tsukihara *et al.*, 1996). It is conceivable that the N-terminus of subunit IV interacting with the carboxy-terminal part of subunit I keeps it in a position, around the entrance of channel D, which promotes the entry of protons from the bulk N phase into the channel. Proteolytic removal of Val5-Val6-Lys7 from the N-terminus of subunit IV might destroy this stabilizing effect of the subunit. Tsukihara *et al.* (1996) have identified another possible proton conduction pathway in the crystal structure of the oxidase consisting of protolytic residues of helices XI and XII, which form with intercalated H_2O molecules a hydrogen-bond net-

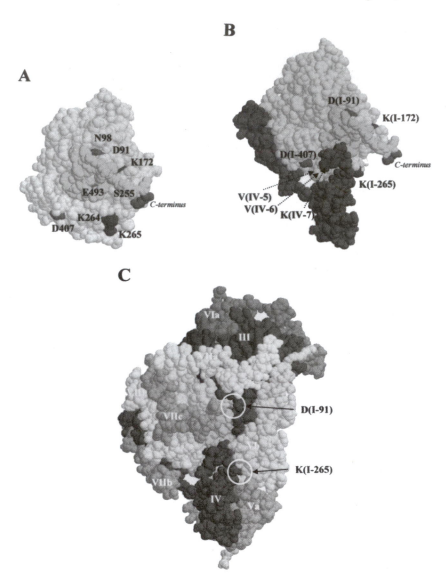

FIGURE 11. Space-fill view of the mass (A, B, and C) subunit I, (B) subunit I and IV, and (C) subunits I, III, IV, Va, Vb, VIa, VIIa, VIIb, VIIc, and VIII of bovine heart cytochrome *c* oxidase protruding at the N side of the membrane. Data from the PDB coordinates of the crystal structure (from Tsukihara *et al.*, 1996) are drawn using the RasMol 2.6 program. The conserved protolytic residues emerging at the N surface, in different colors and with their numbering, of subunits I (yellow) and IV (gray) are shown. The C-terminus of subunit I is also shown in more intensive yellow. For other details, see text. For color representation of this figure, see color tip facing page 69.

work that could transfer pumped protons from the N to the P side of the membrane. The crystal structure of the oxidase shows that the single trans-membrane helix of subunit IV contacts helix XII of subunit I, crossing it at an angle of 50 ° (Fig. 12). Therefore, subunit IV could contribute to stabilize the proposed proton channel, made up by packing of helices XI and XII, through helix–helix interaction in the membrane and interaction of its N-terminus with the C-terminus of subunit I at the N surface of the oxidase (Figs. 11 and 12).

5. A COOPERATIVE PROTON PUMP IN HEME–COPPER OXIDASES

The structural and functional information now available allows us to come back to the cooperative model for proton pumps in redox enzymes (vectorial Bohr mechanism) (Papa *et al.*, 1973; Papa, 1976) and attempt to define it at a molecular level in heme–copper oxidases.

The results presented in Section 3 indicate that transfer via heme *a* of each of the four electrons from Cu_A to the binuclear a_3–Cu_B center for the reduction of O_2 to $2H_2O$ is associated with uptake of protons from the N aqueous phase and their translocation toward the P side of the membrane. Coupling of proton translocation to electron transfer via heme *a* is afforded by the Bohr effect linked to oxidoreduction of this heme. It might be recalled, in this respect, that Raman spectroscopy (Rousseau *et al.*, 1988) as well as X-ray crystallography (Ostermeier *et al.*, 1997) reveal H_2O molecule(s) near heme *a*. It has been proposed that this H_2O could shuttle protons from the N aqueous space to the protolytic group(s), whose pK is governed by the redox state of heme *a* (Rousseau *et al.*, 1988). The crystal structure of the *P. denitrificans* oxidase shows that hydrogen bonds can be formed between the conserved R54 and the formyl substituent of heme *a* and between T50 and the OH group of the farnesyl substituent (Iwata *et al.*, 1995). In the bovine oxidase the threonine is replaced by a serine (S34) and the crystal structure shows that, in addition to the hydrogen bond between the conserved R38 and the formyl substituent of heme *a*, S382 can be hydrogen bonded to the OH group of the farnesyl (Tsukihara *et al.*, 1996). S34 (bovine, T50 in *P. denitrificans* oxidase) and S382 (bovine, S417 in *P. denitrificans*) are situated on the opposite sides of the farnesyl chain. Furthermore, it can be noted that a conserved glutamate (E242 bovine, E278, *P. denitrificans*) of the D proton channel in subunit I can be in connection with both heme *a* and the binuclear center.

The critical role in proton pumping of the Bohr effect linked to heme *a* is substantiated by its vectorial nature (H^+ uptake from the N phase, release in the P phase) as well as by the correspondence of the pH dependence curves of the heme *a* Bohr effect with that of the H^+/e^- ratio for proton ejection associated with ferrocytochrome *c* oxidation in purified cytochrome *c* oxidase

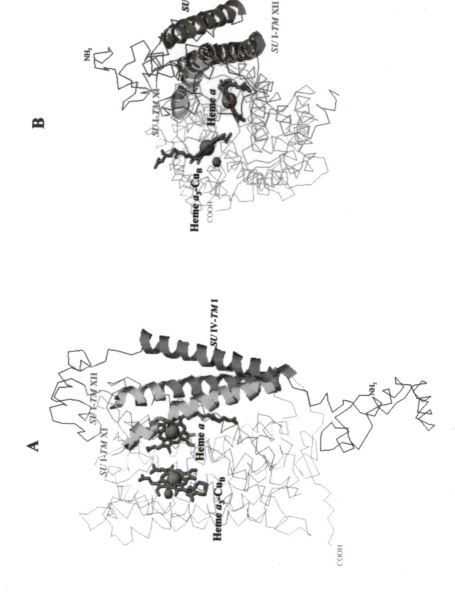

Table IV
Effect of Proteolytic Digestion of Cytochrome c Oxidase on Electron Transfer and Proton Translocation Activity in Reconstituted Cytochrome c Oxidase Vesicles

	Experimental conditions[a]		
	Control	Thermolysin	Trypsin
Km (cyt.c^{2+}, μM)	5.6	6.0	5.6
V_{max} (T.N.·sec^{-1})	226	208	188
R.C.I.	11.7	15.0	9.6
H^+/e^-	0.90	0.81	0.20
H^+ passive conduction (μMH$^+$/min)	63	61	60
Segments digested in subunit IV		(H$_2$N-A-H-G-S-)	(H$_2$N-A-H-G-S-V-V-K-)

[a]For experimental details, see Capitanio et al. (1994).

vesicles (compare Fig. 8B with Fig. 5 in Papa et al., 1987). It can also be recalled that electron flow from heme a to heme a_3–Cu$_B$ is controlled by the aerobic transmembrane ΔpH and in particular by the pH of the N phase (Capitanio et al., 1990). As discussed above, the crystal structure of the bovine oxidase shows, in addition to the residue network providing facile electron transfer from Cu$_A$ to heme a and from the latter to heme a_3–Cu$_B$, another residue network that could mediate direct electron transfer from Cu$_A$ to heme a_3–Cu$_B$. Enhanced contribution of this latter pathway, bypassing heme a, as seems to occur at high electron pressure or when the proton-motive electron transfer via heme a is depressed by the back pressure of aerobic Δμ$_{H^+}$, would also be consistent with a critical role in proton pumping of the Bohr effect linked to heme a (see Section 2).

In the last few months, further progress has been made in the crystallographic elucidation of the structure of cytochrome c oxidase of P. denitrificans (Ostermeier et al., 1997) and bovine heart (Yoshikawa, 1997) and the functional characterization of site-specific mutations in subunit I of bacterial heme–copper oxidases (Konstantinov et al., 1997; Vygodina et al., 1998; Zaslavsky and Gennis, 1998; Adelroth et al., 1997, 1998; Verkhovskaya et al., 1997;

FIGURE 12. (A) Parallel and (B) along the membrane normal from the N side views of subunits I and IV of bovine heart cytochrome c oxidase. Data from the PDB coordinates of the crystal structure (from Tsukihara et al., 1996) are drawn using the RasMol 2.6 program. Transmembrane helices SUI-TMXI, SUI-TMXII, and SUIV-TMI are shown as ribbons to evidentiate their contacts. The contacts of the C-terminal region of subunit I and the N-terminal region of subunit IV at the N surface are also shown. For other details, see text. For color representation of this figure, see color tip facing page 74.

Watmough *et al.*, 1997). The crystal structures of the *P. denitrificans* (Iwata *et al.*, 1995) and bovine oxidase (Tsukihara *et al.*, 1996) revealed two distinct pathways for proton conduction from the N phase to the binuclear center in subunit I: channel D was proposed to be used for pumped protons and channel K for the protons consumed in the chemistry of O_2 reduction to H_2O (Fig. 13). Channel D starts at the N side of subunit I with a conserved cluster of protolytic residues, consisting of (going from the surface to the interior of the protein) D91, N80, N98, N99, N163 (bovine; D124, N113, N131, N132, N199 in *P. denitrificans*) (Ferguson-Miller, 1993; Thomas *et al.*, 1993), which can be connected by a chain of water molecules (Hofacker and Schulten, 1998) to E242 (bovine; E278, *P. denitrificans*). This conserved glutamate located in helix VI (Verkhovskaya *et al.*, 1997) is, on one side, close to ligands of a_3-Cu_B binuclear center (Fann *et al.*, 1995), on the other side it faces a cavity, pointing to heme *a*, contributed by helices VI, X, and II (Iwata *et al.*, 1995; Tsukihara *et al.*, 1996) (see Fig. 14). Molecular dynamics simulation (Hofacker and Schulten, 1998) suggests that E242 (bovine; E278 *P. denitrificans*) can swing during

A   **B**

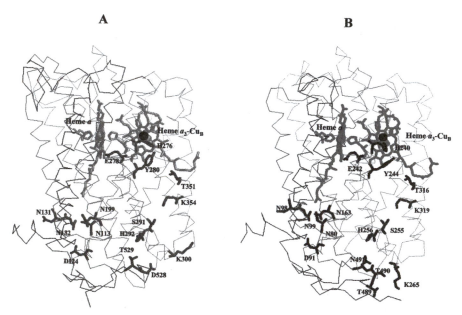

FIGURE 13. View parallel to the membrane of the location in subunit I of (A) *P. denitrificans* and (B) bovine heart cytochrome *c* oxidase of heme *a*, heme a_3–Cu_B, and protolytic residues contributing to proton-conducting pathways. Data from the PDB coordinates of the crystal structure of *P. denitrificans* (Iwata *et al.*, 1995) and bovine heart cytochrome *c* oxidase (Tsukihara *et al.*, 1996) are drawn using the RasMol 2.6 program. For other details, see text.

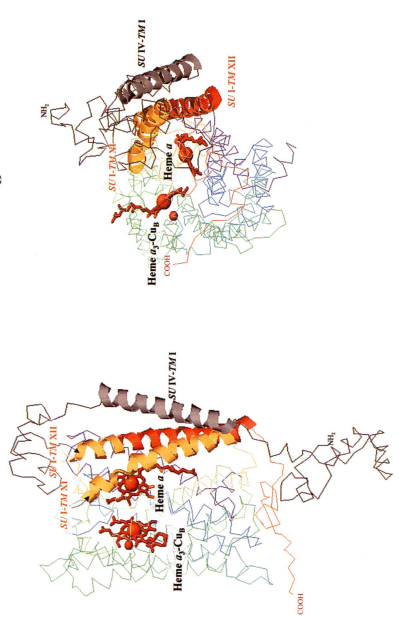

FIGURE 12. (A) Parallel and (B) along the membrane normal from the N side views of subunits I and IV of bovine heart cytochrome *c* oxidase. Data from the PDB coordinates of the crystal structure (from Tsukihara *et al.*, 1996) are drawn using the RasMol 2.6 program. Transmembrane helices SUI-TMXI, SUI-TMXII, and SUIV-TMI are shown as ribbons to evidentiate their contacts. The contacts of the C-terminal region of subunit I and the N-terminal region of subunit IV at the N surface are also shown. For other details, see text.

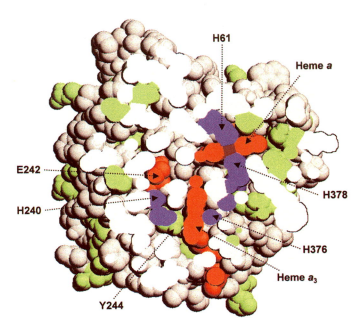

FIGURE 14. Space-fill cross-section view along the membrane normal of subunit I of bovine heart cytochrome *c* oxidase. Data from the PDB coordinates of the crystal structures are drawn using the RasMol 26 program. The hemes *a* and a_3 and the glutamate E242 are shown in red, the histidine ligands to heme *a*, a_3, and Cu_B are in light blue. Polar residues are shown in green. The cavities in the interior of subunit I are in gray, and the cut masses of residues in white are also made visible. For other details, see text.

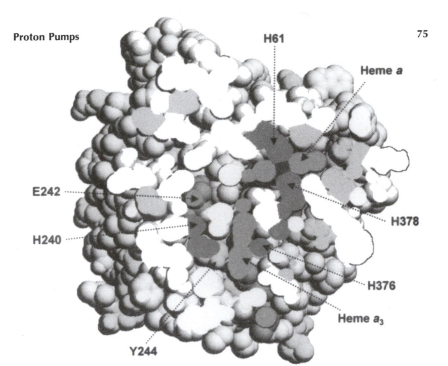

FIGURE 14. Space-fill cross-section view along the membrane normal of subunit I of bovine heart cytochrome c oxidase. Data from the PDB coordinates of the crystal structures are drawn using the RasMol 26 program. The hemes a and a_3 and the glutamate E242 are shown in red, the histidine ligands to heme a, a_3, and Cu_B are in light blue. Polar residues are shown in green. The cavities in the interior of subunit are in gray, and the cut masses of residues in white are also made visible. For other details, see text. For color representation of this figure, see color tip facing this page.

the redox steps of catalysis. On the side of the cavity the glutamate can apparently be connected, by flipping and through H_2O molecules, to the histidine ligands to heme a (Fig. 14).

Evidence has been presented that the redox potential of heme a is lowered by 60 mV upon mutation of N167 of subunit I of *Saccharomyces cerevisiae* cytochrome c oxidase (N163 in the D channel of the bovine and N199 in the (Rich *et al.*, 1996; Chapter 8, this volume). A similar perturbation of the redox Bohr effect in heme a seems to be caused by removal of E242 (bovine numbering; E278 *P. denitrificans*) (Garcia-Horsman *et al.*, 1994). It also been observed that replacement of this glutamate in subunit I of cytochrome c oxidase of *Rhodobacter sphaeroides* blocks electron transfer from Cu_A to heme a (Adelroth *et al.*, 1997). The particular location of this conserved glutamate and its possible alternating connection to heme a and the binuclear center environment may represent the basis for involvement of the D pathway in the

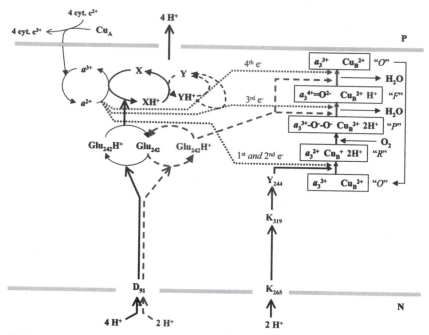

FIGURE 15. A cooperative proton pump model in cytochrome *c* oxidase with the involvement of the vectorial Bohr effect linked to heme *a* and proton conduction pathways identified by mutational analysis of heme–copper oxidases and X-ray crystals of *P. denitrificans* and bovine heart oxidases. Solid black lines denote the pathway of pumped protons, dashed (in channel D), and solid blue lines (in channel K) denote pathways of chemical protons consumed in the four steps of the chemistry of O_2 reduction to H_2O. Dashed black line shows possible cooperative coupling, involving a Y proton trap, between the utilization of the third and fourth electrons and the output pathway of pumped protons. Dotted lines show the transfer of the first and second (in blue) and the third and fourth (in red) electrons from Heme *a* to the binuclear center in the reduction of O_2 to $2 H_2O$. Yellow lines denote the inner (N) and outer (P) surfaces of the membrane. For details, see text.

translocation of both the pumped protons and those consumed in the peroxy→ ferryl and ferryl→oxy intermediate steps of the reduction of O_2 to H_2O (see Fig. 15) (see Konstantinov *et al.*, 1997; Adelroth *et al.*, 1997; Chapter 8, this volume).

Putting together our results with the body of information just summarized, a detailed cooperative model of proton pump (Fig. 15) in cytochrome *c* oxidase and heme–copper oxidases in general (Cu_A is missing in the quinol oxidases), in which proton pumping is essentially provided by the vectorial Bohr effect linked to heme *a* (heme *b* in *bo* oxidases) is proposed. The model we propose incorporates, together with the central role of redox Bohr effect linked to heme

a, cooperative interaction between heme a and the a_3-Cu_B binuclear center (Fig. 15). Upon delivery to the oxidase of the first and second electrons (of the four which reduce O_2 to $2H_2O$), reduction of heme a, through enhancement of the pK of a critical residue of its environment in protonic connection with E242 (bovine numbering) or of E242 itself, will result in the uptake, via channel D, of the first two pumped protons from the N aqueous space. With the transfer of the first and second electrons from heme a to the binuclear center, the first two pumped protons are released in the P aqueous phase directly, or after transient trapping by a second acceptor [Rottenberg, 1998; Rich, this volume] until the third and fourth electrons are utilized in the two-step reduction of the H_2O_2 bound at the binuclear center to $2H_2O$ (see below). The proton release pathway can be contributed by the output structure of D51, which undergoes the redox-linked conformational change described by Yoshikawa et $al.$ (1998). Arrival of the first two electrons at the binuclear center are then associated with uptake, via channel K, of two protons to neutralize the negative charges introduced there, followed, upon oxygen binding, by formation of the peroxy compound (Konstantinov et $al.$, 1997).

Arrival of the third and fourth electrons will result, upon reduction of heme a, in the uptake via channel D of the third and fourth pumped protons. Upon transfer of the third and fourth electron to the binuclear center and their utilization in the P→F and F→O steps the third and fourth pumped protons, together with the first and second if they had been transiently trapped by protolytic residues at the P side, are released in the P aqueous phase. Of the four reductive steps in the catalytic cycle of the oxidase, the reductive cleavage of the peroxy and ferryl compound are, in fact, those associated with the largest ΔG (Babcock & Wikstrom, 1989; Wikstrom, 1989). Response of the redox equilibria of O, F, and P intermediates to mitochondrial energization by ATP (Wikstrom, 1989) as well as direct dynamic measurement of charge transloca-tion in the oxidation of reduced cytochrome c oxidase by O_2 (Verkhovsky et $al.$, 1997), in the isolated P→ and F→O transitions (Konstantinov, 1998), and of proton translocation in the peroxidase reaction (Vygodina et $al.$, 1997) have provided evidence indicating that the P→F and F→O steps are each coupled to pumping of $2H^+/e^-$. If this is indeed the case, one has to think that the P→F and F→O steps induce a conformational change, extending from the binuclear center domain to the proton exit pathway, which results in the release of the pumped protons in the P aqueous phase.

As already pointed out, evidence has been obtained indicating that the D channel, but not the K channel, can also be used for the transfer of the two chemical protons utilized in the P→F and F→O steps (Konstantinov et $al.$, 1997; Adelroth et $al.$, 1997; Konstantinov, 1998). Upon transfer of the third and fourth electrons from the heme a to the binuclear center, the pK of the group linked to heme a returns to the low pK_0 value. At the same time, strong negative charges are generated at the binuclear center so that E242, in the absence of proton delivery by the K channel, now transfers chemical protons to the binuclear center.

The crystal structure of subunit I of bovine heart oxidase shows that the conserved E242 (facing a cavity delimitated by helices VI, II, and X) can, on one side, establish protonic connection, through protolytic residues of helix II and bound water molecules, with H61, ligand to the heme *a* Fe, or with protonable groups of the heme *a* porphyrin. Mutation in *Saccharomyces cerevisiae* of the leucine corresponding to I66 located in the bovine oxidase between E242 and heme *a* has been found to lower the heme *a* redox potential and its pH dependence (Meunier *et al.*, 1998; Rich, this volume). On the other side, E242 is close to the backbone of H240 and could be connected by water molecules to the binuclear center (Puustinen *et al.*, 1997; Riistama *et al.*, 1997; Hofacker & Schulten, 1998).

In the mechanism outlined in Fig. 15, E242 thus plays a central role; disturbance of its sequential switch from the pump to the scalar proton transfer function, as could occur at high rates of electron transfer and/or high transmembrane ΔpH, could represent another cause of decoupling of the proton pump, as is observed under these conditions (Papa *et al.*, 1994; Capitanio *et al.*, 1996). In the model illustrated in Fig. 15, no immediate role is apparent for the redox Bohr effect linked to Cu_B. The Cu_B linked protolytic group with pK_{ox} 6 and pK_{red} 7 (Capitanio *et al.*, 1997a) could only contribute up to a maximum of 0.3 H^+/e^- at pH 6.7; at pH 7.4 and higher, no redox-linked H^+ transfer by the Cu_B-linked group takes place (Capitanio *et al.*, 1997a; contrast Wikström, 1989; Chapter 9, this volume). The possibility of a cooperative or electrostatic interaction of Cu_B ligand in the proton pumping pathway outlined in Fig. 15, however, deserves attention.

The other possible channel for pumped protons identified by Tsukihara *et al.* (1996) in the crystal structure of the bovine heart oxidase (network B in Fig. 10 of Tsukihara *et al.*, 1996), contributed by residues of helices XI and XII of subunit I (see Fig. 12), might be controlled by redox transitions of heme *a*. The OH group of the farnesyl substituent of heme *a* is hydrogen bonded to S382 in this channel (Fig. 12); (see also Papa and Capitanio, 1998). The protolytic residues in this channel, however, are not conserved in other heme–copper oxidases. Furthermore, the H_2O molecules that could be placed by mathematical simulation in this pathway in the crystal structure of *P. denitrificans* oxidase are too few to bridge the gaps between polar residues and the pathway does not contain flexible protonable residues (Hofacker and Schulten, 1998).

It should be noted that in the scheme proposed in Fig. 15, the transfer in the oxidase of each of the four electrons, which reduce O_2 to $2H_2O$, results in the pumping from the N to the P bulk phase of up to 1 H^+/e^- (at pH around 7.4). Results have been produced on which basis it has been proposed that the steps

P→F and F→O are coupled each to the pumping of 2 H^+/e^- from the N to the P phase (Wikström, 1989). If this is the case (see also Vygodina et al., 1998), the two vectorial protons associated with the transfer of the first two electrons, utilized in the O→P step, should transiently accumulate on residues at the P side of subunit I until, upon completion of O_2 reduction in the P→F and F→O steps, all four pumped protons are released in the P phase. The P→F and F→O steps each result in the production of an H_2O molecule. It is conceivable that the release of the pumped protons depends on or even follows the pathway of the release in the P bulk phase of the H_2O molecules along the water channel identified in the crystal structure of the bovine oxidase at the interface between subunit I and II (Tsukihara et al., 1996).

6. POSSIBLE MECHANISMS OF THE PROTON PUMP IN COMPLEXES I AND III

The maximal H^+/e^- and q^+/e^- stoichiometries in complex I and complex III are 2 and 2 and 2 and 1, respectively (see Fig. 1). The proton pump of the bc_1 complex (complex III) is widely explained in terms of the Q-cycle mechanism introduced by Mitchell in 1976 (Mitchell, 1976). This model is based on: (1) the existence in the bc_1 complex of two separate quinone reaction centers, the ubiquinol oxidation site at the P side (Q_P) and the ubiquinol reduction site at the N side of the membrane (Q_N); (2) transfer of the first electron arising from the ubiquinol oxidation at Q_P to the high potential acceptor Fe-S center (E_{m7}, +290 mV) and cycling back of the second electron via the low potential heme b_{566} (E_{m7}, −20 mV) and b_{562} (E_{m7}, +20 mV) to the Q_N site where ubiquinone is re-reduced to ubiquinol (see Fig 2). The existence of the two distinct Q_P and Q_N sites has been definitely substantiated by inhibition studies, mutational analysis, and recently by X-ray crystallographic analysis of the complex (Xia et al., 1997; Chapter 12, this volume). The Q_P site, which has to assure the branching of the first and second electron arising from ubiquinol oxidation to a high- and low-potential acceptor, respectively, is the crucial element of the bc_1 complex. It has been proposed that this site accommodates two quinone molecules in an edge-to-edge (Ding et al., 1995) or a stacked charge transfer complex (Brandt, 1996). Double quinone occupancy of the Q_P site or a mechanism based on one single quinone molecule binding at this site (Chapter 13, this volume) now may be verified by X-ray analysis of the crystals of the bc_1 complex underway in different laboratories (Chapter 12, this volume; Berry et al., 1995; Xia et al., 1997).

On the other hand, Papa et al. (1983, 1989) have proposed a Q-gated proton pump, which is based on a split and not a cyclic electron transfer in the bc_1 complex mediated by a protein stabilized SQ_2^-–QH_2 couple, which departs in this and certain other respects from the Q-cycle formalism. The Q-gated proton pump envisages the existence of a ubiquinone pocket in the bc_1 complex, contributed essentially by domains of cytochrome b and the Rieske FeS protein, containing a SQ_2^-–QH_2 molecule intercalated between the ubiquinone of the pool and the FeS center, which oscillates between the Q_P and the Q_N site. The ubiquinone of the pocket will partition itself between the quinol free state and the SQ_{2N}^- and SQ_{2P}^- state. The Q-gated proton pump model can explain as well as the Q-cycle mechanism the various peculiar features of the electron transfer and proton translocation in the bc_1 complex (Papa *et al.*, 1989). In the Q-gated pump, the Q_P site might also accommodate an additional quinone molecule, as in the case of the Q-cycle. The quinone in the pocket transfers in two successive SQ_2^-–QH_2 cycles the two electrons from the ubiquinol of the pool to the FeS center. The two electrons, however, are not equivalent with respect to proton pumping. The transfer of the first electron of the quinol of the pool is associated with the release in the P phase of two scalar protons. The transfer of the second electron results in vectorial translocation of $2H^+$ and $2q^+$ from the N to the P phase. In the Q-gated proton pump, the central element is represented by the redox cycling of the SQ_2^-–QH_2 couple which results in the transfer of $2\ H^+/e^-$, simply from the differences in the pKs of the SQ_2^- and QH_2. Transmembrane proton pumping is assured by linkage of the proton-motive SQ_2^-–QH_2 cycling to cooperative proton transfer from the N phase to the SQ_2^-–QH_2 couple by redox Bohr affects associated with heme b_{562}, from the SQ_2^-–QH_2 couple to the P phase by redox Bohr effects associated with heme b_{566}, and possibly also with the FeS center. The midpoint redox potentials of these three electron carriers is, in fact, pH dependent (Urban and Klingerberg, 1969; Link *et al.*, 1992).

   In complex I, as in complex III, $2H^+$ are ejected per electron transferred by the enzyme. This stoichiometry allowed Papa *et al.* (1994) to propose that, as in the Q-gated pump of the bc_1 complex, the central element of the pump in complex I is also provided by a protein-stabilized SQ_2^-–QH_2 couple, whose existence in the complex in fact had been discovered by Kotylar *et al.* (1990). Papa *et al.* (1994) proposed that the SQ_2^-–QH_2 couple was intercalated between two FeS centers in the complex. Since all the FeS centers in the complex, however, later have been located on the NADH side of the complex with respect to the protein-bound quinone(s) (see Chapter 14, this volume), one has to envisage that the critical SQ_2^-–QH_2 couple is directly oxidized by the quinol of the pool or by an additional protein-bound Q–SQ_2^- couple (Fig. 16). A number of different detailed models, however, have been proposed by various authors for the proton pump of complex I (for review, see Brandt, 1997). One of these is based on a cooperative linkage between redox Bohr effects at the FeS cluster N-2 and proton-motive redox catalysis of ubiquinone (Brandt, 1997). If this were the case, here we have another example of a pump utilizing a vectorial

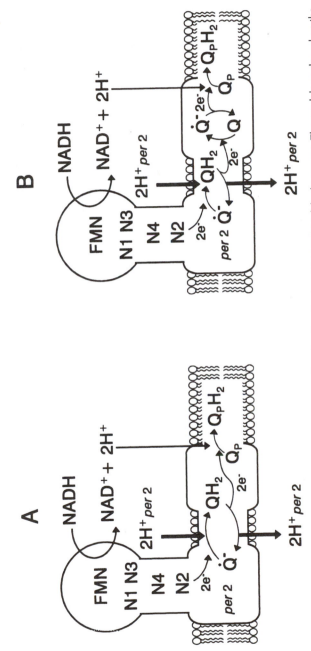

FIGURE 16. Tentative models for redox-linked proton translocation in mitochondrial NADH dehydrogenase. The models are based on the information reviewed by Shulte and Weiss (Chapter 14, this volume) and translocation of 2 H$^+$/e$^-$ by protein-bound Q$_2^-$–QH$_2$ couple redox-connected to the quinone of the pool directly (model A) or via a protein bound Q–Q$_2^-$ couple.

Bohr mechanism. Clearly, more work on the functional properties of complex I needs to be done and hopefully a crystal structure needs to become available to better understand the molecular mechanism of proton pumping in complex I.

ACKNOWLEDGMENTS. The authors are grateful to Prof. Shinya Yoshikawa for discussion and comments and to Drs. Giuseppe Capitanio and Emanuele De Nitto for their contribution to the unpublished experiments presented in this chapter. The research was supported by grant No. 96.03707.CT14 Biotechnology Committee, grant No. 97.01167.PF49 (Biotechnology Oriented Project) Consiglio Nazionale delle Ricerche, Italy, and the National Project of Ministry for University and Scientific and Technological Research (MURST) for Bioenergetics and Membrane Transport.

7. REFERENCES

Abrahams, I. P., Leslie, A. G. W., Lutter, R., and Walker, J., 1994, Structure at 2.8 Å resolution of F1-ATPase from bovine heart mitochondria, *Nature* **370**:621–628.

Adelroth, P., Svensson Ek, M., Mitchell, D. M., Gennis, R., and Brzezinski, P., 1997, Glutamate 286 in cytochrome aa_3 from *Rhodobacter sphaeroides* is involved in proton uptake during the reaction of the fully reduced enzyme with dioxygen, *Biochemistry* **36**:13824–13829.

Adelroth, P., Gennis, R., and Brzezinski, P., 1988, Role of the pathway through K(I-362) in proton transfer in cytochrome *c* oxidase from *R. sphaeroides*, *Biochemistry* **37**:2470–2476.

Azzone, G. F., and Luvisetto, S., 1988, Molecular events in coupling and uncoupling of oxidative phosphorylation, *Ann. NY Acad. Sci.* **550**:277–288.

Babcock, G. T., and Callahan, P. M., 1983, Redox-linked hydrogen bond strength changes in cytochrome *a*: Implications for a cytochrome oxidase proton pump, *Biochemistry* **22**:2314–2319.

Babcock, G. T., and Wikström, M., 1992, Oxygen activation and the conservation of energy in cell respiration, *Nature* **356**:301–309.

Berry, E. A., Shulmeister, V. M., Huang, L., and Kim, S.-H., 1995, A new crystal form of bovine heart ubiquinol:cytochrome *c* oxidoreductase: Determination of space group and unit-cell parameters, *Acta Crystallogr. D* **51**:235–239.

Blair, D. F., Gelles, J., and Chan, S. I., 1986, Redox-linked proton translocation in cytochrome oxidase: The importance of gating the electron flow, *Biophys. J.* **50**:713–733.

Boyer, P. D., Chance, B., Ernster, L., Mitchell, P., Racker, E., and Slater, E. C., 1977, Oxidative phosphorylation and photophosphorylation, *Annu. Rev. Biochem.* **46**:955–1026.

Brandt, U., 1996, Energy conservation by bifurcated electron-transfer in the cytochrome bc_1 complex, *Biochim. Biophys. Acta* **1275**:41–46.

Brandt, U., 1997, Proton-translocation by membrane-bound NADH:ubiquinone–oxidoreductase (complex I) through redox-gated ligand conduction, *Biochim. Biophys. Acta* **1318**:79–91.

Brasseur, R., 1988, Calculation of the three-dimensional structure of *Saccharomyces cerevisiae* cytochrome *b* inserted in a lipid matrix, *J. Biol. Chem.* **263**:12571–12575.

Brown, G. C., 1989, The relative proton stoichiometries of the mitochondrial proton pumps are independent of the proton motive force, *J. Biol. Chem.* **264**:14704–14709.

Brown, G. C., and Brandt, M. D., 1986, Changes in permeability to protons and other cations at high proton motive force in rat-liver mitochondria, *Biochem. J.* **234**:75–81.

Capitanio, N., De Nitto, E., Villani, G., Capitanio, G., and Papa, S., 1990, Proton-motive activity of cytochrome c oxidase: Control of oxidoreduction of the heme centers by the proton-motive force in the reconstituted beef heart enzyme, *Biochemistry* **29:**2939–2945.

Capitanio, N., Capitanio, G., De Nitto, E., Villani, G., and Papa, S., 1991, H^+/e^- stoichiometry of mitochondrial cytochrome complexes reconstituted in liposomes, *FEBS Lett.* **288:**179–182.

Capitanio, N., Peccarisi, R., Capitanio, G., Villani, G., De Nitto, E., Scacco, S., and Papa, S., 1994, Role of nuclear-encoded subunits of mitochondrial cytochrome c oxidase in proton pumping revealed by limited enzymatic proteolysis, *Biochemistry* **33**(41):12521–12526.

Capitanio, N., Capitanio, G., Demarinis, D. A., De Nitto, E., Massari, S., and Papa, S., 1996, Factors affecting the H^+/e^- stoichiometry in mitochondrial cytochrome c oxidase: Influence of the rate of electron flow and transmembrane delta pH, *Biochemistry* **35:**10800–10806.

Capitanio, N., Vygodina, T. V., Capitanio, G., Konstantinov, A. A., Nicholls, P., and Papa, S., 1997a, Redox-linked protolytic reactions in soluble cytochrome c oxidase from beef heart mitochondria, *Biochim. Biophys. Acta* **1318:**255–265.

Capitanio, N., Capitanio, G., De Nitto, E., and Papa, S., 1997b, Vectorial nature of redox Bohr effects in bovine heart cytochrome c oxidase, *FEBS Lett.* **414:**414–418.

Clark, W. M., 1960, *Oxidation–Reduction Potentials of Organic Systems*, Waverley Press, Baltimore, MD.

Cocco, T., Lorusso, M., Di Paola, M., Minuto, M., and Papa, S., 1992, Characteristics of energy-linked proton translocation in liposome reconstituted bovine cytochrome bc_1 complex. Influence of the proton-motive force on the H^+/e^- stoichiometry, *Eur. J. Biochem.* **209:**475–481.

Cocco, T., Di Paola, M., Minuto, M., Carlino, V., Papa, S., and Lorusso, M., 1997, Steady-state proton translocation in bovine heart mitochondrial bc_1 complex reconstituted into liposomes, *J. Bioenerg. Biomembr.* **29:**81–87.

Collinson, I. R., van Raaij, M. J., Runswick, M. J., Fearnley, I. M., Skehel, J. M., Orriss, G. L., Miroux, B., and Walker, J. E., 1994, ATP synthase from bovine heart mitochondria. *In vitro* assembly of a stalk complex in the presence of FI-ATPase and in its absence, *J. Mol. Biol.* **242:**408–421.

Deisenhofer, J., and Michel, H., 1989, The photosynthetic reaction center from the purple bacterium *Rhodopseudomonas viridis*, *Science* **245:**1463–1471.

De Paula, J. C., Peiffer, W. E., Ingle, R. T., Centeno, J. A., Ferguson-Miller, S., and Babcock, G. T., 1990, Hemes a and a_3 environments of plant cytochrome c oxidase, *Biochemistry* **29:**8702–8706.

Ding, H., Moser, C. C., Robertson, D. E., Tokito, M. K., Daldal, F., and Dutton, P. L., 1995, Ubiquinone pair in the Q_0 site central to the primary energy conversion reactions of cytochrome bc_1 complex, *Biochemistry* **34:**15979–15996.

Fann, Y. C., Ahmed, I., Blackburn, N. J., Boswell, J. S., Verkhovskaya, M. L., Hoffman, B. M., and Wikström, M., 1995, Structure of Cu_B in the binuclear heme–copper center of the cytochrome aa_3-type quinol oxidase from *Bacillus subtilis*: An ENDOR and EXAFS study, *Biochemistry* **34:**10245–10255.

Ferguson-Miller, S., 1993, Cytochrome oxidase, *J. Bioenerg. Biomembr.* **25:**167–188.

Fetter, J. R., Qian, J., Shapleigh, J., Thomas, J. W., Garcia-Horsman, A., Schmidt, E., Hosler, J., Babcock, G. T., Gennis, R. B., and Fergusson-Miller, S., 1996, Possible proton relay pathways in cytochrome c oxidase, *Proc. Natl. Acad. Sci. USA* **92:**1604–1608.

Fitton, V., Rigoulet, M., Ouhabi, R., and Guerin, B., 1994, Mechanistic stoichiometry of yeast mitochondrial oxidative phosphorylation, *Biochemistry* **33:**9692–9698.

Fitzgerald, L. D., 1976, Oxygen consumption of mammalian tissues, in: *Biological Handbook, Cell Biology*, vol. 1, (P. L. Altman, and D. Dimmer Katz, eds.), pp. 72–89, Federation of American Societies for Experimental Biology, Bethesda, MD.

Garcia-Horsman, J. A., Barquera, B., Rumbley, J., Ma, J., and Gennis, R., 1994, The superfamily of heme–copper respiratory oxidases, *J. Bacteriol.* **176:**5587–5600.

Hallen, S., and Nilsson, T., 1992, Proton transfer during the reaction between fully reduced cytochrome *c* oxidase and dioxygen: pH and deuterium isotope effects, *Biochemistry* **31:**11853–11859.

Hallen, S., Brzezinski, P., and Malmström, B. G., 1994, Internal electron transfer in cytochrome *c* oxidase is coupled to the protonation of a group close to the bimetallic site, *Biochemistry* **33:**1467–1472.

Haltia, T., Saraste, M., and Wikström, M., 1991, Subunit III of cytochrome *c* oxidase is not involved in proton translocation: A site-directed mutagenesis study, *EMBO J.* **10**(8):2015–2021.

Harris, D. A., and Das, A. M., 1991, Control of mitochondrial ATP synthesis on the heart, *Biochem. J.* **280:**561–573.

Henderson, R., Baldwin, J. M., Ceska, T. A., Zemlin, F., Beckman, E., and Downing, K. H., 1990, Model for the structure of bacteriorhodopsin based on high-resolution electron cryomicroscopy, *J. Mol. Biol.* **213:**899–929.

Hill, B. C., 1994, Modeling the sequence of electron transfer reactions in the single turnover of reduced, mammalian cytochrome *c* oxidase with oxygen, *J. Biol. Chem.* **269:**2419–2425.

Hofacker, I., and Schulten, K., 1998, Oxygen and proton pathways in cytochrome *c* oxidase, *Proteins* **30:**100–107.

Hofhaus, G., Weiss, H., and Leonard, K., 1991, Electron microscopic analysis of the peripheral and membrane parts of mitochondrial NADH dehydrogenase (complex I), *J. Mol. Biol.* **221:**1027–1043.

Iwata, S., Ostermeier, C., Ludwing, B., and Michel, H., 1995, Structure at 2.8 Å resolution of cytochrome *c* oxidase from *Paracoccus denitrificans*, *Nature* **376:**660–669.

Kadenbach, B., Jaraush, S., Hartmass, R., and Merle, P., 1983, Separation of mammalian cytochrome *c* oxidase into 13 polypeptides by a sodium dodecyl sulfate-gel electrophoretic procedure, *Anal. Biochem.* **129:**517–521.

Kennedy, F. G., and Jones, D. P., 1986, Oxygen dependence of mitochondrial function in isolated rat cardiac myocytes, *Am J. Physiol.* **250:**C374–C383.

Khorana, H. G., 1988, Bacteriorhodopsin, a membrane protein that uses light to translocate protons, *J. Biol. Chem.* **263:**7439–7442.

Kilmartin, J. V., and Rossi-Bernardi, L., 1973, Interaction of hemoglobin with hydrogen ions, carbon dioxide, and organic phosphates, *Physiol. Rev.* **53:**836–888.

Konstantinov, A. A., Siletsky, S., Mitchell, D., Kaulen, A., and Gennis, R. B., 1997, The roles of the two proton input channels in cytochrome *c* oxidase from *Rhodobacter sphaeroides* probed by the effects of site-directed mutations in time-resolved electrogenic intraprotein proton transfer, *Proc. Natl. Acad. Sci. USA* **94:**9085–9090.

Link, T. A., Hagen, W. R., Pierik, A. J., Assmann, C., and von Jagow, G., 1992, Determination of the redox properties of the Rieske [2Fe-2S] cluster of bovine heart bc_1 complex by direct electrochemistry of a water-soluble fragment, *Eur. J. Biochem.* **208:**685–691.

Lorusso, M., Cocco, T., Minuto, M., Capitanio, N., and Papa, S., 1995, Proton/electron stoichiometry of mitochondrial bc_1 complex. Influence of pH and transmembrane delta pH, *J. Bioenerg. Biomembr.* **27:**101–108.

Luvisetto, S., Conti, E., Buso, M., and Azzone, G. F., 1991, Flux ratios and pump stoichiometries at sites II and III in liver mitochondria. Effect of slips and leaks, *J. Biol. Chem.* **266:**1034–1042.

Macnab, R. M., 1992, Genetics and biogenesis of bacterial flagella, *Annu. Rev. Genet.* **26:**131–158.

Malmström, B. G., 1989, The mechanism of proton translocation in respiration and photosynthesis, *FEBS Lett.* **250:**9–21.

Markossian, K. A., Poghossian, A. A., Paitian, N. A., and Nalbandyan, R. M., 1978, Superoxide dismutase activity of cytochrome oxidase, *Biochem. Biophys. Res. Commun.* **81:**1336–1343.

Mitchell, P., 1966, *Chemiosmotic Coupling in Oxidative and Photosynthetic Phosphorylation*, Glynn Research Ldt, Bodmin, United Kingdom.

Mitchell, P., 1976, Possible molecular mechanism of the proton-motive function of cytochrome systems, *J. Theor. Biol.* **62**:327–367.

Mitchell, P., 1987a, Respiratory chain systems in theory and practice, in: *Advances in Membrane Biochemistry and Bioenergetics* (C. K. Kim, H. Tedeshi, J. J. Diwan, and J. C. Salerno, eds.), pp. 25–52, Plenum Press, New York.

Mitchell, P., 1987b, A new redox loop formality involving metal-catalysed hydroxide-ion translocation. A hypothetical Cu loop mechanism for cytochrome oxidase, *FEBS Lett.* **222**:235–245.

Mitchell, P., 1988, Possible proton-motive osmochemistry in cytochrome oxidase, *Ann. NY Acad. Sci.* **550**:185–198.

Mitchell, R., and Rich, P. R., 1994, Proton uptake by cytochrome *c* oxidase on reduction and on ligand binding, *Biochim. Biophys. Acta* **1186**:19–26.

Mitchell, P., Moyle, J., and Mitchell, R., 1979, Measurement of → H^+/O in mitochondria and submitochondrial vesicles, *Methods Enzymol.* **55**:627–640.

Mitchell, R., Mitchell, P., and Rich, P. R., 1992, Protonation states of the catalytic intermediates of cytochrome *c* oxidase, *Biochim. Biophys. Acta* **1101**:188–191.

Murphy, M. P., 1989, Slip and leak in mitochondrial oxidative phosphorylation, *Biochim. Biophys. Acta* **977**:123–141.

Murphy, M. P., and Brand, M. D., 1988, The stoichiometry of charge translocation by cytochrome oxidase and the cytochrome bc_1 complex of mitochondria at high membrane potential, *Eur. J. Biochem.* **173**:645–651.

Nicholls, D. G., 1974, The influence of respiration and ATP hydrolysis on the proton-electrochemical gradient across the inner membrane of rat-liver mitochondria as determined by ion distribution, *Eur. J. Biochem.* **50**:305–315.

Nicholls, D. G., and Locke, R. M., 1992, Thermogenic mechanisms in brown fat, *Physiol. Rev.* **64**:1–64.

Nicholls, P., and Wrigglesworth, J. M., 1982, Scalar and vectorial pH effects in cytochrome aa_3: Is there a proton-motive aa_3 cycle? in: *Oxidase and Related Redox Systems* (T. E. King *et al.*, eds.), pp. 1149–1160, Pergamon Press, New York.

Ostermeier, C., Harrenga, A., Ermler, U., and Michel, H., 1997, Structure at 2.7 Å resolution of the *Paracoccus denitrificans* two-subunit cytochrome *c* oxidase complexed with an antibody F_V fragment, *Proc. Natl. Acad. Sci. USA* **94**:10547–10553.

Papa, S., 1976, Proton translocation reactions in the respiratory chain, *Biochim. Biophys. Acta* **456**: 39–84.

Papa, S., and Capitanio, N., 1998, Redox Bohr effects (cooperative coupling) and the role of heme *a* in the proton pump of cytochrome *c* oxidase, *J. Bioenerg. Biomembr.*

Papa, S., and Lorusso, M., 1984, The cytochrome chain of mitochondria: Electron transfer reactions and transmembrane proton translocation, in: *Biomembranes* (R. M. Burton and F. C. Guerra, eds.), pp. 257–290, Plenum Press, London.

Papa, S., Lofrumento, N. E., Quagliarello, E., Meijer, A. J., and Tager, J. M., 1970, Coupling mechanisms in anionic substrate transport across the inner membrane of rat-liver mitochondria, *J. Bioenerg,* **1**:287–307.

Papa, S., Guerrieri, F., Lorusso, M., and Simone, S., 1973, Proton translocation and energy transduction in mitochondria, *Biochemie* **55**:703–716.

Papa, S., Lorusso, M., and Guerrieri, F., 1975, Mechanism of respiration-driven proton translocation in the inner mitochondrial membrane. Analysis of proton translocation associated to oxidation of endogenous ubiquinol, *Biochim. Biophys. Acta* **387**:425–440.

Papa, S., Guerrieri, F., Lorusso, M., Izzo, G., Boffoli, D., and Maida, I., 1981, Redox Bohr effects in the cytochrome system of mitochondria and their role in oxido-reduction and proton translocation, in: *Vectorial Reaction in Electron and Ion Transport in Mitochondria and Bacteria* (F. Palmieri *et al.*, eds.), pp. 57–69, Elsevier/North-Holland Biomedical Press, Amsterdam.

Papa, S., Lorusso, M., Boffoli, D., and Bellomo, E., 1983, Redox-linked proton translocation in the $b-c_1$ complex from beef-heart mitochondria reconstituted in phospholipid vesicles. General characteristics and control of electron flow by $\Delta\mu_{H^+}$, *Eur. J. Biochem.* **137**:405–412.

Papa, S., Guerrieri, F., and Izzo, G., 1986, Cooperative proton-transfer reactions in the respiratory chain: Redox Bohr effects, *Methods Enzymol.* **126**:331–343.

Papa, S., Capitanio, N., and De Nitto, E., 1987, Characteristics of the redox-linked proton ejection in beef heart cytochrome *c* oxidase reconstituted in liposomes, *Eur. J. Biochem.* **164**:507–516.

Papa, S., Capitanio, N., Capitanio, G., De Nitto, E., and Minuto, M., 1991, The cytochrome chain of mitochondria exhibits variable H^+/e^- stoichiometry, *FEBS Lett.* **288**:183–186.

Papa, S., Lorusso, M., and Capitanio, N., 1994a, Mechanistic and phenomenological features of proton pumps in the respiratory chain of mitochondria, *J. Bioenerg. Biomembr.* **26**:609–617.

Papa, S., Capitanio, N., Glaser, P., and Villani, G., 1994b, The proton pump of heme–copper oxidases, *Cell Biol. Int.* **18**(5):345–355.

Pietrobon, D., Zoratti, M., and Azzone, G. F., 1983, Molecular slipping in redox and ATPase H^+ pumps, *Biochim. Biophys. Acta* **723**:317–321.

Proteau, G., Wrigglesworth, J. M., and Nicholls, P., 1983, Proton-motive functions of cytochrome *c* oxidase in reconstituted vesicles, *Biochem. J.* **210**:199–205.

Puustinen, A., Bailey, J. A., Dyer, R. B., Mecklenburg, S. L., Wikström, M., and Woodruff, W. H., 1997, Fourier transform infrared evidence for connectivity between Cu_B and glutamic acid 286 in cytochrome bo_3 from *Escherichia coli*, *Biochemistry* **36**:13195–13200.

Rich, P. R., 1995, Towards an understanding of the chemistry of oxygen reduction and proton translocation in the iron–copper respiratory oxidases, *Aust. J. Plant Physiol.* **22**:479–486.

Rich, P. R., Meunier, B., Mitchell, R., and Moody, A. J., 1996, Coupling of charge and proton movement in cytochrome *c* oxidase, *Biochim. Biophys. Acta* **1275**:91–95.

Rousseau, D. L., Sassaroli, M., Ching, Y.-C., and Dasgupta, S., 1988, The role of water near cytochrome *a* in cytochrome *c* oxidase, *Ann. NY Acad. Sci.* **550**:223–237.

Saiki, K., Nakamura, H., Mogi, T., and Anraku, Y., 1996, Probing a role of subunit IV of the *Escherichia coli* bo-type ubiquinol oxidase by deletion and cross-linking analyses, *J. Biol. Chem.* **271**(76):15336–15340.

Skulachev, V. P., 1994, Decrease of the intracellular concentration of O_2 as a special function of the cellular respiratory system, *Biochemistry (Moscow)* **59**(12):1910–1912.

Thomas, J. W., Puustinen, A., Alben, J. O., Gennis, R. B., and Wikström, M., 1993, Substitution of asparagine for aspartate-135 in subunit I of the cytochrome *bo* ubiquinol oxidase of *Escherichia coli* eliminates proton-pumping activity, *Biochemistry* **32**:10923–10928.

Tittor, J., Soell, C., Oesterhelt, D., Butt, H.-J., and Bamberg, E., 1989, A defective proton pump, point-mutated bacteriorhodopsin asp96→asn is fully reactivated by azide, *EMBO J.* **8**:3477–3482.

Trumpower, B. L., and Gennis, R. B., 1993, Energy transduction by cytochrome complexes in mitochondrial and bacterial respiration: The enzymology of coupling electron transfer reactions to transmembrane proton translocation, *Annu. Rev. Biochem.* **63**:675–716.

Tsukihara, T., Aoyama, H., Yamashila, E., Tomizaki, T., Yamagushi, H., Shinzawa-Itoh, K., Nakashima, R., Yaono, R., and Yoshikawa, S., 1995, Structures of metal sites of oxidized bovine heart cytochrome *c* oxidase at 2.8 Å, *Science* **269**:1069–1074.

Tsukihara, T., Aoyama, T., Yamashila, E., Tomizaki, T., Yamagushi, H., Shinzawa-Itoh, K., Nakashima, R., Yaono, R., and Yoshikawa, S., 1996, The whole structure of the 13-subunit oxidized cytochrome *c* oxidase at 2.8 Å, *Science* **272**:1136–1144.

Urban, P. F., and Klingenberg, M., 1969, On the redox potentials of ubiquinone and cytochrome *b* in the respiratory chain, *Eur. J. Biochem.* **9**:519.

Verkhovskaya, M. L., Garcia-Horsman, A., Puustinen, A., Rigaud, J. L., Morgan, J. E., Verkovsky, M. I., and Wikström, M., 1997, Glutamic acid 286 in subunit I of cytochrome bo_3 is involved in proton translocation, *Proc. Natl. Acad. Sci. USA* **94**:10128–10131.

Villani, G., 1995, Analisi mutazionale e biochimica della fosforilazione ossidativa in procarioti ed eucarioti e sviluppo di approcci diagnostici e terapeutici per le malattie mitocondriali, *Ph.D. Thesis*, Bari, Italy.

Villani, G., and Attardi, G., 1997, *In vivo* control of respiration by cytochrome *c* oxidase in wild-type and mitochondrial DNA mutation-carrying human cells, *Proc. Natl. Acad. Sci. USA* **94:** 1166–1171.

Villani, G., Tattoli, M., Capitanio, N., Glaser, P., Papa, S., and Danchin, A., 1995, Functional analysis of subunits III and IV of *Bacillus subtilis aa$_3$*-600 quinol oxidase by *in vitro* mutagenesis and gene replacement, *Biochim. Biophys. Acta* **1232**(1–2):67–74.

von Jagow, G., and Sebald, W., 1980, *b*-Type cytochromes, *Annu. Rev. Biochem.* **49:**281–314.

Vygodina, T. V., Pecoraro, C., Mitchell, D., Gennis, R., and Konstantinov, A. A., 1998, Mechanism of inhibition of electron transfer by amino acid replacement K362M in a proton channel of *Rhodobacter sphaeroides* cytochrome *c* oxidase, *Biochemistry* **37:**3053–3061.

Walker, J. E., 1992, The NADH:ubiquinone oxidoreductase (complex I) of respiratory chains, *Q. Rev. Biophys.* **25:**253–324.

Watmough, N. J., Katsonouri, A., Little, R. H., Osborne, J. P., Furlong-Nickels, E. F., Gennis, R., Brittain, T., and Greenwood, C., 1997, A conserved glutamic acid in helix VI of cytochrome *bo$_3$* influences a key step in oxygen reduction. *Biochemistry* **36:**13736–13742.

Wikström, M., 1977, Proton pump coupled to cytochrome *c* oxidase in mitochondria, *Nature* **266:**271–273.

Wikström, M., 1989, Identification of of the electron transfers in cytochrome oxidase that are coupled to proton-pumping, *Nature* **338:**776–778.

Wikström, M., and Krab, K., 1979, Proton-pumping cytochrome *c* oxidase, *Biochim. Biophys. Acta* **549:**177–222.

Wikström, M., Krab, K., and Saraste, M., 1981, Proton translocating cytochrome complexes, *Annu. Rev. Biochem.* **50:**623–655.

Wikström, M., Bogachev, A., Finel, M., Morgan, J. E., Puustinen, A., Raitio, M., Verkhovskaya, M., and Verkhovsky, M. I., 1994, Mechanism of proton translocation by the respiratory oxidases. The histidine cycle, *Biochim. Biophys. Acta* **1187:**106–111.

Wilson, D. F., Lindsay, J. G., and Brocklehurst, E. S., 1972, Heme–heme interaction in cytochrome oxidase, *Biochim. Biophys. Acta* **256:**277–286.

Woodruff, W. H., 1993, Coordination dynamics of heme–copper oxidases. The ligand shuttle and the control and coupling of electron transfer and proton translocation, *J. Bioenerg. Biomem.* **25:**177–188.

Woodruff, W. H., Einarsdottir, O., Dyer, R. B., Bagley, K. A., Palmer, G., Atherton, S. J., Goldbeck, R. A., Dawes, T. O., and Kliger, D. S., 1991, Nature and functional implications of the cytochrome *a$_3$* transients after photodissociation of CO-cytochrome oxidase, *Proc. Natl. Acad. Sci. USA* **88:**2588–2592.

Wyman, J., 1968, Regulation in macromolecules as illustrated by haemoglobin, *Q. Rev. Biophys.* **1:**35–80.

Xia, D., Yu, C. A., Kim, H., Xia, J. Z., Kachurin, A. M., Zhang, L., Yu, L., and Deisenhofer, J., 1997, Crystal structure of the cytochrome *bc$_1$* complex from bovine heart mitochondria, *Science* **277:**60–62.

Yoshikawa, S., 1997, The crystal structure of bovine heart cytochrome *c* oxidase, functional implications, *FASEB J.* **11**(9):A858,6.

Yu, C. A., Xia, J.-Z., Kachurin, A. M., Yu, L., Xia, D., Kim, H., and Deisenhofer, J., 1996, Crystallization and preliminary structure of beef heart mitochondrial cytochrome-*bc$_1$* complex, *Biochim. Biophys. Acta* **1275:**47–53.

Zaslavsky, D., and Gennis, R., 1998, Substitution of lysine-362 in a putative proton conduction channel in the cytochrome *c* oxidase from *Rhodobacter sphaeroides* blocks turnover with O$_2$ but not with H$_2$O$_2$, *Biochemistry* **37:**3062–3067.

CHAPTER 4

Uncoupling of Respiration and Phosphorylation

Vladimir P. Skulachev

1. DEFINITIONS

Initially, the term *uncoupling* was introduced to define the respiration processes proceeding without phosphorylation. Such a phenomenon was discovered by Belitser and Tsibakova in 1939. The authors found that the addition of arsenate to a skeletal muscle mince allowed respiration to occur without ATP formation (Belitser and Tsibakova, 1939). However, later, when the role of $\Delta\mu_{H^+}$ and $\Delta\mu_{Na^+}$ as convertible energy currencies was elucidated, the definition of the uncoupling phenomenon was changed. Now, three types of cellular respiration are distinguished:

1. **Energy-coupled respiration.** This is respiration generating $\Delta\overline{\mu}_{H^+}$ or $\Delta\overline{\mu}_{Na^+}$, which are then utilized to perform useful work, namely, chemical work (e.g., ATP synthesis), osmotic work (uphill transport of solutes), or mechanical work (rotation of bacterial flagellum).
2. **Uncoupled respiration.** As in (1), but $\Delta\overline{\mu}_{H^+}$ or $\Delta\overline{\mu}_{Na^+}$ formed are immediately dissipated with no work performed.

Vladimir P. Skulachev • Department of Bioenergetics, A. N. Belozersky Institute of Physico-Chemical Biology, Moscow State University, Moscow 119899, Russia.

Frontiers of Cellular Bioenergetics, edited by Papa *et al.* Kluwer Academic/Plenum Publishers, New York, 1999.

3. **Noncoupled respiration**.   In this case respiration occurs in such a way that neither $\Delta\bar{\mu}_{H^+}$ or $\Delta\bar{\mu}_{Na^+}$ are formed.

Uncoupled and noncoupled respiration, differing in mechanisms, are energy dissipating. In contrast to coupled respiration, both are not controlled by energy consumption in the cell and therefore can be considered as two types of "free respiration" (Skulachev, 1988).

The above classification assumes that the absence of phosphorylation does not always mean uncoupling. Respiration occurring without phosphorylation can be (1) energy coupled, but $\Delta\bar{\mu}_{H^+}$ ($\Delta\bar{\mu}_{Na^+}$) are used to perform a work other than ATP synthesis, or (2) noncoupled. In case (2), the respiratory electron transfer does not generate the proton or sodium ion potentials, since the mechanism of operation of the energy-coupled respiratory chain enzymes is modified in such a way that electrons are transported without formation of the potentials (this phenomenon is defined as *decoupling*), or another respiratory chain initially unable to form $\Delta\bar{\mu}_{H^+}$ ($\Delta\bar{\mu}_{Na^+}$) is employed.

In this chapter, we shall consider uncoupling in its modern meaning when respiratory energy is transiently transduced to membrane potentials to be dissipated as a heat due to downhill flux of H^+ (Na^+). Considering this, we shall principally deal with mitochondria, where uncoupling means an increase in the H^+ conductance of the inner mitochondrial membrane. In uncoupled mitochondria, the rate of the passive (downhill) H^+ influx becomes commensurable to that of active (uphill) H^+ efflux carried out by the respiratory chain $\Delta\bar{\mu}_{H^+}$ generators, i.e., NADH-CoQ reductase, CoQ–cytochrome c reductase and cytochrome c oxidase.

2. ARTIFICIAL UNCOUPLERS

As mentioned above, arsenate was the first uncoupler described (Belitser and Tsibakova, 1939). In this case, uncoupling is due to the fact that ADP-arsenate, instead of ATP, is formed as the final product of oxidative phosphorylation. ADP-arsenate is unstable in water solutions. It spontaneously decomposes to ADP and arsenate. Thus, uncoupling occurs at the very end of energy transduction process.

In 1948, Loomis and Lipmann described uncoupling activity of 2,4-p-dinitrophenol (DNP) (Loomis and Lipmann, 1948). Mitchell (1961) later explained this effect by assuming that DNP increases the H^+ conductance of the coupling membranes and, hence, dissipates $\Delta\bar{\mu}_{H^+}$ which is generated by respiration. In Lehninger's laboratory, it was later found that DNP does increase the H^+ conductance of the planar bilayer phospholipid membrane (BLM) (Bielawski *et al.*, 1966). In our group, it was shown that a large number of hydrophobic weak acids that have a delocalized negative charge in their ionized form are (1) uncouplers in mitochondria and (2) H^+ conductors in BLM. A correlation between efficiencies of uncouplers in mitochondria and BLM was observed, and the term *protonophore* was suggested (Skulachev *et al.*, 1967, 1968). Independently, Chappell and Haarhoff (1967) found that liposome membranes become permeable to H^+ ions in the presence of an uncoupler. Mitchell (1966) reported that this was also the case with mitochondria.

Demonstration of the photonophorous activity of uncouplers proved to be one of the most important pieces of evidence in favor of the Mitchellian chemiosmotic hypothesis of energy coupling. It generally has been accepted that in mitochondria, the DNP anion is electrophoretically expelled from the matrix to be protonated outside mitochondria and to return to the matrix in its neutral (protonated) form. This form is deprotonated in the matrix, regenerating the DNP anion. All these events were assumed to take place in the phospholipid region of the inner mitochondrial membrane with no proteins involved, since they could be reproduced in BLM.

Some doubt of the absolute validity of the last conclusion appeared when we showed that a very small amount of carboxyatractylate (CAtr) specifically inhibiting the ATP–ADP antiporter could cause partial recoupling of the DNP-uncoupled mitochondria. In fact, the level of the mitochondrial electric potential ($\Delta\Psi$) generated by respiration was strongly decreased by DNP and partially increased by CAtr (Andreyev *et al.*, 1989). Then we found that the uncoupling effect of low concentrations of the most potent artificial protonophores (SF6847, FCCP, CCCP, and CCP) on heart or liver mitochondria, bacterial chromatophores, and cytochrome oxidase proteoliposomes is completely abolished by 6-ketocholestanol (kCh). In BLM, kCh caused the opposite effect, potentiating protonophorous activity of uncouplers (Starkov *et al.*, 1994, 1995, 1997a). This phenomenon was later reproduced in plant mitochondria by Vianello *et al.* (1995) and in kidney mitochondria by Chaves *et al.* (1996). Recoupling similar to that done by kCh was shown to be inherent in male sex hormones and progesterone, the effect being specific to animal mitochondria only. Inactive hormone analogues proved to be ineffective. For instance, epiandrosterone (isomer of dihydrotestosterone, lacking hormonal activity) as well as deoxycorticosterone (differing from progesterone by additional hydroxylic group) could not recouple the SF6847-uncoupled mitochondria (Starkov *et al.*, 1997b). In all these cases, no recoupling was possible when high concentrations of protonophores were added (Starkov *et al.*, 1997a,b).

These data were hardly compatible with the classical scheme that explained uncoupling by the circulation of both neutral and anionic uncouplers species via the phospholipid regions of biological membranes. It is more probable that this circulation is assisted, in some way, by membrane protein(s), which facilitate transmembrane diffusion of anionic uncoupler. However, if this is the case, the question arises as to why similar concentrations of uncouplers are effective in mitochondria, where their effect is facilitated by protein(s), and in BLM, where no protein is present.

This problem was solved when we took into account the following fact. BLM occupies such a small part of the experimental system that even a low concentration of uncoupler is sufficient to saturate this membrane. An equal quantity of uncoupler added to a mitochondrial suspension is immediately absorbed by a large amount of mitochondrial membranes, giving a significantly lower uncoupler concentration in the water phase than in the case of BLM. Since the uncoupler concentration in a membrane is equilibrated with that in water, the amount of uncoupler added to mitochondria must be, in the simplest case, much less efficient than the same amount of uncoupler added to BLM. To equalize the acting concentrations of the uncoupler in experiments with mitochondria and BLM, we supplemented the BLM-separated solutions by the same amounts of mitochondria as were used in the study on measuring mitochondrial respiration and $\Delta\Psi$. This was found to strongly increase the concentrations of the uncoupler required to cause a measurable increase in the BLM H^+ conductance. Such an effect was especially large with the most active (and most hydrophobic) uncouplers, such as SF6847. With this uncoupler, the addition of mitochondria to the BLM system decreased the protonophore efficiency of uncoupler in BLM by two orders of magnitude (Starkov et al., 1997a).

Thus, the apparent correlation of absolute values of efficiencies of different protonophores in mitochondria and BLM most probably is a result of the superposition of two oppositely directed effects. On one hand, BLM is in fact much less sensitive to uncouplers due to the absence of uncoupling facilitating proteins. On the other hand, the uncoupler–lipid phase volume ratio is much higher in the BLM experiments than in the studies on mitochondria, a fact that makes BLM much more sensitive to the uncoupler (Starkov et al., 1997a).

Within the framework of the above logic, it seems probable that in mitochondria there are protein(s) that specifically bind uncouplers. Recently reviewing this aspect, Starkov (1997) concluded that there are at least two classes of uncouplers differing in their protein partners.

1. **DNP-like uncouplers**. 2-Azido-4-nitrophenol was shown in the Hatefi group to cause photoaffinity labeling of predominantly a 30-kDa protein. This resulted in (1) inhibition of the state 3 respiration to the level of the state 4 and (2) made the activation of respiration by DNP impossible (Hatefi, 1975; Hanstein, 1976). This looks as if the ATP–ADP-antiporter, which is involved in both (1) ADP import to support ATP synthesis inside mitochondria and (2) uncoupling by DNP, has been inactivated (Andreyev et al., 1989). If this were the case, respiration in the photoinactivated mitochondria would be uncoupled by CCP-like uncouplers or arsenate. On the other hand, it seems

possible that inhibition of the H^+-ATP-synthase by the DNP derivative occurs since, besides the 30-kDa protein, the α-subunit of factor F_1 was also found to be modified. If such an inhibition would take place, arsenate uncoupling would be arrested. In this case, ATPase activity in the inside-out submitochondrial particles would be inhibited. Unfortunately, the authors did not carry out such an analysis.

2. **CCP-like uncouplers.** These uncouplers were studied in Wilson's group. It was found that 2-nitro-4-azidocarbonylcyanide phenylhydrazone (N_3CCP) has a high-affinity binding site in rat heart mitochondria (Katre and Wilson, 1977), rat liver and pigeon heart mitochondria (Katre and Wilson, 1978), and membranes of *Paracoccus denitrificans* and *Tetrahymena pyriformis* (Katre and Wilson, 1980). In contrast to the DNP derivative, N_3CCP, when illuminated, (1) combines with a 10- to 12-kDa polypeptide, and (2) causes irreversible uncoupling. In the same group, the uncoupling activity of substituted 3,5-dichlorosalycylanilides was shown to depend on their steric properties, another feature indicating that a specific binding site on some mitochondrial protein is involved in the uncoupling phenomenon (Storey *et al.*, 1975).

This assumption is in line with the fact that the uncoupling efficiency of most uncouplers appears to be quite different in various types of the coupling membrane. For instance, SF6847 as well as FCCP and other phenylhydrazone derivatives in our experiments were almost 100-fold more efficient in mitochondria and cytochrome oxidase asolectin proteoliposomes than in *Rhodobacter sphaeroides* chromatophores (Starkov *et al.*, 1997a). According to Miyoshi and Fujita (1987), SF6847 efficiency in chloroplasts resembles that found in our study on chromatophores, being much lower than in mitochondria.

Inhibitor analysis revealed some differences between the above-mentioned two groups of uncouplers. In contrast to the effect of the DNP-like uncouplers, which was partially reversed by CAtr, the action of low concentrations of the CCP-like uncouplers and SF6847 was (1) completely reversed by kCh and (2) CAtr resistant. The recoupling effect of kCh was very small with DNP (Starkov *et al.*, 1997a).

All these data indicate that the uncoupling action of low concentrations of both types of uncouplers is somehow mediated by membrane proteins. On the other hand, at high concentrations the same uncouplers operate in an inhibitor-independent fashion, so that this effect can be explained in terms of the classic Mitchellian scheme that postulates the circulation of protonophore molecules in the phospholipid bilayer. The uncoupling effect of gramicidin, which forms H^+-, K^+-, and Na^+-permeable channels in phospholipid membranes, was found to be CAtr and kCh resistant at any concentration of this uncoupler (Starkov *et al.*, 1997a). This fact is consistent with the assumption that the gramicidin effect is protein independent.

Another difference between the usual protonophores and gramicidin is that the latter can uncouple both "protonic"- and "sodium"-coupling membranes. Sodium-coupling membranes use $\Delta\bar{\mu}_{Na^+}$, instead of $\Delta\bar{\mu}_{H^+}$, as a convertible energy currency. They contain primary Na^+ pumps ($\Delta\bar{\mu}_{Na^+}$ generators, e.g., the Na^+-motive respiratory chain) and $\Delta\bar{\mu}_{Na^+}$ consumers (e.g., the Na^+-driven ATP-synthase). This is the case for some marine bacteria. The plasma membrane of the animal cell belongs to the same category: it is energized by Na^+/K^+-ATPase, and $\Delta\bar{\mu}_{Na^+}$ formed is utilized by numerous Na^+, solute-symporters (for reviews, see Skulachev, 1988, 1992, and 1994b).

In contrast to gramicidin, which fails to discriminate between H^+ and Na^+, there are some synthetic Na^+-specific ionophores carrying out Na^+ uniport without simultaneous transport of H^+. One example is ETH 157 (N,N'-dibenzyl-N,N'-diphenyl-1,2-phenilene diacetamide). Compounds of this kind, which can be called *sodiophores*, specifically uncouple the Na^+-coupled oxidative phosphorylation (Avetisyan *et al.*, 1993).

3. NATURAL UNCOUPLERS

3.1. Fatty Acids

Several substances have already been described that cause uncoupling when added to mitochondria or other energy-transducing systems. Among them, fatty acids have been studied in detail.

The uncoupling effect of nonesterified fatty acids was discovered as early as in 1956 by Pressman and Lardy (1956). Later it was found that, in contrast to DNP, FCCP, or gramicidin, fatty acids affect BLM only slightly. Some decrease in the BLM resistance could be observed only if the initial level of this resistance was very high (see, e.g., Walter and Gutknecht, 1984). Low activity of fatty acids in BLM was found to be due to very low permeability of phospholipid membranes to the ionized (anionic) form of fatty acid, although its protonated form easily penetrates through these membranes (Walter and Gutknecht, 1984; Andreyev *et al.*, 1989; Kamp and Hamilton, 1992, 1993). Such relationships are hardly surprising since the ionized fatty acid carboxylate possesses a localized negative charge. This charge is highly hydrated, which prevents diffusion of the fatty acid anion through the hydrophobic membrane core. On the other hand, the negative charge in DNP or FCCP is strongly delocalized over the entire aromatic structure of the protonophore, an effect decreasing hydration and, hence, enhancing the membrane permeability.

Since fatty acids are effective uncouplers in mitochondria, I have assumed that the protein component(s) of the mitochondrial membrane facilitate translocation of the fatty acid anions. The proteins in question were suggested to be thermogenin and the ATP–ADP antiporter (Skulachev, 1988, 1991).

Thermogenin, or the uncoupling protein, is found in mitochondria of brown adipose tissue. This protein can amount to 10–15% of the total protein in these mitochondria (Nicholls and Locke, 1981). It increases the permeability of the inner mitochondrial membrane not only to H^+ (Nicholls, 1979), but also to a large group of anions, including short-chain fatty acids, hexane sulfonate, pyruvate, and Cl^- (reviewed in Skulachev, 1991). As for long-chain fatty acids, their permeability was not directly measured for technical reasons, but it is very probable that they are penetrants, since the permeability of the thermogenin-containing membrane to carboxylate anions increases with an increase in the chain length (Garlid, 1990; Jezek et al., 1990; Jezek and Garlid, 1990). It was also shown that 10^{-6}–10^{-5} M long-chain fatty acids are required for thermogenin to increase the H^+ conductance (Nicholls, 1976; Nicholls and Locke, 1981). I suggested that it is a fatty acid anion rather than H^+ that is translocated by thermogenin (Skulachev, 1988, 1991).

This suggestion is now supported by several lines of evidence from Garlid's and Jezek's groups:

1. Fatty acids compete with Cl^- transported to the brown fat mitochondria or thermogenin proteoliposomes (Jezek et al., 1994). Similar competition takes place between transports of hexane sulfonate and Cl^- (Garlid et al., 1996).
2. The screening of various fatty acids and their derivatives showed that in the thermogenin proteoliposomes, the structural patterns required for the fatty acid uniport, H^+ transport, and competition with Cl^- are identical (Jezek et al., 1996).
3. Isolated thermogenin is labeled by a palmitate derivative, 16-(4-azido-2-nitrophenylamino)[^{3}H]$_4$hexadecanoic acid (N_3-palmitate), with a low stoichiometry (0.75 N_3-palmitate per thermogenin dimer). The labeling is prevented by stearate and hexane sulfonate (Ruzicka et al., 1996).

The question arises of how fatty acids uncouple in tissues other than brown fat, where thermogenin is absent. This author hypothesized that in these cases the ATP–ADP antiporter plays the role of the protein facilitating the translocation of the fatty acid anions (Skulachev, 1988, 1991).

In our group, it was revealed that CAtr, the most effective specific inhibitor of the ATP–ADP antiporter, strongly suppresses respiration uncoupled by fatty acids, increases the $\Delta\Psi$ level, and decreases the fatty-acid-stimulated H^+ conductance in heart muscle and liver mitochondria (Andreyev et al., 1987, 1988, 1989; Brustovetsky et al., 1990). Other inhibitors of the antiporter, namely atractylate, bongkrekic acid, and pyridoxal phosphate, as well as its substrate, ADP, also have some recoupling activity that was lower, however, than that of CAtr (Andreyev et al., 1989). On the other hand, CAtr proved to be inefficient in inside-out submitochondrial particles, whereas bongkrekic acid showed recoupling activity both in mitochondria and in the particles (Dedu-khova *et al.*, 1991). In all cases, the antiporter inhibitors did not recouple when FCCP was used as a protonophore (Andreyev *et al.*, 1989). These data were then confirmed by others on mitochondria (Schönfeld, 1990; Macri *et al.*, 1994; Schönfeld *et al.*, 1996), ATP–ADP antiporter proteoliposomes (Tikhonova *et al.*, 1994; Brustovetsky and Klingenberg, 1994), and digitonin-permeabilized Ehrlich ascites tumor cells (Schönfeld *et al.*, 1996).

Recently, a joint study of Schönfeld's, Jezek's, and Wojtczak's groups (Schönfeld *et al.*, 1996) demonstrated that the ATP–ADP antiporter-mediated translocation of fatty acids in the mitochondrial membrane is required for uncoupling by these compounds. It was found that illumination of the N_3-laurate-treated permeabilized ascites cells strongly suppresses the uncoupling activity of myristate. The control experiments showed that in the dark, N_3-laurate uncouples, as does laurate, heart muscle mitochondria in a CAtr-sensitive fashion. N_3-palmitate was found to label around ten mitochondrial proteins. High labeling was found for a 30-kDa protein that was identified, with a specific antibody, as the ATP–ADP antiporter. N_3-laurate was shown to cause photoinactivation of the main function of the antiporter, i.e., the ATP–ADP exchange.

The above data are hardly consistent with the idea of Winkler and Klingenberg (1994) that fatty acids operate in a "stationary" fashion, either by changing the protein conformation or, when anchored within the protein hydro-phobic core, by facilitating H^+ trafficking.

Comparing thermogenin and the ATP–ADP antiporter, one may conclude that these two proteins are evolutionary closely related. They have a similar amino acid sequence, a similar domain composition, and a similar molecular mass, and they combine with fatty acids and purine nucleotides. The main difference is that thermogenin can bind nucleotides but cannot transport them through the membrane. The binding results in the immersion of the nucleotide in the protein. However, the final event, namely, the release of the nucleotide on the opposite side of the membrane, does not occur. The binding inhibits the fatty-acid-induced uncoupling by thermogenin as it does by that by the antipor-ter. Generally, the ATP–ADP antiporter looks like a bifunctional protein carry-ing out (1) the ATP–ADP exchange and (2) uncoupling by translocation of the fatty acid anions. As for thermogenin, it is specialized in only one function, namely, in uncoupling.

It is interestingly that the glutamate–aspartate antiporter and the phos-phate carrier, two other mitochondrial anion porters belonging to the same family as the ATP–ADP antiporter and thermogenin, also seem to be involved in the uncoupling by fatty acids. Schönfeld (1990) has reported that the degree of the CAtr-induced recoupling of the fatty-acid-uncoupled mitochondria is the highest in heart muscle mitochondria, the lowest in liver mitochondria, with kidney mitochondria occupying the middle position. This corresponds to the relative concentration of the ATP–ADP antiporter in the mitochondria of the above tissues. In our group, it was shown that glutamate, aspartate, and diethyl pyrocarbonate (an inhibitor of the glutamate–aspartate antiporter) cause additional recoupling in liver mitochondria uncoupled by palmitate and partially recoupled by saturating concentration of CAtr (Bodrova *et al.*, 1995; Samartsev *et al.*, 1997). Such an effect was also seen by L. Wojtczak and M. R. Wieckowski (1996, personal communication). Strabergerova and Jezek (1996) reported that micromolar concentrations of N_3-laurate and N_3-palmitate induce photoinhibition of the phosphate transport in mitochondria and combine with the phosphate carrier.

Most probably, all the mitochondrial carriers are composed of (1) surface-located, very specific, substrate-binding sites responsible for the recognition of hydrophilic anions, namely nucleotides, phosphate, or dicarboxylates; and (2) nonspecific anion-binding sites immersed to the hydrophobic core of the protein, which are responsible for the translocation of the recognized anions through the membrane barrier. The hydrophilic substrate cannot reach the type (2) binding sites without assistance of the type (1) binding sites. As for hydrophobic fatty acid anions, they apparently do not need the type (1) binding sites (Skulachev, 1988, 1991).

The mechanism of the action of artificial and natural anionic protonophores is shown on Fig. 1. It is assumed that a natural anionic substrate (A_1^-) combines with a cationic ligand of the anion carrier (C^+) at the inner surface of the inner mitochondrial membrane (Fig. 1A, step 1). The electroneutral com-

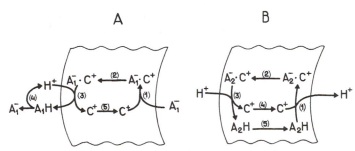

FIGURE 1. The mechanism of translocation of hydrophilic anions and protonophores in mitochondria. It is assumed that efflux of the hydrophilic anion A_1^- (or, e.g., $ATP_{in}^{4-}/ADP_{out}^{3-}$ exchange) is facilitated by combining A_1^- with a cationic ligand C^+ belonging to a protein (anion carrier). (A) Translocation of free C^+ back to the inner membrane surface is $\Delta\Psi$ driven. In the case of the hydrophobic anion A_2^- of an uncoupler, an additional stage of the A_2H influx occurs, which is protein independent. (B) However, the formation of A_2H from A_2^- is catalyzed by the carrier.

plex $A_1^- \cdot C^+$ is translocated to the opposite (outer) membrane side (step 2). Here, A_1^- is protonated, and A_1H releases to the outer medium (step 3) with its subsequent deprotonation to A_1^- (step 4). Then, C^+ electrophoretically returns to the inner membrane surface (step 5).

If an anion belongs to such a hydrophobic compound as a fatty acid, its circulation is restricted to the intramembrane space. In contrast to hydrophilic A_1, protonated fatty acid (A_2H in Fig. 1B) can easily traverse the membrane, and this additional stage results in H^+ translocation from the medium to the intramitochondrial space.

An important feature of this scheme is that the anion carrier catalyzes not only the translocation of an anionic compound, but also its protonation, which is required for decomposition of the $A^- \cdot C^+$ complex (Fig. 1A,B, step 3). This may explain why a strong acid such as dodecyl sulfate uncouples oxidative phosphorylation and increases the H^+ permeability of mitochondria in a CAtr-sensitive fashion (Brustovetsky et al., 1990). Apparently the pK_a of dodecyl sulfate bound to the ATP–ADP carrier is much higher than that of dodecyl sulfate in the water solution. For fatty acids, it was shown that their binding to the liposomal membrane results in a strong alkaline shift of pK_a, which becomes as high as 8.0 (Sankaram et al., 1990). This is not the case for dodecyl sulfate (Yu. Antonenko, unpublished data). Thus, to be protonated at neutral pH, dodecyl sulfate seems to require something more than being located in the water–membrane interface. According to Fig. 1B, this process is catalyzed by the anion carrier (step 3).

The scheme in Fig. 1 does not explain the mechanism of action of cationic uncouplers. There are several publications on the apparent uncoupling effect of hydrophobic weak bases, which, being protonated, acquire a positive charge (see, e.g., Pronevich et al., 1993; Nagamune et al., 1993; Schwaller et al., 1995). All these substances are active at much higher concentrations than SF6847, FCCP, or fatty acids. The possible involvement of carriers in their action has not yet been studied.

3.2. Other Natural Uncouplers and Uncoupling Systems

In addition to the fatty acids, there are some natural protonophores of plant origin. Ravanel and colleagues reported about the uncoupling activity of platanetin (3,5,7,8-tetra-hydroxy-6-dimethylallylflavone) isolated from buds of the plane tree, *Platanus acerifolia*. The half-maximal uncoupling effect varied from 1 to 5×10^{-6} M, depending on the plant species (Ravanel, 1986; Ravanel et al., 1986; Creuzet et al., 1988). We confirmed the uncoupling activity of platanetin on rat liver mitochondria ($C_{1/2} = 1.1 \times 10^{-5}$ M). kCh was shown to recouple the platanetin-uncoupled mitochondria (Starkov et al., 1997a). In the same study, some kCh recoupling was shown for zearalenone, a compound found in plants and fungi ($C_{1/2} = 7 \times 10^{-5}$ M). In plants, this substance seems to be related to the induction of floral buds (Fu *et al.*, 1995). Its uncoupling activity on plant mitochondria was described by Vianello and Macri (1981) (see also Macri and Vianello, 1990).

It remains unclear whether the uncoupling activity of these two compounds is involved in their physiological function. The same uncertainty is inherent in some other cases when a substance of natural origin uncouples *in vitro* at rather high concentrations.

Under some conditions, uncoupling can occur because of the activation of an ion transport futile cycle, e.g., the circulation of K^+ catalyzed by (1) the K^+ uniport (which carries out electrophoretic K^+ influx to mitochondria) and (2) the K^+-H^+ antiporter (which facilitates the K^+ efflux in exchange for H^+). These two porters, operating separately, cannot uncouple. However, their cooperation results in uncoupling (Garlid, 1994, 1996). A similar situation seems to be possible for Ca^{2+} when the Ca^{2+}-uniporter and $Ca^{2+}-2H^+$ antiporter are cooperating (see, e.g., Mikhailova *et al.*, 1996). In bacteria, the circulation of NH_4^+ and NH_3 was suggested as organizing a futile cycle (Neijssel *et al.*, 1990).

4. PHYSIOLOGICAL ASPECTS OF UNCOUPLING

4.1. Endogenous Uncoupling

Isolated mitochondria consume some oxygen even when ADP is not added or is exhausted (the so-called state 4 respiration). The rate of this respiration amounts usually to 5–20% of that in the presence of ADP (state 3). This value can be decreased under some special and obviously nonphysiological conditions (EGTA, sucrose instead of KCl, serum albumin, no added Mg^{2+}, etc.). Toth *et al.* (1990), who studied heart mitochondria from 14- to 21-day-old chicks, reported that ADP caused stimulation of respiration by two orders of magnitude under such conditions.

State 4 respiration is perhaps due to (1) ATP utilization inside mitochondria, (2) uncoupling, (3) decoupling, (4) noncoupled enzymatic oxygen consumption, or (5) noncoupled nonenzymatic oxygen consumption resulting in the formation of superoxide (O_2^-).

The contribution of (1), (4), and (5) can be estimated by inhibitor analysis since endogenous ATP formation is blocked by oligomycin and noncoupled O_2 consumptions are of different sensitivity to the respiratory chain inhibitors than coupled, uncoupled, or decoupled respirations (for details, see Skulachev, 1988, 1996a). On the other hand, it is not so easy to discriminate between uncoupling and decoupling (see Luvisetto *et al.*, 1992; Harper and Brand,

1993). There are serious reasons to assume that the oxygen consumption by state 4 mitochondria is mainly due to uncoupling.

Rolfe and Brand (1994), when studying isolated skeletal muscle and liver cells, concluded that uncoupling accounts for about one fourth and one half, respectively, of the oxygen consumption under resting conditions. The authors assumed that uncoupling contributes 15–33% of oxygen consumption by a rat *in vivo* (Rolfe and Brand, 1997).

4.2. Multiple Functions of Cellular Respiration

On the face of it, uncoupling as a futile cycle is an imperfection of phosphorylating respiration. However, the situation seems more complicated if we take into account the alternative functions of cellular respiration. In fact, ATP synthesis is only one of these functions, although usually the most important quantitatively.

In 1962, I considered four main physiological functions of cellular respiration (Skulachev, 1962):

1. Energy conservation.
2. Energy dissipation (heating).
3. Production of useful substances.
4. Decomposition of harmful substances.

It is obvious that heat production is alternative to energy conservation. This is why we started to study the possible existence of alternative respiratory functions by experiments on cold-exposed, warm-blooded animals.

4.3. Thermoregulatory Uncoupling

Experiments performed by this group showed that when pigeons, previously adapted to cold stress, were exposed to cold for 15 min, their skeletal muscle mitochondria proved to be almost completely uncoupled. The effect was much less pronounced in the nonadapted pigeons (Skulachev and Maslov, 1960; Skulachev, 1963). Similar phenomena (called "thermoregulatory uncoupling") were reproduced in mice. It was also found that an injection of the artificial uncoupler DNP significantly prolonged the survival time for nonadapted mice at their first exposure to cold (Skulachev *et al.*, 1963).

The addition of serum albumin was found to recouple mitochondria from the cold-exposed pigeons. A fraction of free fatty acids, extracted from the mitochondria of cold-treated animals and added to the mitochondria of nontreated animals, caused uncoupling. The total concentration of free fatty acids strongly increased in both muscle tissue and isolated muscle mitochondria of short-term cold-exposed pigeons (Levachev *et al.*, 1965).

A piece of evidence supporting the thermoregulatory uncoupling in intact rat diaphragm muscle was obtained in this group by Zorov and Mokhova (1973).

Later, Grav and Blix (1979) showed that mitochondria isolated from fur seals acclimatized to cold under natural conditions have a much lower respiratory control than those acclimatized to warm conditions., Serum albumin abolished this difference.

The thermoregulatory uncoupling discovered in muscles was then described in brown fat, the mammalian tissue specialized in additional heat production under cold conditions (Nicholls, 1979). Again, fatty acids proved to be natural uncouplers involved in the thermogenic response, the uncoupling being mediated by thermogenin (see Section 3).

It is unclear whether uncoupling plays some role in the regulatory heat production in tissues other than muscles and brown fat. No uncoupling occurred in liver mitochondria of pigeons treated with cold for a short time (Skulachev, 1962). However, cold acclimatization of rats for several weeks resulted in some decrease in the P/O ratio of isolated liver mitochondria (Panagos et al., 1958; Panagos and Beyer, 1960; Smith, 1958). The mechanism of this effect and its possible role in thermoregulation remained obscure. In our group, a partial CAtr-sensitive uncoupling was found in liver mitochondria from ground squirrels waking from hibernation (Amerkhanov et al., 1996). Uncoupling was revealed in flowers of some plants at low ambient temperature (for review, see Skulachev, 1988).

4.4. Noncoupled Respiration: Anabolic and Catabolic Functions

Among the functions of respiration listed in Section 4.2, functions (3) and (4) are related to metabolism rather than to energetics. Formally speaking, they might be carried out by the same energy-coupled respiratory chain that is involved in function (1). However, if it were the case, these functions would be tightly coupled to ATP synthesis and would require ADP to be fulfilled. Such a restriction is apparently sometimes undesirable for the cell. This is why the metabolic functions of respiration are catalyzed, as a rule, by noncoupled respiratory enzymes. For instance, some steps of formation of steroid hormones in adrenal cortex mitochondria, which require oxygenation of the steroid molecule, are mediated by special noncoupled respiratory chain including NADPH-oxidizing flavoprotein, adrenodoxin, and mitochondrial cytochrome P-450. All are localized in the inner mitochondrial membrane together with the usual components of the energy-coupled respiratory chain (Takemori and Kominami, 1984).

The decomposition of xenobiotics is known to be catalyzed by noncoupled respiratory chains in the endoplasmic reticulum. These chains are terminated by microsomal cytochromes P-450 (see, e.g., Archakov, 1975).

4.5. Possible Role of Uncoupling in the Anti-ROS Defense System

The above systems exemplify situations when the contribution of the O_2^--consuming process is relatively small compared to the total respiration rate of the cell. This cannot be the case if we consider such a respiratory function as a decrease in the intracellular O_2 concentration.

Oxygen is a strong oxidant that can chemically oxidize many intracellular compounds with the formation of O_2^-. The latter initiates chain reactions, resulting in the appearance of QH$\cdot$ (redox potential, $+1.35$ V), an oxidant that is even more aggressive than O_2 and O_2^-. Components of the initial and middle parts of the respiratory chain are main O_2 reductants involved in "parasitic" one-electron reduction of O_2 to O_2^-. Among them, semiquinone (CoQH$\cdot$) is apparently employed especially often.

The O_2 concentration and CoQH$\cdot$ lifetime increase under the state 3 $\rightarrow$ state 4 transition. This is because of the strong decrease in the respiration rate and increase in the $\Delta\bar{\mu}_{H^+}$ level, respectively. The latter inhibits the Q-cycle in such a fashion that CoQH$_2$ is still oxidized by the complex III FeS protein, but CoQH$\cdot$ formed cannot be oxidized by cytochrome b_1, which is already completely reduced.

It was postulated that mitochondria possess a special mechanism, called "mild" uncoupling, which prevents a great increase in $\Delta\bar{\mu}_{H^+}$ when ADP is exhausted (Skulachev, 1994a, 1995, 1996a, 1997).

The mechanism of mild uncoupling may be related to the phenomenon of nonohmic resistance of the mitochondrial membrane. It consists of a great increase in the membrane H$^+$ conductance under conditions when $\Delta\bar{\mu}_{H^+}$ increases above a certain threshold. The threshold in question is slightly above the state 3 $\Delta\bar{\mu}_{H^+}$ value (Nicholls, 1974, 1997; Garlid et al., 1986; Garlid, 1989; Murphy, 1989; Wojtczak et al., 1990).

In line with the above reasoning, it was found that (1) addition of either ADP ($+$ P$_i$) or an uncoupler to the state 4 mitochondria strongly suppresses the $O_2^{-\cdot}$ formation (Boveris and Chance, 1973), and (2) the rate of O_2^- formation decreases as the $\Delta\Psi$ lowers. The latter parallels the level of cytochrome b_1 reduction (Liu and Huang, 1996; Liu, 1997).

One of the mechanisms of mild uncoupling might be related to the circulation of superoxide in the mitochondrial membrane. In this connection, I would like to consider the data by McCord and Turrens (1994), who studied the ability of intramitochondrial Mn–superoxide dismutase to catalyze the $O_2^- \rightarrow$ H_2O_2 conversion, using extramitochondrially produced O_2^-. It was found that the consumption of external O_2^- by rat liver mitochondria incubated without a respiratory substrate can be 3.7-fold increased by adding glutamate and malate without ADP. The effect of these respiratory substrates was completely arrested by an uncoupler. The authors speculated that in state 4, a superoxide carrier is activated, which results in fast equilibration of O_2^- across the membrane. This may give rise to uncoupling if one assumes that O_2^- is electrophoretically extruded from matrix with the help of the carrier to come back in protonated form $HO_2^•$. This seems possible, since the pK_a value of O_2^- is about 4.9, i.e., close to that of the majority of artificial uncouplers and fatty acids. Protonation of O_2^- near the outer membrane surface might be facilitated by the O_2^- carrier according to the mechanism described above (see Fig. 1B).

Quite recently, a $HO_2^•$ flux across the inner mitochondrial membrane was postulated by Liu (Liu and Huang, 1996; Liu, 1997). I suggested that the circulation of O_2^- and $HO_2^•$ may explain the mild uncoupling and nonohmic behavior of mitochondrial membrane, presumably involved in the anti-reactive oxygen species (ROS) defense mechanism. If this is the case, we should assume that the mitochondrial O_2^- carrier is activated when $\Delta\bar{\mu}_{H^+}$ exceeds the state 3 level (Skulachev, 1997).

According to Brand and co-workers (Harper and Brand, 1993; Harper *et al.*, 1993), nonohmicity of the mitochondrial membrane is much stronger in a hyperthyroid animal than in a hypothyroid one. This is apparently the reason for higher H^+ conductance of mitochondria when production of thyroid hormones increases (for reviews, see Brand, 1990; Soboll, 1993; Skulachev, 1996a).

The effect of thyroid hormones on the H^+ permeability of mitochondria is different from their action via the nuclear triiodothyronine receptor. It does not require protein synthesis (Horst *et al.*, 1989; Horrum *et al.*, 1992). Moreover, it develops much faster than, for example, induction of the α-glycerophosphate dehydrogenase, the fast effect being mediated by diiodothyronine rather than by triiodothyronine or thyroxine (Horst *et al.*, 1989). There are some indications that the ATP–ADP antiporter possesses a specific thyroid-hormone-binding site of a very high affinity (Sterling, 1986, 1987, 1991; Mowbray and Hardy, 1996).

In our laboratory, it was found that higher (micromolar) concentrations of thyroid hormones added to isolated mitochondria increased the uncoupling activity of low concentrations of SF6847 and FCCP, but they were without effect with DNP. This action was abolished by kCh, male sex hormones, and progesterone. Thyroid hormones were without effect if uncouplers were not added. The uncoupling–stimulating action of thyroid hormones and the re-coupling activity of steroid hormones were arrested by very low concentrations of fatty acids (Starkov *et al.*, 1997b). It was suggested that low amounts of SF6847 and FCCP mimic the action of some natural uncouplers, which is an alternative to the fatty-acid-mediated uncoupling. The above effects might be related to the well-known antagonism of thyroid and steroid hormones as catabolics and anabolics, respectively. One can hypothesize that the thyroid-hormone-induced mild uncoupling decreases the risk of hyperproduction of O_2^-

when the total rate of metabolism is stimulated by these hormones (Skulachev, 1996a, 1997; Starkov *et al.*, 1997b).

If the mild uncoupling, for some reason, is unable to prevent O_2^- formation, the next line of defense is assumed to be actuated. This is the opening of nonspecific pores in the inner mitochondrial membrane, which are permeable by substances of up to 1.5 kDa molecular mass. This results in the collapse of $\Delta\overline{\mu}_{H^+}$, the complete oxidation of CoQH$\cdot$ and the maximal stimulation of oxygen consumption, which is now controlled by respiratory enzyme activities only rather than by $\Delta\overline{\mu}_{H^+}$ or the substrate porter activities (Skulachev, 1994a, 1996a, 1997).

The pore opening is known to be induced by an increase in the level of O_2^- and other ROS. This effect is reversible: the pores close when ROS decreases (for reviews, see Zoratti and Szabo, 1995; Skulachev, 1996a, 1997; Vercesi *et al.*, 1997; Lemasters and Nieminen, 1997; Chernyak, 1997).

If a mitochondrion fails to decrease ROS in spite of the pore opening, it will degrade, since both import and synthesis of proteins in mitochondria require $\Delta\Psi$, which is absent when pores are open (for references, see Skulachev, 1996b). In this way, the mitochondrial population in the cell can be purified of the ROS superproducing organelles ("mitochondrial selection"; see Skulachev, 1996a,b).

Recently, some indications were obtained that the pore opening causes programmed cell death (apoptosis) (Zamzami *et al.*, 1996; Susin *et al.*, 1996; Liu *et al.*, 1996; Petit *et al.*, 1996; Hirsch *et al.*, 1997). In fact, formation of pores results in swelling of the mitochondrial matrix, disruption of the outer mitochondrial membrane, and release of proteins sequestered in the intermembrane space. Among them are mitochondrial cell suicide proteins (MCSP) (Skulachev, 1996b, 1997), namely, a 50-kDa protease [apoptosis inducin factor (AIF)] and cytochrome *c*. It was found by Kroemer and co-workers that the purified AIF, when added to isolated nuclei from HeLa cells, causes typical apoptotic changes (Susin *et al.*, 1996). Independently, Liu *et al.* (1996) showed that a similar effect could be obtained when cytochrome *c*, deoxy-ATP, and a cytosolic protein, Apaf-1 were added to the nuclei from liver cells (for review, see Skulachev, 1997).

The apoptosis initiated by ROS-induced mitochondrial pores may be used by the organism as a mechanism to purify the cell population of the ROS-superproducing cells. This can be defined as cellular selection by analogy with the mitochondrial selection mentioned above (Skulachev, 1996a,b, 1997). Thus, the following chain of events may be involved in the anti-ROS defense system (Skulachev, 1994a):

mild uncoupling $\rightarrow$ strong uncoupling (mitochondrial pore opening)
$\rightarrow$ apoptosis

4.6. Stimulation of ATP Synthesis by Partial Uncoupling

The most paradoxical function of uncoupling consists in stimulating the rate of phosphorylation coupled to respiration. In 1962, I (Skulachev, 1962) suggested that the maximal rate of ATP formation and the maximal thermodynamic efficiency of this process cannot be achieved simultaneously. Partial uncoupling might stimulate the rate-limiting steps of the coupled electron transfer, thus increasing the total oxidative phosphorylation flux.

In 1980, Stucki (1980a,b) presented calculations based on thermodynamic optimising principles which supported the above idea. Later, some observations were published showing that partial uncoupling occurs in liver and muscle mitochondria isolated from animals performing strenuous muscle work (Davies *et al.*, 1982; Klug *et al.*, 1984). In liver, it was found that free fatty acids mediate this uncoupling (Klug *et al.*, 1984). In liver mitochondria, variations in the uncoupling degree between fed and fasted animals were found. These changes were mimicked by perfusion of liver with low concentrations on fatty acids (Soboll and Stucki, 1985). However, in this particular case, uncoupling may be required for metabolism rather than for energetics. As was concluded by Soboll (1995), the

> advantage [of uncoupling] would be a relief of the strong restriction of respiration to the ATP needs of the cell ... to maintain a favorable mitochondrial redox state of NAD^+ during fatty acid oxidation and to generate ketone bodies for peripheral organs during fasting. (p. 578)

The same reasoning may be used to explain why the H^+ permeability of the mitochondrial membrane is rather high in embryonic liver and undergoes a profound reduction during the first postnatal hour (Valcarce *et al.*, 1990).

On the other hand, in all such cases, the most difficult problem is to discriminate between the physiologically useful effect and an *in vitro* artifact or *in vivo* pathology.

5. PATHOLOGICAL ASPECTS OF UNCOUPLING

Hypo- and hyperthyroidism seem to be examples of a situation when endogenous uncoupling is too small or too large, respectively. Unfortunately, both these pathologies are complicated by some other effects of thyroid hormones (Soboll, 1993, 1995; Skulachev, 1997). On the other hand, it is already clear that there are factors, others than thyroid status, that affect the H^+ conductance of mitochondrial membrane.

The first example of a pathology of this kind was described by Ernster, Ikkos, and Luft. This was the so-called Luft's disease (Ernster *et al.*, 1959; Luft *et al.*, 1962). The authors observed the loss of respiratory control in mitochondria isolated from the skeletal muscle of a female patient who suffered from muscle weakness. Special studies revealed that the defect was not a consequence of hyperthyroidism. Later, several cases of the same disease were described (Haydar *et al.*, 1971; Luft, 1994).

According to Cheah *et al.* (1989), malignant hyperthermia is associated with a fatty-acid-mediated uncoupling in mitochondria.

Reye's syndrome, an acute children's disease associated with encephalopathy and fatty infiltration of viscera, was shown to be accompanied by the appearance of a large amount of dicarboxylic fatty acid inducing uncoupling in blood serum (Aprilla and Asimakis, 1977; Tonsgard and Getz, 1985).

Ethanol consumption by rats was shown to result in a less tight attachment of factor F_1 to the membrane, which decreases the coupling (Montgomery *et al.*, 1984). In another study, chronic ethanol consumption by baboons was found to activate phospholipase A_2 in mitochondria and decrease the amount of mitochondrial cardiolipin, phosphatidyl choline, as well as the content and activity of the cytochrome oxidase (Arai *et al.*, 1984).

Ischemia, especially if it is followed by reperfusion, was reported to result in uncoupling in heart mitochondria (for review, see McCord and Turrens, 1994). This might be a result of (1) ROS-induced damage to the mitochondrial membrane and/or (2) the direct uncoupling effect of superoxide that is superproduced under reoxygenation (see Section 4).

Many drugs have been shown to possess an uncoupling activity that apparently was responsible for at least some of their side effects. For example, the anticoagulant phenilin, an analogue of the protonophore dicoumarol, causes uncoupling when added to mitochondria at micromolar concentrations (V. P. Skulachev, unpublished data). N,N'-*bis*-(4-trifluoromethylphenyl)-urea, an impurity in preparations of diuron, proved to be a potent uncoupler (Routaboult *et al.*, 1991).

Sometimes uncoupling explains the therapeutic effect of the drug. This is clearly the case for gramicidin D, which forms H^+-, K^+-, and Na^+-permeable channels in any phospholipid bilayers. So, it is equally dangerous for bacteria, mitochondria, and animal cell outer membranes. This is why gramicidin D is applied for external treatment only.

It is shown that human serum contains a system that kills gram-negative bacteria by means of a channel-forming protein fragment. This fragment is formed because of the specific cleavage of thrombin by bacteria. The process requires a receptor in the outer bacterial membrane and energization of the inner bacterial membrane (Dankert and Esser, 1986; Thaylor and Kroll, 1984). All these properties (proteolysis, the receptor, and $\Delta\bar{\mu}_{H^+}$ requirements) are inherent in some colicines, the bacterial proteins produced by certain *Escherichia coli* strains. *Lactococcus lactis* is found to excrete nisin, a small channel-forming peptide that kill other bacteria (Carneiro de Melo *et al.*, 1996). Uncoupling seems to be involved in the antimitochondrial action of magainins, cationic peptides synthesized by granular glands in the skin of *Xenopus laevis* (de Waal *et al.*, 1991). A protonophoric effect is inherent in some antimalarial drugs (Nissani and Ginsburg, 1989).

Some nongenotoxic carcinogens were shown to possess pronounced uncoupling activity (Keller *et al.*, 1992). On the other hand, the specific killing of some cancer cells by CCCP was described (Newell and Tannock, 1989). These data are in line with recent observation by Kroemer and co-workers (Decaudin *et al.*, 1997) that adriamycin and several other anticancer drugs cause uncoupling, pore opening, and apoptosis.

6. ADDENDUM

In 1997, a new chapter in the uncoupling story was written. It has been found that uncoupling proteins (UCP) similar to thermogenin are present in the great majority of mammalian tissues, as well as in potato.

Fleury *et al.* (1997a) reported that there is a gene in the human genome that codes a protein designated UCP2, which has 59% amino acid identity to the human brown fat UCP (now UCP1). Both UCP1 and UCP2 exhibit three mitochondrial carrier protein motifs and the nucleotide-binding sites. A gene similar to that of human UCP2 (95% identity) was also found in mouse. UCP2 maps to regions of human chromosome 11 and mouse chromosome 7 that have been linked to hyperinsulinemia and obesity. In comparison with UCP1, UCP2 was found to cause stronger *in vivo* lowering of mitochondrial membrane potential when expressed in yeast and stronger inhibition of growth of yeast. Mitochondria isolated from yeast expressing UCP2 showed higher state 4 respiration rate and lower stimulation by FCCP or ADP. In contrast to UCP1, the UCP2 gene is widely expressed in adult human tissues (skeletal muscle, lung, heart, placenta, kidney, spleen, thymus, leukocytes, macrophages, bone marrow, and stomach). In mice, the UCP2 gene expression was found in brown and white fat and at a high level in heart and kidney. A low expression level was observed in liver and brain. The UCP2 expression was up-regulated in white fat in response to fat feeding.

The above finding carried out by the Fleury group initiated a study by Boss and co-workers in Switzerland, where two more representatives of the UCP family were shown to be expressed specifically in the brown fat and skeletal muscle tissues (Boss *et al.*, 1997a,b). They were called $UCP3_L$ and $UCP3_S$, where *L* and *S* are for long and short. $UCP3_S$ is of identical sequence to $UCP3_L$, but contains only 275 amino acids instead of 312 in $UCP3_L$ (amino acids at positions 276 to 312 are absent). $UCP3_L$ has 57% and 73% identical to UCP1 (307 amino acids) and UCP2 (309 amino acids), respectively. The identity to the most closely related mitochondrial anion carrier (α-ketoglutar-ate–malate antiporter) is 32%. $UCP3_S$ contains no purine nucleotide-binding site responsible for inhibition of the fatty-acid-mediated uncoupling in UCP1. Independently, UCP3 was described by Vidal-Puig *et al.* (1997).

Boss *et al.* (1997b) confirmed that the UCP2 mRNA was expressed in various tissues: heart > brown fat > white adipose tissue > skeletal muscle. As for UCP3, its mRNA was most highly expressed in rat brown fat, at high levels in *musculus tensor fascia latae* (fast-twitch glycolytic muscle), *m. tibialis anterior* (fast-twitch oxidative-glycolytic), *m. gastrocnemicus* (mixed), and less in *m. soleus* (slow-twitch oxidative). This suggests that UCP3 is more expressed in glycolytic that in oxidative muscles. The UCP3 mRNA was also detected, although at a much lower level, in rat heart and kidney. In skeletal muscles, the amount of UCP3 mRNA was much higher than that of the UCP2 mRNA (Boss *et al.*, 1997b). Quite recently, Fleury *et al.* (1997b) postulated the existence of UCP4, which is predominantly expressed in neural tissues. A corresponding gene was found in the *X* chromosome.

The UCP3 mRNA level was not affected by cold treatment of rats (Boss *et al.*, 1997b). On the other hand, according to Boss *et al.* (1997a), expression of the UCP2 mRNA increased by factor 2.5 in soleus muscle and brown fat and 4.3 in heart after 48 hr exposure of rats to 6°C. The same cold treatment caused threefold increase in the UCP1 mRNA in brown fat. However, Fleury *et al.* (1997a) failed to observe any effect of cold on the UCP2 mRNA in mice exposed for 10 days to 4°C. This discrepancy may be due to differences in species or in duration of the cold exposure.

In plants, one more uncoupling protein, StUCP, was found by Mueller–Roeber's laboratory in cooperation with the Bouillaud–Fleury group (Laloi *et al.*, 1997). The level of the StUCP mRNA was shown to increase when plants were exposed to cold and decreased when the cold exposure ceased. The effects could be shown both in intact plants (potato) and potato tubers. This finding is in line with the earlier observation of Vercesi *et al.* (1995) about uncoupling in potato tuber mitochondria, which was sensitive to purine nucleotides like that mediated by UCP1. The authors succeeded in isolating a 32-kDa protein from these mitochondria and in reconstituting proteoliposomes that showed an increased H^+ conductance lowered by ATP and GTP by 50% and 35%, respectively. The authors called the protein in question a plant uncoupling mitochondrial protein (PUMP). This name is hardly successful because of H^+ pumps present in the same organelles. As for the name StUCP (*St* from the Latin name of the potato, *Solanum tuberosum*) suggested by the Mueller–Roeber group, it would be adequate only if it would be shown that the protein in question is specific for potato. If that would not be the case, the names pUCP (*p* for plant) or UCP5 seem to be better.

Administration of the β_3-adrenoreceptor agonist Ro-168714 for 32 hr increased in brown fat the level of the UCP2 mRNA by 2.1-fold (Boss *et al.*, 1997a), which was similar to that of the UCP1 mRNA (Muzzin *et al.*, 1989).

Apparently, the effect of cold on UCP2 is mediated by norepinephrine. This is suggested by the fact that stimulation of norepinephrine turnover by cold is severalfold in brown fat (Landsberg and Young, 1978; Young *et al.*, 1982), 48% in soleus muscle, and less than 25% in *m. tibialis anterior* and *m. gastrocnemicus* (Dulloo *et al.*, 1988). Thus, there is a parallel between stimulation of the UCP2 mRNA expression and of the norepinephrine turnover in various tissues (Boss *et al.*, 1997a).

Such a parallel was absent when the effect of 48 hr fasting was studied. It was found that fasting increases UCP2 mRNA expression in *m. soleus*, *m. tibialis anterior*, and *m. gastrocnemicus* by 2.2-, 3.6-, and 2.7-fold, respectively. This was accompanied by 64% lowering of the UCP1 mRNA level in brown fat. No changes in the UCP2 mRNA were found in brown fat and heart (Boss *et al.*, 1997a).

Independently, Gimeno *et al.* (1997) have described UCP2 mRNA in various human and mouse tissues, with predominant expression in white fat and skeletal muscles. In white fat, its amounts were increased five times when mice suffering from obesity (the ob/ob line) as well as mice of the db/db line were studied.

Zhou *et al.* (1997) have demonstrated that protein leptin, the adipocite hormone (mutation in its gene is the reason for obesity of ob/ob mice), is involved in control of the UCP2 mRNA synthesis. It was shown that hyperleptinemia caused by overexpression of leptin due to the leptin gene transfer results in more than tenfold increase in the UCP2 mRNA level in white fat tissues.

Thyroid hormones proved to be one more mechanism controlling UCP2 (as well as UCP3). Lanni *et al.* (1997) and Masaki *et al.* (1997) have shown that thyroid hormones *in vivo* strongly increase level of the UCP2 mRNA. Independently, Gong *et al.* (1997) and Larkin *et al.* (1997) reported similar effects on the UCP3 mRNA level in brown fat and skeletal muscles. Cold exposure, according to Larkin *et al.* (1997), also was effective in the UCP3 mRNA increase in brown fat but not in muscles.

In hepatocytes, where there are no proteins of the UCP family were found, thyroid hormones were shown to also decrease the energy coupling. In our group, it was recently shown in experiments on hepatocytes that energy coupling decreases in the range of hypothyroid > euthyroid > hyperthyroid (Bobyleva *et al.*, 1998). This effect may be due to the thyroid-hormone-induced increase in content of the ATP–ADP antiporter in liver mitochondria (Luciakova and Nelson, 1992; Lunardi *et al.*, 1992; Bobyleva *et al.*, 1997).

Summarizing the above observations, we may conclude that UCP should not now be considered as a brown-fat-specific protein. Besides "classic" UCP1 in brown fat, there are (1) UCP2 expressed in various tissues, and (2) UCP3, which is specific for brown fat and skeletal muscles and exists in long and short forms. UCP2 responds to cold, like UCP1, in a norepinephrine-mediated fashion, whereas UCP3 does not. Even for UCP2, thermoregulatory uncoupling is hardly the only physiological function. Quite recently, indications were obtained (Negre-Salvayre *et al.*, 1997) that UCP2 (and possible other UCPs) are involved in the "mild" uncoupling that prevents fast superoxide production in state 4.

It remains unclear (1) whether both $UCP3_L$ and $UCP3_S$ are competent in uncoupling; (2) what concentrations of UCP are present in mitochondria of different tissues; or (3) whether the already-described uncoupling activity of UCP2 is mediated by fatty acids [according to K. Garlid (personal communication), this is the case; see, however, Fleury *et al.* (1997b)]. To answer these questions, further studies should be done. Already it seems obvious, however, that discoveries of the UCP family open new perspectives for solution of the problem of natural uncoupling.

It is hardly probable that uncoupling activity of fatty acids in tissues other than brown fat is mediated exclusively by UCPs. Rather, UCP2 and -3 are actuated under some specific conditions (state 4, cold stress, fat feeding, fasting), whereas usually the fatty acid uncoupling is mainly assisted by some other mitochondrial proteins. Here, mitochondrial anion carriers, and among them first of all the ATP–ADP and glutamate-aspartate antiporters, should be considered.

7. REFERENCES

Amerkhanov, Z. G., Yegorova, M. V., Markova, O. V., and Mokhova, O. V., 1996, Carboxyatractylate- and cyclosporin A-sensitive uncoupling in liver mitochondria of ground squirrels during hibernation and arousal, *Biochem. Mol. Biol. Int.* **38**:863–870.

Andreyev, A. Yu., Volkov, N. I., Mokhova, E. N., and Skulachev, V. P., 1987, Carboxyatractylate and adenosinediphosphate reduce an uncoupling effect of palmitic acid on skeletal muscle mitochondria, *Biol. Membr.* **4**:474–478. (Russian)

Andreyev, A. Yu., Bondareva, T. O., Dedukhova, V. I., Mokhova, E. N., Skulachev, V. P., and Volkov, N. I., 1988, Carboxyatractylate inhibits the uncoupling effect of free fatty acids, *FEBS Lett.* **226**:265–269.

Andreyev, A. Yu., Bondareva, T. O., Dedukhova, V. I., Mokhova, E. N., Skulachev, V. P., Tsofina, L. M., Volkov, N. I., and Vygodina, T. V., 1989, The ATP/ADP-antiporter is involved in the uncoupling effect of fatty acids on mitochondria, *Eur. J. Biochem.* **182**:585–592.

Aprilla, J. R., and Asimakis, G. K., 1977, Reye's syndrome: The effect of patient serum on mitochondrial respiration *in vitro*, *Biochem. Biophys. Res. Commun.* **79**:1122–1129.

Arai, M., Gordon, E. R., and Lieber, C. S., 1984, Decreased cytochrome oxidase activity in hepatic mitochondria after chronic ethanol consumption and the possible role of decreased cytochrome aa_3 content and changes in phosphilipids, *Biochim. Biophys. Acta* **797**:320–327.

Archakov, A. I., 1975, *Microsomal Oxidation*, Nauka, Moscow. (Russian)

Avetisyan, A. V., Bogachev, A. V., Murtasina, R. A., and Skulachev, V. P., 1993, ATP-driven Na^+ transport and Na^+-dependent ATP synthesis in *Escherichia coli* grown at low $\Delta\bar{\mu}_{H^+}$, *FEBS Lett.* **317**:267–270.

Belitser, V. A., and Tsibakova, E. T., 1939, On the coupling mechanism of the respiration and phosphorylation, *Biokhimiya* **4**:516–521. (Russian)

Bielawski, J., Thompson, T. E., and Lehninger, A., 1966, The effect of 2,4-dinitrophenol on the electric resistance of phospholipid bilayer membranes, *Biochim. Biophys. Res. Commun.* **24**: 948–954.

Bobyleva, V., Bellei, M., Kneer, N., and Lardy, H., 1997, The effects of the ergosteroid 7-oxo-dehydroepiandrosterone on mitochondrial membrane potential: Possible relathionship, *Arch. Biochem. Biophys.* **341**:122–128.

Bobyleva, V., Pazienza, T. L., Maseroli, R., Salviolu, S., Cossarizza, A., Franceschi, C., and Skulachev, V. P., 1998, Decrease in mitochondrial energy coupling by thyroid hormones: A physiological effect rather than a pathological hyperthyroidism consequence, *FEBS Lett.* **430**:409–413.

Bodrova, M. E., Markova, O. V., Mokhova, E. N., and Samartsev, V. N., 1995, Involvement of ATP/ADP-antiporter in fatty acid-induced uncoupling in rat liver mitochondria, *Biochemistry (Moscow)* **60**:1027–1033.

Boss, O., Samec, S., Dulloo, A., Seydoux, J., Muzzin, P., and Giacobino, J.-P., 1997a, Tissue-dependent up-regulation of rat uncoupling protein-2 expression in response of fasting or cold, *FEBS Lett.* **412**:111–114.

Boss, O., Samec, S., Paoloni-Giacobino, A., Rossier, P., Dulloo, A., Seydoux, J., Muzzin, P., and Giacobino, J.-P., 1997b, Uncoupling protein-3: A new member of the mitochondrial carrier family with tissue-specific expression, *FEBS Lett.* **408**:39–42.

Boveris, A., and Chance, B., 1973, The mitochondrial generation of hydrogen peroxide. General properties and effect of hyperbaric oxygen, *Biochem. J.* **134**:707–716.

Brand, M. D., 1990, The proton leak across the mitochondrial inner membrane, *Biochim. Biophys. Acta* **1018**:128–133.

Brustovetsky, N. N., and Klingenberg, M., 1994, The reconstituted ADP/ATP carrier can mediate H^+ transport by free fatty acids, which is further stimulated by mersalyl, *J. Biol. Chem.* **269**: 27329–27336.

Brustovetsky, N. N., Dedukhova, V. I., Yegorova, M. V., Mokhova, E. N., and Skulachev, V. P., 1990, Inhibitors of the ATP/ADP antiporter suppress stimulation of mitochondrial respiration and H^+ permeability by palmitate and anionic detergents, *FEBS Lett.* **272**:187–189.

Carneiro de Melo, A. M. S., Cook, G. M., Miles, R. J., and Poole, R. K., 1996, Nisin stimulates oxygen consumption by *Staphylococcus aureus* and *Escherichia coli*, *Appl. Environ. Microbiol.* **62**:1831–1834.

Chappel, J. B., and Haarhoff, K. N., 1967, The penetration of the mitochondrial membrane by anions and cations, in: *Biochemistry of Mitochondria* (E. C. Slater *et al.*, eds.), pp. 75–91, Academic Press, London.

Chavez, E., Moreno-Sanchez, R., Zazueta, C., Cuellar, A., Ramirez, J., Reyes-Vivas, H., Bravo, C., and Rodriguez-Enriquez, S., 1996, On the mechanism by which 6-ketocholestanol protects mitochondria against uncoupling-induced Ca^{2+} efflux, *FEBS Lett.* **379**:305–308.

Cheah, K. S., Cheah, A. M., Fletcher, J. E., and Rosenberg, H., 1989, Skeletal muscle mitochondrial respiration of malignant hyperthermia-susceptible patients. Ca^{2+}-induced uncoupling and free fatty acids, *Int. J. Biochem.* **21**:913–920.

Chernyak, B. V., 1997, Redox regulation of the mitochondrial permeability transition pore, *Biosci. Rep.* **17**:293–302.

Creuzet, S., Ravanel, P., Tissut, M., and Kaouadji, M., 1988, Uncoupling properties of three flavonols from plane-tree buds, *Phytochemistry* **27**:3093–3099.

Dankert, J. P., and Esser, A. F., 1986, Complement-mediated killing of *Escherichia coli*: Dissipation of membrane potential by a C9-derived peptide, *Biochemistry* **25**:1094–1100.

Davies, K. J. A., Quantanilha, A. T., Brooks, G. A., and Packer, K., 1982, Free radicals and tissue damage produced by exercise, *Biochem. Biophys. Res. Commun.* **107**:1198–1205.

Decaudin, D., Geley, S., Hirsch, T., Castedo, M., Marchetti, P., Macho, A., Kofler, R., and Kroemer, G., 1997, Bcl-2 and Bcl-X_L antagonize the mitochondrial dysfunction preceding nuclear apoptosis induced by chemotherapeutic agents, *Cancer Res.* **57**:62–67.

Dedukhova, V. I., Mokhova, E. N., Skulachev, V. P., Starkov, A. A., Arrigoni-Martelli, E., and Bobyleva, V. A., 1991, Uncoupling effect of fatty acids on heart muscle mitochondria and submitochondrial particles, *FEBS Lett.* **295**:51–54.

Dulloo, A. G., Young, J. B., and Landsberg, L., 1988, Sympathetic nervous system responses to cold exposure and diet in rat skeletal muscle, *Am. J. Physiol.* **255**:E180–E188.

Ernster, L., Ikkos, D., and Luft, R., 1959, Enzymatic activities of human skeletal muscle mitochondria: A tool in clinical metabolic research, *Nature* **184**:1851–1854.

Fleury, C., Neverova, M., Collins, S., Raimbault, S., Champigny, O., Levi-Meyrueis, C., Bouillaud, F., Seldin, M. F., Surwit, R. S., Ricquier, D., and Warden, C. H., 1997a, Uncoupling protein-2: A novel gene linked to obesity and hyperinsulinemia, *Nature Genet.* **15**:269–272.

Fleury, G., Pecquer, C., Raimbault, S., Levi-Meyrues, C., Vacher, D., Ricquier, D., and Bouillaud, F., 1997b, Mitochondrial uncoupling proteins, in: *New Perspectives in Mitochondrial Research* (Meeting Abstracts), p. C5, Padova University Press, Padova, Italy.

Fu, Y., Li, H., and Meng, F., 1995, The possible role of zearalenone in the floral gradient in *Nicotiana tabacum, J. Plant Physiol.* **347**:197–202.

Garlid, K. D., 1989, On the nature of ion leaks in energy-transducing membranes, *Biochim. Biophys. Acta* **976**:109–120.

Garlid, K. D., 1990, New insights into mechanisms of anion uniport through the uncoupling protein of brown adipose tissue mitochondria, *Biochim. Biophys. Acta* **1018**:151–154.

Garlid, K. D., 1994, Mitochondrial cation transport: A progress report, *J. Bioenerg. Biomembr.* **26**:537–542.

Garlid, K. D., 1996, Cation transport in mitochondria—The potassium cycle, *Biochim. Biophys. Acta* **1275**:123–126.

Garlid, K. D., Orosz, D. E., Modriansky, M., Vassanelli, S., and Jezek, P., 1996, On the mechanism of fatty acid-induced proton transport by mitochondrial uncoupling protein, *J. Biol. Chem.* **271**:2615–2620.

Gimeno, R. E., Dembski, M., Weng, X., Deng, N., Shyjan, A. W., Gimeno, C. J., Iris, F., Ellis, S. J., Woolf, E. A., and Tartaglia, L. A., 1997, Cloning and characterization of an uncoupling protein homolog: A potential molecular mediator of human thermogenesis, *Diabetes* **46**:900–906.

Gong, D. W., He, Y., Karas, M., and Reitman, M., 1997, Uncoupling protein-3 is a mediator of thermogenesis regulated by thyroid hormone, β-3 adrenergic agonists, and leptin, *J. Biol. Chem.* **272**:24129–24132.

Grav, H. J., and Blix, A. S., 1979, A source of nonshivering thermogenesis in fur seal skeletal muscle, *Science* **294**:87–89.

Hanstein, W. G., 1976, Uncoupling of oxidative phosphorylation, *Biochim. Biophys. Acta* **456**:129–148.

Harper, M.-E., and Brand, M. D., 1993, The quantitative contributions of mitochondrial proton leak and ATP turnover reactions to the changed respiration rates of hepatocytes from rats of different thyroid status, *J. Biol. Chem.* **268**:14850–14860.

Harper, M.-E., Ballantyne, J. S., Leach, M., and Brand, M. D., 1993, Effects of thyroid hormones on oxidative phosphorylation, *Biochem. Soc. Trans.* **21**:785–792.

Hatefi, Y., 1975, Energy conservation and uncoupling in mitochondria, *J. Supramol. Str.* **3**:210–213.

Haydar, N. A., Conn, H. L., Afifi, A., Nakid, N., Bellas, S., and Fawaz, K., 1971, Severe hyper-metabolism with primary abnormality of skeletal muscle mitochondria, *Ann. Intern. Med.* **74:** 548–558.

Hirsch, T., Marzo, I., and Kroemer, G., 1997, Role of the mitochondrial permeability transition pore in apoptosis, *Biosci. Rep.* **17:**67–76.

Horrum, M. A., Tobin, R. B., and Ecklund, R. E., 1992, The early triodothyronine-induced changes in state IV respiration is not regulated by the proton permeability of the mitochondrial inner membrane, *Biochem. Int.* **28:**813–821.

Horst, C., Rokos, H., and Seitz, H. J., 1989, Rapid stimulation of hepatic oxygen consumption by 3,5-iodo-L-thyronine, *Biochem. J.* **261:**945–950.

Jezek, P., and Garlid, K. D., 1990, New substrates and competitive inhibitors of the Cl⁻ translocat-ing pathway of the uncoupling protein of brown adipose tissue mitochondria, *J. Biol. Chem.* **265:**19303–19311.

Jezek, P., Orosz, D. E., and Garlid, K. D., 1990, Reconstitution of the uncoupling protein of brown adipose tissue mitochondria. Demonstration of GDT-sensitive halide anion uniport, *J. Biol. Chem.* **265:**19296–19302.

Jezek, P., Orosz, D. E., Modriansky, M., and Garlid, K. D., 1994, Transport of anions and protons by the mitochondrial uncoupling protein and its regulation by nucleotides and fatty acids, *J. Biol. Chem.* **269:**26184–26190.

Jezek, P., Modriansky, M., and Garlid, K. D., 1996. A structure–activity study of fatty acid intercation with bilayer and mitochondrial uncoupling protein, *Europ. Bioenerg. Conf.* **9:**50.

Kamp, F., and Hamilton, J. A., 1992, pH gradients across phospholipid membranes caused by fast flip-flop of un-ionized fatty acids, *Proc. Natl. Acad. Sci. USA* **89:**11367–11370.

Kamp, F., and Hamilton, J. A., 1993, Movement of fatty acids, fatty analogues, and bike acids across phospholipid bilayers, *Biochemistry* **32:**11074–11086.

Katre, N. V., and Wilson, D. F., 1977, Interaction of uncouplers with the mitochondrial membrane: A high-affinity binding site, *Arch. Biochem. Biophys.* **174:**578–585.

Katre, N. V., and Wilson, D. F., 1978, Interaction of uncouplers with the mitochondrial membrane: Identification of the high affinity binding site, *Arch. Biochem. Biophys.* **191:**647–656.

Katre, N. V., and Wilson, D. F., 1980, A specific uncoupler-binding protein in *Tetrahymena pyriformis*, *Biochim. Biophys. Acta* **593:**224–229.

Keller, B. J., Marsman, D. S., Popp, J. A., and Thurman, R. G., 1992, Several nongenotoxic carcinogens uncouple mitochondrial oxidative phosphorylation, *Biochim. Biophys. Acta* **1102:**237–244.

Klug, G. A., Krause, J., Ostlung, A.-K., Knoll, G., and Brdiczka, D., 1984, Alterations in liver mitochondria function as a result of fasting and exhaustive exercise, *Biochim. Biophys. Acta* **764:**272–282.

Laloi, M., Klein, M., Riesmeier, J. W., Müller-Röber, B., Fleury, C., Bouillaud, F., and Ricquier, D., 1997, A plant cold-induced uncoupling protein, *Nature* **389:**135–136.

Landsberg, L., and Young, J. B., 1978, Fasting, feeding and regulation of the sympathetic nervous system, *N. Engl. J. Med.* **298:**1295–1301.

Lanni, A., De Felice, M., Lombardi, A., Moreno, M., Fleury, C., Ricquier, D., and Goglia, F., 1997, Induction of UCP2 mRNA by thyroid hormones in rat heart, *FEBS Lett.* **418:**171–174.

Larkin, S., Mull, E., Miao, W., Pittner, R., Albrandt, K., Moore, C., Young, A., Denaro, M., and Beaumont, K., 1997, Regulation of the third member of the uncoupling protein family, UCP3, by cold and thyroid hormone, *Biochem. Biophys. Res. Commun.* **240:**222–227.

Lemasters, J. J., and Nieminen, A.-L., 1997, Mitochondrial oxygen radical formation during reductive and oxidative stress to intact hepatocytes, *Biosci. Rep.* **17:**281–292.

Levachev, M. M., Mishukova, E. A., Sivkova, V. G., and Skulachev, V. P., 1965, Energetics of pigeon at self-warming after hypothermia, *Biokhimiya* **30:**864–874. (Russian)

Liu, S.-S., 1997, Generating, partitioning, targeting and functioning of superoxide in mitochondria, *Biosci. Rep.* **17:**259–272.

Liu, S.-S., and Huang, J. P., 1996, Co-existence of reactive oxygen cycle with Q-cycle in respiratory chain—A hypothesis for generation, partitioning and functioning of O_2^- in mitochondria, in *Proceedings of International Symposium on Natural Antioxidants: Molecular Mechanisms and Health Effects* (D. Moores, ed.), pp. 513–529, AOCS Press, Champaign, IL.

Liu, X., Naekyung, C., Yang, J., Jemmerson, R., and Wang, X., 1996, Induction of apoptotic program in cell-free extracts: Requirement for dADP and cytochrome *c*, *Cell* **86:**147–157.

Loomis, W. L., and Lipmann, F., 1948, Reversible inhibition of the coupling between phosphorylation and oxidation, *J. Biol. Chem.* **173:**807–808.

Luft, R., 1994, The development of mitochondrial medicine, *Proc. Natl. Acad. Sci. USA* **91:**8731–8738.

Luft, R., Ikkos, D., Palmieri, G., Ernster, L., and Afzelius, B., 1962, A case of severe hypermetabolism of non-thyroid origin with a defect in the maintenance of mitochondrial respiratory control: A correlated clinical, biochemical and morphological study, *J. Clin. Invest.* **41:**1776–1804.

Luciakova, K., and Nelson, D., 1992, Transcript levels for nuclear-encoded mammalian mitochondrial respiratory-chain components are regulated by thyroid hormone in an uncoordinated fashion, *Eur. J. Biochem.* **207:**247–251.

Lunardi, J., Kurko, O., King Engel, W., and Attardi, G., 1992, The multiple ADP/ATP translocase genes are differentially expressed during human muscle development, *J. Biol. Chem.* **267:** 15267–15270.

Luvisetto, S., Schmehl, I., Intravaia, E., Conti, E., and Azzone, G. F., 1992, Mechanisms of loss of thermodynamic control in mitochondria due to hyperthyroidism and temperature, *J. Biol. Chem.* **267:**15348–15355.

Macri, F., and Vianello, A., 1990, Zearalenone dissipated electrochemical gradient in higher plant mitochondria and microsomes, *J. Plant Physiol.* **136:**754–757.

Macri, F., Vianello, A., Petrussa, E., and Mokhova, E. N., 1994, Effect of carboxyatractylate on transmembrane electrical potential of plant mitochondria in different metabolic states, *Biochem. Mol. Biol. Int.* **34:**217–224.

Masaki, T., Yoshimatsu, H., Kakuma, T., Hidaka, S., Kurokawa, M., and Sakata, T., 1997, Enhanced expression of uncoupling protein 2 gene in rat white adipose tissue and skeletal muscle following chronic treatment with thyroid hormone, *FEBS Lett.* **418:**323–326.

McCord, J. M., and Turrens, J. F., 1994, Mitochondrial injury by eschemia and reperfusion, *Curr. Top. Bioenerg.* **17:**173–195.

Mikhailova, L. M., Kushnareva, Yu. E., and Andreyev, A. Yu., 1996. Inhibitors of ADP/ATP-antiporter induce two Ca^{2+}-dependent uncoupling pathways in rat liver mitochondria, *Biochemistry* (Moscow) **61:**905–910.

Mitchell, P., 1961, Coupling of phosphorylation to electron and hydrogen transfer by a chemiosmotic type of mechanism, *Nature* **191:**144–148.

Mitchell, P., 1966, Chemiosmotic coupling and photosynthetic phosphorylation, *Biol. Rev.* **41:** 445–502.

Miyoshi, H., and Fuijita, T., 1987, Quantitative analyses on uncoupling activity of SF6847 (2,6-di-*t*-butyl-4-(2,2-diacynovinyl)phenol) and its analogs with spinach chloroplasts, *Biochim. Biophys. Acta* **894:**339–345.

Montgomery, R. I., Spach, P. I., and Cunningham, C. C., 1984, Effect of chronic ethanol consumption on the liver mitochondrial ATP synthase, *Fed. Proc.* **43:**1878.

Mowbray, J., and Hardy, D. L., 1996, Direct thyroid hormone signalling via ADP-ribosylation controls mitochondrial nucleotide transport and membrane leakiness by changing the conformation of the adenine nucleotide transporter, *FEBS Lett.* **394:**61–65.

Murphy, M. P., 1989, Slip and leak in mitochondrial oxidative phosphorylation, *Biochim. Biophys. Acta* **977:**123–141.

Muzzin, P., Revelli, J.-P., Ricquier, D., Meier, M. K., Assimacopoulos-Jeannet, F., and Giacobino, J.-P., 1989, The novel thermogenic β-adrenergic agonist Ro 16-8714 increases the interscapular brown-fat β-receptor adenylate cyclase and the uncoupling protein mRNA level in obese (fa/fa) Zucker rats, *Biochem. J.* **261:**721–724.

Nagamune, H., Fukushima, Y., Takada, J., Yoshida, K., Unami, A., Shimooka, T., and Terada, H., 1993, The lipophilic weak base (Z)-5-methyl-2-[2-(1-naphthyl)ethenyl]-4-piperidine (AU-142) is a potent protonophore type cationic uncoupler of oxidative phosphorylation in mitochondria, *Biochim. Biophys. Acta* **1141:**231–237.

Negre-Salvayre, A., Hirtz, C., Carrera, G., Cazenave, G., Troly, M., Salvayre, R., Penicaud, L., and Casteilla, L., 1997, A role for uncoupling protein-2 as a regulator of mitochondrial hydrogen peroxide generation, *FASEB J.* **11**(10):809–815.

Neijssel, O. M., Buurman, E. T., and de Mattos, M. J. T., 1990, The role of futile cycles in the energetics of bacterial growth, *Biochim. Biophys. Acta* **1018:**252–255.

Newell, K. J., and Tannock, I. F., 1989, Reduction of intracellular pH as a possible mechanism for killing cells in acidic regions of solid tumors: Effects of carbonylcyanide-3-chlorophenylhydrazone, *Cancer Res.* **49:**4477–4482.

Nicholls, D. G., 1974, The influence of respiration and ATP hydrolysis on the proton–electrochemical gradient across the inner membrane of rat-liver mitochondria as determined by ion distributin, *Eur. J. Biochem.* **50:**305–315.

Nicholls, D. G., 1976, The bioenergetics of brown adipose tissue mitochondria, *FEBS Lett.* **61:** 103–110.

Nicholls, D. G., 1979, Brown adipose tissue mitochondria, *Biochem. Biophys. Acta* **549:**1–22.

Nicholls, D. G., 1997, The non-ohmic proton leak—25 years on, *Biosci. Rep.* **17:**252–258.

Nicholls, D. G., and Locke, R. M., 1984, Thermogenic mechanisms in brown fat, *Physiol. Rev.* **64:**1–64.

Nissani, E., and Ginsburg, H., 1989, Protonophoric effects of antimalarial drugs and alkylamines in *Escherichia coli* membranes, *Biochim. Biophys. Acta* **978:**293–298.

Panagos, S., and Beyer, R. E., 1960, Oxidative phosphorylation in liver mitochondria from cold-exposed rats, *Am. J. Physiol.* **199:**836–839.

Panagos, S., Beyer, R. E., and Masoro, E. J., 1958, Oxidative phosphorylation in liver mitochondria prepared from cold-exposed rats, *Biochim. Biophys. Acta* **29:**204–210.

Petit, P. X., Susin, S.-A., Zamzami, N., Mignotte, B., and Kroemer, G., 1996. Mitochondria and programmed cell death: Back to the future, *FEBS Lett.* **396:**7–13.

Pressman, B. C., and Lardy, H. A., 1956, Effect of surface activity agents on the latent ATPase of mitochondria, *Biochim. Biophys. Acta* **21:**458–466.

Pronevich, L. A., Topaly, E. E., Kirpichenok, M. A., Grandberg, I. I., and Topaly, V. P., 1993, A benzopyrylium dye as a novel effective cationic uncoupler of oxidative phosphorylation, *FEBS Lett.* **323:**179–182.

Ravanel, P., 1986, Uncoupling activity of a series of flavones and flavonols on isolated plant mitochondria, *Phytochemistry* **25:**1015–1020.

Ravanel, P., Tissut, M., and Douce, R., 1986, Platanetin: A potent natural uncoupler and inhibitor of the exogenous NADH dehydrogenase in intact plant mitochondria, *Plant Physiol.* **80:** 500–504.

Rolfe, D. F. S., and Brand, M. D., 1994, The contribution of mitochondrial proton leak to basal metabolic rate in ther at, in: *Europ. Bioenerg. Conf.* **8:**101.

Rolfe, D. F. S., and Brand, M. D., 1997, The physiological significance of mitochondrial proton leak in animal cells and tissues, *Biosci. Rep.* **17:**9–16.

Routaboult, J.-M., Mougin, C., Ravanel, R., Tissut, M., Mrlina, G., and Calmon, J.-P., 1991, Effects of *N,N'-bis*-(4-trifluoromethylphenyl)-urea on isolated plant mitochondria and thylakoid membranes, *Phytochemistry* **30**:733–738.

Ruzicka, M., Borecky, J., Hanus, J., and Jezek, P., 1996, Photoaffinity labelling of the mitochondrial uncoupling protein by ^{3}H-azido-fatty acid affects the anion channel, in: *Europ. Bioenerg. Conf.* **9**:190.

Samartsev, V. N., Smirnov, A. V., Zeldi, I. P., Markova, O. V., Mokhova, E. N., and Skulachev, V. P., 1997, Involvement of aspartate/glutamate antiporter in fatty acid-induced uncoupling of liver mitochondria, *Biochim. Biophys. Acta* **1319**:251–257.

Sankaram, M. B., Brophy, P. J., Jordi, W., and Marsh, D., 1990, Fatty acid pH titration and the selectivity of interaction with extrinsic proteins in dimyristoylphosphatidylglycerol dispersions. Spin label ESR studies, *Biochim. Biophys. Acta* **1021**:63–69.

Schönfeld, P., 1990, Does the function of adenine nucleotide translocase in fatty acid uncoupling depend on the type of mitochondria? *FEBS Lett.* **264**:246–248.

Schönfeld, P., Jezek, P., Belyaeva, E. A., Borecky, J., Slyshenkov, V. S., Wieckowski, M. R., and Wojtczak, L., 1996, Photomodification of mitochondrial proteins by azido fatty acids and its effect on mitochondrial energetics. Further evidence for the role of the ADP/ATP carrier in fatty-acid-mediated uncoupling, *Eur. J. Biochem.* **240**:387–393.

Schwaller, M.-A., Allard, B., Lescot, E., and Moreau, F., 1995, Protonophoric activity of ellipticine and isomers across the energy-transducing membrane of mitochondria, *J. Biol. Chem.* **270**: 22709–22713.

Skulachev, V. P., 1962, *Interrelations of the Respiratory Chain Oxidation and Phosphorylation*, Akademiya Nauk SSSR, Moscow. (Russian)

Skulachev, V. P., 1963, Regulation of the coupling oxidation and phosphorylation, *Int. Congr. Biochem.* **5**:365–375.

Skulachev, V. P., 1988, *Membrane Bioenergetics*, Springer, Berlin.

Skulachev, V. P., 1991, Fatty acid circuit as a physiological mechanism of uncoupling of oxidative phosphorylation, *FEBS Lett.* **294**:158–162.

Skulachev, V. P., 1992, Chemiosmotic systems and the basic principles of cell energetics, in: *Molecular Mechanisms in Bioenergetics* (L. Ernster, ed.), pp. 37–73, Elsevier, Amsterdam.

Skulachev, V. P., 1994a, Lowering of the intracellular O_2 concentration as a special function of respiratory systems of the cell, *Biochemistry* (Moscow) **59**:1910–1912.

Skulachev, V. P., 1994b, The latest news from the sodium world, *Biochim. Biophys. Acta* **1187**: 216–221.

Skulachev, V. P., 1995, Non-phosphorylating respiration as a mechanism to minimize formation of reactive oxygen species in the cell, *Mol. Biol.* **29**:709–715. (Russian)

Skulachev, V. P., 1996a, Role of uncoupled and non-coupled oxidations in maintenance of safely low levels of oxygen and its one-electron reductants, *Q. Rev. Biophys.* **29**:169–202.

Skulachev, V. P., 1996b, Why are mitochondria involved in apoptosis? Permeability transition pores and apoptosis as a selective mechanisms to eliminate superoxide-producing mitochondria and cell, *FEBS Lett.* **397**:7–10.

Skulachev, V. P., 1997, Membrane-linked systems preventing superoxide formation, *Biosci. Rep.* **17**:347–366.

Skulachev, V. P., and Maslov, S. P., 1960, Role of non-phosphorylating oxidation in thermoregulation, *Biokhimiya* **25**:1058–1064. (Russian)

Skulachev, V. P., Maslov, S. P., Sivkova, V. G., Kalinichenko, L. P., and Maslova, G. M., 1963, Cold-induced uncoupling of oxidation and phosphorylation in the muscles of white mice, *Biokhimiya* **28**:70–79. (Russian)

Skulachev, V. P., Sharaff, A. A., and Liberman, E. A., 1967 Proton conductors in the respiratory chain and artificial phospholipid membranes, *Nature* **216**:718–719.

Skulachev, V. P., Sharaff, A. A., Jaguzhinsky, L. S., Jasaitis, A. A., Liberman, E. A., and Topali,

V. P., 1968, The effect of uncouplers on mitochondria, respiratory enzyme complexes and artificial phospholipid membranes, *Curr. Mod. Biol.* **2**:98–105.

Smith, R. E., 1958, Symposium: "Metabolism and Cold-Adaptation" (in Discussion), *Fed. Proc.* **17**:1069.

Soboll, S., 1993, Thyroid hormone action on mitochondrial energy transfer, *Biochim. Biophys. Acta* **1144**:1–16.

Soboll, S., 1995, Regulation of energy metabolism in liver, *J. Bioenerg. Biomembr.* **27**: 571–582.

Soboll, S., 1985, Regulation of the degree of coupling of oxidative phosphorylation in intact rat liver, *Biochim. Biophys. Acta* **807**:245–254.

Starkov, A. A., 1997, "Mild" uncoupling of mitochondria, *Biosci. Rep.* **17**:273–280.

Starkov, A. A., Dedukhova, V. I., and Skulachev, V. P., 1994, 6-Ketocholestanol abolishes the effect of the most potent uncouplers of oxidative phosphorylation in mitochondria, *FEBS Lett.* **355**:305–308.

Starkov, A. A., Dedukhova, V. I., Bloch, D. A., Severina, I. I., and Skulachev, V. P., 1995, Some male sex hormones, progesterone and 6-ketocholestanol, counteract uncoupling effects of low concentrations of the most active protonophores, in: *Thirty Years in Mitochondrial Bioenergetics and Molecular Biology* (F. Palmieri *et al.*, eds.), pp. 51–55, Elsevier, Amsterdam.

Starkov, A. A., Bloch, D. A., Chernyak, B. V., Dedukhova, V. I., Mansurova, S. E., Severina, I. I., Simonyan, R. A., Vygodina, T. V., and Skulachev, V. P., 1997, 6-Ketocholestanol is a recoupler for mitochondria, chromatophores and cytochrome oxidase proteoliposomes, *Biochim. Biophys. Acta* **1318**:159–172.

Starkov, A. A., Simonyan, R. A., Dedukhova, V. I., Mansurova, S. E., Palamarchuk, L. A., and Skulachev, V. P., 1997, Regulation of the energy coupling in mitochondria by some steroid and thyroid hormones, *Biochim. Biophys. Acta* **1318**:173–183.

Sterling, K., 1986, Direct thyroid hormone activation of mitochondria: The role of adenine nucleotide translocase, *Endocrinology* **119**:292–295.

Sterling, K., 1987, Direct thyroid hormone activation of mitochondria: Identification of adenine nucleotide translocase (AdNT) as the hormone receptor, *Trans. Assoc. Am. Physicians* **100**: 284–293.

Sterling, K., 1991, Thyroid hormone action: Identification of the mitochondrial thyroid hormone receptor as adenine nucleotide translocase, *Thyroid* **1**:167–171.

Storey, B. T., Wilson, D. F., Bracey, A., Rosen, S. L., and Stephenson, S., 1975, Steric and electronic effects on the uncoupling activity of substituted 3,5 dichlorosalicylanilides, *FEBS Lett.* **49**:338–341.

Strabergerova, H., and Jezek, P., 1996, Azido fatty acid interaction with mitochondrial phosphate carrier, in: *Europ. Bioenerg. Conf.* **9**:191.

Stucki, J., 1980a, The optimal efficiency and the economic degree of coupling of oxidative phosphorylation, *Eur. J. Biochem.* **109**:269–283.

Stucki, J., 1980b, The thermodynamic-buffer enzymes, *Eur. J. Biochem.* **109**:257–267.

Susin, S. A., Zamzami, N., Castedo, M., Hirsch, T., Macho, A., Daugas, E., Geuskens, M., and Kroemer, G., 1996, Bcl-2 inhibits the mitochondrial release of an apoptogenic protease, *J. Exp. Med.* **184**:1331–1341.

Takemori, S., and Kominami, S., 1984, The role of cytochrome P-450 in adrenal steroidogenesis, *Trends Biochem. Sci.* **9**:393–396.

Thaylor, P. W., and Kroll, H.-P., 1984, Interaction of human complement proteins with serum-sensitive and serum-resistant strains of *Escherichia coli*, *Mol. Immunol.* **21**:609–620.

Tikhonova, I. M., Andreyev, A. Yu., Antonenko, Yu. N., Kaulen, A. D., Komrakov, A.Yu., and Skulachev, V. P., 1994, Ion permeability induced in artificial membranes by the ATP/ADP antiporter, *FEBS Lett.* **337**:231–234.

Tonsgard, J. H., and Getz, G. S., 1985, Effect of Reye's syndrome serum on isolated chinchilla liver mitochondria, *J. Clin. Invest.* **76**:816–825.

Toth, P. P., Sumerix, K. J., Ferguson-Miller, S., and Suelter, C. H., 1990, Respiratory control and ADP:O coupling ratios of isolated chick heart mitochondria, *Arch. Biochem. Biophys.* **276**: 199–211.

Valcarce, C., Vitorica, J., Satrustegui, J., and Cuezva, J. M., 1990, Rapid postnatal developmental changes in the passive proton permeability of the inner membrane in rat liver mitochondria, *J. Biochem.* **108**:642–645.

Vercesi, A. E., Martins, I. S., Silva, M. P., and Leite, H. M., 1995, PUMPing plants, *Nature* **375**:24.

Vercesi, A. E., Kowaltowski, A. J., Grijalba, M. T., Meinicke, A. R., and Castilho, R. F., 1997, The role of reactive oxygen species in mitochondrial permeability transition. *Biosci. Rep.* **17**: 43–52.

Vianello, A., and Macri, F., 1981, Effect of zearalenone (F-2) on pea stem, maize root and rat liver mitochondria, *Planta* **153**:443–446.

Vianello, A., Macri, F., Braidot, E., and Mokhova, E. N., 1995, Effect of 6-ketocholestanol on FCCP- and DNP-induced uncoupling in plant mitochondria, *FEBS Lett.* **365**:7–9.

Vidal-Puig, A., Solanes, G., Grujic, D., Flier, J. S., and Lowell, B. B., 1997, UCP3: An uncoupling protein homologue expressed preferentially and abundantly in skeletal muscle and brown adipose tissue, *Biochem. Biophys. Res. Commun.* **235**:79–82.

de Waal, A., Gomes, A. V., Mensink, A., Grootegoed, J. A., and Westerhoff, H. V., 1991, Magainins affect respiratory control, membrane potential and motility of hamster spermatozoa, *FEBS Lett.* **293**:219–223.

Walter, A., and Gutknecht, J., 1984, Monocarboxylic acid permeation through lipid bilayer membranes, *J. Membr. Biol.* **77**:255–264.

Winkler, E., and Klingenberg, M., 1994, Effect of fatty acids on H^+ transport activity of the reconstituted uncoupling protein, *J. Biol. Chem.* **269**:2508–2515.

Wojtczak, L., Bogucka, K., Duszynski, J., and Zablocka, B., 1990, Regulation of mitochondrial resting state respiration: Slip, leak, heterogeneity? *Biochim. Biophys. Acta* **1018**:177–180.

Young, J. B., Savill, E., Rothwell, N. J., Stock, M. J., and Landsberg, L., 1982, Effect of diet and cold exposure on norepinephrine turnover in brown adipose tissue of the rat, *J. Clin. Invest.* **69**:1061–1071.

Zamzami, N., Susin, S. A., Marchetti, P., Hirsch, T., Gomez-Monterrey, I., Castedo, M., and Kroemer, G., 1996, Mitochondrial control of nuclear apoptosis, *J. Exp. Med.* **183**:1533–1544.

Zhou, Y.-T., Shimabukuro, M., Koyama, K., Lee, Y., Wang, M.-Y., Trieu, F., Newgard, C. B., and Unger, R. H., 1997, Induction by leptin of uncoupling protein-2 and enzymes of fatty acid oxidation, *Proc. Natl. Acad. Sci. USA* **94**:6386–6390.

Zoratti, M., and Szabo, I., 1995, The mitochondrial permeability transition, *Biochim. Biophys. Acta* **1241**:139–176.

Zorov, D. B., and Mokhova, E. N., 1973, Effect of a short-term cold exposure on respiration of the diaphragm muscle, *Biol. Nauki* **7**:45–53. (Russian)

CHAPTER 5

Crystallization, Structure, and Possible Mechanism of Action of Cytochrome *c* Oxidase from the Soil Bacterium *Paracoccus denitrificans*

Hartmut Michel, So Iwata, and Christian Ostermeier

1. INTRODUCTION

Cellular respiration is one of the most fundamental processes of life. Most of the energy available to animals is generated by it. In the so-called respiratory chain, four large membrane protein complexes act together to oxidize substrates and finally to reduce oxygen. In the respiratory chains of mitochondria and in many bacteria, either NADH or succinate, both formed preferentially in the citric acid cycle, are oxidized by complex I or complex II, respectively, and ubiquinol is generated. Ubiquinol is oxidized by complex III, also known as the cytochrome bc_1 complex, and the electrons are transferred to cytochrome *c*. Cytochrome *c* is oxidized by complex IV, the cytochrome *c* oxidase (see Fig. 1). The electrons of cytochrome *c* are used to reduce molecular oxygen, and water is formed. Complexes I, III, and IV are able to transport (or pump) protons

Hartmut Michel, So Iwata, and Christian Ostermeier • Max-Planck-Institut für Biophysik, D-60528 Frankfurt am Main, Germany.

Frontiers of Cellular Bioenergetics, edited by Papa *et al.* Kluwer Academic/Plenum Publishers, New York, 1999.

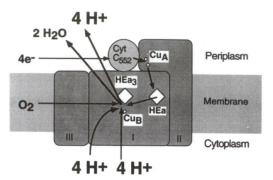

FIGURE 1. Schematic presentation of the cytochrome *c* oxidase from *Paracoccus denitrificans*. Subunits I, II, and III are shown in dark gray, the membrane in light gray. Cytochrome c_{552}, also shown in light gray, donates electrons to Cu_A. From there, the electrons are transfered first to heme *a* (HEa) and then to heme a_3 (HEa$_3$) and Cu_B. Oxygen diffuses via subunit III to the heme a_3–Cu_B binuclear site. Protons are taken up by two different pathways, some of them to be used in water formation, and some of them to be pumped across the membrane.

across the membrane in addition to those protons that are released from ubiquinol on the periplasmic side of the bacterial membrane (or in the intermembrane space of mitochondria) by complex III, or consumed on the cytoplasmic (or matrix) side in complex IV upon water formation. The electrochemical potential difference of protons is used to drive ATP synthesis by the H^+-translocating ATPase. In some sense, the respiratory chain catalyzes the detonating gas reaction, but it has to make certain that energy is stored in the electrochemical proton gradient and that no dangerous side products are formed, especially in the reaction catalyzed by the cytochrome *c* oxidase. Generation of superoxides or peroxides would be dangerous.

The cytochrome *c* oxidases are members of a large superfamily of heme–copper-containing terminal oxidases (see Garcia-Horsman *et al.*, 1994; Calhoun *et al.*, 1994, for review; for general reviews, see Saraste, 1990; Malatesta *et al.*, 1995), which, for example, also includes the cytochrome *bo* ubiquinol oxidase from *Escherichia coli*. Especially, the sequences of subunits I and II are well conserved. Subunit II of cytochrome *c* oxidases contains the binuclear Cu_A center that receives the electrons from cytochrome *c* and transfers them to heme *a* and finally to the binuclear heme a_3-Cu_B center. The Cu_A center is absent in the ubiquinol oxidases, but could be "restored" by genetic engineering (Wilmanns *et al.*, 1995). The heme groups, whose chemical identity can vary (hemes A, B, or O have been found), and Cu_B are bound to subunit I. The total number of subunits varies from two or three in some bacteria to 13 in mammalian mitochondria. During the last 10 years, site-directed mutagenesis studies combined with spectroscopy provided much structural information (Calhoun *et al.*, 1994) that was mostly correct.

In this chapter, we describe the structure of the cytochrome c oxidase from the soil bacterium *Paracoccus denitrificans* as determined by X-ray crystallography (Iwata *et al.*, 1995). This cytochrome c oxidase has the advantage of being well-suited for site-directed mutagenesis. Also quite surprising, in 1995, the structure of the metal site of cytochrome c oxidase from bovine heart mitochondria and, in 1996, its complete protein structure were published (Tsukihara *et al.*, 1995, 1996).

2. CRYSTALLIZATION

The "bottleneck" of membrane protein structure determination is the unavailability of well-ordered crystals. Up to now, atomic models based on X-ray or electron crystallographic structure determinations are available for members of only seven membrane protein families, namely, bacterial photosynthetic reaction centers, bacteriorhodopsin, bacterial porins, prostaglandin H_2 synthase-1, plant light-harvesting complex II, bacterial light-harvesting complexes, and now cytochrome c oxidases.

It is remarkable that it took about 20 years to get suitable crystals of the bovine cytochrome c oxidase using a conventional crystallization strategy. As was the case for photosynthetic reaction centers, the choice of the detergent was critical. Crystallization trials with the bacterial enzyme proceeded for about 6 years, and finally succeeded within 2 years, when a novel strategy, namely, the cocrystallization with an F_V fragment of a conformation-specific monoclonal antibody, was used (Ostermeier *et al.*, 1995). The antibody fragment binds to a discontinuous epitope at the periplasmic side of subunit II, affecting neither cytochrome c oxidation nor proton pumping (Kannt and Michel, unpublished data).

The F_V fragment is the only part of the cytochrome c oxidase–F_V fragment complex involved in protein–protein contacts in the a,b-plane of the crystal lattice. It acts by enhancing the polar surfaces of cytochrome c oxidase, but it does not cover any hydrophobic surfaces. The antibody fragment could also be used to isolate the cytochrome c oxidase in a rapid and mild way (Kleymann *et al.*, 1995).

3. THE STRUCTURE OF THE *PARACOCCUS* CYTOCHROME c OXIDASE

A view parallel to the membrane of the entire cytochrome c oxidase is shown in Fig. 2A. The part integrated into the membrane has a trapezelike

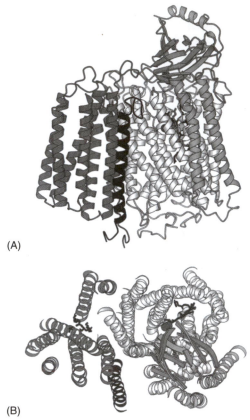

(A)

(B)

FIGURE 2. (A) Ribbon representation of the cytochrome *c* oxidase from *Paracoccus denitrificans* in a view parallel to the membrane. Subunit I is shown in light gray, subunit II in medium gray, subunit III in dark gray, and subunit IV in black. Heme *a* (black), heme a_3 (gray), and Cu_B bound to subunit I, as well as Cu_A bound to subunit II are barely visible. (B) Presentation of the major secondary structure elements of the cytochrome *c* oxidase in a view from the periplasmic space. Color code as in (A). The lipid (black) bound into the V-shaped cleft of subunit III is clearly visible. The β-sheet-rich polar domain of subunit II, binding the Cu_A center, covers most of heme a_3 and Cu_B.

appearance from the direction shown. The width at the cytoplasmic surface is about 90 Å; at the periplasmic surface it is approximately 75 Å. The height of the trapezoid is 75 Å, and 22 transmembrane helices form the trapezoid. The globular domain of subunit II is attached to the trapezoid from the periplasmic side. The central part of the complex is composed of subunit I, which binds both hemes and Cu_B. Subunit I is associated with subunit II at one side and subunit III at the other. The amino- and carboxy-termini of subunit II protrude into the periplasmic space and form a globular domain that contains Cu_A. The antibody F_V fragment binds to this globular domain. In a view perpendicular to the membrane (Fig. 2B), cytochrome *c* oxidase has an oval shape, with the largest dimensions of 90 and 60 Å.

3.1. The Structure of the Protein Subunits

The structure of the protein subunits is briefly described in the following sections.

3.1.1. Subunit I

Subunit I consists mainly of 12 transmembrane helices. In general, the loops on the periplasmic side are longer than those on the cytoplasmic side. The helices show a fascinating and unexpected arrangement: The 12 closely packed helices that are in a simple anticlockwise sequential arrangement when viewed from the periplasmic side (Fig. 2B) can be described to form three symmetry-related semicircular arcs. Due to this remarkable architecture, three *pores* are formed. Two of them are blocked by heme *a*, including its hydroxyethyl farnesyl side chain, or heme a_3, the third one by mostly conserved aromatic residues. Subunit I forms the core of the whole oxidase complex and is responsible for the oxidation of molecular oxygen to water as well as for the redox-coupled pumping of protons.

3.1.2. Subunit II

Subunit II has only two transmembrane helices, which are firmly bound to subunit I (Haltia *et al.*, 1994). It is the only subunit possessing a polar domain. This domain consists of a ten-stranded β-barrel with similarities to the class I copper proteins like plastocyanin. It is bound to the periplasmic side of subunit I and contains the binuclear Cu_A center. The Cu_A center of the corresponding domain in the quinol oxidases is thought to have been lost during evolution.

3.1.3. Subunits III and IV

Subunit III possesses seven transmembrane helices. They are arranged in an irregular manner. They form two bundles, one consisting of the first two helices, and the other of helices III to VII. Both bundles are in a V-shaped arrangement. The association of subunit III with subunit I is weak (Haltia *et al.*, 1994), and subunits III and IV are removed from the core complex by many detergents. In some members of the oxidase family, the first two helices seem to be fused to subunit I, and subunit III possesses only five helices (Castresana *et al.*, 1994). Subunit I and II can be fused at the DNA level. The resulting construct of the *E. coli* ubiquinol oxidase was found to be active (Ma *et al.*, 1993). It has been ruled out that subunit III is involved directly in proton pumping; it may play a more indirect role during assembly or as a linker to other proteins. Furthermore, the structure of the bacterial oxidase indicated a possible channel for diffusion of oxygen leading from the cleft between the two helix bundles of subunit III directly to the binuclear center (Riistama *et al.*, 1996). In the cleft, subunit III contains at least one firmly bound lipid molecule (Iwata *et al.*, 1995). The role of subunit III therefore might be to prevent blockage of the channel entrance by other proteins. Due to the high solubility of oxygen in the hydrophobic interior of the membrane generated by the alkyl chains of the lipids, it is likely that the lipids bound to subunit III might help to allow rapid diffusion of oxygen to the heme a_3–Cu_B center.

The crystal structure of the *Paracoccus* cytochrome *c* oxidase provided the final proof for the existence of a fourth subunit. It consists of only one transmembrane helix with a small *N*-terminal extension at the cytoplasmic side. Its function is unknown; deletion of its gene does not cause any phenotype (Witt and Ludwig, 1997). A number of additional lipid and detergent molecules could be localized around subunit IV.

3.2. The Structure of the Metal Centers

3.2.1. Cu_A Center

As an important difference among the type I copper proteins, Cu_A has been suggested to be a mixed-valence [Cu(1.5)–Cu(1.5)] complex (Blackburn *et al.*, 1994; Larsson *et al.*, 1995; Farrar *et al.*, 1995; Henkel *et al.*, 1995). This suggestion agrees with the crystal structure. The ligands of the two copper atoms are the N_δ-atoms of two histidine residues, one methionine sulfur, one backbone carbonyl oxygen from a glutamate residue (which could not be predicted to be a ligand of Cu_A), and two cysteine thiolates. The latter bridge the two copper atoms (Fig. 3). Distances between the two copper atoms of 2.6 Å in the bacterial (Iwata *et al.*, 1995), 2.7 Å in the beef heart (Tsukihara *et al.*, 1995), and 2.5 Å in the reengineered *bo* oxidase (Wilmanns *et al.*, 1995) have been published.

In the interface region between subunit I and II, a non–redox-active metal can be assigned. An Mg site was modeled in the bovine oxidase structure (Tsukihara *et al.*, 1995), and an Mg–Mn site is in a similar location in the bacterial oxidase. The function of this metal center is still unclear.

3.2.2. Heme *a* and Heme a_3–Cu_B Center

Heme *a* is a low-spin heme with two axial histidine ligands, whereas heme a_3 is a high-spin heme with one histidine ligand. The shortest distance between

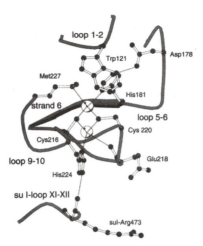

FIGURE 3. The Cu_A center. The two Cu_A atoms are shown as white spheres. They are bridged by two thiolate–sulfurs from cysteine residues. All four atoms lie in one plane. The further Cu_A ligands, all from subunit II, are shown. Tryptophan 121 may form an entry pathway for electrons. Histidine 224 forms a hydrogen bond to the backbone carbonyl oxygen atom of arginine 473 from subunit I and may form the exit pathway for electrons. Taken from Ostermeier *et al.* (1996)

the two hemes is only 4.7 Å. Both are located 15 Å from the periplasmic surface in the hydrophobic core of subunit I. The heme planes are perpendicular to the membrane, and the interplanar angle between the two heme groups is 108° in the bacterial cytochrome *c* oxidase. The electronic coupling between heme *a* and the binuclear heme a_3–Cu_B center is very strong, and fast electron transfer between these two redox centers was recently used to explain the high operational oxygen affinity of the oxidase by "kinetic trapping" of bound oxygen (Verkhovsky *et al.*, 1996).

The heme a_3–Cu_B center is the catalytic core for O_2 reduction. The high-spin iron of heme a_3 is coordinated by histidine 411 as an axial ligand. The free coordination site of the iron points toward the free coordination site of Cu_B, which is ligated by the three histidines 276, 325, and 326. In the bacterial oxidase, one of the three histidines, His 325, seems to be disordered when the oxidized form is crystallized in the presence of azide (Iwata *et al.*, 1995). There is spectroscopic evidence that a bridging ligand may exist between the two metals (Fann *et al.*, 1995), but there was no evidence for a ligand in the bovine cytochrome *c* oxidase structure (Tsukihara *et al.*, 1995). In contrast, electron density between iron and copper was observed in the bacterial enzyme. The identity of this electron density is unclear. It might be caused by the presence of one or two water molecules, or one water plus one hydroxy anion, which would be bound to Cu_B. The distance between iron and copper is 5.2 Å in the bacterial

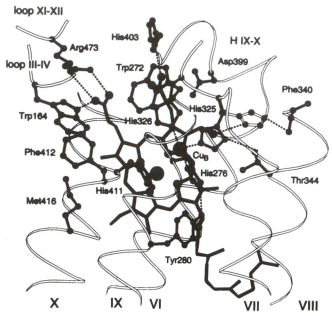

FIGURE 4. The binuclear center in subunit I. Heme a_3 and some important neighboring residues are shown as black ball-and-stick models. The transmembrane helices of subunit I involved are indicated. The heme a_3 iron atom and the Cu_B atom are shown as black spheres. Histidine 276 and histidine 326 are clear ligands of Cu_B. There is no electron density for the side chain of histidine 326. It can be modeled in two different conformations. In one, it is a Cu_B ligand, and not in the other (shown in white). Switching of the side chain between both positions might be an essential part of the proton pump mechanism.

cytochrome c oxidase. However, the central heme a_3 iron atom was found to be 0.7 Å out of plane. The question of whether a hydroxyl group exists as a ligand to Cu_B has to remain unanswered at the present resolution of the structure determination.

4. POSSIBLE MECHANISMS

Protein structures are not determined to provide beautiful textbook figures but to understand the function and mechanism of action of the protein under investigation. In the case of cytochrome c oxidase, three important mechanistic questions have to be answered: (1) the precise pathway for electron transfer, (2) the proton transfer pathways, and (3) the mechanism of coupling proton transfer and proton pumping to oxygen reduction. Cu_A is the first acceptor for electrons delivered from cytochrome c. The electrons are then transferred to heme a, which then passes them further to the binuclear center (Hill, 1994). It is very likely that the endergonic proton pumping reaction is directly coupled to the exergonic redox reactions at the binuclear heme a_3–Cu_B center (Babcock and Wikström, 1992). Many models and hypotheses have been put forward to explain this coupling (see Babcock and Wikström, 1992; Einarsdóttir, 1995; Einarsdóttir et al., 1995; Hallen et al., 1994; Mitchell and Rich, 1994; Morgan et al., 1994; Musser and Chan, 1995; Rich, 1995; Verkhovsky et al., 1995; Wikström, 1989).

Inspection of the structure of the bacterial cytochrome c oxidase shows two possible proton transfer pathways that are in agreement with the results of site-directed mutagenesis experiments. Pathway (1) leads along the conserved polar face of transmembrane helix VIII (Thomas et al., 1993a). A prominent residue at this face is lysine 354. It is therefore called the K-pathway. In addition, it includes the hydroxyl group of the hydroxyethylfarnesyl side chain of heme a_3 and tyrosine 280. Replacement of the residues of the polar face of transmembrane helix VIII leads to a loss of the enzymatic activity. Pathway (2) starts at a conserved aspartate (124); it is therefore called the D-pathway, at the entrance of the "pore" formed by helices II–VI (see Fig. 2B), and leads into a partly polar cavity containing several solvent molecules. The conserved glutamate residue 278 is found at the end of the cavity. Beyond this glutamate, the proton pathway becomes rather speculative. It may lead to histidine 325, which is not visible in the electron density map. Replacing the conserved Asp 124 by Asn or two subsequent Asn residues (113,131) by hydrophobic residues abolishes proton pumping, but oxygen reduction and water formation still occur, albeit at a reduced rate (Thomas et al., 1993b; Garcia-Horsman et al., 1995; Fetter et al., 1995). These findings suggest at a first glance that the D-pathway is the one used for protons to be pumped and the K-pathway for protons consumed in water formation. However, such an assignment depends on the precise mechanism of coupling proton pumping and redox reactions.

The papers by Rich (1995) and Morgan et al. (1994) seem to be of special importance for formulating a mechanism of redox-linked proton pumping. Rich postulates electroneutrality of redox changes around the heme–copper center. It is an attractive hypothesis, considering the low polarity of the surrounding hydrophobic membrane environment. This means that upon a single reduction of one of the metals a proton has to be taken up for charge compensation and two protons for double reduction. In the catalytic cycle when protons are taken up from the cytoplasmic side later and consumed in water formation, those protons taken up first are expelled to the periplasmic side by electrostatic interactions, and thus pumped. Rich (1995) does not present chemical details.

Morgan et al. (1994) postulate that a histidine ligand of Cu_B "cycles" between the imidazolate, imidazole, and imidazolium states twice upon reduction of one molecule of dioxygen. The protons of the imidazolium are those being pumped. Also in this model the uptake of the protons to be consumed upon water formation leads to the expulsion of the protons to be pumped. The histidine shuttle mechanism presented by Iwata *et al.* (1995) is compatible with the structure and strictly obeys the electroneutrality principle. The key residue is histidine 325, which would be a Cu_B ligand in the imidazolate and imidazole states but not in the imidazolium state. In the latter, it might assume a position suited for proton transfer to the periplasmic side, which is realized upon arrival of protons at the binuclear site needed for water formation. However, if a hydroxy group were a Cu_B ligand [a special case of the finding of one oxygen with an exchangable proton(s) as a Cu_B ligand (Fann *et al.*, 1995)], the first proton taken up upon reduction of the binuclear center would unavoidably lead to the formation of water, and the histidine cycle–shuttle mechanism would have to be reformulated.

The idea of having at least one hydroxy anion between the heme a_3 iron and Cu_B in the oxidized form of the enzyme is attractive, since otherwise a strong electrostatic repulsion between the iron atom with one positive charge and Cu_B with two positive charges would exist. Knowledge of the protonation states of the histidine ligands to Cu_B, especially of histidine 325, and the pK values for protonation–deprotonation is required to support or to exclude a histidine cycle mechanism. If the histidine ligands stay neutral during the catalytic cycle, alternative mechanisms have to be sought. Of particular interest is the answer to the question of what the proton acceptors for those protons are that are taken up upon reduction of cytochrome *c* oxidase (Mitchell and Rich, 1994).

What is also clearly needed is the determination of the precise structure of the intermediates (see Babcock and Wikström, 1992) of the redox reactions of the heme–copper oxidases. There is good reason to believe that the structures of some of these intermediates can be determined by X-ray crystallography using the available cytochrome *c* oxidase crystals.

NOTE ADDED IN PROOF. Since the submission of the manuscript serious doubts have been raised against the postulate (Wikström, 1989) that the oxidative part of cytochrome *c* oxidase catalytic cycle is exclusively coupled to proton pumping, and a new model of proton pumping, based on the electroneutrality principle, has been proposed (Michel, H., 1998, The mechanism of proton pumping by cytochrome *c* oxidase, *Proc. Natl. Acad. Sci. USA* **95**:12819–12824).

ACKNOWLEDGMENTS. We thank Dr. C. R. D. Lancaster and A. Harrenga for preparing some of the figures.

5. REFERENCES

Babcock, G. T., and Wikström, M., 1992, Oxygen activation and the conservation of energy in cell respiration, *Nature* **356:**301–309.

Blackburn, N. J., Barr, M. E., Woodruff, W. H., von der Ooost, J., and deVries, S., 1994, Metal–metal bonding in biology: EXAFS evidence for a 2.5 Å copper–copper bond in the Cu_A center of cytochrome oxidase, *Biochemistry* **33:**10401–10407.

Calhoun, M. W., Thomas, J. W., and Gennis, R. B., 1994, The cytochrome oxidase superfamily of redox-driven proton pumps, *Trends Biochem. Sci.* **19:**325–330.

Castresana, J., and Saraste, M., 1995, Evolution of energetic metabolism: The respiration-early hypothesis, *Trends Biochem. Sci.* **20:**443–448.

Castresana, J., Lübben, M., Saraste, M., and Higgins, D. G., 1994, Evolution of cytochrome oxidase, an enzyme older than atmospheric oxygen, *EMBO J.* **13:**2516–2525.

Einarsdóttir, Ó., 1995, Fast reactions of cytochrome oxidase, *Biochim. Biophys. Acta* **1229:**129–147.

Einarsdóttir, Ó., Geogiadis, K. E., and Sucheta, A., 1995, Intramolecular electron transfer and conformational changes in cytochrome *c* oxidase, *Biochemistry* **34:**496–508.

Fann, Y. C., Ahmed, I., Blackburn, N. J., Boswell, J. S., Verkhovskaya, M. L., Hoffman, B. M., and Wikström, M., 1995, Structure of Cu_B in the binuclear heme–copper center of the cytochrome aa_3-type quinol oxidase from *Bacillus subtilis*: An ENDOR and EXAFS study, *Biochemistry* **34:**10245–10255.

Farrar, J. A., Lappalainen, P., Zumft, W. G., Saraste, M., and Thomson, A. J., 1995, Spectroscopic and mutagenesis studies on the Cu_A centre from the cytochrome *c* oxidase complex of *Paracoccus denitrificans*, *Eur. J. Biochem.* **232:**294–303.

Fetter, J. R., Qian, J., Shapleigh, J., Thomas, J. W., García-Horsman, A., Schmidt, E., Hosler, J., Babcock, G. T., and Gennis, R. B., 1995, Possible proton relay pathways in cytochrome *c* oxidase, *Proc. Natl. Acad. Sci. USA* **92:**1604–1608.

Garcia-Horsman, J. A., Barquera, B., Rumbley, J., Ma, J., and Gennis, R. B., 1994, The superfamily of heme–copper respiratory oxidases, *J. Bacteriol.* **176:**5587–5600.

Garcia-Horsman, J. A., Puustinen, A., Gennis, R. B., and Wikström, M., 1995, Proton transfer in cytochrome bo_3 ubiquinol oxidase of *Escherichia coli*: Second-site mutations in subunit I that restore proton pumping in the mutant Asp 135 → Asn, *Biochemistry* **34:**4428–4433.

Hallén, S., Brzezinski, P., and Malmström, B. G., 1994, Internal electron transfer in cytochrome *c* oxidase is coupled to the protonation of a group close to the bimetallic site, *Biochemistry* **33:** 1467–1472.

Haltia, T., Semo, N., Arrondo, J. L. R., Goñi, F. M., and Freire, E., 1994, Thermodynamic and structural stability of cytochrome *c* oxidase from *Paracoccus denitrificans*, *Biochemistry* **33:** 9731–9740.

Henkel, G., Müller, A., Weissgräber, S., Buse, G., Soulimane, T., Steffens, G. C. M., and Nolting, H.-F., 1995, The active sites of the native cytochrome *c* oxidase from bovine heart mito-chondria: EXAFS-spectroscopic characterization of a novel homobinuclear copper center (Cu_A) and of the heterobinuclear Fe_{a3}–Cu_B center, *Angew. Chem. Int. Ed. Engl.* **34:**1488–1492.

Hill, B. C., 1994, Modeling the sequence of electron transfer reactions in the single turnover of reduced, mammalian cytochrome *c* oxidase, *J. Biol. Chem.* **269:**2419–2425.

Iwata, S., Ostermeier, C., Ludwig, B., and Michel, H., 1995, Structure at 2.8 Å resolution of cytochrome *c* oxidase from *Paracoccus denitrificans*, *Nature* **376:**660–669.

Kleymann, G., Ostermeier, C., Ludwig, B., Skerra, A., and Michel, H., 1995, Engineered Fv fragments as a tool for the one-step purification of integral multisubunit membrane protein complexes, *Biotechnology* **13:**155–160.

Larsson, S., Källebring, B., Wittung, P., and Malmström, B. G., 1995, The Cu_A center of cytochrome *c* oxidase: Electronic structure and spectra of models compared to the properties of Cu$_A$ domains, *Proc. Natl. Acad. Sci. USA* **92**:7167–7171.

Ma, J., Lemieux, L., and Gennis, R. B., 1993, Genetic fusion of subunits I, II, and III of the cytochrome *bo* ubiquinol oxidase from *Escherichia coli* results in a fully assembled and active enzyme, *Biochemistry* **32**:7692–7697.

Malatesta, F., Antonini, G., Sarti, P., and Brunori, M., 1995, Structure and function of a molecular machine: cytochrome *c* oxidase, *Biophys. Chem.* **54**:1–33.

Mitchell, R., and Rich, P. R., 1994, Proton uptake by cytochrome *c* oxidase on reduction and on ligand binding, *Biochim. Biophys. Acta* **1186**:19–26.

Morgan, J. E., Verkhovsky, M. I., and Wikström, M., 1994, The histidine cycle: A new model for proton translocation in the respiratory heme–copper oxidases, *J. Bioenerg. Biomembr.* **26**: 699–608.

Musser, S. M., and Chan, S. I., 1995, Understanding the cytochrome *c* oxidase proton pump: Thermodynamics of redox linkage, *Biophys. J.* **68**:2543–2555.

Ostermeier, C., Iwata, S., Ludwig, B., and Michel, H., 1995, F$_V$ fragment-mediated crystallization of the membrane protein bacterial cytochrome *c* oxidase, *Nature Struct. Biol.* **2**:842–846.

Ostermeier, C., Iwata, S., and Michel, H., 1996, Cytochrome *c* oxidase, *Curr. Opin. Struct. Biol.* **6**:460–466.

Rich, P. R., 1995, Towards an understanding of the chemistry of oxygen reduction and proton translocation in the iron–copper respiratory oxidases, *Aust. J. Plant Physiol.* **22**:479–486.

Riistama, S., Puustinen, A., Garcia-Horsman, A., Iwata, S., Michel, H., and Wikström, M., 1996, Chanelling of dioxygen into the respiratory enzyme, *Biochim. Biophys. Acta* **1275**:1–4.

Saraste, M., 1990, Structural features of cytochrome oxidase, *Q. Rev. Biophys.* **23**:331–366.

Thomas, J. W., Lemieux, L. J., Alben, J. O., and Gennis, R. B., 1993a. Site-directed mutagenesis of highly conserved residues in helix VIII of subunit I of the *bo* ubiquinol oxidase from *Escherichia coli*. An amphipathic transmembrane helix that may be important in conveying protons to the binuclear center, *Biochemistry* **32**:11173–11180.

Thomas, J. W., Puustinen, A., Alben, J. O., Gennis, R. B., Wikström, M., 1993b. Substitution of asparagine-135 in subunit I of the cytochrome *bo* ubiquinol oxidase of *Escherichia coli* eliminates proton-pumping activity, *Biochemistry* **32**:10923–10928.

Tsukihara, T., Aoyama, H., Yamashita, E., Tomizaki, T., Yamaguchi, H., Shinzawa-Itoh, K., Nakashima, R., Yaono, R., and Yoshikawa, S., 1995, Structure of metal sites of oxidized bovine heart cytochrome *c* oxidase at 2.8 Å, *Science* **269**:1069–1074.

Tsukihara, T., Aoyama, H., Yamashita, E., Tomizaki, T., Yamaguchi, H., Shinzawa-Itoh, K., Nakashima, R., Yaono, R., and Yoshikawa, S., 1996, The whole structure of the 13-subunit oxidized cytochrome *c* oxidase at 2.8 Å, *Science* **272**:1136–1144.

Verkhovsky, M. I., Morgan, J. E., and Wikström, M., 1995, Control of electron delivery to the oxygen reduction site of cytochrome *c* oxidase: A role for protons, *Biochemistry* **34**:7483–7491.

Verkhovsky, M. I., Morgan, J. E., Puustinen, A., and Wikström, M., 1996, Kinetic trapping of oxygen in cell respiration, *Nature* **380**:268–270.

Wikström, M., 1989, Identification of the electron transfers in cytochrome oxidase that are coupled to proton-pumping, *Nature* **338**:776–778.

Wilmanns, M., Lappalainen, P., Kelly, M., Sauer-Eriksson, E., and Saraste, M., 1995, Crystal structure of the membrane-exposed domain from a respiratory quinol oxidase complex with an engineered dinuclear copper center, *Proc. Natl. Acad. Sci. USA* **92**:11955–11959.

Witt, H., Ludwig, M., 1997, Isolation, analysis and deletion of the gene coding for subunit IV of cytochrome *c* oxidase in *Paracoccus denitrificans*, *J. Biol. Chem.* **272**:5514–5517.

CHAPTER 6

The Structure of Crystalline Bovine Heart Cytochrome *c* Oxidase

Shinya Yoshikawa, Kyoko Shinzawa-Itoh, and Tomitake Tsukihara

1. INTRODUCTION

Cytochrome *c* oxidase, the terminal enzyme in cell respiration, reduces molecular oxygen (O_2) to water. The enzyme couples oxygen reduction with the active transport of protons (Ferguson-Miller and Babcock, 1996; Malmström, 1990). Since its discovery (Warburg, 1924), this enzyme has been one of the most extensively investigated systems in bioenergetics, because of its physiological importance. The redox-active centers of the enzyme contain the transition metals, Fe and Cu, and these active-site metals have been the focus of a variety of spectroscopic investigations on the intriguing O_2 reduction process (Ferguson-Miller and Babcock, 1996; Malmström, 1990). So far, the enzyme has been isolated from about 80 different organisms, and most of the amino acid sequences from each of these organisms have been determined. Buse and co-workers determined the amino acid sequence of the bovine enzyme, and this pioneering work has been the basis for subsequent structure–function investi-

Shinya Yoshikawa and Kyoko Shinzawa-Itoh • Department of Life Science, Himeji Institute of Technology, Kamigohri, Akoh Hyogo 678-12, Japan. **Tomitake Tsukihara** • Institute for Protein Science, Osaka University, Suita, Osaka 565, Japan.

Frontiers of Cellular Bioenergetics, edited by Papa *et al.* Kluwer Academic/Plenum Publishers, New York, 1999.

gations on the enzyme (Hensel and Buse, 1990; Meinecke and Buse, 1986). An ingenious investigation of the subunit composition of the enzyme performed by Kadenback is also a benchmark for the structural characterization of the enzyme (Kadenbach *et al.*, 1983). On the other hand, time-resolved resonance Raman spectroscopy has been a powerful experimental technique for identifying the structure of the intermediate species during the O_2 reduction, such as the dioxygen (Fe^{2+}-O_2), the ferryl oxide ($Fe^{4+} = O$), and the hydroxide (Fe^{3+}-OH) intermediates (Ogura *et al.*, 1991; Kitagawa and Ogura, 1997). Substantial effort also has been directed toward elucidating a coupling mechanism between the O_2 reduction and the proton pumping (Ferguson-Miller and Babcock, 1996; Malmström, 1990; Wikström and Morgan, 1992). However, these past investigations were performed without knowledge of the three-dimensional structure of the enzyme at the atomic resolution. Without this detailed structural knowledge, contributions from these studies for understanding the reaction mechanism of this enzyme at the atomic level were limited. Thus, the two papers on the crystal structure of the enzyme isolated from two different species (Tsukihara *et al.*, 1995; Iwata *et al.*, 1995) marked the beginning of a new era in the history of research on cytochrome *c* oxidase.

In this chapter, the crystal structure of beef heart cytochrome *c* oxidase at its reported current atomic resolution, 2.8 Å, will be summarized and the mechanism of the enzyme reaction will be discussed.

2. AN OVERVIEW OF THE CRYSTAL STRUCTURE OF BEEF HEART CYTOCHROME *c* OXIDASE

Cytochrome *c* oxidase, isolated from bovine heart muscle, is dimeric within crystals. These crystals, prepared in the presence of decyl maltoside, are shown in Fig. 1 (Tsukihara *et al.*, 1996). The middle portion of the structure is composed of 28 transmembrane α-helices. The largest three subunits, subunits I, II, and III (Kadenbach *et al.*, 1983), which are encoded by mitochondrial genes, form the core of each monomer. The other ten nuclear-encoded subunits surround the core. This enzyme has four redox-active metal sites, hemes *a* and a_3, Cu_A and Cu_B, and two redox-inactive metals, Mg and Zn. Cu_A is a dinuclear copper site in subunit II that receives electrons from cytochrome *c*. A nuclear-encoded subunit, subunit Vb, on the matrix surface holds a zinc atom. The magnesium atom is bound at the interface between subunits I and II. The other three metals are in subunit I. The distance between the iron atom of heme a_3 and Cu_B, which is a mononuclear copper site, is 4.5 Å in the fully oxidized state. These two metals form the O_2 binding and reduction site. The Cu_A site is in the extramembrane domain of subunit II on the cytosolic side. The distances from the center of the two copper atoms of Cu_A to hemes *a* and a_3 irons are 19 Å

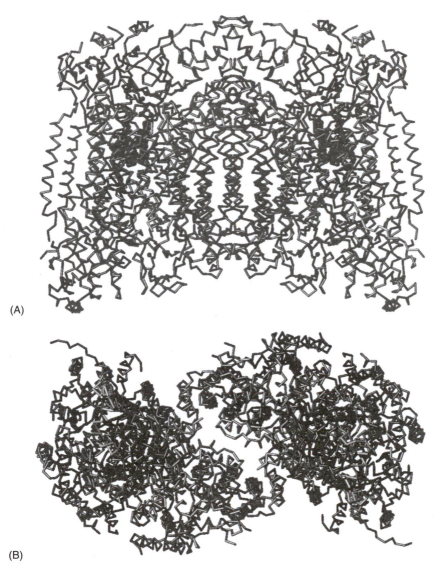

(A)

(B)

FIGURE 1. Cα backbone trace of dimer of bovine heart cytochrome *c* oxidase. Each monomer consists of 13 different subunits. (A) A side view against the transmembrane helices and (B) a top view from the cytosolic side.

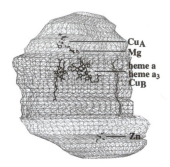

FIGURE 2. A schematic representation of metal site location in bovine heart cytochrome *c* oxidase. Molecular surface shown by the cage is defined with the electron density at 5 Å resolution. A dark region in the cage shows transmembrane part. The thickness of the transmembrane part is 48 Å. The hydrophilic parts protruding to the cytosolic (upper) and matrix (lower) sides are 37 Å and 32 Å in thickness, respectively. The structures with and without a ball denote hemes *a* and a_3, respectively. The heme ball near heme a_3 is Cu_B. Two small balls at the membrane surface level are a magnesium atom and a water oxygen bound to the magnesium. The two balls, close to the molecular surface and 8 Å above the membrane surface level, are the two copper atoms in Cu_A site. Zinc is located in the hydrophilic part in the matrix side.

and 22 Å, respectively. The magnesium site is between Cu_A and heme a_3 (Fig. 2). In addition to these metals, five phosphatidylethanolamines, three phosphatidylglycerols, and two cholates have been detected in the crystal structure (Tsukihara *et al.*, 1996).

3. STRUCTURE OF THE REDOX ACTIVE METAL SITES AND THE ELECTRON TRANSFER BETWEEN THEM

3.1. Cu_A Site

The electron density peak at the Cu_A site of beef heart cytochrome *c* oxidase at the current level of atomic resolution is spherical in shape (Tsukihara *et al.*, 1995). However, the electron density value at the center of this spherical peak is 1.5 times higher than the value expected for a single copper atom. Furthermore, the six functional groups of amino acid residues surrounding the copper sphere do not provide a reasonable coordination structure for a single copper atom. On the other hand, two copper atoms placed 2.7 Å apart provide a dinuclear metal coordination structure similar to the structure of the iron–sulfur cluster of ferredoxin (Fig. 3). The two iron atoms in the iron–sulfur cluster are replaced with two copper atoms, and the two inorganic sulfur atoms

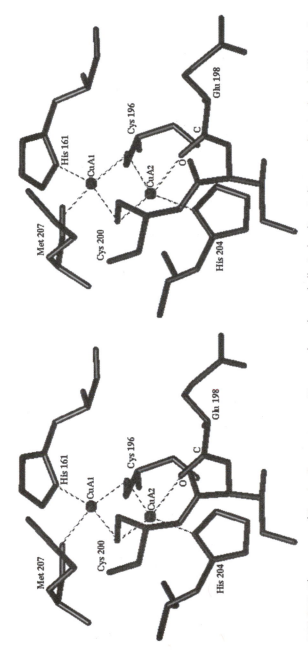

FIGURE 3. The structure of Cu_A site. Copper atom positions are shown by two balls. Broken lines indicate coordination bonds. All the amino acids given here belong to subunit II. The peptide carbonyl group of Glu[198] is illustrated with a bar marked with C and O.

are replaced with two Sγ atoms of cysteine residues. The two sulfur atoms and two copper atoms form a rhombus, with the interatomic distances of 2.7 Å for the Cu–Cu distance and of 3.8 Å for the Sγ–Sγ distance. One of the copper atoms forms a tetrahedral coordination complex that includes the two Sγ atoms of Cys[196] and Cys[200], the imidazole nitrogen of His[161], and the sulfur atom of Met[207]. The peptide carbonyl of Glu[198], the imidazole nitrogen of His[204], and the two Sγ atoms are ligands that also form a tetrahedral complex to the other copper atom.

The electron paramagnetic resonance spectrum of Cu$_A$ suggests that the oxidation state of the dinuclear center is [Cu$^{1.5+}$ ... Cu$^{1.5+}$] in which one electron is delocalized between the two Cu^{2+} ions, since a dinuclear copper enzyme, N$_2$O reductase, shows a similar spectrum in which the dinuclear center is assigned to the [Cu$^{1.5+}$ ... Cu$^{1.5+}$] "half-reduced" state (Kröneck *et al.*, 1988). Thus, the Cu$_A$ site is dinuclear but is a one electron reduction site. Then, why is the Cu$_A$ site dinuclear? One possible reason for the existence of a dinuclear copper site is to provide separate sites for electron entry and exit. Small electron transfer proteins like plastocyanin and cytochrome *c* need only one site for both electron entry and exit, because these proteins are mobile electron carriers in the aqueous environment of the cell. Cu$_A$ is fixed in cytochrome *c* oxidase. Thus, one of the copper atoms is the site for receiving electrons from cytochrome *c* and the other for donating the electrons to heme *a*.

The redox equilibrium behavior of Cu$_A$ indicates that the midpoint potential of Cu$_A$ as well as those of hemes *a* and *a*$_3$ and possibly Cu$_B$ are influenced by the oxidation states at other redox sites. These interactions make the potentials of the two hemes essentially identical in any given overall oxidation state, while the absolute potential is influenced by the overall oxidation state (Wikström *et al.*, 1981). Later, this model was modified to some extent by more accurate measurements of the redox behavior. However, the interactions between the redox centers has not been eliminated from the model (Blair *et al.*, 1986; Wang *et al.*, 1986; Nichols and Wrigglesworth, 1988, and references therein). A dinuclear copper center with conformational flexibility would be better suited to influence the midpoint potential than a mononuclear center. This is consistent with investigations on the polynuclear iron–sulfur proteins that demonstrate a wide range of midpoint potentials (Jensen, 1986). Furthermore, the dinuclear center has a greater number of ligands, compared with the mononuclear center, which could relay changes in the oxidation state at the metal site to other regions of the protein. Thus, the larger dinuclear metal site is better suited for the redox coupling between the metal sites. In order to elucidate the mechanism of this coupling interaction, the redox-dependent changes in the conformation at the Cu$_A$ site need to be examined at high resolution.

3.2. Hydroxylfarnesylethyl Groups

The two hemes share the same common structural framework, which includes a hydroxylfarnesylethyl group attached to position 2 of the porphyrin ring and a formyl group at position 8. Three possible roles have been proposed for the long alkyl side chain of hemes a and a_3. According to one proposal, this long side chain functions as an anchor to stabilize the interaction between the heme and the protein within the hydrophobic interior of the protein. However, in many types of hemoproteins, heme B, which does not have the large side chain, interacts strongly with the protein. In 1971, Caughey (1971) suggested a folding of the side chain in which the double bonds of the three isoprenoid units are aligned in parallel and in close proximity for the formation of a conjugated π election system. Furthermore, for this particular side chain folding, the π election system of the isoprenoid side chain is conjugated with the π electron system of the porphyrin to provide an electron transfer pathway. According to this proposal, the electron transfer rate is regulated by the conformation of the alkyl side chain. The third proposal, a ligand shuttle mechanism, was formulated by Woodruff et al. (1991). In this mechanism, the unsaturated group of the long alkyl side chain reversibly coordinates to either Cu_B or Fe_{a3}. They proposed that this shuttle movement of the unsaturated group between the two metal sites pumps protons.

The hydroxylfarnesylethyl group has been discovered in the heme O of the O_2 reduction sites of the bacterial cytochrome bo complex (Wu et al., 1992). Furthermore, proton pumping capability has been detected for this terminal oxidase (Puustinen et al., 1989). Thus, these findings support the Woodruff mechanism. Previous investigations indicate that the long alkyl side chain is an indispensable feature for proper functioning of the dinuclear O_2 reduction site. Recently, however, it has been shown that the O_2 reduction site of a bacterial terminal oxidase, cytochrome cbb_3, contains Cu_B and heme B, not heme O. The various spectral properties of cytochrome c oxidase and its O_2 reduction ability are essentially identical with those of heme $O–Cu_B$ center (Gray et al., 1994; Garcia-Horsman et al., 1994). The structural and functional properties of cytochrome cbb_3 indicates that the second and third proposals for the role of the hydroxylfarnesylethyl side chain are unlikely.

In the current crystal structure, the plane of heme a is positioned perpendicular to the membrane surface, with the long alkyl side chain directed toward the membrane surface at the matrix side. This structure is consistent with the first proposal described above. The alkyl side chain of heme a_3 is in a U-shaped conformation, with the chain terminus close to the transmembrane surface of subunit I. Loosely packed hydrophobic and aromatic amino acid side chains surround the isoprenoid side chain. Thus, the alkyl side chain of heme a_3 serves as an important structural element of a proposed channel for O_2, which leads to the reduction site. (As described in Section 3.4, two other possible O_2 channels have been identified in the crystal structure.) Interestingly, this function for the long side chain of heme a_3 had never been proposed, before the crystal structure appeared.

All the enzymes in the terminal oxidase family couple the reduction of O_2 with proton pumping. It has been widely accepted that this coupling follows a common reaction mechanism (Ferguson-Miller and Babcock, 1996; Malmström, 1990). However, insertion of the large alkyl groups of hemes a and a_3 into the transmembrane region of the protein is expected to induce large conformational changes. Thus, compensating conformational changes must occur in other regions of the protein if the reaction mechanism is identical regardless of the structure of hemes. Alternatively, the mechanism of the enzyme reaction in different organisms could be altered without the occurrence of large conformational changes to compensate for placement of the long side chain within the protein core. For a better understanding of the role of the alkyl group, it is important to solve the crystal structure of a terminal oxidase that does not contain the hydroxylfarnesylethyl group.

3.3. Magnesium Site

One of the magnesium ligands is a subunit II carboxyl side chain of Glu[198], the peptide carbonyl of which is coordinated to a copper atom at the Cu_A site. In addition to Glu[198], Asp[369], and His[368] of subunit I and a bound water molecule form a tetrahedral coordination complex at the magnesium site. At the resolution of the crystal structure determination, electron density for an isolated magnesium atom is difficult to identify. Fortunately, mutagenesis results had previously implicated the amino acid residues listed above as ligands to the magnesium site (Hosler *et al.*, 1995; Espe *et al.*, 1995). Thus, assignment of electron density surrounding the four ligands to the magnesium site is quite reasonable because of the mutagenesis results. One of the ligands to magnesium, His[368], is hydrogen bonded to one of the propionate groups of heme a_3. Thus, Glu[198], the magnesium site, and His[368] provide a possible electron transfer pathway directly from Cu_A to heme a_3.

3.4. Heme *a* and the O_2 Reduction Site

Heme *a* is coordinated symmetrically by two histidine imidazoles, and the iron atom is in the plane of the porphyrin. Thus, this prosthetic group is in a typical ferric low-spin heme coordination (Tsukihara *et al.*, 1995). Heme *a* and heme a_3 are in close proximity. The two histidines, one an axial ligand to heme *a* and the other an axial ligand to heme a_3, are separated by only one residue in

the amino acid sequence. The shortest distance between the two heme periph-
eries is only 4 Å, and the close positioning of the two hemes suggests extremely
rapid electron transfer between them.

The Cu_B is coordinated by three histidines: His[290], His[291], and His[240]
(Tsukihara *et al.*, 1995). The three coordinating imidazole nitrogen atoms form
an equilateral triangle with the copper atom at the center. This histidine
triangle, with Cu_B at the center, is in parallel with the heme a_3 porphyrin plane.
The Cu_B is 4.7 Å from heme a_3 iron and 1.0 Å away from a line, which is
normal to the heme a_3 plane and passes through the heme a_3 iron (Fig. 4).
Electron density maps indicate the absence of larger atoms such as chloride
between the two metals. However, smaller-sized atoms such as oxygen and
nitrogen may be present. At higher resolution, careful calculation of the electron density difference should reveal the existence of atoms bridging heme a_3
iron and Cu_B. This electron density difference is based on the subtraction of
model structure factor amplitudes from experimental X-ray diffraction amplitudes. The model structure factors are calculated from the refined model
coordinates without the inclusion of a bridging ligand. Trigonal planar coordination for cupric copper (Cu^{2+}) is unstable. Thus, Cu_B^{2+} in cytochrome c
oxidase is likely to coordinate the fourth ligand, and this fourth ligand is
located between the two metals. However, one important point is that the
metal–metal distance—4.7 Å—is too long for a bridging coordination by a
single oxygen atom.

This $Fe_{a3}–Cu_B$ distance from the crystal structure (4.7 Å) and distance
measured from extended X-ray absorption fine structure (EXAFS) (3.5 Å)

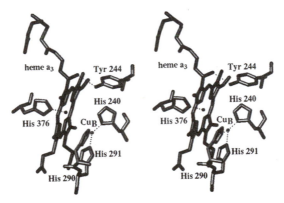

FIGURE 4. Structure of the O_2 binding and reduction site. The broken lines are coordination
bonds, and a dotted line is a hydrogen bond between Tyr[244] and the hydroxyl group of the
hydroxylfarnesylethyl group of heme a_3.

(Scott, 1989) disagree. The EXAFS experiments were performed at liquid nitrogen temperatures, while the X-ray diffraction intensities were measured at 8°C. This temperature difference could influence the pH of the microenviron- ment at the O_2 reduction site, ultimately changing the coordination state of the metals. Such a significant conformational change suggests that different con- figurations, which determine the Cu^{2+}–Fe_{a3} interatomic distance, are involved in the enzyme under physiological conditions. To further identify and charac- terize temperature-dependent conformational changes at the active site, we are planing to collect X-ray diffraction data at the same cryogenic temperatures that the EXAFS experiments were performed at.

Tyr^{244}, a residue near the O_2 reduction site, is hydrogen bonded to a hydroxyl group on the isoprenoid chain of heme a_3. This tyrosine is likely to be a proton donor to the intermediate species during the O_2 reduction, since a hydrogen bond network is located between Tyr^{244} and Lys^{265} (Fig. 5). The latter residue is at the matrix surface side of the protein. The hydrogen bond network includes hydrogen bonds between hydrogen-bond-forming residues and fixed waters. Ser^{255} and His^{256} in the network are connected by a cavity. Cavities

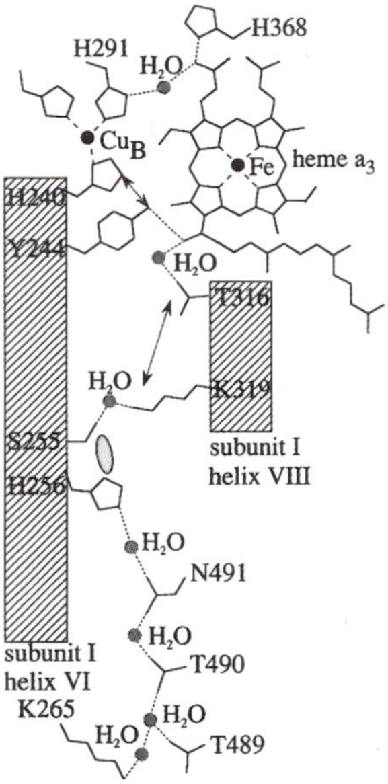

FIGURE 5. A schematic representation of a possible channel for chemical protons. A dark oval, dotted lines, and a dotted line with arrows denote an internal cavity, hydrogen bonds, and a possible hydrogen-bond configuration, respectively. Dark balls and black balls are fixed waters and metals, respectively. Starting from Lys^{265} of subunit I placed on the matrix surface, the hydrogen bond network ends at Tyr^{244} near the heme a_3 and Cu_B site.

in the protein are spaces without any significant electron density but are large enough to contain mobile or unordered water molecules. Thus, protons can be transferred through the cavity. There is no hydrogen bond between Lys[319] and Thr[316] in the network. However, a small conformational change in the amino acid side chains of either residue would form a new hydrogen bond between the two residues without affecting the conformation of the peptide backbone. Such an alternate positioning of the two amino acid side chains is referred to in this chapter as a possible hydrogen bond configuration. The conformational change could be induced by the change in the oxidation state or the ligand-binding state of the enzyme. Alternatively, the conformation forming the hydrogen bond could exist only for a transient period of time too short to detect by X-ray diffraction. These conformational changes may also regulate the rate of proton donation.

In addition to the proton-donating site, a water channel leads to the O_2 reduction site. The channel begins at one of the propionates of heme a_3, which forms a hydrogen bond to His[368], one of the magnesium ligands. Two opposing arrays of hydrophilic amino acids, one belonging to subunit I and the other to subunit II, are situated at the interface between subunits I and II and form a hydrophilic channel from the propionate of heme a_3 up to the cytosolic surface. This channel is too narrow to accept even a single water molecule, suggesting a tightly controlled movement of water from the active site to the cytosolic surface side. This regulated movement prevents the back flow of water from the cytosolic side. Such a back flow could result in the influx of protons and short circuit the proton pumping.

Some redox states of cytochrome c oxidase do not contain enough electron equivalents for reducing O_2 completely to water. This would produce a partially reduced O_2 intermediate or a high valence state of the heme iron such as $Fe^{4+} = O$, if O_2 is freely accessible to the O_2 reduction site. Thus, the supply of O_2 to the O_2 reduction site of cytochrome c oxidase must be strictly regulated. In fact, from close inspection of the atomic coordinates of subunit I, there is no channel wide enough for the passage of O_2 to its reduction site. Most of the amino acid residues of subunit I in the immediate vicinity of the Cu_B side of heme a_3 are hydrophobic or aromatic, and they are loosely packed. Three small channel openings at the surface of subunit I connect to the O_2 reduction site. The channels are too narrow to accept even a single O_2 molecule. However, small conformational changes could widen the channels to allow for O_2 transfer. These conformational changes are likely to be controlled by the changes in oxidation or the binding of ligands at the redox active metal sites. One of these possible O_2 channels passes along the shortest pathway from Cu_B to the surface of subunit I. The entrance of this channel is on the surface of the protein in the space between two monomers where, as described below, two cardiolipin molecules are located.

Another possible channel, significantly longer than the one described above, leads from the surface of subunit I to the O_2 reduction site. A crevice of subunit III, which contains three phospholipids, as described in Section 7.1, is located near the entrance of this possible channel. Both of the channels contain bound phospholipids at their entrances. The third possible O_2 channel, which passes along the long alkyl side chain of heme a_3, as described above, opens at the hydrophobic surface of the protein and is exposed to the "bulk phospholipid" in the mitochondrial membrane.

3.5. Electron Transfers between the Three Redox Centers

Electron transfer within cytochrome c oxidase has been examined for more than 30 years, since the pioneering work of Gibson and Greenwood (1963), who used the flow-flash method to follow changes in the absorption spectrum. Recently, the direction of electron flow within the enzyme has been identified as cytochrome c–Cu_A–heme a–heme a_3 (Hill, 1994). A hydrogen bond network between Cu_A and heme a includes one of the ligands of Cu_A—His[204]—a peptide bond between Arg[438] and Arg[439], and a propionate group of heme a. This hydrogen bond network provides a facile electron transfer path from Cu_A to heme a (Fig. 6), and the double-bond character of peptide bond promotes the efficiency of electron transfer. Another network connecting Cu_A and heme a_3 through the magnesium site, as described in Section 3.3, could also be an effective electron transfer path directly from Cu_A to heme a_3. However, the direct electron transfer from Cu_A to heme a_3 has never been detected in the rapid kinetic experiments (Hill, 1994), which suggests that the rate of electron transfer from Cu_A to heme a is at least two orders of magnitude higher than that from Cu_A to heme a_3. The difference in the electron transfer rate is expected because of the shorter distance between Cu_A and heme a compared with one between Cu_A and heme a_3 (19 Å vs. 22 Å). Why then does this enzyme have a direct electron transfer path from Cu_A to heme a_3?

4. MECHANISM OF THE O_2 REDUCTION AND THE ROLE OF Cu_B

It is well known that a one-electron reduction of O_2, the triplet dioxygen, is not favorable energetically, while two-electron reduction is quite favorable (Caughey *et al.*, 1976). This property of O_2 provides the stability for the oxygenated forms of hemoglobin and myoglobin, in which the Fe^{2+}-O_2 is isolated in the protein, and therefore, the second electron is not easily accessible to the bound O_2. In fact, some artificial reductants like hydroquinone can stimulate the auto-oxidation of these globins. Thus, Cu_B placed near the O_2 binding site must stimulate the O_2 reduction in a two-electron process to form

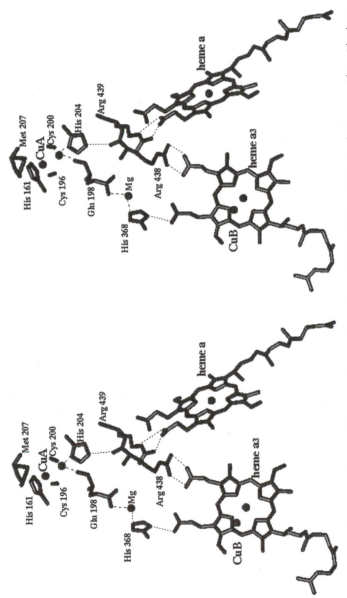

FIGURE 6. Hydrogen bond network between three redox active metal sites bovine heart cytochrome *c* oxidase. The heme structures with and without a small ball (Cu$_B$) are hemes *a* and *a$_3$*, respectively. Cu$_A$ site is shown by two small blue balls and six amino acids coordinated to the coppers. The other three amino acids that belong to subunit I are placed between Cu$_A$ site and the two hemes. Dotted lines indicate hydrogen bonds and broken lines are coordination bonds.

peroxide. This process is likely to be limited by electron transfer across a short distance of about 3 Å from Cu_B^{1+} to O_2 on Fe_{a3}^{2+}. The process could be very rapid, in the order of a picosecond. On the other hand, formation of Fe^{2+}-O_2 is limited by the O_2 transfer through one or more of the proposed O_2 channels described above. Even the shortest channel is 8 Å in length. Furthermore, the O_2 transfer is most likely accompanied by conformational changes along the channel. Thus, the O_2 transfer process should be much slower than the electron transfer rate from Cu_B to the bound O_2. This indicates that the oxygenated form of cytochrome c oxidase is unlikely to be detected under physiological conditions.

Unexpectedly, the oxygenated form of the enzyme has been observed during the reaction of the fully reduced enzyme with O_2, using time-resolved resonance Raman spectroscopy (Ogura *et al.*, 1990; Han *et al.*, 1990; Varotsis *et al.*, 1990). The isotropic shifts of the Fe-O_2 vibration observed by the exchange of $^{16}O_2$ with $^{18}O_2$ and $^{16}O^{18}O$ indicate that O_2 binds to Fe^{2+} in an end-on fashion in the same manner as in hemoglobin (Ogura *et al.*, 1993). Furthermore, the lifetime of the oxygenated form is "unusually" long, with a $t_{1/2}$ of about 0.1 msec at 4°C (Ogura *et al.*, 1993, 1991; Kitagawa and Ogura, 1997). The resonance Raman results suggest a structure that diminishes the electron transfer rate from Cu_B to the bound O_2.

At present, the crystal structure of bovine heart cytochrome c oxidase has been reported only for the fully oxidized state (Tsukihara *et al.*, 1995, 1996). Known crystal structures of many metalloproteins indicate that the conformational perturbations induced by changes in the oxidation state at the metal site are very small. Thus, the structure of the O_2 reduction site in the fully reduced state is likely to be essentially identical with that in the fully oxidized state. When O_2 binds to Fe_{a3}^{2+} in an end-on fashion without forming any interaction between Cu_B and one of the oxygen atoms, the coordination state of Cu_B^{1+} is trigonal planer with three histidine imidazole side chains as ligands. The Cu^{1+} in such a coordination is very stable. Thus, this coordination stability lowers the rate of the electron transfer from Cu_B^{1+} to the O_2 bound to Fe_{a3}^{2+}. Furthermore, formation of a hydrogen bond between Tyr[244] and the O_2 at Fe_{a3}^{2+} lengthens Cu_B-O_2 distance. Such a configuration is less favorable for the electron transfer from Cu_B to the O_2 at Fe_{a3}. However, electron transfer from the fifth ligand side of heme a_3 to the bound O_2 would be accelerated by the hydrogen bond formation, which decreases the negative charge on the O_2. Thus, the O_2 reduction is likely to be limited by this hydrogen bond formation. The current crystal structure indicates that the hydrogen bond formation must be accompanied by a small change in the conformation of Tyr[244], and this conformational change may not be very rapid, which is consistent with the resonance Raman results.

Recent Raman evidence (Kitagawa and Ogura, 1997) shows that the hydroperoxo intermediate formed in cytochrome c oxidase (Fe^{3+}-OOH) is very rapidly decomposed to a ferryloxo intermediate ($Fe^{4+}=O$), with an extra oxidation equivalent releasing H_2O, as in the case of peroxidase (Kitagawa and Mizutani, 1994). The extra oxidation equivalent at the iron atom, giving $Fe^{5+}=O$, has been proposed based on the raman measurements (Kitagawa and Ogura, 1997). Cu_B, with an extremely high redox potential, will reduce the Fe_{a3} at the high oxidation state during the enzymic turnover. Also, the Cu_B^{1+} could be a scavenger of any activated oxygen accidentally formed.

It has long been widely accepted that Cu_B^{1+} promotes the two electron reduction of O_2 bound at Fe_{a3}^{2+} (Caughey $et\ al.$, 1976; Malmström, 1990; Ferguson-Miller and Babcock, 1996). However, the crystal structure suggests that heme a, not Cu_B, triggers the normal O_2 reduction, and that Cu_B protects heme a_3 against the potential damages of activated oxygen intermediates by scavenging for free radicals and by controlling the electron entry to Fe_{a3}. These functions of Cu_B are in addition to the role that Cu_B serves for the reduction of high oxidation states Fe_{a3} during enzymatic turnover.

Two well-known but peculiar properties of bovine heart cytochrome c oxidase in the fully oxidized state are the unusually slow rate of cyanide binding and the unusually slow reduction of heme a_3 when ferrocytochrome c is used as the reductant (Gibson $et\ al.$, 1965). [Here, the fully oxidized state denotes the state of the enzyme as isolated from beef heart muscle under aerobic conditions. Four electron equivalents are required for attaining the fully reduced state from the fully oxidized state (Steffens $et\ al.$, 1993). The strong magnetic coupling between Fe_{a3}^{+3} and Cu_B^{2+} is also a well-known property of the enzyme in the fully oxidized state (Van Gelder and Beinert, 1969; Malmström, 1990). Furthermore, heme a_3 of the enzyme under normal turnover conditions receives electron equivalents and binds cyanide much faster than the Fe_{a3}^{3+} in the fully oxidized state (Antonini $et\ al.$, 1977). Thus, the fully oxidized enzyme is unlikely to be involved in the enzymic turnover.] The crystal structure of this enzyme in the fully oxidized state (Tsukihara $et\ al.$, 1995, 1996) does not indicate that access for cyanide to heme a_3 is obstructed. The three possible O_2 channels could be very effective pathways for HCN, which is the predominant species of cyanide at neutral pH (Tsukihara $et\ al.$, 1996). Both the proximity of heme a to heme a_3 and the networks of bonds between Cu_A and heme a_3 including hydrogen bonds and coordination bonds to the Mg ion could support very facile electron transfer to heme a_3. Thus, the slow reduction of heme a_3 and the slow cyanide binding to the Fe_{a3}^{3+} are likely to be due to the structural and functional characteristics of the O_2 reduction site.

A preliminary analysis of the electron density between the two metals in the O_2 reduction site shows the presence of a bridging ligand, suggesting that Cu_B^{2+} is in a tetracoordinated state and Fe_{a3}^{3+} is in a hexacoordinated state. Thus, the reduction of Fe_{a3}^{3+} is accompanied by breaking the magnetic coupling between Fe_{a3}^{3+} and Cu_B^{2+} as well as removal of the bridging ligand. Similarly, cyanide binding would be obstructed by the presence of a bridging ligand. Thus, the peculiar properties of the fully oxidized form are most likely to be induced by Cu_B and the bridging ligand.

5. MECHANISM OF PROTON PUMPING

Cytochrome c oxidase creates an electrochemical potential across the mitochondrial membrane by pumping protons from the matrix side to the cytosolic side, together with the proton uptake of "chemical" protons from the matrix side and the electron transfer from the cytochrome c at the cytosolic side (Ferguson-Miller and Babcock, 1996; Malmström, 1990). The redox-coupled proton pumping has been an intriguing and extensively investigated aspect in cytochrome c oxidase research, since Wikström's (1977) discovery of the proton pumping function.

The coupling between the O_2 reduction and the proton translocation has been carefully and ingeniously investigated by examining the effect of the electrochemical potential on the reverse reaction of the O_2 reduction, the water oxidation (Wikström and Morgan, 1992). These investigations demonstrated that only two electron transfer steps out of four during the O_2 reduction to water were driven to the reverse direction by applying an electrochemical potential opposite in polarity to the polarity under normal physiological conditions. Papa and co-workers (Capitanio *et al.*, 1996) reported that the efficiency of proton uptake was dependent on the electron transfer rate, which suggests an indirect coupling. Most of the redox-dependent structural and functional changes for this enzyme have been associated with the proton pumping function. Also, the effects of pH and proton–deuterium exchange on reaction rates were examined to identify which steps during the O_2 reduction process were coupled with the proton pumping (Malmström, 1990). However, the results of such experiments have not provided further information about the mechanism of the proton pumping. These experimental results may identify the O_2 reduction steps coupled with the proton pumping. It is difficult to determine whether this coupling between the pumping and electron transfer is direct or indirect.

Mechanisms of proton pumping by this enzyme have been proposed without knowledge of the crystal structure. Chan and co-workers proposed (Gelles *et al.*, 1986) that redox changes at the copper ion of Cu_A induce ligand exchange, and this ligand exchange accounts for the redox-coupled unidirectional proton transfer. This was the first proposal for the mechanism of the redox coupled proton pumping. However, this proposal has been downplayed after a proton-pumping function coupled with the O_2 reduction to water was discovered in a bacterial terminal oxidase that lacks Cu_A (Puustinen *et al.*,

1989). It has been well accepted that the proton-pumping mechanism of the bacterial terminal oxidase is identical to that of beef heart enzyme (Malmström, 1990; Ferguson-Miller and Babcock, 1996).

Recently, more sophisticated mechanisms for redox-coupled proton pumping have been proposed (Wikström et al., 1994), based on the coordination structure of the O_2 reduction site deduced from mutagenesis results. In this mechanism, one of the three histidine ligands to Cu_B acts as a unidirectional proton shuttle. According to this model, this histidine is either liganded or dissociated from Cu_B, depending on the protonation state of the imidazole side chain as well as on the oxidation and ligand binding states of Fe_{a3}. In this mechanism, the histidine imidazole is the proton-pumping site as well as a key component of the O_2 reduction site. Thus, one of the most important functions of this proton-pumping site must be to prevent the pumping protons from entering the dioxygen reduction site. Otherwise, the pumping protons will be taken up by activated intermediates of O_2 reduction that have an extremely high affinity for protons. This proton trapping would short-circuit the proton pumping.

This mechanism seems consistent with the crystal structure determination of the azide-bound, fully oxidized bacterial cytochrome c oxidase in which electron density for one of the three histidine imidazole ligands of Cu_B is missing (Iwata et al., 1995). This structure suggests that one of the three histidine imidazole ligands is mobile. Thus, it seems very difficult for this mobile imidazole to form a barrier between pumping protons and O_2 reduction intermediates bound to Fe_{a3}. Furthermore, the Fe_{a3}–Cu_A distance of the azide-inhibited bacterial enzyme is 5.2 Å, clearly longer than that of the fully oxidized bovine heart enzyme, 4.7 Å. This difference suggests the Cu_B site is flexible, which is also likely to induce proton leakage into the O_2 reduction site.

Two hydrogen bond networks from the matrix side to the cytosolic side are detected in the crystal structure (Fig. 7) (Tsukihara et al., 1996). These networks include hydrogen bonds, cavities, and possible hydrogen bond configurations. Neither network contains any redox-active metals as the primary component. In network B, heme a lies on a side branch and could regulate the proton transfer in the main branch through changes in oxidation state. However, this side branch is not the site for pumping protons. The hydrogen bond network from the matrix side to Tyr[244] is unlikely to be used for the proton pumping, since it terminates at the O_2 reduction site (Fig. 5). The O_2 reduction site includes a water channel, as described above, which could be a pathway for hydronium ions (H_3^+O). However, no bypass between Tyr[244] and the water channel, which is well segregated from the O_2 reduction site, has been detected in the crystal structure. Thus, the crystal structure of bovine heart cytochrome c oxidase indicates that redox-active metal sites are not involved directly in the proton pumping. No dead-end branches are shown in the figures except for heme a. Some dead-end branches reach fairly close to the cytosolic surface.

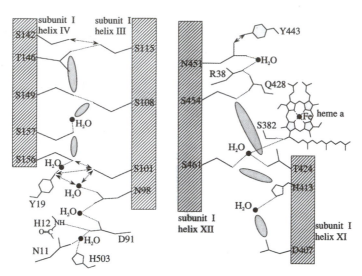

FIGURE 7. Schematic representations of possible channels for vectorial protons. (A) A possible channel for pumping protons, placed between helices III and IV of subunit I. Two amino acids, His[503] and Asn[11], are placed at the entrance of the channel on the matrix side surface. Ser[142] and Ser[115] constitute the exit on the cytosolic surface. (B) The second possible channel for proton pumping between helices III and IV. This network connects Asp[407] on the matrix surface with Tyr[443] on the cytosolic surface. This figure includes a dead-end branch reaching to heme *a*. This branch could control the proton transfer in the main stream, depending on the oxidation state of heme *a*.

Thus, the redox and ligand binding states of the metal sites could induce a conformation change, accompanying changes in the peptide backbone, that was not very large but large enough to provide a new hydrogen bond network from the dead-end branches to the cytosolic side. But we do not assume any conformational change accompanying any movement of the peptide backbone.

A pair of amino acid side chains in a possible hydrogen bond configuration as defined above is not connected by a hydrogen bond, in the crystal structure in the oxidation state. A small conformational change that does not affect the configuration of peptide backbone, depending on the oxidation state and ligand binding, would form a new hydrogen bond between them. The crystal structure shows a small space near the amino acid side chains to allow the conformational change. The conformation depicted in the crystal structure is the most stable one, and other discrete but less stable conformations are possible. However, the population of these less-stable conformations are too low to detect in the crystal structure. Some could be the conformation in which the hydrogen bond is placed between the amino acid residues. The oxidation state and the ligand binding could change the relative stability of these conformations to make the hydrogen-bonded conformation most stable and detectable in the crystal structure. Therefore, even in the fully oxidized state, these possible hydrogen bonding configurations in the networks could be in the hydrogen-bonded conformation, though for a very limited period.

Network A in Fig. 7 contains two points along the network connected by possible hydrogen bonding configurations, while the network B has only a single possible hydrogen bonding configuration near the cytosolic surface. Thus, the network B could be a leaky point for protons. Such a structure could discharge the proton gradient when it is too high for the physiological conditions.

Mutation of the homologous residue of Asp^{91} in the bacterial enzyme to asparagine eliminates the proton-pumping activity (Thomas *et al.*, 1993). As seen in the Fig. 7, Asp^{91} is hydrogen bonded to several functional groups. Thus, exchanging the Asp with Asn would induce significant changes in the hydrogen bonding configuration. This mutagenesis result is consistent with the involvement of network A in the proton pumping. Most of the amino acid residues in network A are conserved, which also supports the possible importance of this network for proton pumping. However, the amino acid sequence of a proton-pumping apparatus that does not include a metal site could show higher amino acid sequence variability among different organisms compared with the sequence variability of a proton-pumping apparatus that includes a metal site. The function of a metal coordination compound is influenced critically by the structure of the ligands, and these ligands are expected to be highly conserved. However, many combinations of amino acids can transfer protons quite effectively. Thus, sets of amino acid residues not completely conserved could be the proton-pumping site. For elucidation of proton-pumping mechanism, the most important task is to identify the redox-coupled conformational changes.

6. PHOSPHOLIPIDS AND CHOLATE

The electron density is not strong enough for identifying the unsaturated bonds in the hydrocarbon tails of the eight phospholipid molecules detected in the crystal structure. The head groups of the phospholipids are near the surfaces of the mitochondrial inner membrane at both the cytosolic side and the matrix side. The location of the membrane surfaces are indicated by the transmembrane (or α-helical) region of subunits I and III. Two cholate molecules are detected near the membrane surface levels. Each has a negatively charged head group on one end. The cholate molecules are likely to be a contaminants, since the enzyme from beef heart muscle for the crystallization is exposed to cholate during its purification. The enzyme is extracted with cholate from beef heart muscle and fractionated with ammonium sulfate in the presence of cholate (Yoshikawa *et al.*, 1977). Then, the enzyme preparation is washed several times with a neutral sodium phosphate buffer containing decyl maltoside. The critical micellar concentration of cholate anion is about one order of magnitude higher than that of decyl maltoside. Thus, most of the cholate attached to the transmembrane surface should be replaced with decyl maltoside. However, the presence of the two cholates in the crystal structure indicates sites that have an extremely high binding affinity.

7. ROLES OF SUBUNITS OTHER THAN SUBUNITS I AND II

As stated above, Cu_A is located in subunit II and the other three redox-active metals are in subunit I. It has been established that a cytochrome c oxidase preparation containing only subunits I and II, isolated from a bacteria, *Paracoccus denitrificans*, has the basic functional characteristics of cytochrome c oxidase: reduction of O_2 to water and redox-coupled proton pumping (Ludwig and Schatz, 1980). Thus, one of the biggest mysteries concerning the structure of mammalian cytochrome c oxidase is the role of the other 11 subunits. The crystal structure of the beef heart enzyme suggests various possible functions of these subunits.

7.1. Subunit III

The second biggest subunit, subunit III, holds three phospholipids in the fairly large V-shaped crevice between two helix bundles. One bundle consists of helices I and II and the other of helices III–VII. The hydrocarbon tails of these phospholipids bound within the subunit are located at the entrance of a possible O_2 channel that leads to the O_2 reduction site of subunit I. Thus, subunit III could provide a pool of O_2 molecules for the O_2 reduction site. The O_2 pool could result in an extremely high effective O_2 concentration. In addition, the O_2 pool (or the site of O_2 storage) may bind O_2 molecules tightly. Thus, complete removal of O_2 from cytochrome c oxidase preparation may not be easily achieved. Any residual O_2 could influence the results of experiments designed to be under anaerobic conditions.

7.2. Nuclear-Encoded Subunits

Seven of ten nuclear-encoded subunits possess a single transmembrane helix. All these helices as well as those of the core subunits are in close contact and the helices intersect at angles of either O°, 20°, or 50°, which results in a very stable arrangement of helix contacts (Chotia *et al.*, 1987). Thus, at least one of the roles of these subunits is to stabilize the conformation of the enzyme. The transmembrane helices of the nuclear-encoded subunits are distributed evenly on the core subunits, with the exception of two regions: the transmembrane region facing the intermonomer space in the dimer and the crevice of subunit III containing three phospholipids.

Perhaps subunit Va has one of the most peculiar structures of nuclear-encoded subunits. The peptide of subunit Va is in an α-helix conformation, and this α-helix is coiled into a superhelix. This subunit is located along with subunits IV and VIc in the matrix space and has no direct contact with the core subunits. The shortest distance between this subunit and subunit I, the nearest core subunit, is 6 Å. Thus, water molecules can pass freely between the two subunits. Subunit Va and the extramembrane regions of subunits IV and VIc are the largest extramembrane domains of this enzyme in the matrix side. A possible role for this moiety is to control the movement of the enzyme in the mitochondrial membrane, which depends on the viscosity of the medium in the matrix space. The effect of the absence of subunit Va on activity may provide clues for achieving a better understanding of its physiological role.

7.3. Subunits for Dimer Formation and ADP Binding

The extramembrane domain of subunit VIa on the cytosolic side is attached to subunit III. This subunit may serve as a lid for the phospholipid pool of subunit III. The amino-terminal end of subunit VIa is in tight contact with helices VII and VIII of subunit I on the other monomer. The extramembrane subunit, VIb, stabilizes the conformation of the extramembrane domain of subunit II, which includes the Cu_A site. The subunit VIb on the cytosolic side is placed very near the pseudo-twofold symmetry axis relating the two monomers, so that this subunit on each monomer is in close contact. The three contact points between the two monomers serve to effectively stabilize the dimeric state of the enzyme. Thus, any monomer–dimer transition is unlikely. Breaking the strong interactions between the monomers without irreversible denaturation of these subunits is unlikely. A monomer–dimer transition has been proposed based on previous hydrodynamic analyses using techniques such as analytical ultracentrifugation and gel filtration (Suarez *et al.*, 1984). However, the current crystal structure suggests a reexamination of these hydrodynamic analyses.

This enzyme is likely to function in the dimer state under physiological conditions. Then, what functional advantage over the monomer does the dimer serve in the physiological functioning? One possibility is that formation of the dimer regulates the function of the enzyme through contacts between subunit VIa and helices VII and VIII of subunit I of the twofold-related monomer. The two helices are placed near the O_2 reduction site. This subunit could communicate structural changes between the two monomers. Furthermore, one of the cholate molecules in the crystal structure binds near the amino-terminal of subunit VIa. The size and shape of a cholate ion is very similar to that of ADP. Both molecules have a negatively charged group at one end. Actually, from model building, an ADP molecule fits quite well to the cholate binding site (Tsukihara *et al.*, 1996). On the other hand, Frank and Kadenbach (1996) have established that ADP–ATP ratio in liposomes affects the efficiency of the proton pumping. Furthermore, they have established immunochemically the importance of the ADP–ATP binding site on subunit VIa for regulation of the enzyme activity (Anthony *et al.*, 1993). Thus, the ADP or ATP binding to subunit VIa could influence the conformation of both monomers, which in turn would affect the physiological function of the enzyme.

Our preliminary results indicate that our crystals of cytochrome *c* oxidase, in addition to eight phospholipids detectable in the crystal structure as described above, contain one cardiolipin per monomer of the enzyme. It has been previously proposed that cardiolipin is indispensable for the activity of this enzyme (Robinson, 1982). However, no cardiolipin has been discovered in the electron density of the enzyme. This suggests that the cardiolipin binds the enzyme, but with enough mobility to flatten the electron density. The intermonomer space between the two monomers contains enough space to position two cardiolipin molecules. Perhaps these two cardiolipin molecules, with their head groups on either of the membrane surfaces and with their hydrocarbon tails projecting inside of the membrane, stabilize the dimer state. Thus, even without the existence of nuclear-encoded subunits such as in the bacterial enzyme, the core complex could form the dimer. These phospholipids could also serve as the O_2 pool placed near one of the O_2 channels of subunit I. These phospholipids will be released when the dimer is split into two monomers. The release of phospholipids will lower the possibility for regeneration of the dimer.

7.4. Cytochrome *c* Binding Site

The extramembrane domain of subunit II, the carboxyl-terminal moiety of subunit VIa, subunit VIb, and a part of the cytosolic surface of subunit I create a concave surface that can accept a cytochrome *c* molecule (about 25 Å in diameter). Steady-state kinetics of cytochrome *c* oxidation has been examined extensively for many years (Malmström, 1990). These experiments demonstrated that the rate of cytochrome *c* oxidation at very low concentrations of ferrocytochrome *c* is significantly higher than that expected from a simple Michaelis–Menten type kinetics at physiologically relevant concentrations of reduced cytochrome. These kinetics have been interpreted as indicating the presence of an extra binding site for cytochrome c with an extremely high affinity (Ferguson-Miller *et al.*, 1976). Another interpretation for the unusual kinetics is the presence of two conformational states in equilibrium, each having a single binding site for reduced cytochrome c with different affinity. The concave surface observed in the crystal structure of this enzyme does not provide a space large enough for accepting two molecules of cytochrome c. The stable dimer structure as described above suggests that the unusual steady-state kinetics is due to a repulsive interaction between the two cytochrome c binding sites, each in different monomers.

ACKNOWLEDGMENT. The authors wish to thank Dr. Herbert L. Axelrod for valuable discussion and assistance in the correction of syntax and grammatical errors.

8. REFERENCES

Anthony, G., Reimann, A., and Kadenbach, B., 1993, Tissue specific regulation of bovine heart cytochrome-c oxidase activity by ADP via interaction with subunit VIa, *Proc. Natl. Acad. Sci. USA* **90**:1652–1656.

Antonini, E., Brunori, M., Colosimo, A., Greenwood, C., and Wilson, M. T., 1977, Oxygen "pulsed" cytochrome c oxidase: Functional properties and catalytic relevance, *Proc. Natl. Acad. Sci. USA* **74**:3128–3125.

Blair, D. F., Ellis, W. R., Wang, H., Gray, H. B., and Chan, S. I., 1986, Spectroelectrochemical study of cytochrome c oxidase: pH and temperature dependencies of the cytochrome potentials, *J. Biol. Chem.* **261**:11524–11537.

Capitanio, N., Capitanio, G., Demarinis, D. A., De Nitto, E., Massari, S., and Papa, S., 1996, Factors affecting the H^+/e^- stoichiometry in mitochondrial cytochrome c oxidase: Influence of the rate of electron flow and transmembrane ΔpH, *Biochemistry* **35**:10800–10806.

Caughey, W. S., 1971, Structur–function relationship in cytochrome c oxidase and other hemo-proteins. *Adv. Chem. Ser.* **100**:248–270.

Caughey, W. S., Wallace, W. J., Volpe, J. A., and Yoshikawa S., 1976, Cytochrome c oxidase, in: *The Enzymes*, vol. 13C (P. D. Boyer, ed.), pp. 299–344, Academic Press, New York.

Chotia, C., Levitt, M., and Richardson, D., 1987, Structure of proteins: Packing of α-helices and pleated sheets, *Proc. Natl. Acad. Sci. USA* **74**:4130–4134.

Espe, M. P., Hosler, J. P., Ferguson-Miller, S., Babcock, T., and McCracken, J., 1995, A continuous wave and pulsed EPR characterization of the Mu^{2+} binding site in *Rhodobacter sphaeroides* cytochrome c oxidase, *Biochemistry* **34**:7593–7602.

Ferguson-Miller, S., and Babcock, G. T., 1996, Heme/copper terminal oxidases, *Chem. Rev.* **96**:2889–2907.

Ferguson-Miller, S., Brautigen, D. L., and Margoliash, E., 1976, Correlation of the kinetics of electron transfer activity of various eukaryotic cytochromes with binding to mitochondrial cytochrome c oxidase, *J. Biol. Chem.* **251**:1104–1115.

Frank, V., and Kadenbach, B., 1996, Regulation of the H^+/e^- stoichiometry of cytochrome c oxidase from bovine heart by intermitochondrial ATP/ADP ratios, *FEBS Lett.* **382**:121–124.

Garcia-Horsman, J. A., Berry, E., Shapleigh, J. P., Alben, J. O., and Gennis, R. B., 1994, A novel cytochrome c oxidase from *Rhodobacter sphaeroides* that lacks Cu_A, *Biochemistry* **33**:3113–3119.

Gelles, J., Blair, D. F., and Chan, S. I., 1986, The proton-pumping site of cytochrome c oxidase: A model of its structure and mechanism, *Biochim. Biophys. Acta* **853**:205–236.

Gibson, Q. H., and Greenwood, C., 1963, Reactions of cytochrome oxidase with oxygen and carbon monoxide, *Biochem. J.* **86**:54–554.

Gibson, Q. H., Greenwood, C., Wharton, D. C., and Palmer, G., 1965, The reaction of cytochrome oxidase with cytochrome c, *J. Biol. Chem.* **240**:888–894.

Gray, K. A., Grooms, M., Mylhykallio, H., Moomaw, C., Slaughter, C., and Daldal, F., 1994, *Rhodobacter capsulatus* contains a novel cb-type cytochrome c oxidase without a Cu_A center, *Biochemistry* **33**:3120–3127.

Han, S., Ching, Y., and Rousseau, D. L., 1990, Primary intermediate in the reaction of oxygen with fully reduced cytochrome c oxidase, *Proc. Natl. Acad. Sci. USA* **87**:2491–2495.

Hensel, S., and Buse, G., 1990, Studies on cytochrome-c oxidase, XIV, *Biol. Chem. Hoppe-Seyler* **371**:411–422.

Hill, B. C., 1994, Modeling the sequence of electron transfer reaction in the single turnover of reduced, mammalian cytochrome c oxidase, *J. Biol. Chem.* **269**:2419–2425.

Hosler, J. P., Espe, M. P., Zhen, Y., Babcock, G. T., and Ferguson-Miller, S., 1995, Analysis of site-directed mutants locates a non-redox-active metal near the active site of cytochrome c oxidase of *Rhodobacter sphaeroides*, *Biochemistry* **34**:7586–7592.

Iwata, S., Ostermeier, C., Ludwig, B., and Michel, H., 1995, Structure at 2.8 Å resolution of cytochrome c oxidase from *Paracoccus denitrificans*, *Nature* **376**:660–669.

Jensen, L. H., 1986, The iron–sulfur proteins: An overview, in: *Iron–Sulfur Protein Research* (H. Matsubara, Y. Katsube, and K. Wada, eds.), pp. 3–21, Springer-Verlag, Berlin.

Kadenbach, B., Ungibauer, M., Jarausch, J., Büge, U., and Kuhn-Nentwig, L., 1983, The complexity of respiratory complexes, *Trends Biochem. Sci.* **8**:398–400.

Kitagawa, T., and Mizutani, Y., 1994, Resonance Raman spectra of highly oxidized metalloporphyrins and heme proteins, *Coordination Chem. Rev.* **135/136**:685–735.

Kitagawa, T., and Ogura, T., 1997, Oxygen activation mechanism at the binuclear site of heme–copper oxidase superfamily as revealed by time-resolved resonance Raman spectroscopy, *Prog. Inorg. Chem.* **45**:431–479.

Kröneck, P. M. H., Antholine, W. A., Riester, J., and Zumft, W. G., 1988, The cupric site in nitrous oxide reductase contains a mixed-valence [Cu(II), Cu(I)] binuclear center: A multifrequency electron paramagnetic resonance investigation, *FEBS Lett.* **242**:70–74.

Ludwig, B., and Schatz, G., 1980, A two-subunit cytochrome c oxidase (cytochrome a a_3) from *Paracoccus denitrificans*, *Proc. Natl. Acad. Sci. USA* **77**:196–200.

Malmström, B. G., 1990, Cytochrome c oxidase as a redox-linked proton pump, *Chem. Rev.* **90**:1247–1260.

Meinecke, L., and Buse, G., 1986, Studies on cytochrome-c oxidase, XIII, *Biol. Chem. Hoppe-Seyler* **367**:67–73.

Nichols, P., and Wrigglesworth, J. M., 1988, Routes of cytochrome a_3 reduction, *Ann. NY Acad. Sci.* **550**:59–67.

Ogura, T., Takahashi, T., Shinzawa-Itoh, K., Yoshikawa, S., and Kitagawa, T., 1990, Observation of the Fe^{II}-O_2 stretching Raman band for cytochrome oxidase compound A at ambient temperature, *J. Am. Chem. Soc.* **112**:5630–5631.

Ogura, T., Takahashi, T., Shinzawa-Itoh, K., Yoshikawa, S., and Kitagawa, T., 1991, Time-resolved resonance Raman investigation of cytochrome oxidase catalysis: Observation of a new oxygen-isotope sensitive Raman band, *Bull. Chem. Soc. Jpn.* **64**:2901–2907.

Ogura, T., Takahashi, S., Hirota, S., Shinzawa-Itoh, K., Yoshikawa, S., Appelman, E. H., and Kitagawa, T., 1993, Time-resolved resonance Raman elucidation of the pathway for dioxygen reduction by cytochrome c oxidase, *J. Am. Chem. Soc.* **115:**8527–8536.

Puustinen, A., Finel, M., Virkki, M., and Wikström, M., 1989, Cytochrome $o(bo)$ is a proton pump in *Paracoccus denitrificans* and *Escherichia coli*, *FEBS Lett.* **249:**163–167.

Robinson, N. C., 1982, Specificity and binding affinity of phospholipids to the high-affinity cardiolipin sites of beef heart cytochrome c oxidase, *Biochemistry* **21:**184–188.

Scott, R. A., 1989, X-ray absorption spectroscopic investigations of cytochrome c oxidase structure and function, *Annu. Rev. Biophys. Biophys. Chem.* **18:**137–158.

Shinzawa-Itoh, K., Ueda, H., Yoshikawa, S., Aoyama, H., Yamashita, E., and Tsukihara, T., 1995, Effects of ethyleneglycol chain length of dodecyl polyethyleneglycol monoether on the crystallization of bovine heart cytochrome c oxidase, *J. Mol. Biol.* **246:**572–575.

Steffens, G. C. M., Souliname, T., Wolff, G., and Buse, G., 1993, Stoichiometry and redox behavior of metals in cytochrome-c oxidase, *Eur. J. Biochem.* **213:**1149–1157.

Suarez, M. D., Revzin, A., Narlock, R., Kempner, E. S., Thomson, D. A., and Ferguson-Miller, S., 1984, The functional and physical form of mammalian cytochrome c oxidase determined by gel filtration, radiation inactivation, and sedimentation equilibrium analysis, *J. Biol. Chem.* **259:**13791–13799.

Thomas, J. W., Puustinen, A., Alben, J. O., Gennis, R. B., and Wikström, M., 1993, Substitution of asparagine for aspartate-135 in subunit I of the cytochrome bo ubiquinol oxidase of *Escherichia coli* eliminates proton-pumping activity, *Biochemistry* **32:**10923–10928.

Tsukihara, T., Aoyama, H., Yamashita, E., Tomizaki, T., Yamaguchi, H., Shinzawa-Itoh, K., Nakashima, R., Yaono, R., and Yoshikawa, S., 1995, Structures of metal sites of oxidized bovine heart cytochrome c oxidase at 2.8 Å, *Science* **269:**1069–1074.

Tsukihara, T., Aoyama, H., Yamashita, E., Tomizaki, T., Yamaguchi, H., Shinzawa-Itoh, K., Nakashima, R., Yaono, R., and Yoshikawa, S., 1996, The whole structure of the 13-subunit oxidized cytochrome c oxidase at 2.8 Å, *Science* **272:**1136–1144.

Van Gelder, B. F., and Beinert, H., 1969, Studies of the heme component of cytochrome c oxidase by EPR spectroscopy, *Biochim. Biophys. Acta* **189:**1–24.

Varotsis, C., Woodruff, W. H., and Babcock, G. T., 1990, Time resolved Raman detection of v(Fe-O) in an early intermediate in the reaction of O_2 by cytochrome oxidase, *J. Am. Chem. Soc.* **112:**1297.

Wang, H., Blair, D. F., Ellis, W. R., Gray, H. B., and Chan, S. I., 1986, Temperature dependence of the reduction potential of Cu_A in carbon monoxide inhibited cytochrome c oxidase, *Biochemistry* **25:**167–171.

Warburg, O., 1924, Üben Eisen, den sauerstoffubentragenden Bestandteil des Atmungsfermentes, *Biochem. Z.* **152:**479–494.

Wikström, M. K. F., 1977, Proton pump coupled to cytochrome c oxidase in mitochondria, *Nature* **266:**271–273.

Wikström, M., and Morgan, J. E., 1992, The dioxygen cycle, *J. Biol. Chem.* **267:**10266–10273.

Wikström, M., Krab, S., and Saraste, M., 1981, *Cytochrome Oxidase-α Synthesis*, Academic Press, London.

Wikström, M., Bogachev, A., Finel, M., Morgan, J. E., Puustinen, A., Raitio, M., Verkhovskaya, M., and Verkhovsky, M. I., 1994, Mechanism of proton translocation by the respiratory oxidases. The histidine cycle, *Biochim. Biophys. Acta* **1187:**106–111.

Woodruff, W. H., Einersdóttir, Ó., Dyer, R. B., Bagley, K. A., Palmer, G., Atherton, S. J., Goldbeck, R. A., Dawes, T. D., and Kliger, D. S., 1991, Nature and functional implications of the cytochrome a_3 transients after photodissociation of CO-cytochrome oxidase, *Proc. Natl. Acad. Sci. USA* **88:**2588–2592.

Wu, W., Chang, C. K., Varotsis, C., Babcock, G. T., Puustinen, A., and Wikström, M., 1992, Structure of the heme O prosthetic group from the terminal quinol oxidase of *Escherichia coli, J. Am. Chem. Soc.* **114:**1182–1187.

Yoshikawa, S., Choc, M. G., O'Toole, M. C., and Caughey, W. S., 1977, An infrared study of CO binding to heart cytochrome *c* oxidase and hemoglobin A, *J. Biol. Chem.* **252:**5498–5508.

CHAPTER 7

Electron and Proton Transfer in Heme–Copper Oxidases

Yuejun Zhen, Denise A. Mills, Curtis W. Hoganson,
Rebecca Lucas, Wenjun Shi, Gerald T. Babcock,
and Shelagh Ferguson-Miller

1. INTRODUCTION

In eukaryotes, the process of energy transduction coupled to electron transfer occurs in mitochondria, where cytochrome c oxidase catalyzes the transfer of electrons derived from foodstuffs to oxygen, the final electron sink. The reduction of oxygen to water and the concomitant translocation of protons are carried out by a member of a family of enzymes: the heme–copper oxidases (Saraste, 1990; García-Horsman *et al.*, 1994). A number of these oxidases have now been identified in bacterial systems; all use a heme–copper center to carry out the oxygen chemistry, but not all have similar auxiliary metal centers. The eukaryotic and some prokaryotic enzymes contain an additional heme a, an additional bimetallic copper center, and a magnesium ion. These oxidases use

Yuejun Zhen, Denise A. Mills, Rebecca Lucas, and Shelagh Ferguson-Miller • Department of Chemistry, Michigan State University, East Lansing, Michigan 48824. **Curtis W. Hoganson, Wenjun Shi, and Gerald T. Babcock** • Department of Biochemistry, Michigan State University, East Lansing, Michigan 48824.

Frontiers of Cellular Bioenergetics, edited by Papa *et al.* Kluwer Academic/Plenum Publishers, New York, 1999.

cytochrome c as their immediate electron donor. Another group of oxidases in the family use quinol as a substrate and contain neither an extra copper center nor magnesium. Although the two groups are likely to have similar energy transduction mechanisms, this discussion will focus on the cytochrome c oxidases with the additional copper and magnesium and on our efforts to define the roles of these auxiliary metals in controlling electron input, proton output, and coupling efficiency.

The cytochrome c oxidase from *Rhodobacter sphaeroides* has been shown to be an excellent model for the mammalian enzyme and approachable by genetic, biochemical, and spectroscopic techniques (Cao *et al.*, 1992; Hosler *et al.*, 1992; Shapleigh and Gennis, 1992). With the recent solution of high-resolution three-dimensional crystal structures of the enzymes from the closely related *P. denitrificans* (Iwata *et al.*, 1995) and from beef heart (Tsukihara *et al.*, 1995, 1996), more incisive analysis of the mechanism of electron and proton translocation is now possible. In particular, proton channels previously suggested by mutational analysis (Thomas *et al.*, 1993; Fetter *et al.*, 1995; García-Horsman *et al.*, 1995) have been further substantiated and defined by examination of the crystal structure. However, it is still not clear precisely how and where the protons move in the protein to allow uptake of both "substrate" and "pumped" protons. Nor is there any evidence regarding the route of proton exit, a critical question with respect to whether a direct coupling mechanism (Wikström *et al.*, 1994; Iwata *et al.*, 1995) or an indirect mechanism (Tsukihara *et al.*, 1996) is operative.

Equally controversial is the question of how, where, and what controls electron movement through the protein (Gray and Malmström, 1989; Moser *et al.*, 1995). The fundamental issue that rests on defining these processes is how the rate and efficiency of energy transduction is regulated.

2. ELECTRON PATHWAYS AND CONTROL OF ELECTRON TRANSFER

How electrons move through proteins has been debated for many years. Although it is clear that electron transfer rates are dependent on the driving force (redox potential difference), the reorganizational energies of the centers involved and the distance between them (Marcus and Sutin, 1985), the role of the structure of the intervening protein is not as apparent. Can it, or does it, determine the rate or direction of electron flow by providing preferred pathways? In the case of cytochrome c oxidase, the entry of electrons from cytochrome c appears to occur at the fastest rate through the unique dinuclear Cu_A center located in the C-terminal domain of subunit II (Hill, 1991). The Cu_A site is close to the outside surface of the protein where cytochrome c must approach, and very rapid rates of electron transfer from cytochrome c to Cu_A have been measured ($> 40,000$/sec) (Geren *et al.*, 1995). It is therefore of interest to determine the exact position of the cytochrome *c* binding site in order to understand the relationship between rates, distance, and protein structure.

Similarly, Cu_A is close to the interface of subunits I and II, but still is at a significant distance from the heme groups to which it must pass on the electrons (metal center to metal center distance: Cu_A to heme *a*, 19 Å; Cu_A to heme a_3, 22 Å). The similar distances from copper to the two heme groups lead to the question of whether there are features of the Cu_A center itself, or the intervening protein structure, that contribute to apparent selectivity for transfer to heme *a*. The ability of the dinuclear Cu_A center to delocalize electrons over a large area and the existence of a shorter "through-bond" route between Cu_A and heme *a* have been postulated to account for the rapid transfer between these two centers (Ramirez *et al.*, 1995), as well as different reorganizational energies of the two hemes (Brzezinski, 1996).

The data discussed here relate to three specific questions regarding electron transfer in cytochrome *c* oxidase:

1. Where does cytochrome *c* sit when transferring electrons into the oxidase, and is there more than one docking position or entry site?
2. Is the unique dinuclear character of the Cu_A center essential for rapid, directed electron flow or proton translocation?
3. Is there a through-bond pathway between Cu_A and heme *a* that contributes to the speed and direction of electron flow?

To address these issues, a number of mutants have been made in subunit II of *R. sphaeroides* cytochrome *c* oxidase. These mutants, in the vicinity of Cu_A, were designed to alter the site of electron donation from cytochrome *c*, the dinuclear character of Cu_A, or the predicted pathway to heme *a*.

Four examples of mutants that address these goals will be discussed: E157Q, predicted to be involved in cytochrome *c* binding; D214N, predicted to be involved in cytochrome *c* binding and hydrogen bonded to a Cu_A ligand; M263L, a direct but weak ligand to Cu_A; and H260N, a direct strong ligand to Cu_A and in a predicted pathway to heme *a*.

2.1. General Characteristics of Subunit II Mutants

Figure 1 shows the location of the residues under consideration. Their general characteristics are summarized in Table I. Each mutant was introduced into an overproducing plasmid (Zhen *et al.*, 1998a) and purified by a *his*-tag method as described previously (Mitchell and Gennis, 1995; Qian *et al.*, 1997). Analyses were carried out by UV/visible, resonance Raman, electron paramagnetic resonance (EPR), and inductively coupled plasma atomic emission spectroscopy (ICP-AES), and activities were determined by steady-state (Hosler *et al.*, 1992) and rapid kinetic methods (Pan *et al.*, 1993; Geren *et al.*, 1995).

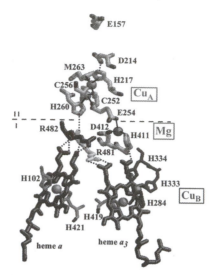

FIGURE 1. Structure and arrangement of the metal centers and associated amino acids in cytochrome *c* oxidase. The solid lines indicate the bonds between metals and their ligands. The dashed lines indicate hydrogen bonds. Model was produced in Rasmol from coordinates of the bovine oxidase (pdb1OCC.ent) *Rhodobacter sphaeroides* numbering. (Tsukihara *et al.*, 1995, 1996).

The mutant forms show an array of modified characteristics from no change in Cu_A properties (E157Q), to a minor change in the Cu_A redox potential (D214N), to different degrees of severe disruption of the Cu_A center (M263L and H260N). In all the mutants, except H260N, the other metal centers are essentially unaltered, indicating localized effects of the subunit II mutations

Table I
Subunit II Mutants from *R. sphaeroides*

Enzymes	α-Band (nm)	Turnover (e/s)	CO binding	Cu_A EPR	Proton pumping[a] H^+/e^- ratio
Wild type	606	1700	100%	Normal	0.7
E157Q	606	600	100%	Normal	0.7
D214N	606	400	100%	Normal	0.7
H260N	605	15	100%	Distorted	0.3
M263L	606	150	100%	Distorted	0.5

[a]Proton translocation was measured following reconstitution of the purified enzymes into phospholipid vesicles at 22°C by a spectroscopic method in 2.4 ml of 50 μM NaHCO$_3$-KOH, pH7.4, 45 mM KCl, 44 mM sucrose, 50 μM phenol red, pH 7.4 using an Aminco DW-2 spectrophotometer and a 556.8–504.7 nm wavelength pair.

and allowing for meaningful interpretation of the results. In H260N, the non-redox-active Mg/Mn site and the heme a center are slightly disturbed.

Optical spectra of the four purified mutants are similar to that of wild type, and the CO-binding capacities for all of them are 100% wild type. Resonance Raman spectroscopy is a sensitive technique to probe any changes associated with the two heme groups, and the Raman spectra of the four mutants did not show any perturbations, which is consistent with the optical and CO-binding results and argues against any changes in enzyme activity being caused by the disruption of heme a or heme a_3. The EPR spectra of D214N and E157Q did not reveal any change in the Cu_A center and the Cu/Fe ratios determined by ICP-AES are equivalent to those of the wild-type enzyme (data not shown).

Activities of all the mutants decrease dramatically, as shown in Table I. H260N has about 1% wild-type activity, which is not due to any exogenous activity and can be inhibited by CN^-. The other three mutants retain different amounts of wild-type activity, ranging from 10 to 30%.

2.2. Surface Charge Changes Close to Cu_A: Effects on Cytochrome c Interaction

The electrostatic interaction between cytochrome c and cytochrome c oxidase has been shown to be important in binding soluble cytochrome c to the enzyme (Ferguson-Miller *et al.*, 1978). Previous studies indicated that the C-terminal domain of subunit II of cytochrome c oxidase is the major binding site for cytochrome c, while the crystal structure suggests that subunit I and some of the small subunits of nuclear origin in the mammalian enzyme may also contribute to the binding. In the subunit II C-terminal domain, there are seven highly conserved negatively charged residues (Asp or Glu) present in all the known sequences. Beside the one that is involved in ligating the Mg/Mn site (Glu 254 in *R. sphaeroides*), there is evidence for involvement of some of these residues in cytochrome c binding (Millett *et al.*, 1983; Witt *et al.*, 1995).

The two carboxylate residues whose mutant forms are described here are both conserved and located at the surface of subunit II in a likely region for cytochrome c binding (Fig. 1). One approach to estimating effects on binding is steady-state kinetic analysis, but this method is complicated to interpret in the case of cytochrome c oxidase, because of nonlinear reciprocal- and Eadie–Hofstee-type plots. Apparent low and high K_m phases have been shown to relate to high- and low-affinity binding reactions, respectively (Ferguson-Miller *et al.*, 1976; Cooper, 1990), but the significance of these phases for electron input remains unclear. This type of study with D214N and E157Q shows increased K_m values for both high- and low-affinity phases, suggesting decreased binding affinity. These results are consistent with data from rapid kinetic analysis showing weaker binding of cytochrome c (Zhen *et al.*, 1997;

1998b). Since EPR and Raman spectra of D214N and E157Q show no structural change in the Cu_A and heme centers in these two mutants, the decreased activities appear largely due to the weakened interaction between cytochrome c and cytochrome c oxidase. However, additional studies will be necessary to quantitate the binding changes.

The D214N mutation, besides weakening the cytochrome c binding, also disrupts the hydrogen bond with the Cu_A ligand, His-217, which would be predicted to increase the redox potential of the Cu_A center due to loss of a negatively charged residue close to it. The redox potential difference between Cu_A and heme a can be estimated from the time-resolved ruthenium–cytochrome c kinetic assay (Pan *et al.*, 1993); in fact, a decrease of the redox potential difference between the two redox partners in D214N is observed (Zhen *et al.*, 1997). Based on the native spectral characteristics of heme a in this mutant, its redox potential is assumed to be unaltered and the change is attributed to an increase in the redox potential of the Cu_A center. This would be expected to decrease the steady-state reduction of heme a, and hence to lower the steady-state activity, as is observed. The redox difference between Cu_A and heme a in E157Q appears unaltered, indicating that the lower enzyme activity is due solely to the weak interaction between cytochrome c and oxidase (Pan *et al.*, 1993; Zhen *et al.*, 1997).

We conclude that both residues are part of the cytochrome c interaction domain; a more detailed definition of which is currently underway.

2.3. Decoupling of the Binuclear Cu_A Site: Effects on Electron Transfer and Proton Pumping

While the EPR spectra of D214N and E157Q did not reveal any change in the Cu_A center in these two mutants, the spectra of M263L and H260N are significantly changed compared to wild type (Fig. 2). The signal at $g = 2.19$ is replaced in M263L by four hyperfine splittings at the $g_{\parallel}$ region, with a splitting constant ($A_{\parallel}$) of about 50 G, characteristic of type 1 copper proteins. This spectrum is similar to the EPR spectrum of M227I, the homologous residue in *P. denitrificans*, especially in the $g_{\parallel}$ region (Zickermann *et al.*, 1995). Metal analysis indicates that the Cu/Fe ratio is about 1.5 in M263L (as in wild type), suggesting no loss of copper and spectral characteristics due to a decoupled binuclear center.

In H260N, the mutational effect on the Cu_A center is even more dramatic than M263L, with the amplitude of the signal at the g_1 region significantly decreased. The effect of the alteration of Cu_A has spread to the Mg/Mn site and to heme a, evidenced by the changes in the hyperfine structures of the Mn EPR spectrum and the slight shift of the α-peak in the optical spectrum. These results can be understood on the basis of the crystal structure (Ostermeier *et al.*,

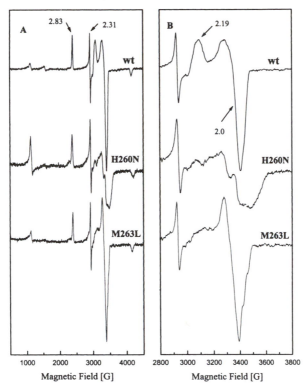

FIGURE 2. EPR spectra of wild-type and subunit II mutants. EPR spectra were recorded at 10 K, 2 mW microwave power at 9.48 GHz. The spectra were normalized by adjusting the low-spin heme a signals at $g = 2.31$ to the same height. Related g values are indicated by arrows. (B) The enlarged Cu_A spectra from A. All the enzyme samples were obtained from cells grown in Sistrom's medium with low Mn content (0.5 μM) and high Mg (1200 μM) to minimize Mn contamination.

1997), which shows a shared ligand between Cu_A and the Mg site and a wide-ranging hydrogen-bonded network that links the Cu_A, the Mg/Mn site, and heme a center together. Thus, the H260N mutation, by altering Cu_A, causes a disturbance of the Mg/Mn and heme a centers (Fig. 3). In contrast, the Mn EPR spectra of M263L did not show any change, suggesting much more localized effects.

The redox potential difference between Cu_A and heme a, estimated by time-resolved kinetic measurements, indicates that the redox potential of Cu_A has become higher than heme a in both H260N and M263L, rather than lower as in wild type (Zhen *et al.*, 1997). This would be expected to result in inhibition

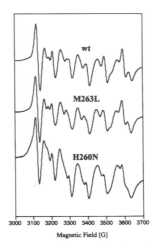

FIGURE 3. Mn EPR spectra of wild-type and mutants. Both wild-type and mutants were grown in Sistrom's medium with 700 μM Mn, 50 μM Mg. Spectra were recorded at X-band using a Bruker EP-300E spectrometer at 110 K with 20 mW microwave power at 9.44 GHz. The modulation amplitude was 12.6 G.

of electron transfer from Cu_A to heme *a* and to contribute to the low catalytic activities of these two mutants.

Beside being an initial electron acceptor, Cu_A was originally proposed to be involved in proton pumping (Gelles *et al.*, 1987). Proton-pumping assays of all four mutants were done after reconstitution into phospholipid vesicles. All the mutants were capable of pumping protons, even H260N, which has a very low turnover rate (Table I). The proton-to-electron ratios for D214N, E157Q, and M263L are comparable to that of wild-type enzyme. The low proton-to-electron ratio in H260N may be due to its very low turnover rate, resulting in more significant competition from the back-leak rate of protons into the vesicle.

2.4. Disruption of a Putative "Through-Bond" Pathway

The crystal structures of cytochrome *c* oxidase reveal that His-260 is hydrogen bonded to a highly conserved arginine pair in subunit I, which is proposed to form a part of a through-bond electron transfer pathway from Cu_A to heme *a* (Iwata *et al.*, 1995; Ramirez *et al.*, 1995; Tsukihara *et al.*, 1996). The H260N mutation will undoubtedly disrupt this hydrogen-bond interaction, and this could account for a lower enzymatic activity than that expected for the disruption of the Cu_A center and alteration in its redox potential, as compared to M263L. However, given the severe changes in structure and redox potential in this mutant, it is not possible to draw any conclusions regarding the pathway model based on this mutant alone.

2.5. Conclusions

The modification of residues in the vicinity of Cu_A results in a variety of functional changes. Mutation of two carboxylates, Glu-157 and Asp-214, causes altered steady-state and time-resolved kinetics with cytochrome c, indicative of their involvement in the cytochrome c binding domain. Alteration of two direct Cu_A ligands, His-260 and Met-263, gave mutants that appear to have lost the electronic coupling between the two coppers. The resulting changes in structure and redox potential caused severe inhibition of overall electron transfer rates but retention of proton pumping. The results strongly support the model that Cu_A is the sole input site for electrons from cytochrome c, that the electronic coupling between the two copper atoms is important for rapid transfer to heme a, but that its unique structure is not required for the proton pump.

3. PROTON PATHWAYS AND CONTROL OF PUMPING EFFICIENCY

The majority of the energy for pumping protons through cytochrome oxidase comes from the redox energy released by reduction of oxygen at the binuclear center. Protons are required for this conversion of oxygen to water (substrate protons), as well as for translocation across the membrane ("pumped" protons), but it is not known whether both types of protons move through the protein via the binuclear center (a direct coupling mechanism) or whether the pumped protons take a separate pathway (an indirect coupling mechanism). Mutagenesis studies and the X-ray crystal structures of cytochrome c oxidase provide mutually supportive evidence for at least two separate input sites for protons, but the subsequent routes through the protein are less well defined. Additional pathways are also suggested by the structural data (Tsukihara et al., 1996).

Tsukihara and colleagues recognized a possible proton channel in the crystal structure involving residues in helix III and IV. Since the pathway does not involve the a_3–Cu_B center, it presumably would be indirectly linked to the active site by conformational changes. Other proposals suggest that the pumped protons are directly linked to the active site chemistry by ligand-switching mechanisms (Rousseau et al., 1993; Woodruff, 1993; Wikström et al., 1994). Suggested mechanisms include a substitution of the heme a_3 ligand with a tyrosine (Tyr-422), whose involvement was subsequently disproved

(Mitchell *et al.*, 1996), or cycling on and off of a histidine ligand of Cu_B (Wikström *et al.*, 1994). It is envisioned that the histidine switches from the imidazolate form to imidazolium, and subsequently dissociates from the Cu_B and releases two protons. This mechanism makes use of the protons to be translocated as part of the charge neutralization at the active site, required to compensate for electron input (Rich *et al.*, 1996), and hence necessary for electron transfer activity.

The direct coupling mechanisms suggest that there can be no "slip" between proton translocation and activity. However, a number of studies have indicated that there are conditions under which the stoichiometry of protons pumped per electron (H^+/e^-) can differ from the normal 1:1. In mammalian systems, the ability to produce heat and to respond to signals indicating too much stored energy are important physiological functions that could be accomplished by less efficient proton pumping. Some clues as to how this may be achieved, even with a direct coupling process, come from studies of mutant forms of the bacterial enzyme, as well as from comparison of the responses of the mammalian and bacterial enzymes to possible regulatory metabolites. Indeed, evidence from mutant bacterial oxidases suggests that inefficient (uncoupled) activity can be supported by reversal of the normal exit pathway for protons, and that the more complex structure of the mammalian enzyme could allow regulatory metabolites to affect the control of this process.

3.1. Mutants at Aspartate 132 in *Rhodobacter sphaeroides* Cytochrome *c* Oxidase

An aspartate residue (Asp-132 in *R. sphaeroides*) in an interior loop between helices II and III of subunit I has been shown to be essential for proton pumping. Mutation of this aspartate to a noncarboxylate residue (alanine or asparagine) significantly decreases enzyme activity and abolishes proton pumping, but does not perturb the spectral characteristics of the metal centers (Fetter *et al.*, 1995). Its position on the matrix side of the membrane buried about 5 Å into the protein suggests that it is required for the uptake of protons from the interior side of the membrane.

3.1.1. Reverse Respiratory Control

When the mutant (D132A or D132N) enzyme is reconstituted into vesicles [cytochrome oxidase vesicles (COV)], it shows abnormal behavior with ionophores, demonstrating "reverse" respiratory control (Fetter *et al.*, 1996). Since D132A COV are still able to generate a normal membrane potential, the enzyme is correctly inserted and chemical protons are taken up on the interior. The reversal of respiratory control is seen by monitoring oxygen consumption

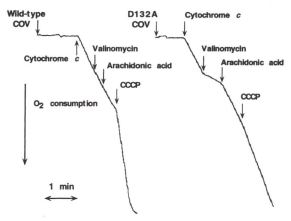

FIGURE 4. Effect of ionophores and arachidonate on steady-state activity of wild-type and D132A COV. Oxygen consumption measured as in Tables II and III with 0.02 nmoles aa_3 wild-type COV and 0.03 nmoles aa_3 mutant D132A COV and the addition of 28 μM arachidonate.

rates (Fig. 4) or measuring the initial rate of oxidation of cytochrome c by rapid-scanning, stopped-flow spectrophotometry (Fig. 5). The observed rate of cytochrome c oxidation is highest in the absence of ionophores, that is, in the presence of a membrane potential and/or pH gradient (Fig. 5B). In contrast, the wild-type COV have substantially decreased activity when the membrane potential is present (Fig. 5A). These data suggest that protons driven from the exterior by the positive-outside membrane potential are being used by the mutant oxidase to support electron transfer, instead of the protons normally supplied from the interior by the Asp-132-linked channel. In support of this idea, decreased activity was observed with D132A COV made with a high internal potassium concentration, setting up a more positive-inside membrane potential (Table II). Similarly, when the interior of the COV is made acidic, activity of the mutant is low but respiratory control is closer to normal, indicating that protons can be driven from the interior at low enough pH (Table II and Fig. 5D).

3.1.2. Chemical Rescue

It is sometimes observed that single-site mutations can be chemically "rescued" by supplying the missing chemical group in a soluble form, usually at very high concentration (Tittor *et al.*, 1989). The activity of the D132A enzyme is substantially increased (three- to sevenfold) by adding micromolar levels of unsaturated long-chain fatty acids (Fetter *et al.*, 1996). Among the

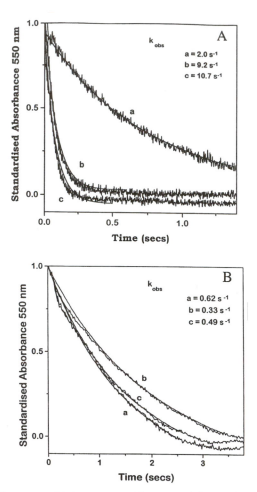

FIGURE 5. Cytochrome *c* oxidation measured by stopped flow. COV were made by the cholate dialysis method (Fetter *et al.*, 1995), with 20 mg/ml asolectin lipids and 2 μM *aa₃*. These were diluted to 80 nM *aa₃* in 50 μM NaHCO₃, 45 mM KCl, 44 mM sucrose pH 7.4. Cytochrome c^{2+} was prereduced with dithionite and desalted through Sephadex G-75. Final cytochrome c^{2+} concentration was 1.5 μM (equivalent to about five turnovers). A kinetic trace of difference spectra (fully oxidized cytochrome *c* subtracted) at 550 nm, (a) with no ionophores, (b) with valinomycin (2 μM), or (c) valinomycin + CCCP (10 μM). (A) Wild-type COV, (B) D132A COV, both with global fit overlaid; (C) D132A COV with 140 μM arachidonic acid + 7 μM BSA; and (D) D132A COV with pH 6.0 buffered interior, global fit only shown. Observed first-order rate constants (k_{obs}) from global fitting inset. Measurements were made with an OLIS rapid-scanning stopped-flow spectrophotometer.

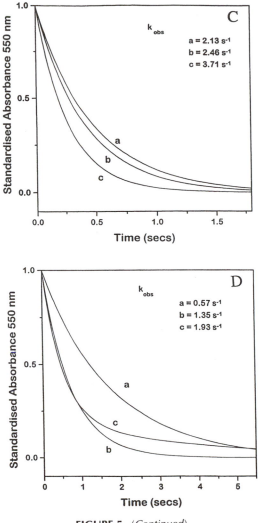

FIGURE 5. (Continued)

most effective was arachidonate, which is also able to restore normal respiratory control to D132A COV (Fetter *et al.*, 1996) (Fig. 5C), indicating that the "repair" is facilitating proton access from the inside of the membrane. In the presence of an appropriate concentration of bovine serum albumin (BSA) to buffer the added fatty acids and combat their uncoupling activity, there is still an increase in D132A activity (sixfold); but no proton pumping can be detected

Table II
Steady-State Activities for D132A and Wild-Type COVs
with Intraliposomal Low pH or High Ionic Strength

Conditions[a]	Rates of oxygen consumption (e^-/sec/aa_3)				
	No ionophores	+ Valinomycin	+ Val. + CCCP	RCRΨ[b]	RCR[c]
Wild type					
Control	133	163	936	1.2	7.0
High [K$^+$] in	137	197	683	1.4	5.0
D132A					
Control	41	19	29	.46	.71
High [K$^+$] in	15	7	22	.47	1.5
pH 6.0 in	27	27	34	1.0	1.3

[a]COV were reconstituted with 75 mM MES-NaOH buffer at pH 6.0 or with 200 mM KCl or as normal (see Fig. 4). Turnover was measured in a Gibson oxygraph in 10 mM Hepes-KOH, 41 mM KCl, 38 mM sucrose, pH 7.4, with 5–10 μl COVs.
[b]RCRΨ, activity with valinomycin divided by coupled activity.
[c]RCR, respiratory control ratio; uncoupled activity with valinomycin (1.2 μM) and CCCP (12 μM) divided by coupled activity with no ionophores.

even by stopped-flow measurements (Fig. 6), which should detect any initial pH change that would have been missed in the slower time-scale measurements (Fetter *et al.*, 1995).

3.1.3. A Model: Reversal of the Proton Exit Pathway

A model was developed to explain the behavior of Asp-132 mutants: low activity, reverse respiratory control, and fatty acid stimulation (Fetter *et al.*, 1996) (Fig. 7D). The presence of a membrane potential (negative inside) or proton gradient (alkaline inside) promotes the flow of protons through the outlet path in the reverse direction, opposite to the normal flow (Fig. 7C). In the wild-type enzyme (Fig. 7A), this flow is insignificant—2% the normal rate—but in the Asp-132 mutants these protons are able support a low activity by substituting for those supplied by the normal uptake route (Fig. 7B) and required for charge neutralization. Absence of a membrane potential (upon addition of valinomycin) decreases activity, because reverse flow is not promoted. Reversing the membrane potential, by high salt inside, reduces the activity by further removing the driving force for protons to enter via the exit pathway.

The addition of arachidonate or other long-chain unsaturated fatty acids is able to enhance activity by inserting a carboxyl group into the vicinity of the missing aspartate. This stimulates the flow of protons through the normal proton channel and hence restores normal respiratory control (Fig. 7D). De-

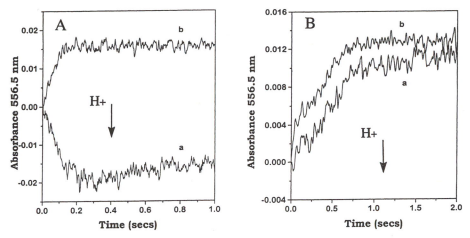

FIGURE 6. Phenol red absorbance changes with COV. Measurements in the stopped flow under the same conditions as in Fig. 5 but with the addition of 50 μM phenol red, 140 μM arachidonic acid, and 7 μM BSA and with (a) valinomycin and (b) valinomycin + CCCP. Kinetic trace at the isosbestic point for cytochrome c at 556.5 nm. (A) Wild-type COV. (B) D132A COV.

creasing the pH of the interior of the vesicles is also able to stimulate activity, presumably by driving the flow of protons through the proton channel in the correct direction, and again restoring normal respiratory control. However, no proton pumping has been measured for the D132A COV even by stopped-flow methods in the presence of arachidonate and BSA (Fig. 6), suggesting that there is still some reversal of proton flow under all conditions, in the modified inlet channel as well as the outlet route.

3.1.4. Evidence from Mutants of Cytochrome bo_3

The role of the aspartate in proton uptake has been studied extensively with mutations in the equivalent region in cytochrome bo_3 ubiquinol oxidase from *Escherichia coli* (Thomas *et al.*, 1993; García-Horsman *et al.*, 1995). Moving the aspartate to two other nearby sites (139 or 142: *P. denitrificans* residues 131 or 134) or substituting a glutamate for the aspartate allowed restoration of proton pumping, but the mutants still had lowered activity compared to wild type and respiratory control was not measurable in this system. The results demonstrate the importance of the carboxyl moiety in a particular location, but indicate that precise positioning is not essential.

It is interesting to note that in cytochrome bo_3, the asparagine substitution

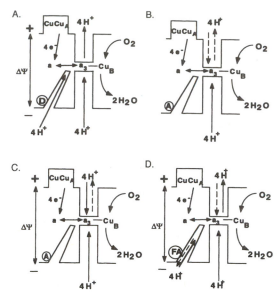

FIGURE 7. Model of proton pathways in wild-type and D132A COV. Taken from Fetter *et al.* (1996). (A) Wild-type cytochrome *c* oxidase showing position of aspartate 132 (D). (B) D132A mutant with no membrane potential showing reversibility of exit pathway. (C) D132A mutant with membrane potential showing facilitation of proton entry into exit pathway. (D) D132A mutant with fatty acid at position of the missing aspartate, facilitating proton entry through the normal uptake pathway

for aspartate gives much higher uncoupled activities (45% of wild type) than the equivalent mutation in cytochrome aa_3 (Thomas *et al.*, 1993; García-Horsman *et al.*, 1995). The bo_3 oxidase lacks both the Cu_A and the magnesium sites, which are located at the interface between subunits I and II, where there is a proposed H_2O channel directly above the a_3–Cu_B center. This channel provides an obvious route for proton exit (Ferguson-Miller and Babcock, 1996), but the lack of metal centers in the bo_3 enzyme could make this region less well structured, leading to less stringent control of proton back-flow through the exit channel and explaining the difficulty in achieving coupled, reconstituted bo_3 COV.

3.1.5. Significance

The behavior of the Asp-132 mutants in the proton uptake pathway on the inside of the membrane draws attention to the nature and regulation of the proton exit pathway on the outside of the membrane and its potential for controlling the efficiency of proton pumping. The actual involvement of the proton exit route in the regulation of efficiency remains to be established, as does the location of this pathway. But the control of efficiency is a critical physiological function, and results from studies on modification of this property in the mammalian oxidase are interesting to consider in the light of the bacterial enzyme behavior.

3.2. Regulation of the Efficiency of Bovine Oxidase by ATP

The allosteric regulation of the activity of cytochrome c oxidase has been the subject of considerable investigation and speculation, with much effort going into attempts to define the role of adenine nucleotides as reporters of the energy status of the cell. There is evidence for several binding sites for ATP in eukaryotic oxidases (Taanman *et al.*, 1994; Kadenbach *et al.*, 1995). Renewed interest in the issue of nucleotide binding has come from the bovine crystal structure that shows two likely nucleotide binding sites, occupied by cholate, one of which is associated with subunit VIa, as suggested by other analyses (Anthony *et al.*, 1993; Taanman *et al.*, 1994; Kadenbach *et al.*, 1995). This site on the interior side of the oxidase has been implicated in regulating efficiency (Belyanovich *et al.*, 1996; Frank and Kadenbach, 1996).

The *R. sphaeroides* cytochrome c oxidase has only three subunits, although it behaves very similarly to the bovine heart enzyme except for considerably faster turnover (Hosler *et al.*, 1992). Because it is lacking subunit VIa, it makes an excellent control for nucleotide effects. Reconstitution of wild-type *R. sphaeroides* or bovine heart cytochrome c oxidase with either nucleotide (ADP or ATP) or phosphate in the interior was performed. Steady-state measurements of oxygen consumption show that 5 mM ATP or phosphate, but not ADP, causes a significant decrease in the respiratory control ratio (RCR) only in bovine heart COV (Table III). The decrease is mainly due to a higher coupled rate of activity rather than a decreased uncoupled activity. Measurements of proton pumping show that internal ATP does not abolish the ability of the bovine heart COV to pump protons, but it affects the efficiency as previously determined (Anthony *et al.*, 1993; Kadenbach *et al.*, 1995). ADP shows no effect on the RCR of any COV. However, previous data (Anthony *et al.*, 1993; Belyanovich *et al.*, 1996) has indicated a stimulation of activity with intraliposomal ADP with bovine heart oxidase.

It is not unreasonable to consider ATP as an allosteric modifier of cytochrome oxidase activity, since its levels (as well as the ADP.Pi/ATP ratio) are likely to vary with the energy state, while Pi levels are more constant. If the observed effects are in fact mediated by subunit VIa, as results from Kadenbach and colleagues suggest, it is also possible that they are transmitted through the transmembrane region from the proposed ATP binding site to the exterior

Table III
Effect of Nucleotides on Steady-State Activity
of Bovine Heart or *R. sphaeroides* COV

Conditions[a]	Rates of oxygen consumption ($e^-/sec/aa_3$)		
	Coupled[b]	Uncoupled[c]	RCR
Bovine heart COV	29	162	5.6
5 mM Pi_{in}	89	200	2.3
5 mM ATP_{in}	69	176	2.6
5 mM ADP_{in}	22	143	6.6
R. sphaeroides COV	65	643	9.9
5 mM Pi_{in}	59	547	9.4
5 mM ATP_{in}	68	602	8.8

[a]5 mM ATP, ADP, or NaH_2PO_4 was added to the reconstitution mixture and to the first dialysis during COV preparation. All external nucleotides were removed by extensive dialysis. Other details as Table II.
[b]No ionophores added.
[c]With valinomycin and CCCP.

surface of the oxidase. Subunit VIa has a substantial external domain in contact with subunit I that could conformationally affect the reversibility of an external exit route for protons.

3.3. Conclusion

Our studies on the mutant enzyme (D132A) reveal the possibility that activity can be maintained in the absence of net proton translocation (zero efficiency). This appears to occur by the inward flow of protons to maintain electroneutrality at the active site, probably via the proton exit channel. This finding raises the possibility that the mammalian enzyme could normally use variable reversal of the exit channel to control efficiency.

In the mammalian enzyme, ATP binding to an interior site on the enzyme appears to act as a negative effector of efficiency (Kadenbach *et al.*, 1995; Belyanovich *et al.*, 1996); our results support this finding. There is further evidence to indicate that the nucleotide binding site involved is in the interior portion of subunit VIa. Since this small subunit traverses the membrane and loops extensively over the outside surface of subunits I and III, it may directly interact with as yet unidentified residues that form a proton exit channel or indirectly induce conformational changes in this region. This provides a specific testable hypothesis for the regulation of efficiency in cytochrome oxidase.

4. SUMMARY

The function of cytochrome c oxidase is to transfer electrons from cytochrome c to oxygen, with conservation of the energy released in a proton gradient. Understanding how and where electrons and protons move through the protein is key to determining the mechanism of coupling between the two and how energy transduction is regulated.

Results discussed in this chapter show that the rate of electron entry and transfer in oxidase is determined by the unique structure of the bimetallic copper center, Cu_A. Mutations of copper ligands that lead to loss of its dinuclear spectral characteristics but retention of two copper ions and no change in the other redox-active metal centers cause a dramatic decrease in overall activity (1/10 to 1/100 of wild type). The data indicate that Cu_A is the major electron entry pathway, and its delocalized electronic structure and redox potential are critical for rapid electron transfer from cytochrome c to heme a. Mutation of two surface carboxyls close to Cu_A, which does not alter the unique bimetallic structure, identifies them as part of the cytochrome c binding domain. All the mutants retain proton-pumping activity, ruling out a critical role of the Cu_A structure in proton translocation.

Protons are required in cytochrome oxidase to convert oxygen to water (substrate protons) and to be translocated across the membrane (pumped protons). A direct coupling mechanism utilizes both kinds of protons to maintain charge neutrality during electron transfer into the active site. Results discussed here suggest that charge neutrality can also be maintained by reversal of the exit channel for protons, when the normal route for uptake of pumped protons from the interior is blocked. The result is uncoupled electron transfer that is accelerated rather than inhibited by an externally positive membrane potential.

In the mammalian oxidase, but not the bacterial form, uncoupled electron transfer is observed in the presence of high ATP concentrations in the interior of reconstituted vesicles. A likely ATP binding site on the inside of the membrane on subunit VIa is observed in the crystal structure, and its possible involvement in controlling coupling efficiency by allosteric effects on the outside of the membrane are discussed.

ACKNOWLEDGMENT. Supported by NIH AM26916 (SF-M).

5. REFERENCES

Anthony, G., Reimann, A., and Kadenbach, B., 1993, Tissue-specific regulation of bovine heart cytochrome c oxidase activity by ADP via interaction with subunit VIa, *Proc. Natl. Acad. Sci. USA* **90:**1652 1656.

Belyanovich, L., Arnold, S., Kohnke, D., and Kadenbach, B., 1996, A molecular switch in cytochrome *c* oxidase turns on thermogenesis in heart at low work load, *Biochem. Biophys. Res. Commun.* **229**:485–487.

Brzezinski, P., 1996, Internal electron-transfer reactions in cytochrome *c* oxidase, *Biochemistry* **35**: 5612–5615.

Cao, J., Hosler, J., Shapleigh, J., Gennis, R., Revzin, A., and Ferguson-Miller, S., 1992, Cytochrome aa_3 of *Rhodobacter sphaeroides* as a model for mitochondrial cytochrome *c* oxidase, *J. Biol. Chem.* **267**:24273–24278.

Cooper, C. E., 1990, The steady-state kinetics of cytochrome *c* oxidation by cytochrome oxidase, *Biochim. Biophys. Acta* **1017**:187–203.

Ferguson-Miller, S., and Babcock, G., 1996, Heme/copper terminal oxidases, *Chem. Rev.* **96**:2889–2907.

Ferguson-Miller, S., Brautigan, D., and Margoliash, E., 1976, Correlation of the kinetics of electron transfer activity of various eukaryotic cytochrome *c* with binding to mitochondrial cytochrome *c* oxidase, *J. Biol. Chem.* **251**:1104–1115.

Ferguson-Miller, S., Brautigan, D., and Margoliash, E., 1978, Definition of cytochrome *c* binding domains by chemical modification, *J. Biol. Chem.* **253**:149–159.

Fetter, J. R., Qian, J., Shapleigh, J., Thomas, J. W., García-Horsman, J. A., Schmidt, E., Hosler, J., Babcock, G. T., Gennis, R. B., and Ferguson-Miller, S., 1995, Possible proton relay pathways in cytochrome *c* oxidase, *Proc. Natl. Acad. Sci. USA* **92**:1604–1608.

Fetter, J. R., Sharpe, M., Qian, J., Mills, D., Ferguson-Miller, S., and Nicholls, P., 1996, Fatty acids stimulate activity and restore respiratory control in a proton channel mutant of cytochrome *c* oxidase, *FEBS Lett.* **393**:155–160.

Frank, V., and Kadenbach, B., 1996, Regulation of the H^+/e^- stoichiometry of cytochrome *c* oxidase from bovine heart by intramitochondrial ATP/ADP ratios, *FEBS Lett.* **382**:121–124.

García-Horsman, J. A., Barquera, B., Rumbley, J., Ma, J., and Gennis, R. B., 1994, The superfamily of heme-copper respiratory oxidases, *J. Bacteriol.* **176**:5587–5600.

García-Horsman, J. A., Puustinen, A., Gennis, R. B., and Wikström, M., 1995, Proton transfer in cytochrome bo_3 ubiquinol oxidase of *Escherichia coli*: Second-site mutations in subunit I that restore proton pumping in the mutant Asp135→Asn, *Biochemistry* **34**:4428–4433.

Gelles, J., Blair, D. F., and Chan, S. I., 1987, The proton-pumping site of cytochrome *c* oxidase: A model of its structure and mechanism, *Biochim. Biophys. Acta* **853**:205–236.

Geren, L. M., Beasley, J. R., Fine, B. R., Saunders, A. J., Hibdon, S., Pielak, G. J., Durham, B., and Millett, F., 1995, Design of a ruthenium–cytochrome *c* derivative to measure electron transfer to the initial acceptor in cytochrome *c* oxidase, *J. Biol. Chem.* **270**:2466–2662.

Gray, H. B., and Malmström, B. G., 1989, Long-range electron transfer in multisite metalloproteins, *Biochemistry* **28**:7499–7505.

Hill, B. C., 1991, The reaction of the electrostatic cytochrome *c*–cytochrome oxidase complex with oxygen, *J. Biol. Chem.* **266**:2219–2226.

Hosler, J. P., Fetter, J., Tecklenburg, M. M. J., Espe, M., Lerma, C., and Ferguson-Miller, S., 1992, Cytochrome aa_3 of *Rhodobacter sphaeroides* as a model for mitochondrial cytochrome *c* oxidase, *J. Biol. Chem.* **267**:24264–24272.

Iwata, S., Ostermeier, C., Ludwig, B., and Michel, H., 1995, Structure at 2.8 Å resolution of cytochrome *c* oxidase from *Paracoccus denitrificans*, *Nature* **376**:660–669.

Kadenbach, B., Barth, J., Akgun, R., Freund, R., Linder, D., and Possekel, S., 1995, Regulation of mitochondrial energy generation in health and disease, *Biochim. Biophys. Acta* **1271**:103–109.

Marcus, R. A., and Sutin, N., 1985, Electron transfers in chemistry and biology, *Biochim. Biophys. Acta* **811**:265–322.

Millett, F., de Jong, C., Paulson, L., and Capaldi, R. A., 1983, Identification of specific carboxylate groups on cytochrome *c* oxidase that are involved in binding cytochrome *c*, *Biochemistry* **22:**546–552.

Mitchell, D. M., and Gennis, R. B., 1995, Rapid purification of wild-type and mutant cytochrome *c* oxidase from *Rhodobacter sphaeroides* by Ni^{2+}-NTA affinity chromatography, *FEBS Lett.* **368:**148–150.

Mitchell, D. M., Ädelroth, P., Hosler, J. P., Fetter, J. R., Brzezinski, P., Pressler, M. A., Aasa, R., Malmström, B. G., Alben, J. O., Babcock, G. T., Gennis, R. B., and Ferguson-Miller, S., 1996, A ligand-exchange mechanism of proton pumping involving tyrosine-422 of subunit I of cytochrome oxidase is ruled out, *Biochemistry* **35:**824–828.

Moser, C. C., Page, C. C., Farid, R., and Dutton, P. L., 1995, Biological electron transfer, *J. Bioenerg. Biomembr.* **27:**263–274.

Ostermeier, C., Harrenga, A., Ermler, U., and Michel, H., 1997, Structure at 2.7 Å resolution of the *Paracoccus denitrificans* two-subunit cytochrome *c* oxidase complexed with an antibody FV fragment, *Proc. Natl. Acad. Sci. USA* **94:**10547–10553.

Pan, L. P., Hibdon, S., Liu, R.-Q., Durham, B., and Millett, F., 1993, Intracomplex electron transfer between ruthenium–cytochrome *c* derivatives and cytochrome *c* oxidase, *Biochemistry* **32:** 8492–8498.

Qian, J., Shi, W., Pressler, M., Hoganson, C., Mills, D., Babcock, G. T., and Ferguson-Miller, S., 1997, Aspartate-407 in *Rhodobacter sphaeroides* cytochrome *c* oxidase is not required for proton pumping or Mn binding, *Biochemistry* **36:**2539–2543.

Ramirez, B. E., Malmström, B. G., Winkler, J. R., and Gray, H. B., 1995, The currents of life: The terminal electron-transfer complex of respiration, *Proc. Natl. Acad. Sci. USA* **92:**11949–11951.

Rich, P., Meunier, B., Mitchell, R., and Moody, R., 1996, Coupling of charge and proton movement in cytochrome *c* oxidase, *Biochim. Biophys. Acta* **1275:**91–95.

Rousseau, D. L., Ching, Y.-C., and Wang, J., 1993, Proton translocation in cytochrome *c* oxidase: Redox linkage through proximal ligand exchange on cytochrome a_3, *J. Bioenerg. Biomembr.* **25:**165–176.

Saraste, M., 1990, Structural features of cytochrome oxidase, *Q. Rev. Biophys.* **23:**331–366.

Shapleigh, J. P., and Gennis, R. B., 1992, Cloning, sequencing, and deletion from the chromosome of the gene encoding subunit I of the aa_3-type cytochrome *c* oxidase of *Rhodobacter sphaeroides, Mol. Microbiol.* **6:**635–642.

Taanman, J.-W., Turina, P., and Capaldi, R., 1994, Regulation of cytochrome *c* oxidase by interaction of ATP at two binding sites, one on subunit VIa, *Biochemistry* **33:**11833–11841.

Thomas, J. W., Puustinen, A., Alben, J. O., Gennis, R. B., and Wikström, M., 1993, Substitution of asparagine for aspartate-135 in subunit I of the cytochrome *bo* ubiquinol oxidase of *Escherichia coli* eliminates proton-pumping activity, *Biochemistry* **32:**10923–10928.

Tittor, J., Soell, C., Oesterhelt, Butt, H.-J., and Bamberg, E., 1989, A defective proton pump, point-mutated bacteriorhodopsin Asp 96 → Asn is fully reactivated by azide, *EMBO J.* **8:**3477–3482.

Tsukihara, T., Aoyama, H., Yamashita, E., Tomizaki, T., Yamaguchi, H., Shinzawa-itoh, K., Nakashima, R., Yaono, R., and Yoshikawa, S., 1995, Structure of metal sites of oxidized bovine heart cytochrome *c* oxidase at 2.8 Å, *Science* **269:**1069–1074.

Tsukihara, T., Aoyama, H., Yamashita, E., Tomizaki, T., Yamaguchi, H., Shinzawa-itoh, K., Nakashima, R., Yaono, R., and Yoshikawa, S., 1996, The whole structure of the 13-subunit oxidized cytochrome *c* oxidase at 2.8 Å, *Science* **272:**1136–1144.

Wikström, M., Bogachev, A., Finel, M., Morgan, J. E., Puustinen, A., Raitio, M., Verkhovskaya, M. L., and Verkhovsky, M. I., 1994, Mechanism of proton translocation by the respiratory oxidases. The histidine cycle, *Biochim. Biophys. Acta* **1187:**106–111.

Witt, H., Zickermann, V., and Ludwig, B., 1995, Site-directed mutagenesis of cytochrome *c* oxidase reveals two acidic residues involved in the binding of cytochrome *c*, *Biochim. Biophys. Acta* **1230:**74–76.

Woodruff, W. H., 1993, Coordination dynamics of heme–copper oxidases. The ligand shuttle and the control and coupling of electron transfer and proton translocation, *J. Bioenerg. Biomembr.* **25:**177–188.

Zhen, Y., Wang, K., Mills, D., Ferguson-Miller, S., and Millett, F., 1997, The binuclear character of Cu_A in cytochrome *c* oxidase is important for electron transfer, not for proton pumping, *Biophys. J.* **72:**A93.

Zhen, Y., Qian, J., Follmann, K., Hosler, J., Haywary, T., Nilsson, T., and Ferguson-Miller, S., 1998a, Overexpression and purification of cytochrome *c* oxidase from *Rhodobacter sphaeroides*, *Protein Exp. Purif.* **13:**326–336.

Zhen, Y., Wang, K., Sadoski, R., Grinnell, S., Geren, L., Ferguson-Miller, S., Durham, B., and Millett, F., 1998b, Definition of interaction domain for the reaction of cytochrome *c* with *Rb. sphaeroides* cytochrome *c* oxidase, *Biophys. J.* **74:**A77.

Zickermann, V., Verkhovsky, M., Morgan, J., Wikström, M., Anemuller, S., Bill, E., Steffens, G. C., and Ludwig, B., 1995, Perturbation of the Cu_A site in cytochrome-*c* oxidase of *Paracoccus denitrificans* by replacement of Met227 with isoleucine, *Eur. J. Biochem.* **234:**686–693.

CHAPTER 8

Mechanism of Proton-Motive Activity of Heme–Copper Oxidases

Peter R. Rich

1. INTRODUCTION

Mammalian mitochondrial cytochrome c oxidase is responsible for the terminal step in the respiratory electron transfer chain of oxidation of cytochrome c and reduction of molecular oxygen. Energy from these reactions is conserved both because of the vectorial nature of the redox reactions and because of an associated mechanism that results in translocation of additional protons across the membrane. It has become clear that the mammalian enzyme is one member of a large and diverse superfamily of homologous oxidases that is widely distributed and includes both cytochrome c and quinol-oxidizing forms (Saraste *et al.*, 1991). The influences of molecular biology and successes in solving two oxidase structures to atomic resolution (Tsukihara *et al.*, 1995, 1996; Iwata *et al.*, 1995) have clarified many issues and allow consideration of mechanistic questions at the atomic level. Of particular interest is the way in which electron transfer and oxygen reduction chemistry are coupled to the processes that result in the net transfer of protons across the membrane and the conservation of energy for subsequent use in endergonic processes such as ATP synthesis. The purpose of the present chapter is to review recent ideas on some of the critical

Peter R. Rich • Department of Biology, University College, London WC1E 6BT, England.

Frontiers of Cellular Bioenergetics, edited by Papa *et al.* Kluwer Academic/Plenum Publishers, New York, 1999.

features that likely underlie the mechanism of coupling of proton and electron transfers and to relate them to the available structural information.

2. PRINCIPLES OF COUPLING

The central features and underlying principles of the chemiosmotic mechanism by which membrane-located exergonic electron transfer processes can be coupled to those that require energy were described by Mitchell in the 1960s (Mitchell, 1961, 1966, 1968). Mitchell realized that electron transfer chains are coupled to the movement of protons across the membranes in which they are situated so that electron transfer results in a proton-motive force that is then used to drive the endergonic processes. Importantly, he also proposed a realistic and testable model, termed the vectorial redox loop, of the chemistry that would be required in a series of membrane-bound electron transfer components so that electron transfer would necessarily result in coupled proton translocation across the membrane. A central feature concerned the postulate that the components of the redox chain alternated between electron carriers and hydrogen atom carriers and that these were spatially organized in the membrane. In the case of cytochrome oxidase, electrons are donated from cytochrome c (the electron carrier) from the intermembrane space side of the membrane (the positive, or P phase), whereas protons required for water formation from oxygen (which can be considered to be the hydrogen atom carrier) are taken up from the matrix (the negative, or N phase). This organization is as envisaged in the redox loop proposal and automatically leads both to electrical charging of the membrane (since charges move across the membrane) and to formation of a proton gradient (since protons are consumed from the N phase).

The simple redox loop is usually complicated by details of the protein structures that surround the catalytic reaction sites. Much of the key chemical and redox transformations may take place in a relatively small part of the protein structure, with channels in the protein allowing access of ions to this core. In the case of cytochrome oxidase, it is now clear that electron transfer only proceeds across part of the membrane and that channels must exist to allow the protons required for water production to move through considerable distances. However, a further complication in understanding coupling in the oxidases is that of the chemistry associated with the additional process of translocation of protons from the N phase to the P phase. This process is tightly coupled to the oxygen reduction chemistry, and a description in terms of the redox loop formulation has always proved elusive. Clearly, it required proton transfer pathways through the protein, and these may overlap with or may be quite separate from the pathway for those "substrate" protons that are required for water formation.

3. THE IMPORTANCE OF LOCAL ELECTRONEUTRALITY WITHIN THE REACTION CORE

I have emphasized (Rich, 1995, 1996) a further important consideration when exploring the detailed mechanism of coupling of ion and electron transfer within the reaction cycle of the protein. This concerns the need for local change neutralization of those catalytic intermediates that are produced in a region of low dielectric strength, a condition likely to be the case in a number of important ion-motive enzymes, including the proton-motive oxidases (Rich, 1996). In the absence of sufficient solvent or protein charge rearrangements, electron transfer into such systems can provide a strong driving force for uptake of a charge-compensating proton or possibly another cation. The notion of a need for local electroneutrality within regions of proteins of low dielectric strength may have relevance not only to cytochrome oxidase and cytochrome *bo* (Mitchell and Rich, 1994; Moody and Rich, 1994; Mitchell *et al.*, 1992) and other ion-motive systems (Rich, 1996; Rich *et al.*, 1995), but also to soluble enzymes with occluded reaction cores. For example, we have discussed this notion in relation to the ligand-associated proton chemistry of horseradish peroxidase (Meunier *et al.*, 1997; Meunier and Rich, 1997; Meunier *et al.*, 1995).

From a physical point of view, the requirement for charge compensation originates in free energy considerations and the factors that control redox potentials. A variety of factors affect the redox potentials of groups in proteins, so that there is no single dominating factor. Moore and co-workers (Moore, 1996; Moore and Pettigrew, 1990; Moore *et al.*, 1986; Gunner and Honig, 1991; Churg and Warshel, 1986) have divided the factors into energetic changes associated with electrostatic interactions, bonding interactions, and protein conformational changes. In the present context, it is the first of these which is of importance. Electrostatic interactions can involve several components: the ions in solution (nonspecific screening and any specific ion–protein complex formation at the protein surface), the surface charges on the protein, the solvent itself, and the charges in the protein around the redox center. For proteins in which the catalytic center is buried in a region of relatively low dielectric strength, as is the case for cytochrome oxidase, it is this latter factor that can become particularly important.

I have utilized a simple equation, derived from the Born model in which the ion is treated as a rigid sphere and the medium as a structureless continuum, in order to illustrate the energy cost of movement of a charge onto a specific redox center within a protein (Rich, 1996). The energetic cost of the movement of a single electrical charge is dependent on the effective radius of the charged group, r_i, and on the dielectric constant, D, of the surroundings. Hence, the energy cost increases as the charge becomes more localized and as the dielectric constant of the surrounding medium decreases. A quantitative estimate of the difference in electrostatic energy between a reduced (with one net charge) and oxidized (with zero net charge) species can be expressed by the equation:

$$\Delta G = N \cdot e^2 \cdot (1/4\pi\varepsilon_0) \cdot q^2/Dr_i \qquad \text{J} \cdot \text{mole}^{-1}$$

where N is Avogadro's number; e is the electronic charge of a single electron (1.6×10^{-19} C); ε_0 is the permittivity of free space (8.8542×10^{-12} $C^2 \cdot N^{-1} \cdot m^{-2}$); q is the number of charges; r_i is the radius of the redox group; and D the dielectric strength of the surrounding medium. For a redox center of radius 3 Å, initially with a zero net charge, and embedded in a protein of dielectric strength 3.5 [an assumed average value for the interior of proteins (Moore and Pettigrew, 1990)], this indicates that a single electron transfer into it would contribute more than 130 kJ·mole^{-1} (>1350 mV) in electrostatic energy. Although this is a crude simplification of any real microscopic situation, it does illustrate the point that the introduction of a charged species into a protein of low dielectric strength would, in the absence of other factors, be energetically very costly.

Any of a number of energy-minimizing processes could raise the effective dielectric constant, and hence reduce the energy cost, their balance being dependent on the protein structure and external conditions. If the metal center is exposed to the aqueous medium, both specific and nonspecific ion redistributions can occur in the medium in the vicinity of the charge. If the metal center is buried inside the protein, then solvents, local charges, and hydrogen bonds within the protein can redistribute so as to minimize the charge imbalance. The greater the extent of these processes, the more effective is the energy minimization of introduction of the charge.

If the extent of such processes in the surrounding protein structure is limited, then the energetic cost of introduction of the charge becomes high. Delocalization of the charge itself can help to decrease the energy cost, and such delocalization is quite feasible within larger cofactor structures such as the porphyrin ring macrocycle or in a multimetal redox center such as an iron–sulfur cluster. However, if there is an appropriate site in the protein structure, a specific counter-ion may bind in response to the introduction of the charge, i.e., the affinity of the site for the counter-ion is increased in response to the charge. In the special case of a protonation, the result is redox-linked pK shifts (Moore *et al.*, 1984).

Such charge-associated protonations can be important factors that are additional to the microscopic electron transfer rate constants in controlling equilibrium constants and observed rates of reaction. For example, we have shown the importance of the co-binding of a proton in controlling the binding and rate constants of anionic ligands in the distal heme pocket of horseradish peroxidase under some conditions (Meunier *et al.*, 1997). Control of reaction rates through limitations of proton movement rates might be particularly important if the proton has to move through the protein structure in order to arrive at its binding site, as is likely to be the case for compensation of charge changes in the deeply buried binuclear center of the proton-motive oxidases (Verkhovsky *et al.*, 1995; Hallén *et al.*, 1994).

Below, I assess the importance for oxidase mechanism of redox-linked protonations internal balance of charges, and structures that might raise dielectric strength in localized regions of the protein.

4. INTERNAL CHARGE BALANCING AND PROTON-MOTIVE MECHANISM

Local charge neutralization is likely to form a central feature of the coupling mechanism of several of the more complex ion-motive enzymes. In a general form (Fig. 1), electron transfer into a region of locally low dielectric strength is electrically counterbalanced by the uptake of a cation (in the case of oxidases, this is a proton). Subsequent electron transfer to an acceptor, A (in the case of oxidases, this acceptor is oxygen), at later stages in the reaction cycle provides a strong driving force for protonation of the reduced acceptor product (in this case, an oxide that becomes protonated to form water). However, the charge-counterbalancing cations are spatially separated from the reduced acceptor and cannot themselves be used to form the final product; the system may be said to be "electrostatically locked." Uptake of substrate protons is therefore required to complete the reaction. As these are taken up from the N phase,

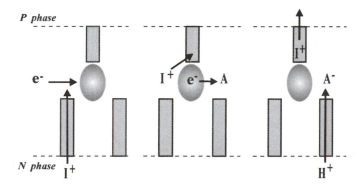

FIGURE 1. Coupling of ion translocation to electron transfer in a proton-motive enzyme of low internal dielectric strength. Details are in the text and the model is an extension of that presented in Rich (1996).

they cause loss of the counterbalancing cations into the aqueous P phase. In Fig. 1, the cations, I^+, and protons are shown to remain separate as they enter by separate channels, which seems quite likely if I^+ is not itself a proton. However, when I^+ is itself a proton, as in the case of the oxidases, a single shared entry channel is possible, with separation of I^+ from substrate by ligand rearrangements in the catalytic site.

5. GATING OF THE PROTON TRANSFER ROUTE

The nature of the gating process by which the charge-counterbalancing cations are taken up from the N phase in the initial reduction step but are repulsed into the P phase during the final chemical transformation is critical to the mechanism outlined in Fig. 1. This is essential since exit back through the N phase channel would prevent any net vectorial proton translocation. Uptake of cations from the N phase during reduction is likely to be ensured by a large permanent activation barrier for passage of protons through the P phase channel, the energy required to overcome this barrier at later steps being supplied by chemical transformation of substrate. In addition, for complex enzymes whose electron transfer pathway involves multiple steps, correct sidedness of proton uptake may be governed by the spatial position of the electron entry route.

A number of possible mechanisms may be envisaged to ensure exit of cations into the P phase. In the simplest case, electrostatic interaction with the incoming substrate protons itself may be enough, provided that the N phase route of the translocated protons is blocked by entry of the substrate protons. This could occur if at least part of the N phase route were shared by translocated and substrate protons (although such a common route would presumably require additional mechanisms to prevent access of the translocated protons to the chemical transformation site). For multistep enzymes, other possibilities include the physical movement of protons away from the N phase channel during an intermediate reaction step, or a physical closing of the N phase channel by amino acid and hydrogen bond rearrangements, in response to internal changes in electrostatic potential fields. Some of these are considered below in more specific detail in relation to the oxidases.

6. CHARGE-LINKED PROTONATION SITES AND STRUCTURAL FEATURES

We have empirically tested the degree to which the binuclear center balances charge changes by protonation changes. To date, we have found no exception to the general rule that all stable redox or ligand-binding changes in the binuclear center are counterbalanced by protonation changes (Rich *et al.*,

1996). The pH dependencies of midpoint potentials of the metals during classical redox potentiometry are also roughly consistent with proton-linked charge counterbalance, although the redox-linked phenomena are complicated by an intricate electrostatic linkage between several of the metal centers and multiple protonation sites (reviewed in Wikström et al., 1981). It should be stressed that it is quite feasible that conditions may be found where other factors outweigh the protonation changes in minimizing energy costs. Indeed, some conditions for cytochrome oxidase where strict charge counterbalance may not hold have been described recently (Capitanio et al., 1997). If the affected protonations are key to proton-motive mechanism, then it becomes possible that the proton-motive mechanism may become impaired under such conditions.

The structures of the four subunit cytochrome oxidase from *Paracoccus denitrificans* (Iwata et al., 1995) and the 13 subunit enzyme from beef heart mitochondria (Tsukihara et al., 1995, 1996) have been solved by X-ray crystallography to 2.8 Å resolution, and coordinates for the latter are available. Possible protonation sites and routes in subunit I of the protein can be identified (Brown et al., 1993; Babcock and Wikström, 1992; Hosler et al., 1993). The most obvious candidates for stable sites of charged-linked protonation in subunit I include the histidine ligands to the metal centers, the heme propionates, and residues close to them (such as D364, D-369, and H-368), and conserved protonatable residues K-319, E-242, and Y-244 (unless otherwise stated, all numbering refers to the bovine sequence). Figure 2 illustrates the positions of some of the residues that will be discussed.

It has already been pointed out (Iwata et al., 1995) that two possible pores

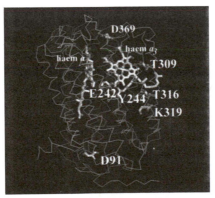

FIGURE 2. Positions of residues that might be involved in proton transfer or binding. The structure has been drawn with QUANTA from the coordinates of the bovine enzyme published by Tsukihara et al. (1996), and numbering of residues is also according to the bovine subunit I sequence. Pore A may link the N phase to E242 through D91 and other residues, and pore B may link the N phase to the binuclear center through K319, Y244, T316, T309, and other residues.

of relatively hydrophilic residues can be identified in the crystal structures. One of these (pore B) appears to lead from the N phase via K-319 to the binuclear center, whereas a second (pore A) may lead from the N phase to the conserved E-242 (Fig. 2). A putative route for proton exit into the P phase (Iwata *et al.*, 1995) and possible multiple routes for entry of oxygen to the binuclear center from the hydrophobic membrane phase (Tsukihara *et al.*, 1996) have also been described. In addition, a third possible proton transfer route from the N phase to the P phase has been noted (Tsukihara *et al.*, 1996). It seems likely that at least some of these pathways play essential roles in the catalytic reaction cycle and its coupled proton transfer mechanism.

7. CHEMICAL MODELS FOR PROTON TRANSLOCATION

There have been a wide range of ideas on possible coupling mechanisms in proton-motive oxidases, and most of these were formulated before specific structural information was available. These have been well reviewed elsewhere recently (see, for example, Ferguson-Miller and Babcock, 1996; Trumpower and Gennis, 1994; Malmström, 1993; Brunori and Wilson, 1995; Papa *et al.*, 1994), and will not be reiterated here. Instead, I will describe two mechanisms that are valid in the light of, and have been related to, the structural information. Two protons are known to be bound when the peroxy state is formed from the oxidized enzyme (Mitchell and Rich, 1994) and both models assume that these protons are destined to be translocated, in response to uptake of substrate protons for water generation, rather than acting as the substrate protons themselves. One model, introduced by Wikström and termed the "histidine cycle," assumes that the two protons are bound to one of the histidine ligands of Cu_B (Wikström *et al.*, 1994). Hence, they are quite close to the oxygen intermediates and could reach their binding sites either through the same channel used by substrate protons or perhaps through a separate route. The second model assumes that the translocated protons are accumulated in the region of the conserved glutamate, E-242, and other moieties in its surroundings (for example, the heme propionates and closely associated protonatable groups such as H-368, D364, D369), and are taken up into this region through channel A (Fig. 2), which is physically separated both from the binuclear center and possibly from its substrate proton channel B. We have called this model the "glutamate trap" (Meunier and Rich, 1997a).

7.1. The Histidine Cycle

In this mechanism, which was suggested by Wikström and colleagues (Wikström *et al.*, 1994) before the crystal structure information was available, a histidine ligand to the Cu_B is proposed to be the binding site for protons that are destined to be translocated across the membrane. A double protonation of this histidine from imidazolate to imidazolium was envisaged in response to changes in the oxygen intermediates, followed by a conformational change of the residue so that the protons were unavailable to the oxygen intermediates themselves and instead were ultimately ejected into the aqueous P phase at later stages in the catalytic cycle.

Some support for the histidine cycle has come from the subsequent observation in the bacterial crystal structure (Iwata *et al.*, 1995) that one of the histidine ligands to Cu_B (equivalent to H-290 in the beef heart structure) was disordered and therefore might be able to adopt several conformations (although such disorder in the same histidine residue was not observed in the beef heart enzyme structure). The histidine cycle was elaborated (Iwata *et al.*, 1995) by combination with the charge-counterbalancing restrictions of the binuclear center intermediates (Rich, 1996), and separate routes for substrate and translocated protons (through pore B and pore A, respectively) were suggested.

7.2. The Glutamate Trap

This proposal in more general form was also first developed (Rich, 1995) before the detailed structural information became available and is centered on the necessity for charge balancing by protonation changes of the stable catalytic intermediates. As in Fig. 1, the protons required for charge balancing are physically separated from the oxygen intermediates and ultimately are translocated into the P phase when the oxide products are protonated to form water. This proposal can be made more explicit in light of the structural information suggesting that electron transfer to heme *a* is linked to protonation, through channel B in Fig. 2, of the conserved glutamate residue, E-242. Beside allowing heme *a* to be more easily reduced, this protonation may be necessary before electron transfer to the binuclear center can proceed, if this same proton can charge compensate the reduction of the binuclear center. Indeed, a shift of the electron density from the heme *a* to the binuclear center may well cause a relocation of this proton to a site more strongly redox-linked to the binuclear center, for example, to one of the heme a_3 propionate residues or nearby residues such as H-368, D364, or D369 (Fig. 3) or even to ligands of the magnesium atom. Such electron transfer-linked relocation of the proton, for which there is some experimental support (Hallén *et al.*, 1994), may form the basis of the gating mechanism by which the P phase exit route is favored. Several protons may be accumulated in the protein through this glutamate-trapping mechanism, as is required by the model (Rich, 1995). The positions of the residues that might hold these protons indicate that they are unlikely to be available subsequently for protonation of the oxide products of oxygen reduc

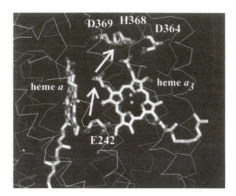

FIGURE 3. Details of the glutamate trap model for proton coupling. The structures have been drawn with QUANTA from the coordinates of the bovine enzyme published by Tsukihara *et al.* (1996).

tion, a requirement of the general mechanism. Hence, when an oxide product is formed (presumably during the $P \rightarrow F$ and the $F \rightarrow O$ reaction steps), additional protons are required for oxide conversion to water, and these must enter the protein by a route which gives access to the oxygen intermediates. The most obvious possible pathway for the passage of substrate protons from the N phase onto oxygen intermediates in the binuclear center is via the separate channel A depicted in Fig. 2. At both the $P \rightarrow F$ and the $F \rightarrow O$ steps, two protons are expected to be required for water formation, so that two are ejected into the P phase, an expectation consistent with measurements (Wikström, 1989).

7.3. Experimental Progress

The two possible channels, A and B in Fig. 2, were already recognized in predicted models before the crystal structures appeared (Fetter *et al.*, 1995). Furthermore, a range of studies of site-directed and random (Meunier and Colson, 1994; Hosler *et al.*, 1993) mutants also highlighted the importance of residues in these regions. For example, mutagenesis of the lysine and threonine residues in channel B, or the glutamate residue in channel A, has dramatic inhibitory effects on catalytic turnover. In addition, mutagenesis of a conserved aspartate at the predicted N phase opening of channel A led to an enzyme that was apparently devoid of the associated proton-translocating function (Fetter *et al.*, 1995; Thomas *et al.*, 1993). Mutagenesis of the glutamate residue associated with channel B leads to a catalytically inactive enzyme that has no peroxidatic partial reaction cycle (Konstantinov *et al.*, 1997) and can be re-oxidized from the reduced state only partially, perhaps to the peroxy form, and without the capability of any fast proton uptake from the medium (Svensson-Ek *et al.*, 1996). Given that the reduced binuclear center already has bound protons, this result appears consistent with a role for the glutamate in the proton translocation functions. We have studied a mutant form of cytochrome *c* oxidase that we have isolated from *Saccharomyces cerevisiae* (Meunier and Colson, 1994), which has a single mutation of 167N in subunit I (equivalent to 166 in the bovine structure in Fig. 1), a position in the crystallographic data that is between heme *a* and the conserved glutamate residue in channel A. It seems clear that this change perturbs the redox-linked protonations associated with heme *a*, as does removal of the glutamate itself (Garcia-Horsman *et al.*, 1994). Overall, such studies reinforce the notion of two separate channels for proton input: channel A for translocated protons and channel B for substrate protons.

However, recent and more detailed analyses of the effects of some of these mutations have complicated this relatively simple view. For example, mutation of the lysine residue in channel A produces a protein that is inactive in catalytic turnover and yet the reduced enzyme can be reoxidized rapidly by oxygen after photolysis of carbon monoxide (Svensson *et al.*, 1995) and retains a peroxidatic section of the catalytic cycle (Konstantinov *et al.*, 1997). Both functions should be expected to require uptake of both substrate and translocated protons from the N phase. In this same mutant, reduction of the binuclear center to the doubly reduced form, which can react with oxygen, is no longer possible (S. Jünemann *et al.*, unpublished data). It might have been expected that the catalytic cycle might be able to proceed relatively normally from the oxidized to the peroxy form in this mutant, since the required proton uptake to reach this stage should be associated only with the proton-translocating channel A.

Overall, much of these recent data are not easy to reconcile with a simple view of two separate proton channels and have led to a suggestion that the structures may even change their roles at different stages in the catalytic cycle (Konstantinov *et al.*, 1997). Although the proposed channel regions are critical, a major question must be raised concerning the proposed function of channel B. Clearly, channel B is an essential structural element and we are now actively considering whether it might play a role other than proton transfer, for example, one in which it acts as a "dielectric well" in order to decrease the energy cost of production of some of the transient and unstably charged intermediates of the binuclear center. Such a dielectric well may be contrasted with a proton well (Mitchell, 1968) by being proton impermeable but with a high effective dielectric constant (e.g., by having an extended hydrogen bonding network between the binuclear center and the N phase) that can decrease the energy cost of charging the binuclear center. Further studies will be required to test these and the other ideas that are rapidly unfolding as this major class of enzymes reveals new aspects of important protein structural chemistry.

ACKNOWLEDGMENTS. I am grateful to HFSP (RG-464/95 M) and The Wellcome Trust (Project Grant 049722) for financial support, and to my colleagues Drs. Brigitte Meunier and Susanne Jünemann for useful comments and discussions.

8. REFERENCES

Babcock, G. T., and Wikström, M., 1992, Oxygen activation and the conservation of energy in cell respiration, *Nature* **356**:301–309.

Brown, S., Moody, A. J., Mitchell, R., and Rich, P. R., 1993, Binuclear centre structure of terminal protonmotive oxidases, *FEBS Lett.* **316**:216–223.

Brunori, M., and Wilson, M. T., 1995, Electron transfer and proton pumping in cytochrome oxidase, *Biochimie* **77**:668–676.

Capitanio, N., Vygodina, T. V., Capitanio, G., Konstantinov, A. A., Nichols, P., and Papa, S., 1997, Redox-linked protolytic reactions in soluble cytochrome-c oxidase from beef-heart mitochondria: redox Bohr effects, *Biochim. Biophys. Acta* **1318**:255–265.

Churg, A. K., and Warshel, A., 1986, Control of the redox potential of cytochrome *c* and microscopic dielectric effects in proteins, *Biochemistry* **25**:1675–1681.

Ferguson-Miller, S., and Babcock, G. T., 1996, Heme/Copper terminal oxidases, *Chem. Rev.* **96**: 2889–2907.

Fetter, J. R., Qian, J., Shapleigh, J., Thomas, J. W., García-Horsman, A., Schmidt, E., Hosler, J., Babcock, G. T., Gennis, R. B., and Ferguson-Miller, S., 1995, Possible proton relay pathways in cytochrome *c* oxidase, *Proc. Natl. Acad. Sci. USA* **92**:1604–1608.

Garcia-Horsman, J. A., Barquera, B., Rumbley, J., Ma, J., and Gennis, R. B., 1994, The superfamily of heme-copper oxidases, *J. Bacteriol.* **176**:5587–5600.

Gunner, M. R., and Honig, B., 1991, Electrostatic control of midpoint potentials in the cytochrome subunit of the *Rhodopseudomonas viridis* reaction center, *Proc. Natl. Acad. Sci. USA* **88**: 9151–9155.

Hallén, S., Brzezinski, P., and Malmström, B. G., 1994, Internal electron transfer in cytochrome *c* oxidase is coupled to the protonation of a group close to the bimetallic site, *Biochemistry* **33**: 1467–1472.

Hosler, J. P., Ferguson-Miller, S., Calhoun, M. W., Thomas, J. W., Hill, J., Lemieux, L., Ma, J., Georgiou, C., Fetter, J., Shapleigh, J., Tecklenburg, M. M. J., Babcock, G. T., and Gennis, R. B., 1993, Insight into the active-site structure and function of cytochrome oxidase by analysis of site-directed mutants of bacterial cytochrome aa_3 and cytochrome *bo*, *J. Bioenerg. Biomemb.* **25**:121–136.

Iwata, S., Ostermeier, C., Ludwig, B., and Michael, H., 1995, Structure at 2.8 Å resolution of cytochrome *c* oxidase from *Paracoccus denitrificans*, *Nature* **376**:660–669.

Konstantinov, A. A., Siletsky, S., Mitchell, D., Kaulen, A., and Gennis, R. B., 1997, The roles of the two proton input channels in cytochrome *c* oxidase from *Rhodobacter sphaeroides* probed by the effects of site-directed mutations on time-resolved electrogenic intraprotein proton transfer, *Proc. Natl. Acad. Sci. USA* **94**:9085–9090.

Malmström, B. G., 1993, Vectorial chemistry in bioenergetics: Cytochrome *c* oxidase as a redox-linked proton pump, *Acc. Chem. Res.* **26**:332–338.

Meunier, B., Rodriguez-Lopez, J. N., Smith, A. T., Thorneley, R. N. F., and Rich, P. R., 1995, Laser photolysis behaviour of ferrous horseradish peroxidase with carbon monoxide and cyanide: effects of mutations in the distal heme pocket, *Biochemistry* **34**:14687–14692.

Meunier, B., Rodriguez-Lopez, J. N., Smith, A. T., Thorneley, R. N. F., and Rich, P. R., 1998, Redox- and anion-linked protonation sites in horseradish peroxidase, *Biochem. J.* **330**:303–309.

Meunier, B., and Colson, A.-M., 1994, Random deficiency mutations and reversions in the cytochrome *c* oxidase subunits I, II and III of *Saccharomyces cerevisiae*, *Biochim. Biophys. Acta* **1187**:112–115.

Meunier, B., and Rich, P. R., 1997a, Coupling of protons transfer to oxygen chemistry in cyto- chrome oxidase; the roles of residues I67 and E243. *Oxygen Homeostasis and Its Dynamics* (Y. Ishimura, H. Shimada, and M. Suematsu, eds.), Springer-Verlag, Tokyo, pp. 106–111.

Meunier, B, and Rich, P. R., 1997b, Photolysis of the cyanide adduct of the ferrous horseradish peroxidase, *Biochim. Biophys. Acta* **1318**:235–245.

Mitchell, P., 1961, Coupling of phosphorylation to electron and proton transfer by a chemi-osmotic type of mechanism, *Nature* **191**:144–148.

Mitchell, P., 1966, *Chemiosmotic coupling in oxidative and photosynthetic phosphorylation*, Glynn Research Ltd., Bodmin, United Kingdom.

Mitchell, P., 1968, *Chemiosmotic coupling and energy transduction*, Glynn Research Ltd., Bodmin, United Kingdom.

Mitchell, R., and Rich, P. R., 1994, Proton uptake by cytochrome *c* oxidase on reduction and on ligand binding, *Biochim. Biophys. Acta* **1186**:19–26.

Mitchell, R., Mitchell, P., and Rich, P. R., 1992, Protonation states of the catalytic cycle intermedi- ates of cytochrome *c* oxidase, *Biochim. Biophys. Acta* **1101**:188–191.

Moody, A. J., and Rich, P. R., 1994, The reaction of hydrogen peroxide with pulsed cytochrome *bo* from *Escherichia coli*, *Eur. J. Biochem.* **226**:731–737.

Moore, G. R., 1996, Haemoproteins, in: *Protein Electron Transfer* (D. S. Bendall, ed.), BIOS Scientific Publishers Ltd., Oxford, England, pp. 189–216.

Moore, G. R., and Pettigrew, G. W., 1990, Redox potentials, in: *Cytochromes c: Evolutionary, Structural and Physicochemical Aspects*, Springer-Verlag, Berlin, chapter 7, pp. 309–362.

Moore, G. R., Harris, D. E., Leitch, F. A., and Pettigrew, G. W., 1984, Characterisations of ionisations that influence the redox potential of mitochondrial cytochrome *c* and photosyn- thetic bacterial cytochrome c_2, *Biochim. Biophys. Acta* **764**:331–342.

Moore, G. R., Pettigrew, G. W., and Rogers, N. K., 1986, Factors influencing redox potentials of electron transfer proteins, *Proc. Natl. Acad. Sci. USA* **83**:4998–4999.

Papa, S., Capitanio, N., Glaser, P., and Villani, G., 1994, The proton pump of heme-copper oxidases, *Cell Biol. Int.* **18**:345–356.

Rich, P. R., 1995, Towards an understanding of the chemistry of oxygen reduction and proton translocation in the iron-copper respiratory oxidases, *Aust. J. Plant. Physiol.* **22**:479–486.

Rich, P. R., 1996, Electron transfer complexes coupled to ion translocation, in: *Protein Electron Transfer* (D. S. Bendall, ed.), BIOS Scientific Publishers Ltd., Oxford, England, pp. 217–248.

Rich, P. R., Meunier, B., and Ward, F. B., 1995, Predicted structure and possible ionmotive mechanism of the sodium-linked NADH-ubiquinone oxidoreductase of *Vibrio alginolyticus*, *FEBS Lett.* **375**:5–10.

Rich, P. R., Meunier, B., Mitchell, R. M., and Moody, A. J., 1996, Coupling of charge and proton movement in cytochrome *c* oxidase, *Biochim. Biophys. Acta* **1275**:91–95.

Saraste, M., Holm, L., Lemieux, L., Lübben, M., and van der Oost, J., 1991, The happy family of cytochrome oxidases, *Biochem. Soc. Trans.* **19**:608–612.

Svensson, M., Hallén, S., Thomas, J. W., Lemieux, L. J., Gennis, R. B., and Nilsson, T., 1995, Oxygen reaction and proton uptake in helix VIII mutants of cytochrome bo_3, *Biochemistry* **34**: 5252–5258.

Svensson-Ek, M., Thomas, J. W., Gennis, R. B., Nilsson, T., and Brzezinski, P., 1996, Kinetics of electron and proton transfer during the reaction of wild type and helix VI mutants of cytochrome bo_3 with oxygen, *Biochemistry* **35**:13673–13680.

Thomas, J. W., Puustinen, A., Alben, J. O., Gennis, R. B., and Wikström, M., 1993, Substitution of asparagine for aspartate-135 in subunit I of the cytochrome *bo* ubiquinol oxidase of *Escherichia coli* eliminates proton-pumping activity, *Biochemistry* **32**:10923–10928.

Trumpower, B. L., and Gennis, R. B, 1994, Energy transduction by cytochrome complexes in mitochondrial and bacterial respiration: The enzymology of coupling electron transfer reactions to transmembrane proton translocation, *Annu. Rev. Biochem.* **63**:675–716.

Tsukihara, T., Aoyama, H., Yamashita, E., Tomizaki, T., Yamaguchi, H., Shinzawa-Itoh, K., Nakashima, R., Yaono, R., and Yoshikawa, S., 1995, Structures of metal sites of oxidized bovine heart cytochrome *c* oxidase at 2.8 Å, *Science* **269**:1069–1074.

Tsukihara, T., Aoyama, H., Yamashita, E., Tomizaki, T., Yamaguchi, H., Shinzawa-Itoh, K., Nakashima, R., Yaono, R., and Yoshikawa, S., 1996, The whole structure of the 13-subunit oxidized cytochrome *c* oxidase at 2.8 Å, *Science* **272**:1136–1144.

Verkhovsky, M. I., Morgan, J. E., and Wikström, M., 1995, Control of electron delivery to the oxygen reduction site of cytochrome *c* oxidase: A role for protons, *Biochemistry* **34**:7483–7491.

Wikström, M., 1989, Identification of the electron transfers in cytochrome oxidase that are coupled to proton-pumping, *Nature* **338**:776–778.

Wikström, M., Krab, K., and Saraste, M., 1981, *Cytochrome Oxidase A Synthesis*, Academic Press, London, pp. 1–198.

Wikström, M., Bogachev, A., Finel, M., Morgan, J. E., Puustinen, A., Raitio, M., Verkhovskaya, M., and Verkhovsky, M. I., 1994, Mechanism of proton translocation by the respiratory oxidases. The histidine cycle, *Biochim. Biophys. Acta* **1187**:106–111.

CHAPTER 9

Oxygen Reduction and Proton Translocation by the Heme–Copper Oxidases

Mårten Wikström, Joel E. Morgan, Gerhard Hummer,
William H. Woodruff, and Michael I. Verkhovsky

1. INTRODUCTION

Cytochrome c oxidase (EC 1.9.3.1), the oxygen-reducing member of the respiratory chains of mitochondria and several aerobic bacteria, catalyzes flux of reducing equivalents to the highly oxidizing O_2–H_2O couple and conserves energy as a proton-motive force. These enzymes are members of a large family of related heme–copper oxidases; others are the hydroquinone oxidases, of which cytochrome bo_3 of *Escherichia coli* is the best characterized. Apart from the intriguing catalysis of O_2 reduction to water, this class of enzymes conserves energy by vectorial functioning as proton and electrical charge trans-

Mårten Wikström, Joel E. Morgan, and Michael I. Verkhovsky • Helsinki Bioenergetics Group, Department of Medical Chemistry, Institute of Biomedical Sciences and Biocentrum Helsinki, University of Helsinki, FI-00014 Helsinki, Finland. **Gerhard Hummer** • Theoretical Biology and Biophysics Group, Los Alamos National Laboratory, Los Alamos, New Mexico 87545. **William H. Woodruff** • Chemical Science and Technology Division, Los Alamos National Laboratory, Los Alamos, New Mexico 87545.

Frontiers of Cellular Bioenergetics, edited by Papa *et al.* Kluwer Academic/Plenum Publishers, New York, 1999.

locators. This leads to primary conservation of respiratory redox energy as an electrochemical proton gradient across the mitochondrial or bacterial membrane, which is secondarily utilized as the driving force for the oxidative synthesis of ATP (Mitchell, 1966). Before the elucidation of the crystal structures of cytochrome *c* oxidase from *Paracoccus denitrificans* (Iwata *et al.*, 1995) and bovine heart mitochondria (Tsukihara *et al.*, 1995, 1996), the heme–copper oxidases have been the subject of many review articles, which have highlighted their functional (Malmström, 1990a,b; Cooper, 1990; Chan and Li, 1990; Wilson and Bickar, 1991; Rich, 1991; Babcock and Wikström, 1992), structural (Capaldi, 1990; Saraste, 1990; Saraste *et al.*, 1991; Gennis, 1992; Haltia and Wikström, 1992; Hosler *et al.*, 1993), and evolutionary aspects (Kadenbach *et al.*, 1987, 1991; Castresana *et al.*, 1995; Saraste *et al.*, 1996). The present account is not intended as a review. Instead, we will set a rather ambitious focus on the most recent knowledge pertinent to the function of these enzymes as molecular machines.

2. STRUCTURE

The impact of the three-dimensional structures at 2.8 Å resolution of cytochrome *c* oxidase from *P. denitrificans* (Iwata *et al.*, 1995) and bovine heart mitochondria (Tsukihara *et al.*, 1995, 1996) on the elucidation of function cannot be overstated, which we hope to make clear in this account. The heme–copper oxidases consist of up to 13 different subunit polypeptides, of which three (COI, COII, and COIII) always appear to be present in the cytochrome *c* and quinol oxidases. In fact, the bacterial enzymes just contain these three subunits, sometimes with one additional small subunit that does not seem to have homologous counterparts in other members of the family.

2.1. Subunit III

COIII, while nearly always present, has no bound redox center. It is highly hydrophobic, with either seven (in cytochrome *c* oxidases) or five (in quinol oxidases) transmembrane helices, and binds phospholipid molecules tightly. Its functional role is uncertain. Deletion of the corresponding gene results in poor assembly of the enzyme (Haltia *et al.*, 1989). Recently, it has been suggested that it forms an "oxygen storage device" (Tsukihara *et al.*, 1996), or that it is part of a channel for O_2 into the enzyme's active site (Riistama *et al.*, 1996).

2.2. Subunit II

COII has two transmembrane helices that are in contact with COI. Its main mass is outside the membrane, forming a β-sheet structure above the domain of

COI that contains the heme groups. In the cytochrome c oxidases, this β-structure binds the Cu_A center, which consists of two copper ions lying closely together (Iwata *et al.*, 1995; Tsukihara *et al.*, 1995; Wilmanns *et al.*, 1995). The Cu_A center serves as an immediate one-electron acceptor from cytochrome c and as the electron donor for heme a (Hill, 1991). Although there are speculations about direct electron transfer from Cu_A to the binuclear site (bypassing heme a; compare Chapter 3, this volume), all the present kinetic data suggest that if this occurs at all, it would have to be insignificantly slow.

2.3. Subunit I and the Binuclear Site

COI is the best-conserved subunit, and in cytochrome c oxidase it consists of 12 transmembrane helices, which when viewed perpendicularly to the membrane are arranged symmetrically in three semicircular "arcs" consisting of four helices each (Iwata *et al.*, 1995; Tsukihara *et al.*, 1996). The three arcs form three "pores" (A, B, C), in which transmembrane helices II, VI, and X have key structural roles in the center, while the other helices are more peripheral (Fig. 1). COI is the functionally most important subunit and contains two heme groups and a copper site in all known members of the enzyme family (Fig. 2), although the structure of the heme moieties may vary.

The low-spin heme group of cytochrome c oxidase (heme a; Fe_a) is coordinated to fully conserved histidine imidazoles of helices II and X (Hosler

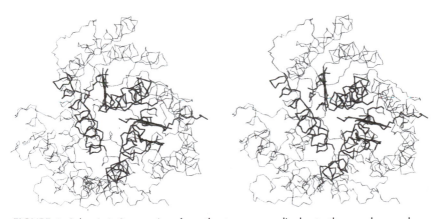

FIGURE 1. Subunit I. Stereo view from the top, perpendicular to the membrane plane. Hemes a (left) and a_3 (right), and Cu_B with its three histidine ligands (below heme a_3) are highlighted. The structure may be described as three "arcs" of four helices each, i.e., numbers XI–**II** (upper left), III–**VI** (lower left), and VII–**X** (right), of which the last helix that takes a central position is highlighted in each case. Data are from the bovine cytochrome aa_3 structure (Tsukihara *et al.*, 1996), using the HyperChem program. For color representation of this figure, see color tip facing page 199.

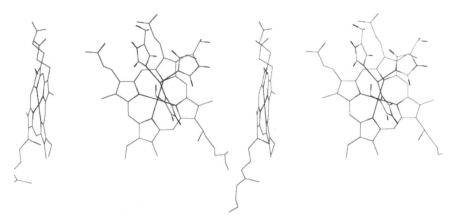

FIGURE 2. The binuclear site and heme *a*. Stereo view in the membrane plane toward heme a_3 (right) from a point behind Cu_B. The three histidine ligands of Cu_B are shown: H291 (upper left), H290 (upper right), and H240 (lower right). Data are from the bovine cytochrome aa_3 structure (Tsukihara *et al.*, 1996), using HyperChem software. For color representation of this figure, see color tip facing page 199.

et al., 1993; Iwata *et al.*, 1995; Tsukihara *et al.*, 1995); it resides within pore C with its plane perpendicular to the membrane and the heme iron about one third across the membrane thickness from the outside. Its two propionate groups point toward the outside (Figs. 1, 2). In the cytochrome *c* oxidases where this heme has the heme A structure, the hydroxyethylfarnesyl side chain points downward and helps to block pore C from access from the inside of the membrane. In the quinol oxidase of *E. coli* (cytochrome bo_3), the low-spin heme is usually a protoheme (heme B) without the long side chain, although heme O may alternatively be inserted in cases of enzyme overexpression (Puustinen *et al.*, 1992).

The other heme group (a_3) forms part of the binuclear oxygen-reducing heme–copper center. It has either a heme A or a heme O (heme o_3) structure, except in the most distant members of the heme–copper oxidase family (the cytochrome cbb_3 oxidases) (de Gier *et al.*, 1996). This heme is located in pore B, again perpendicular to the membrane, almost at right angles to heme *a* (108°) and with the heme iron at roughly the same depth in the membrane as Fe_a. The assumption that the heme planes are both exactly perpendicular to the membrane puts Fe_{a3} at most about 0.7 Å closer to the *o*-side of the membrane than Fe_a (Fig. 2), largely due to the tilting of helix X relative to the membrane normal (Fig. 1). The propionate side chains are again pointing upward, but this time the hydroxyethylfarnesyl side chain is tilted away from pore B into the surrounding phospholipid bilayer. As a result, and in contrast to the case for heme a, this allows access to heme a_3 from the inside of the membrane through pore B. Heme a_3 is coordinated on the proximal side to a conserved histidine imidazole of helix X, which is only one residue (phenylalanine) remote from one of the Fe$_a$ ligand histidines. This highly conserved his-phe-his motif yields a potential electron transfer path of 16 covalent bonds between the two heme irons (Woodruff, 1993), and places the phenylalanine near the heme edges, their closest approach being only some 5 Å.

An early copper electronuclear double resonance (ENDOR) study indicated that Cu$_B$ has three nitrogenous (histidine imidazole) ligands in a trapped intermediate state of the catalytic cycle (Cline *et al.*, 1983). More recently, a combination of extended X-ray absorption fine structure (EXAFS) and ENDOR spectroscopy of Cu$_B$ in cytochrome aa_3-600 from *Bacillus subtilis* showed that Cu$_B$ has three histidine ligands in the oxidized enzyme, all with distinct hyperfine coupling constants, and at 1.98 ($\pm$ 0.02) Å from the copper (Fann *et al.*, 1995). Site-directed mutagenesis already had suggested that three specific histidines in COI might have such a role, based on the absence of the Fourier transform infrared spectroscopy (FTIR) signature of Cu$_B^+$-CO upon their mutation (Hosler *et al.*, 1993): two adjacent ones (H290 and H291)* in helix VII (which actually loses its helical structure in this domain), and one in helix VI (H240). One of the two adjacent histidine ligands of helix VII (H290) was not resolved in the *Paracoccus* X-ray structure, presumably due to azide in the binuclear pocket (Iwata *et al.*, 1995), but all three are resolved in the structure of the mitochondrial enzyme (Tsukihara *et al.*, 1995).

A combination of EXAFS and ENDOR spectroscopy and heavy water substitution revealed that Cu$_B$ has a fourth oxygenous ligand in the oxidized enzyme, with exangeable proton(s) (Fann *et al.*, 1995). This is most likely a water or a hydroxide ion. The Cu$_B$ site is thus tetragonal but with low symmetry. The oxygenous Cu$_B$ ligand is not resolved in the X-ray structures, however. The relatively low resolution combined with the difficulty of resolving a small nonproteinaceous ligand at a short distance from a heavy metal ion may be among the reasons for this.

Cu$_B$ lies about 4–5 Å from Fe$_{a3}$ on the distal side of the heme, thus being part of pore B (Fig. 2). Since the early work of van Gelder and Beinert (1969) (see also Wikström *et al.*, 1981), it is known that Fe$_{a3}$ and Cu$_B$ are tightly coupled magnetically in the oxidized enzyme, the former being in a ferric high ($S = \frac{5}{2}$), or possibly intermediate spin ($S = \frac{3}{2}$) state, and the latter a cupric ion ($S = \frac{1}{2}$). The magnetic coupling, which usually quenches any potential electron paramagnetic resonance (EPR) signatures from both metals, has been thought

---

*The amino acid numbering in this chapter is that of the cytochrome aa_3 enzyme from bovine heart mitochondria.

to require a bridging ligand, but such ligands are not seen in the X-ray structure. Modeling the Cu_B site suggests that the water or OH^- ligand must point toward the distal face of heme a_3. An OH^- bridge between the metals would be compatible with the EXAFS data on cytochrome aa_3-600 (Fann *et al.*, 1995). A model compound of the binuclear center with a bridging OH^- ligand has an Fe–Cu distance of 3.66 Å (Fox *et al.*, 1996). Although the Fe_{a3}–Cu_B distance is 4.5 Å in the X-ray structure of the mitochondrial cytochrome aa_3, EXAFS distances often tend to be shorter than those from X-ray crystallography. Another possible bridging structure could be made up by an Fe_{a3}-bound OH^-, hydrogen-bonded to a water ligand of Cu_B.

3. THE CATALYTIC CYCLE

Cytochrome *c* oxidase from mitochondria has been the main object of dynamic studies, and this work has been reviewed extensively. Here, we will briefly describe the catalytic cycle, with emphasis on some recent findings, as well as on some still open questions.

3.1. Kinetics and Reaction Intermediates

The binuclear site of the oxidized enzyme is in the ferric–cupric **O** state. Under equilibrium conditions, the first electron will reduce Cu_B, giving rise to distinct high-spin ferric EPR signals from Fe_{a3} due to the loss of magnetic coupling (Fig. 3, state **E**) (see Wikström *et al.*, 1981). O_2 binds to the Fe_{a3} after a second electron has reduced it (mixed-valence enzyme), and a primary "oxy" intermediate is formed (Chance *et al.*, 1975) with $\tau \sim 5$ μsec at room temperature (see Babcock and Wikström, 1992), and with absorption maxima in the difference spectrum at about 595 and 430 nm (Chance *et al.*, 1975; Verkhovsky *et al.*, 1996a). The oxy intermediate (**A**, Fig. 3) has also been characterized in the cytochrome bo_3 quinol oxidase from *E. coli* where the spectrum is different due to the different heme structure (Verkhovsky *et al.*, 1996a). Prior to the binding of the O_2 to Fe_{a3}, it interacts weakly with Cu_B ($K_D \sim 8$ mM) (Verkhovsky *et al.*, 1994; Bailey *et al.*, 1996). Also, the O_2 binding to Fe_{a3} is weak ($K_D \sim 0.35$ mM) (Chance *et al.*, 1975; Verkhovsky *et al.*, 1994), especially in relation to the high operative oxygen affinity for cytochrome *c* oxidase ($K_{M,app.} \sim 0.1$ μM).

In the absence of a further electron source, the oxy species **A** is converted at about 5000 sec^{-1} [$\tau \sim 200$ μsec; room temperature (RT)] into a new species (not shown in Fig. 3), originally called compound C by Chance and co-workers (1975), who described it in low-temperature studies. Subsequently, this state has also been called **P** on the basis of its presumed peroxy structure. The **P** state

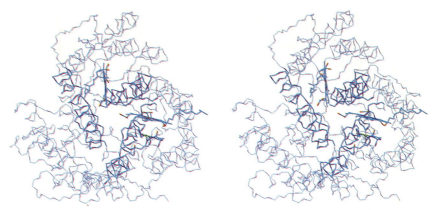

FIGURE 1. Subunit I. Stereo view from the top, perpendicular to the membrane plane. Hemes *a* (left) and *a*₃ (right), and Cu$_B$ with its three histidine ligands (below heme *a*₃) are highlighted. The structure may be described as three "arcs" of four helices each, i.e., numbers XI–**II** (upper left), III–**VI** (lower left), and VII–**X** (right), of which the last helix that takes a central position is highlighted in each case. Data are from the bovine cytochrome *aa*₃ structure (Tsukihara *et al.*, 1996), using the HyperChem program.

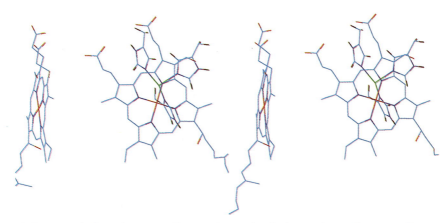

FIGURE 2. The binuclear site and heme *a*. Stereo view in the membrane plane toward heme *a*₃ (right) from a point behind Cu$_B$. The three histidine ligands of Cu$_B$ are shown: H291 (upper left), H290 (upper right), and H240 (lower right). Data are from the bovine cytochrome *aa*₃ structure (Tsukihara *et al.*, 1996), using HyperChem software.

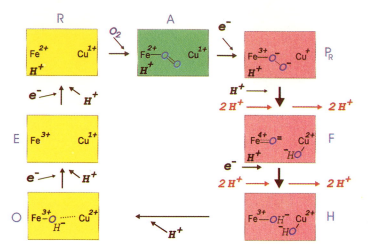

FIGURE 3. The catalytic cycle. Major intermediates of the binuclear heme iron–copper site are described in the text. The left side amounts to the "electron-filling" phase of the cycle, and the right to the "power stroke," which is associated with proton translocation.

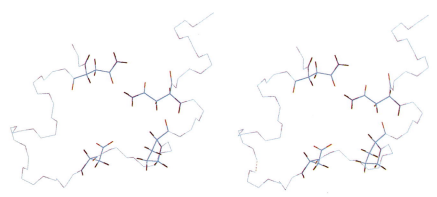

FIGURE 4. Proton input domain of the D channel. Stereo view of the end of helix II (left), the beginning of helix III (right), and the loop that connects them. Key residues mentioned in the text are shown highlighted, that is, N80 in helix II, D91 and P95 in the loop, and N91 in helix III. Only the polypeptide backbone is shown otherwise. Data are from the bovine cytochrome aa_3 structure (Tsukihara *et al.*, 1996) using HyperChem software.

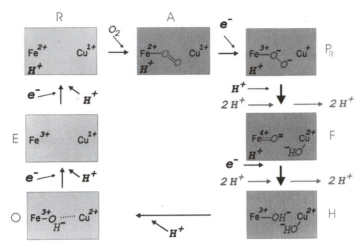

FIGURE 3. The catalytic cycle. Major intermediates of the binuclear heme iron–copper site are described in the text. The left side amounts to the "electron-filling" phase of the cycle, and the right to the "power stroke," which is associated with proton translocation. For color representation of this figure, see color tip on facing page.

is characterized by a strong band at 607 nm ($\varepsilon \sim 11$ mM^{-1}cm^{-1}) in the difference spectrum versus oxidized enzyme, and a red-shifted Soret band.

In the reaction of the fully reduced enzyme with O_2, compound **A** is again formed first and with the same kinetics, but now the subsequent reaction step is much faster [$\tau \sim 32$ μsec (RT) at 1 mM O_2]. This step includes electron transfer from the low spin heme (Fe$_a$) to the oxygen-bound binuclear site (Fig. 3) (Han et al., 1990a; Hill, 1991), which obviously cannot occur in the mixed-valence enzyme. It is this fast electron transfer that traps the weakly bound O_2 irreversibly to the binuclear site, which is largely responsible for the low apparent K_M for O_2 (Verkhovsky et al., 1996b). The fact that the rate of this 32-μsec phase depends on both the O_2 concentration and on the eT rate constant shows that the preceding reactions of weak O_2 binding to Cu$_B$ and Fe$_{a3}$ are essentially at equilibrium prior to this kinetic "trapping" step (Verkhovsky et al., 1994). The product of the 32-μsec phase thus has three electrons in the binuclear site (**P$_R$**, Fig. 3). The structure of this state is presently under discussion, the major issue being whether or not the O–O bond has been broken. Recently, it was shown that the product of the 32-μsec phase exhibits the same optical spectral characteristics as the product of the corresponding 200-μsec phase for the mixed-valence enzyme, i.e., the characteristics of the **P** state (Morgan et al., 1996; Sucheta et al., 1997). This finding is important insofar as it shows that Fe$_{a3}$ is initially the same optically (and hence has the same or a very similar structure) whether the oxygen reaction site contains two or three electrons. We have called these two P states $\mathbf{P_R}$ and $\mathbf{P_M}$, referring to their origins in the fully reduced and mixed valence enzyme, respectively (Morgan *et al.*, 1996).

The $\mathbf{P_R}$ state is subsequently converted into the so-called F state at 5000 $\sec^{-1}$ ($\tau \sim 200$ μsec) at RT without further electron input into the binuclear site (Fig. 3). F has a characteristic absorption maximum at about 582 nm in the difference spectrum versus oxidized enzyme ($\varepsilon \sim 5.2$ mM^{-1}cm^{-1}). It was initially proposed to have a ferryl iron–cupric structure (Wikström, 1981), and this has received strong support from resonance Raman (RR) data (Varotsis and Babcock, 1990; Ogura *et al.*, 1990; Han *et al.*, 1990a). Similarly, it was recently shown that discrete input of a single electron into the $\mathbf{P_M}$ state by means of a photoreductant yields the F state, and that one-electron reduction of F yields the oxidized ferric–cupric enzyme (O) (Verkhovsky *et al.*, 1996c). The $\mathbf{P_M}$ and F states are thus two and one electron, respectively, more oxidized than the oxidized enzyme in state O, and $\mathbf{P_R}$ has the same overall number of electrons as F in the binuclear site.

Roughly simultaneously with the conversion of $\mathbf{P_R}$ into F, the electron "hole" at Fe$_a$ created by the 32-μsec phase is equilibrated with Cu$_A$ with about 50% oxidation of the latter (Morgan *et al.*, 1996; Sucheta *et al.*, 1997). Thus, after this phase, the fourth electron remains about 50–50 equilibrated between the Fe$_a$ and Cu$_A$ centers (not shown in Fig. 3).

In the final step, F receives the fourth electron (from being shared between Cu$_A$ and Fe$_a$) and is converted into O, with $\tau \sim 1$–2 msec at RT. This step includes a ferric-hydroxy intermediate (H, Fig. 3), which has been detected by RR spectroscopy (Han *et al.*, 1990b). The oxidized enzyme is regenerated and a new cycle can begin. It should be noted that the reductive part of the cycle (Fig. 3; left) is largely rate-determining for enzyme turnover, and that the reduction of the binuclear site is limited kinetically by the simultaneous uptake of protons (Verkhovsky *et al.*, 1995).

3.2. The Structures of F and P

Having established the separate identity of F and P and their redox relationships to one another and to state O, we may consider their structures. But first we should recall that these compounds can be produced by yet another technique: by addition of H_2O_2 to the oxidized enzyme (Bickar *et al.*, 1982; Konstantinov *et al.*, 1992; for a review, see Fabian and Palmer, 1995). Briefly, low concentrations of peroxide produce P, whereas high concentrations produce F. However, these reactions are complicated; there are several exceptions to this rule of thumb that will not be considered here (but see for example, Fabian and Palmer, 1995). We should also recall that there are two forms of

P, P_R and P_M, of which the latter is the one usually encountered, except in the reaction of the fully reduced enzyme with O_2, as discovered only recently (Morgan *et al.*, 1996; Sucheta *et al.*, 1997).

The fact that P_M is produced by reacting the two-electron reduced enzyme with O_2, by addition of H_2O_2 to the oxidized enzyme, as well as by energy-dependent two-electron oxidation of the oxidized enzyme plus water, and that the controlled reduction of P_M by two electrons yields the oxidized enzyme, all support the conclusion that the P_M state must have a structure that is electronically equivalent to the peroxy species:

$$Fe^{3+}-O^--O^- \quad Cu^{2+} \tag{1}$$

The alternative structure

$$R(+) \quad Fe^{4+}=O \quad Cu^{2+} \tag{2}$$

where $R(+)$ is a porphyrin radical, has been excluded (Fabian and Palmer, 1995). Structure (2) where $R(+)$ is a protein radical, has been deemed highly unlikely (Morgan *et al.*, 1996).

The structure

$$Fe^{4+}=O \quad \quad Cu^{3+} \tag{3}$$

is another possibility for P_M. One-electron reduction of structure (3) would be expected to yield

$$Fe^{4+}=O \quad Cu^{2+} \tag{4}$$

which is the relatively well-established structure of the intermediate **F**. P_M has the same optical characteristics as P_R, but one less electron in the binuclear site. Since P_R cannot be the same as **F**, but precedes it kinetically, we conclude that structure (3) is also very unlikely for P_M. Hence, the O–O bond is likely to be intact in the P_R and P_M species.

We may then further consider electronic equivalents of structure (1). Noteworthy alternative structures of P_M are the ferrous peroxy species

$$Fe^{2+}-O^--O^- \quad Cu^{3+} \tag{5}$$

and the ferrous superoxy form

$$Fe^{2+}-O_2^- \quad Cu^{2+} \tag{6}$$

Recalling that the optical spectrum reflects the Fe_{a3}, the corresponding P_R states should then have the same Fe_{a3} structure, but an additional reducing equivalent in Cu_B, so that structure (5) would have Cu^{2+} and structure (6) would have Cu^+. The P_R intermediate is interesting insofar as it is expected to be EPR-active. In fact, Blair *et al.* (1985) have described an oxygen intermediate in low-temperature experiments with an unusual Cu_B EPR spectrum. If the identity of this EPR-active intermediate is the P_R state, its structure can be unequivocally assigned, because the addition of one electron to the different remaining P_M structures (1), (5), and (6), in such a way as to keep Cu_B in the cupric EPR-active state, has one unique solution, namely, the peroxy structure

$$Fe^{2+} - O^- - O^- \quad Cu^{2+} \tag{7}$$

which was the structure proposed by Blair *et al.* (1985) for the EPR-active intermediate. We emphasize, however, that it has not been established whether the EPR-active intermediate is equivalent to P_R. It could alternatively be a form of the oxyferryl state, as suggested by Hansson *et al.* (1982). Clearly, it would be important to find out whether the EPR-active state has a "P-like" optical spectrum.

Kitagawa and co-workers (see Kitagawa and Ogura, 1997, for a review) have performed detailed RR studies on the oxygen intermediates. RR modes from the oxy intermediate **A**, from a ferric-hydroxy state (**H**), and from a ferryl intermediate (**F**) at 785 cm^{-1} are well established, also from work in two other laboratories (see Section 3.1). Two additional modes, at 804 (Kitagawa and Ogura, 1997) and at 356 cm^{-1} (Babcock and Varotsis, 1993; Kitagawa and Ogura, 1997) are more controversial. The 804-cm^{-1} mode (like that at 785 cm^{-1}) was shown unequivocally to involve an Fe=O bond stretch. Kitagawa *et al.* have argued that this species is equivalent to P_R, i.e., the 607-nm compound observed as an intermediate of the reaction of the fully reduced enzyme with O_2. According to this view, therefore, the O−O bond must already be broken in P_R, which disagrees with our conclusion above. However, we think that both the 804- and the 356-cm^{-1} compounds must arise from species in the catalytic cycle *beyond* P_R, but before the final 785-cm^{-1} **F** state. It is especially noteworthy that all the above RR modes, except that at 804 cm^{-1}, have been observed in cytochrome bo_3 (Hirota *et al.*, 1994), although the oxygen reaction proceeds through the same type of intermediates as in cytochrome aa_3 (Morgan *et al.*, 1995).

3.3. Proton Consumption

A net total of four protons will be consumed during a complete turnover of the enzyme

$$4e^- + 4\,H^+ + O_2 \rightarrow 2\,H_2O$$

From work of Mitchell and Rich (1994) and Capitanio *et al.* (1997), we know that two protons will be bound by the enzyme upon two-electron reduction of the binuclear site (but see Section 4). Two further protons are taken up during the oxidative phase of the catalytic cycle, one linked to the conversion of **P** to **F**, and the other to the **F** → **O** step (Fig. 3). It is especially noteworthy that neither the binding of O_2 to the enzyme nor the conversion of the **A** state into **P** (whether $\mathbf{P_M}$ or $\mathbf{P_R}$) is associated with net uptake of protons (Oliveberg *et al.*, 1991; Mitchell and Rich, 1994).

The net proton uptake by the enzyme is shown schematically in Fig. 3. This occurs vectorially from the inside of the membrane, which is important from the point of view of energy conservation. The four electrons are taken up from the outside (cytochrome *c*), and eventually the electrical charges of these electrons and protons annihilate in the formation of two water molecules from O_2. Thus, the catalytic chemistry itself is vectorially orientated relative to the membrane, as originally suggested by Mitchell (1966), and leads to the translocation of four electrical charge equivalents per catalytic O_2 cycle and to generation of both $\Delta\psi$ and ΔpH (i.e., of $\Delta\mu_H{}^+$).

3.4. Proton Translocation (Pumping)

An additional four protons per O_2 are translocated across the membrane from the aqueous *i* phase to the aqueous *o* phase linked to the catalytic cycle, which doubles the energy conservation efficiency. Thus, the complete cytochrome *c* oxidase reaction may be written

$$4e_o^- + 8\,H_i^+ + O_2 \rightarrow 2\,H_2O + 4\,H_o^+$$

where subindices *i* and *o* denote the input (inside) and output (outside) domains on each side of the coupling membrane. Notice that from an energetic viewpoint, it makes no difference from which side of the membrane O_2 enters or water leaves, since both are uncharged electrically.

4. MECHANISTIC LINKAGE OF PROTON TRANSLOCATION TO CATALYSIS

The linkage between the events of O_2 reduction and proton translocation was established by the discovery of the heme–copper oxidases as proton pumps (Wikström, 1977; van Verseveld *et al.*, 1981; Solioz *et al.*, 1982; Puustinen *et al.*, 1989; Raitio and Wikström, 1994). To understand this linkage in molecular detail, we first must understand the mechanism of O_2 reduction. Interestingly, much of the knowledge on intermediates of this reaction stems originally from observations in isolated mitochondria where the proton–electron linkage was utilized to perturb the catalytic cycle and to identify some of its intermediate (Wikström, 1981). It may also be of some historical interest that this work was actually initiated by projects aimed at finding energy-rich intermediates of the respiratory chain predicted by the chemical hypothesis of oxidative phosphorylation (Slater, 1966). The observation of an energy-dependent spectral shift in mitochondrial heme a_3 (Erecinska *et al.*, 1972) prompted further studies, which eventually led to the discovery of proton translocation by cytochrome *c* oxidase (see Wikström, 1977).

4.1. The Reactions that Are Linked to Proton Translocation

Analysis of the dependence of the **P–F** and **F–O** equilibria on phosphorylation potential (Wikström, 1989) revealed that these reaction steps are coupled to the synthesis of 1 and 0.75 ATP molecules, respectively, during reversal of the reaction in intact mitochondria. Since the ATP/4e$^-$ ratio for the overall oxidase reaction is 2 (Chamalaun and Tager, 1969), this shows that it is the reactions converting the **P** intermediate to **O** via **F** that are associated with the main energy conservation events. The energy requirement of translocation of 4 H$^+$ ions by the pump, per O$_2$ reduced, is met by the **P** $\to$ **F** $\to$ **O** reactions, while the reduction phase is much less exergonic. The subsequent binding of O$_2$ and the formation of **P$_R$** are not coupled to energy conservation (Babcock and Wikström, 1992). While this is a thermodynamic argument demonstrating linkage between a specific subset of catalytic interconversions and the synthesis of ATP, it is suggestive of a mechanistic linkage of these partial reactions with proton pumping.

However, proof of mechanistic linkage can only come from dynamic studies. This was recently achieved by time-resolved measurements of the generation of $\Delta\psi$ across the membrane of liposomes, into which cytochrome *c* oxidase was incorporated (Verkhovsky *et al.*, 1997). This study showed unequivocally that energy conservation is associated specifically, and to an equal extent, to the **P** $\to$ **F** and **F** $\to$ **O** transitions, while O$_2$ binding to the reduced enzyme and formation of **P$_R$** are not associated with net charge translocation. The energetics of the reductive half of the catalytic cycle is limited to translocation of electrons from cytochrome *c* to the binuclear site and to the associated net uptake of protons from the *i*-side into the enzyme.

4.2. Transfer versus Translocation of Protons

Proton pumping includes three basic events: uptake of proton(s) from the *i*-side, translocation to the opposite side of the osmotic barrier, and release on the *o*-side. It seems likely that uptake and release of protons occur by passive transfer along proton-conducting pathways that connect the aqueous *i*- and *o*-phases with the translocation machinery inside the membrane. This avoids large conformational changes associated with the translocation event, and has stimulated research on possible proton "channels." In addition, proton uptake occurs due to the protonation of reduced dioxygen to form water at the binuclear site.

4.3. Proton Channels and Water Chains

The first experimental evidence for the existence of discrete pathways for proton uptake came from mutagenesis work where it was shown that changing a conserved aspartic acid in subunit I (D91) into asparagine abolished proton pumping in multiturnover experiments with cytochrome bo_3 from *E. coli*, but with only about 50% inhibition of respiration (Thomas *et al.*, 1993; see also Fetter *et al.*, 1995). The crystal structures show that D91 lies in a "loop" on the *i*-side of the membrane, between transmembrane helices II and III (Fig. 4), as modeling had already predicted (Hosler *et al.*, 1993). Mutagenesis of two nearby asparagines (N80 and N98) had the same effect, and the acidic side chain of D91 could be moved into two other unique positions in this domain (replacing P95 and N98) with retention of proton translocation (García-Horsman *et al.*, 1995). The latter three loci are indeed close to one another in the bovine aa_3 structure, and were suggested to form the inlet of a proton channel (Fig. 4). With these findings and the crystal structure as guides, Iwata *et al.* (1995) proposed a channel for "pumped" protons leading from the D91 domain close to the the *i*-side of the membrane to the residue S193 near the middle (D channel), and the same potential proton transfer structure was noted by Tsukihara *et al.* (1996).

Iwata *et al.* (1995) suggested linkage of the top of the D channel to E242 near the binuclear center via solvent (water) molecules, while Tsukihara *et al.* (1996) did not find such connectivity in the bovine structure. Simulation of bound water positions in the bovine cytochrome aa_3 structure using the poten-

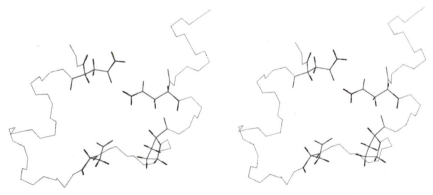

FIGURE 4. Proton input domain of the D channel. Stereo view of the end of helix II (left), the beginning of helix III (right), and the loop that connects them. Key residues mentioned in the text are shown highlighted, that is, N80 in helix II, D91 and P95 in the loop, and N91 in helix III. Only the polypeptide backbone is shown otherwise. Data are from the bovine cytochrome aa_3 structure (Tsukihara *et al.*, 1996) using HyperChem software. For color representation of this figure, see color tip facing page 199.

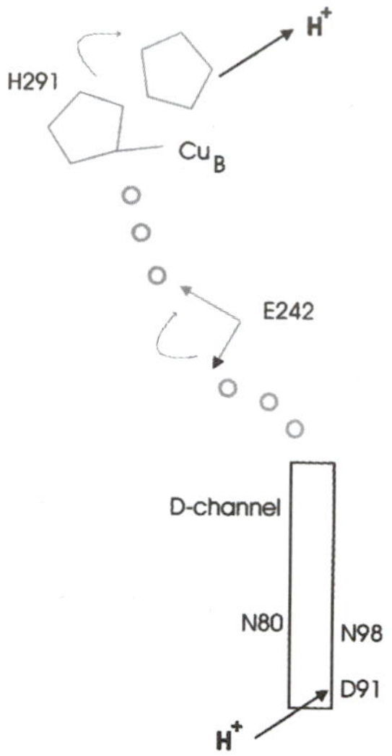

FIGURE 5. Schematic view of the proton translocation pathway. The D channel is shown with its proton input domain (lower right). The upper end of the D channel connects to E-242 through hydrogen bonding via three bound water molecules (circles). The side chain of E-242 can "flip" (arrows) into an upper position, where it can hydrogen bond to another array of three water molecules (circles). The latter leads to the Cu_B ligand, H291, which can accept protons, dissociate from copper, and release protons into an output channel (see text for details). For color representation of this figure, see color tip on facing page.

tial of mean force (PMF) (Hummer *et al.*, 1995, 1996) formalism nicely links the upper part of the D channel to E242 via three bound water molecules (A. Puustinen *et al.*, unpublished data) (Fig. 5), which is consistent with the report by Iwata *et al.* (1995). Furthermore, according to the PMF calculations, there are several hydrogen-bonded water molecules in the D channel itself, in good agreement with both crystal structures.

In addition, these PMF calculations predict the presence of at least three water molecules that may form a chain that starts within H-bonding distance from the Nϵ of the H291 copper ligand and goes toward E242 (A. Puustinen *et al.*, unpublished data) (Fig. 5). Hydrogen bonding of this network to E242 does not seem possible, however, unless the glutamic acid side chain swings upward from its position in the crystal structure into a second position where one of its oxygens is within H-bonding distance from the backbone oxygen of G239. Such mobility seems quite possible, because the carboxylic acid side chain shows no strong neighbor interactions, apart from those with water molecules. E242 might thus shuttle protons from the D channel into this latter water network, which leads to H291 (see Section 4.6).

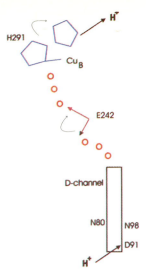

FIGURE 5. Schematic view of the proton translocation pathway. The D channel is shown with its proton input domain (lower right). The upper end of the D channel connects to E-242 through hydrogen bonding via three bound water molecules (circles). The side chain of E-242 can "flip" (arrows) into an upper position, where it can hydrogen bond to another array of three water molecules (circles). The latter leads to the Cu_B ligand, H291, which can accept protons, dissociate from copper, and release protons into an output channel (see text for details).

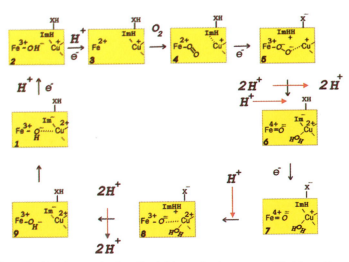

FIGURE 6. The histidine cycle. Details of this mechanism are modified from Morgan *et al.* (1994) as described in the text. Two of the Cu_B histidine ligands are only indicated by bonds to the copper, while a third is denoted Im^-, ImH, or $ImHH^+$, depending on its state of protonation. $XH-X^-$ denotes a closely lying proton donor to ImH, possibly a water cluster. State 1 is the oxidized binuclear site (**O**), state 2 the half-reduced, and state 3 the reduced form (**R**). Number 4 denotes the oxy compound (**A**) and 5 the P_R intermediate. Number 6 is the ferryl intermediate (**F**), which after receiving the fourth electron turns into 7. Number 8 is a transition state and 9 is the hydroxy intermediate (**H**).

Iwata *et al.* (1995) and Tsukihara *et al.* (1996) also identified another channel in the vicinity of transmembrane helix VIII, with K319 as one key residue (K channel). In accordance with the proposal of Hosler *et al.* (1993), they suggested that this channel may be used for uptake of those protons that are consumed in the formation of water. More recent experiments with the K319M mutant (K362M in the enzymes from *Rhodobacter* and *E. coli*) Siletsky *et al.* (1996) have shown, however, that reduction of Fe_{a3} is inhibited, while the $F \rightarrow O$ step (Fig. 3) and its associated charge translocation are unaffected. Enzyme reduction is limited by the associated uptake of about two protons (see Section 3.3). On the other hand, the $F \rightarrow O$ step is associated with both proton pumping and uptake of one proton to form water (Fig. 3). The picture thus emerges that the K channel may be used exclusively for proton uptake associated with reduction of Fe_{a3}, whereas the D channel may be used otherwise. This view is consistent with the notion that proton uptake coupled to the oxygen chemistry is mechanistically related to the events of proton translocation, whether the proton will be consumed or translocated.

Capitanio *et al.* (1997) have recently shown that there are three redox-linked acid–base groups that are associated to the binuclear site metals. Two of these ($pK_{ox} \sim 7.2$; $pK_{red} > 12$) may be due to an OH^- ligand of Fe_{a3} and of a histidine imidazolate ligand of Cu_B, respectively, in the oxidized enzyme.

4.4. Proton Translocation and the Histidine Cycle

In 1994, we presented a concrete chemical model of the proton-motive function of the heme–copper oxidases: the "histidine cycle" (Morgan *et al.*, 1994). Based on the crystal structure, Iwata *et al.* (1995) proposed a related model, which specifically involves the copper ligand H290 (H325 in *aa₃* from *Paracoccus*). Tsukihara *et al.* (1996) also discussed proton translocation, but their proposals are difficult to assess due to lack of detail. Rich (1995) has presented a more schematic model that is based on the principle of electron-eutrality in the binuclear site (Mitchell and Rich, 1994). According to this, the uptake of the protons for the oxygen chemistry forces the release of protons to the *o*-side of the membrane for electrostatic reasons, which is also a key principle in the histidine cycle (with the obvious exception of transition states) (Morgan *et al.*, 1994). This means that two of the four protons taken up to be consumed in the oxygen chemistry are actually an intrinsic part of the proton pump mechanism. On the other hand, the finding by Capitanio *et al.* (1997) that the number of protons taken up on reduction depends on pH does not allow a generalization of the electroneutrality principle, so that uptake of negative charge on reduction of the binuclear site must always be accompanied by proton uptake. In some cases, proton translocation can be decoupled from electron transfer, by mutagenesis (Thomas *et al.*, 1993; García-Horsman *et al.*,

1995) and at high pH (Verkhovskaya *et al.*, 1992), provided that the protonation deficiency still allows the chemistry to occur at a reasonable rate.

Reduction of the binuclear site may lead to protonation of a bridging hydroxide ligand and of a histidine imidazolate ligand of Cu_B (Mitchell *et al.*, 1992; Mitchell and Rich, 1994; Morgan *et al.*, 1994; Capitanio *et al.*, 1997). The two protons may be taken up via the K and D channels, respectively (states 1–3, Fig. 6). If so, it is unlikely that both will be translocated across the membrane in later reaction steps (contrast Iwata *et al.*, 1995; Rich, 1995).

4.5. Formation of the P_R State

When the binuclear site is reduced, O_2 binds to form the oxy intermediate **A** (state 4, Fig. 6). If an electron is available at Fe_a, the **A** species relaxes, coupled to electron transfer from Fe_a to the binuclear site, which forms the P_R state. This is not linked to net translocation of electrical charge across the dielectric (Verkhovsky *et al.*, 1997), but may be regarded as an initiation step of proton translocation. Hallén and Nilsson (1992) found a D_2O isotope effect of

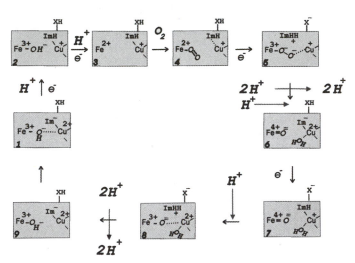

FIGURE 6. The histidine cycle. Details of this mechanism are modified from Morgan *et al.* (1994) as described in the text. Two of the Cu_B histidine ligands are only indicated by bonds to the copper, while a third is denoted Im⁻, ImH, or ImHH⁺, depending on its state of protonation. XH–X⁻ denotes a closely lying proton donor to ImH, possibly a water cluster. State 1 is the oxidized binuclear site (**O**), state 2 the half-reduced, and state 3 the reduced form (**R**). Number 4 denotes the oxy compound (**A**) and 5 the P_R intermediate. Number 6 is the ferryl intermediate (**F**), which after receiving the fourth electron turns into 7. Number 8 is a transition state and 9 is the hydroxy intermediate (**H**). For color representation of this figure, see color tip facing page 206.

1.4 on the rate of $\mathbf{P_R}$ formation, and we have found an effect of 2.0 (M. Wikström *et al.*, unpublished data). Even though this reaction is not accompanied by net proton uptake (Oliveberg *et al.*, 1991), we suggest that it is associated with proton transfer to a histidine ligand of Cu_B and dissociation of ImH_2^+ from the copper (state 5). $\mathbf{P_R}$ is a key state where the dissociated ImH_2^+ ligand is stabilized by an electrostatic nonbonding interaction with the oxygenous ligand on the iron (Morgan *et al.*, 1994). The peroxy ligand on Fe_{a3} might bridge to the copper (state 5, Fig. 6).

The imidazole residue receives the proton locally (from XH, Fig. 6) and ImH_2^+ remains near the copper. If the proton were taken from the D channel proper, and/or if the ImH_2^+ would move to its output position, we would have expected an electrogenic response. Since electron transfer between the hemes associated with the formation of $\mathbf{P_R}$ appears to take place along the membrane plane (Fig. 2), there is no obvious way to charge compensate a movement of ImH_2^+ to an output position in this reaction step.

4.6. Discharge of the P_R State

$\mathbf{P_R}$ has the same overall number of electrons as the ferryl state ($\mathbf{F}$) into which it relaxes with translocation of protons (states $5 \rightarrow 6$, Fig. 6). In order for this reaction to work as a proton pump, the ImH_2^+ must first move to a proton output position, swinging out from its initial location near the copper. We suggest that this first step is rate-limiting for the $\mathbf{P_R} \rightarrow \mathbf{F}$ transition, so that $\mathbf{P_R}$ will be occupied experimentally. The major reason why the reaction can proceed only via the output conformation of ImH_2^+ is that protonation of the oxygenous ligand is impossible, or very slow, in the absence of connectivity to a proton-uptake network. This connectivity is established only when ImH_2^+ has moved away. However, if normal proton flux has been impeded, e.g., by mutations in the D channel, the requirement of protons to complete the oxygen chemistry may have to be satisfied by proton transfer from ImH_2^+, which will "decouple" the pump. The probability (rate constant) of such erroneous proton transfer appears to be much higher in the bo_3 enzyme from *E. coli* (Thomas *et al.*, 1993; García-Horsman *et al.*, 1995) than in cytochrome aa_3 from *Rhodobacter* (Fetter *et al.*, 1995), because analogous mutations in the D channel inhibit enzyme turnover much more efficiently in the latter case.

The D channel should serve the histidine ligand with protons, which may occur via the residue E242 and the waters linking the two (Fig. 5). This proposal is supported by the finding that while the E242C mutant retains reasonable turnover in the *E. coli* enzyme, proton pumping is abolished (M. Verkhovskaya *et al.*, unpublished observations).

In order for the H291 ligand to become protonated in the $\mathbf{P_R}$ state without this step being electrogenic, the proton should be taken initially from water molecules close to H291 (XH, Fig. 6; compare with Fig. 5). The coordination change at copper and the swing out of H291 into its output position now bring this water network into H-bonding contact to the distal oxygenous ligand on the iron and might include the water molecule that was formed on reduction of Fe_{a3} (state $2 \rightarrow 3$). These events initiate protonation of the oxygen ligand via the D channel. Thus, the access of protons is gated to either the histidine or to the oxygen ligand.

As soon as the ImH_2^+ output state of $\mathbf{P_R}$ is reached, the $O-O$ bond is broken by fast protonation of the peroxy ligand and electron transfer from iron and copper. These events lead to changes at the binuclear site that initiate back movement of the imidazolate anion to bind Cu_B, leaving two protons in the output domain (see Section 4.7).

The entire reaction converting $\mathbf{P_R}$ into $\mathbf{F}$ has a time constant of about 200 μsec at RT and is associated with the first half of charge translocation across the dielectric, mainly due to proton translocation (Verkhovsky *et al.*, 1997). This finding excludes the possibility that the reaction of the fully reduced enzyme with O_2 would be partially uncoupled from proton pumping (contrast Nilsson *et al.*, 1990; Babcock and Wikström, 1992)., However, time-resolved measurements of proton release from cytochrome oxidase vesicles into the outer aqueous medium do not show a fast 200-μsec phase (Nilsson *et al.*, 1990; Oliveberg *et al.*, 1991). This discrepancy with charge translocation may be due to translocated protons dwelling close to the membrane surface for a substantial time, before they are eventually released into the aqueous medium, as Heberle and Dencher (1992) showed for bacteriorhodopsin.

4.7. The Shuttling of Imidazole

The histidine cycle postulates protonation and deprotonation of a single histidine imidazole ligand of Cu_B. The predicted water networks connecting the side chain of E242 to the D channel and to H291, respectively (Fig. 5), suggest that H291 is the key histidine (H326 in the *Paracoccus aa₃*, and H334 and *E. coli bo₃* enzymes). Interestingly, the bond between the Nε of H291 and Cu_B is out of the imidazole plane by about 30° in the bovine aa_3 crystal structure (Fig. 2), which is unusual for metal-bonded histidines. This weakens the $Cu-N$ bond, facilitating dissociation of the histidine, and may favor bonding as the Im^- anion. The crystal structure allows movement of the H291 side chain "upward" (toward the protonic output side), in an arc that is constrained sterically between heme a_3 and the highly conserved W236, to reach an output position near the two heme a_3 propionates. Since this movement need not be extensive (the ε nitrogen of H291 may move less than 2 Å only along the membrane normal), the plane of H291 remains parallel to that of W236 and at van der Waals distance, as in the crystal structure. This feature may be of special interest.

A large cluster of water molecules is predicted by the PMF approach just above the histidine in the output position (G. Hummer and M. Wikström, unpublished observations), and waters are also seen here in the crystal structure (Tsukihara *et al.*, 1996). The tightly bound Mg ion is just above this water cluster, at about 6–8 Å from the imidazole. Tsukihara *et al.* (1996) suggested that this domain might be the beginning of an output channel for water. In our view this may rather represent the output pathway for the proton pump, as also considered by Iwata *et al.* (1995). This means that the immediate surroundings of the imidazole are quite different in the output position from what they are in the input state near Cu_B.

A major apparent problem with the histidine cycle is that the imidazole side chain must take up and release two protons on each shuttle motion. In aqueous solution the pK_a values for dissociation of ImH_2^+ and ImH are about 6.4 and 14, respectively. However, these pK_a values are very sensitive to a number of parameters, e.g., the local electric potential (and the effective dielectric constant), metal ligation (Cu_B), and hydrogen bonding to adjacent amino acid or water residues. The buildup of excess negative charge at the binuclear site stabilizes the ImH_2^+ state in intermediates 5 and 8 (Fig. 6), both in its input and output positions. Movement of ImH_2^+ to the output position is the prerequisite for establishing contact between the water chain connected to the D channel and the oxygenous ligand, and thus for the chemistry to proceed. Hence, the chemistry is gated by this movement.

In the output position, ImH_2^+ may be stabilized by interactions with a propionate carboxylate of heme a_3 and the backbone carbonyl of the histidine itself. The closeness and parallel orientation of H291 and W236 may be important in this connection, because it may be expected to further stabilize the ImH_2^+ state due to a cation–π interaction between the two (Dougherty, 1996).

Deprotonation of ImH_2^+ may be initiated by the positive charge formed at Cu_B due to its oxidation to Cu_B^{2+} in steps 5 → 6 and 7 → 9 (Fig. 6). Both protons must be stripped off from the imidazolium cation when the imidazole moves in the electric field between the negative charges stabilizing ImH_2^+ and the positive charge formed at Cu_B. This field relaxes by the back movement of Im^- to bind the copper. Solvation of the released protons by the nearby water cluster in the output site and the binding of Im^- to copper should both favor this transition.

4.8. Conversion of the Ferryl State to the Oxidized Enzyme

The second cycle of proton translocation is coupled to the reduction of **F**. Input of the fourth electron into the binuclear site causes reduction of Cu_B coupled to protonation of Im^- via the D channel (state 7). A transition state follows (state 8), which initiates the subsequent events of proton translocation. Transfer of the fourth electron directly to $Fe_{a3}^{4+}=O$, rather than to Cu_B^{2+}, may not be plausible thermodynamically when the oxygen ligand on Fe_{a3} cannot be protonated (Morgan *et al.*, 1994). Electron transfer to Cu_B is also not especially favorable, so that the occupancies of states 6 and 7 (Fig. 6) may both be reasonably high. Hence, two different ferryl species may be observable experimentally, which could correspond to the 804 and 785 cm^{-1} modes observed by RR spectroscopy (Kitagawa and Ogura, 1997).

As before, the transition state (state 8, Fig. 6) is reached by proton transfer to ImH. ImH_2^+ leaves the copper, facilitating protonation of the oxygen ligand and electron transfer from Cu_B, as discussed in Section 4.7, and state 9 is formed with uptake and release of two protons.

5. EPILOGUE

The function of the heme–copper oxidases as molecular machines is only beginning to be understood. The newly gained structural information may be the end of one arduous road (Beinert, 1995), but will certainly be the beginning of at least another. We envisage new structural information at higher resolution and in different catalytic states of the enzyme in the near future. Analogously, the advance on the atomic level will be accelerated especially by RR, FTIR (Hellwig *et al.*, 1996; Lübben and Gerwert, 1996), and EXAFS spectroscopy, which will serve as important complements to the crystallographic efforts. This structural work will be accompanied by dynamic studies using both spectroscopic and time-resolved electrometric techniques. The combination of all these approaches will eventually lead to a deeper understanding of function. Consequently, the admittedly speculative model proposed here can and will soon be rigorously tested experimentally. At best, this will lead to its further modifications. But even if the model will be disproved entirely, we believe that it may still have served as a stimulus for the search for correct answers.

6. REFERENCES

Babcock, G. T., and Varotsis, C., 1993, Discrete steps in dioxygen activation—The cytochrome oxidase/O₂ reaction, *J. Bioenerg. Biomembr.* **25**:71–80.

Babcock, G. T., and Wikström, M., 1992, Oxygen activation and the conservation of energy in cell respiration, *Nature* **356**:301–309.

Bailey, J. A., James, C. A., and Woodruff, W. H., 1996, Flow-flash kinetics of O₂ binding to cytochrome *c* oxidase at elevated [O₂]: Observations using high pressure stopped flow for gaseous reactants, *Biochem. Biophys. Res. Commun.* **220**:1055–1060.

Beinert, H., 1995, Crystals and structures of cytochrome *c* oxidase—The end of an arduous road, *Curr. Biol.* **2**:781–785.

Bickar, D., Bonaventura, J., and Bonaventura, C., 1982, Cytochrome *c* oxidase binding of hydrogen peroxide, *Biochemistry* **24**:2661–2666.

Blair, D. F., Witt, S. N., and Chan, S. I., 1985, Mechanism of cytochrome c oxidase-catalyzed dioxygen reduction at low temperatures. Evidence for two intermediates at the three-electron level and entropic promotion of the bond-breaking step, *J. Am. Chem. Soc.* **107:**7389–7399.

Capaldi, R. A., 1990, Structure and function of cytochrome c oxidase, *Annu. Rev. Biochem.* **59:** 569–596.

Capitanio, N., Vygodina, T. V., Capitanio, G., Konstantinov, A. A., Nicholls, P., and Papa, S., 1997, Redox-linked protolytic reactions in soluble cytochrome c oxidase from beef heart mitochondria: Redox Bohr effects, *Biochim. Biophys. Acta* **1318:**255–265.

Castresana, J., Lübben, M., and Saraste, M., 1995, New archaebacterial genes coding for redox proteins: Implications for the evolution of aerobic metabolism, *J. Mol. Biol.* **250:**202–210.

Chamalaun, R. A. F. M., and Tager, J. M., 1969, Stoicheiometry of oxidative phosphorylation with tetramethyl-p-phenylenediamine in rat liver mitochondria, *Biochim. Biophys. Acta* **180:** 204–206.

Chan, S. I., and Li, P. M., 1990, Cytochrome c oxidase: Understanding nature's design of a proton pump, *Biochemistry* **29:**1–12.

Chance, B., Saronio, C., and Leigh, J. S., Jr., 1975, Functional intermediates in the reaction of membrane-bound cytochrome oxidase with oxygen, *J. Biol. Chem.* **250:**9226–9237.

Cline, J., Reinhammar, B., Jensen, P., Venters, R., and Hoffman, B. M., 1983, Coordination environment for the type 3 copper center of tree laccase and Cu_B of cytochrome c oxidase as determined by electron nuclear double resonance, *J. Biol. Chem.* **258:**5124–5128.

Cooper, C. E., 1990, The steady-state kinetics of cytochrome c oxidation by cytochrome oxidase, *Biochim. Biophys. Acta* **1017:**187–203.

de Gier, J.-W. L., Schepper, M., Reijnders, W. N. M., van Dyck, S. J., Slotboom, D. J., Warne, A., Saraste, M., Krab, K., Finel, M., Stouthamer, A. H., van Spanning, R. J. M., and van der Oost, J., 1996, Structural and functional analysis of aa_3-type and cbb_3-type cytochrome c oxidases of *Paracoccus denitrificans* reveals significant differences in proton pump design, *Mol. Microbiol.* **20:**1247–1260.

Dougherty, D. A., 1996, Cation–π interactions in chemistry and biology: A new view of benzene, Phe, Tyr, and Trp, *Science* **271:**163–168.

Erecinska, M., Wilson, D. F., Sato, N., and Nicholls, D., 1972, The energy dependence of the chemical properties of cytochrome c oxidase, *Arch. Biochem. Biophys.* **151:**188–193.

Fabian, M., and Palmer, G., 1995, The interaction of cytochrome oxidase with hydrogen peroxide: The relationship of compounds P and F, *Biochemistry* **34:**13802–13810.

Fann, Y. C., Ahmed, I., Blackburn, N. J., Boswell, J. S., Verkovskaya, M. L., Hoffman, B. M., and Wikström, M., 1995, Structure of Cu_B in the binuclear heme–copper center of the cytochrome aa_3-type quinol oxidase from *Bacillus subtilis*: An ENDOR and EXAFS study, *Biochemistry* **34:**10245–10255.

Fetter, J. R., Qian, J., Shapleigh, J., Thomas, J. W., García-Horsman, J. A., Schmidt, E., Hosler, J., Babcock, G. T., Gennis, R. B., and Ferguson-Miller, S., 1995, Possible proton relay pathways in cytochrome c oxidase, *Proc. Natl. Acad. Sci. USA* **92:**1604–1608.

Fox, S., Nanthakumar, A., Wikström, M., Karlin, K. D., and Blackburn, N. J., 1996, XAS structural comparisons of reversibly interconvertible oxo- and hydroxo-bridged heme–copper oxidase model compounds, *J. Am. Chem. Soc.* **118:**24–34.

García-Horsman, J. A., Puustinen, A., Gennis, R. B., and Wikström, M., 1995, Proton transfer in cytochrome bo_3 ubiquinol oxidase of *Escherichia coli*: Second-site mutations in subunit I that restore proton pumping in the mutant Asp135 → Asn, *Biochemistry* **34:**4428–4433.

Gennis, B., 1992, Site-directed mutagenesis studies on subunit I of the aa_3-type cytochrome c oxidase of *Rhodobacter sphaeroides*: A brief review of progress to date, *Biochim. Biophys. Acta* **1101:**184–187.

Hallèn, S., and Nilsson, T., 1992, Proton transfer during the reaction between fully reduced cytochrome *c* oxidase and dioxygen: pH and deuterium isotope effects, *Biochemistry* **31**:11853–11859.

Haltia, T., and Wikström, M., 1992, Cytochrome oxidase: Notes on structure and mechanism, in: *Molecular Mechanisms in Bioenergetics* (L. Ernster, ed.), pp. 217–239, Elsevier, Amsterdam.

Haltia, T., Finel, M., Harms, N., Nakari, T., Raitio, M., Wikström, M., and Saraste, M., 1989, Deletion of the gene for subunit III leads to defective assembly of bacterial cytochrome oxidase, *EMBO J.* **8**:3571–3579.

Han, S., Ching, Y.-C., and Rousseau, D. L., 1990a, Cytochrome *c* oxidase: Decay of the primary oxygen intermediate involves direct electron transfer from cytochrome *a*, *Proc. Natl. Acad. Sci. USA* **87**:8408–8412.

Han, S., Ching, Y. C., and Rousseau, D. L., 1990b, Ferryl and hydroxy intermediates in the reaction of oxygen with reduced cytochrome *c* oxidase, *Nature* **348**:89–90.

Hansson, O., Karlsson, B., Aasa, R., Vänngård, T., and Malmström, B. G., 1982, The structure of the paramagnetic oxygen intermediate in the cytochrome *c* oxidase reaction, *EMBO J.* **1**:1295–1297.

Heberle, J., and Dencher, N. A., 1992, Surface-bound optical probes monitor protein translocation and surface potential changes during the bacteriorhodopsin photocycle, *Proc. Natl. Acad. Sci. USA* **89**:5996–6000.

Hellwig, P., Rost, B., Kaiser, U., Ostermeier, C., Michel, H., and Mäntele, W., 1996, Carboxyl group protonation upon reduction of the *Paracoccus denitrificans* cytochrome *c* oxidase: Direct evidence by FTIR spectroscopy, *FEBS Lett.* **385**:53–57.

Hill, B. C., 1991, The reaction of the electrostatic cytochrome *c*–cytochrome oxidase complex with oxygen, *J. Biol. Chem.* **266**:2219–2226.

Hirota, S., Mogi, T., Ogura, T., Hirano, T., Anraku, Y., and Kitagawa, T., 1994, Observation of the $Fe-O_2$ and $FeIV=O$ stretching Raman bands for dioxygen reduction intermediates of cytochrome *bo* isolated from *Escherichia coli*, *FEBS Lett.* **352**:67–70.

Hosler, J. P., Ferguson-Miller, S., Calhoun, M. W., Thomas, J. W., Hill, J., Lemieux, L., Ma, J., Georgiou, C., Fetter, J., Shapleigh, J., Tecklenburg, M. M. J., Babcock, G. T., and Gennis, R. B., 1993, Insight into the active-site structure and function of cytochrome oxidase by analysis of site-directed mutants of bacterial cytochrome aa_3 and cytochrome *bo*, *J. Bioenerg. Biomembr.* **25**:121–136.

Hummer, G., García, A. E., and Soumpasis, D. M., 1995, Hydration of nucleic acid fragments: Comparison of theory and experiment for high-resolution crystal structures of RNA, DNA, and DNA–drug complexes, *Biophys. J.* **68**:1639–1652.

Hummer, G., García, A. E., and Soumpasis, D. M., 1996, A statistical mechanical description of biomolecular hydration, *Faraday Disc.* **103**:175–189.

Iwata, S., Ostermeier, C., Ludwig, B., and Michel, H., 1995, Structure at 2.8 Å resolution of cytochrome *c* oxidase from *Paracoccus denitrificans*, *Nature* **376**:660–669.

Kadenbach, B., Kuhn-Nentwig, L., and Buge, U., 1987, Evolution of a regulatory enzyme: Cytochrome *c* oxidase (complex IV), *Curr. Top. Bioenerg.* **15**:113–161.

Kadenbach, B., Stroh, A., Hüther, F. J., Reimann, A., and Steverding, D., 1991, Evolutionary aspects of cytochrome *c* oxidase, *J. Bioenerg. Biomembr.* **23**:321–334.

Kitagawa, T., and Ogura, T., 1997, Oxygen activation mechanism at the binuclear site of heme–copper oxidase superfamily as revealed by time-resolved resonance Raman spectroscopy, *Progr. Inorg. Chem.* **45**:431–479.

Konstantinov, A. A., Capitanio, N., Vygodina, T. V., and Papa, S., 1992, pH changes associated with cytochrome *c* oxidase reaction with H_2O_2. Protonation state of the peroxy and oxoferryl intermediates, *FEBS Lett.* **312**:71–74.

Lübben, M., and Gerwert, K., 1996, Redox FTIR difference spectroscopy using caged electrons reveals contributions of carboxyl groups to the catalytic mechanism of heme–copper oxidases, *FEBS Lett.* **397**:303–307.

Malmström, B. G., 1990a, Cytochrome *c* oxidase as a redox-linked proton pump, *Chem. Rev.* **90**: 1247–1260.

Malmström, B. G., 1990b, Cytochrome oxidase: Some unsolved problems and controversial issues, *Arch. Biochem. Biophys.* **280**:233–241.

Mitchell, P., 1966, *Chemiosmotic Coupling in Oxidative and Photosynthetic Phosphorylation*, Glynn Research Ltd., Bodmin, United Kingdom.

Mitchell, R., and Rich, P. R., 1994, Proton uptake by cytochrome *c* oxidase on reduction and on ligand binding, *Biochim. Biophys. Acta* **1186**:19–26.

Mitchell, R., Mitchell, P., and Rich, P. R., 1992, Protonation states of the catalytic intermediates of cytochrome *c* oxidase, *Biochim. Biophys. Acta* **1101**:188–191.

Morgan, J. E., Verkhovsky, M. I., and Wikström, M., 1994, The histidine cycle: A new model for proton translocation in the respiratory heme–copper oxidases, *J. Bioenerg. Biomembr.* **26**: 599–608.

Morgan, J. E., Verkhovsky, M. I., and Wikström, M., 1995, Identification of a "peroxy" intermediate in cytochrome bo_3 of *Escherichia coli*, *Biochemistry* **34**:15633–15637.

Morgan, J. E., Verkhovsky, M. I., and Wikström, M., 1996, Observation and assigment of peroxy and ferryl intermediates in a reduction of dioxygen to water by cytochrome *c* oxidase, *Biochemistry* **35**:12235–12240.

Nilsson, T., Hallén, S., and Oliveberg, M., 1990, Rapid proton release during flash-induced oxidation of cytochrome *c* oxidase, *FEBS Lett.* **260**:45–47.

Ogura, T., Takahashi, S., Shinzawa-Itoh, K., Yoshikawa, S., and Kitagawa, T., 1990, Observation of the $Fe^{4+}=O$ stretching Raman band for cytochrome oxidase compound F at ambient temperature, *J. Biol. Chem.* **265**:14721–14723.

Oliveberg, M., Hallén, S., and Nilsson, T., 1991 Uptake and release of protons during the reaction between cytochrome *c* oxidase and molecular oxygen: A flow-flash investigation, *Biochemistry* **30**:436–440.

Puustinen, A., Finel, M., Virkki, M., and Wikström, M., 1989, Cytochrome *o* (*bo*) is a proton pump in *Paracoccus denitrificans* and *Escherichia coli*, *FEBS Lett.* **249**:163–169.

Puustinen, A., Morgan, J. E., Verkhovsky, M., Thomas, J. W., Gennis, R. B., and Wikström, M., 1992, The low spin heme site of cytochrome *o* from *Escherichia coli* is promiscuous with respect to heme type, *Biochemistry* **31**:10363–10368.

Raitio, M., and Wikström, M., 1994, An alternative cytochrome oxidase of *Paracoccus denitrificans* functions as a proton pump, *Biochim. Biophys. Acta* **1186**:100–106.

Rich, P. R., 1991, The osmochemistry of electron-transfer complexes, *Biosci. Rep.* **11**:539–571.

Rich, P. R., 1995, Towards an understanding of the chemistry of oxigen reduction and proton translocation in the iron–copper respiratory oxidases, *Aust. J. Plant. Physiol.* **22**:479–486.

Riistama, S., Puustinen, A., García-Horsman, A., Iwata, S., Michel, H., and Wikström, M., 1996, Channeling of dioxygen into the respiratory enzyme, *Biochim. Biophys. Acta* **1275**:1–4.

Saraste, M., 1990, Structural features of cytochrome oxidase, *Q. Rev. Biophys.* **23**:331–366.

Saraste, M., Holm, L., Lemieux, L., Lübben, M., and van der Oost, J., 1991, The happy family of cytochrome oxidases, *Biochem. Soc. Trans.* **19**:608–612.

Saraste, M., Castresana, J., Higgins, D., Lübben, M., and Wilmanns, M., 1996, Evolution of cytochrome oxidase, in: *Origin and Evolution of Biological Energy Conservation*, (H. Baltscheffsky, ed.), pp. 255–289, VCH Publishers, New York.

Siletsky, S. A., Kaulen, A. D., Mitchell, D., Gennis, R. B., and Konstantinov, A. A., 1996, Resolution of two proton conducting pathways in cytochrome *c* oxidase, *EBEC Short Rep.* **9**:90.

Slater, E. C., 1966, Oxidative phosphorylation, in: *Comprehensive Biochemistry*, vol. 14 (M. Florkin and E. H. Stotz, eds.), pp. 327–396, Elsevier, Amsterdam.

Solioz, M., Carafoli, E., and Ludwig, B., 1982, The cytochrome *c* oxidase of *Paracoccus denitrificans* pumps protons in a reconstituted system, *J. Biol. Chem.* **257:**1579–1582.

Sucheta, A., Georgiadis, K. E., and Einarsdottir, Ó., 1997, Mechanism of cytochrome *c* oxidase-catalyzed reduction of dioxygen to water: Evidence for peroxy and ferryl intermediates at room temperature, **36:**554–565.

Thomas, J. W., Puustinen, A., Alben, J. O., Gennis, R. B., and Wikström, M., 1993, Aspartate-135 in subunit I of cytochrome *bo* from *Escherichia coli* may be involved in proton-pumping, *Biochemistry* **32:**10923–10928.

Tsukihara, T., Aoyama, H., Yamashita, E., Tomizaki, T., Yamaguchi, H., Shinzawa-Itoh, K., Nakashima, R., Yaono, R., and Yoshikawa, S., 1995, Structure of metal sites of oxidized bovine heart cytochrome *c* oxidase at 2.8 Å, *Science* **269:**1069–1074.

Tsukihara, T., Aoyama, H., Yamashita, E., Tomizaki, T., Tamaguchi, H., Shinzawa-Itoh, K., Nakashima, R., Yaono, R., and Yoshikawa, S., 1996, The whole structure of the 13-subunit oxidized cytochrome c oxidase at 2.8 Å, *Science* **272:**1136–1144.

van Gelder, B. F., and Beinert, H., 1969, Studies of the heme components of cytochrome *c* oxidase by EPR spectroscopy, *Biochim. Biophys. Acta* **189:**1–24.

van Verseveld, H. W., Krab, K., and Stouthamer, A. H., 1981, Proton pump coupled to cytochrome *c* oxidase in *Paracoccus denitrificans*, *Biochim. Biophys. Acta* **635:**525–534.

Varotsis, C., and Babcock, G. T., 1990, Nanosecond time-resolved resonance Raman spectroscopy, *Biochemistry* **29:**7357–7362.

Verkhovskaya, M., Verkhovsky, M., and Wikström, M., 1992, pH-dependence of proton translocation in cells from *Escherichia coli*, *J. Biol. Chem.* **267:**14559–14562.

Verkovsky, M. I., Morgan, J. E., and Wikström, M., 1994, Oxygen binding and activation: Early steps in the reaction of oxygen with cytochrome *c* oxidase, *Biochemistry* **33:**3079–3086.

Verkhovsky, M. I., Morgan, J. E., and Wikström, M., 1995, Control of electron delivery to the oxygen reduction site of cytochrome *c* oxidase: A role for protons, *Biochemistry* **34:**7483–7491.

Verkhovsky, M. I., Morgan, J. E., Puustinen, A., and Wikström, M., 1996a, The "ferrous-oxy" intermediate in the reaction of dioxygen with fully reduced cytochromes aa_3 and bo_3, *Biochemistry* **35:**16241–16246.

Verkhovsky, M. I., Morgan, J. E., Puustinen, A., and Wikström, M., 1996b, Kinetic trapping of oxygen in cell respiration, *Nature* **380:**268–270.

Verkovsky, M. I., Morgan, J. E., and Wikström, M., 1996c, Redox transitions between oxygen intermediates in cytochrome *c* oxidase, **93:**12235–12239.

Verkovsky, M. I., Morgan, J. E., Verkhovskaya, M. L., and Wikström, M., 1997, Translocation of electrical charge during a single turnover of cytochrome *c* oxidase, *Biochim. Biophys. Acta* **1318:**6–10.

Wikström, M., 1977, Proton pump coupled to cytochrome *c* oxidase in mitochondria, *Nature* **266:** 271–273.

Wikström, M., 1981, Energy-dependent reversal of the cytochrome oxidase reaction, *Proc. Natl. Acad. Sci. USA* **78:**4051–4054.

Wikström, M., 1989, Identification of the electron transfers in cytochrome oxidase that are coupled to proton-pumping, *Nature* **338:**776–778.

Wikström, M., Krab, K., and Saraste, M., 1981, *Cytochrome Oxidase—A Synthesis*, Academic Press, London.

Wilmanns, M., Lappalainen, P., Kelly, M., Sauer-Eriksoon, E., and Saraste, M., 1995, Crystal structure of the membrane-exposed domain from a respiratory quinol oxidase complex with an engineered dinuclear copper center, *Proc. Natl. Acad. Sci. USA* **92:**11955–11959.

Wilson, M. T., and Bickar, D., 1991, Cytochrome oxidase as a proton pump, *J. Bioenerg. Biomembr.* **23:**755–771.

Woodruff, W. H., 1993, Coordination dynamics of heme–copper oxidases. The ligand shuttle and the control and coupling of electron transfer and proton translocation, *J. Bioenerg. Biomembr.* **25:**177–187.

CHAPTER 10

Transient Spectroscopy of the Reaction between Cytochrome *c* Oxidase and Nitric Oxide

A New Hypothesis on the Mechanism of Inhibition and Oxygen Competition

Alessandro Giuffrè, Paolo Sarti, Emilio D'Itri, Gerhard Buse, Tewfik Soulimane, and Maurizio Brunori

1. INTRODUCTION

It is of great interest, though somewhat puzzling, that nitric oxide (NO) can play simultaneously the dual role of messenger (Moncada *et al.*, 1991), and inhibitor (Bredt and Snider, 1994; Nathan, 1992). The balance between these two functions will depend on a number of parameters, ranging from the cell–

Alessandro Giuffrè, Paolo Sarti, Emilio D'Itri, and Maurizio Brunori • Department of Biochemical Sciences and CNR Center of Molecular Biology, University of Roma "La Sapienza," 00185 Rome, Italy. **Gerhard Buse and Tewfik Soulimane** • Institut für Biochemie, Rheinisch-Westfälische Technische Hochschule, 52057 Aachen, Germany.

Frontiers of Cellular Bioenergetics, edited by Papa *et al.* Kluwer Academic/Plenum Publishers, New York, 1999.

tissues pathophysiological state to the relative concentration *in vivo* of O_2 and NO.

The prime role of NO as a messenger is attributed to activation of guanylyl cyclase (Moncada and Higgs, 1993; Moncada *et al.*, 1991). This enzyme, which controls many cellular functions by cyclic GMP protein–kinase-dependent reactions, is activated upon binding of NO to its heme. The mechanisms of NO toxicity are largely unknown, and several biological macromolecules (including some mitochondrial enzymes) have been proposed as possible targets. Cleeter *et al.* (1994) first reported the reversible inhibition of respiration by NO, and proposed that binding of endogenous NO to cytochrome *c* oxidase may play a role in some neurodegenerative diseases such as Parkinson's disease; thus, understanding the mechanism by which NO inhibits cell respiration via cytochrome *c* oxidase acquires biochemical as well as medical interest.

In a recent paper, Brown (1995) discussed the quantitative aspect of regulation of cell respiration by endogenous NO. By raising the points (1) is the control of cytochrome oxidase relevant to a substantial modulation of cell respiration and (2) are the levels of NO synthesized *in vivo* compatible with the inhibition–modulation of this enzyme, he came to the conclusion that NO could control respiration by inhibiting cytochrome oxidase. Thus, intracellular 0.1–1.0 μM NO, in the presence of 25–50 μM O_2 (average concentration levels of the two gases found *in vivo*) (Brown, 1995), may lead to a substantial inhibition of cytochrome oxidase and thereby of respiration.

Brown and Cooper (1994), working with brain synaptosomes, found that NO is a powerful inhibitor of cytochrome *c* oxidase ($K_i = 270$ nM at $[O_2] = 140$ μM), yet inhibition is quickly reversible in the presence of O_2. These observations, which have been reproduced in other laboratories, form the basis of a new hypothesis suggesting that pathophysiology of NO basically resides on its ability to inhibit cytochrome oxidase (Brown, 1995).

Interestingly, Brown presented both direct and indirect evidence, all converging toward the NO-dependent control of cytochrome oxidase and thereby of ATP synthesis. It may be difficult to attribute to the same interaction (NO–oxidase) the dual role of a (potent) inhibitor and a (mild) modulator, particularly if the K_i is in the submicromolar concentration range. In this regard, Brown (1995) speculates that NO inhibition of cytochrome oxidase can be reasonably thought of as a self-limiting interaction linked to an increase in the local concentration of oxygen following inhibition of oxidase, and thereby destruction of NO by direct reaction in bulk or *in situ* (at the binding site of the enzyme). Convincing though indirect evidence that *in vivo* a fraction of oxidase is inhibited by NO is that the apparent K_m for oxygen measured in intact living tissues–cells (> 1 μM) is greater than that determined *in vitro* (< 1 μM) (Brown, 1995).

In summary, the ability of NO to compete very efficiently with O_2 and the rapid reversal of inhibition are the crucial features of the hypothesis suggesting cytochrome oxidase as a prime NO target.

These observations are not easily understood on the basis of what is known about the kinetics and thermodynamics of the reaction of these two gases with the reduced cytochrome a_3–Cu_B center (Malatesta *et al.*, 1995; Blackmore *et al.*, 1991). Binding of NO to fully reduced cytochrome c oxidase was shown to involve formation of a complex with the Fe^{2+} of cytochrome a_3 (Blackmore *et al.*, 1991; Gibson and Greenwood, 1963) with a second-order rate constant of 0.4–1×10^8 M^{-1} sec^{-1}. NO binding to the oxidized enzyme was used to break the magnetic coupling between cytochrome a_3^{3+} and Cu_B^{2+} in the binuclear center, with detection of the electron paramagnetic resonance (EPR) signal of cytochrome a_3 (Stevens *et al.*, 1979). Stopped-flow experiments carried out with the solubilized enzyme indicated that inhibited oxidase is a complex of NO with reduced cytochrome a_3 and postulated that Cu_B is crucial to the mechanism of NO inhibition (Torres *et al.*, 1995).

In this chapter, we report rapid mixing experiments that provide a clue to the mechanism of inhibition of beef heart cytochrome c oxidase by NO and its reversal in the presence of O_2. Using a sequential-mixing stopped-flow, we have shown that inhibition by NO demands (at least) one turnover; moreover, the onset of inhibition by NO as well as its release are both sensitive to the O_2/NO ratio. Photodiode array stopped-flow measurements allowed us to observe the spectrum of the species inhibited by NO under turnover conditions and the decay of the nitrosyl derivative into a NO-free species, depending on the O_2/NO concentration ratio. We have discovered that displacement of NO in the presence of O_2 occurs with a rate constant, $k = 0.13$ sec^{-1}, much faster than estimated for other hemeproteins assigned to the dissociation of NO from reduced cytochrome a_3.

2. MATERIALS AND EXPERIMENTAL DESIGNS

2.1. General Information

All the experiments were carried out using cytochrome c oxidase from beef heart, purified according to Soulimane and Buse (1995). The purification protocol yields an enzyme that is homogeneously "fast." All throughout the chapter, the concentration of cytochrome oxidase is expressed as that of the functional unit (cytochrome aa_3.

Lauryl maltoside was from Biomol (Hamburg, Germany). Horse heart cytochrome c (type VI) and ascorbate were from Sigma Chemical Co. (St. Louis, MO). Ruthenium hexamine was from Aldrich (Milwaukee, WI).

Reduced cytochrome c was freshly prepared before each experiment using dithionite; excess reductant was removed by gel filtration. Cytochrome *c* concentration was determined spectrophotometrically using the extinction coefficient (reduced-oxidized) $\varepsilon = 19$ mM^{-1} cm^{-1} at 550 nm.

NO (gas) was from Air Liquide (France). Stock solutions of NO were freshly prepared before use by equilibrating with NO gas in a volume (typically 10 ml) of exhaustively degassed buffer; equilibration was carried out using a NO (gas)-containing 2 liters tonometer. From solubility data, the concentration of NO in solution was taken to be 2 mM at 20°C. It should be kept in mind, however, that the actual (nominal) NO concentration in the experiments can be overestimated since it may drop with manipulations, due to contaminating O_2. Oxygen is known to react with NO to generate NO_2^- following the stoichiometry (Goldstein and Czapski, 1995):

$$4 \; NO + O_2 + 2 \; H_2O \rightarrow 4 \; NO_2^- + 4 \; H^+$$

The rate law for this complex reaction in water is second order in NO and first order in O_2, and the calculated rate constant is $k = 2.3 \times 10^6$ M^{-2} sec^{-1} (Goldstein and Czapski, 1995).

In some experiments, the NO dissociation rate from reduced cytochrome oxidase has been measured directly, using deoxyhemoglobin as NO scavenger. Owing to the high reactivity of NO, however, we have checked the NO stability independently under experimental conditions similar to those used to measure its rate of dissociation. By photodiode array spectroscopy and using deoxyhemoglobin as NO indicator, we have shown that in the presence of ascorbate and ruthenium hexamine the concentration of NO does not change over 1 hr. On the contrary and in agreement with the literature (Brudwig *et al.*, 1980), 200 μM NO in buffered solution, upon addition of 50 mM dithionite, rapidly ($t \ll 50$ sec) disappears from solution.

When degassed oxidized oxidase is mixed with a solution of NO previously incubated with excess dithionite (a few tens of seconds), the enzyme is initially reduced ($t_{1/2} \approx 1$ sec), and is thereafter very slowly (many minutes to half an hour) nitrosylated. This observation confirms that NO reacts with dithionite and that some product of this reaction, when incubated with dithionite reduced oxidase, very slowly regenerates the nitrosylated enzyme (as already reported by Yonetani *et al.* [1972]).

2.2. The Photodiode Array Experiments

These are rapid mixing stopped-flow experiments carried out with a Durrum-Gibson instrument equipped with a photodiode array detector (TN6500, Tracor Northern, Madison, WI). The diode array stopped-flow spectrometer can acquire with 3 msec dead time up to 80 spectra of 1024 elements as a function of time. It is equipped with a 2-cm observation chamber. The acquisition time of a single spectrum depends on the range of wavelengths selected. In all our experiments, the acquisition time was 10 msec.

2.3. The Sequential Mixing Experiments

The sequential-mixing stopped-flow spectrophotometer from Applied Photophysics (DX.17MV, Applied Photophysics, Leatherhead, England) is a single wavelength instrument that allows one to carry out in sequence two 1:1 mixing events separated by a delay time whose duration can be fixed, from 10 msec to several tens of minutes. We have used the sequential mixing device to first mix cytochrome oxidase with NO, and then, after the preset delay time, mix the product of the first mixing with reduced cytochrome c (experimental mixing conditions specified in the text and figure legends).

Data analysis was carried out by means of the software MATLAB (Math-Works, South Natick, MA) running on an Intel 486-based computer. Time-resolved optical spectra were analyzed by singular value decomposition (SVD) according to Henry and Hofrichter (1992).

Kinetic simulations were carried out using a differential equations solver algorithm, implemented by Dr. E. Henry (NIH, Bethesda, MD), running on a Silicon Graphics Workstation.

3. RESULTS AND DISCUSSION

In order to better understand the rationale underlying the experiments herein presented, it may be worth reconsidering some recent findings on the interaction between cytochrome oxidase and NO, namely, those from Brown and Cooper (1994), Takehara et al. (1995), and Torres et al. (1995). To our knowledge, their work provides the strongest evidence supporting the original finding of Cleeter et al. (1994) that inhibition of cytochrome oxidase by NO and its reversibility depend on oxygen concentration.

3.1. The Oxygen-Dependent Reversibility of Cytochrome Oxidase Inhibition by NO

Brown and Cooper (1994) measured oxygraphically the respiratory activity of both synaptosomes and purified cytochrome oxidase in the presence and absence of NO. The experiments were such that NO, either produced by light-activated degradation of nitroprusside in the reaction medium or supplemented as a gas in solution, was added to the respiring system when some 20–40% oxygen was already consumed. The experimental setup allowed them to measure simultaneously the O_2 and NO concentrations. The polarographic profiles

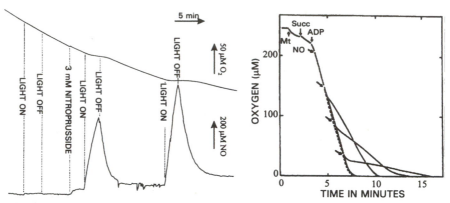

FIGURE 1. NO inibition and its reversal in the presence of oxygen: (Left panel) Light-activated NO release from nitroprusside causes reversible inhibition of synaptosomal respiration. Recordings indicate O_2 concentration (upper trace) and NO concentration (lower trace) determined by specific electrodes. From Brown and Cooper (1994) with permission. (Right panel) Oxygen-dependent inhibition of phosphorylating mitochondrial respiration by NO ≈ 0.8 μM, added to the incubation medium at different oxygen concentrations as indicated by the arrows. Mt, 1 mg mitochondrial protein; Succ, 5 mM succinate; ADP, 600 μM. Modified from Takehara *et al.* (1995).

of respiring synaptosomes presented by Brown and Cooper (1994) are shown in Fig. 1 (left panel). The data show that (1) a single pulse of micromolar NO strongly inhibits respiration when oxygen is in the range 100 to 200 μM, concentration values far above those usually detected in tissues/cells ($[O_2]$ ≈ some μM), and (2) inhibition is under those conditions removed within about 60–80 sec, at least judging from the time course of NO disappearance from solution. The inhibition of respiration is O_2-concentration dependent as proved by the K_i values of 270 and 60 nM NO estimated at approximately 150 and 30 μM O_2, respectively (Brown and Cooper, 1994).

These findings are consistent with those reported by Takehara *et al.* (1995) and by Torres *et al.* (1995). Following oxygraphically the respiration of mitochondria, Takehara *et al.* (1995) observed recovery of respiration after complete inhibition induced by the addition of NO, provided oxygen was present in excess over NO. As shown in Fig. 1 (right panel), the degree of inhibition induced by the addition of a constant submicromolar amount of NO, and estimated as a decrease in slope of oxygen consumption, was greater at lower O_2, i.e. on increasing the NO/O_2 ratio. On lowering the oxygen tension, the time required to recover activity was also significantly longer (see Fig. 1).

Using soluble cytochrome oxidase and monitoring under turnover conditions the formation and degradation of the nitrosyl derivative as a function of time, Torres *et al.* (1995) also observed oxygen concentration-dependent reversal of inhibition. In their hands, oxidase activity was about 50% inhibited at 1.3 μM NO, when O_2 concentration was about 50 μM.

These results show that inhibition of cytochrome oxidase by NO, though highly effective, is sensitive to the concentration of oxygen and activity can be quickly recovered.

In this context, we have carried out experiments designed to provide a clue to the mechanism of inhibition of mammalian cytochrome oxidase and its reversal in the presence of oxygen. In particular, we have investigated the onset of enzymatic inhibition observed by following the oxidation of cytochrome *c* oxidation, thus also collecting information on the minimum number of electrons necessary to bind NO.

3.2. The Onset of Inhibition

The experiment was designed as follows: Degassed oxidized oxidase is first mixed anaerobically with degassed buffer containing or not containing NO. After a prefixed delay, the first mixing product is mixed with an air-equilibrated solution of cytochrome c^{2+} (second mixing). As shown in Fig. 2 (trace 1), in the absence of NO, the oxidation of cytochrome c^{2+} is complete in about 1 sec following a monoexponential decay ($k = 2.3 \text{ sec}^{-1}$). In the presence of NO, at various concentrations from 30 to 250 μM, cytochrome *c* is oxidized biphasically. Moreover, the total amount of cytochrome *c* oxidized at constant O_2 decreases progressively with the increase of NO concentration (see Fig. 2). This finding confirms that inhibition depends on the NO/O_2 ratio.

The kinetic analysis of the time courses of cytochrome *c* oxidation observed at the highest NO–O_2 ratio shows that the fast phase ($k_f = 55 \text{ sec}^{-1}$) and the slow phase ($k_s = 6.3 \text{ sec}^{-1}$) together account for the oxidation of approximately 3 mole of cytochrome *c* per mole of aa_3, half delivered to oxidase in the burst phase ($k = 7 \times 10^6 \text{ M}^{-1} \text{ sec}^{-1}$) and half in the slow phase, before the reaction stops. From the data shown in Fig. 2, we may conclude that when $[O_2]$ = 125 μM and [NO] = 140 μM, inhibition is set within about 0.5 sec, when the amount of cytochrome *c* oxidized normalized to cytochrome oxidase accounts for maximally one turnover. Thus, at the highest NO concentrations, the onset of inhibition is complete within the first turnover, and catalysis is totally prevented. The onset of inhibition was also found to be sensitive to the NO/O_2 ratio, a finding consistent with the dependence on O_2 concentration of the extent of inhibition of respiration, previously found by others (Takehara *et al.*, 1995; Brown and Cooper, 1994).

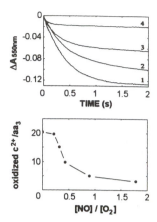

FIGURE 2. Inhibition of cytochrome c^{2+} oxidation by NO. Double mixing experiments. First mixing: degassed cytochrome oxidase, 1.5 μM (aa_3), is mixed with degassed buffer $\pm$ NO; second mixing: previous mixture is mixed with cytochrome c^{2+}, 15 μM, in air-equilibrated buffer. O_2, 140 μM (final concentration in the observation chamber). (Top panel) Time courses of cytochrome c^{2+} oxidation. NO (final concentration in the observation chamber) = 0 μM (1); 45 μM (2); 60 μM (3); and 250 μM (4). (Bottom panel) Dependence on the NO/O_2 ratio of the number of equivalents of cytochrome c oxidized, normalized for the concentration of cytochrome aa_3. Experimental conditions: Cytochrome oxidase in 0.1 M K–phosphate pH 7.3 + 0.1% lauryl maltoside at 20°C. Wavelength, 550 nm. Light path, 1 cm. Delay time, 10 msec.

In order to investigate a (possible) dependence of enzymatic activity on the anaerobic incubation of oxidized cytochrome oxidase with NO (first mixing), the delay time before the second mixing was varied systematically from milliseconds to several minutes. The observed time courses (not shown) were almost superimposable, indicating that anaerobic preincubation of the oxidized enzyme with NO is irrelevant with respect to inhibition, which therefore requires electron flux through the enzyme, i.e., turnover.

3.3. The Spectral Intermediates Populated during Turnover and Nitrosylation

Cytochrome oxidase in air was mixed with an anaerobic solution of ascorbate and ruthenium hexamine, to sustain turnover, in the presence and absence of NO. Under these conditions, one may predict that the reduction of the enzyme will be associated with binding of either O_2 or NO. As demanded by the results of the sequential mixing experiment, at some point, and indeed in the presence of NO, inhibition of cytochrome oxidase activity by complexation with NO will occur. Since inhibition by NO is known to be reversible (Brown and Cooper, 1994; Cleeter *et al.*, 1994), it is expected that under excess O_2 over NO, enzymatic activity will be eventually restored. Figure 3 shows the relevant optical species detected as a function of time (panels labeled as A, B, C), and the relative time courses at 444 *minus* 433 nm (panels labeled as a, b, c). As shown in the figure, three different experiments were carried out. Oxidized degassed oxidase was mixed with reductants in the presence of (1) 140 μM O_2 without NO (panels A, a); (2) 140 μM O_2 and 50 μM NO (panels B, b), and (3) without O_2, NO 50 μM (panels C, c). Cytochrome oxidase was initially oxidized and under the above conditions decayed in about 0.1 sec to a first intermediate (spectrum 1) displaying a peak rising at 444 nm (partial reduction of the hemes). Depending on conditions, this intermediate had the following

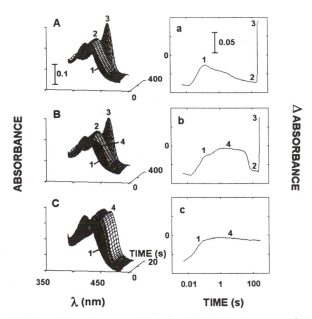

FIGURE 3. Inhibition of cytochrome oxidase by NO during turnover. Absolute spectra collected in the 380- to 480-nm region (left panels) and corresponding time courses at 444 *minus* 433 nm. Spectral changes (from 10 msec to approximately 140 sec) observed upon mixing oxidized oxidase, 1.25 μM (aa_3), with ascorbate, 20 mM, and ruthenium hexamine, 300 μM, under the following conditions: (A,a) O_2 = 140 μM in the absence of NO; (B,b) O_2 = 140 μM, and NO = 50 μM; and (C,c) NO = 50 μM in the absence of O_2. The numbers indicated at different times correspond to the species identified within the spectra reported in panels A, B, and C (see text). Other experimental conditions as in Fig. 2. Light path, 2 cm.

fate: In the presence of O_2 (panels A, a) the enzyme entered turnover and shifted to a second species (spectrum 2), with a main peak at 428 nm; after O_2 exhaustion, the fully reduced oxidase was accumulated (spectrum 3). In the presence of both NO and O_2 (panels B, b) it relaxed, $t_{1/2} = 0.5$ sec, to the nitrosylated fully reduced species, identical to that observed in the absence of O_2 (see spectrum 4); this derivative, however, was unstable and decayed itself to the turnover intermediate, i.e., the main species observed in the absence of NO (spectrum 2). Also under this condition, once oxygen was exhausted, the enzyme became fully reduced, with spectral features identical to spectrum 3 observed in the absence of NO (confront panels A, a). This finding strongly suggests that over this period of time NO has disappeared from solution most likely because of reaction with O_2 (Goldstein and Czapski, 1995). In the presence of NO without oxygen (panels C, c), the first intermediate relaxed with a $t_{1/2} = 0.2$ sec to a stable, fully reduced nitrosylated species (spectrum 4), identical to the one just described in panel B.

Thus, in the presence of excess O_2, inhibition can be reversed and the function recovered. The decay of the nitrosyl derivative into the turnover intermediate can be also investigated directly, by mixing the NO-bound reduced cytochrome oxidase with a large excess of O_2 (Giuffrè *et al.*, 1996). In this experiment, the nitrosylated fully reduced enzyme decays within a few seconds biphasically into the turnover intermediate, and the major kinetic component ($k = 0.13$ sec^{-1}) corresponds to NO dissociation from cytochrome a_3.

3.4. Determining the NO Off Rate

A NO dissociation rate constant of 0.13 sec^{-1} is much greater than typical values reported for dissociation of NO from other reduced hemeproteins (such as hemoglobin) (Antonini and Brunori, 1971). This finding poses the question whether O_2 reacts directly with NO while bound to the enzyme.

This question is crucial to the understanding of activity recovery, and thus of the regulatory role of NO at the mitochondrial level. Therefore, we have determined independently the rate constant for NO dissociation from cytochrome oxidase in the absence of O_2.

Deoxyhemoglobin in molar excess over NO was used to scavenge from solution NO dissociating from cytochrome oxidase. The overall reaction can be summarized as follows:

$$COX^{red} - NO + Hb^{deoxy} \rightarrow COX^{red} + Hb - NO$$

In this rapid-mixing photodiode array experiment, the nitrosyl derivative of reduced cytochrome oxidase is mixed under anaerobic conditions with excess deoxyhemoglobin in the presence of a small amount of dithionite. The SVD analysis allowed us to sort out the kinetics of formation of the main spectral

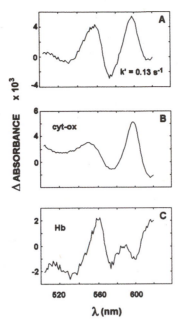

FIGURE 4. NO dissociation from reduced cytochrome c oxidase in the absence of O_2 and presence of hemoglobin. Fully reduced oxidase (1.25 μM aa_3) incubated with 5 μM NO, 20 mM ascorbate, and 300 μM ruthenium hexamine was mixed with 12.5 μM deoxyhemoglobin in the presence of a small amount of dithionite. (A) Optical component deconvoluted by SVD analysis of the raw data (not shown); this optical transition occurs at $k = 0.13$ sec^{-1}, and is contributed by the dissociation of NO from (B) reduced cytochrome a_3 and binding to (C) deoxyhemoglobin. Other experimental conditions as in Fig. 2.

components, despite the unfavorable signal-to-noise ratio. The most relevant optical component, which accumulates at $k = 0.13$ sec^{-1} (top spectrum), is shown in Fig. 4. It combines the spectral contributions of NO dissociation from reduced cytochrome a_3 (middle spectrum) and NO binding to deoxy hemoglobin (lower spectrum).

This indicates that NO dissociates from reduced cytochrome a_3 with the same rate constant, both in the presence and absence of O_2. Therefore, we also conclude that reversal of NO inhibition does not demand some chemical reaction between O_2 and bound NO, but rather it occurs via NO dissociation from the fully reduced cytochrome a_3–Cu$_B$ center.

In conclusion, time-resolved optical spectroscopy confirmed that the inhibited state of the enzyme corresponds to the fully reduced nitrosyl-derivative of cytochrome a_3, in agreement with Torres *et al.* (1995). The inhibition was found to be rapid and reversible. In the presence of O_2, the nitrosyl derivative, promptly generated, decays to a fully active turnover intermediate optically indistinguishable from the one generated at steady-state in the absence of NO. These observations account for the recovery of respiration observed by others (Takehara *et al.*, 1995; Brown and Cooper, 1994) in the presence of O_2, based on experiments in the time range of minutes.

The rate at which NO is displaced by O_2 from the cytochrome a_3^{2+}–NO adduct (with recovery of function), measured by mixing the fully reduced NO-derivative of oxidase with excess O_2, was found to be faster than expected based on comparison with other hemeproteins; for example the rate constant for NO dissociation from hemoglobin R-state is 10^{-3} to 10^{-4} s^{-1} (Antonini and Brunori, 1971). This is a new important finding consistent with the prompt release of NO inhibition in the presence of oxygen.

3.5. The Mechanism of Interaction between NO and Cytochrome Oxidase

By low-temperature EPR spectroscopy, Brudwig *et al.* (1980) showed that NO can bind to oxidized Cu_B, and proposed also that NO can be "metabolized" by oxidase. The inhibitory relevance of this binding, however, was not evident in their work. Evidence collected in the last few years indicates on the other hand that cytochrome oxidase inhibition depends on the binding of NO to the fully or half-reduced binuclear site (cytochrome a_3–Cu_B).

As far as the mechanism of NO inhibition, given that the rate constants for O_2 and NO binding to reduced cytochrome c oxidase are somewhat similar (k_{on} = 2×10^8 and 4×10^7 M^{-1} sec^{-1}, respectively), a scheme involving a simple kinetic competition between O_2 and NO for the fully reduced cytochrome a_3–Cu_B center cannot explain the rapid onset of inhibition and the competition between the two gases. For this reason, Torres *et al.* (1995) postulated the existance of a turnover intermediate with an affinity for NO much higher than that for O_2. We have recently extended this hypothesis by proposing a possible kinetic mechanism that may account for the onset of inhibition and for its time- and oxygen-dependent evolution (Giuffrè *et al.*, 1996), and that represents an alternative to that of Torres *et al.* (1995). These authors outlined that the different reactivity of O_2 and NO with the half-reduced states of the binuclear center is crucial to inhibition, and indicated reduced Cu_B as the prime NO target. We propose on the other hand that reduced cytochrome a_3 is the preferential target. According to simulations of a possible mechanism (Giuffrè *et al.*, 1996), we believe that nitrosylation of cytochrome a_3 in the half-reduced binuclear center better explains both the potent inhibitory activity of NO and its competition with oxygen, and thus accounts for the idea that NO can exert either an inhibitory or a regulatory role of respiration *in vivo*.

4. CONCLUDING REMARKS

Using a model based on our new kinetic data, we were able to simulate the fast onset of inhibition of cytochrome c oxidase activity, the high inhibitory effect, and the competition between O_2 and NO seen in presteady and steady-state experiments (Giuffrè et al., 1996). This model predicts that under in vivo-like conditions, i.e., in the presence of micromolar O_2 and micromolar NO, the latter will inhibit the reaction with oxygen by efficiently trapping 1 e⁻ on cytochrome a_3 ("inhibition"). However, the partition of electrons between cytochrome a_3 and Cu_B, by favoring the reduction of copper (assumed not accessible to NO), thermodynamically favors the oxygen-dependent pathway, thus subtracting the enzyme from the nitrosylation reaction ("competition"). In other words, inhibition of cytochrome oxidase is strictly linked to NO binding to cytochrome a_3^{2+}. The degree of inhibition, in turn, will depend on (1) the redox equilibrium in the binuclear center [favoring the oxygen reaction by one to three orders of magnitude (Giuffrè et al., 1996)], and (2) the ratio between the affinity constant of NO for reduced cytochrome a_3 ($k_{aff} = 4 \times 10^8$ M⁻¹) and that of O_2 for reduced Cu_B (k_{aff} ranging from 1×10^2 and 7×10^3 M⁻¹), a ratio that favors the NO reaction by five to six order of magnitudes. Last, the presence of oxygen and its well-known reactivity with free NO, together with the high NO off rate from cytochrome a_3, [$k = 0.13$ sec⁻¹ (Giuffrè et al., 1996)], would favor the destruction of bulk NO, and therefore release of inhibition ("reversal").

It is not surprising therefore that Brown and Cooper (1994), working on brain synaptosomes, reported that in the presence of 30 μM O_2 the oxidase activity was half-inhibited at 60 nM NO. In conclusion, we provide new evidence that, because of its peculiar reactivity with a partially reduced cytochrome a_3–Cu_B binuclear center, NO is a potent but also reversible inhibitor of oxidase. As discussed by Brown (1995), NO levels synthesized by different tissues (endothelium, smooth muscle, brain, etc.), challenged with different stimuli, are in the range 0.1 to 5.0 μM. Given that O_2 concentration in tissues is about 30 μM, these NO levels should be sufficient to inhibit (partially or totally) cytochrome c oxidase. Thus, our data support the view that oxidase inhibition by NO may indeed occur in some tissues under conditions of cellular stress, with effects that in the extreme may lead to serious cellular damage and onset of pathology.

ACKNOWLEDGMENTS. We wish to particularly thank Dr. Eric Henry (NIH, Bethesda, MD) for stimulating discussions about the kinetic model and its computation. Our work is partially supported by M.U.R.S.T., Italy PRIN "Bioenergetica e Trasporto di membrane" to P. S., and by EU Grant SCl-CT91-0698 (CAESAR Project) to M.B. and G.B.

5. REFERENCES

Antonini, E., and Brunori, M., 1971, Hemoglobin and myoglobin in their reactions with ligands, in: A. Neuberger and E. L. Tatum, *Frontiers in Biology*, North-Holland Publishing Company, Amsterdam, pp. 31–33.

Blackmore, R. S., Greenwood, C., and Gibson, Q. H., 1991, Studies of the primary oxygen intermediate in the reaction of fully reduced cytochrome oxidase, *J. Biol. Chem.* **266:**19245–19249.

Bredt, O. S., and Snyder, O. H., 1994, Nitric oxide: A physiological messenger molecule, *Annu. Rev. Biochem.* **63:**175–195.

Brown, G. C., 1995, Nitric oxide regulates mitochondrial respiration and cell functions by inhibiting cytochrome oxidase, *FEBS Lett.* **369:**136–139.

Brown, G. C., and Cooper, C. E., 1994, Nanomolar concentrations of nitric oxide reversibly inhibit synaptosomal respiration by competing with oxygen at cytochrome oxidase, *FEBS Lett.* **356:** 295–298.

Brudwig, G. W., Stevens, T. H., and Chan, S. I., 1980, Reactions of nitric oxide with cytochrome oxidase, *Biochemistry* **19:**5275–5285.

Cleeter, M. W. J., Cooper, J. M., Darley-Usmar, V. M., Moncada, S., and Schapira, A. H. V., 1994, Reversible inhibition of cytochrome *c* oxidase, the terminal enzyme of the mitochondrial respiratory chain, by nitric oxide, *FEBS Lett.* **345:**50–54.

Gibson, Q. H., and Greenwood, C., 1963, Reactions of cytochrome oxidase with oxygen and carbon monoxide, *Biochem. J.* **86:**541–555.

Giuffrè, A., Sarti, P., D'Itri, E., Buse, G., Soulimane, T., and Brunori, M., 1996, On the mechanism of inhibition of cytochrome *c* oxidase by nitric oxide, *J. Biol. Chem.* **271:**33404–33408.

Goldstein, S., and Czapski, G., 1995, Kinetics of nitric oxide autoxidation in aqueous solution in the absence and presence of various reductants. The nature of the oxidizing intermediates, *J. Am. Chem. Soc.* **117:**12078–12084.

Henry, E. R., and Hofrichter, J., 1992, Singular value decomposition: Application to analysis of experimental data, *Methods Enzymol.* **210:**129–192.

Malatesta, F., Antonini, G., Sarti, P., and Brunori, M., Structure and function of a molecular machine: Cytochrome *c* oxidase, 1995, *Biophys. Chem.* **54:**1–33.

Moncada, S., and Higgs, E. A., 1993, The L-arginine-nitric oxide pathway, *N. Engl. J. Med.* **329:** 2002–2012.

Moncada, S., Palmer, R. M., and Higgs, E. A., 1991, Nitric oxide: Physiology, pathophysiology and pharmacology, *Pharmacol. Rev.* **43:**109–142.

Nathan, C., 1992, Nitric oxide as a secretory product of mammalian cells, *FASEB J.* **6:**3051–3064.

Soulimane, T., and Buse, G., 1995, Integral cytochrome *c* oxidase. Preparation and progress towards a three-dimensional crystallization, *Eur. J. Biochem.* **227:**588–595.

Stevens, T. H., Brudwig, G. W., Bocian, D. F., and Chan, S. I., 1979, Structure of cytochrome a_3–Cu_{a3} couple in cytochrome *c* oxidase as revealed by nitric oxide binding studies, *Proc. Natl. Acad. Sci. USA:* **76:**3320–3324.

Takehara, Y., Kanno, T., Yoshioka, T., Inoue, M., and Utsumi, K., 1995, Oxygen-dependent regulation of mitochondrial energy metabolism by nitric oxide, *Arch. Biochem. Biophys.* **323:**27–32.

Torres, J., Darley-Usmar, V. M., and Wilson, M. T., 1995, Inhibition of cytochrome *c* oxidase in turnover by nitric oxide: Mechanism and implications for control of respiration, *Biochem. J.* **312:**169–173.

Wilson, M. T., Antonini, G., Malatesta, F., Sarti, P., and Brunori, M., 1994, Probing the oxygen binding site of cytochrome oxidase by cyanide, *J. Biol. Chem.* **269:**24114–24119.

Yonetani, T., Yamamoto, H., Erman, J. E., Jr., and Reed, G. H., 1972, Electromagnetic properties of hemeproteins. Optical and electron paramagnetic resonance characteristics of nitric oxide derivatives of metalloporphyrin-apohemoprotein complexes, *J. Biol. Chem.* **247:**2447–2455.

CHAPTER 11

Energy Transduction in Mitochondrial Respiration by the Proton-Motive Q-Cycle Mechanism of the Cytochrome bc_1 Complex

Bernard L. Trumpower

The cytochrome bc_1 complex is an oligomeric electron transfer enzyme located in the inner membrane of mitochondria, where it participates in respiration, and the plasma membrane of bacteria, where it participates in respiration, denitrification, and nitrogen fixation (Trumpower and Gennis, 1994). The cytochrome bc_1 complex transfers electrons from ubiquinol to cytochrome c and links this electron transfer to translocation of protons across the membrane in which it resides, thus converting the available free energy of the oxidation–reduction reaction into an electrochemical proton gradient. The relationship between the cytochrome bc_1 complex, the cytochrome c oxidase complex, and some of the dehydrogenases that form the ubiquinol substrate for the cytochrome bc_1 complex is shown in Fig. 1.

The mechanism by which the cytochrome bc_1 complex converts energy from the oxidation–reduction reaction into an electrochemical proton gradient

Bernard L. Trumpower • Department of Biochemistry, Dartmouth Medical School, Hanover, New Hampshire 03755.

Frontiers of Cellular Bioenergetics, edited by Papa *et al*. Kluwer Academic/Plenum Publishers, New York, 1999.

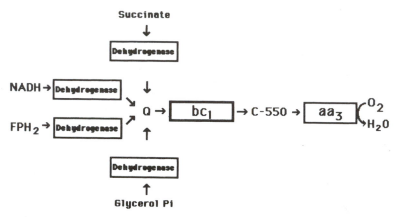

FIGURE 1. The cytochrome bc_1 complex in mitochondrial respiration. The mitochondrial bc_1 complex is a confluence point for reducing equivalents from substrate-specific dehydrogenases that oxidize hydrogen-containing metabolites, such as glycerol–phosphate, succinate, NADH, and reduced flavin, and reduce ubiquinone to ubiquinol. The bc_1 complex oxidizes ubiquinol and reduces cytochrome c and uses the energy from this oxidation–reduction to translocate protons outward across the inner mitochondrial membrane, contributing to the mitochondrial proton-motive force.

is generally accepted to be the proton-motive Q cycle (Trumpower, 1990; Brandt and Trumpower, 1994). Here, I will review some of the experimental data from which the Q cycle was deduced and explain some features of this mechanism that are sometimes misunderstood.

In mitochondria, the cytochrome bc_1 complex consists of 10 or 11 subunit polypeptides (Schägger et al., 1986, Brandt et al., 1994), and these are displayed across the inner mitochondrial membrane approximately as shown in Fig. 2. The cytochrome bc_1 complex has now been crystallized in several laboratories, and hopefully before not too long, we will be able to see the structure of this enzyme in atomic detail (Yue et al., 1991; Berry et al., 1992). In mammalian mitochondria the enzyme contains 11 subunits, while in yeast subunit number 10 is missing. The reason for this difference is that the presequence of the iron–sulfur protein is cleaved in one step and retained as an 11th subunit in mammalian mitochondria (Brandt et al., 1993), but is cleaved in two steps and degraded in yeast (Fu et al., 1990) and Neurospora mitochondria (Hartl et al., 1986). However, only three of these subunits—cytochrome b, cytochrome c_1, and the iron–sulfur protein—are necessary for the electron transfer and energy-transducing functions of the cytochrome bc_1 complex.

The reason we know that only three subunits are required for the activity of the cytochrome bc_1 complex is that only three subunits are present when this

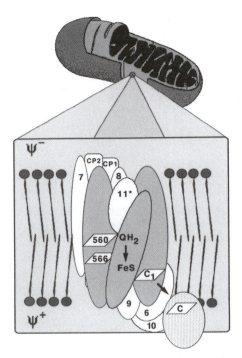

FIGURE 2. Arrangement of subunits in the cytochrome bc_1 complex. The figure shows a model for the topographical disposition of subunits of the bc_1 complex across the inner mitochondrial membrane. The three redox-containing subunits are shaded and the supernumerary subunits, for which there are apparently no counterparts in prokaryotes, are in white. The asterisk on subunit 11 indicates that one of the mitochondrial subunits, formed by cleavage and retention of the iron–sulfur protein presequence in the complex, is present in mammalian mitochondria but absent from *S. cerevisiae* and *N. crassa* mitochondria. The model is based on indirect methods, such as chemical cross-linking and topographically directed chemical and immunological probing of subunit location, combined with predictions of protein structures. This hypothetical model should soon be displaced by a more exact structure, when the crystal structure of the enzyme is solved.

enzyme is purified from bacteria (Yang and Trumpower, 1986), and the three-subunit bacterial enzyme appears to be functionally identical to the mito-chondrial enzyme (Yang and Trumpower, 1988). The sodium dodecyl sulfate–polyacrylamide gel electrophoresis (SDS-PAGE) gels in Fig. 3 show the subunit compositions of the bc_1 complexes from mitochondria and bacteria. These two samples were run on different occasions on different SDS-PAGE gels and then were photographed together, so the mobilities of subunits in the two samples

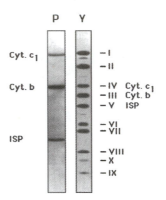

FIGURE 3. SDS-PAGE showing subunit compositions of the cytochrome bc_1 complexes from *Saccharomyces cerevisiae* and *Paracoccus denitrificans*. The yeast complex consists of ten subunits, numbered to the right. Only three of the yeast subunits—cytochrome *b*, cytochrome c_1, and iron–sulfur protein—contain redox prosthetic groups. These three redox-containing subunits are the only subunits present in the purified complex from *P. denitrificans*, shown on the left. The functions of the non-redox-containing subunits in the yeast complex are not known. The photograph is a composite of two different electrophoresis gels; consequently, the migration positions of subunits in the two complexes cannot be compared.

are not directly comparable. The point of this comparison is the number of subunits rather than their sizes.

The gel on the right in Fig. 3 is of the enzyme from yeast, which contains ten subunits (Brandt *et al.*, 1994). The gel on the left is of the enzyme from a gram-negative bacteria, *Paracoccus denitrificans*. In all bacteria from which the enzyme has been purified, it contains only three or, in some cases, four subunits (Ljungdahl *et al.*, 1987). In all mitochondria from which the enzyme has been purified to date, the enzyme contains 10 or 11 subunits. Since the three-subunit bacterial enzyme is functionally equivalent to the mitochondrial enzyme (Yang and Trumpower, 1988), some of the following experiments were carried out with *Paracoccus* enzyme, and some were carried out with beef heart enzyme; the results are, for all practical purposes, completely interchangeable.

Cytochrome *b* is a transmembrane protein with two heme groups. The low-potential cytochrome *b* heme, referred to as cytochrome b_L or cytochrome b_{566}, is located close to the electropositive outer surface of the membrane (Fig. 2). The higher-potential cytochrome *b* heme, referred to as cytochrome b_H or cytochrome b_{560}, is located more interiorly, toward the negative surface of the membrane. The heme of cytochrome c_1 is located near the electrically positive surface of the complex, and the iron–sulfur cluster of the iron–sulfur protein likewise is located in this region. The iron–sulfur protein has now been crystallized, and there is a 1.5 Å crystal structure available (Iwata *et al.*, 1996).

The enzyme oxidizes ubiquinol and transfers electrons through the iron–sulfur protein and cytochrome c_1, eventually to cytochrome c (Fig. 2). The cytochrome bc_1 complex spans the membrane, and there are two ubiquinone–ubiquinol reaction domains in the enzyme. One of these domains is referred to as center P, because it is at the electropositive surface of the membrane and involves the iron–sulfur protein and the low-potential heme group of cytochrome b, the so-called b_L heme group; this is a ubiquinol oxidation site. The other site for ubiquinol–ubiquinone reaction is on the negative side of the membrane and is referred to as center N. This is a ubiquinone–semiquinone reduction site under catalytic conditions. These two sites are spatially isolated and electrically insulated so that the semiquinone at the center N site cannot react with the semiquinone that is formed at the center P site.

The Q cycle mechanism is described in Fig. 4. Ubiquinol is oxidized at center P in a reaction in which the two electrons take two different paths. One electron is transferred through the iron–sulfur cluster of the iron–sulfur protein to cytochrome c_1 and then on to cytochrome c and oxygen. The other electron, which remains with the semiquinone that was formed as a result of the one electron oxidation of ubiquinol, rapidly reduces the cytochrome b_L heme, after

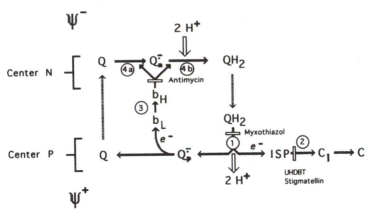

FIGURE 4. The proton-motive Q cycle mechanism of electron transfer and energy transduction by the cytochrome bc_1 complex. The path of electron transfer from ubiquinol to cytochrome c through the redox prosthetic groups of the cytochrome bc_1 complex is depicted as a series of numbered reactions, shown by solid arrows. Dashed arrows represent movement of ubiquinol and ubiquinone between the site where ubiquinol is oxidized at the positive side of the membrane ("center P") and the site where ubiquinone and ubisemiquinone are reduced at the negative side of the membrane ("center N"). Open arrows show the reactions in which protons are released during oxidation of ubiquinol and taken up during reduction of ubiquinone. Open rectangles show the reactions that are blocked by antimycin, myxothiazol, stigmatellin, and UHDBT.

which the electron flows from the b_L heme endergonically to the higher-potential cytochrome b_H heme. That electron residing on the cytochrome b_H then rereduces the quinone that was formed at center P back to semiquinone at center N. During the oxidation step at center P, two protons are released to the intermembrane space of the mitochondria or the periplasmic space of the bacteria. When the quinone is reduced to semiquinone at center N, there is no proton uptake due to the pKa of the semiquinone.

Up to this point in the Q cycle, one electron has been transferred from ubiquinol to cytochrome c_1. This released two protons at the positive surface of the membrane. The second electron from ubiquinol oxidation went through the b cytochromes and equilibrates between the b_H heme and the semiquinone at center N. At this point the Q cycle mechanism is only half complete. To complete the reaction mechanism, a second ubiquinol oxidation occurs, almost exactly as the first time. An electron goes to cytochrome c_1 and eventually downstream to oxygen through cytochrome c. The oxidation of ubiquinol again forms a semiquinone at center P, again sending an electron through the b hemes. However, this time, rather than reducing quinone to semiquinone at center N, there is already a semiquinone at center N that was formed during the first turnover of the enzyme. When the second electron passes through the b hemes, it reduces the semiquinone at center N to ubiquinol. The reduction of ubisemiquinone to ubiquinol at center N takes up two protons from the negative side of the membrane, and this completes the cycle (Fig. 4).

During one complete Q cycle, two molecules of ubiquinol are oxidized to ubiquinones, but one molecule of ubiquinol is regenerated by rereduction of one of these ubiquinones. One complete Q cycle requires that the iron–sulfur protein, cytochrome c_1, and the two hemes of cytochrome b undergo two redox turnovers, in which steps 1, 2, and 3 in Fig. 4 are duplicated. Cytochrome b_H first reduces ubiquinone to ubisemiquinone anion (step 4a) and then reduces the ubisemiquinone anion to ubiquinol (step 4b).

As a result of the net oxidation of one ubiquinol, two molecules of cytochrome c are reduced, four protons are deposited on the positive side of the membrane coincident with the two divergent center P oxidations, and two protons are consumed on the negative side of the membrane as b_H rereduces a ubiquinone to ubiquinol. This stoichiometry of proton translocation implies that for each one molecule of cytochrome c reduced, a pH meter will detect two protons appearing outside of the mitochondrial membrane, while measurements of potassium uptake in the presence of valinomycin will detect only one compensating positive charge electrophoresed through the membrane.

To demonstrate proton translocation with the purified cytochrome bc_1 complex, it is necessary to reconstitute it into phospholipid vesicles (Yang and Trumpower, 1988), which is shown in Fig. 5. The purified cytochrome bc_1 complex is mixed with phospholipid vesicles in the presence of dodecyl

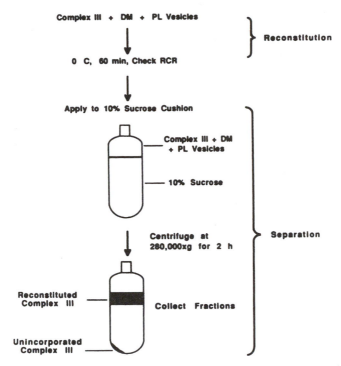

FIGURE 5. Incorporation of purified cytochrome bc_1 complex into phospholipid vesicles. The figure depicts the method used to incorporate the three-subunit bc_1 complex from *P. denitrificans* into phospholipid vesicles (PL) formed by sonication of egg yolk phospholipids, using dodecyl maltoside (DM) to facilitate fusion of the lipoprotein complex with the phospholipid bilayer of the vesicles.

maltoside, allowed to stand at 4°C for some time, then overlaid onto a sucrose cushion and centrifuged. The enzyme that has not incorporated into the vesicles goes to the bottom of the centrifuge tube. The enzyme that is incorporated into the phospholipid has a buoyant density less than the 10% sucrose and stays on top of the sucrose cushion from where it is recovered.

The vesicular reconstituted cytochrome bc_1 complex is then reduced with ubiquinol. If a pulse of oxidant such as ferricyanide is then added, there is a rapid extrusion of protons from the vesicles into the medium, as shown in Fig. 6. If one calculates how many protons have been extruded per molecule of oxidant added, the proton-to-electron stoichiometry is close to four protons extruded for two electrons transferred; in the example shown, this ratio was 3.9. If an uncoupler is added, two protons are released into the medium when ubiquinol is oxidized; however, there is no proton translocation by the proton-

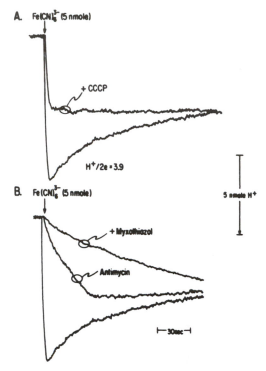

FIGURE 6. Proton translocation by the purified *P. denitrificans* cytochrome bc_1 complex reconstituted into phospholipid vesicles. The bc_1 complex was incorporated into phospholipid vesicles as shown in Fig. 5 and then diluted to 1 μM in pH 7.2 buffer at 15°C. After adding 0.1 mM DBH to reduce the bc_1 complex, 3.6 μM valinomycin was added to dissipate the membrane potential and 1 μM cytochrome *c* was added to enhance the oxidation rate. Proton extrusion was initiated by oxidation with 5 μM ferricyanide. Where indicated, 5 μM CCCP, 6 μM antimycin, or 10 μM myxothiazol were added.

motive Q cycle. Consequently, in the presence of an uncoupler, the $H^+/2\ e^-$ ratio drops from four to two, shown by the upper trace in Fig. 6a. The traces in Fig. 6b are controls, showing that the burst of acidification is inhibited by inhibitors of the bc_1 complex.

Early in our work with the cytochrome bc_1 complex, Carol Edwards discovered that she could extract the iron–sulfur protein from the cytochrome bc_1 complex, and that eliminated the catalytic activity of the complex. When she added the iron–sulfur protein back to the enzyme, it would bind and remain with the enzyme through washing with buffer and recentrifugation, and it would restore the activity (Trumpower and Edwards, 1979; Trumpower *et al.*, 1980). The SDS-PAGE gels in Fig. 7 show the purified iron–sulfur protein, the

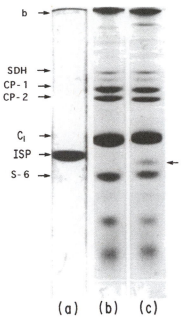

b →

SDH →
CP-1 →
CP-2 →

c_1 →
ISP →
S-6 →

(a)   (b)   (c)

FIGURE 7. SDS-PAGE showing purified iron–sulfur protein, succinate–cytochrome c reductase complex depleted of iron–sulfur protein, and depleted reductase complex after reconstitution with iron–sulfur protein. Arrows on the left show the migration positions of cytochrome b, which aggregated during denaturation and is at the top of the gel, the flavoprotein subunit of succinate dehydrogenase (SDH), core protein 1 (CP-1) and core protein 2 (CP-2), cytochrome c_1, iron–sulfur protein (ISP), and subunit 6 of the bc_1 complex (S-6). The low molecular weight subunits of the bc_1 complex are not resolved by this gel system. The arrow on the right shows the iron–sulfur protein after reconstitution to the depleted reductase complex.

cytochrome bc_1 complex from which iron–sulfur protein has been extracted, and the depleted cytochrome bc_1 complex after reconstitution with the purified protein. In this experiment, about 0.2 equivalent of iron–sulfur protein was reconstituted to the depleted complex. By adding back varying amounts of iron–sulfur protein, we could restore about 40% of the cytochrome c reductase activity of the enzyme.

In the following experiments, we looked at the pre–steady-state reduction of the cytochromes in the enzyme in order to ascertain which cytochromes can be reduced in the absence of iron–sulfur protein. The top trace in Fig. 8 shows that when the iron–sulfur protein is extracted from the cytochrome bc_1 complex, cytochrome b is still rapidly reduced by succinate in the presence of succinate dehydrogenase. On the right is an absorption spectrum taken at the

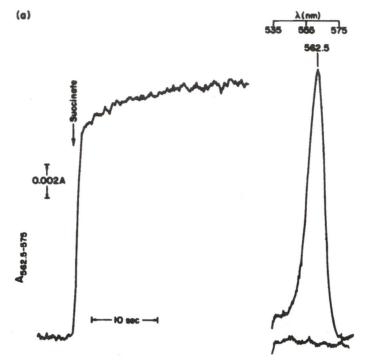

FIGURE 8. Pre-steady-state reduction of cytochromes b and c_1 showing that iron–sulfur protein is required for reduction of cytochrome c_1, but is not required for reduction of cytochrome b. The tracing on the left in (a) shows reduction of cytochrome b in succinate–cytochrome c reductase complex depleted of iron–sulfur protein. The tracing on the left in (b) shows reduction of cytochrome c_1 in depleted reductase complex after reconstitution with iron–sulfur protein. Optical spectra taken at the end of the reactions are shown on the right.

end of the reduction, confirming that cytochrome b is reduced. Although traces for cytochrome c_1 reduction are not shown, from the spectrum in Fig. 8a one can see there is no cytochrome c_1 reduced at the end of the reaction. If c_1 had been reduced, there would be an absorbance peak at approximately 553 nm.

The data in Fig. 8b show the behavior of the enzyme after it is reconstituted with iron–sulfur protein. Addition of the iron–sulfur protein restores the rapid reduction of cytochrome c_1. The optical spectrum taken at the end of the reaction in Fig. 8b confirms that reduction of cytochrome c_1 is restored in addition to reduction of cytochrome b. This summarizes an extensive series of experiments showing that the iron–sulfur protein is required for reduction of cytochrome c_1 but is not required of cytochrome b (Trumpower *et al.*, 1980; Edwards *et al.*, 1981, 1982).

(b)

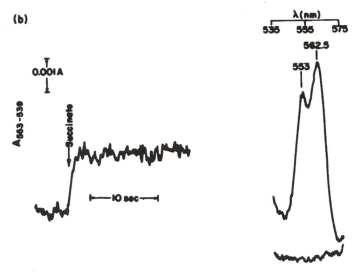

FIGURE 8. (*Continued*)

The reason the iron–sulfur protein is required for reduction of cytochrome c_1 is self-evident from Fig. 4; it is the direct redox protein that transfers an electron from ubiquinol to cytochrome c_1. The reason that the iron–sulfur protein is not required for reduction of cytochrome b, however, requires a bit more explanation. Under normal catalytic conditions, electrons are flowing through the enzyme in the clockwise direction (Fig. 4). However, under pre–steady-state conditions, if one blocks electron transfer through center P, electrons can enter the b cytochromes by a reverse of the reaction at center N. If the iron–sulfur protein is extracted so that the b cytochromes cannot be reduced by the clockwise pathway through center P (Fig. 4), they can be reduced by counterclockwise electron flow through center N. That is, in the absence of the iron–sulfur protein, c_1 reduction could not proceed because center P is inactive, but the b cytochromes could be reduced because center N remains active.

Antimycin inhibits oxidation–reduction of cytochrome b_H. There is a wealth of data indicating that this inhibitor binds to cytochrome b_H. It changes the b_H optical spectrum, and there is a collection of yeast mutants that are resistant to antimycin and all those mutations are on the mitochondrial encoded cytochrome b gene. The mapping of such inhibitor-resistant mutants to confined domains in cytochrome b has allowed the folding pattern of cytochrome b to be inferred prior to solving the crystal structure of the bc_1 complex (Colson, 1993; Gennis *et al.*, 1993; Brasseur, 1988).

If antimycin inhibits oxidation–reduction of cytochrome b_H, and if the iron–sulfur protein is removed so that the b cytochromes cannot be reduced through center P, then one would expect that the combination of removing the iron–sulfur protein and adding antimycin would block reduction of the b cytochromes in addition to blocking reduction of cytochrome c_1. In other words, in the presence of antimycin, the iron–sulfur protein should be required for reduction of both cytochrome c_1 and the b cytochromes. This is shown in Fig. 9. The top traces show that there is no reduction of either cytochrome b or c_1 in a cytochrome bc_1 complex where the iron–sulfur protein has been extracted and antimycin has been added. On the right is an optical spectrum taken at the end of these two reactions to confirm that there has been no reduction of those cytochromes. The bottom traces show that if iron–sulfur protein is added back to the complex, it restores the rapid reduction of cytochrome b and cytochrome c_1. The optical spectrum on the right shows the reduction of the cytochromes at the end of the reaction.

Since antimycin inhibits oxidation of cytochrome b_H, this inhibits electron transfer through the cytochrome bc_1 complex to cytochrome c under catalytic conditions. However, antimycin does not directly inhibit reduction of cytochrome c_1 during a single turnover of the enzyme under pre–steady-state conditions (Bowyer and Trumpower, 1981). When John Bowyer did the experiments shown in Fig. 10, this result was somewhat surprising, because the thinking at that time was that antimycin inhibits electron transfer between cytochromes b and c_1. The top two traces in Fig. 10 show reduction of cytochrome c_1, followed at 553 nm, after addition of succinate and succinate dehydrogenase in the absence or presence of antimycin. If one adds antimycin and then adds succinate, cytochrome c_1 reduction proceeds as a monophasic, rapid reaction. The decrease in absorbance change between the trace in the presence of antimycin (Fig. 10b) and in the absence of antimycin (Fig. 10a) is due to the fact that, at this wavelength pair, some of the absorbance change is due to reduction of cytochrome b and overlap from the cytochrome b spectrum. Antimycin shifts the optical spectrum of cytochrome b about 1 nm to the red and consequently diminishes the overlap from the cytochrome b (Bowyer and Trumpower 1981).

Although antimycin does not inhibit reduction of cytochrome c_1 in a first, pre–steady-state turnover, one would predict that it would inhibit the reduction of cytochrome c_1 if the enzyme is forced to turn over more than once. In other words, cytochrome c_1 can be reduced once without reoxidizing the b cytochromes, but in order to reduce cytochrome c_1 a second time, it is necessary to reoxidize the b cytochromes. In the middle traces in Fig. 10, we added approximately one equivalent of cytochrome c so that the enzyme has to turn over twice. In the absence of antimycin, the cytochrome c plus c_1 goes reduced in less than a second. However, in the presence of antimycin, there is rapid reduction of one equivalent of c type cytochrome, and then the reaction is inhibited about three orders of magnitude. The second electron enters the c

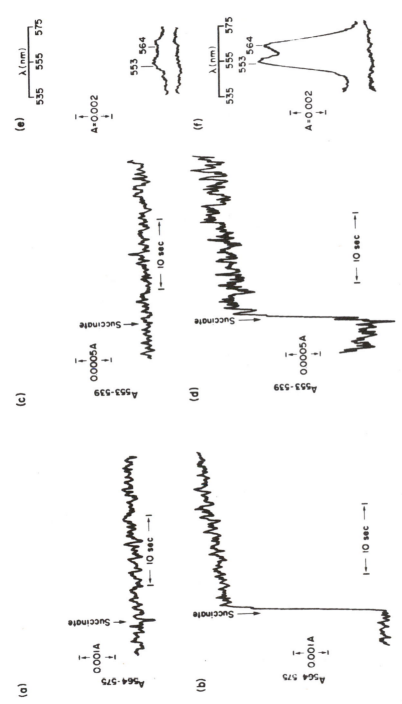

FIGURE 9. Pre-steady-state reduction of cytochromes b and c_1 showing that, in the presence of antimycin, iron–sulfur protein is required for reduction of both cytochromes b and c_1. The traces at the top show lack of reduction of (a) cytochrome b and (c) cytochrome c_1 in succinate–cytochrome c reductase complex depleted of iron–sulfur protein in the presence of antimycin. The traces at the bottom show reduction of (b) cytochrome b and (d) cytochrome c_1 in depleted reductase complex after reconstitution with iron–sulfur protein in the presence of antimycin. Optical spectra taken at the end of the reactions are shown on the right.

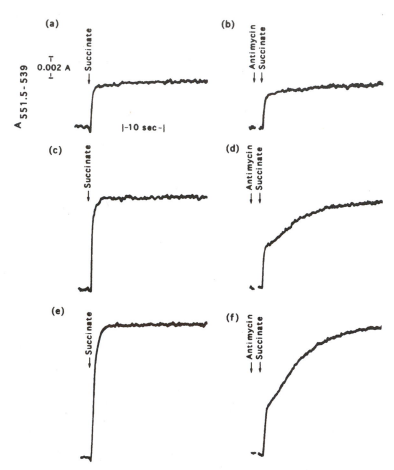

FIGURE 10. Pre-steady-state reduction of cytochromes $c_1 + c$, showing that antimycin does not inhibit pre-steady-state reduction of cytochrome c_1 during a single turnover of the bc_1 complex, but does inhibit reduction of cytochrome $c_1 +$ cytochrome c during multiple turnovers of the complex. Isolated succinate–cytochrome c reductase complex was suspended at 0.37 μM in (a) and (b). In (c) and (d), 0.22 μM cytochrome c was added, and in (e) and (f), 0.44 μM cytochrome c was added. The reductase complex with or without additional cytochrome c was then reduced with succinate. Antimycin was added where indicated. The absorbance and time scales shown in (a) are identical for all of the traces.

cytochromes at a markedly diminished rate. The bottom traces show the result if 1.8 equivalents of cytochrome c are added, so that there is a total of 2.8 equivalents of c type cytochrome. Again the cytochrome c_1 plus c are reduced rapidly and monotonically in the absence of the inhibitor. When antimycin is added, only the first turnover proceeds in an uninhibited manner; antimycin inhibits subsequent turnovers.

In addition to antimycin, which inhibits at center N, there is a group of inhibitors that inhibit at center P. These inhibitors all block the oxidation of ubiquinol, although there are differences in how they block that reaction. One of these is called myxothiazol, a toxin produced by myxobacteria. Myxothiazol inhibits oxidation of ubiquinol by binding to cytochrome b in the vicinity of the b_L heme, and it displaces or prevents ubiquinol from gaining access to the center P quinol oxidase (Von Jagow *et al.*, 1984). The other inhibitor is a compound called stigmatellin, which is thought to bind to the iron–sulfur protein in the cytochrome bc_1 complex (Von Jagow and Ohnishi, 1985).

Earlier we saw that if the iron–sulfur protein is removed, antimycin will block reduction of the b cytochromes. By that same reasoning—if there are two routes of cytochrome b reduction, in order to block b reduction you have to block both of those routes—one would expect that myxothiazol or stigmatellin would not block b reduction because b reduction should occur by the center N pathway. Likewise, antimycin should not block b reduction, because b reduction should be able to proceed by the center P pathway. However, if you add both inhibitors together, you would expect to fully block the reduction of the b cytochromes.

If myxothiazol and stigmatellin are acting at center P, they should block reduction of cytochrome c_1, and that is shown in Fig. 11. Trace (a) shows reduction of cytochrome c_1 with DBH, a ubiquinol analogue, in the absence of any inhibitors, and trace (b) shows that this is not inhibited by antimycin, as also shown before. However, c_1 reduction is inhibited by myxothiazol, and it is also inhibited by stigmatellin, as shown by the traces in (c) and (d).

In the experiments shown in Fig. 12, menaquinol was used to reduce the b cytochromes, since this quinol has a sufficiently low potential that it reduces approximately 80% of the total b cytochromes, compared to the less strongly reducing DBH, which reduces only 40% of the cytochrome b. Antimycin, myxothiazol, and stigmatellin added alone do not inhibit reduction of cytochrome b. However, if either of the two center P inhibitors, myxothiazol or stigmatellin, is added in combination with antimycin, the combination of inhibitors blocks reduction of the b cytochromes, since they block both center N and center P pathways.

There is also a class of center P inhibitors that resemble ubiquinone, and one of the best-studied examples of these is a benzoxythiazol called UHDBT. This hydroxy-benzoquinone inhibits electron transfer by binding to center P,

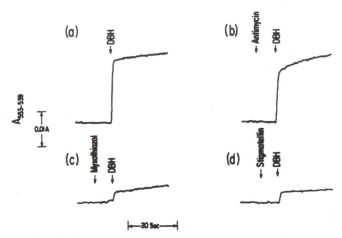

FIGURE 11. Effect of inhibitors on pre-steady-state reduction of cytochrome c_1. DBH, an analogue of ubiquinol, was used to reduce cytochrome c_1 in purified *P. denitrificans* bc_1 complex. Inhibitors were added where indicated.

either very near to or on the iron–sulfur protein (Bowyer *et al.*, 1982). This inhibitor was used to investigate an unusual feature of the bc_1 complex, which is elicited by antimycin and which can be inferred from Fig. 4. When ubiquinol is rapidly oxidized, for example, by adding a pulse of oxygen in the presence of cytochrome c and cytochrome oxidase, it rapidly forms ubisemiquinone, which reduces cytochrome b. This means that rapid oxidation at the high potential end of the respiratory chain will drive the reduction of the b cytochrome. It is difficult to observe this oxidant-induced b reduction under normal turnover conditions, because the b cytochromes are reoxidized through center N so fast that at room temperature their reduction is transient and is missed. However, if antimycin is added to prevent reoxidation of the b cytochromes and then an oxidant is added to generate the ubisemiquinone through oxidation of the c cytochromes and iron–sulfur protein, the oxidant will drive the reduction of the b cytochromes and antimycin will trap them in the reduced state.

If myxothiazol and UHDBT act at center P, and if the iron–sulfur protein participates in forming ubisemiquinone at center P as shown in Fig. 4, then adding these inhibitors or removing the iron–sulfur protein should block the oxidant-induced reduction of cytochrome b (Bowyer and Trumpower, 1980; Bowyer *et al.*, 1981). The effects of the inhibitors are shown in Fig. 13. Decyl ubiquinol (DBH), the substrate analogue of ubiquinol, is used to partially reduce the b cytochromes. The midpoint potential of DBH at pH 7 is about 60

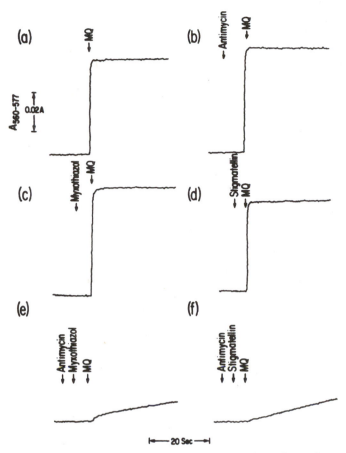

FIGURE 12. Effect of inhibitors on pre-steady-state reduction of cytochrome *b*. Menaquinol was used to reduce cytochrome *b* in purified *P. denitrificans* bc_1 complex. Inhibitors were added where indicated.

mV; consequently, DBH will reduce about 40% of the total cytochrome *b* heme, as shown in trace (a). If the reduction of *b* by DBH is allowed to come to completion and then an oxidant is added, at room temperature one does not see any oxidation-driven reduction of cytochrome *b*, because of their rapid reoxidation through center N.

However, if antimycin is added after reduction of the cytochrome *b* with the quinol and then the oxidant is added, antimycin elicits oxidant-induced reduction of cytochrome *b* by slowing down or inhibiting the reoxidation as

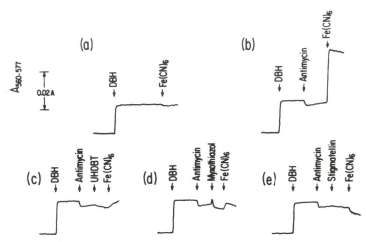

FIGURE 13. Effect of inhibitors on oxidant-induced reduction of cytochrome *b*. Purified *P. denitrificans bc₁* complex was reduced with DBH, an analogue of ubiquinol, after which ferricyanide was to elicit oxidant-induced reduction of cytochrome. Inhibitors were added where indicated.

shown in trace (b). UHDBT, myxothiazol, and stigmatellin all inhibit the oxidant-induced reduction (Fig. 13c, d, and e), as expected if these are center P inhibitors.

As noted earlier, the iron–sulfur protein should be required for oxidant-induced reduction if it is the oxidant for ubiquinol, and that is shown in Fig. 14. The data in the top panel were obtained with cytochrome bc_1 complex from which the iron–sulfur protein was removed. Ascorbate was added to reduce cytochrome c_1 and antimycin was added to allow observation of any oxidant-induced reduction of cytochrome *b*. When succinate is then added along with succinate dehydrogenase, there is no reduction of cytochrome *b* by the succinate. If cytochrome c plus cytochrome *c* oxidase is then added, there is no oxidant-induced reduction in the bc_1 complex where the iron–sulfur protein is missing.

On the right in Fig. 14 are optical spectra taken at the beginning and end

--→

FIGURE 14. Involvement of iron–sulfur protein in oxidant-induced reduction of cytochrome *b*. Ascorbate, antimycin, and succinate were added to isolated succinate–cytochrome c reductase complex depleted of iron–sulfur protein (top trace) or to depleted reductase complex reconstituted with iron–sulfur protein (bottom trace). Cytochrome c plus cytochrome c oxidase were then added to elicit oxidant-induced reduction of cytochrome *b*. The middle trace is a control in which the reductase complex was omitted.

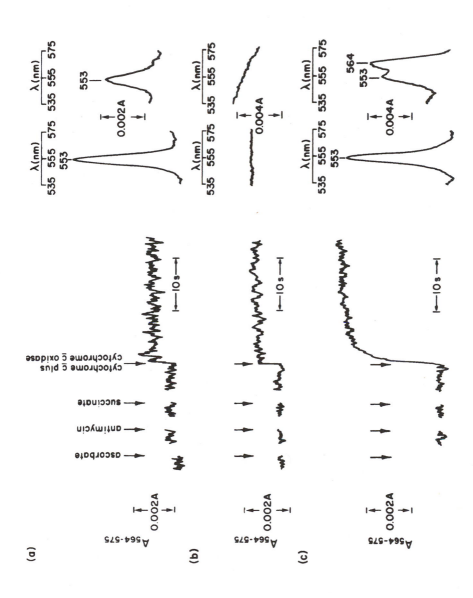

of the reaction. The spectra in the top panel show that the cytochrome c_1 is fully reduced at the beginning of the reaction and at the end of the reaction it has been partially oxidized. During the course of this reaction, the ascorbate is slowly rereducing the c_1, so that if we could scan the spectrum on a much faster time scale, we presumably could see that the c_1 would have been fully oxidized, but during the time interval of several seconds, there has been partial rereduction of the cytochrome c_1. The small upward deflection in the kinetic trace in Fig. 14a is not reduction of cytochrome b, but a turbidity artifact caused by adding the cytochrome c oxidase. That is shown in the set of traces in Fig. 14b, which are like those in (a) except there is no cytochrome bc_1 complex. The same additions were made: ascorbate, antimycin, succinate, and cytochrome c oxidase, but no cytochrome bc_1 complex was added. We see the same jump in the kinetics trace in (b) as in (a), confirming that it is a turbidity change. The changes in the baseline of the absorbance traces on the right in Fig. 14b further confirm the turbidity change due to diluting the cytochrome c oxidase out of detergent into the aqueous reaction mixture.

If this experiment is repeated with cytochrome bc_1 complex that has been reconstituted with iron–sulfur protein, shown in the kinetics trace in Fig. 14c, there is oxidant-induced reduction of cytochrome b, resulting from the addition of cytochrome c and cytochrome c oxidase. The upward deflection of the kinetics traces suggests that cytochrome b has been reduced. On the right are the absorption traces before adding the oxidant, showing that initially the cytochrome c_1 is reduced, but there is no cytochrome b reduced. After adding the oxidant, there is oxidation of cytochrome c_1 and reduction of cytochrome b. This experiment established that the iron–sulfur protein is required for the oxidant-induced reduction of cytochrome b.

The cytochrome bc_1 complex exhibits an unusual redox response during pre–steady-state reduction, known as the triphasic reduction (Jin $et\ al.$, 1981). The kinetics trace in Fig. 15 shows the pre–steady-state reduction of cytochrome c_1 and cytochrome b on a millisecond time scale (Tang and Trumpower, 1986). Reduction of cytochrome c_1 is a monophasic reaction, but the cytochrome b initially goes partially reduced, then it goes partially reoxidized, and finally it goes rereduced to completion in about a second. In this and similar experiments where triphasic reduction is observed, the rate of b reduction is first order with respect to the ubiquinol substrate.

Examination of the traces in Fig. 15 reveals that the rate of the first phase of b reduction closely matches c_1 reduction, whereas the rate of the second reduction is much slower than the rate of c_1 reduction. This suggests that the initial phase of b reduction is mechanistically linked to c_1 reduction. The intermediate reoxidation phase starts at about the time c_1 reduction reaches completion. The interpretation that has been given for the redox behavior of the two cytochromes is that cytochrome b and c_1 are initially reduced through

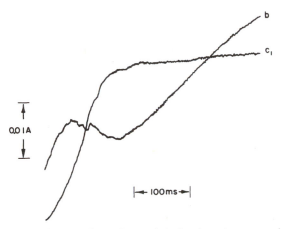

FIGURE 15. Triphasic reduction of cytochrome *b*. Isolated succinate–cytochrome *c* reductase complex was reduced with succinate, reduction of the cytochromes was monitored with a stopped-flow spectrophotometer.

the center P pathway (Tang and Trumpower, 1986). The reason cytochrome *b* does not go fully reduced is that as ubiquinol is oxidized to ubiquinone at center P, the ubiquinone then reoxidizes the *b* cytochromes through center N. Following the initial burst of *b* reduction and during its reoxidation, the reaction reaches a point where cytochrome c_1 and the iron–sulfur protein are fully reduced and can no longer act as an oxidant for ubiquinol. At that point, ubiquinol rereduces the *b* cytochromes through center N, and the reduction proceeds more slowly through center N than the initial reduction through the high potential, iron–sulfur-protein-dependent center P pathway. As noted above, this behavior is most clearly observed when the rate is first order with respect to ubiquinol. As the ubiquinol concentration is increased, the triphasic reaction is compressed in time to a point where it is not observable (Tang and Trumpower, 1986). However, we assume that a similar redox behavior ensues under conditions where the rate is zero order with respect to ubiquinol, but that the response is merely hidden by the limited time resolution of the spectrophotometer and mixing chamber.

The triphasic reduction of cytochrome *b* is consistent with the oxidation–reduction potentials of the redox centers of the bc_1 complex, since the initial phase of *b* reduction follows the energetically favored pathway through center P; when that pathway is blocked by the reduction of cytochrome c_1 and the iron–sulfur protein, reduction of cytochrome *b* proceeds through the lower potential center P pathway. Fig. 16 shows a thermodynamic profile of the cytochrome bc_1 complex disposed across the intermitochondrial membrane and

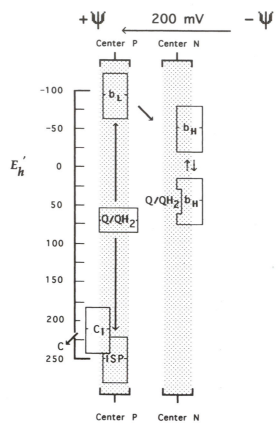

FIGURE 16. Thermodynamic profile of the cytochrome bc_1 complex. In this depiction of the Q cycle, the redox prosthetic groups are arranged vertically according to their oxidation–reduction potentials and displayed horizontally according to their electrochemical disposition across the inner mitochondrial membrane or bacterial plasma membrane. The open boxes delineate the approximate range along the vertical axis of oxidation–reduction potentials spanned by the redox components as their oxidation–reduction status varies in response to changes in rates of electron transfer through the cytochrome bc_1 complex. Cytochrome b_H is a mixture of two potentiometric species, in which reversible binding of ubiquinol raises the potential of a variable portion of the b_H from approximately −50 to +40 mV. The potentials of the ubisemiquinone couples at centers N and P are not shown.

spanning an electrochemical potential distance that approaches 200 mV at the extremes (Brandt and Trumpower, 1994). The iron–sulfur protein and the low potential b_L heme are at center P, and the higher potential b_H heme is at center N. The potential of the latter heme group changes slightly, dependent on whether ubiquinone or ubiquinol is bound at center N, proximal to the heme group (Salerno *et al.*, 1989).

The thermodynamic profile illustrates that at center P the oxidation of ubiquinol drops one electron to a potential of about 250 or 280 mV, and sends a second electron, in a thermodynamically linked reaction, to a potential as low as -100 mV. The midpoint potential of b_L is somewhere between -60 and -100 mV, differing to that extent in mitochondria and photosynthetic bacteria, so that at center P the quinone–quinol couple reacts across a potential span of about 350 mV. At center N, where the b_H cytochrome must reduce quinone to semiquinone and semiquinone to quinol, these two reactions must take place at approximately the same potential.

The relationship between the midpoint potentials of the quinone–semiquinone and semiquinone–quinol couples and the equilibrium constant for formation of semiquinone is shown in Fig. 17a. The midpoint potential for the two electron oxidation–reduction of ubiquinol–ubiquinone must be the arithmetic average of the two midpoint potentials for the oxidation–reduction of the constituent semiquinone couples. In other words, the midpoint potentials of the two semiquinone couples must be equidistant above and below the midpoint potential of the quinone–quinol couple. As the midpoint potentials of the two semiquinone couples move progressively further apart and the midpoint potential of the semiquinone–quinol couple approaches $+250$ mV, as it is at center P, then the equilibrium constant for formation of semiquinone at center P is somewhere in the neighborhood of 1×10^{-6} or less. One would thus expect that the semiquinone at center P never reaches detectable concentrations. On the other hand, at center N, where the two semiquinone couples are approximately equal potential, one would expect that the semiquinone would approach the concentration of quinone and quinol.

This is shown graphically in Fig. 17b, which illustrates how the concentration of semiquinone drops as the potentials of the two semiquinone couples move apart. On the other hand, as the potentials of the two semiquinone couples become approximately equal, the concentration of semiquinone approaches a third of the total quinone redox species in the pool. This predicts that a semiquinone should be detectable at center N. A thermodynamically stable ubisemiquinone with properties expected for the ubisemiquinone at center N was first detected in purified succinate–cytochrome reductase complex (Ohnishi and Trumpower, 1980). Titration of the electron paramagnetic resonance (EPR) detectable semiquinone signal indicated a midpoint potential of 30–40 mV at pH 8.5, and the EPR signal from the semiquinone was eliminated by antimycin.

As noted earlier, UHDBT is a structural analogue of ubiquinone that inhibits electron transfer in the bc_1 complex. This inhibitor ionizes and is inhibitory only in the protonated form, so high-affinity binding to the bc_1 complex is observed only below pH 6.5, the pK_a of the hydroxy group on the inhibitor (Trumpower and Haggerty, 1980). There is one high-affinity binding site for UHDBT in the cytochrome bc_1 complex, and if iron–sulfur protein is extracted

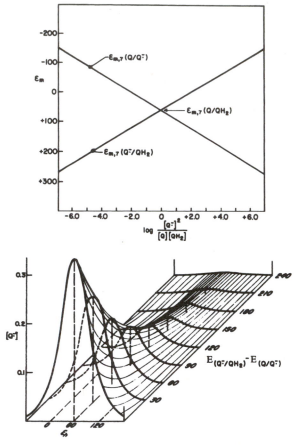

FIGURE 17. (a) Relationship between the midpoint potentials of the quinone–semiquinone and semiquinone–quinol couples and the equilibrium constant for formation of semiquinone. (b) Relationship between the concentration of ubisemiquinone anion and the difference in midpoint potentials of the quinone–semiquinone and semiquinone–quinol couples.

from the bc_1 complex and added back in incremental amounts, the amount of UHDBT required to inhibit electron transfer is proportional to the amount of iron–sulfur protein reconstituted (Bowyer *et al.*, 1982), indicating that UHDBT binds with high affinity only in the presence of the iron–sulfur protein. Furthermore, when UHDBT binds to the complex, it changes the midpoint potential of the iron–sulfur protein by 70–90 mV. These results are suggestive that UHDBT inhibits by binding to the iron–sulfur protein and that it binds approximateley 15-fold more tightly to reduced iron–sulfur protein. Strictly speaking, however, these results indicate that the inhibitor binds preferentially to some site in the bc_1 complex when the iron–sulfur protein is present and in the reduced form, but they do not establish conclusively that the inhibitor binds directly to the iron–sulfur protein.

The results in Fig. 18 provide more convincing evidence that UHDBT binds to the iron–sulfur protein and also show that ubiquinol interacts with the iron–sulfur protein (Meinhardt *et al.*, 1987). These traces show the EPR spectrum of the iron–sulfur protein, which has a diagnostic central g_Y resonance at 1.90 and a g_x resonance at 1.80. The top two traces show that the g_x resonance is shifted to 1.76 by UHDBT, indicating that this hydroxy-quinone is changing the electronic environment around the iron–sulfur cluster. In the left panels we varied the redox potential applied to the cytochrome bc_1 complex and looked at the g_x 1.80 signal in the absence of UHDBT. At +220 mV, the iron–sulfur protein is reduced, but ubiquinone is in the oxidized form. Under these conditions, the g_x resonance is at 1.80. When the redox potential is lowered to +20 mV, the iron–sulfur protein remains fully reduced, and ubiquinone is reduced to ubiquinol. Coincident with reduction to ubiquinol, the g_x signal shifts from 1.80 to 1.79. This change in signal resonance indicates that ubiquinone and/or

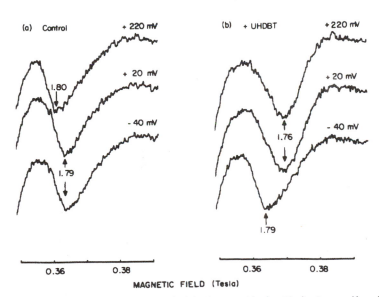

FIGURE 18. Interaction of UHDBT and ubiquinone with the Rieske iron–sulfur cluster through a common binding site. Isolated succinate–cytochrome c reductase complex was poised at the indicated potentials in the absence or presence of UHDBT.

ubiquinol are electronically affecting the iron–sulfur cluster. If the potential is lowered to −40 mV, there is no further change in the g_x resonance beyond that resulting from reduction of the quinone to quinol.

The results in the right panel (Fig. 18) show that UHDBT shifts the g_x resonance from 1.80 to 1.76 under conditions where ubiquinone is in the oxidized form, and from this we conclude that UHDBT displaces ubiquinone from the vicinity of the iron–sulfur cluster. When the quinone is reduced to quinol at a potential of +20 mV, UHDBT still shifts the g_x resonance to 1.76. In other words, UHDBT displaces both ubiquinone and ubiquinol from the electronic environment of the iron–sulfur cluster. The bottom trace in Fig. 18b shows that if the potential is lowered to −40 mV, so that UHDBT undergoes reduction, the spectrum is the same as was obtained with ubiquinol in the absence of UHDBT, because the inhibitor in the reduced form has not displaced ubiquinol.

Further evidence that ubiquinone affects the electronic environment of the iron–sulfur cluster is shown in Fig. 19, which shows a titration curve of the change in g_x 1.80 resonance of the iron–sulfur protein as the redox potential is varied. The signal amplitude increases as the redox potential is increased from 0 to 100 mV, with an apparent midpoint at approximately +30 mV, the midpoint potential of the ubiquinone–ubiquinol couple at pH 8.5. The fact that dependence of the iron–sulfur cluster g_x 1.80 resonance on redox potential

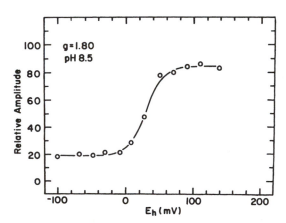

FIGURE 19. Interaction of ubiquinone with the Rieske iron-sulfur cluster. Relative amplitude of the $g = 1.80$ resonance in the EPR spectrum of the Rieske iron–sulfur protein in purified *P. denitrificans* bc_1 complex was measured at various potentials. A theoretical two-electron reduction curve with a midpoint potential of +30 mV is drawn through the experimentally determined data points.

matches the midpoint potential of ubiquinone–ubiquinol is persuasive evidence that ubiquinone and/or ubiquinol interact electronically with the iron–sulfur cluster.

ACKNOWLEDGMENT. The research described in this chapter was supported by NIH Research Grant GM 20379.

REFERENCES

Berry, E. A., Huang, L. S., Earnest, T. N., and Jap, B. K. (1992) X-Ray diffraction by crystals of beef heart ubiquinol–cytochrome-c oxidoreductase, *J. Mol. Biol.* **224**:1161–1166.

Bowyer, J. R., and Trumpower, B. L., 1980, Inhibition of the oxidant-induced reduction of cytochrome b by a synthetic analog of ubiquinone, *FEBS Lett.* **115**:171–174.

Bowyer, J. R., and Trumpower, B. L., 1981, Rapid reduction of cytochrome c_1 in the presence of antimycin and its relevance to the mechanism of electron transfer in the cytochrome bc_1 segment of the mitochondrial respiratory chain, *J. Biol. Chem.* **256**:2245–2251.

Bowyer, J. R., Edwards, C. A., and Trumpower, B. L., 1981, Involvement of the iron–sulfur protein of the cytochrome bc_1 segment of the mitochondrial respiratory chain in the oxidant-induced reduction of cytochrome b, *FEBS Lett.* **126**:93–97.

Bowyer, J. R., Edwards, C. A., Ohnishi, T., and Trumpower, B. L., 1982, An analogue of ubiquinone which inhibits respiration by binding to the iron–sulfur protein of the cytochrome bc_1 segment of the mitochondrial respiratory chain, *J. Biol. Chem.* **257**:8321–8330.

Brandt, U., and Trumpower, B. L., 1994, The protonmotive Q cycle in mitochondria and bacteria, *Crit. Rev. Biochem. Mol. Biol.* **29**:165–197.

Brandt, U., Yu, L., Yu, C. A., and Trumpower, B. L., 1993, The mitochondrial targeting presequence of the Rieske iron-sulfur protein is processed in a single step after insertion into the cytochrome bc_1 complex in mammals and retained as a subunit in the complex, *J. Biol. Chem.* **268**:8387–8390.

Brandt, U., Uribe, S., and Trumpower, B. L., 1994, Isolation and characterization of QCR10, the nuclear gene encoding the 8.5 kDa subunit 10 of the *Saccharomyces cerevisiae* cytochrome bc_1 complex, *J. Biol. Chem.* **269**:12947–12953.

Brasseur, R., 1988, Calculation of the three-dimensional structure of *Saccharomyces cerevisiae* cytochrome b inserted in a lipid matrix, *J. Biol. Chem.* **263**:12571–12575.

Colson, A. M., 1993, Random mutant generation and its utility in uncovering structural and functional features of cytochrome b in *Saccharomyces cerevisiae*, *J. Bioenerg. Biomembr.* **3**: 211–220.

Edwards, C. A., Bowyer, J. R., and Trumpower, B. L., 1981, Pathways of electron transfer in the cytochrome bc_1 segment of the mitochondrial respiratory chain as deduced by resolution and reconstitution of the iron–sulfur protein, in: *Vectorial Reactions in Electron and Ion Transport in Mitochondria and Bacteria* (F. Palmieri, E. Quagliariella, N. Siliprandi, and E. C. Slater, eds.) pp. 139–148, Elsevier/North Holland, Amsterdam.

Edwards, C. A., Bowyer, J. R., and Trumpower, B. L., 1982, Function of the iron–sulfur protein of the cytochrome bc_1 segment in electron transfer reactions of the mitochondrial respiratory chain, *J. Biol. Chem.* **257**:3705–3713.

Fu, W., Japa, S., and Beattie, D. S., 1990, Import of the iron–sulfur protein of the cytochrome bc_1 complex into yeast mitochondria, *J. Biol. Chem.* **265**:16541–16547.

Gennis, R. B., Barquera, B., Hacker, B., Van Doren, S. R., Arnaud, S., Crofts, A. R., Davidson, E., Gray, K. A., and Daldal, F., 1993, The bc_1 complexes of *Rhodobacter sphaeroides* and *Rhodobacter capsulatus*, *J. Bioenerg. Biomembr.* **25**:195–209.

Hartl, F.-U., Schmidt, B., Wachter, E., Weiss, H., and Neupert, W., 1986, Transport into mitochondria and intramitochondrial sorting of the Fe/S protein of ubiquinol–cytochrome c reductase, *Cell* **47**:939–951.

Iwata, S., Saynovits, M., Link, T. A., and Michel, H., 1996, Structure of a water soluble fragment of the "Rieske" iron sulfur protein of the bovine heart mitochondrial cytochrome bc_1 complex determined by MAD phasing at 1.5 Angstrom resolution, *Structure* **4**:567–579.

Jin, Y. Z., Tang, H. L., Li, S. L., and Tsou, C. L., 1981, The triphasic reduction of cytochrome b in the succinate–cytochrome c reductase, *Biochim. Biophys. Acta* **637**:551–554.

Ljungdahl, P. O., Pennoyer, J. D., Robertson, D., and Trumpower, B. L., 1987, Purification of highly active cytochrome bc_1 complexes from phylogenetically diverse species by a single chromatographic procedure, *Biochim. Biophys. Acta* **891**:227–242.

Meinhardt, S. W., Yang, X., Trumpower, B. L., and Ohnishi, T. O., 1987, Identification of a stable ubisemiquinone and characterization of the effects of ubiquinone oxidation–reduction status on the Rieske iron–sulfur protein in the three subunit ubiquinol–cytochrome c oxidoreductase complex from *Paracoccus denitrificans*, *J. Biol. Chem.* **262**:8702–8706.

Ohnishi, T., and Trumpower, B. L., 1980, Differential effects of antimycin on ubisemiquinone bound in different environments in isolated succinate–cytochrome c reductase complex, *J. Biol. Chem.* **255**:3278–3284.

Salerno, J. C., Xu, Y., Osgood, M. P., Kim, C. H., and King, T. E., 1989, Thermodynamic and spectroscopic characterization of the cytochrome bc_1 complex: Role of quinone in the behavior of cytochrome b-562, *J. Biol. Chem.* **264**:15398–15403.

Schägger, H., Link, T. A., Engel, W. D., and Von Jagow, G., 1986, Isolation of the eleven protein subunits of the bc_1 complex from beef heart, *Methods Enzymol.* **126**:224–237.

Tang, H. L., and Trumpower, B. L., 1986, Triphasic reduction of cytochrome b and the protonmotive Q cycle, *J. Biol. Chem.* **261**:6209–6215.

Trumpower, B. L., 1990, The protonmotive Q cycle: Coupling of proton translocation to electron transfer by the cytochrome bc_1 complex, *J. Biol. Chem.* **265**:11409–11412.

Trumpower, B. L., and Edwards, C. A., 1979, Purification of a reconstitutively active iron–sulfur protein (oxidation factor) from succinate–cytochrome c reductase complex of bovine heart mitochondria, *J. Biol. Chem.* **254**:8697–8706.

Trumpower, B. L., and Gennis, R., 1994, Energy transduction by cytochrome complexes in mitochondrial and bacterial respiration: The enzymology of coupling electron transfer reactions to transmembrane proton translocation, *Annu. Rev. Biochem.* **63**:675–716.

Trumpower, B. L., and Haggerty, J. G., 1980, Inhibition of electron transfer in the cytochrome bc_1 segment of the mitochondrial respiratory chain by a synthetic analogue of ubiquinone, *J. Bioenerg. Biomembr.* **12**:151–164.

Trumpower, B. L., Edwards, C. A., and Ohnishi, T., 1980, Reconstitution of the iron–sulfur protein responsible for the $g = 1.90$ EPR signal and associated cytochrome c reductase activities to depleted succinate–cytochrome c reductase complex, *J. Biol. Chem.* **255**:7487–7493.

Von Jagow, G., and Ohnishi, T., 1985, The chromone inhibitor stigmatellin—Binding to the ubiquinol oxidation center at the C-side of the mitochondrial membrane, *FEBS Lett.* **185**:311–315.

Von Jagow, G., Ljungdahl, P. O., Graf, P., Ohnishi, T., and Trumpower, B. L., 1984, An inhibitor of mitochondrial respiration which binds to cytochrome b and displaces quinone from the iron–sulfur protein of the cytochrome bc_1 complex, *J. Biol. Chem.* **259**:6318–6326.

Yang, X., and Trumpower, B. L., 1986, Purification of a three subunit ubiquinol–cytochrome *c* oxidoreductase complex from *Paracoccus denitrificans, J. Biol. Chem.* **261:**12282–12289.

Yang, X., and Trumpower, B. L., 1988, Protonmotive Q cycle pathway of electron tranfer and energy transduction in the three-subunit ubiquinol–cytochrome *c* oxidoreductase complex of *Paracoccus denitrificans, J. Biol. Chem.* **263:**11962–11970.

Yue, W. H., Zou, Y. P., Yu, L., and Yu, C. A., 1991, Crystallization of mitochondrial ubiquinol–cytochrome *c* reductase, *Biochemistry* **30:**2303–2306.

CHAPTER 12

The Crystal Structure of Mitochondrial Cytochrome bc_1 Complex

Chang-An Yu, Li Zhang, Anatoly M. Kachurin,
Sudha K. Shenoy, Kai-Ping Deng, Linda Yu, Di Xia,
Hoeon Kim, and Johann Deisenhofer

1. INTRODUCTION

The cytochrome bc_1 complex (commonly known as ubiquinol–cytochrome c reductase, or complex III) is a segment of the mitochondrial respiratory chain that catalyzes antimycin-sensitive electron transfer from ubiquinol to cytochrome c (Rieske, 1967; Hatefi et al., 1962). The reaction is coupled to the translocation of protons across the mitochondrial inner membrane to generate a proton gradient and membrane potential for ATP synthesis.

Bovine heart mitochondrial cytochrome bc_1 complex was first isolated in 1962 (Hatefi et al., 1962); since then, several purification methods have been introduced (Ljungdahl et al., 1987; Engel et al., 1980; Yu and Yu, 1980; Hatefi, 1978; Rieske, 1967). The purified cytochrome bc_1 complex contains 11 protein

Chang-An Yu, Li Zhang, Anatoly M. Kachurin, Sudha K. Shenoy, Kai-Ping Deng, and Linda Yu • Department of Biochemistry and Molecular Biology, Oklahoma State University, Stillwater, Oklahoma 74078.   **Di Xia, Hoeon Kim, and Johann Deisenhofer** • Howard Hughes Medical Institute and Department of Biochemistry, University of Texas, Southwestern Medical Center, Dallas, Texas 75235.

Frontiers of Cellular Bioenergetics, edited by Papa et al. Kluwer Academic/Plenum Publishers, New York, 1999.

subunits, as revealed by high-resolution sodium dodecyl sulfate–polyacryl-amide gel electrophoresis (SDS-PAGE) (Gonzales-Halphen et al., 1988; Schägger et al., 1986). The complex consists of 2165 amino acid residues and four prosthetic groups, with a total molecular mass of 248 kDa, without counting the bound phospholipids. The amino acid sequences of all subunits are available either by peptide (Schägger et al., 1985, 1987; Borchart et al., 1985, 1986; Wakabayashi et al., 1982, 1985) or nucleotide (Yu et al., 1995; Gencic et al., 1991; Usui et al., 1990) sequencing. The essential redox compo-nents of the cytochrome bc_1 complex are two b cytochromes (b_{565} and b_{562}), one c-type cytochrome (c_1), one high-potential iron–sulfur cluster (FeS), and a ubiquinone.

Based on recent intensive biochemical and biophysical investigations of the electron transfer and proton translocation mechanisms, investigators in the field now generally favor the proton-motive Q-cycle hypothesis (Mitchell, 1976; Trumpower, 1990a). The key feature of the Q-cycle hypothesis is the involvement of two separate binding sites for ubiquinone and ubiquinol: Ubiq-uinol is first oxidized at the Q_o site near the P side of the inner mitochondrial membrane, and ubiquinone is reduced at the Q_i site near the N side of the membrane. According to the Q-cycle model, one electron is transferred from ubiquinol to the Rieske iron–sulfur center, and then to cytochrome c via cytochrome c_1. The newly generated reactive ubisemiquinol then reduces the low-potential cytochrome b_{566} heme (b_L). Reduced b_L rapidly transfers an electron to the high-potential cytochrome b_{562} heme (b_H), which is located on the opposite side of the membrane. A ubiquinone or ubisemiquinone bound at the Q_i site then oxidizes the reduced b_H. Proton translocation is the result of deprotonation of ubiquinol at the Q_o site and protonation of ubiquinol at the Q_i site. The Q-cycle mechanism is supported by several experimental observa-tions, such as the oxidant-induced reduction of cytochrome b (Erecinska et al., 1972; Wikström and Berden, 1972), the detection of antimycin-sensitive and -insensitive transient Q radicals (Nagaoka et al., 1981; De Vries et al., 1981; Ohnishi and Trumpower, 1980), the ejection of two protons per one electron transfer in cytochrome bc_1 complexes from many sources (Hauska et al., 1983), and the binding of specific inhibitors to the Q_i or Q_o site (Von Jagow and Link, 1986).

In addition to the redox-active protein subunits, mitochondrial cyto-chrome bc_1 complex also carries non–redox-active proteins, the so-called supernumeral subunits (Trumpower, 1990b). Since these supernumeral sub-units are not present in bacterial cytochrome bc_1 complexes, their role in the complex has long been assumed to be structural rather than catalytic. The failure to isolate viable mitochondrial bc_1 without the supernumeral subunits suggests that the vital role of these subunits is in maintaining the structural integrity of the complex. Recent studies showed that plant mitochondrial bc_1

complexes from wheat, potato, and spinach have mitochondrial-processing peptidase (MPP) activity in addition to electron transfer activity (Glaser *et al.*, 1994). MPP activity was associated with the core subunits (Eriksson *et al.*, 1994, 1996; Braun and Schmitz, 1995; Braun *et al.*, 1992) of the plant complex. However, a similar peptidase activity is not detected in the bovine bc_1 complex, even though the sequences of core protein subunits 1 and 2 are highly homologous with the β and α subunits of MPP (Braun and Schmitz, 1995), respectively. The crystal structure of bovine bc_1 has a putative MPP active site and a zinc-binding motif in core subunits (Xia *et al.*, 1997), even though no detectable MPP activity is found in the active bc_1 complex. The three-dimensional structure of the bc_1 complex also revealed an unidentified polypeptide (subunit 9) bound to the putative active site of MPP in the core proteins. Probably, the lack of MPP activity in the bovine cytochrome bc_1 complex is due to the binding of this unidentified polypeptide to the MPP active site at core subunits 1 and 2. If this were the case, one would expect to find MPP activity when this bound inhibitor polypeptide is released or when its binding to core proteins is weakened. Indeed, when crystalline cytochrome bc_1 complex is treated with nonionic detergents, such as Triton X-100 or zwitergen, the electron transfer activity is impaired and MPP activity is observed (Deng *et al.*, 1997).

In this chapter, we summarize our work on the determination of the three-dimensional structure of beef heart mitochondrial cytochrome bc_1 complex and the activation of MPP activity in this complex. For other work on this and related complexes, excellent reviews are available (Gennis and Trumpower, 1994; Hauska *et al.*, 1983).

2. PREPARATION AND CRYSTALLIZATION OF THE CYTOCHROME bc_1 COMPLEX

2.1. Preparation and Properties

The cytochrome bc_1 complex used for crystallization is prepared from a highly purified succinate–cytochrome c reductase preparation (Yu and Yu, 1980). Succinate dehydrogenase in extensively dialyzed succinate–cytochrome c reductase is solubilized and removed by alkaline treatment (pH 10) in the presence of 10 mM succinate under anaerobic conditions. The resulting cytochrome bc_1 particle is suspended in 50 mM TrisCl buffer, pH 8.0, containing 0.66 M sucrose. This cytochrome bc_1 particle contains the cytochrome bc_1 complex and a protein fraction (QPs) that converts succinate dehydrogenase into succinate–Q reductase and is capable of reconstitution with pure succinate dehydrogenase to form succinate–cytochrome c reductase. Highly purified cytochrome bc_1 complex is obtained from the oxidized bc_1 particle by deoxycholate solubilizating followed by a 14-step ammonium acetate fractionation (Yu and Yu, 1980). A trace amount of cytochrome c oxidase and cytochrome c is added to the cytochrome bc_1 particle to ensure the fully oxidized state of the bc_1 complex. Purified complex contains (in nmoles per mg protein) 8.3 of cytochrome b, 4.7 of cytochrome c_1, 3.5 of ubiquinone, and 250 of phospholipid. The specific activity of purified complex is 11 μmole cytochrome c reduced per min per nmole of cytochrome b at 23°C. When the purified complex is subjected to high-resolution SDS-PAGE (Gonzalez-Halphen *et al.*, 1988; Schägger *et al.*, 1986), 11 protein bands are observed.

2.2. Crystallization and Data Collection

The cytochrome bc_1 complex was diluted with an equal volume of 0.5 M sucrose and precipitated with an equal volume of 50% saturated ammonium acetate. The precipitates were collected by centrifugation and redissolved in 20 mM ammonium acetate buffer, pH 7.2, containing 20% glycerol, to a protein concentration of 20 mg/ml. To this cytochrome bc_1 complex was added 0.5 M MOPS buffer, pH 7.2, and 2% detergent solution [decanoyl-N-methylglucamide (DMG), or short (diheptanoyl) side-chained phosphatidyl choline (SPC)], to give a final concentration of 50 mM MOPS and 0.1% detergent. A 0.57 : 1 volume of 12% PEG-4000 in 50 mM MOPS buffer containing 0.5 M KCl, 20% glycerol, and 0.1% detergent was slowly added to the cytochrome bc_1 complex solution with constant stirring. After 2 hr incubation at 0°C, the mixture was centrifuged for 10 min at 40,000×g to remove precipitates formed, if any. The clear solution was used for crystallization. For a quick test, 4-μl aliquots of clear supernatant solution were used for the hanging drop setup and incubated with 200 μl of equilibrating solution containing 18% PEG, 0.5 M KCl, and 20 mM Tris-Cl buffer, pH 7.0. Under these conditions, crystals formed overnight. To grow crystals for X-ray diffraction, 30- to 50-μl aliquots of clear supernatant solution were placed in small test tubes (6 × 50 mm) and incubated with equilibrating solution at 0 to 4°C. Crystals were mature in 3–4 weeks; they were rectangular in shape, ranging in size from 0.5 to 1 mm (Yu *et al.*, 1995). Figure 1 shows crystals of cytochrome bc_1 complex grown in 20% glycerol. Following our first report of crystallization of cytochrome bc_1 complex in 1991 (Yue *et al.*, 1991), several other investigators have also reported crystallization of this complex (Yu *et al.*, 1996; Lee *et al.*, 1995; Berry *et al.*, 1992, 1995; Kawamoto *et al.*, 1994; Kubota *et al.*, 1991).

Crystals of the cytochrome bc_1 complex are very stable in the cold and show full enzymatic activity when redissolved in aqueous solution. Crystalline cytochrome bc_1 complex is in the oxidized state and contains (in nmoles per mg protein) 2.5 of ubiquinone, 8.4 of cytochrome b, 4.2 of cytochrome c_1, and 140 of phospholipids. About 36% of the phospholipid associated with crystalline

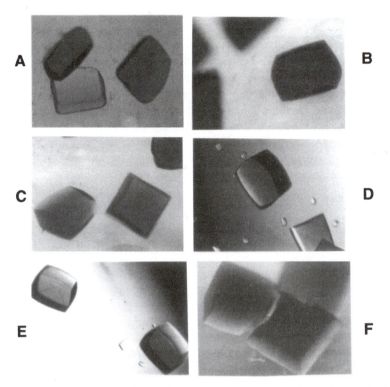

FIGURE 1. Crystals of the cytochrome bc_1 complex grown in 20% glycerol. Crystals of the cytochrome bc_1 complex grown under different conditions. (A) Native cytochrome bc_1 complex. (B) Cytochrome c–cytochrome bc_1 complex. (C) UHDBT–cytochrome bc_1 complex. (D) Stigmatellin–cytochrome bc_1 complex. (E) Stigmatellin–antimycin–cytochrome bc_1 complex. (F) Antimycin–myxothiazole–cytochrome bc_1 complex.

cytochrome bc_1 complex is diphosphatidylglycerol. Absorption spectra of the redissolved crystals show a Soret: UV ratio of 0.88 and 1.01 in the oxidized and the reduced forms, respectively.

Although crystals of the cytochrome bc_1 complex are very stable at 4°C, they are highly sensitive to X rays. However, the crystals can be frozen in the presence of a high concentration of glycerol. Under cryogenic conditions (100°K), these crystals diffract X rays to 2.6 Å resolution using a synchrotron X-ray source. The crystals have the symmetry of space group I4$_1$22, with unit cell dimensions of $a = b = 153.5$ Å and $c = 597.7$ Å, with one bc_1 monomer in the crystallographic asymmetric unit; each monomer consists of 11 different protein subunits.

2.3. Cocrystallization of the Cytochrome bc_1 Complex with Inhibitors

The crystallization conditions described for native cytochrome bc_1 complex were also used for crystallization of the cytochrome bc_1–inhibitor complexes. The cytochrome bc_1 complex was incubated with electron transfer inhibitors (Von Jagow and Link, 1986), individually or in combination, at 4°C for 30 min before being mixed with the precipitating solution (0.57 volume). The molar ratios of inhibitors to the cytochrome bc_1 complex used were 1.3, for antimycin, MOA-Stilbene, myxothiazole, and stigmatellin, and 2.1 for 5-undecyl-6-hydroxy-4,7-dioxobenzothiazole (UHDBT). The shapes and sizes of crystals of the bc_1–inhibitor complex were very similar to those of the native cytochrome bc_1 complex, and they diffracted X rays to a resolution comparable to that of the native crystals.

2.4. Cocrystallization of the Cytochrome bc_1 Complex with Cytochrome *c*

Cocrystallization of the cytochrome bc_1 complex with cytochrome *c* was achieved by incubating the complex with a slight molar excess (1.1 : 1) of beef heart cytochrome *c* for 1 hr before mixing with the precipitating buffer solution containing no KCl. Crystals of the cytochrome bc_1–cytochrome *c* complex formed within 4 weeks. The presence of cytochrome *c* in the crystals was confirmed by spectral analysis of washed crystals or crystalline solution obtained by dissolving washed crystals in Tris-Cl buffer, pH 8.0, containing 0.1% deoxycholate. The molar ratio of the cytochrome *c*–cytochrome bc_1 complex was approximately 1. Crystals of the cytochrome bc_1–cytochrome *c* complex also diffracted X rays to a resolution comparable to that of native crystals.

3. STRUCTURE ANALYSIS OF THE CYTOCHROME bc_1 COMPLEX BY X-RAY DIFFRACTION

3.1. Structure Overview

Crystals grown in the presence of glycerol diffract X rays to 2.6 Å resolution under cryogenic conditions and a native data set is completed at 2.9 Å resolution. The diffraction quality of heavy-metal-derivatized crystals is comparable to that of native crystals. Initial multiple isomorphous replacement anomalous scattering (MIRAS) phases were determined with seven heavy metal derivatives. Electron density maps obtained with the MIRAS phases were subsequently improved with cyclic density modification procedures. Better phases were then used to improve further the initial MIRAS phases.

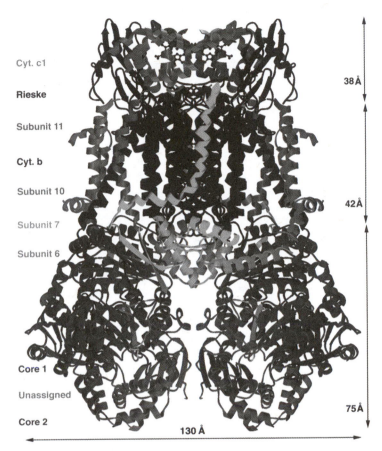

Cyt. c1

Rieske

Subunit 11

Cyt. b

Subunit 10

Subunit 7

Subunit 6

Core 1

Unassigned

Core 2

38Å

42Å

75Å

130 Å

FIGURE 2. Ribbon diagram of dimeric mitochondrial cytochrome bc_1 complex. Color code for each subunit is indicated. The unassigned peptide in between subunits core 1 and core 2 is most likely subunit 9. The diagram is produced with program Setor.

The electron density map at 2.9 Å resolution clearly shows close interaction between two crystallographic symmetry-related monomers (Fig. 2). The dimer is pear-shaped, with a maximal diameter of 130 Å and a height of 155 Å (Yu *et al.*, 1996), similar to the dimensions observed by electron microscopic studies (Akiba *et al.*, 1996; Leonard *et al.*, 1981). Subunits core 2 and cytochrome *b* contribute major dimer interactions across the twofold symmetry axis. Subunit 6 interacts with cytochrome *b* of one monomer and with core 1 and core 2 of the other monomer.

The cytochrome bc_1 complex can be divided into three regions: the transmembrane helix region, the matrix region, and the intermembrane space region. More than half of the molecular mass is located in the matrix region of the molecule, extending from the transmembrane helix region by 75 Å. This region consists of core 1, core 2, subunit 6, part of subunit 7, cytochrome c_1 (C-terminal part), N-terminal portion of the iron–sulfur protein (ISP), and a sequence-unassigned polypeptide of about 50 residues, most likely to be subunit 9. The intermembrane space region, which consists mainly of the water soluble parts of cytochrome c_1 and ISP, extends 38 Å into the cytoplasm from the membrane surface. The transmembrane helix region is about 42 Å thick, with 13 transmembrane helices in each monomer.

Based on the electron density map at 2.9 Å resolution, amino acid sequences to the following protein subunits of the cytochrome bc_1 complex have been assigned: core 1 (445 residues), core 2 (422 residues), cytochrome b (378 residues), subunit 6 (104 residues), subunit 7 (74 residues), ISP (196 residues), subunit 10 (49 residues), subunit 11 (48 residues), and parts of cytochrome c_1 (73 of 241 residues). Including sequence-unassigned polyalanine models, nearly 2000 residues (about 14,500 nonhydrogen atoms) have been assigned, accounting for about 92% of the residues in the cytochrome bc_1 complex.

In the crystal, the transmembrane regions of neighboring molecules form planes normal to the crystallographic 4_1 axis, stacked on top of each other with alternating orientations of dimeric bc_1 complexes. Most crystal contacts are made by the two core subunits in the matrix regions of the molecule, a few by the transmembrane helix region, and none by the proteins that extend into the intermembrane space. This distribution of crystal contacts between bc_1 dimers is correlated with and may be responsible in part for the nonuniform quality of the electron density.

3.2. Localization of Redox Centers

A difference map, based on the anomalous scattering effect in the diffraction data from native crystals, showed four strong peaks in the crystallographic asymmetric unit that are at least 8 SD above the mean values of the electron density maps (Fig. 3). When the X-ray wavelength used in the data collection was shifted to the absorption edge of the iron atoms in the crystal, the heights of these peaks increased while the noise remained unchanged, indicating that the anomalous scattering signal indeed comes from the iron atoms of the bc_1 crystal. Two of these four peaks are located in the transmembrane region: one near the matrix side (N side) surface and one near the intermembrane space (P side) surface; the other two peaks reside in the intermembrane space region. Based on the known biochemical and biophysical data (Tolkatchev *et al.*, 1996; Miki *et al.*, 1994; Degli Esposti *et al.*, 1993; Robertson and Dutton, 1988; Von Jagow *et al.*, 1986; Widger *et al.*, 1984; Glaser and Crofts, 1984), the two peaks in the transmembrane region are assigned to heme irons of b_L and b_H, with b_L

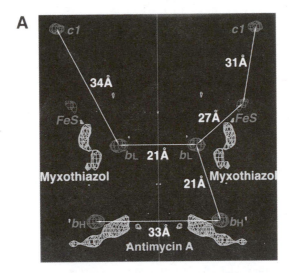

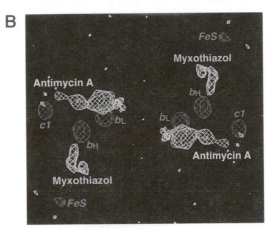

FIGURE 3. Anomalous difference Fourier map shows eight maxima related by twofold molecular axis. The assignment of these peaks and the distance between the peaks are indicated. The difference in electron density between antimycin A bound and native, and between myxothiazole bound and native is also shown. (A) Side view. (B) Top view.

being closer to the P side and b_H to the N side. The two peaks in the intermembrane space near the P side membrane surface are assigned to the iron–sulfur cluster and the heme iron of cytochrome c_1, with the iron–sulfur cluster being closer to b_L. The relative location of these four redox centers are indicated in Fig. 3.

The distances between heme irons in b_H, b_L, c_1, and the iron–sulfur cluster

(FeS) are listed in Fig. 3. The angle FeS–b_L–b_H is 95°, the angle c_1–FeS–b_L is 61°, and the angle c_1–b_L–b_H is 155°. The distance of 21 Å between the two b-type hemes inside the membrane is in good agreement with the prediction of 22 Å for the distance between the putative heme ligands in the primary sequence (Widger *et al.*, 1984) and with the spin relaxation of the b_H and b_L from electron paramagnetic resonance (EPR) studies (Von Jagow *et al.*, 1986). The distance of 31 Å between the iron–sulfur cluster and the c_1 heme iron is longer than that expected based on the observed high electron transfer rate (Meinhardt and Crofts, 1982; Tsai *et al.*, 1983). The distance between the b_L of the two monomers is less than 21 Å. This indicates close interaction between these two b_L cytochromes and suggests a functional dimer in the cytochrome bc_1 complex.

3.3. Localization of Inhibitor Binding Sites

Biochemical and biophysical studies of the interaction between the cytochrome bc_1 complex and electron transfer inhibitors have made significant contributions to our knowledge of the electron transfer pathway in this complex. Antimycin blocks electron exit from cytochrome b_H and causes a spectral red shift. Binding of antimycin to the cytochrome bc_1 complex destabilizes ubisemiquinone at the Q_i site, indicating that antimycin binds near cytochrome b_H. Inhibitor-resistant mutant analysis suggests that antimycin binds on the matrix side of the inner membrane (Brasseur *et al.*, 1996; Gennis *et al.*, 1993; Howell *et al.*, 1989). In a difference electron density map between antimycin A-bound and native crystals, a well-defined feature with the shape of antimycin A appears in a pocket in the transmembrane region of cytochrome b next to the b_H heme iron site (Fig. 4). An extended negative electron density close to the density of antimycin A is also observed. The negative electron density could represent part of a ubiquinone molecule that is bound in the native crystal but displaced by bound antimycin A. This suggests that the antimycin A binding site partly overlaps the quinone reduction site, Q_i. Mutations in cytochrome b that influence the Q_i site involve residues lining the wall of the antimycin A binding pocket (Brasseur *et al.*, 1996).

Myxothiazol stops the electron flow from ubiquinol to the iron–sulfur center and, indirectly, to b_L. The difference map between myxothiazol-bound and native crystals showed a well-fitting myxothiazol density in the Q_o pocket within cytochrome b, midway between the iron positions of b_L and the iron–sulfur center (Fig. 5). The myxothiazol binding pocket is underneath the cd helices of cytochrome b, and mutations in the cd helices lead to loss of quinol oxidation (Brasseur *et al.*, 1996). Thus, the myxothiazol binding site is probably the quinol oxidation site, Q_o, of the cytochrome bc_1 complex. Unlike the Q_i site, this site is not in direct contact with either the b_L heme or the iron–sulfur

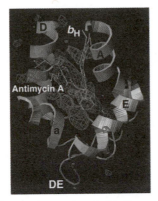

FIGURE 4. Antimycin binding pocket. Antimycin binding pocket is made of part of membrane spanning helices D, A, and E, and DE loop and heme b_H. The blue density is the positive difference density between antimycin A bound and native and the coral color is the negative difference density between antimycin A bound and native, which may represent bound ubiquinone in the native crystal being replaced upon antimycin A binding. The antimycin A density is in direct contact with the edge of heme b_H. This diagram is produced using program O.

cluster; it is surrounded by aromatic residues that could mediate electron transfer.

Another potent and specific electron transfer inhibitor of the cytochrome bc_1 complex is UHDBT, which blocks electron transfer between the iron–sulfur protein and cytochrome c_1 (Trumpower and Haggerty, 1980) and inhibits

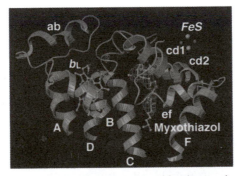

FIGURE 5. Myxothiazol binding pocket. Myxothiazol binding pocket is located between group A and group B helices (see text), and is made of transmembrane helices C, F, and cd and ef loops. The myxothiazol density is not in direct contact with either FeS or heme b_L.

oxidant-induced cytochrome *b* reduction (Bowyer and Trumpower, 1980). UHDBT binds near the iron–sulfur cluster, as indicated by the modification of EPR signals in its presence. Preliminary analysis of the difference electron density maps between the UHDBT-bound and the native crystals indicates that the dioxobenzothiazole part of the UHDBT molecule is bound near the iron–sulfur cluster. Binding of UHDBT also enhances the electron density of the anomalous scattering peak of the iron–sulfur cluster, suggesting that this inhibitor reduces the mobility of ISP in the crystal. The electron density of UHDBT is in the same pocket and partially overlap with that of myxothiazol.

E-β-methoxyacrylate-stilbene (MOAS) is another potent Q_o site inhibitor. The binding affinity of MOAS to the cytochrome bc_1 complex decreases when ISP and cytochrome c_1 are in the reduced states. A difference electron density map between MOAS-bound and native crystals shows a well-fitting MOAS density in the Q_o pocket, close to heme b_L, but not in direct contact with it. Binding of MOAS to the cytochrome bc_1 complex greatly diminishes the electron density of the anomalous scattering peak of the iron–sulfur cluster, suggesting that this inhibitor also increases the mobility of ISP in the crystal. The electron density of MOAS does not overlap with that of UHDBT, but does overlap a great deal with that of myxothiazol.

Stigmatellin inhibits electron transfer between the iron–sulfur cluster and cytochrome c_1 by increasing the midpoint potential of ISP. Binding of stigmatellin to the cytochrome bc_1 complex also exerts an immobilization effect on ISP. This effect is slightly stronger than that observed with UHDBT, judging from their ability to enhance the anomalous scattering peak of the iron–sulfur cluster. Biochemical analysis also suggests that stigmatellin is bound to the cytochrome bc_1 complex more tightly than UHDBT. The electron density of stigmatellin largely overlaps with that of UHDBT and partly with that of myxothiazol.

3.4. Intermembrane Space Region: Cytochrome c_1 and ISP

The electron density is less well-defined for the intermembrane space region of the molecule as compared with that for the matrix region. In spite of the compact and rigid structure of the extramembrane domain of ISP (Iwata *et al.*, 1996), the electron density of this domain is particularly poor, indicating that ISP is highly mobile in the crystal. The amino acid sequence assignment of ISP in the crystal was possible due to the aid of reported coordinates for the extramembrane domain of ISP. The 196-amino-acid residue of ISP can be divided into three domains: the NH_2-terminal tail of 70 residues that begins with a NH_2-terminal random coil interacting with core subunits and ends with a membrane-spanning helix; a COOH-terminal compact "head" that is a globular soluble domain of about 120 residues; and a flexible linker connecting the two. The head domain of ISP in one monomer is connected to the tail domain of the protein in the other monomer, suggesting a functional dimer for the cytochrome bc_1 complex. The current model for the extramembrane regions of cytochrome c_1 has several fragments; only the sequence of the COOH-terminal transmembrane helix had been assigned. Part of the uninterpreted electron density near the cytochrome c_1 heme may belong to subunit 8 (hinge protein), which mediates the interaction between cytochromes c_1 and c.

3.5. Membrane-Spanning Region: Cytochrome b

Eight of the 13 transmembrane helices, labeled A to H, belong to cytochrome b protein (Fig. 6). The rest, one each, belong to cytochrome c_1, ISP, subunits 7, 10, and 11. Most of them are tilted with respect to the plane of the membrane. In the bc_1 dimer, the 16 transmembrane helices of cytochromes b form the core of the transmembrane helix region, with the other ten transmembrane helices bound to the periphery (Fig. 7). Viewed from either side of the membrane, the overall shape of the bc_1 dimer's transmembrane helix region is approximately elliptical, with long and short axes of about 90 Å and 80 Å (Fig. 8).

Topographically, cytochrome b is a rather simple molecule, with four long loops (named AB, CD, DE, and EF) and three short loops (BC, FG, and GH) connecting the eight transmembrane helices. The DE loop is on the matrix side

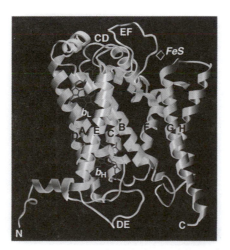

FIGURE 6. Ribbon diagram of cytochrome b. Cytochrome b has eight membrane-spanning helices. Both N- and C-termini are on the matrix side of the membrane. The group A helices are made of helices A, B, C, D, and E, and the group B helices are made of helices F, G, and H.

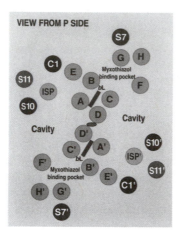
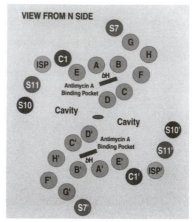

FIGURE 7. Schematic diagram of the arrangement of the membrane-spanning helices of cytochrome bc_1 complex, as shown in the views of (A) P side and (B) N side.

and the other three long loops are on the intermembrane space side of the membrane. Both the AB and EF loops each contain one helix: the ab and ef helices, respectively. The CD loop has two short helices, named cd1 and cd2, which form a hairpin structure. All the extramembrane helices run along the membrane surface, except for the ef helix, which is buried inside the membrane. The DE loop has no regular secondary structure. Both the NH_2- and COOH-termini of the cytochrome b subunit are located in the mitochondrial matrix. The eight transmembrane helices of the cytochrome b subunit are approximately where predicted from hydropathy analysis of its primary protein sequence and from mutagenesis studies (Brasseur *et al.*, 1996). They associate

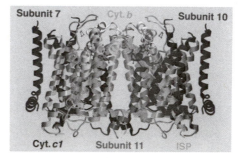

FIGURE 8. Membrane-spanning region of cytochrome bc_1 complex showing 26 helices in the dimer.

into two groups: helices A, B, C, D, and E in one group (group A), and helices F, G, and H in the other (group B) (Fig. 6). These two groups are in close contact on the matrix side and separated on the intermembrane space side (Fig. 7). The two b-type hemes are bound within the four-helix bundle made of helices A, B, C, and D. The axial ligands of both hemes are histidines (Widger *et al.*, 1984): H83 and H182 for bL, and H97 and H196 for bH. These histidines are located in transmembrane helices B and D, and are absolutely invariant in the sequences of b-type cytochromes from over 900 different species (Degli Esposti *et al.*, 1993). Other invariant residues as well as residues conferring inhibitor-resistant mutations (Brasseur *et al.*, 1996) in cytochrome b are around the myxothiazol and antimycin A binding sites.

The cytochrome b dimer forms two large cavities, related by twofold symmetry (Fig. 8), which are accessible from the membrane bilayer. The cavities are made of the transmembrane helices D, C, F, and H in one monomer and helices D' and E' from the other monomer of cytochrome b and the transmembrane helices of c_1, ISP, and subunits 10 and 11. The Q_i pocket of one monomer and the Q_o pocket of the other monomer are accessible through the same cavity. We assume that hydrophobic molecules, such as ubiquinol, antimycin A, or myxothiazol, can reach their binding sites in the enzyme complex through these cavities, and that ubiquinol made at the Q_i site of one monomer can proceed to the nearby Q_o site of the other monomer without having to leave the bc_1 complex. This explains well the reported observation that isolated succinate–cytochrome c reductase needs only 1 mole of Q per mole of enzyme to show maximum activity (Yu and Yu, 1981). The electron transfer activity of succinate–cytochrome c reductase can only be stimulated by the addition of exogenous Q when the Q content in the preparation is less than 1 mole per mole of cytochrome c_1. The cavities appear tightly sealed on the surface of the intermembrane space by the surface helices cd1 and cd2 of cytochrome b. However, they are accessible by solvent from the matrix side through channels leading to subunits core 1, core 2, and subunit 6. This arrangement allows proton uptake at the Q_i site from the matrix and prevents leakage of protons through the bc_1 complex.

3.6. Matrix Region and Core Proteins

More than half of the total molecular mass of the cytochrome bc_1 complex, including the two core proteins, subunit 6, and part of subunit 7, is in the matrix region. Subunit 7 contributes to both the matrix and transmembrane regions. The COOH-terminal 50 residues of subunit 7 form a long, bent transmembrane helix, and the NH_2-terminal part of the protein associates with the core 1 protein as part of a β-sheet which serves as one of the membrane anchors for the subunits in the matrix region (see Fig. 2). Subunit 6 consists of four helices and connecting loops. It is attached to the matrix side of exposed residues of the helices F, G, and H of cytochrome *b* of one monomer and contacts the subunits core 1 and core 2 of the other monomer.

Subunits core 1 and core 2 are synthesized in the cytosol as precursor proteins of 480 and 453 amino acid residues, respectively. Proteolytic processing in the mitochondrial matrix removes 34 residues from the NH_2-terminus of precursor core 1 protein and 14 residues from the NH_2-terminus of precursor core 2 protein. In our electron density maps, density for core 1 starts at residue 36 and for core 2 at residue 33. Core 1 and core 2 have 21% sequence identity, and their three-dimensional structures are remarkably similar (Fig. 9). A rotation around an approximate twofold symmetry axis superimposes 384 α-carbon atoms from each subunit with an root mean square (rms) deviation of 1.7 Å.

Each of the core subunits consists of two structural domains of about equal size and almost identical folding topology, which are also related by approximate twofold rotational symmetry. Both domains are folded into one mixed β-sheet of five or six β-strands, flanked by three α-helices on one side and one α-helix, from the other domain, on the other side. Rotation of the COOH-terminal domain on the NH_2-terminal domain superimposes 134 α-carbons from each domain of core 1 with an rms deviation of 2.0 Å and 124 α-carbons from each domain of core 2 with an rms deviation of 2.1 Å. Only about 10% of the superimposed residues are chemically identical. Searches with the program DALI (Holm and Sander, 1996) did not identify known protein structures closely related to the core subunit domains.

The overall shape of each core protein resembles a bowl. In the assembled bc_1 complex, the NH_2-terminal domain of core 1 is interacting with the COOH-terminal domain of core 2, and vice versa. Core 1 and core 2 enclose a large cavity that was also observed by electron microscopy (Akiba *et al.*, 1996;

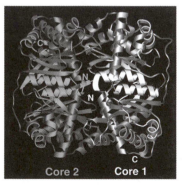

FIGURE 9. Structure of core subunits. Ribbon diagram of subunits core 1 and 2 heterodimer.

Leonard *et al.*, 1981). Because the two internal approximate twofold rotation axes of core 1 and core 2 differ in direction by 14.5°, the two bowls representing these subunits come together in the form of a ball with a crack leading to the internal cavity. The crack is filled with part of the 50-residue sequence-unassigned polypeptide. The amino acid residues lining the wall of the cavity are mostly hydrophilic.

The protein sequences of core subunits of beef bc_1 complex are homologous with the subunits of the general mitochondrial matrix processing peptidase (MPP), which cleaves signal peptides of nucleus-encoded proteins after import into the mitochondrion. MPP belongs to the pitrilysin family of zinc metalloproteases with an inverted zinc binding motif whose members include insulin-degrading enzymes from mammals and protease III from bacteria (Rawlings and Barrett, 1991). Mammalian MPP is a soluble heterodimer of α- and β-subunits localized in the mitochondrial matrix. Both subunits of MPP are required for protease activity (Paces *et al.*, 1993), with the active site being part of the β-subunit. The core 1 protein of the bovine bc_1 complex has 56% sequence identity to the β-subunit of rat MPP, 38% to yeast, and 42% to potato, while the core 2 protein is 27% identical to the α-subunit of rat MPP, 28% to yeast, and 30% to potato. We therefore can expect that the MPPs of these species are very similar in structure to the bovine core 1 and core 2 subunits. In plant mitochondria, the MPP activity is membrane bound and is an integral part (core proteins) of the bc_1 complex (Glaser *et al.*, 1994).

4. PROTEIN-PROCESSING PEPTIDASE ACTIVITY

4.1. MPP Activity of the Cytochrome bc_1 Complex

In addition to the sequence homology between core subunits of bovine bc_1 complex and subunits of MPP, a sequence of $Y_{91}XXE_{94}H_{95}(X)_{76}E_{171}$ (Fig. 10), similar to the zinc binding motif, $HXXEH(X)_{76}E$, in the β-subunit of MPP, is present in the core 1 subunit of bovine enzyme. Furthermore, residues K_{300} and R_{301} of bovine core 2 subunit are structurally close to the putative zinc binding motif of core 1 and may contribute to the active site. However, no MPP activity is detected in the active bovine cytochrome bc_1 complex. Based on three-dimensional structural information of the bovine complex, the lack of MPP activity was attributed to the binding of an unidentified polypeptide at the active site of MPP in the core subunits (Fig. 2). This is indeed the case, because when crystalline bc_1 complex is treated with nonionic detergents such as Triton X100, electron transfer activity is lost, while MPP activity appears (Fig. 11). Triton X-100 disrupts the structural integrity of the complex, causing the loss of electron transfer activity and a weakening of the binding of inhibitory polypep-

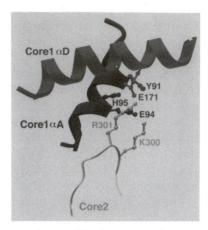

FIGURE 10. Putative zinc binding motif for the core 1 and core 2 heterodimer. In the crystal structure, no electron density for zinc atom is found; instead, R_{301} of core 2 and E_{171} of core 1 forms a salt bridge.

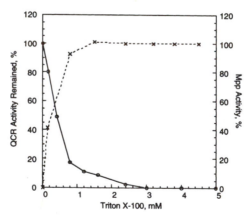

FIGURE 11. Reactivation of MPP from crystalline cytochrome bc_1 complex by Triton X-100. Aliquots of the crystalline cytochrome bc_1 complex (30 μg in 20 μl) in 15 mM Tris-Cl buffer, pH 8.0, were mixed with 40 μg of substrate peptide, VPASVRY**SHTDIK** ($_{-7}V_{+6}$), and indicated concentrations of Triton X100 and incubated at 37°C for 24 hr. At the end of incubation, ubiquinol–cytochrome c reductase activity (QCR) and MPP activity were assayed. The QCR activity (○) was determined by the reduction of cytochrome c spectrophotometrically using reduced Q2 as substrate. The 100% QCR activity equals 5.5 μmole cytochrome c reduced per minute per nmole cytochrome b, at room temperature. The MPP activity (×) was determined by product formation using high-pressure liquid chromatography. The highest product peak was taken as a 100% of MPP activity.

tide to core subunits, and thus activates MPP. When synthetic polypeptides composed of various lengths of the C-terminal end of presequences and the N-terminal end of mature subunits are used as substrate, the specificity of MPP is governed more by the length of the N-terminal end of the mature subunit than by the length of the C-terminal end of the presequence. Although the MPP activity is not stimulated by the addition of divalent cations, it is inhibited by the addition of EDTA. The EDTA-inhibited activity is restored by the addition of divalent cations, indicating that divalent cations are essential.

4.2. Cloning, Overexpressing, and *in Vitro* Reconstitution of MPP from Core Proteins

The cDNAs encoding core 1 and core 2 subunits of the bovine bc_1 complex were amplified from bovine heart cDNA library using polymerase chain reaction, cloned into the expression vector, pET30^{a+} and overexpressed as *His-tag* fusion proteins in *Escherichia coli* (Yu *et al.*, 1997). Recombinant core 1 and core 2 proteins were purified to homogeneity from soluble cell-free extract by a Ni-NTA resin. Purified recombinant core 1 and core 2 alone have no MPP activity. However, when these two recombinant proteins are mixed together, the MPP activity appears (Yu *et al.*, 1997). Maximum activity is observed when the molar ratio of these two core proteins reaches one. Like the MPP activity in Triton X100-treated bc_1 complex, the reconstituted MPP activity is stimulated by the addition of divalent cations and inhibited by EDTA.

5. STRUCTURAL BASIS OF THE ELECTRON TRANSFER REACTION

5.1. Does the Cytochrome bc_1 Complex Function as a Dimer or a Monomer

Most reports in the literature favor the dimeric association of the cytochrome bc_1 complex, even though isolation of monomeric complex has been documented (Nalecz and Azzi, 1985; Musatov and Robinson, 1994). Crystal structures show the bc_1 complex as a dimer, but whether or not the enzyme functions as a dimer is still subject to debate. The Q-cycle hypothesis does not require a functional dimer, nor do most biochemical and biophysical data discriminate one monomer from the other. However, recent crystallographic information suggests the dimer as a functional unit. First, the distance between two b_L hemes from each monomer, 21 Å, is the same as that between b_L and b_H hemes within a monomer. Since many electron transfer systems operate with a free energy difference close to zero (Moser *et al.*, 1992), electron transfer between these two b_L hemes should be possible, especially when the membrane is highly energized. Second, dimeric cytochrome bc_1 complex forms two symmetry-related cavities in the membrane-spanning region. Although each monomer contains both Q_i and Q_o sites, they cannot communicate with each other. It is the cavity that brings the Q_o site of one monomer close to the Q_i site of the other. Thus, it is possible for a quinone molecule reduced at the Q_i site to be used in the nearby Q_o site without having to leave the bc_1 complex. This is consistent with the kinetic data showing that 1 mole of ubiquinone per mole of c_1 is sufficient for a maximal rate of electron transfer. Third, in the crystal structure, the head domain of ISP of one monomer interacts with cytochrome b of the other monomer. In summary, crystallographic studies not only show a physical dimeric association in the bc_1 complex, but also suggest that the dimer is functional.

5.2. Ubiquinone Binding Sites

Biochemical studies strongly suggest that oxidation of ubiquinol by the cytochrome bc_1 complex is via the Q cycle. The key feature of the Q-cycle mechanism is the involvement of two conceptually separated ubiquinone–ubiquinol binding sites: one for ubiquinone reduction and one for ubiquinol oxidation. Structural analysis of the bc_1–inhibitor complexes reveals two separate inhibitor binding pockets, located at opposite sides of the membrane, for Q_o and Q_i site inhibitors. Binding of the Q_i site inhibitor, antimycin, may cause dislocation of bound Q at the Q_i site, as suggested by negative electron density in a difference electron density map between inhibitor-bound and native crystals (Fig. 4). Binding of the Q_o site inhibitors, however, does not generate such a negative electron density in the difference maps, indicating that ubiquinone may not be bound in the Q_o pocket in the native crystal. The failure to detect bound ubiquinone at the Q_o pocket may be due to the low binding affinity of the Q_o site for oxidized Q or to the fact that the cytochrome bc_1 complex used for crystallization contains substoichiometric amounts of Q (~0.6 mole per mole complex) (Yue *et al.*, 1991). How many Q molecules are bound to the Q_o site is a question to be answered in the future. Two Q per Q_o site in bacterial cytochrome bc_1 complex has been indicated by EPR characteristics of the iron–sulfur cluster (Ding *et al.*, 1995; Brandt, 1996).

The observations that different Q_o site inhibitors bind at slightly different locations in the Q_o pocket and show different effects on ISP encourage us to speculate that Q may bind to different locations within the Q_o pocket, depending on its redox states, with the fully reduced form being closer to ISP and the half-reduced form being closer to b_L. In other words, the Q binding site closer to the ISP in the Q_o pocket has higher binding affinity for fully reduced Q and the site closer to b_L has higher binding affinity for ubisemiquinone. How cytochrome b senses such a shift in Q binding positions and subsequently releases ISP requires further structural investigation.

5.3. Electron Transfer Rate and Distances between the Redox Centers

The distances 21 Å between b_L and b_H and 27 Å between b_L and FeS accommodate well the fast electron transfers observed between the two involved redox centers, assuming ubiquinol is bound somewhere between b_L and FeS. However, the distance of 31 Å between heme c_1 and FeS is difficult to explain in view of the fast electron transfer rate between these two redox centers. Perhaps electron transfer between these two redox centers is facilitated by an as yet unidentified electron mediator, such as an aromatic or sulfhydryl amino acid residue located between these two centers. It has been reported (Yu *et al.*, 1975) that one of the cysteines in cytochrome c_1 is able to reduce heme c_1 upon illumination with visible light. Alternatively, movement of ISP may facilitate the observed fast electron transfer in this region. The iron–sulfur cluster (FeS) is reduced at a position 27 Å from b_L and 31 Å from cytochrome c_1. Once reduced, it moves closer to cytochrome c_1 and may form a transient complex with cytochrome c_1 to donate its electron. The iron–sulfur cluster reduced by the first electron of ubiquinol cannot donate an electron to cytochrome c_1 before the second electron of ubiquinol is transferred to cytochrome b_L. It is tempting to speculate that the change of the Q-binding position and reduction of b_L causes a conformational change in cytochrome b, which moves ISP close enough to heme c_1 to have fast electron transfer (Meinhardt and Crofts, 1982; Tsai *et al.*, 1983). This also would explain why a more powerful reductant, ubisemiquinone, reduces b_L but not FeS during quinol oxidation.

5.4. Movement of Iron–Sulfur Protein

Among the three regions of the bc_1 molecule in the crystalline form, the intermembrane space region, which includes ISP and cytochrome c_1, is the most flexible and perhaps is under constant motion, judging from the low electron density of FeS compared to the heme irons of b_H or b_L. Binding of various Q_o site inhibitors produces different effects on the anomalous signal of FeS, suggesting that these inhibitors are capable of affecting the mobility of FeS and, most likely, of the whole extramembrane domain of ISP (Kim *et al.*, 1997). Since inhibitors are bound to transmembrane helices of cytochrome b protein, the inhibition of electron transfer between FeS and heme c_1 must be indirect and a consequence of the conformational change exerted on the part of cytochrome b near the surface of cytoplasmic side, probably the CD and EF loops. These loops form a cap that seals the Q_o pocket and blocks the exposure of FeS to the pocket. Binding of the Q_i site inhibitor, antimycin, causes no detectable change in ISP mobility, supporting the speculation that ISP movement is caused by a conformational change in cytochrome b near the Q_o pocket. Apparently, the presence of the CD and EF loops between the transmembrane helices of cytochrome b and ISP provides the conformational change that makes the synchronized motion of ISP possible.

5.5. Electron Transfer Events at the Q_0 Site

One of the most important features of the Q-cycle hypothesis is the bifurcation of electron transfer at the Q_o site, the mechanism of which is still under intensive investigation. The information from crystal structure and crystallographic studies of inhibitor-complex suggests two transient quinone binding sites in the inhibitor binding pocket: one Q binding site is likely to be where UHDBT binds, closer to ISP, and designated as P1; the other Q binding site is where MOAS binds, closer to b_L heme and designated as P2. Analogous to inhibitor binding, binding of Q to P1 will cause ISP to be less mobile, and binding of Q to P2 will release ISP. Therefore, the electron transfer events at the Q_o site can be described as follows: A ubiquinol molecule comes in, first binds to the P1 site, ISP is immobilized, and one proton is released, followed by transfer of one electron to ISP. After the second proton is released, the ubisemiquinone radical moves from the P2 site, which causes the ISP to be released, and prevents the remaining electron from going to ISP. The second electron of ubiquinol then reduces the b_L heme. This completes the cycle in a manner similar to that in the proposed catalytic switch model (Brandt and Von Jagow, 1991). This model requires that the P1 site have a higher affinity for quinol when FeS is in the oxidized state and that the P2 site have a higher affinity for ubisemiquinone when b_L is in the oxidized state. The lack of detectable ubisemiquinone radical at the Q_o site, however, is inconsistent with this proposed electron transfer. The reported detection of a transient, antimycin-A-insensitive ubisemiquinone radical at the Q_o site was recently questioned, as it was not sensitive to the Q_o site inhibitors such as myxothiazol, MOA-Stilbene, or stigmatellin.

An alternative hypothesis for the electron transfer event at the Q_o site is that the electron donor of FeS is the ubiquinol heme b_L^{3+} complex and not the ubiquinol alone. Once the first electron of ubiquinol in the complex transfers to FeS, the second electron immediately transfers to heme b_L and then to heme b_H; thus, no semiubiquinone is generated. The electron transfer from heme b_L to b_H results in a conformational change of cytochrome b protein, which makes (or allows) the reduced ISP move to a position closer to heme c_1 to allow electron transfer to take place. In this way, the second electron of ubiquinol may reach its destination (the Q_i site in the symmetric-related monomer) before or at the same time as the first electron reaches heme c_1. Since the oxidation of the reduced FeS depends on the transfer of the second electron of ubiquinol from heme b_L to heme b_H, bifurcation is obligatory for the oxidation of ubiquinol in the bc_1 complex. The intermonomer electron transfer, as supported by the three-dimensional structure of the cytochrome bc_1 complex showing the connection of the Q_i pocket of one monomer to the Q_o pocket of the other, would be more efficient than the electron transfer within the same monomer, because electron transfer between the hemes b_L and b_H is much faster than the movement of Q from Q_o pocket to Q_i pocket.

5.6. Proton Translocation: Pumping/Gating

Although the Q-cycle hypothesis explains well the $2H^+$/electron stoichiometry of proton transfer and the structural information obtained so far generally supports this hypothesis, the proton transfer path in the bc_1 molecule during ubiquinol oxidation is still not clear. No obvious proton channel was found in the cytochrome bc_1 complex at the current level of structure resolution. This, of course, does not exclude the participation of bound water in proton movement. Since the simplest bacterial bc_1 complex has no supernumeral subunits and has full proton translocation ability, proton translocation must be achieved through transmembrane helices of cytochrome b, c_1, and ISP. Van der Waals surface of the matrix side of these transmembrane helices, particularly in cytochrome b, shows several openings for proton uptake from the matrix. Since the Van der Waals surface of the cytoplasmic side is totally sealed, proton exit must be accompanied by some sort of conformational change or gating device. The available biochemical data indicate that the iron–sulfur cluster of ISP plays an essential role in proton exit. It was reported that the iron–sulfur cluster of ISP in the bc_1 complex can be destroyed by the destruction of one of the histidine ligands during illumination of the hematoporphyrin-treated complex. The resulting complex leaks protons when reconstituted into phospholipid vesicles. A strong redox-state-dependent pKa of the histidyl ligands of FeS cluster also suggests that they may be involved in proton release (Link and Iwata, 1996).

5.7. Assembly of Mitochondrial Cytochrome bc_1 Complex

The crystal structure and the experiments on MPP activity of the core subunits provide some understanding of the assembly of process of this large multisubunit cytochrome complex. From the structural point of view, the cytochrome b dimer should be formed first, followed by the association of cytochrome c_1, and then ISP. Other membrane-spanning subunits, such as subunits 7, 10, and 11, are needed to assist the precise positioning of cytochrome c_1 and ISP. For example, subunit 10 of yeast bc_1 was found to be essential for ISP assembly (Brandt *et al.*, 1994). The core subunits most likely function as stabilizing factors for the entire complex. In order to do so, they may act as a suicidal enzyme by clipping subunit 9 off the ISP precursor, then forming a close interaction with subunits c_1 and 7.

ACKNOWLEDGMENTS. The work described in this chapter was supported in part by a grant from the National Institutes of Health (GM 3072 to C. A. Y.) and from the Welch Foundation (to J. D.). J. D. is an investigator in the Howard Hughes Medical Institute.

6. REFERENCES

Akiba, T., Toyoshima, C., Matsunaga, T., Kawamoto, M., Kubota, T., Fukuyama, K., Namba, K., and Matsubara, H., 1996, Three-dimensional structure of bovine cytochrome bc_1 complex by electron cryomicroscopy and helical image reconstruction, *Nat. Struct. Biol.* **3**:553–561.

Berry, E. A., Huang, L., Earnest, T. N., and Jap, B. K., 1992, X-ray diffraction by crystals of beef heart ubiquinol–cytochrome *c* oxidoreductase, *J. Mol. Biol.* **224**:1161–1166.

Berry, E. A., Shulmeister, V. M., Huang, L., and Kim, S-H., 1995, A new crystal form of bovine heart ubiquinol–cytochrome *c* oxidoreductase: Determination of space group and unit-cell parameters, *Acta Crystallog.* **D51**:235–239.

Borchart, U., Machleidt, W., Schagger, H., Link, T. A., and von Jagow, G., 1985, Isolation and amino acid sequence of the 8 kda DCCD-binding protein of beef heart ubiquinol–cytochrome *c* reductase, *FEBS Lett.* **191**:125–129.

Borchart, U., Machleidt, W., Schagger, H., Link, T. A., and Von Jagow, G., 1986, Isolation and amino-acid sequence of the 9.5-kilodalton protein of beef heart ubiquinol–cytochrome *c* reductase, *FEBS Lett.* **200**:81–86.

Bowyer, J. R., and Trumpower, B. L., 1980, Inhibition of the oxidant-induced reduction of cytochrome *b* by a synthetic analogue of ubiquinone, *FEBS Lett.* **115**:171–174.

Brandt, U., 1996, Energy conservation by bifurcated electron-transfer in the cytochrome bc_1 complex, *Biochim. Biophys. Acta* **1275**:41–46.

Brandt, U., and Von Jagow, G., 1991, Analysis of inhibitor binding of the mitochondrial cytochrome *c* reductase by fluorescence quench titration, *Eur. J. Biochem.* **195**:163–170.

Brandt, U., Uribe, S., Schagger, H., and Trumpower, B. L., 1994, Isolation and characterization of QCR10, the nuclear gene encoding the 8.5 kda subunit of the *Saccharomyces cerevisiae* cytochrome bc_1 complex, *J. Biol. Chem.* **269**:12947–12953.

Brasseur, G., Saribas, A. S., and Daldal, F., 1996, A compilation of mutations located in the cytochrome *b* subunit of the bacterial and mitochondrial bc_1 complex, *Biochim. Biophys. Acta* **1275**:61–69.

Braun, H.-P., and Schmitz, U. K., 1995, Are the "core" proteins of the mitochondrial bc_1 complex evolutionary relics of a processing protease, *Trends Biochem. Sci.* **20**:171–175.

Braun, H.-P., Emmermann, M., Kraft, V., and Schmitz, U. K., 1992, The general mitochondrial processing peptidase from potato is an integral part of cytochrome *c* reductase of the respiratory chain, *EMBO J.* **11**:3219–3227.

Degli Esposti, M., de Vries, S., Crimi, M., Ghelli, A., Patarnello, T., and Myer, A., 1993, Mitochondrial cytochrome *b*: Evolution and structure of the protein, *Biochim. Biophys. Acta* **1143**:243–271.

Deng, P-P., Xia, D., Kachurin, A. M., Kim, H., Deisenhofer, J., Yu, L., and Yu, C. A., 1997, Detection of a matrix processing peptidase activity in the cytochrome bc_1 complex from bovine heart mitochondria, *Biophys. J.* **72**:319a.

de Vries, S., Albracht, S. P. J., Berden, J. A., and Slater, E. C., 1981, A new species of bound ubisemiquinone anion in QH_2:cytochrome c oxidoreductase, *J. Biol. Chem.* **256**:11996–11998.

Ding, H., Moser, C. C., Robertson, D. E., Tokito, M. K., Daldal, F., and Dutton, L. P., 1995, Ubiquinone pair in the Q_o site central to the primary energy conversion reactions of cytochrome bc_1 complex, *Biochemistry* **34**:15979–15996.

Engel, W. D., Schagger, H., and Von Jagrow, W., 1980, Ubiquinol–cytochrome c reductase: Isolation in Triton X-100 by hydroxyapatite and gel chromatography: Structural and functional properties, *Biochim. Biophys. Acta* **592**:211–222.

Erecinska, M., Chance, B., Wilson, D. F., and Dutton, P. L., 1972, Aerobic reduction of cytochrome b_{566} in pigeon heart mitochondria, *Proc. Natl. Acad. Sci. USA* **69**:50–54.

Eriksson, A. C., Sjoling, S., and Glaser, E., 1994, The ubiquinol–cytochrome c oxidoreductase complex of spinach leaf mitochondria is involved in both respiration and protein processing, *Biochim. Biophys. Acta* **1186**:221–231.

Eriksson, A. C., Sjoling, S., and Glaser, E., 1996, Characterization of the bifunctional mitochondrial processing peptidase (mpp)/bc_1 complex in spinach oleracea, *J. Bioenerg. Biomembr.* **28**:283–290.

Gencic, S., Schagger, H., and von Jagow, G., 1991, Core I protein of bovine ubiquinol–cytochrome c reductase: An additional member of the mitochondrial-protein-processing family: Cloning of bovine core I and core II cDNAs and primary structure of the proteins, *Eur. J. Biochem.* **199**:123–131.

Gennis, R. B., and Trumpower, B. L., 1994, Energy transduction by cytochrome complexes in mitochondrial and bacterial respiration: The enzymology of coupling electron transfer reactions to transmembrane proton translocation, *Annu. Rev. Biochem.* **63**:675–716.

Gennis, R. B., Barquera, B., Hacker, B., Van Doren, S. R., Arnaud, S., Crofts, A. R., Davdson, E., Gray, K., and Daldal, F., 1993, The bc_1 complexes of *Rhodobacter sphaeroides* and *Rhodobacter capsulatus*, *J. Bioenerg. Biomembr.* **25**:195–210.

Glaser, E. G., and Crofts, A. R., 1984, A new electrogenic step in the ubiquinol: Cytochrome c_2 oxidoreductase complex of *Rhodopseudomonas*, *Biochim. Biophys. Acta* **766**:322–333.

Glaser, E., Eriksson, A., and Sjöling, S., 1994, Bifunctional role of the bc_1 complex in plants: Mitochondrial bc_1 complex catalyses both electron transport and protein processing, *FEBS Lett.* **346**:83–87.

Gonzalez-Halphen, D., Lindorfer, M. A., and Capaldi, R. A., 1988, Subunit arrangement in beef heart complex III, *Biochemistry* **27**:7021–7031.

Hatefi, Y., 1978, Preparation and properties of dihydroubiquinone: Cytochrome c oxidoreducatse (complex III), *Methods Enzymol.* **53**:35–47.

Hatefi, Y., Haavik, A. G., and Griffiths, D. E., 1962, Studies on the electron transfer system. XLI: Reduced coenzyme Q (QH_2)-cytochrome c reductase, *J. Biol. Chem.* **237**:1681–1685.

Hauska, G., Hurt, E., Gabellini, N., and Lockau, W., 1983, Comparative aspects of quinol–cytochrome c/plastocyanin oxidoreductase, *Biochim. Biophys. Acta* **726**:97–133.

Holm, L., and Sander, C., 1996, Mapping the protein universe, *Science* **273**:595–603.

Howell, N., Appel, J., Cook, J. P., Howell, B., and Hauswirth, W. W., 1989, Mutational analysis of the mouse mitochondrial cytochrome b gene, *J. Mol. Biol.* **203**:607–618.

Iwata, S., Saynovits, M., Link, T. A., and Michel, H., 1996, Structure of a water soluble fragment of the "Rieske" iron–sulfur protein of the bovine heart mitochondrial cytochrome bc_1 complex determined by MAD phasing at 1.5 A resolution, *Structure* **4**:567–579.

Kawamoto, M., Kubota, T., Matsunaga, T., Fukuyama, K., Matsubara, H., Shinzawa-Itoh, K., and Yoshikawa, S., 1994, New crystal forms and preliminary x-ray diffraction studies of mitochondrial cytochrome bc_1 complex from bovine heart, *J. Mol. Biol.* **244**:238–241.

Kim, H., Xia, D., Deisenhofer, J., Yu, C. A., Kachurin, A., Zhang, L., and Yu, L., 1997, X-ray crystallographic studies on an inhibitor-induced conformational change of the iron–sulfur protein in mitochondrial bc_1 complex, *FASEB J.* **11**:1084a.

Kubota, T., Kawamoto, M., Fukuyama, K., Shinzawa-Itoh, K., Yashikawa, S., and Matsubara, H., 1991, Crystallization and preliminary X-ray crystallographic studies of bovine heart mitochondrial cytochrome bc_1 complex, *J. Mol. Biol.* **221**:379–382.

Lee, J. W., Chan, M., Law, T. V., Kwon, H. J., and Jap, B. K., 1995, Preliminary cryocrystallographic study of the mitochondrial cytochrome bc_1 complex: Improved crystallization and flash-cooling of a large membrane protein, *J. Mol. Biol.* **252**:15–19.

Leonard, K., Wingfield, P., Arad, T., and Weiss, H., 1981, Three-dimensional structure of ubiquinol: cytochrome *c* reductase from *Neurospora* mitochondria determined by electron microscopy of membrane crystals, *J. Mol. Biol.* **149**:259–274.

Link, T. A., and Iwata, S., 1996, Functional implications of the structure of the "Rieske" iron–sulfur protein of bovine heart mitochondrial cytochrome bc, *Biochim. Biophys. Acta* **1275**: 54–60.

Ljungdahl, P. O., Pennoyer, J. D., Robertson, D. E., and Trumpower, B. L., 1987, Purification of highly active cytochrome bc_1 complexes from phylogenetically diverse species by a single chromatographic procedure, *Biochim. Biophys. Acta* **891**:227–241.

Meinhardt, S. W., and Crofts, A. R., 1982, Kinetic and thermodynamic resolution of cytochrome *c*-1 and cytochrome *c*-2 from *Rhodopseudomonas sphaeroides*, *FEBS Lett.* **149**:223–227.

Miki, T., Miki, M., and Orii, Y., 1994, Membrane potential-linked reversed electron transfer in the beef heart cytochrome bc_1 complex reconstituted into potassium-loaded phospholipid, *J. Biol. Chem.* **269**:1827–1839.

Mitchell, P., 1976, Possible molecular mechanisms of the protonmotive function of cytochrome systems, *J. Theor. Biol.* **62**:327–367.

Moser, C. C., Keske, J. M., Warncke, K., Farid, R. S., and Dutton, P. L., 1992, Nature of biological electron transfer, *Nature* **355**:796–802.

Musatov, A., and Robinson, N. C., 1994, Detergent-solublized monomeric and dimeric cytochrome bc_1 isolated from bovine heart, *Biochemistry* **33**:13005–13012.

Nagaoka, S., Yu, L., and King, T. E., 1981, Evidence of ubisemiquinone radicals in electron transfer at the cytochrome *b* and c_1 region of the cardial respiratory chain, *Arch. Biochem. Biophys.* **205**:334–343.

Nalecz, M. J., and Azzi, A., 1985, Molecular conversion between monomeric and dimeric states of the mitochondrial cytochrome *b*-*c*-1 complex: Isolation of active monomers, *Arch. Biochem. Biophys.* **236**:619–628.

Ohnishi, T., and Trumpower, B. L., 1980, Differential effects of antimycin on ubisemiquinone bound in different environments in isolated succinate–cytochrome *c* reductase complex, *J. Biol. Chem.* **255**:3278–3284.

Paces, V., Rosenberg, L. E., Fenton, W. A., and Kalousek, F., 1993, The beta subunit of the mitochondrial processing peptidase from rat liver: Cloning and sequencing of a cDNA and comparison with a proposed family of metallopeptidases, *Proc. Natl. Acad. Sci. USA* **90**: 5355–5358.

Rawlings, N. D., and Barrett, A. J., 1991, Homologues of insulinase: A new superfamily of metalloendopeptidases, *Biochem. J.* **275**:389–391.

Rieske, J. S., 1967, Preparation and properties of reduced coenzyme Q-cytochrome *c* reductase (complex III of the respiratory chain), *Methods Enzymol.* **10**:239–245.

Robertson, D. E., and Dutton, P. L., 1988, The nature and magnitude of the charge–separation reactions of the ubiquinol cytochrome c_2 oxidoreductase, *Biochim. Biophys. Acta* **935**:273–291.

Schägger, H., Borchart, U., Aquila, H., Link, T. A., and Von Jagow, G., 1985, Isolation and amino acid sequence of the smallest subunit of beef heart bc_1 complex, *FEBS Lett.* **190**:89–94.

Schägger, H., Link, T. A., Engel, W. D., and Von Jagow, G., 1986, Isolation of the eleven protein subunits of the bc_1 complex from beef heart, *Methods Enzymol.* **126:**224–237.

Schägger, H., Borchart, U., Machleidt, W., Link, T. A., and Von Jagow, G., 1987, Isolation and amino acid sequence of the Rieske's iron–sulfur protein of beef heart ubiquinol:cytochrome c reductase, *FEBS Lett.* **219:**161–168.

Tolkatchev, D., Yu, L., and Yu, C. A., 1996, Potential induced redox reactions in mitochondrial and bacterial cytochrome b-c_1 complexes, *J. Biol. Chem.* **271:**12356–12363.

Trumpower, B. L., 1990a, The protonmotive Q cycle: Energy transduction by coupling of proton translocation to electron transfer by the cytochrome bc_1 complex, *J. Biol. Chem.* **265:**11409–11412.

Trumpower, B. L., 1990b, Cytochrome bc_1 complexes of microorganism, *Microbiol. Rev.* **54:** 101–129.

Trumpower, B. L., and Haggerty, J. G., 1980, Inhibition of electron transfer in the cytochrome bc_1 segment of the mitochondrial respiratory chain by a synthetic analogue of ubiquinone, *J. Bioenerg. Biomembr.* **12:**151–157.

Tsai, A. L., Olson, J. S., and Palmer, G., 1983, The oxidation of yeast complex III, *J. Biol. Chem.* **258:**2122–2125.

Usui, S., Yu, L., and Yu, C. A., 1990, Cloning and sequencing of a cDNA encoding the rieske iron–sulfur protein of bovine heart mitochondrial ubiquinol–cytochrome c reductase, *Biochem. Biophys. Res. Commun.* **167:**575–579.

Von Jagow, G., and Link, T. A., 1986, Use of specific inhibitors on the mitochondrial bc_1 complex, *Methods Enzymol.* **126:**253–271.

Von Jagow, G., Link, T. A., and Ohnishi, T., 1986, Organization and function of cytochrome b and ubiquinone in the cristae membrane of beef mitochondria, *J. Bioenerg. Biomembr.* **18:** 157–179.

Wakabayashi, S., Tekada, H., Matsubara, H., Kim, C. H., and King, T. E., 1982, Identity of the heme-not-containing protein in bovine heart cytochrome c_1 preparation with the protein mediating c_1–c complex formation: A protein with high glutamic acid content, *J. Biochem.* **91:**2077–2085.

Wakabayashi, S., Tekao, T., Shimonish, Y., Kuramitsu, S., Matsubara, H., Wang, T-Y., Zhang, Z-P., and King, T. E., 1985, Complete amino acid sequence of the ubiquinone binding protein (QP-C): A protein similar to the 14,000-dalton subunit of the yeast ubiquinol–cytochrome c reductase complex, *J. Biol. Chem.* **260:**3

Widger, W. R., Cramer, W. A., Herrmann, R. G., and Trebst, A., 1984, Sequence homology and structural similarity between cytochrome b of mitochondrial complex iii and the chloroplast b_6f complex: Position of the cytochrome b hemes in the membrane, *Proc. Natl. Acad. Sci. USA* **81:**674–678.

Wikström, M., and Berden, J. A., 1972, Oxireduction of cytochrome b in the presence of antimycin, *Biochim. Biophys. Acta* **283:**403–420.

Xia, D., Yu, C. A., Kim, H., Xia, J-Z, Kachurin, A. M., Zhang, L., Yu, L., and Deisenhofer, J., 1997, Crystal structure of the cytochrome bc_1 complex from bovine heart mitochondria, *Science* **277:**60–66.

Yu, C. A., and Yu, L., 1980, Resolution and reconstitution of succinate–cytochrome c reductase: Preparations and properties of high purity succinate dehydrogenase and ubiquinol–cyto-chrome c reductase, *Biochim. Biophys. Acta* **591:**409–420.

Yu, C. A., and Yu, L., 1981, Ubiquinone protein, *Biochim. Biophys. Acta* **639:**99–128.

Yu, C.-A., Chiang, Y. L., Yu, L., and King, T. E., 1975, Photoreduction of cytochrome c_1, *J. Biol. Chem.* **250:**6218–6221.

Yu, C. A., Xia, D., Deisenhofer, J., and Yu, L., 1994, Crystallization of mitochondrial cytochrome b-c_1 complex from gel with or without reduced pressure, *J. Mol. Biol.* **243:**802–805.

CHAPTER 13

Structural Aspects of the Cytochrome bc_1 Complex

Thomas A. Link

1. INTRODUCTION

The ubiquitous cytochrome bc_1 complexes form the middle part of electron transfer chains of mitochondria and many bacteria. The related $b_6 f$ complexes are found in chloroplasts, algae, and some gram-positive bacteria. Together, they form the family of bc complexes; they all contain two heme b centers, a high-potential heme center (heme c_1 or heme f), and a "Rieske" iron–sulfur protein comprising a high-potential [2Fe-2S] cluster.

The bc complexes oxidize hydroquinones (ubiquinone, plastoquinone, or menaquinone) and transfer electrons to a water-soluble redox protein, cytochrome c (for bc_1 complexes) or plastocyanin (for $b_6 f$ complexes). It is generally accepted that electron transfer is coupled to proton translocation by the proton-motive Q-cycle mechanism first proposed by Peter Mitchell (1976). According to this mechanism, hydroquinone is oxidized at the Q_p site; the first electron is transferred to the "Rieske" [2Fe-2S] cluster and from there via cytochrome c_1 to cytochrome c. The second electron is transferred to the low-potential heme b_L and from there across the membrane dielectric to heme b_H. A

Thomas A. Link • Universitätsklinikum Frankfurt, ZBC, Institut für Biochemie I, Molekulare Bioenergetik, D-60590 Frankfurt/Main, Germany.

Frontiers of Cellular Bioenergetics, edited by Papa *et al.* Kluwer Academic/Plenum Publishers, New York, 1999.

second molecule of hydroquinone is then oxidized at the Q_P site, the first electron again transferred to the "Rieske" [2Fe-2S] cluster and on to cytochrome c, and the second electron again to cytochrome b. Two electrons are then transferred from heme b_H to a quinone molecule bound in the Q_N site, which is reduced to hydroquinone. In total, half the electrons are recycled back to quinone; therefore, this mechanism leads to a doubling of the proton-motive efficiency of the bc_1 complex. The Q-cycle mechanism is discussed in detail by Trumpower (Chapter 11, this volume). Therefore, I will not discuss these reactions any further, but will refer to those aspects of the structure of the bc_1 complex that are a prerequisite for its function.

The minimal structural requirements for the Q-cycle mechanism are:

1. The presence of two independent quinone binding sites that are in contact with different sides of the membrane: the hydroquinone oxidation (Q_P) site at the positive P side of the membrane, and the quinone reduction (Q_N) site at the negative N side of the membrane.
2. The existence of two independent electron transfer pathways: a high-potential pathway along the P side of the membrane, formed by the "Rieske" iron–sulfur cluster and heme c_1–f and catalyzing electron transfer from hydroquinone to cytochrome c or plastoquinone, respectively; and a low-potential pathway across the membrane where electrons are driven against the membrane potential from the P side (b_L) to the N side (b_H).
3. An obligatory branching of the electrons during hydroquinone oxidation at the Q_P site. Although electron transfer from semiquinone to the "Rieske" center ($E_m \approx +300\,mV$) and cytochrome c_1 ($E_m \approx +250\,mV$) is energetically favorable compared to electron transfer to heme b_L ($E_m \approx 0\,mV$) by 25–30 kJ/mole, the second electron must not be allowed to go into the high-potential chain after the first electron has been delivered to the "Rieske" center, even under conditions where the second electron acceptor (heme b_L) is already reduced. Otherwise, uncoupling of electron transfer and proton translocation would occur, which is not observed experimentally.

2. SUBUNITS AND THEIR PHYLOGENETIC RELATIONSHIP

2.1. Subunits Containing Redox Centers

The bc complexes from mitochondria, chloroplasts, and bacteria all contain three catalytic subunits comprising the four redox centers:

1. Cytochrome b, a hydrophobic membrane protein containing both heme b centers in a transmembrane arrangement. In bc_1 complexes of mito-

chondria and bacteria like *Paracoccus* or *Rhodobacter*, cytochrome b has around 400 amino acid residues and contains eight membrane-spanning α-helices; in $b_6 f$ complexes, the protein is split into cytochrome b_6 containing only the first four transmembrane helices coordinating both hemes and subunit IV which is homologous to the rest of cytochrome b. Exceptions from this general pattern have been observed: e.g., the menaquinone oxidizing bc complex of *Bacillus subtilis* contains a split cytochrome b; subunit IV is fused to the c-type cytochrome.

2. Cytochrome c_1 or cytochrome f, both consisting of a N-terminal water-soluble domain and a C-terminal membrane anchor. The water-soluble domain contains a high-potential heme center; the core structure of cytochrome c_1 is related to the structure of cytochrome c (Zhang *et al.*, 1998), while cytochrome f shows no homology to cytochrome c_1 or to other known cytochromes. The water-soluble part of cytochrome f consists of a N-terminal heme-binding peptide (residues 1–25), a large domain, and a small domain (Martinez *et al.*, 1994). The fold of the large domain is like that of the type III domain of fibronectin (FnIII) or the bacterial chaperone protein PapD. As there is no detectable sequence homology, the relevance of this structural homology is unclear.

3. A "Rieske" iron–sulfur protein that consists of a C-terminal water-soluble domain coordinating the [2Fe-2S] cluster that is attached to the membrane and to the rest of the complex through a N-terminal membrane anchor. The "Rieske" cluster is coordinated within a 45-residue cluster-binding fold that is structurally related to rubredoxins, small (45–54 amino acid residues) mononuclear electron transport proteins containing a single iron atom coordinated by four cysteine residues (Iwata *et al.*, 1996). The general topology of the cluster-binding fold of the "Rieske" protein and of rubredoxins is similar; the metal-binding loops of the two proteins superpose with a root mean square deviation of the Cα atoms of 0.66 Å. The two coordinating cysteine residues and iron atom Fe-1 of the "Rieske" protein superpose with two of the coordinating cysteine residues and the iron atom of rubredoxin, while the acid-labile sulfur atoms of the iron–sulfur fragment superpose with the Sγ atoms of the other two cysteine residues of rubredoxin.

An analogous homology to a water-soluble electron transfer protein as in the "Rieske" iron sulfur protein is found in subunit II of cytochrome oxidase: both proteins have their redox centers in water-soluble domains that are connected to the rest of the complexes through N-terminal membrane anchors. Subunit II contains a dicopper Cu_A site, but is structurally related to plastocyanin, a water-soluble mononuclear copper protein. One of the metal ions of the dinuclear center superposes on the metal ion of the respective mononuclear counterpart. Moreover, the peripheral domains of the complexes I and II of the respiratory chain also show some relationship to water-soluble redox proteins (Table I). Therefore, it appears to be a general pattern that respiratory electron

Table I

Subunits of Electron Transfer Complexes Related to Water-Soluble Electron Transport Proteins

Respiratory complex	Subunit	Redox centers	Related to	Redox centers
Complex I	24 kDa	[2Fe-2S]	NAD$^+$ reducing H$_2$ase (A. eutrophus), α subunit	[2Fe-2S]
Complex I	51 kDa	[4Fe-4S]	NAD$^+$ reducing H$_2$ase (A. eutrophus), α subunit	[4Fe-4S]
Complex I	75 kDa	[2Fe-2S], 2 × [4Fe-4S]	NAD$^+$ reducing H$_2$ase (A. eutrophus), γ subunit	[xFe-4S]
Complex II	Iron–sulfur protein	[2Fe-2S]	Plant type ferredoxins	[2Fe-2S]
Complex II	Iron–sulfur protein	[3Fe-4S], [4Fe-4S]	Various ferredoxins	2 × [4Fe-4S]; [3Fe-4S], [4Fe-4S]
Complex III	"Rieske" iron–sulfur protein	"Rieske" [2Fe-2S]	Rubredoxin	[Fe]
Complex III	Cytochrome c_1	Heme c	Cytochrome c	Heme c
Complex IV	Subunit II	Cu$_A$ [2Cu]	Plastocyanin	"Blue" [Cu]

transport complexes have recruited soluble redox proteins by attaching them to the hydrophobic core of the machinery; the water-soluble redox domains are essential for electron transfer to aqueous reaction partners.

2.2. Subunits without Redox Centers

Mitochondrial bc_1 complexes contain subunits, which are related to the family of matrix-processing peptidases (MPPs) and to pitrilysin, a zinc-containing endoprotease found in *Escherichia coli* that cleaves small peptides (reviewed in Braun and Schmitz, 1995). MPPs contain two subunits of approximately 50 kDa each, which are homologous to each other—α-MPP and β-MPP—and cleave targeting sequences, which direct mitochondrial proteins synthesized in the cytosol to the mitochondrion. The β-subunit of MPP contains an "inverted zinc binding site" characterized by the sequence motif His-Phe-Leu-Glu-His, while the α-subunit contains an incomplete inverted zinc binding site; therefore, it has been suggested that the β-subunit is the catalytically active subunit.

The core proteins are the largest subunits of the respective complexes and have their names for historial reasons; they are located peripherally and can be released from the isolated bc_1 complexes by salt treatment in some species (e.g., *Neurospora crassa*; Hovmöller *et al.*, 1981). In most bc_1 complexes, the core proteins differ from the subunits of MPP and the bc_1 complexes show no processing activity. In mammalian mitochondria, both core proteins have an incomplete "inverted zinc binding" sequence; in core protein I, the first histidine is replaced by tyrosine, while the essential residues are completely absent in core protein II. The bc_1 complex from *Neurospora crassa* exhibits processing activity if combined with α-MPP; core protein I is β-MPP, containing a complete inverted zinc binding site. In plants, both subunits of MPP are integrated into the bc_1 complex as core proteins and the isolated bc_1 complex is a highly active processing peptidase.

The bc_1 complexes from mitochondria and $b_6 f$ complexes from chloroplasts contain additional small subunits (molecular mass between 6 and 15 kDa) without redox centers (Schägger *et al.*, 1994, 1995) (Table II) the presence of a fourth subunit has been shown in some bacterial bc_1 complexes, but its significance is unclear (Yu and Yu, 1991).

Subunit IX of bovine bc_1 complex was found to be the 78-amino acid presequence of the "Rieske" iron–sulfur protein (Brandt *et al.*, 1993). The iron–sulfur protein is nuclear coded and requires the presequence for mitochondrial targeting. The protein is processed and the presequence is cleaved in a single proteolytic step after it is inserted into the bc_1 complex, apparently by the core proteins that have retained some processing activity despite the modified inverted zinc binding site. This is the first instance in which a cleaved presequence has been shown to be retained as a subunit of a mature complex.

Table II
Peptide Subunits of the bc_1 Complexes of Beef Heart and of S. *cerevisiae*

Bovine					Yeast	
	Subunit	M_r	TMH[a]	Homology	Subunit	M_r
I	Core I	49,212	0	35%	UCR1	47,409
II	Core II	46,524	0	32%	UCR2	38,705
III	Cytochrome b	43,851	8	53%	Cytochrome b	44,920
IV	Cytochrome c_1	27,903	1	56%	Cytochrome c_1	28,388
V	Iron–sulfur protein	21,781	1	55%	UCR1	20,271
VI	Cyt. b-associated	13,388	0	36%	UCR7	14,565
VII	Core-associated	9,589	1	25%	UCRQ	10,843
VIII	"Hinge" protein	9,171	0	41%	UCRH	14,528
IX	ISP targeting peptide	7,997	0	—	?	—
X	Cyt. c_1 associated	7,326	1	37%	UCR9	7,345
XI	ISP associated	6,520	1	27%	UCRX	8,462
Sum		243,262	13			235,436

[a]TMH, transmembrane helices.

Subunit XI of bovine bc_1 complex is a small hydrophobic subunit containing a single transmembrane helix; it can be removed by detergent (Schägger *et al.*, 1990). The subunit is not required for electron transfer activity in isolated bovine bc_1 complex. When the corresponding yeast subunit (QCR10) was deleted, ubihydroquinone–cytochrome c oxidoreductase activity was decreased by 40% and the "Rieske" iron–sulfur protein was lost when the bc_1 complex was purified from the strain (Brandt *et al.*, 1994).

The other subunits were found to be associated with large subunits upon dissociation of bovine bc_1 complex; they were either coisolated or they form subcomplexes, e.g., the c_1 subcomplex formed by cytochrome c_1 and subunits VIII and X. Bovine subunit VIII was named "hinge protein," as it appears to be involved in the interaction between the bc_1 complex and the basic cytochrome c (Wakabayashi *et al.*, 1982). The yeast homologue was named the "17-kDa subunit," according to its migration behavior in sodium dodecyl sulfate-polyacrylamide gel electrophoresis (SDS-PAGE); the true molecular mass is 14.5 kDa. Both the bovine and yeast subunits have a great excess of acidic residues within their N-terminal parts (25 consecutive acidic residues in yeast, 8 consecutive glutamic acid residues in bovine heart). The yeast subunit contains an incomplete "EF hand" calcium binding motif, where two residues providing calcium coordinating oxygens are replaced by basic residues. The incomplete calcium binding motif is not conserved in the bovine subunit. The isolated yeast subunit but not the isolated homologous bovine subunit was found to bind calcium (M. E. Schmitt and B. L. Trumpower, unpublished results). Deletion of the yeast subunit blocked maturation of cytochrome c_1 and caused temperature-sensitive petite growth of the yeast (Yang and Trumpower, 1994). In a strain where the effect of the deletion was covered by a suppressor, the yeast subunit was found to regulate ionic strength-dependent half-of-the-sites reactivity (Schmitt and Trumpower, 1990).

2.3. Other Constituents

An additional important structural component of the bc_1 complex are phospholipids. Eight to nine molecules of cardiolipin are bound to each bc_1 monomer; their removal by extended detergent washes results in irreversible inactivation. In addition, approximately 100 molecules of neutral phospholipids (phosphatidylethanolamine and/or phosphatidylcholine) per bc_1 complex dimer are required for full electron transfer activity; this corresponds to one complete phospholipid annulus (Schägger *et al.*, 1990). In comparison, all phospholipids but approximately nine molecules of cardiolipin can be removed from bovine heart cytochrome c oxidase without an effect on the activity (M. Boese and T. A. Link, unpublished results).

3. STRUCTURES OF THE CATALYTIC SUBUNITS AND THE REDOX CENTERS

3.1. Cytochrome b

Cytochrome b contains eight membrane-spanning helices. The existence of a ninth transmembrane helix had been initially proposed from hydropathy analyses and from sequence comparison between cytochrome b and b_6 (Widger *et al.*, 1984); however, studies of mutations conferring resistance to either Q_P- or Q_N-site inhibitors as well as gene fusion experiments indicated that the fourth helix (now called *cd* helix) is not transmembranous but peripherally located at the P side of the membrane. The transmembrane location of the heme b centers as well as their assigment has been confirmed by site-directed mutagenesis (Yun *et al.*, 1991). The constraints implied by the following have allowed the construction of models of the three-dimensional structure of the helices A–F of cytochrome b (Crofts *et al.*, 1990; Link *et al.*, 1994):

1. The arrangement of the heme ligands (two pairs of histidines, which are 13 residues apart in the helices B and D).
2. The location of inhibitor-resistant mutants affecting either the Q_P- or the Q_N-reaction sites (reviewed in Degli Esposti *et al.*, 1993; Brasseur *et al.*, 1996).

3. Nearest-neighbor relationships deduced from the analysis of second-site pseudorevertants of cytochrome b (di Rago *et al.*, 1990).

Recently, the crystal structure of bovine heart bc_1 complex has been reported by three groups (Xia *et al.*, 1997; Zhang *et al.*, 1998; Iwata *et al.*, 1998). In general, there is good agreement between the models and the X-ray structures that gives support to the conclusions drawn from the biochemical experiments.

3.1.1. Structural Models of Cytochrome b

The core of cytochrome b is a right-handed all-antiparallel 4-α-helical bundle formed by the helices A–D (Figs. 1, 2). Helices B and D each contain a pair of heme-coordinating histidine residues 13 residues apart in the sequence. In the case of cytochrome b_6, an additional residue is inserted between the histidine ligands in helix D; this will lead to altered steric constraints within the 4-α-helical bundle and to changes in the geometry of the heme–ligand interaction (see Section 3.1.2.), providing a rationale for the spectroscopically observed differences between the cytochrome b and b_6. Helices A and C each contain a pair of conserved glycine residues spaced, like the histidine ligands, 13 residues apart; these residues have been predicted to be critical for heme packing within the tightly constrained 4-α-helical bundle (Tron *et al.*, 1991). Mutational analysis of the first glycine in the bc_1 complex of *Rhodobacter sphaeroides* proved that a large (valine) or even a charged residue is not tolerated in this position (Yun *et al.*, 1992).

The 4-α-helical bundles are the simplest helical domain structures found in many proteins, including hemoproteins; geometric analysis indicated that an angle of 20° between the axes of the helices is favorable for side chain packing. The all-antiparallel arrangement provides electrostatically favorable helix–dipole interactions. Examples of hemoproteins of this type are the well-charac-

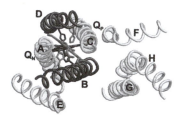

FIGURE 1. Arrangement of the eight transmembrane helices of cytochrome b in the membrane. The view is from the P side of the membrane; helices B and D containing the histidines that coordinate heme are shown in dark grey. Heme b_L is shown in light gray (top), heme b_H in dark gray (bottom). The location of the Q_P and of the Q_N site is indicated.

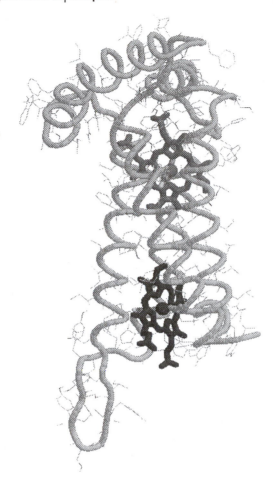

FIGURE 2. Model structure of the 4-α-helical bundle forming the center of cytochrome b. Heme b_L (medium gray) and the *cd* helix at the P side are shown at the top, heme b_H at the N side (dark gray) at the bottom of the figure. The *cd* helix is modeled as a single bent helix, while X-ray crystallography shows a hairpin structure (*cd1*-turn-*cd2*).

terized water-soluble cytochrome b_{562} from *E. coli*, which has a single heme center with histidine–methionine coordination, and the heme binding domain of the microsomal cytochrome b_5, which has a heme center with *bis*-histidine coordination. The packing of two heme centers on top of each other within the 4-α-helical bundle in cytochrome b provides additional constraints on the geometry and is the cause of the "steric strain" of cytochrome b (see Section 3.1.2.).

Within the 4-α-helical bundle, the heme centers are arranged on top of each other with an iron–iron distance of approximately 21 Å; the edge-to-edge distance of the heme centers is approximately 12 Å. This geometry enables rapid electron transfer between the heme centers across the membrane dielectric (the low-potential electron pathway).

The location of the helices E and F as well as the position of the quinone reaction sites shown in Fig. 1 could be deduced from the analysis of mutations conferring resistance to either the Q_P- or Q_N-site inhibitors (for a recent compilation, see Brasseur et al., 1996). Crystal structures obtained in the presence of myxothiazol and antimycin, which represent Q_P- or Q_N-site inhibitors, respectively, show two independent quinone reaction sites on opposite sides of the 4-α-helical bundle. This is in accordance with the requirements of the Q-cycle mechanism and is consistent with the fact that the binding of Q_P- or Q_N-site inhibitors is independent and additive (von Jagow and Link, 1986).

The helices have long interhelical connections on the P side, while the connections on the N side are short except for the de loop between helices D and E; this loop is involved in the formation of the Q_N site. The connection between the helices C and D, which was modeled as a long continuous helix, forms a split helix (cd1 and cd2) with a hairpinlike structure; the first part of helix cd1 and the connecting part between the transmembrane helix C and cd1 is involved in the formation of the Q_P site (Link et al., 1993). Since the helices E and F are located on opposite sides of the 4-α-helical bundle, the long connecting ef loop crosses over the bundle; residues in this loop are involved in the formation of the Q_P site and in the interaction between cytochrome b and the "Rieske" iron–sulfur protein at the P side of the membrane (Geier et al., 1992).

3.1.2. The Environment of the Heme Centers and the Steric Strain of Cytochrome b

As discussed above, packing of two heme centers within the 4-α-helical bundle imposes steric constraints on the protein as well as on the environment of the heme centers. This steric strain of the heme centers is essential for their spectroscopic and electrochemical properties. The distortion of the porphyrin ring systems leads to a splitting of the Q_{ox} and Q_{oy} components of the α (Q) band of the reduced cytochrome b; this splitting is 8 nm for heme b_L of bovine bc_1 complex (566 and 558 nm for Q_{ox} and Q_{oy}, respectively) and approximately 2–3 nm for heme b_H.

A different type of strain is observed in the electron paramagnetic resonance (EPR) spectra of the oxidized cytochrome b. The EPR spectra of oxidized low-spin (d^5) hemes reflect the distribution of the electron hole between the d orbitals belonging to the t_{2g} set (Taylor, 1977). Highly axial spectra with a limiting g_z value of 4 are expected if the electron hole is completely delocalized between the d_{xz} and d_{yz} orbitals, i.e., for a complete x–y degeneracy of the d orbitals of the central iron atom, while lower g_z values correspond to a situation where the d_{yz} orbital (by convention) is energetically higher than the d_{xz} orbital, so that the hole is partially localized on d_{yz}. The g_z values for the heme centers in bovine bc_1 complex are observed at 3.78 (b_L) and 3.44 (b_H), corresponding to a highly delocalized situation; the g_z values for unconstrained hemes are close to 3. High g_z values are supposed to originate from nonzero dihedral angles between the coordinating imidazole rings; they have been observed in sterically hindered model compounds, while low g_z values correspond to a parallel orientation of the imidazole rings (Palmer, 1985). Therefore, in cytochrome b we have the paradoxical situation that optical spectra of the reduced hemes indicate lower symmetry of the porphyrin ring compared to unconstrained cytochromes, while EPR spectra of the oxidized hemes indicate higher symmetry around the iron atoms. A transition to low g_z values (2.93) is induced by partial denaturation, e.g., upon freezing–thawing or after addition of moderate amounts of detergent; this partial denaturation of cytochrome b is accompanied by a shift of the absorption maximum to 560 nm and by a decrease of the redox potential (Link *et al.*, 1986).

The close proximity of the heme centers gives rise to exciton coupling between their Soret band transitions, resulting in intense circular dichroism (CD) spectra. From the analysis of the CD spectra of reduced cytochrome b, a dihedral angle between the heme planes of approximately 45° has been deduced (Palmer and Degli Esposti, 1994); this value is similar to that observed in the X-ray structures. The CD spectra of oxidized cytochrome b differ significantly. In the oxidized state, the heme planes were assumed to be almost parallel. This would require a major conformational to occur upon oxidation–reduction, which we can exclude as the X-ray structures were obtained from the oxidized complex. As an alternative explanation, we suggest that the differences observed in the CD spectra are due to the electronic nature of the oxidized low-spin heme and the interference of additional (charge transfer) bands, which are not present in closed-shell (low-spin d^6) hemes (T. A. Link and O. S. Tautu, unpublished results).

3.2. Cytochrome c_1

Before the determination of the crystal structure of the bc_1 complex, little was known about the three-dimensional structure of cytochrome c_1, since it shows no significant sequence homology to other cytochromes or even to cytochrome f, its counterpart in b_6f complexes. The crystal structure (Zhang *et al.*, 1998) revealed significant structural homology between the core of cytochrome c_1 and helices 1, 3, and 5 of cytochrome c; however, cytochrome c_1 contains long insertions between these conserved parts surrounding the heme group. These insertions are responsible for the fact that the water-soluble domain of cytochrome c_1 is approximately two times the size of ordinary cytochromes c; in addition, cytochrome c_1 contains a transmembrane helix at its carboxy-terminus. Parts of the insertions constitute acidic surfaces that appear to form the docking site for the basic cytochrome c, together with the "hinge" protein.

There is no homology between cytochromes c_1 and f, except for the heme binding pentapeptide Cys-X-X-Cys-His characteristic for cytochromes having covalently bound heme c. Surprisingly, in Western blots following SDS-PAGE, cytochrome f reacted with antibodies against native cytochrome c_1 (Link, 1988).

Like cytochrome b, oxidized cytochrome c_1 shows a highly anisotropic EPR spectrum with a g_z value of 3.37 (for the bovine system). For comparison, cytochrome c shows a "normal" g_z value of 3.03. The 695-nm band, which is assigned to the charge transfer transition from the highest occupied porphyin π orbital into the iron d_z2 orbital, is shifted by 500 cm^{-1} to 13,900 cm^{-1} compared to 14,400 cm^{-1} in cytochrome c. These spectroscopic properties indicate steric strain also in cytochrome c_1, which might be correlated with the apparently lower stability of the ligand environment compared to cytochrome c. This is indicated by the fact that a transition that resembles the alkaline transition of cytochrome c, i.e., a ligand exchange from methionine to a nearby nitrogen ligand, is facilitated in isolated cytochrome c_1 (Finnegan et al., 1996). Isolated cytochromes c_1 were shown to contain at least a fraction with histidine–histidine coordination, which increased at alkaline pH; this fraction had a g_z value of 2.96.

3.3. "Rieske" Iron–Sulfur Protein

The "Rieske iron–sulfur protein was first described and isolated by Rieske et al. (1964) and later was identified as the "oxidation factor" of the bc_1 complex (Trumpower and Edwards, 1979). The protein contains a [2Fe-2S] cluster showing a distinct EPR spectrum with $g_{av} = 1.91$ compared to $g_{av} = 1.96$ for plant-type ferredoxins. Its redox potential (+300 mV) is approximately 700 mV more positive than that of plant-type [2Fe-2S] ferredoxins. From the analysis of the EPR, electron nuclear double resonance (ENDOR), and electron spin echo envelope modulation (ESEEM) spectra in combination with mutational analyses, it was predicted that the [2Fe-2S] cluster should have two histidine ligands coordinating through their Nδ atoms to one iron atom [Fe(II) of the reduced cluster] (Gurbiel et al., 1989, 1991; Britt et al., 1991).

3.3.1. Structure of the Water-Soluble Catalytic Domain

The water-soluble domain of the protein containing the [2Fe-2S] cluster is formed by the C-terminal two thirds of the protein, while the N-terminal part forms the membrane anchor having a large content of α-helix structure (Link *et al.*, 1996a). The two parts are connected by a flexible linker that is easily attacked by proteases. After proteolytic cleavage of the bc_1 complex, we have isolated a water-soluble fragment [iron–sulfur fragment (ISF)] of bovine heart mitochondrial bc_1 complex containing the intact [2Fe-2S] cluster; this fragment has a molecular mass of 14,592 Da and has been found to be highly stable and suited for electrochemical and spectroscopic studies. The reduced fragment could be crystallized (Link *et al.*, 1996a) and its structure could be solved at 1.5 Å resolution using the multiwavelength anomalous dispersion (MAD) technique (Iwata *et al.*, 1996).

The ISF is a flat spherical molecule with dimensions of 45 × 40 × 25 Å. The structure contains three layers of antiparallel β-sheets (Fig. 3). Sheet 1 is formed by the strands β9, β10, and β1; sheet 2 by strands β2, β3, and β4; and sheet 3 by the strands β5–β8. The central sheet 2 contains longer strands than the other sheets and may be regarded as the "spine" of the protein. The only α-helix and a long loop are inserted between the strands β3 and β4; both are lacking in "Rieske" proteins from b_6f complexes. These parts of the structure have contact with the sheets 2 and 3, mostly through a salt bridge–hydrogen bond network that connects the bottom part of the metal binding fold (β6, β7, and β8) to the helix α1, the loop α1-β4, and the lower part of strand β3. Five arginine residues (Arg 99, Arg 101, Arg 126, Arg 170, and Arg 172) and four acidic residues (Glu 105, Glu 109, Asp 123, and Asp 166) are involved in the network. We have suggested that the complex salt bridge network is essential for the stability of the protein. This is demonstrated by the fact that in a yeast mutant where Asp 166 had been replaced by Asn, no [2Fe-2S] cluster could be

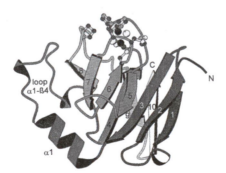

FIGURE 3. Schematic ribbon diagram of the structure of the water-soluble fragment of the "Rieske" iron–sulfur protein. Atoms in the iron–sulfur cluster, its ligands, the disulfide bond, and residue Pro 175 are also shown. The figure was produced using the program MOLSCRIPT (Kraulis, 1991).

observed (Gatti *et al.*, 1989). Other random mutants where the salt bridge network was affected showed 20–30% of the wild-type activity most likely due to instability of the protein: This included the mutants Glu 105 → Gly, Asp 123 → Asn, and Asp 123 → Gly (Graham *et al.*, 1992).

The β-sheet 3 is the bottom part of the metal binding fold, which can be separated from the rest of the structure and forms a small domainlike structure of its own; the cluster is at the tip of the fold. The [2Fe-2S] cluster is coordinated by two cysteine and two histidine residues (Fig. 4). The ligands coordinating the cluster originate from loops β4–β5 and β6–β7 (loops are named according to the secondary structure elements at the both ends). Both loops contribute one cysteine and one histidine residue each: loop β4–β5 contains Cys 139 and His 141, and loop β6–β7 contains Cys 158 and His 161. His 141 and His 161 are ligands for Fe-2, while Cys 139 and Cys 158 are ligands for Fe-1; thus, the coordination pattern is 2 + 2 compared to the 3 + 1 pattern (Cys-X$_4$-Cys-X$_2$-Cys ... Cys) observed in plant type ferredoxins.

Both loops β4–β5 and β6–β7 contain an additional cysteine residue (Cys 144 and Cys 160, respectively); these cysteines form a disulfide bond connecting the two loops. The presence of this disulfide bond provides a rationale for the observation that mutation of each of the four totally conserved cysteine residues results in the loss of the cluster (Graham and Trumpower, 1991; Davidson *et al.*, 1992; Van Doren *et al.*, 1993); the disulfide bridge seems to be

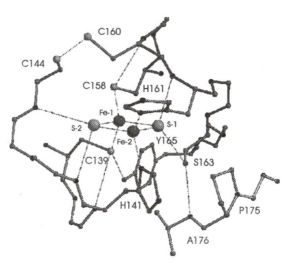

FIGURE 4. Structure of the "Rieske" [2Fe-2S] cluster. The view is from the top of Fig. 3; the histidine-coordinated iron (Fe-2) is in the front. The disulfide bridge and the hydrogen bonds of the sulfur atoms of the cluster are indicated by dotted bonds.

important for the stabilization of the fold around the cluster, as the two loops are not shielded by other parts of the protein. The "Pro loop" containing the fully conserved sequence Gly-Pro175-Ala-Pro 177 (a part of the loop β8–β9) covers the cluster from the other side. Mutations in this loop have been found to be critical for cluster stability. Using a random mutagenesis approach in yeast, respiratory-deficient (Gatti *et al.*, 1989) and temperature-sensitive (Beckmann *et al.*, 1989; Graham *et al.*, 1992) mutations of the "Rieske" protein leading to complete loss of activity were found in positions 168 (Ser → Pro), 169 (Gly → Asp), 170 (Arg → Gly), 176 (Ala → Val), and 179 (Asn → Lys); in mutants where either Pro 175 was converted to serine or leucine, the activity was less than 20% of the wild type. In the mutant Pro 175 → Ser, this was due to the instability of the cluster, while the specific activity per [2Fe-2S] cluster was not reduced (Gatti *et al.*, 1989). The same result was found for the mutant Ala 176 → Thr, while the mutant Ala 176 → Val had lost activity completely (Graham *et al.*, 1992).

3.3.2. Structure of the [2Fe-2S] Cluster

In the crystal structure, one iron atom of the [2Fe-2S] cluster (Fe-1) is coordinated by the Sγ atoms of two cysteine residues, the other iron (Fe-2) by the Nδ atoms of two imidazoles; this essentially confirms the predictions from the analysis of ENDOR and ESEEM spectroscopy. The fact that two imidazoles are directly coordinated to the [2Fe-2S] cluster is essential for the electrochemical properties of the cluster. However, the coordination geometry around the histidine-coordinated Fe-2 is much more asymmetric than anticipated; the angle Nδ–(Fe-2)–Nδ between the two histidine ligands is 91°, a value expected for octahedral but not for tetrahedral geometry.

Both iron atoms of the [2Fe-2S] cluster are in the high-spin state; in the oxidized cluster, coupling between the two iron atoms gives an EPR-silent $S = 0$ ground state. The reduced cluster has a localized mixed valence state, i.e., one iron atom has the (formal) oxidation state Fe(II), while the other iron remains as Fe(III) (Fee *et al.*, 1984). The reduced iron Fe(II) (Fe-2) is close to the surface of the protein with its histidine ligands fully exposed to the solvent, while Fe(III) (Fe-1) is buried within the protein and surrounded by the three loops forming the metal binding fold.

All sulfur atoms participate in two NH-S or OH-S hydrogen bonds in addition to the iron and carbon bonds, except the Sγ of Cys 158, which has only a single hydrogen bond to the nitrogen atom of Cys 160. This pattern contributes to the stability of the cluster as well as to the high redox potential. Two OH-S hydrogen bonds are observed: S-1–Ser 163 Oγ and Cys 139 Sγ–Tyr 165 Oη. Both Ser 163 and Tyr 165 are conserved in all known "Rieske" iron–sulfur proteins of ubiquinone-oxidizing bc_1 complexes; they are important for the high redox potential observed. In "Rieske" clusters of menaquinone-oxidizing *bc* complexes, the residue that corresponds to Ser 163 is replaced by an alanine, which cannot form a hydrogen bond to the cluster; the redox potential of these "Rieske" clusters is approximately 150 mV lower than in those "Rieske" clusters containing serine in this position. Similar shifts of the redox potential that were accompanied by a decrease of the electron transfer activity were obtained when the residues corresponding to Ser 163 and Tyr 165 were exchanged by site-directed mutagenesis of the bc_1 complexes in yeast (Denke *et al.*, 1998) and in *Paracoccus denitrificans* (Schröter *et al.*, 1998), demonstrating the importance of these hydrogen bonds.

Multiple factors contribute to the high redox potential of the cluster: (1) The presence of hydrogen bonds; (2) the overall charge of the cluster, $0/-1$ for the oxidized and reduced state, respectively, compared to $-2/-3$ for [2Fe-2S] clusters with four cysteine coordination; (3) the electronegativity of the histidine ligands; and (4) the solvent exposure of the Fe(II). The last factor was found to be largely responsible for the observed differences of the redox potential between the "Rieske" center of the bc_1 complex and "Rieske"-type centers in bacterial dioxygenases, which have a much more negative redox potential (Link *et al.*, 1996b).

The hydrophobic surface around His 141 and His 161 is likely to be a part of the hydroquinone oxidation (Q_P-) site of the bc_1 complex, which is located at the interface between the "Rieske" protein and cytochrome *b* within or close to the membrane (see Section 5.2).

3.3.3. Redox-Dependent Protonation–Deprotonation

The redox potential of the "Rieske" [2Fe-2S] cluster is pH dependent, both in the bc_1 complex and in the isolated "Rieske" protein. From the pH dependence, two groups with redox dependent pK_a values could be determined (Link *et al.*, 1992). The pK_a values of 7.6 and 9.2 of the oxidized protein shift to above 11 on the reduced protein. Two redox-dependent pK_a values or a pH dependence of $\Delta E_m/\Delta pH < -60$ mV/pH, which cannot be explained with a single deprotonation step, have been found in all "Rieske" proteins involved in hydroquinone oxidation, including those from plastohydroquinone-oxidizing $b_6 f$ complexes, menahydroquinone-oxidizing *bc* complexes of gram-positive bacteria, and "Rieske" proteins from archaebacteria, where they are most likely part of hydroquinone-oxidizing oxidases. Therefore, the existence of two redox-dependent pK_a values can be considered as an intrinsic feature of "Rieske" clusters, which should be associated with conserved residues of these clusters.

The deprotonation of the oxidized "Rieske" protein could also be observed in the CD spectrum of the [2Fe-2S] cluster, indicating that this is not a general electrostatic effect. From these data, it has been proposed that deprotonation should occur on the histidine ligands of the "Rieske" cluster (Link, 1994). The electron density at the histidine ligands will be decreased and therefore their pK_a lowered due to the electron-withdrawing effect of the Fe(III) of the oxidized cluster compared to the Fe(II) of the reduced cluster.

This proposal is strongly supported by the structure of the ISF. The hydrogens at the $N\varepsilon$ atoms of the histidine ligands are fully exposed and are not involved in any hydrogen bond. In addition, there are no other residues in the vicinity of the cluster that are likely to undergo redox-dependent protonation–deprotonation; the fully conserved Tyr 165 is not exposed to solvent and the $O\eta$ is hydrogen bonded to the $S\gamma$ of Cys 139 (see Section 3.3.2.). Deprotonation of aspartate, glutamate, or tyrosine could be excluded by Fourier transform infrared (FTIR) spectroscopy (F. Baymann, T. A. Link, D. E. Robertson, and W. Mäntele, manuscript in preparation).

The histidine ligands of the cluster are most likely the binding site for quinones and inhibitors within the hydroquinone oxidation (Q_P-) site (see Section 5.2). Since the cluster is exposed at the tip of the structure, it can interact with the electron donor hydroquinone within the Q_P site. Due to the redox dependence of the pK_a values of the histidine ligands, they will bind protons more tightly upon reduction. The pK_a shift of the histidines reflects the increase of electron density upon reduction. The accompanying increase of the strength of hydrogen bonds between the $N\varepsilon$ atoms of the histidines and bound acceptors is considered to be essential for hydroquinone oxidation in the Q_P site (see Section 5.2).

4. GENERAL TOPOLOGY AND ORIENTATION OF THE REDOX CENTERS

Both mitochondrial bc_1 and chloroplast b_6f complexes are present in the membrane as structural dimers (Schägger et al., 1994) with a twofold axis of symmetry, while the bacterial complexes have been found to be monomers. Cytochrome b forms the central part of the complex as well as, in the case of dimers, the contact site between monomers. The other subunits are arranged peripherally, both within the membrane as well as in the aqueous phase. A low-resolution picture of the bc_1 complex from *Neurospora crassa* was obtained by electron microscopy of two-dimensional crystals (Leonard et al., 1981). The water-soluble domains of the "Rieske" iron–sulfur protein and of cytochrome c_1 extend into the periplasmic space or the mitochondrial intermembrane space, respectively, while the core proteins are bound at the inner side of the membranes. Insight into the general topological relationship of the subunits has been provided by proteolysis and specific labeling experiments (González-Halphen et al., 1988; Link, 1988).

The topographical distribution of the redox centers has been analyzed by the spin relaxation technique using a water-soluble dysprosium probe (Ohnishi *et al.*, 1989). The location of the "Rieske" cluster and the heme c_1 was consistent with the proposed orientation of the subunits at the P side of the membrane; the heme b centers were found to be asymmetrically oriented within the membrane, with the heme b_L center being located closer to the P side and heme b_H center in the middle of the membrane.

Several groups have reported crystals of the bc_1 complex (Yu *et al.*, 1994; Kawamoto *et al.*, 1994; Berry *et al.*, 1995; Lee *et al.*, 1995). At the moment this chapter was revised (April 1998), two of these groups have reported crystal structures (Xia *et al.*, 1997; Zhang *et al.*, 1998). Unfortunately, the "Rieske" protein was disordered in the first structure reported by Xia and co-workers. In the structure of the chicken heart enzyme reported by Zhang and co-workers, the structure of the water soluble "Rieske" fragment (Iwata *et al.*, 1996) was fitted to the electron density as a rigid body. In addition, both groups determined distances between the redox centers by analyzing anomalous scattering data of the crystals. Figure 5 shows a schematic representation of the relative locations obtained by these groups.

There is excellent agreement between the groups and with the distances obtained from spectroscopy and from model building as far as the distances within cytochrome b are concerned. However, the position of the "Rieske"

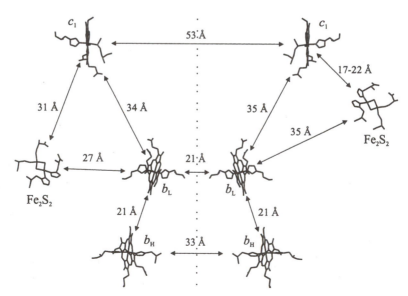

FIGURE 5. Schematic representation of the relative distances of the redox centers within the bc_1 complex as determined by Yu *et al.* (1996) (left) and Zhang *et al.* (1998) (right).

[2Fe-2S] cluster is at variance; while both structures show a fairly long distance between the iron atom of heme b_L and the "Rieske" cluster (27 and 35 Å, respectively), the "Rieske" cluster is distant from heme c_1 (31 Å) in one crystal form but it is close to heme c_1 in the other crystal form (17–21 Å). Only the latter distance is compatible with the fast electron transfer observed between these redox centers ($k_{ET} > 10^5$ sec^{-1}) (Crofts and Wang, 1989) but not the long distance. Therefore, the different crystal forms seem to represent different functional states of the "Rieske" protein during catalytic turnover. Apparently, the "Rieske" protein moves between a position closer to heme b_L ("b position"), where it interacts with hydroquinone and semiquinone, respectively, and a position close to heme c_1 ("c_1 position"), where it can transfer electrons to cytochrome c_1 (Zhang et al., 1998). The movement of the "Rieske" protein during catalysis is in agreement with conclusions drawn from inhibitor binding experiments ("catalytic switch" mechanism) (Brandt et al., 1991).

An unresolved issue has been the interaction between the monomers of a dimer. A cooperative catalytic function had been proposed for the Q_N site in the dimeric Q-cycle mechanism (de Vries et al., 1982), but this remained controversial. Rapid exchange of Q_N site inhibitors between dimers has been proposed to explain hyperbolic inhibition kinetics (Bechmann et al., 1992). Rapid exchange of inhibitors of the Q_N but not of the Q_P site would fit a model in which the heme b_H centers should be closer to each other than the heme b_L centers. However, the distance between the two heme b_H centers is 12 Å larger than that between the two heme b_L centers (33 vs. 21 Å, respectively). This makes a dimeric function of the Q_N site rather unlikely. On the other hand, slow electron transfer between the heme b_L centers of a dimer could be conceivable at least in inhibitor titration experiments where the intramonomer electron transfer is blocked. This could also lead to a redox equilibration between the heme b_H centers of a dimer via the route $b_H(A)–b_L(A)–b_L(B)–b_H(B)$.

5. QUINONE REACTION AND INHIBITOR BINDING SITES

The parts of cytochrome b that are involved in the formation of the two (Q_P-, Q_N-) quinone reaction sites have been identified by exhaustive studies of mutants conferring either resistance or hypersensitivity to specific inhibitors of either quinone reaction site (for a recent compilation, see Brasseur et al., 1996). All residues where relevant mutations have been detected cluster in loops at the membrane surfaces or in the adjacent ends of transmembrane helices. Two regions at the P side are involved in the formation of the Q_P site: Q_PI, the C-terminal part of helix C and the cd loop; and Q_PII, the ef loop. Two regions at the N side are involved in the formation of the Q_N site: Q_NI, the N-terminal part preceding helix A; and Q_NII, the de loop and the N-terminal part of helix E.

Three of these regions ($Q_N I$, $Q_P I$, $Q_P II$) contain sequences with high helical amphipathic and mutability moments that form surface-oriented amphipathic helices (*a*, *cd*1, *cd*2, and *ef*); this indicates that both peripheral and trans-membrane helices contribute to the quinone reaction sites. This predicted topology of the quinone binding sites of the bc_1 complex strongly resembles the topology of the quinone binding sites of the bacterial photosynthetic reaction centers (Fig. 6): the quinone head group is bound between a transmembrane and a membrane-attached helix through hydrogen bonds, while the isoprenoid tail seems to contribute to the binding through van der Waals contacts with the

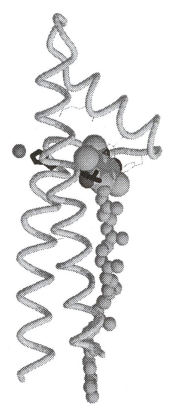

FIGURE 6. The structure of the secondary quinone (Q_B) site of the bacterial photosynthetic reaction center from *Rhodobacter sphaeroides* (from PDB file 1PCR) (Ermler *et al.*, 1994). The quinone head group is shown in a space-filling representation, the quinone tail as ball-and-stick model, and residues Leu L 189 and His L 190 in black. His L 190 is coordinated to the iron atom (gray ball, left) as well as to the proximal carbonyl oxygen of quinone; the other ligands to the iron atom have been omitted.

transmembrane helix. This suggests that the type of quinone binding site depicted in Fig. 6 might constitute an archetypical structural module that could be expected to occur in many if not all enzymes using quinone as a substrate.

5.1. Quinone Reduction (Q_N) Site

The function of the quinone reduction (Q_N) site is equivalent to that of the secondary quinone binding (Q_B) site of bacterial reaction centers: quinone is bound and reduced to hydroquinone in two consecutive one-electron transfer steps where both electrons are delivered by the same donor. The intermediate semiquinone is stabilized within the Q_N site; it can be detected using EPR spectroscopy (de Vries *et al.*, 1980). However, there is no detectable sequence homology between the Q_B and the Q_N site; despite their functional equivalence and their presumed topological similarity (see this section), both sites appear to be structurally diverse. This is reflected, for example, in their affinity for inhibitors: at the Q_B site, several inhibitors of the Q_N site (diuron, which binds to yeast but not to mammalian bc_1 complexes, and hydroxyquinoline-N-oxides) and of the Q_P site are bound with high affinity, but not antimycin, which is a specific inhibitor of the Q_N site

Antimycin binds in the vicinity of heme b_H and induces spectral shifts of this heme center; binding of antimycin displaces the Q_N^- semiquinone radical. In difference electron density maps of antimycin-bound bc_1 complex, antimycin showed a strong area of density close to heme b_H, while some negative peaks may indicate the removal of ubi(semi)quinone as well as conformational changes of side chains in the site (Yu *et al.*, 1996).

Besides heme b_H, two fully conserved residues appear to have central roles in the Q_N site: histidine 202 and aspartate 229 (mitochondrial numbering). His 202 is located in the beginning of the Q_NII region close to the end of helix D. Mutations of the bacterial His 217 (the corresponding residue in *Rhodobacter capsulatus*) and in other residues nearby [Thr 219, Gly 220, Asn 221 (corresponding to the mitochondrial Ser 206), Asn 222, Asn 223] altered the kinetics of the Q_N site or the redox potential of b_H. Mutations in His 217/202 strongly affected the stability of the bound semiquinone Q_N^-: the stability of Q_N^- was increased upon mutation of His to the more basic arginine and decreased upon mutation to the acidic aspartate or to the hydrophobic leucine (Gray *et al.*, 1994). These effects are consistent with a direct interaction between one of the carbonyl groups of the semiquinone and His 202 in a similar manner as Q_A and Q_B of the photosynthetic reaction center interact with His M217/219 and His L190, respectively.

Asp 229 is part of the amphipathic N-terminal end of helix E; on the same face of the helix are the residues (Met,Phe) 221, Phe 225, and Lys 228. Mutations of all of these residues have been reported to cause functional defects of the Q_N site (Hacker *et al.*, 1993) and/or inhibitor resistance toward antimycin, HQNO, or diuron. From the analysis of site-directed mutants it can be concluded that the presence of an acidic residue in this part of the reaction pocket is essential; therefore, Asp 229 is most likely involved in electrogenic charge compensation and/or proton transfer at the Q_N site (Crofts *et al.*, 1995).

5.2. Hydroquinone Oxidation (Q_P) Site

The function of the Q_P site differs from that of the Q_N and Q_B sites: hydroquinone is oxidized and the two electrons are transferred to two different acceptors. The first electron goes to the "Rieske" iron–sulfur cluster, the second electron to heme b_L; the coupling between the two electron transfer steps is obligatory. The semiquinone is activated and no stable intermediate semiquinone (Q_P^-) is generally observed even under conditions where the second electron transfer step is blocked, although a kinetically trapped free radical has been associated with the semiquinone Q_P^- (de Vries *et al.*, 1981). It is not clear, however, whether the semiquinone described is a trapped intermediate of the catalytic reaction or a species that is generated only in the inhibited state.

As discussed in Section 1, the two redox centers accepting one electron each are located in different protein subunits. Domains both of cytochrome *b* and the "Rieske" iron–sulfur protein contribute to the formation of the Q_P site: the tip of the "Rieske" protein, including the exposed histidine ligands and residue Leu 142, the environment of heme b_L, and the Q_PI and Q_PII regions (*cd* and *ef* loops, respectively) of cytochrome *b*. The interaction between cytochrome *b* and the "Rieske" iron–sulfur protein is mediated by the N-terminal membrane anchor of the "Rieske" protein (interacting with the transmembrane part of cytochrome *b*) and by the *ef* loop of cytochrome *b* (Geier *et al.*, 1992; Giessler *et al.*, 1994).

Despite these differences, the Q_P and the Q_B site appear to have significant structural homology. This refers not only to the general topology and to the binding of inhibitors like stigmatellin and undecyl-hydroxy-dioxobenzothiazole (UHDBT) but there is also statistically relevant homology between the sequence Trp 142-Gly-Ala-Thr-Val/Ile in the *cd* loop (Q_PI region) of cytochrome *b* and the sequence His M 217/219-Gly-Ala-Thr-Val/Ile of the M subunit of the reaction center (Degli Esposti *et al.*, 1993); the reaction center sequence is part of the Q_A site where His M 217/219 interacts with the C(4)=O carbonyl of Q_A. A homologous sequence is also present in the Q_B site (Fig. 6). This supports the proposal of a common quinone binding motif (see Section 5). Additional evidence for a model of the Q_P site homologous to the Q_A and Q_B sites has been discussed by Robertson *et al.* (1990).

The structure of the ISF has allowed us to build a molecular model of the

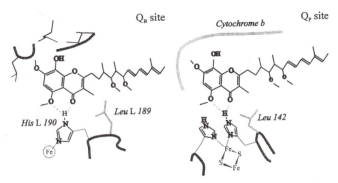

FIGURE 7. Binding of the inhibitor stigmatellin in the secondary quinone (Q_B) site of the bacterial reaction center from *Rhodopseudomonas viridis* (left; X-ray structure) (Lancaster and Michel, 1997) and in the hydroquinone oxidation (Q_P) site of the cytochrome bc_1 complex (right; model).

binding of quinones and quinonoid inhibitors within the Q_P site (Fig. 7) (Link and Iwata, 1996). This model is based on the proposal that quinones and inhibitors like stigmatellin and hydroxyquinones, which show strong structural homology to the natural substrate ubiquinone, bind directly to the [2Fe-2S] cluster. This is indicated by the observations that (1) the affinity of the bc_1 complex for stigmatellin is largely decreased when the iron–sulfur protein is depleted (Brandt *et al.*, 1991); (2) stigmatellin binds 5 orders of magnitude more tightly to the reduced [2Fe-2S] cluster than to the oxidized and shifts the EPR spectrum of the (reduced) cluster (von Jagow and Ohnishi, 1985); and that (3) that the g_x signal of the EPR spectrum is sensitive to the redox state of bound quinones (Siedow *et al.*, 1978; de Vries *et al.*, 1979). However, neither inhibitors nor substrates will bind tightly to the isolated ISF (T. A. Link, unpublished results), indicating that cytochrome *b* forms the major part of the quinone binding site.

In the structure of the ISF, the Nε atoms of both histidine ligands to Fe-2 [Fe(II) in the reduced cluster] are completely exposed to solvent within a hydrophobic surface at the tip of the structure (Fig. 3). The redox potential of the isolated ISF depends on the polarity of the environment (Link *et al.*, 1992, 1996b); it is identical within 10 mV to that of the "Rieske" cluster in the bc_1 complex (Link *et al.*, 1996a). These data indicate that the histidines are exposed in the Q_P site and that the polarity within the site is high. The Q_B site of the photosynthetic reaction center is filled with six ordered water molecules when quinones are not present (Lancaster and Michel, 1997). The mode of binding of stigmatellin in the Q_B site has been determined by X-ray crystallography (Lancaster and Michel, 1997): it binds in the same position as ubiquinone via a three-center hydrogen bond to residue His L 190, which is a ligand to the Fe^{2+} (Fig. 7).

The model of the Q_P site contains a three-center hydrogen bond between stigmatellin and one of the histidine ligands of the cluster (Fig. 7). This interaction appears to be essential for the high affinity of stigmatellin and could explain the experimental results described above, in particular, the strong redox dependence of stigmatellin binding. The higher affinity of stigmatellin to the reduced "Rieske" cluster reflects the high pK_a values of the histidine ligands and the increased hydrogen bond strength. Since both histidines are almost equally exposed in the structure of the ISF, we cannot yet decide which of the histidine ligands forms the hydrogen bond with quinone and/or stigmatellin or whether both histidines may interact either simultaneously or consecutively during a reaction cycle.

The analogy between the Q_B and the Q_P site can be extended to the hydrophobic Leu L 189 in the Q_3 site, which is adjacent to the His L 190, forming the hydrogen bond to stigmatellin. Leu L 189 is in van der Waals contact with the side chain of stigmatellin or with the side chains of quinones or other inhibitors (Lancaster and Michel, 1997). In the structure of the ISF, the corresponding Leu 142 is exposed at the hydrophobic tip adjacent to the histidine ligands. Mutation of this residue in *Rhodobacter capsulatus* (Leu 136) to Gly, Asp, His, or Arg completely blocked quinone binding, while the "Rieske" cluster was still present (Liebl *et al.*, 1997). Therefore, it appears that a residue that can provide hydrophobic interactions with quinones and inhibitors is essential for quinone binding in the Q_B as well as in the Q_P site.

On the distal side of the pocket, stigmatellin (and quinones) will interact with parts of cytochrome *b*. Several residues in the Q_P site have been identified where mutations weakened the binding of stigmatellin (reviewed in Link *et al.*, 1993). However, each single mutation had a rather weak effect on the affinity of stigmatellin (less than 7 kJ/mole change in binding energy), which could indicate that either the backbone interacts specifically with stigmatellin so that the effect of the mutations is through an altered backbone conformation or that the distal side of the pocket interacts unspecifically through van der Waals interactions.

A controversial issue is the number of quinone molecules bound in the Q_P site. From an analysis of the effect of the size and redox state of the quinone pool on the g_x signal of the EPR spectrum of the reduced "Rieske" cluster in membranes of *Rhodobacter capsulatus*, Ding *et al.* (1992) proposed that the Q_P site can accommodate two ubiquinone molecules: a strongly bound species (Q_{os}) binding with three- to fourfold higher affinity than the secondary quinone Q_B in the reaction center, and a weakly bound species (Q_{ow}) with 20-fold lower affinity than Q_{os} (i.e., weaker than Q_B). Both Q_{os} and Q_{ow} showed equal affinity within a factor of two in the oxidized (Q) and reduced (QH_2) state. From a subsequent kinetic analysis of mutants in the Q_P site, Ding *et al.* (1995) suggested that Q_{os} exchanges with the quinone pool at a rate that is much slower than the time scale of turnover, while Q_{ow} exchanges at a rate that is much faster than the time scale of turnover. Ding *et al.* (1995) have also discussed several possible modes of quinone binding in the Q_P site and proposed a kinetic mechanism: a linear edge-to-edge arrangement of Q_{os} and Q_{ow} and fast electron transfer from hydroquinone in the Q_{ow} site to semiquinone in the Q_{os} site (formed by the first electron transfer to the "Rieske" cluster) should allow charge separation and move the semiquinone away from the "Rieske" cluster, thus preventing transfer of the second electron to the high-potential chain. Since Q_{os} is supposed to bind closer to the "Rieske" cluster, the quinone binding site shown in Fig. 7 should be equated with Q_{os}, which acts as a prosthetic group. Brandt (1996) has suggested a proton-gated charge-transfer mechanism that involves a stacked quinone/hydroquinone pair ($Q_{os}/Q_{ow}H_2$); after activation by deprotonation ($Q_{os}/Q_{ow}H^-$), symproportionation of the two quinones occurs and the double semiquinone (Q_{os}^-/Q_{ow}^-) simultaneously transfers two electrons to the "Rieske" cluster and to heme b_L. The symproportionation is driven by the oxidized "Rieske" center acting as a Lewis acid.

From the crystal structures determined so far, it cannot be determined whether the Q_P site can bind two quinone molecules simultaneously or perhaps only a single quinone molecule in alternate positions. However, while there is good evidence for the binding of quinone to the histidine ligands of the "Rieske" cluster, binding of quinones to the axial histidine of either heme b can be excluded since the histidines are buried within the 4-α-helical bundle. If the axial histidines were accessible, one would expect a redox-dependent deprotonation of the histidines to occur on the oxidized heme; this was observed, for example, for the water soluble cytochrome b_{562} from *E. coli* (Moore *et al.*, 1985) but not for cytochrome b (Link *et al.*, 1995). An edge-to-edge orientation of two quinones within the Q_P site is also difficult to reconcile with the observation that mutations in different residues affected both quinones to the same extent. The Q_P site appears to be one contiguous reaction site with respect to the binding of quinones, while mutations in different parts of the Q_P site show specific effects on the binding of different inhibitors (reviewed in Link *et al.*, 1993).

A mechanism that follows directly from the electrochemical behavior of the "Rieske" protein and that requires only a single quinone molecule is the proton-gated affinity change mechanism (Fig. 8) (Link, 1997). The mechanism is based on the observation that the affinity of the reduced "Rieske" cluster for quinone intermediates (e.g., the inhibitor stigmatellin) is several orders of magnitude larger than that of the oxidized cluster (Section 3.3.3). In the reaction cycle (Fig. 8), deprotonation of hydroquinone (step 1) forms the

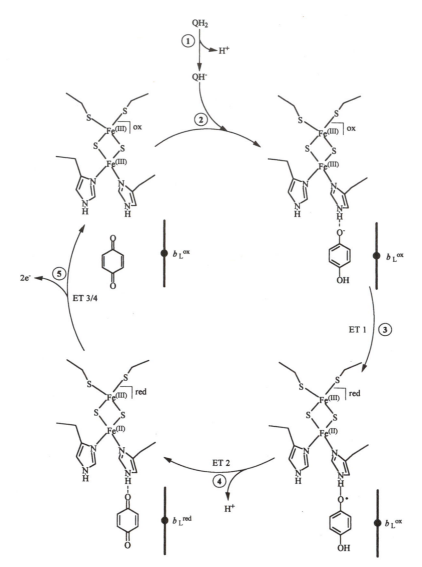

FIGURE 8. Proton-gated affinity change mechanism of hydroquinone oxidation in the Q_P site. (Step 1) Hydroquinone must first be deprotonated, before it can bind to the oxidized "Rieske" cluster (step 2). (Step 3) The first electron is transferred to the "Rieske" cluster (ET 1); thereby, a semiquinone is formed that is bound and magnetically coupled to the "Rieske" cluster (bottom right). (Step 4) After the second deprotonation step, the second electron is transferred to heme b_L (ET 2). (Step 5) After electron transfer from the "Rieske" cluster to cytochrome c_1 (ET 3) and from heme b_L to heme b_H (ET 4), quinone is released and the "Rieske" cluster can bind the next molecule of deprotonated hydroquinone. The different line types between the imidazole of the "Rieske" cluster and the quinone molecule indicate the change of the binding constant in different redox states: broken line, weak binding; full line, strong binding.

activation barrier (Brandt and Okun, 1997; Brandt, 1998); the deprotonated hydroquinone interacts with one of the histidine ligands of the oxidized "Rieske" cluster (step 2). Electron transfer (ET 1) from the deprotonated hydroquinone to the "Rieske" cluster generates a semiquinone bound tightly to the reduced "Rieske" cluster (step 3); the high affinity of the reduced "Rieske" cluster for semiquinone will catalyze the reaction and prevent dissociation of the semiquinone. After the second deprotonation step, the bound semiquinone will reduce heme b_L and oxidized quinone can be released. In the final step, electrons are transferred from heme b_L to b_H and from the reduced "Rieske" cluster to cytochrome c_1; this reaction seems to require movement of the "Rieske" protein from a position close to heme b_L (b position) to a position close to heme c_1 (c_1 position). A modified version of the reaction scheme fully exploiting the catalytic mobility of the "Rieske" protein has been developed (Iwata et al., 1998).

The key step in this mechanism is the change of the affinity of the "Rieske" cluster toward the substrate upon reduction and the stabilization of the semiquinone on the reduced "Rieske" cluster, which not only will prevent dissociation of the semiquinone but prevent the reduced "Rieske" protein from transferring its electron to cytochrome c_1 before the bound semiquinone has transferred the second electron to heme b_L. The bound semiquinone would not be observed in EPR spectroscopy, since it is magnetically coupled to the reduced paramagnetic "Rieske" cluster. The distance of 27 Å between the center of the "Rieske" cluster and the iron atom of heme b_L in the b position determined by X-ray crystallography is compatible with this mechanism: the distance between the center of the "Rieske" cluster and the distal oxygen is 13 Å, and the distance between a heme edge and the central iron atom 6 Å. Therefore, electron transfer between semiquinone bound to the "Rieske" cluster and heme b_L would occur over a distance of less than 12 Å, which is compatible with electron transfer rates $>10^6 \text{ sec}^{-1}$.

6. SUMMARY AND PERSPECTIVE

The bc_1 complex is one of the best-studied membrane proteins. The wealth of information obtained from research in many laboratories and using enzymology, protein chemistry, electrochemistry, spectroscopy, and mutagenesis has allowed the construction of fairly detailed models discussed above. Although the emerging X-ray structures are consistent with the models in many aspects, they have shown unexpected features of the protein; in particular, the large movement of the "Rieske" protein required for the reaction cycle. These findings will set the stage for new kinetic and spectroscopic experiments at a higher level.

Paramagnetic resonance techniques have been extremely powerful tools over the past 20 years. Their main disadvantage is that they are "blind" for one of the two redox states of each metal center of the bc_1 complex and for two of the three redox states of quinones. This limitation has been evident, for example, in the analysis of the "double-occupancy" model of the Q_P site, since quinone or hydroquinone could not be observed directly but only through their effect on the reduced "Rieske" cluster, and since binding of quinones to the oxidized "Rieske" cluster could not be monitored at all using EPR techniques. Vibrational (FTIR and resonance Raman) spectroscopy is an obvious choice for the next generation of experiments, since it can provide local and specific structural information in each redox state. In particular, kinetically resolved vibrational spectroscopy should allow the identification of reaction intermediates on a realistic time scale, although these studies are difficult in the case of the bc_1 complex since quinones are hydophobic weak binding substrates with dissociation constants in the 30-mM range.

A review in this rapidly evolving field is partially outdated at the time the book is printed. Any reader interested in the latest information on the bc_1 complex is referred to the Web page of the bc_1 complex site at http://arc-gen1.life.uiuc.edu/bc-complex_site/, maintained by A. R. Crofts at the University of Illinois at Urbana-Champaign (UIUC). Thanks to the efforts of Dr. Crofts, the reader will find news concerning the structure(s) of the bc_1 complex, among other useful information and links.

ACKNOWLEDGMENT. I thank Drs. E. A. Berry, J. Deisenhofer, and C.-A. Yu for generously releasing details of their crystallographic analyses of crystals of the bc_1 complex prior to publication; Drs. C. R. D. Lancaster and H. Michel for the making the structure of the stigmatellin complex of the bacterial reaction center available prior to publication; and Dr. B. L. Trumpower for providing unpublished information on calcium binding. The author's work on the structure of the "Rieske" iron–sulfur protein was supported by the Deutsche Forschungsgemeinschaft (DFG) through the Priority Programme "Metal Ions in Biology and their Coordination Chemistry," Grant Li 474/2.

7. REFERENCES

Bechmann, G., Weiss, H., and Rich, P. R., 1992, Non-linear inhibition curves for tight-binding inhibitors of dimeric ubiquinol-cytochrome c oxidoreductases. Evidence for rapid mobility, *Eur. J. Biochem.* **208**:315–325.

Beckmann J. D., Ljungdahl, P. O., and Trumpower, B. L., 1989, Mutational analysis of the mitochondrial Rieske iron–sulfur protein of *Saccharomyces cerevisiae*. I. Construction of a RIP1 deletion strain and isolation of temperature-sensitive mutants, *J. Biol. Chem.* **264**:3713–3722.

Berry, E. A., Shulmeister, V. M., Huang, L., and Kim, S.-H., 1995, A new crystal form of bovine heart ubiquinol:cytochrome *c* oxidoreductase: Determination of space group and unit-cell parameters, *Acta Crystallogr. D* **51**:235–239.

Brandt, U., 1996, Bifurcated ubihydroquinone oxidation in the cytochrome bc_1 complex by proton-gated charge transfer, *FEBS Lett.* **387**:1–6.

Brandt, U., 1998, The chemistry and mechanics of ubihydroquinone oxidation at center P (Q_o) of the cytochrome bc_1 complex, *Biochim. Biophys. Acta* **1365**:261–268.

Brandt, U., and Okun, J., 1997, Role of deprotonation events in ubihydroquinone:cytochrome *c* oxidoreductase from bovine heart and yeast mitochondria, *Biochemistry* **36**:11234–11240.

Brandt, U., Haase, U., Schägger, H., and von Jagow, G., 1991, Significance of the "Rieske" iron–sulfur protein for formation and function of the ubiquinol oxidation pocket of mitochondrial cytochrome *c* reductase (bc_1 complex), *J. Biol. Chem.* **266**:19958–19964.

Brandt, U., Yu, L., Yu, C.-A., and Trumpower, B. L., 1993, The mitochondrial targeting pre-sequence of the Rieske iron–sulfur protein is processed in a single step after insertion into the cytochrome bc_1 complex in mammals and retained as a subunit in the complex, *J. Biol. Chem.* **268**:8387–8390.

Brandt, U., Uribe, S., Schägger, H., and Trumpower, B. L., 1994, Isolation and characterization of QCR10, the nuclear gene encoding the 8,5-kDa subunit 10 of the *Saccharomyces cerevisiae* cytochrome bc_1 complex, *J. Biol. Chem.* **269**:12947–12953.

Brasseur, G., Saribas, A. S., and Daldal, F., 1996, A compilation of mutations located in the cytochrome *b* subunit of the bacterial and mitochondrial bc_1 complex, *Biochim. Biophys. Acta* **1275**:61–69.

Braun, H.-P., and Schmitz, U. K., 1995, Are the "core" proteins of the mitochondrial bc_1 complex evolutionary relics of a processing peptidase? *Trends Biochem. Sci.* **20**:171–175.

Britt, R. D., Sauer, K., Klein, M. P., Knaff, D. B., Kriauciunas, A., Yu, C.-A., Yu, L., and Malkin, R., 1991, Electron spin echo envelope modulation spectroscopy supports the suggested coordination of two histidine ligands to the Rieske Fe-S centers of the cytochrome b_6f complex of spinach and the cytochrome bc_1 complexes of *Rhodospirillum rubrum*, *Rhodobacter sphaeroides* R-26, and bovine heart mitochondria, *Biochemistry* **30**:1892–1901.

Crofts, A. R., and Wang, Z., 1989, How rapid are the internal reactions of the ubiquinol:cytochrome c_2 oxidoreductase? *Photosynthesis Res.* **22**:69–87.

Crofts, A. R., Hacker, B., Barquera, B., Yun, C.-H., and Gennis, R. B., 1992, Structure and function of the *bc*-complex of *Rhodobacter sphaeroides*, in: *Research in Photosynthesis*, vol. II (N. Murata, ed.), pp. 463–470, Kluwer Academic Publishers, Dordrecht.

Crofts, A. R., Barquera, B., Bechmann, G., Guergova, M., Salcedo-Hernandez, R., Hacker, B., Jong, S., and Gennis, R. B., 1995, Structure and function of the bc_1-complex of *Rb. sphaeroides*, in: *Photosynthesis: From Light to Biosphere*, vol. II (P. Mathis, ed.), pp. 749–752, Kluwer Academic Publishers, Dordrecht.

Davidson, E., Ohnishi, T., Atta-Asafo-Adjei, E., and Daldal, F., 1992, Potential ligands to the [2Fe-2S] Rieske cluster of the cytochrome bc_1 complex of *Rhodobacter capsulatus* probed by site-directed mutagenesis, *Biochemistry* **31**:3342–3351.

Degli Esposti, M., de Vries, S., Crimi, M., Ghelli, A., Patarnello, T., and Meyer, A., 1993, Mitochondrial cytochrome *b*: Evolution and structure of the protein, *Biochim. Biophys. Acta* **1143**:243–271.

Denke, E., Merbitz-Zahradnik, T., Hatzfeld, O. M., Snyder, C. H., Link, T. A., and Trumpower, B. L., 1998, Alteration of the midpoint potential and catalytic activity of the Rieske iron–sulfur protein by changes of amino acids forming hydrogen bonds to the iron–sulfur cluster, *J. Biol. Chem.* **273**:9085–9093.

de Vries, S., Albracht, S. P. J., and Leeuwerik, F. J., 1979, The multiplicity and stoichiometry of the prosthetic groups in QH₂:cytochrome *c* oxidoreductase as studied by EPR, *Biochim. Biophys. Acta* **546**:316–333.

de Vries, S., Berden, J. A., and Slater, E. C., 1980, Properties of a semiquinone anion located in the QH$_2$:cytochrome c oxidoreductase segment of the mitochondrial respiratory chain, *FEBS Lett.* **122**:143–148.

de Vries, S., Albracht, S. P. J., Berden, J. A., and Slater, E. C., 1981, A new species of bound ubisemiquinone anion in QH2:cytochrome c oxidoreductase, *J. Biol. Chem.* **256**:11996–11998.

de Vries, S., Albracht, S. P. J., Berden, J. A., and Slater, E. C., 1982, The pathway of electrons through QH$_2$:cytochrome c oxidoreductase studied by pre–steady-state kinetics, *Biochim. Biophys. Acta* **681**:41–53.

Ding, H., Robertson, D. E., Daldal, F., and Dutton, P. L., 1992, Cytochrome bc_1 complex [2Fe-2S] cluster and its interaction with ubiquinone and ubihydroquinone at the Q$_o$ site: A double-occupancy Q$_o$ site model, *Biochemistry* **31**:3144–3158.

Ding, H., Moser, C. C., Robertson, D. E., Tokito, M. K., Daldal, F., and Dutton, P. L., 1995, Ubiquinone pair in the Q$_o$ site central to the primary energy conversion reactions of cytochrome bc_1 complex, *Biochemistry* **34**:15979–15996.

di Rago, J.-P., Netter, P., and Slonimski, P. P., 1990, Intragenic suppressors reveal long distance interactions between inactivating and reactivating amino acid replacements generating three-dimensional constraints in the structure of mitochondrial cytochrome b, *J. Biol. Chem.* **265**:15750–15757.

Ermler, U., Fritzsch, G., Buchanan, S. K., and Michel, H., 1994, Structure of the photosynthetic reaction centre from *Rhodobacter sphaeroides* at 2.65 Angstroms resolution: Cofactors and protein–cofactor interactions, *Structure* **2**:925–936.

Fee, J. A., Findling, K. L., Yoshida, T., Hille, R., Tarr, G. E., Hearshen, D. O., Dunham, W. R., Day, E. P., Kent, T. A., and Münck, E., 1984, Purification and characterization of the Rieske iron-sulfur protein from *Thermus thermophilus*. Evidence for a [2Fe-2S] cluster having non-cysteine ligands, *J. Biol. Chem.* **259**:124–133.

Finnegan, M. G., Knaff, D. B., Qin, H., Gray, K. A., Daldal, F., Yu, L., Yu, C.-A., Kleis-San-Francisco, S., and Johnson, M. K., 1996, Axial heme ligation in the cytochrome bc_1 complexes of mitochondrial and photosynthetic membranes. A near-infrared magnetic circular dichroism and electron paramagnetic resonance study, *Biochim. Biophys. Acta* **1274**:9–20.

Gatti, D. L., Meinhardt, S. W., Ohnishi, T., and Tzagoloff, A., 1989, Structure and function of the mitochondrial bc_1 complex. A mutational analysis of the yeast Rieske iron–sulfur protein, *J. Mol. Biol.* **205**:421–435.

Geier, B. M., Schägger, H., Brandt, U., and von Jagow, G., 1992, Point mutation in cytochrome b of yeast ubihydroquinone:cytochrome-c oxidoreductase causing myxothiazol resistance and facilitated dissociation of the iron–sulfur subunit, *Eur. J. Biochem.* **208**:375–380.

Giessler, A., Geier, B. M., di Rago, J.-P., Slonimski, P. P., and von Jagow, G., 1994, Analysis of cytochrome-b amino acid residues forming the contact face with the iron–sulfur subunit of ubiquinol:cytochrome-c reductase in *Saccharomyces cerevisiae*, *Eur. J. Biochem.* **222**:147–154.

González-Halphen, D. M., Lindorfer, M. A., and Capaldi, R., 1988, Subunit arrangement in beef heart complex III, *Biochemistry* **27**:7021–7031.

Graham, L. A., and Trumpower, B. L., 1991, Mutational analysis of the mitochondrial Rieske iron–sulfur protein of *Saccharomyces cerevisiae*. III. Import, protease processing, and assembly into the cytochrome bc_1 complex of iron–sulfur protein lacking the iron–sulfur cluster, *J. Biol. Chem.* **266**:22485–22492.

Graham, L. A., Brandt, U., Sargent, J. S., and Trumpower, B. L., 1992, Mutational analysis of assembly and function of the iron–sulfur protein of the cytochrome bc_1 complex in *Saccharomyces cerevisiae*, *J. Bioenerg. Biomembr.* **25**:245–257.

Gray, K. A., Dutton, P. L., and Daldal, F., 1994, Requirement of histidine 217 for ubiquinone reductase activity (Q_i site) in the cytochrome bc_1 complex, *Biochemistry* **33**:723–733.

Gurbiel, R. J., Batie, C. J., Sivaraja, M., True, A. E., Fee, J. A., Hoffman, B. M., and Ballou, D. P., 1989, Electron-nuclear double resonance spectroscopy of ^{15}N-enriched phthalate dioxygenase from *Pseudomonas cepacia* proves that two histidines are coordinated to the [2Fe-2S] Rieske-type clusters, *Biochemistry* **28**:4861–4871.

Gurbiel, R. J., Ohnishi, T., Robertson, D., Daldal, F., and Hoffman, B. M., 1991, Q-band ENDOR spectra of the Rieske protein from *Rhodobacter capsulatus* ubiquinol–cytochrome *c* oxidoreductase show two histidines coordinated to the [2Fe-2S] cluster, *Biochemistry* **30**:11579–11584.

Hacker, B., Barquera, B., Crofts, A. R., and Gennis, R. B., 1993, Characterization of mutations in the cytochrome *b* subunit of the bc_1 complex of *Rhodobacter sphaeroides* that affect the quinone reductase site (Q_c), *Biochemistry* **32**:4403–4410.

Hovmöller, S., Leonard, K., and Weiss, H., 1981, Membrane crystals of a subunit complex of mitochondrial cytochrome reductase containing the cytochromes *b* and c_1, *FEBS Lett.* **123**:118–122.

Iwata, S., Lee, J. W., Okada, K., Lee, J. K., Iwata, M., Rasmussen, B., Link, T. A., Jap, B. K., 1998, Complete structure of the 11-subunit bovine mitochondrial cytochrome bc_1 complex, *Science* **281**:64–71.

Iwata, S., Saynovits, M., Link, T. A., and Michel, H., 1996, Structure of a water soluble fragment of the "Rieske" iron–sulfur protein of the bovine heart mitochondrial cytochrome bc_1 complex determined by MAD phasing at 1.5 Å resolution, *Structure* **4**:567–579.

Kawamoto, M., Kubota, T., Matsunaga, T., Fukuyama, K., Matsubara, H., Shinzawa-Itoh, K., and Yoshikawa, S., 1994, New crystal forms and preliminary X-ray diffraction studies of mitochondrial cytochrome bc_1 complex from bovine heart, *J. Mol. Biol.* **244**:238–241.

Kraulis, P. J., 1991, MOLSCRIPT: A program to produce both detailed and schematic plots of protein structures, *J. Appl. Crystallogr.* **24**:946–950.

Lancaster, C. R. D., and Michel, H., 1997, The coupling of light-induced electron transfer and proton uptake as deduced from crystal structures of reaction centres from *Rhodopseudomonas viridis* modified at the binding site of the secondary quinone, Q_B, *Structure* **5**:1339–1359.

Lee, J. W., Chan, M., Law, T. V., Kwon, H. J., and Jap, B. K., 1995, Preliminary cryocrystallographic study of the mitochondrial cytochrome bc_1 complex: Improved crystallization and flash-cooling of a large membrane protein, *J. Mol. Biol.* **252**:15–19.

Leonard, K., Wingfield, P., Arad, T., and Weiss, H., 1981, Three-dimensional structure of ubiquinol: cytochrome *c* reductase from *Neurospora* mitochondria determined by electron microscopy of membrane crystals, *J. Mol. Biol.* **149**:259–274.

Liebl, U., Sled, V., Brasseur, G., Ohnishi, T., and Daldal, F., 1997, Conserved nonliganding residues of *Rhodobacter capsulatus* Rieske iron-sulfur protein are essential for [2Fe-2S] cluster properties and communication with the quinone pool protein structure properties of the [2Fe-2S] cluster, *Biochemistry* **36**:11675–11684.

Link, T. A., 1988, Zur Struktur des mitochondrialen bc_1-Komplexes, PhD thesis, Ludwig-Maximilians-Universität, München.

Link, T. A., 1994, Two pK values of the oxidised "Rieske" [2Fe-2S] cluster observed by CD spectroscopy, *Biochim. Biophys. Acta* **1185**:81–84.

Link, T. A., 1997, The role of the "Rieske" iron–sulfur protein in the hydroquinone oxidation (Q_p) site of the cytochrome bc_1 complex—The proton-gated affinity change mechanism, *FEBS Lett.* **412**:257–264.

Link, T. A., and Iwata, S., 1996, Functional implications of the structure of the "Rieske" iron–

sulfur protein of bovine heart mitochondrial cytochrome bc_1 complex, *Biochim. Biophys. Acta* **1275**:54–60.

Link, T. A., Schägger, H., and von Jagow, G., 1986, Analysis of the structures of the subunits of the cytochrome bc_1 complex from beef heart mitochondria, *FEBS Lett.* **204**:9–15.

Link, T. A., Hagen, W. R., Pierik, A. J., Assmann, C., and von Jagow, G., 1992, Determination of the redox properties of the Rieske [2Fe-2S] cluster of bovine heart bc_1 complex by direct electrochemistry of a water-soluble fragment, *Eur. J. Biochem.* **208**:685–691.

Link, T. A., Haase, U., Brandt, U., and von Jagow, G., 1993, What information do inhibitors provide about the structure of the hydroquinone oxidation site of ubihydroquinone:cytochrome c oxidoreductase? *J. Bioenerg. Biomembr.* **25**:221–232.

Link, T. A., Wallmeier, H., and von Jagow, G., 1994, Modelling the three-dimensional structure of cytochrome *b*, *Biochem. Soc. Trans.* **1994**:197–203.

Link, T. A., Cheesman, M. R., and Thomson, A. J., 1995, CD and MCD spectroscopy of the heme centers of the bc_1 complex, *J. Inorg. Biochem.* **59**:436.

Link, T. A., Saynovits, M., Assmann, C., Iwata, S., Ohnishi, T., and von Jagow, G., 1996a, Isolation, characterisation, and crystallisation of a water-soluble fragment of the Rieske iron–sulfur protein of bovine heart mitochondrial bc_1 complex, *Eur. J. Biochem.* **237**:71–75.

Link, T. A., Hatzfeld, O. M., Unalkat, P., Shergill, J. K., Cammack, R., and Mason, J. R., 1996b, Comparison of the Rieske 2-iron–2-sulfur centre in bc_1 complex and in bacterial dioxygenases by CD spectroscopy and cyclic voltammetry, *Biochemistry* **35**:7546–7552.

Martinez, S. E., Huang, D., Szczepaniak, A., Cramer, W. A., and Smith, J. L., 1994, Crystal structure of chloroplast cytochrome *f* reveals a novel cytochrome fold and unexpected heme ligation, *Structure* **2**:95–105.

Mitchell, P., 1976, Possible molecular mechanisms of the protonmotive force function of cytochrome systems, *J. Theor. Biol.* **62**:327–367.

Moore, G. R., Williams, R. J. P., Peterson, J., Thomson, A. J., and Mathews, F. S., 1985, A spectroscopic investigation of the structure and redox properties of *Escherichia coli* cytochrome *b*-562, *Biochim. Biophys. Acta* **829**:83–96.

Ohnishi, T., Schägger, H., Meinhardt, S. W., LoBrutto, R., Link, T. A., and von Jagow, G., 1989, Spatial organization of the redox active centers in the bovine heart ubiquinol–cytochrome c oxidoreductase, *J. Biol. Chem.* **264**:735–744.

Palmer, G., 1985, The electron paramagnetic resonance of metalloproteins, *Biochem. Soc. Trans.* **1985**:548–560.

Palmer, G., and Degli Esposti, M., 1994, Application of exciton coupling theory to the structure of mitochondrial cytochrome *b*, *Biochemistry* **33**:176–185.

Rieske, J. S., MacLennan, D. H., and Coleman, R., 1964, Isolation and properties of an iron-protein from the (reduced coenzyme Q)–cytochrome c reductase complex of the respiratory chain, *Biochem. Biophys. Res. Commun.* **15**:338–344.

Robertson, D. E., Daldal, F., and Dutton, P. L., 1990, Mutants of ubiquinol–cytochrome c_2 oxidoreductase resistant to Q_o site inhibitors: Consequences for ubiquinone and ubiquinol affinity and catalysis, *Biochemistry* **29**:11249–11260.

Schägger, H., Hagen, T., Roth, B., Brandt, U., Link, T. A., and von Jagow, G., 1990, Phospholipid specificity of bovine heart bc_1 complex, *Eur. J. Biochem.* **190**:123–130.

Schägger, H., Cramer, W. A., and von Jagow, G., 1994, Analysis of molecular masses and oligomeric states of protein complexes by blue native electrophoresis and isolation of membrane protein complexes by two-dimensional native electrophoresis, *Anal. Biochem.* **217**:220–230.

Schägger, H., Brandt, U., Gencic, S., and von Jagow, G., 1995, Ubiquinol–cytochrome-c reductase from human and bovine mitochondria, *Methods Enzymol.* **260**:82–96.

Schmitt, M. E., and Trumpower, B. L., 1990, Subunit 6 regulates half-of-the-sites reactivity of the dimeric cytochrome bc_1 complex in *Saccharomyces cerevisiae*, *J. Biol. Chem.* **265**:17005–17011.

Schröter, T., Hatzfeld, O. M., Gemeinhardt, S., Korn, M., Friedrich, T., Ludwig, B., and Link, T. A., 1998, Mutational analysis of residues forming hydrogen bonds in the "Rieske" [2Fe-2S] cluster of the cytochrome bc_1 complex in *Paracoccus denitrificans*, *Eur. J. Biochem.* **255**: 100–106.

Siedow, J. N., Power, S., De La Rosa, F. F., and Palmer, G., 1978, The preparation and characterization of highly purified, enzymically active complex III from baker's yeast, *J. Biol. Chem.* **253**:2392–2399.

Taylor, C. P. S., 1977, The EPR of low spin heme complexes, *Biochim. Biophys. Acta* **491**:137–149.

Tron, T., Crimi, M., Colson, A.-M., and Degli Esposti, M., 1991, Structure/function relationships in mitochondrial cytochrome *b* revealed by the kinetic and circular dichroic properties of two yeast inhibitor-resistant mutants, *Eur. J. Biochem.* **199**:753–760.

Trumpower, B. L., 1981, Function of the iron–sulfur protein of the cytochrome b-c_1 segment in electron-transfer and energy-conserving reactions of the mitochondrial respiratory chain, *Biochim. Biophys. Acta* **639**:129–155.

Trumpower, B. L., and Edwards, C. A., 1979a, Identification of oxidation factor as a reconstitutively active form of the iron–sulfur protein of the cytochrome b-c_1 segment of the respiratory chain, *FEBS Lett.* **100**:13–16.

Trumpower, B. L., and Edwards, C. A., 1979b, Purification of a reconstitutively active iron–sulfur protein (oxidation factor) from succinate–cytochrome *c* reductase complex of bovine heart mitochondria, *J. Biol. Chem.* **254**:8697–8706.

Van Doren, S. R., Gennis, R. B., Barquera, B., and Crofts, A. R., 1993, Site-directed mutations of conserved residues of the Rieske iron–sulfur subunit of the cytochrome bc_1 complex of *Rhodobacter sphaeroides* blocking or impairing quinol oxidation, *Biochemistry* **32**:8083–8091.

von Jagow, G., and Link, T. A., 1986, Use of specific inhibitors on the mitochondrial bc_1 complex, *Methods Enzymol.* **126**:253–271.

von Jagow, G., and Ohnishi, T., 1985, The chromone inhibitor stigmatellin–binding to the ubiquinol oxidation center at the C-side of the mitochondrial membrane, *FEBS Lett.* **185**:311–315.

Wakabayashi, S., Takeda, H., Matsubara, H., Kim, C. H., and King, T. E., 1982, Identity of the heme-not-containing protein in bovine heart cytochrome c_1 preparation with the protein mediating c_1–c complex formation—A protein with high glutamic acid content, *J. Biochem.* **91**:2077–2085.

Widger, W. R., Cramer, W. A., Herrmann, R. G., and Trebst, A., 1984, Sequence homology and structural similarity between cytochrome *b* of mitochondrial complex III and the chloroplast b_6-f complex: Position of the cytochrome *b* hemes in the membrane, *Proc. Natl. Acad. Sci. USA* **81**:674–678.

Xia, D., Yu, C.-A., Kim, H., Xia, J.-Z., Kachurin, A. M., Zhang, L., Yu, L., and Deisenhofer, J., 1997, Crystal structure of the cytochrome bc_1 complex from bovine heart mitochondria, *Science* **277**:60–66.

Yang, M., and Trumpower, B. L., 1994, Deletion of QCR6, the gene encoding subunit six of the mitochondrial cytochrome bc_1 complex, blocks maturation of cytochrome c_1, and causes temperature-sensitive petite growth in *Saccharomyces cerevisiae*, *J. Biol. Chem.* **269**:1270–1275.

Yu, C. A., Xia, D., Deisenhofer, J., and Yu, L., 1994, Crystallization of mitochondrial cytochrome b–c_1 complex from gel with or without reduced pressure, *J. Mol. Biol.* **243**:802–805.

Yu, C. A., Xia, J.-Z., Kachurin, A. M., Yu, L., Xia, D., Kim, H., and Deisenhofer, J., 1996, Crystallization and preliminary structure of beef heart mitochondrial cytochrome-bc_1 complex, *Biochim. Biophys. Acta* **1275**:47–53.

Yu, L., and Yu, C.-A., 1991, Essentiality of the molecular weight 15000 protein (subunit IV) in the cytochrome $b-c_1$ complex of *Rhodobacter sphaeroides*, *Biochemistry* **30:**4934–4939.

Yun, C.-H., Crofts, A. R., and Gennis, R. B., 1991, Assignment of the histidine axial ligands to the cytochrome b_H and cytochrome b_L components of the bc_1 complex from *Rhodobacter sphaeroides* by site-directed mutagenesis, *Biochemistry* **30:**6747–6754.

Yun, C.-H., Wang, Z., Crofts, A. R., and Gennis, R. B., 1992, Examination of the functional roles of five highly conserved residues in the cytochrome *b* subunit of the bc_1 complex of *Rhodobacter sphaeroides*, *J. Biol. Chem.* **267:**5901–5909.

Zhang, Z., Huang, L., Shulmeister, V. M., Chi, Y.-I., Kim, K. K., Hung, L.-W., Crofts, A. R., Berry, E. A., and Kim, S.-H., 1998, Electron transfer by domain movement in cytochrome bc_1, *Nature* **392:**677–684.

CHAPTER 14

Structure, Function, and Biogenesis of Respiratory Complex I

Ulrich Schulte and Hanns Weiss

1. INTRODUCTION

Respiratory complex I is the proton pumping NADH–ubiquinone oxidoreductase (E.C. 1.6.99.3). Characteristic features of the complex are the large number of protein subunits, one FMN and several iron–sulfur (FeS) clusters as redox groups, and the sensitivity of the electron transfer reaction to a number of naturally occurring compounds like rotenone and piericidin (Weiss *et al.*, 1991; Singer and Ramsay, 1992; Walker, 1992). The complex is found in many bacteria, often referred to as NADH dehydrogenase I, and in mitochondria of most eukaryotes (Friedrich and Weiss, 1997). Eukaryotes known to lack complex I are the fermentative yeast *Saccharomyces*, *Schizosaccharomyces*, and *Kluyveromyces* (Nosek and Fukuhara, 1994; Büschges *et al.*, 1994). In many bacteria and in mitochondria of plants and fungi a second NADH–quinone oxidoreductase is present, which is a non-proton-pumping, single-subunit FAD enzyme referred to as NADH dehydrogenase 2 or alternative NADH dehydrogenase (Young *et al.*, 1981; Møller and Palmer, 1982; Matsushita *et al.*, 1987; de Vries and Grivell, 1988).

Ulrich Schulte and Hanns Weiss • Institut für Biochemie, Heinrich-Heine-Universität, D 40225 Düsseldorf, Germany.

Frontiers of Cellular Bioenergetics, edited by Papa *et al.* Kluwer Academic/Plenum Publishers, New York, 1999.

Biochemical studies of bacterial complex I are most advanced with *Escherichia coli* (Leif *et al.*, 1995) and *Paracoccus denitrificans* (Yagi *et al.*, 1992). Genetic and biochemical studies of complex I from *Thermus thermophilus* and *Rhodobacter capsulatus* have also been reported (Yagi *et al.*, 1988; Dupuis, 1992). The mitochondrial complex from bovine heart is the best characterized and is the reference enzyme in this chapter (Walker, 1992). *Neurospora crassa* has been established as a suitable eukaryotic microorganism for studying complex I (Weiss *et al.*, 1991). Mitochondrial complex I has also been studied in various plants (Laterme and Boutry, 1993; Rasmussen *et al.*, 1994; Herz *et al.*, 1994).

Several enzymes related in part to complex I have been identified. In chloroplasts and cyanobacteria, an "alien" complex I has been proposed to be involved in cyclic electron transport in photosynthesis (Friedrich *et al.*, 1995). Prokaryotic hydrogenases as well as formiate hydrogenlyase are other relatives of complex I (Friedrich and Weiss, 1997).

In this chapter we give an overview of the current knowledge about complex I and provide some new approaches initiated recently in our laboratory to gain new insights into this complicated machinery. Medical aspects of complex I, which have been reviewed recently (Cooper *et al.*, 1992; Brown and Wallace, 1994), are not dealt with in this chapter.

2. STRUCTURE OF COMPLEX I

2.1. Genes and Subunits

The prokaryotic complex I is composed of 14 different subunits (Leif *et al.*, 1995). In *E. coli*, the 14 genes coding for these subunits are arranged in an operon (Weidner *et al.*, 1993). In *P. denitrificans* (Xu *et al.*, 1993) and *R. capsulatus* (Dupuis *et al.*, 1996a), the 14 complex I genes are clustered with several unidentified reading frames (URFs) interspersed. The URFs are not conserved between the species and their inactivation does not affect complex I assembly or function (Dupuis *et al.*, 1996b). The order of the complex I genes is highly conserved among these bacteria.

Homologues of the 14 subunits of the bacterial complex are found in all organisms analyzed so far and appear to form the minimal complex (Friedrich *et al.*, 1995). According to their primary structure, seven subunits are located peripherally with regard to the phospholipid bilayer and the other seven subunits are supposed to be membrane spanning. The binding sites for NADH and all known prosthetic groups have been identified on the subunits considered to be peripheral.

The subunits of the mitochondrial complex are of dual genetic origin. The seven membrane-spanning minimal subunits are encoded on the mitochondrial genome (Chomyn *et al.*, 1985; Ise *et al.*, 1985). The seven peripheral subunits are nuclear encoded and are imported from the cytosol. In plants, however, homologues to two peripheral subunits (49- and 30-kDa subunits) (see Section 2.3) are encoded in the mitochondrial genome (Kubo *et al.*, 1993; Gabler *et al.*, 1994).

In addition to the 14 minimal subunits, up to 28 accessory subunits are found in mitochondrial complexes. Some of the accessory subunits are conserved between fungi and animals, but many appear to be species specific (Walker, 1992; Azevedo and Videira, 1994). The sequences of 39 of 43 different subunits of the bovine enzyme have been published (Walker *et al.*, 1992; Buchanan and Walker, 1996) and 26 subunits have been sequenced of complex I from *N. crassa*, with some 10 subunits still awaiting their sequencing (Weiss *et al.*, 1991; Azevedo and Videira, 1994; Azevedo *et al.*, 1994; Duarte *et al.*, 1996; Videira, Schulte, and Weiss, unpublished data). For both the mammalian and the fungal complex, the subunit composition is virtually independent of the isolation procedure used, and there are as yet no indications of subunits being isolation artifacts rather than bona fide parts of the enzyme. Of the many complex I genes in *N. crassa* that have been inactivated, all had profound impacts on the enzyme (see Table II).

Unfortunately, a general nomenclature for complex I subunits has not yet been adopted. For now, we follow the nomenclature suggested by Walker (1992). The mitochondrially coded subunits are designated ND1 to ND6 and ND4L. For historical reasons the subunits of the flavoprotein fragment (51, 24, and 10 kDa), six subunits of the iron–sulfur protein fragment (75, 49, 30, 18, 15, and 13 kDa), and two subunits with apparent molecular weights of 42 and 39 kDa, respectively, are named according to their apparent molecular mass determined by gel electrophoresis. The other nuclear-coded subunits that were identified during large-scale sequencing efforts are named by the four N-terminal amino acids of the bovine subunit in one-letter code (e.g., TYKY). Subunits that failed to yield a N-terminal sequence by Edman degradation are named by their apparent molecular mass on gels preceded by the letter B (e.g., B13). Subunits of *N. crassa* with no apparent bovine homologue are designated by their calculated molecular mass.

The function of accessory subunits is mostly unknown. It has been suggested that many may serve in making the structure more rigid and compact. This might promote a more efficient energy coupling due to the highly fixed position of the redox groups. In addition, the cover provided by these subunits would shield the redox groups against reactive substances like oxygen, thereby preventing the formation of potentially hazardous compounds. Three subunits, however, may have a more specific function. It has been suggested that PGIV, a 20-kDa subunit conserved in bovine and *N. crassa*, displays a binding site for an iron–sulfur cluster due to its many conserved cysteines (Dupuis *et al.*, 1991a). Two other subunits have been implicated in a biosynthetic pathway still to be characterized. Subunit SDAP, which is an acyl-carrier protein, has been identified in complex I of bovine (Runswick *et al.*, 1991), *N. crassa* (Sackmann *et al.*, 1991), and *Arabidopsis thaliana* (Shintani and Ohlrogge, 1994). The 39-kDa subunit shows a NAD(P)$^+$ binding site and is related to different dehydrogenases (Fearnley and Walker, 1992). The possible function of these two subunits will be discussed in Section 4.2.

2.2. Prosthetic Groups

Prosthetic groups identified in complex I are one FMN and several FeS clusters. Seven different clusters have been detected by electron paramagnetic resonance (EPR) (Sled' *et al.*, 1993). Binuclear clusters [2Fe-2S] are designated N1a, N1b, and N1c (Ohnishi, 1979) (a slightly different nomenclature has been proposed by Beinert and Albracht, 1982). Cluster N1a has a very low midpoint potential around -370 mV and is not reduced by NADH. In *N. crassa*, it has not yet been identified (Wang *et al.*, 1991). As with most of the other clusters, N1b has a midpoint potential around -250 mV. The ratio of cluster N1b to FMN and the other clusters has been reported by Albracht and co-workers to be 0.5 in bovine (van Belzen *et al.*, 1990, 1992). In general, however, an equimolar ratio is observed for all clusters and the FMN. Cluster N1c so far has been observed only in *E. coli* (Leif *et al.*, 1995). Four tetranuclear FeS clusters [4Fe-4S], designated N2, N3, N4, and N5, have been identified in complex I. Cluster N2 has the highest midpoint potential of all clusters and shows a pronounced pH dependence of its redox potential (Ingledew and Ohnishi, 1980). These characteristic features have prompted suggestions of a prominent role of this cluster in the various mechanistic models of complex I (see Section 3.2). There has been some debate about cluster N5 being a bona fide component of complex I. It has been detected only in bovine and only in substoichiometric amounts (Ohnishi, 1979; Beinert and Albracht, 1982). In addition to the EPR-detectable clusters, complex I can be expected to contain FeS clusters that might be diamagnetic or coupled and therefore EPR silent, or that have so far escaped detection (Ragan, 1987). For example, the presence of a spin $\frac{3}{2}$ cluster has been suspected. A total of eight to nine binding motifs have been found in the primary structures of the subunits (Fearnley and Walker, 1992). The iron and acid-labile sulfur content of the various complex I preparations give 20–25 moles iron and sulfur per mole of complex I, which would accommodate, for example, two binuclear and five tetranuclear centers.

The analysis of primary structures of complex I subunits has revealed binding sites for the substrates as well as the prosthetic groups. An assignment of the FeS clusters detected by EPR to the known binding sites has been obtained mainly by the resolution of the enzyme into subcomplexes, the analysis of single overexpressed subunits, and the characterization of mutant complexes lacking single subunits. However, the assignment is not yet unambiguous (Table I).

The NADH oxidation site is located on the 51-kDa subunit, which shows a conserved sequence motif known to form the ADP binding pocket (Pilkington *et al.*, 1991). Photoaffinity labeling with NAD analogues also identified the 51-kDa subunit as NADH binding site (Chen and Guillory, 1981; Deng *et al.*, 1990; Yagi and Dinh, 1990). A second sequence motif of a potential nucleotide binding site is found in the 39-kDa subunit (Fearnley and Walker, 1992). This subunit is present only in mitochondrial enzymes and is most likely not involved in the general electron transfer (see Section 4.2).

A binding site for ubiquinone has been assigned to subunit ND1. This subunit was labeled by dihydrorotenone (Earley *et al.*, 1987) and a photoactivatable analogue of rotenone (Earley and Ragan, 1984), which is known to act as a noncompetitive inhibitor with respect to ubiquinone (see Section 3.3). The assignment to subunit ND1 is further supported by homology of part of this subunit to bacterial glucose–ubiquinone oxidoreductase. This single polypeptide enzyme oxidizes glucose to gluconolactone in the periplasm of many bacteria and delivers the electrons via pyrroloquinoline–quinone to ubiquinone (Matsushita *et al.*, 1989). Like complex I, it is inhibited by a number of naturally occurring compounds in a partially competitive manner with regard to ubiquinone (Friedrich *et al.*, 1990, 1994). The significance of the sequence similarity and the relatedness of the inhibitor binding site, however, have been questioned (Fearnley and Walker, 1992; Sakamoto *et al.*, 1995). Attempts to label the ubiquinone-binding subunit by ubiquinone analogues have led to

Table I
FeS Protein Subunits of Complex I

Mitochondrial complex I[a]	Bacterial complex I[b]	Binding motifs	FeS cluster[c]
75 kDa	nuoG	1 [2Fe-2S][d]	N1b (N1a)
		2 [4Fe-4S]	N4, N5 (?)
51 kDa	nuoF	1 [4Fe-4S]	N3
24 kDa	nuoE	1 [2Fe-2S]	N1a (N1b)
TYKY	nuoI	2 [4Fe-4S]	? (N2)
PSST	nuoB	1 [4Fe-4S]	N2 (?)

[a]Nomenclature according to Walker (1992).
[b]Nomenclature according to Weidner *et al.* (1993).
[c]Nomenclature according to Ohnishi (1979).
[d]Subunit of *E. coli* binds an additional [2Fe-2S] cluster designated N1c.

inconsistent results. The most recent experimental results have yielded a specific labeling of a 9.5-kDa subunit of the *N. crassa* complex I (Heinrich and Werner, 1992). This subunit, however, is not found in prokaryotic complexes.

In addition to the NADH binding site, the 51-kDa subunit provides a binding site for a tetranuclear FeS cluster (Table I) and most likely also for the FMN. This is in agreement with the presence of this subunit in the flavoprotein fragment and the absence of flavin and cluster N3 in a mutant complex I lacking this subunit (Fecke *et al.*, 1994). A close proximity of FMN and cluster N3 has also been suggested by the magnetic interaction of both (Salerno *et al.*, 1977). Binding motifs for a binuclear cluster are found in the 24-kDa subunit (Pilkington and Walker, 1989) and the 75-kDa subunit (Walker, 1992). Since the flavoprotein fragment is thought to comprise cluster N1b, this cluster has been attributed to the 24-kDa subunit leaving N1a for the 75-kDa subunit. EPR spectra of the *P. denitrificans* subunit homologous to the 75-kDa subunit overexpressed in *E. coli*, however, suggest the location of N1b as well as cluster N4 on this subunit (Yano *et al.*, 1994). In that case, N1a would be located on the 24-kDa subunit, an assignment that is further supported by EPR analysis of a heterodimer of *P. denitrificans* subunits homologous to the 51- and 24-kDa subunit expressed in *E. coli* (Yano *et al.*, 1996). The *E. coli* homologue of the 75-kDa subunit (nuoG) shows a second motif for a binuclear center, which most likely binds the *E. coli*-specific cluster N1c (Leif *et al.*, 1995). As already mentioned, the 51-kDa subunit binds cluster N3. Cluster N4, which is present in the iron protein fragment of bovine as well as the NADH dehydrogenase fragment of *E. coli* (see Section 2.3), is most likely located on the 75-kDa subunit, the only subunit displaying the 4Fe-4S sequence motive, which is present in both fragments.

A second potential binding site for a tetranuclear cluster has been identified in the 75-kDa subunit (Fearnley and Walker, 1992; Weidner *et al.*, 1993). It could harbor cluster N5 or a cluster not yet identified by EPR. The subunits TYKY and PSST are candidates to harbor cluster N2. TYKY is a typical 2[4Fe-4S] protein (Dupuis *et al.*, 1991b) and has been suggested to bind two cluster N2 in a heterodimeric complex (van Belzen *et al.*, 1992). The sequence motif found in PSST differs from the typical consensus pattern in that it lacks the fourth cysteine, which might be replaced by a glutamate (Arizmendi *et al.*, 1992; Ohnishi, 1993). This unusual arrangement would correspond to the high midpoint potential of cluster N2 and its pH dependence.

All the above-mentioned subunits involved in the binding of prosthetic groups are minimal subunits found in complex I of all species. As mentioned, one of the accessory subunits of mitochondrial complex I, subunit PGIV, is discussed as a potential FeS protein. As many as eight tandemly arranged cysteines are strictly conserved in the subunits of bovine and *N. crassa* (Dupuis *et al.*, 1991a). The subunit is part of the bovine subcomplex Ia (Finel *et al.*,

1992). In *N. crassa*, it is found in the membrane part (see Section 2.3). Since all EPR detectable FeS clusters have been assigned to minimal subunits and the membrane arm of the *N. crassa* complex I does not yield any appreciable EPR-visible FeS signal, the significance of the conserved cysteines remains obscure.

2.3. Three-Dimensional Structure and Subunit Arrangement

Isolation procedures yielding a monodisperse enzyme in detergent solution have been described for bovine, *N. crassa*, and *E. coli* (Finel *et al.*, 1992; Ise *et al.*, 1985; Leif *et al.*, 1995). A molecular mass of more than 900 kDa has been calculated from sequence data for bovine complex I (Buchanan and Walker, 1996). The complex from *N. crassa* might be slightly smaller (Leonard *et al.*, 1987). The molecular mass of *E. coli* complex I as determined by sedimentation velocity sucrose gradient centrifugation is approximately 550 kDa, which is in good agreement with 524 kDa obtained by the addition of the masses of the single subunits (Leif *et al.*, 1995).

The overall structure of mitochondrial complex I has been determined by electron microscopy (Fig. 1). Analysis of two-dimensional crystals of whole complex I from *N. crassa* and a membrane-intrinsic fragment revealed an

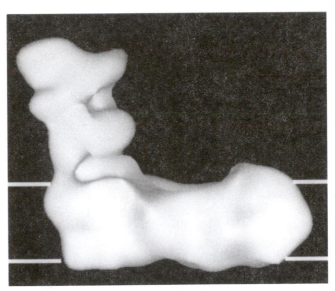

FIGURE 1. Three-dimensional model of complex I from *N. crassa* as reconstructed from electron microscopic analysis. The membrane is represented by two lines. The resolution is about 3.5 nm. Taken from Guenebaut *et al.* (1996).

L-shaped structure, which was confirmed by single-particle analysis (Hofhaus *et al.*, 1991; Guenebaut *et al.*, 1996). One arm of the L forms the membrane-intrinsic domain and the other protrudes into the mitochondrial matrix. Support for this L-shaped structure is provided by the resolution of isolated complex I into subcomplexes. Subcomplexes can also be obtained *in vivo* from mutants of *N. crassa* and *E. coli* lacking single subunits (Schulte and Weiss, 1995; A. Berger and T. Friedrich, unpublished data).

The peripheral arm of complex I has been analyzed in considerable detail. A flavoprotein fragment and an iron–protein fragment, both water soluble, can be extracted from bovine complex I by perchlorate (Galante and Hatefi, 1979; Ragan *et al.*, 1982). The two subcomplexes together appear to contain all the known prosthetic groups of complex I, but with more or less altered properties. The flavoprotein fragment formed by the 51-, 24-, and 10-kDa subunits harbors the FMN and two FeS clusters supposed to be N1b and N3 (Ohnishi *et al.*, 1985).

A flavoprotein subcomplex of *P. denitrificans* comprising the homologues to the 51- and 24-kDa subunits has been obtained by coexpression in *E. coli*. It also yielded EPR signals from two FeS clusters that were assigned, however, to cluster N1a and N3 (Yano *et al.*, 1996). Subcomplex Iα, obtained by treatment of the bovine complex I with lauryldimethylaminoxide, includes most if not all peripheral subunits and possibly also a small membrane-intrinsic segment (Finel *et al.*, 1992; Finel, 1993). All known prosthetic groups have been identified in subcomplex Iα. Subcomplex Iλ, comprising some 15 different subunits still including all prosthetic groups, has been obtained by a slightly modified procedure (Finel *et al.*, 1994).

A peripheral arm can be prepared from *N. crassa* grown in the presence of chloramphenicol, which inhibits mitochondrial protein synthesis (Friedrich *et al.*, 1989), or from *N. crassa* mutants lacking single subunits of the membrane arm (Schulte and Weiss, 1995) (see Section 4.1). This subcomplex differs from the bovine peripheral subcomplexes in that it lacks cluster N2. While the homologue to TYKY has been shown to be present in the peripheral arm of *N. crassa* (Duarte *et al.*, 1997), the localization of PSST being the most likely subunit for binding cluster N2 has not yet been determined.

The peripheral arm of complex I from *E. coli* falls into two fragments upon pH shift: the NADH dehydrogenase fragment and the connecting fragment (Leif *et al.*, 1995). The water-soluble NADH dehydrogenase fragment comprises the homologues to the 75-, 51-, and 24-kDa subunits and includes the FMN and the EPR-detectable FeS clusters N1b, N1c, N3, and N4. The presence of cluster N1a, which was not analyzed by EPR, and of a third tetranuclear cluster is suggested from the binding motifs of the three subunits and the iron content of the fragment (Braun *et al.*, 1998). The connecting fragment is formed by the homologues to TYKY, PSST, and the 49- and 30-

kDa subunits. It harbors one tetranuclear FeS cluster detected by EPR, which most likely is cluster N2.

The membrane intrinsic part of complex I is less well characterized. The hydrophobic fragment obtained by treatment of bovine complex I with perchlorate is merely the insoluble residue resulting from this treatment. Fragment Iβ, resulting from treatment of bovine complex I with lauryldimethylaminoxide, is supposed to include some but not all membrane-spanning subunits (Finel *et al.*, 1992). A membrane arm of the *N. crassa* complex I can be obtained either by dissociation of the peripheral subunits by chaotropic salts (Hofhaus *et al.*, 1991; Zensen *et al.*, 1992) or from a mutant of *N. crassa* lacking the homologue to the 49-kDa subunit of the peripheral arm (Schulter ans Weiss, 1995). Cluster N2, which was previously assigned to the membrane arm of *N. crassa* (Schmidt *et al.*, 1992), could not be detected in the isolated membrane arm. When complex I of *E. coli* was cleaved by pH-shift, a fragment including all membrane-spanning subunits was isolated (Leif *et al.*, 1995). So far, no prosthetic group has been unequivocally located in the membrane part.

3. FUNCTION OF COMPLEX I

3.1. Electron Transfer

The electron transfer activity of complex I is best measured in its native membrane surrounding as rotenone-sensitive electron transfer from NADH to ubiquinone or quinol analogues. The apparent K_M value for NADH is in the range of 2 to 14 μM (Møller and Palmer, 1982; Yagi, 1986; Friedrich *et al.*, 1989, 1994). NADPH is only poorly used as a substrate due to a much higher K_M value of about 500 μM and a much lower turnover (Rydström *et al.*, 1978). Deamino-NADH with a K_M value of about 100 μM has been used to discriminate between complex I activity and the activity of the non–proton-pumping alternative NADH dehydrogenase present in fungi, plants, and several bacteria, which shows a reduced activity with deamino-NADH (Matsushita *et al.*, 1987; Friedrich *et al.*, 1994). Due to its low solubility in water, the native electron acceptor ubiquinone-10 cannot be added to the assay mixture, and a number of quinone analogues with shorter side chains, namely ubiquinone-1, ubiquinone-2, and decylubiquinone (2,3-dimethoxy-5-methyl-6-decyl-1,4-benzoquinone) have been used (Hatefi *et al.*, 1960). K_M values for ubiquinone-2 are about 10–20 μM (Friedrich *et al.*, 1989; Fato *et al.*, 1996). In general, the longer the isoprenyl or alkyl side chain, the lower is the apparent K_M value (Degli Esposti *et al.*, 1996). This is not only due to a higher affinity of complex I to the long-chain quinones but it also reflects the better partition into the lipid phase. Inhibitor sensitivity of the reaction with quinone analogues decreases with reduced hydrophobicity, because short-chain quinones also react at artificial binding sites (Degli Esposti et al., 1996).

In detergent solution, no inhibitor-sensitive NADH–Q redox activity of complex I can be measured. Reconstitution of the isolated enzyme into phospholipid membranes restores some of the activity with hydrophobic quinones, but generally with reduced sensitivity to inhibitors (Ise et al., 1985; Finel et al., 1992; Leif et al., 1995). A prominent exception has been the highly active preparation of bovine complex I established by Hatefi et al. (1962). This preparation, however, is a polydisperse suspension of complex I that contains considerable amounts of lipids. More recently, complex I from bovine and N. crassa has been isolated in monodisperse form in dodecylmaltoside, yielding a high rotenone sensitive NADH–decylubiquinone oxidoreductase activity when reconstituted in phospholipid membranes (Buchanan and Walker, 1996, and Schulte and Weiss, unpublished data).

Subcomplexes of complex I show electron transfer activity from NADH to artificial acceptors, like ferricyanide, as long as the 51-kDa subunit is included. The peripheral subcomplexes also catalyze electron transfer to short-chain quinones. These activities are invariably insensitive to rotenone and piericidin A, and the K_M values for quinones are much higher compared to the intact enzyme (Friedrich et al., 1989; Finel et al., 1992). It is therefore generally accepted that artificial Q binding sites are involved in these reactions. It has been suggested, however, that these binding sites stem from internal binding sites involved in some kind of Q cycle (Weiss and Friedrich, 1991).

Data on the exact electron route through complex I are scarce. A tentative sequence of electron transfer steps can be deduced from the redox potentials of the involved redox groups and their putative spatial arrangement. Flavin as a two-electron acceptor most likely acts as direct oxidant for NADH. From the flavin, the electrons can be transferred one at a time to the isopotential FeS clusters N1b, N3, and N4. EPR studies suggested that N1 is reduced first (Krishnamoorthy and Hinkle, 1988). It has been reported that complex I has to be activated by several turnover cycles before steady-state activity is reached (Burbaev et al., 1989; Kotlyar and Vinogradov, 1990). Assuming that the activation is a reduction of an electron pool formed by clusters N1b, N3, and N4, a branched electron pathway is favored. This view is further supported by the fact that loss of cluster N4 does not significantly affect the electron transfer rate (Krishnamoorthy and Hinkle, 1988). The very low redox potential of cluster N1a is puzzling, making it difficult, for thermodynamic reasons, to integrate it in an electron pathway leading from flavin to quinone. Cluster N2, with a rather high midpoint potential, is supposed to reduce quinone in one-electron steps. Ubisemiquinones formed transiently have been detected by EPR (Suzuki and King, 1983; Kotlyar et al., 1990; de Jong and Albracht, 1994).

No time resolution of the fast internal redox reactions has been possible so far. EPR spectroscopy in combination with the freeze–quench technique could not yet resolve the fast electron transfer among the FeS centers (van Belzen and Albracht, 1989). Resolution of the redox reactions by time-resolved UV/VIS absorption spectroscopy has been hampered by the difficulty to differentiate between the redox components involved.

FeS clusters as well as flavin show broad and overlapping absorption maxima (Fig. 2A). Upon reduction with equimolar amounts of NADH, slight changes in the spectrum are detected, which are revealed in detail in the difference of the spectrum of the complex reduced by NADH minus the

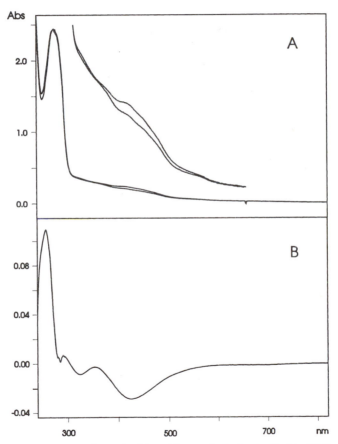

FIGURE 2. UV/VIS spectra of complex I from *N. crassa* in dodecylmaltoside. (A) Spectrum of complex I (2 μM) before and 5 sec after addition of NADH (3 μM). (B) Difference of the spectrum taken 5 sec after addition of NADH minus the spectrum taken before addition of NADH. The diode array spectrometer Zeiss, Specord S10 was used

spectrum of the air-oxidized complex (Fig. 2B). The added NADH gets immediately oxidized and gives rise to the maximum at 260 nm. Since both flavin and FeS clusters have smaller extinction coefficients in the reduced state, the difference spectrum shows minima at 425 and 320 nm. In addition, a small maximum around 295 nm is visible. The chromophore yielding this spectral change has not yet been identified. Neither flavin nor FeS clusters is known to display such a maximum in its redox difference spectra. No ubiquinone could be extracted from the enzyme preparation, and EPR spectra did not show a ubisemiquinone. The reduced enzyme is rapidly oxidized by dissolved oxygen. Interestingly, the 295-nm absorbance has a longer lifetime than the minima at 425 and 320 nm.

From an *N. crassa* mutant lacking the gene for the homologue of the 39-kDa subunit (Table II, nuo40), we have obtained a complex I containing all subunits and detectable prosthetic groups, except the 39-kDa homologue. Nevertheless, the complex is unable to transfer electrons from NADH to decylubiquinone. Reduction of the mutant complex with NADH results in a redox difference spectrum showing the two minima at 425 nm and 320 nm, but no maximum at 295 nm. Except for a moderate broadening of the N2 signal, EPR spectroscopy showed no difference between mutant and wild-type complex I. At this point, we cannot decide whether the 295-nm chromophore is not present in the mutant complex, resulting in a disrupted electron pathway, or whether the electron pathway is disrupted, preventing a reduction of the chromophore. As discussed later (Section 4.3), the 39-kDa subunit might be involved in a not yet characterized biosynthetic pathway modifying an unidentified component of complex I.

3.2. Proton Translocation

Proton translocation by complex I has been measured in coupled submitochondrial membranes. Coupling ratios between 3 $H^+/2e^-$ (De Jonge and Westerhoff, 1982) and 5 $H^+/2e^-$ (Lemasters *et al.*, 1984) have been reported. As a consensus the ratio, 4 $H^+/2e^-$ is generally accepted (Weiss and Friedrich, 1991). Various mechanistic models have been proposed to yield this coupling in complex I.

Proton translocation has been suggested to be linked to redox reactions at two sites, namely, FMN and the cluster N2–ubiquinone couple. Assuming that flavin is fully protonated and deprotonated during a redox cycle, a directed uptake of protons from the negative site and release of protons on the positive site would yield two transferred protons per two electrons. Since the flavin has no immediate access to protons of the positive site, this mechanism would require a gated proton channel for which no experimental evidence so far is available. Full protonation and deprotonation, on the other hand, would require an appropriate shift in pK_a values of the involved redox states, which could

Table II
Incompletely Assembled Parts of Complex I in *N. crassa* Mutants

Designation of mutant[a]	Accumulated parts[b]	Number of subunits encoded in		Redox groups
		Nucleus	Mitochondria	
nuo75[c]	Membrane arm	~12	7	None known
nuo51[d]	Complex I Δ51 kDa	~27	7	N1, N2, N4
nuo49	Membrane arm	~12	7	None known
nuo40	Complex I Δ39 kDa	~27	7	FMN, N1–N4
nuo29.9[e]	Membrane arm	n.d.	n.d.	n.d.
nuo21.3a[f]	Peripheral arm	~12	0	FMN, N1, N3, N4
	Large intermediate	6 + 2	5	None known
	Small intermediate Δ21.3a kDa	3	2	None known
nuo21.3b[g]	Complex I Δ 21.3b kDa	n.d.	n.d.	n.d.
	Peripheral arm	n.d.	n.d.	n.d.
nuo20.9[h]	Peripheral arm	~12	0	FMN, N1, N3, N4
	Small intermediate	4	2	None known
nuo12.3[e]	Peripheral arm	n.d.	n.d.	n.d.
	Membrane arm Δ12.3 kDa	n.d.	n.d.	n.d.
acp-1[i]	Membrane arm	~12	7	None known

[a]Nuo refers to NADH:ubiquinone oxidoreductase and the number refers to the calculated molecular mass of the inactivated subunit.
[b]Large and small intermediate refer to the respective membrane arm intermediates; Δ indicates a subunit missing in the assembled part.
[c]Harkness *et al.* (1995) and unpublished results.
[d]Fecke *et al.* (1994).
[e]Duarte *et al.* (1995).
[f]Nehls *et al.* (1992).
[g]Alves and Videira (1994).
[h]Identical phenotype in nuo 10.4.
[i]Schneider *et al.* (1995).

not be confirmed (Sled' *et al.*, 1994). A proton-motive ubiquinone cycle was already proposed by Suzuki and King in 1983, and later on was advanced and modified in several ways (Ragan, 1990; Kotlyar *et al.*, 1990; Weiss and Friedrich, 1991; Vinogradov, 1993). The pH dependence of the midpoint potential of cluster N2 has been of particular interest, suggesting a role in proton translocation (Ohnishi and Salerno, 1982). This cluster has also been shown to be involved in energy-induced structural changes in the enzyme (de Jong *et al.*, 1994). A spin interaction of ubisemiquinone with cluster N2 has recently been reported (Vinogradov *et al.*, 1995). In this study, two distinct species of ubisemiquinones were detected, one of which was identified as causing the split of the $g = 2.05$ signal of cluster N2. A mutual distance between N2 and the semiquinone of about 1 nm was calculated from the data. However, the split N2 signal was attributed to a spin interaction of two N2 clusters in another study (de Jong *et al.*, 1994; van Belzen *et al.*, 1996). In summary, there is no mechanistic model for redox-linked proton translocation available, which could experimentally be tested.

3.3. Inhibitors

A large number of naturally occurring as well as chemically synthesized compounds with different chemical structures have been described as inhibitors of complex I. Few inhibit oxidation of NADH, whereas the large majority inhibits reduction of ubiquinone (Singer and Ramsey, 1992; Friedrich *et al.*, 1994; Degli Esposti *et al.*, 1994).

Rhein (4,5-dihydroxyanthraquinone-2-carboxylic acid) has long been known to competitively inhibit NADH (Kean *et al.*, 1971). Diphenyleneiodonium irreversibly blocks reoxidation of the flavin (Majander *et al.*, 1994).

Rotenone and piericidin A are the most widely used inhibitors at the ubiquinone reduction site. The stoichiometry of inhibition of both is most likely 1:1, although titration is hampered by the hydrophobicity of the inhibitors (Singer and Ramsey, 1992). A classification of inhibitors of the ubiquinone reduction site according to the mode of action has been proposed (Friedrich *et al.*, 1994). Piericidin-A-type inhibitors that act partially competitively include annonin VI, phenalamid A_2, aurachins, and thiangazol. Noncompetitive inhibitors like phenoxan, aureothin, and benzimidazole belong to the rotenone type. The two classes of inhibitors apparently do not bind to the same site. This can be interpreted either as two inhibitor binding sites close to the same ubiquinone reduction site or as two different ubiquinone binding sites blocked by one class of inhibitors each (Friedrich *et al.*, 1994; Singer and Ramsay, 1994).

Another group of inhibitors, namely, derivatives of *N*-methyl-4-phenylpyridinium (MPP$^+$), has received much attention recently (Singer and Ramsay, 1992, and references cited therein). They are neurotoxic compounds formed in the brain through the oxidation of tetrahydropyridines by monoamine oxidases A and B. Symptoms caused by MPP$^+$ resemble those of Parkinson's disease patients. As the toxic effect of these compounds has been correlated with the inhibition of complex I, complex I deficiency is discussed as a possible cause of Parkinson's disease (Mizuno *et al.*, 1989; Schapira *et al.*, 1990; Shoffner *et al.*, 1991; Benecke *et al.*, 1992).

4. BIOGENESIS OF COMPLEX I

4.1. Assembly Intermediates

Assembly of complex I has been followed by pulse labeling of growing *N. crassa* cultures with [^{35}S]methionine (Tuschen *et al.*, 1990). Most of the added labeled amino acid is taken up by the hyphae in a few minutes, and endogenously synthesized methionine quickly dilutes the intracellular pool of labeled methionine, resulting in a pulse–chase course. After labeled mitochondrial proteins solubilized in detergent are fractionated by sucrose-gradient centrifugation, complex I and assembly intermediates of it can be identified by immunoprecipitation, sodium dodecyl sulfate–polyarylamide gel electrophoresis (SDS-PAGE), and autoradiography (Schulte and Weiss, 1995). Whole, detergent-solubilized complex I is quite stable and sediments like a 800-kDa protein in a well-defined band through the gradient. After pulse labeling of wild type, it takes about 1 hr for all the labeled subunits to appear in the mature complex. Smaller subcomplexes labeled only transiently can be precipitated after shorter periods of incorporation of radioactive methionine. Distinct assembly intermediates are revealed by immunoprecipitation using subunit specific antisera.

Assembly of the peripheral arm and of the membrane arm progresses independently (Tuschen et al., 1990). Both arms finally join in the formation of mature complex I. If formation of the membrane arm is retarded by chloramphenicol, which inhibits synthesis of the mitochondrially encoded subunits, excess peripheral arm is accumulated in the mitochondria, allowing its isolation (Friedrich et al., 1989). A so-called large membrane arm assembly intermediate comprising five of the seven mitochondrially coded subunits, which is probably caused by an uneven effect of chloramphenicol on the synthesis of the different mitochondrially coded subunits, is also accumulated (Tuschen et al., 1990).

A complete block in the assembly pathway is caused by the lack of single subunits in mutants obtained by gene disruption (Nehls et al., 1992). Loss of subunits of the peripheral arm results in either the assembly of a complex lacking basically that single subunit (nuo21.3b, nuo51, nuo40) or formation of the free membrane arm without detectable peripheral arm or part thereof (nuo29.9, nuo49, nuo75) (Table II). A most remarkable effect is caused by the loss of the acyl carrier protein subunit, which is part of the peripheral arm. Not only formation of the peripheral arm is prevented, but assembly of the membrane arm is also impaired, with only small amounts of it accumulating (Schneider et al., 1995). Mutants lacking a subunit of the membrane arm accumulate the peripheral arm and two different assembly intermediates of the membrane arm (nuo10.4, nuo12.3, nuo20.9, nuo21.3) (Table II). The so-called large and small membrane arm intermediates are complementary to each other with regard to the subunit composition.

The subcomplexes listed in Table II are the final products of the disrupted assembly pathways and the term "intermediates" is misleading in this respect. Four subcomplexes are classified as intermediates, characterized by their transient labeling in pulse-labeling experiments. The membrane arm and its large assembly intermediate have been identified in wild type. In the presence of chloramphenicol, the peripheral arm is also accumulated in wild type in detectable amounts. The small membrane arm intermediate is transiently detectable in mitochondria of mutants nuo40 and nuo49.

The large membrane arm intermediate is tightly associated with two proteins not found in mature complex I (Schulte et al., 1994). The two proteins are present in the intermediate immunoprecipitated from wild-type mitochondria with various subunit specific antisera as well as in the intermediate isolated by chromatographic steps from mitochondria of mutant nuo21, in which the intermediate accumulated (Table II) (Schulte and Weiss, 1995). The proteins of molecular masses of 84 and 30 kDa, respectively, are found in about equimolar amounts to the other subunits.

Antisera raised against the two proteins immunoprecipitate the entire large membrane arm intermediate. Neither any other assembly intermediate nor mature complex I nor any other mitochondrial protein is coprecipitated. Therefore, the proteins were termed "complex I intermediate-associated proteins" (CIA84 and CIA30). In wild-type mitochondria, the CIA proteins are found in roughly the same amounts in the free state and bound to the assembly intermediate. The accumulation of the intermediate in mutant nuo21 results in a shift of most of the CIA proteins to the bound form. Exclusively free CIA proteins are found in mutant nuo20.9, which cannot assemble a stable large membrane arm intermediate.

Pulse-labeling experiments indicate that the CIA proteins cycle between the free and the bound state and are involved in the assembly of many molecules of complex I. While the labeled complex I subunits are only transiently present in the intermediate and are replaced by more recently synthesized nonradioactive proteins, the CIA proteins bound to the intermediate are permanently labeled.

We have cloned the genes of the CIA proteins from *N. crassa*. The deduced amino acid sequence of CIA84 reveals a globular protein preceded by a typical import sequence of 24 amino acids. The import sequence of CIA30 comprises 12 amino acids. So far, no sequence similarity is found to other known proteins.

The proteins appear to be specifically involved in a distinct step in the assembly of complex I. A conceivable function of the two proteins would be to keep the large membrane arm intermediate in an assembly competent state preventing aggregation or misfolding, thus working as a complex-I-specific chaperone. Since in wild type the large membrane arm intermediate is present in excess of the small intermediate, it may have to be stabilized. A scheme depicting the assembly pathway of complex I is shown in Fig. 3.

4.2. Subunits Involved in a Biosynthetic Pathway

Two of the so-called accessory subunits of the mitochondrial complex I, i.e., subunits not found in the minimal bacterial complex (see Section 2.1),

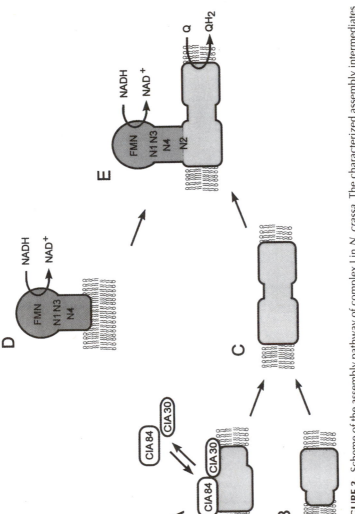

FIGURE 3. Scheme of the assembly pathway of complex I in *N. crassa*. The characterized assembly intermediates and the proposed stepwise association are shown. (A) Large assembly intermediate of the membrane arm. (B) Small assembly intermediate of the membrane arm. (C) Peripheral arm. (D) Membrane arm. (E) Complex I. CIA refers to complex I intermediate-associated protein; N1–4 refers to the EPR visible FeS clusters.

show remarkable sequence similarity to proteins involved in biosynthetic pathways.

The 10-kDa subunit SDAP is closely related to prokaryotic acyl-carrier proteins (ACPs) participating in fatty acid synthesis. The mitochondrial ACP has been found in bovine (Runswick *et al.*, 1991), *N. crassa* (Brody and Mikolajczyk, 1988; Sackmann *et al.*, 1991), and *Arabidopsis thaliana* (Shintani and Ohlrogge, 1994). *S. cerevisiae*, which lacks respiratory complex I, also has a mitochondrial ACP. Disruption of the *acp* gene results in different phenotypes in *N. crassa* and *S. cerevisiae* (Schneider *et al.*, 1995). In the *N. crassa* ACP mutant assembly of the membrane arm of complex I is rather ineffective, leaving a large fraction of membrane arm subunits in ill-defined aggregated state. No stable assembly of the subunits of the peripheral arm is found. Other complexes of the respiratory chain are not affected by the loss of ACP. A further phenotypic feature of the mutant is a fourfold increase of the lysophospholipid content. Other complex I mutants of *N. crassa* have the same low lysophospholipid content as wild type. In *S. cerevisiae*, loss of the *acp* gene gives a pleiotropic respiratory-deficient *pet* phenotype. The mutant is unable to grow on nonfermentable substrates and the mitochondria are essentially devoid of cytochromes. The same phenotype has been reported for a yeast mutant, in which the gene for a protein homologous to β-ketoacyl ACP synthase, the condensing enzyme of bacterial fatty acid synthetase (FAS type II), has been disrupted (Harington *et al.*, 1993). The *S. cerevisiae* mutant lacking the ACP has recently been shown to be devoid of lipoic acid (Brody *et al.*, 1997).

The pantothenate of the mitochondrial ACP from *N. crassa* is predominantly loaded with hydroxymyristate and shorter hydroxy acids, e.g., hydroxylaureate (Mikolajczyk and Brody, 1990; M. Massow, unpublished data). A single acyl group yielding a mass difference of 302 between acylated and deacylated ACP has been reported for the ACP from bovine complex I (Runswick *et al.*, 1991).

Involvement of the ACP in a mitochondrial de novo fatty acid synthesis would be a plausible explanation. Mitochondria are known to import the bulk of phospholipids from the endoplasmic reticulum, and long-chain fatty acids for cardiolipin synthesis are imported from the cytosol (Daum, 1985; Voelker, 1991); however, synthesis of lipoic acid is at least one possible function. As indicated by the increased amount of lysophospholipids in the *N. crassa asp*-1 mutant, another function might be the repair of damaged or hydrolyzed phospholipids. Still unresolved is the reason for the ACP being a subunit of complex I. The severe impact of the loss of ACP on the assembly of complex I in *N. crassa* suggests that beyond synthesis of fatty acids, the ACP might have another function in the biosynthesis of complex I.

The other biosynthetic subunit of the mitochondrial complex I is the 39-kDa subunit. The most prominent feature of the primary structure is the presence of 9–10 of 11 conserved sequence positions in the adenine binding pocket of a NADH binding site (Wierenga *et al.*, 1985; Fearnley and Walker, 1992). An insertion required between the first two conserved glycines is common to many other dehydrogenases. The function of this binding site is unknown. Fearnley and Walker (1992) have suggested that the 39-kDa subunit might be involved in a NADH-NAD$^+$ transhydrogenase activity associated with complex I or that it provides an alternative NADH oxidation site used under certain circumstances. However, no experimental data have yet been obtained to support any of these suggestions.

Regarding the entire sequence, the 39-kDa subunit shows significant sequence similarity to β-hydroxysteroid dehydrogenases (Fearnley and Walker, 1992), dihydroflavonol reductases (R. Schneider, personal communication), nucleotide sugar epimerases and dehydratases, isoflavone reductases, and some unassigned open reading frames (ORFs) in the genomes of chloroplasts and cyanobacteria (Fig. 4). β-Hydroxysteroid dehydrogenases catalyze oxidation and isomerization steps of the biosynthesis of steroids. Dihydroflavonol and isoflavone reductases are part of the biosynthesis of flavonoids (Heller and Forkmann, 1993). Dihydroflavonol reductase catalyzes the stereospecific reduction of dihydroflavonols to the respective flavandiols. Isoflavone reductase catalyzes the NADPH-dependent reduction of a double bond in isoflavones to form isoflavanones. Synthesis of flavonoids so far has been attributed to plants. However, in the genomes of yeast and nematodes, genes homologous to dihydroflavonol reductase have been found. UDP-glucose epimerase (galactowaldenase) catalyzes the isomerization of UDP-glucose to UDP-galactose by oxidation of a hydroxyl group to a keto group followed by rereduction to a hydroxyl group in the changed conformation, while UDP-glucose dehydratase is involved in polysaccharide biosynthesis.

A sequence alignment of the enzymes named above shows conserved amino acids throughout the sequence, with the adenine binding pocket showing the highest and the C-terminal part the lowest similarity (Fig. 4). At least some of the sequences contain a YXXXK motif characteristic of the family of short-chain alcohol dehydrogenases. However, sequence similarity of the 39-kDa subunit to other dehydrogenases including the ketoacyl reductase of fatty acid synthesis is essentially restricted to the NAD binding pocket.

Phylogenetic analysis reveals an almost equal distance between the enzyme classes included (Fig. 4). Bootstrap analysis indicates that besides the grouping of the plastidal and cyanobacterial ORFs with isoflavone reductases, no clear grouping of the different enzyme classes can be concluded.

The enzymes related to the 39-kDa subunit are involved in diverse reactions, and common features are not readily apparent. The enzymes catalyze NAD(P)H-dependent reactions of biosynthetic pathways. The catalyzed redox reactions in most cases include the conversion of a keto group into a hydroxyl

A

```
Dhfr P.h.    mplhlrcsatvcvtgaagfigswlvmrllergynvhatvrdpenkkkvkhl----lelpkadtn
Dhfr S.c.    mttektvvfvsgatgfialhvvddllktgykvigsgrsqekndgll-------kkfksnpnl
Dhfr C.e.    (39)nnnetkvlvtgasgfigthcveillkngyrvrgtvrdlnnkakvqp-----ikkldkknh
Gluepi M.j.       milvtggagfigshivdkliennydviildnltt------------gnknninpk
Gluepi S.l.       msgkylvtggagyvgsvvaqhlveagnevvvlhnlst-----------gfragvpag
Gludhy S.f.       mrvlvtggagfigshftgqlltgaypdlgatrtvvldkltyagn(3)lehvaghpd
Hsdh F.v.    mrtlvyvvtggcgflgrhiinnlilfesslkevrvydl-drpmlld----vek-cniik
Hsdh H.s.    mgwsclvtgaggllgqriv-rllveekelkeiraldkafrpelre-----fsklqnrtk
Cdh N.s.     (5)lttdlgcvlvtggsgfvganlvtelldrgyavrsfdra--------------psplgdhag
39 kDa B.t.  (49)svsgivatvfgatgflgryvvnhlgrmgsqvivphrcepydtmh-------lrpmgdlgq
39 kDa H.s.  (47)svsgivatvfgatgflgryvvnhlgrmgsqviipyrcdkydimh-------lrpmgdlgq
40 kDa N.c.  (47)slgghtatvfgatgqlgryivnrlarqgctvvipfrde-ynkrh-------lkvtgdlgk
Orf 39 P.p.        mtllvigatgtlgrqivrraldegynvkcmvrnlr-----------ksaflkewg
Ycf 39 C.p.        msilvigatgtlgrqivrsaldegyqvrclvrnlr-----------kaaflkewg
Hyppro S.s.       mnrkilvtgatgsngteivkrlaaknvqvramvrdfd-----------rakkiafpn
Ifr A.t.     matekskilviggtgyigkflveasakaghstfalvreatlsdpvkgk----tvqsfkdlg
Ifr P.s.     maten-kililgatgaigrhivwasikagnptyalvrk-tsdnvnkpk(12)llknyqasg
                  lv gatgf g     v l   g v       r
```

```
Dhfr P.h.    ltllkadltvegsfdeaiqg----cqgvfhvatpmdfeskdp-----enevikptvrgmlsiie
Dhfr S.c.    smeivediaapnafdkvfqkhgkeikvvlhiaspvhfnttdf------ekdllipavngtksile
Dhfr C.e.    lelveadlldstcwkkavag----cdyvlhvaspfpivs--------dercittavegtmnvlk
Gluepi M.j.  aefvnadirdkdldekinfk---dvevvihqaaqinvrnsven----pvydgdinvlgtinile
Gluepi S.l.  asfyrgdirdqdfmrkvfrg-rlsfdgvlhfaafsqvgesvvk----pekywdnnvggtmalle
Gludhy S.f.  lefvrgdiadhgwwrrlmeg----vglvvhfaaeshvdrsies----seafvrtnvegtrvllq
Hsdh F.v.    ivpvigdvrnkstldealrs----advvihiasindvagkft-----ndsimdvvningtknvvd
Hsdh H.s.    ltvlegdildepflkracqd----vsvvihtaciidvfgvth-----resimnvnvkgtqllle
Cdh N.s.     leviegdicdketvaaavkd----idtvihtaaiidlmggasvteayrqrsfavnvegtknlvh
39 kDa B.t.  iifmdwngrdkdsirraveh----ssvvinlvgrewetq-------nfdfedvfvkipqaiaq
39 kDa H.s.  llflewdardkdsirrvvqh----snvvinligrdwetk--------nfdfedvfvkipqaiaq
40 kDa N.c.  vvmiefdlrntqsieeesvrh----sdvvynligrdyptk--------nfsfedvhiegaeriae
Orf 39 P.p.  aelvygdlklpesilqsfcg----vtavidastsrpsd---------pynteqidldgkialie
Ycf 39 C.p.  akliwgdlsqpesllpaltg----irviidtstsrptd---------pagvyqvdlkgkkalid
Hyppro S.s.  vevvegnfdrpetllealae----vdraflltnstera-----------eaqqlafvd
Ifr A.t.     vtilhgdlndheslvkaikq----vdvvistvgsmq------------------ildqtkiis
Ifr P.s.     villegdindhetlvnaikq----vdtvicaagrll-----------------iedqvkvik
                   gd  d      a        vvih a                    v g
```

```
Dhfr P.h.    s--cakantvkrlvftssagtldvqeqqklfyd(4)sdldfiyakkmtgwmyfaskilaekaam
Dhfr S.c.    aiknyaadtvekvvitssvaalaspgdmkdtsf(7)nkdtwescqanavsaycgskkfaektaw
Dhfr C.e.    a--iaedgnvrklvltsscaavnglwsliqnns(11)edswsnlesdmvdcyiksktlaekaaw
Gluepi M.j.  m--mrkyd-idkivfassggavygepnylpvde---------nhpinplspyglskyvgeeyik
Gluepi S.l.  a--mrgag-vrrlvfsst-aatygepeqvpive----------saptrptnpygasklavdhmit
Gludhy S.f.  a--avdag-vgrfvhist-devygsiaegswpe----------dhpvapnspyaatkaasdllal
Hsdh F.v.    s--clyng-vrvlvytssysavgpnflgdamir----gnentyyqsnhkeaylskqlsekyil
Hsdh H.s.    a--cvqas-vpvfiytssievagpnsykeiiqn----gheeeplentwptpypyskklaekavl
Cdh N.s.     a--sqeag-vkrfvytasnsvvmggqdivngde-------tmpyttrfndlytetkvvaekfvl
39 kDa B.t.  v--skeag-vekfihishln----------------------adikssskylrskavgekevr
39 kDa H.s.  l--skeag-vekfihvshln----------------------aniksssrylrnkavgekvvr
40 kDa N.c.  r--vakyd-vdrfihvssyn----------------------adpnseceffatkargeqvvr
Orf 39 P.p.  a--akaak-vqrfiffsiln----------------------adqypkvplmnlksqvvnylq
Ycf 39 C.p.  a--akamk-iekfiffsiln----------------------sekysqvplmriktvteellk
Hyppro S.s.  a--arqnq-vkhivklsqfa----------------------adahspvrflryhaaveaaiq
Ifr A.t.     a--ikeagnvkrflp-sefgvdvdrt----------------savepaksafagkiqirrtie
Ifr P.s.     a--ikeagnvkrffp-sefgldvdrh----------------davepvrqvfeekasirrvve
                  a    ag v  fv  s                            y  k   e
```

FIGURE 4. (A) Sequence alignment of the 39-kDa subunit of complex I and related proteins. Dhfr P.h., dihydroflavonol reductase from *Petunia hybrida* (Beld *et al.*, 1989); Dhfr S.c., ORF from *Saccharomyces cerevisiae* homologous to dihydroflavonol reductase (submitted to Genbank by U. Hebling, B. Hofmann, and H. Delius); Dhfr C.e., ORF from *Caenorhabditis elegans* homologous to dihydroflavonol reductase (submitted to Genbank by C. Hembry); Gluepi M.j., UDP-glucose epimerase from *Methanococcus jannaschii* (Bult *et al.*, 1996); Gluepi S.l., UDP-glucose epimerase from *Streptomyces lividans* (Adams *et al.*, 1988); Gludhy S.f., dTDP-glucose dehydratase from *Streptomyces fradiae* (Merson-Davies and Cundliffe, 1994); Hsdh F.v., 3-β-hydroxysteroid dehydrogenase/steroid isomerase from fowlpox virus (submitted to Genbank by M. A. Skinner, J. B. Moore, M. M. Binns, and M. E. Boursnell); Hsdh H.s., 3-β-hydroxysteroid dehydrogenase/delta 5-delta 4-isomerase from human

```
Dhfr P.h.   eeakkkni-----dfisiipplvvgpfitptfppslitalslitgneahyciik---qgqyvhl
Dhfr S.c.   dfleenqssikf-tlstinpgfvfgpqlfadslrnginsssaiianlvsyklgd(5)sgpfidv
Dhfr C.e.   dfierlpedkkf-pmtvinptlvfgpayiteqgasitlmrkfmngempaappln----mpivdv
Gluepi M.j. lynrlygi-----eyailrysnvygerqdpkgeagvisifidkmlknqspiifg(4)trdfvyv
Gluepi S.l. geaaahgl-----gavsvpyfnvaganrgvrlvhdpeshliplvlqvaqgrrea-----isvyg
Gludhy S.f. ayhrtygl-----dvrvtrcsnnygprqypekavplfttnlldglp-------------vplyg
Hsdh F.v.   eangtmsniglrlctcalrplgvfgeycpvletlyrrsyksrkmykyaddkvfh-----srvya
Hsdh H.s.   aangwnlkngdtlytcalrptyiygeggpflsasinealnnngilssvgkfstv-----npvyv
Cdh N.s.    aengkhdm-----ltcairpsgiwgrgdqtmfrkvfenvlaghvkvlvgnknik----ldnsyv
39 kDa B.t. etfpeat---------iikpaeifgredrflnyfanirwfggvplislgkktvk-----qpvyi
39 kDa H.s. dafpeai---------ivkpsdifgredrflnsfasmhrfgpiplgslgwktvk-----qpvyv
40 kDa N.c. sifpett---------ivrpapmfgfedrllhklas----vkniltsngmqeky-----npvhv
Orf 39 P.p. kssisyt--------vfslggffqglisqyaipild-----kksvwvtgestpi-----ayidt
Ycf 39 C.p. esglnyt--------ifklcgffqgliggyavpild-----qqtvwittestsi-----aymdt
Hyppro S.s. gsgmtyt--------flrpnlfmqgll-nfqstits-----qnafyaaisdakv------svdv
Ifr A.t.    aegipyt---------yavtgcfggyylptlvqfepgltspprdkvtilgdgna---kavinke
Ifr P.s.    segvpyt--------ylcchaftgyflrnlaqida--tdpprdkvvilgdgnv---rgayvte
                          p         g                                    v

Dhfr P.h.   ddlceahiflyehpkadgrficsshhaiiydvak--mvrekw---------peyyvptefkgid
Dhfr S.c.   rdvskahllafekpecaggrlflcedmfcsgealdilneef----------pqlkgkiatgepg
Dhfr C.e.   rdvalahfeamrrpesdnerilvikmvkcywiprftapyff-------vrlyalfdpetkaslp
Gluepi M.j. gdvakanlmaln----wkneivnigtgketsvnelfdii--------kheigfrgeaiydkpre
Gluepi S.l. ddyptpdtcvrdyihvadlaeahllavrrrpgnehlicnl-----gngngfsvrevvetvrrvt
Gludhy S.f. -dggntrewlhvddhcrgvalvgaggrpgviyni-gggte-----ltnaeltdrilelcgadtk
Hsdh F.v.   gnvawmhilaarnmiengqhsplcknvyycydtsptehyhdf(7)lgmdlrnt-clplwclrfi
Hsdh H.s.   gnvawahilalralrdpkkapsvrgqfyyisddtphqsydnl(7)fglrldsrwslpltlmywi
Cdh N.s.    hnllfngfilagqdlvpggtap---gqayfindgepinmfefa----rpvlaacgrplptf-yvsg
39 kDa B.t. vdvtkgiinaikdpdargktfafvgpsryl-lfdlvqyvfav----ahrpflpyplphfayrwi
39 kDa H.s. vdvskgivnavkdpdangksfafvgpsryl-lfhlvkyifav---ahrlflpfplplfayrwv
40 kDa N.c. idvgqaleqmlwddntasetfelygpktyt-taeisemvdre----iykrrrhvnvpk-kilkp
Orf 39 P.p. qdaaklvikslgvpstenrilplvgnkawt-saeiitlcekl----sgqktqisqipl-sllka
Ycf 39 C.p. idiarftlrslvlketnnrvfplvgtrswn-sadiiqlcerl----sgqnakvtrvpi-aflel
Hyppro S.s. rdiadvavvaltetehegkiynltgpqalt-hae---maeql----saalnr-----qi-afvdi
Ifr A.t.    ediaaytikavddprtlnkilyikpsnntlsmneivtlwekk----igkslekthlpeeqllks
Ifr P.s.    advgtytiraandpntlnkavhirlpnnyltanevialwekk----igktlektyvseeqvlkd
                  dv       a         d          g            e          p

Dhfr P.h.   kdlpvvsfsskk--------ltdmgfqfkytledmykgaidtcrqkqllpfstrsaedng(35)
Dhfr S.c.   sgstfltkncck------cdnrktknllgfqfnkfrdcivdtasqllevqsks
Dhfr C.e.   rlcqevkfdns-----------kaqrllgmtmrdskealidmahslidlgiierk
Gluepi M.j. gevyriyldikk----------------------aeslgwkpeidlkegikrvvnwm(6)
Gluepi S.l. ghpipeimaprr-----------grdpavlvasagtareklgwnpsradlaivsdawew(12)
Gludhy S.f. alrrvadrpghd-----------rrysvdttkireelgyaprtgiteglgtvaw(23)
Hsdh F.v.   aninkglrvlls(7)pllnpytlikecttftie-tdkafkdfgyvplytweesrsktqlw(15)
Hsdh H.s.   gfllevvsflls(7)ppfnrhtvtlsnsvftfs-ykkaqrdlaykplysweeakqktvew(17)
Cdh N.s.    rlvhkvmmawqw(10)lieplaverlylnnyfs-iakakrdlgyeplftteqamaecmpy(21)
39 kDa B.t. grlfeispfepw-----ttrdkverihttdkilphlpgledlgveatplelkaievlrrh(24)
39 kDa H.s. arvfeispfepw-----itrdkvermhitdmklphlpgledlgiqatplelkaievlrrh(21)
40 kDa N.c. iagvlnkalwwp----imsadeierefhdqvidpeaktfkdlgiepadianftyhylqsy(26)
Orf 39 P.p. lrkitktlqwtw-----------nisdrlafaevltsgeifmapmnevyeilsidksevisl(31)
Ycf 39 C.p. arntscffewgw---------niadrlaftevlsksqffnssmdevykifkieststtil(33)
Hyppro S.s. psevmrdqllni---------------gmppwqadgviedyahyrrneaaavs(26)
Ifr A.t.    iqespipinvvl---------sinhavfvngdtnisiepsfgveaselypd-vkytsvde(6)
Ifr P.s.    iqtssfphnyll---------alyhsqqikgdavyeidpakdveaydaypd-vkyttade(6)
                    dv    g
```

(Rheaume *et al.*, 1991); Cdh N.s., cholesterol dehydrogenase from *Nocardia* sp. (Horinouchi *et al.*, 1991); 39-kDa B.t., 39-kDa subunit of complex I from bovine (Fearnley *et al.*, 1991); 39-kDa H.s., 39-kDa subunit of complex I from human (Baens *et al.*, 1993); 40-kDa N.c., homologue of 39-kDa subunit of complex I from *N. crassa* (Röhlen *et al.*, 1991); ORF39 P.p., ORF 39 from chloroplast genome of *Porphyra purpurea* (Reith and Munholland, 1995); ycf39 C.p., ORF from *Cyanophora paradoxa* (submitted to Genbank by V. L. Stirewalt, C. B. Michalowski, W. Luffelhardt, H. J. Bohnert, and D. A. Bryant); hyppro S.s., ORF from *Synechocystis* sp. (Kaneko *et al.*, 1995); Ifr A.t., isoflavonoid reductase from *Arabidopsis thaliana* (submitted to Genbank by E. Babiychuk, S. Kushnir, E. Belles-Boix, M. van Montagu, and D. Inze); Ifr P.s., isoflavone reductase from *Pisum sativum* (Paiva *et al.*, 1994). In view of the large number of sequences of closely related enzymes from different organisms available in the data banks, two to three sequences were arbitrarily chosen from each class; sequences

```
dhfr-1   ..mplhlrcsatvcvtgaagfigswlvmrllergynvhatvrdpenkkkvkhllelpk..adtnlt
dhfr-2   ..mgsq...setvcvtgasgfigswlvmrllerrltvratvrdptnvkkvkhlldlpk..aethlt
   ifr   .....matekskilviggtgyigkflveasakaghstfalvreatlsdpvkgktvqsf.kdlg.vt
hsdh-1   ......mrtlvyvvtggcgflgrhiinnlilfesslke.vrvyd..ldrpmlldlvekcniikiv
hsdh-2   ........mgwsclvtgaggllgqriv.rllveekelke.iraldkafrpelreefsklqnrtklt
   cdh   mgdaslttdlgcvlvtggsgfvganlvtelldrgya....vrsfdrapsp........lgdhagle
39kda-1  (47)rssvsgivatvfgatgflgryvvnhlgrmgsqvivphrcepydtm.....hlrpmgdlgqii
39kda-2  (45)rssvsgivatvfgatgflgryvvnhlgrmgsqviipyrcdkydim.....hlrpmgdlgqll
40kda    (45)rsslgghtatvfgatgqlgryivnrlarqgctvvipfrde.ynkr.....hlkvtgdlgkvv
orf39    .........mtllvigatgtlgrqivrraldegynvkcmvrnl........r.ksaflkewg.ae
ycf39    .........msilvigatgtlgrqivrsaldegyqvrclvrnl........r.kaaflkewg.ak
hyppro   ........mnrkilvtgatgsngteivkrlaaknvqvramvrdf........d.rakkiafpn.ve
                  v gatg lg    v  l    g  v        vr

dhfr-1   llkadltvegsfdeaiqgcqgvfhvatpmdfeskdp    enevikptvrgmlsiiescakantvkr
dhfr-2   lwkadladegsfdeaikgctgvfhvatpmdfeskdp    enevikptiegmlgimkscaaaktvrr
   ifr   ilhgdlndheslvkaikqvdvvistvgsm.......    .......qildqtkiisaikeagnvkr
hsdh-1   pvigdvrnkstldealrsadvvihiasindvagkft    ndsimdvningtknvvdsclyng.vrv
hsdh-2   vlegdildepflkracqdvsvvihtaciidvfgvth    resimnvnvkgtqlllleacvqas.vpv
   cdh   viegdicdketvaaavkdidtvihtaaiidlmggas(5)rqrsfavnvegtknlvhasqeag.vkr
39kda-1  fmdwngrdkdsirravehssvvinlvgr.ewetq..    nfdfedvfvkipgaiaqvskeag.vek
39kda-2  flewdardkdsirrvvqhsnvvinligr.dwetk..    nfdfedvfvkipgaiaqlskeag.vek
40kda    miefdlrntqsieesvrhsdvvynligr.dyptk..    nfsfedvhiegaeriaervakyd.vdr
orf39    lvygdlklpesilqsfcgvtavidastsrpsd....    pynteqidldgkialieaakaak.vqr
ycf39    liwgdlsqpesllpaltgirviidtstsrptd....    pagvyqvdlkgkkalidaakamk.iek
hyppro   vvegnfdrpetllealaevdraflltnstera....    ........eaqqlafvdaarqng.vkh
             gd    s    a    vi    d              v    g          a  v

dhfr-1   lvftssagtldvqeqqklfydqtswsdldfiyakkmtgwmyfaskilaekaameeakkknidfis.
dhfr-2   lvftssagtvniqehqlpvydescwsdmefcrakkmtawmyfvsktlaeqaawkyakennidfit.
   ifr   flp.sefgvdvd.................rtsavepaksafagkiqirrtieaegipytyavtgc
hsdh-1   lvytssysavgp.....nflgdamirgnentyyqsnhkeayplskqlsekyileangtmsniglrl
hsdh-2   fiytssievagp.....nsykeiiqngheeeplentwptpypyskklaekavlaangwnlkngdtl
   cdh   fvytasnsvvmg........gqdivngdetmpyttrfndlytetkvvaekfvlaengk.....hdm
39kda-1  fihishln...................adikssskylrskavgekevretfpeat......
39kda-2  fihvshln...................aniksssrylrnkavgekvvrdafpeai......
40kda    fihvssyn...................adpnseceffatkargeqevvrsifpett......
orf39    fiffsiln...................adqypkvplmnlksqvvnylqkssisyt......
ycf39    fiffsiln...................sekysqvplmriktvteellkesglnyt......
hyppro   ivklsqfa...................adahspvrflryhaaveaaiqgsgmtyt......
             f   s                         y    k   e                .

dhfr-1   ....iipplvvgpfitptfppslitalslitgneahyciik.qggqyvhlddlceahiflyehpkad
dhfr-2   ....iiptlvvgpfimssmppslitalspitgneahysiir.qggfvhlddlcnahiylfenpkae
   ifr   fggyylptlvqfepglt.spprdkvtilgd.gn.akavink.......eediaaytikavddprtl
hsdh-1   ctcalrplgvfgeycpvletlyrrsyksrkmykyaddkvfh.srvyagnvaw...mhilaarnmien
hsdh-2   ytcalrptyiygeggpflsasinealnnngilssvgkfstv.npvyvgnvaw...ahilalralrdp
   cdh   ltcairpsgiwgrgdqtmfrkvfenvlaghvkvlvgnknikldnsyvhnlih..gfilagqdlvpg
39kda-1  ...iikpaeifgredrflnyfanirwfggvplislgkktvk.qpvyivdvtk..giinaikdpdan
39kda-2  ...ivkpsdifgredrflnsfasmhrfgpiplgslgwktvk.qpvyvvdvsk..givnavkdpdan
40kda    ...ivrpapmfgfedrllhklasvkni....ltsngmqeky.npvhvidvgq..aleqmlwddnta
orf39    .. vfslggffqglisqyaipildkksvvrtpgest.......piayidtqdaaklvikslgvpste
ycf39    ...ifklcgffqgligqyavpildqqtvvwittest.......siaymdtidiarftlrslvlketn
hyppro   ...flrpnlfmqgll.nfqstitsqnafyaaisda.......kvs.vdvrdiadvavvaltetehe
             p    g                        yv

dhfr-1   gr...ficsshhaiiydvakmvre...........kwpeyyvptefkgidkdlpvvsfsskk.lt
dhfr-2   gr...yicsshdciilldlakmlre...........kypeyniptefkgvdenlksvcfsskk.lt
   ifr   nkil.yikpsnntlsmneivtlwe.......kkigkslekthlpeeql..lksiqespipinvvls
hsdh-1   gqhsplcknvvygcvdtsptehyhdfnmhffnqlgmdlrnt.clplwcl...rfianinkglrvlls
hsdh-2   kkapsvrgqfyyisddtphqsydnlnyilskefglrldsrwslpltlm...ywigfllevvsflls
   cdh   gtap...gqayfindgepinmf.e......farpvlaacgrplptfyvs.grlvhkvmmawqwlhf
39kda-1  gktfafvgpsryllfdlvqyvfav.........ahrpflpyplphfayrwigrlfeispfepwttr
39kda-2  gksfafvgpsryllfhlvkyifav.........ahrlflpfplplfayrwvarvfeispfepwitr
40kda    setfelygpktyttaeisemvdre.........iykrrrhvnvpkkilkpiagvlnkalwwpimsa
orf39    nrilplvgnkawtsaeiitlcekl.........sgqktqisqipls.l..lkalrkitktlqwtwn
ycf39    nrvfplvgtrswnsadiiqlcerl.........sgqnakvtrvpia.f..lelarntscffewgwn
hyppro   gkiynltgpqalthae...maeql.........saalnr....qia.f..vdipsevmrdqllnig
             g    g                             p
```

Figure 4. (*Continued*) were aligned by pile up and adjusted for highest overall sequence identity; numbers in brackets give additional residues; residues conserved in the majority of sequences are indicated. (B) Phylogenetic tree of enzymes related to the 39-kDa subunit. Tree construction is based on the above alignment using the program package Phylip by J. Felsenstein.

```
dhfr-1   dmgfqf............kytledmykgaidtcrqkqllpfstrsaednghnreaiaisaqn(22)
dhfr-2   dlgfef............kysledmftgavdtcrakgllrpshekpvdgkt
   ifr   inhavf............vngdtnisiepsfgveaselypd.vkyt.svdeylsyfa
hsdh-1   picsytpllnpytlikecttftietdkafkdfgyvp..lytweesrsktqlwireleak.ss(6)
hsdh-2   piysyqppfnrhtvtlsnsvftfsykkaqrdlaykp..lysweeakqktvewvgslvdrhke(7)
   cdh   kfalpeplieplaverlylnnyfsiakakrdlgyep..lftteqamaecmpyyvemfhqmes(11)
39kda-1   dkveri..........httdkilphlpgledlgveatplelkaievlrrhrtyrwlsseied(12)
39kda-2   dkverm..........hitdmklphlpgledlgiqatplelkaievlrrhrtyrwlsaeied(9)
 40kda   deiere..........fhdqvidpeaktfkdlgiepadianftyhyl...qsyr..snayyd(19)
 orf39   isdrlafaevltsgeifmapmn.....evyeilsidksevislekyf...qeyfgkilkvlk(18)
 ycf39   iadrlaftevlsksqffnssmd.....evykifkieststtilesyl...qeyfsrilkrlk(14)
hyppro   mppwqa..dgvi............edyahyrrneaaavssgiqd...aigkeprsfntf(9)
                                d       l                   y
```

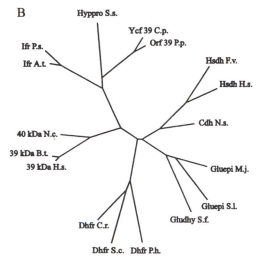

FIGURE 4. (*Continued*)

group or vice versa. Isoflavone reductases, however, catalyze the reduction of a double bond. With the exception of the sugar-modifying enzymes, all enzymes catalyze the modification of complex polycyclic compounds, which have been synthesized from precursors bound to coenzyme A.

We consider the 39-kDa subunit to be most likely involved in some kind of biosynthetic pathway rather than respiratory electron transfer. The ACP might well be involved in the same pathway at an earlier stage. No indications, however, point to a participation of the 39-kDa subunit in fatty acid synthesis.

In continuation of the discussion of the function of the ACP, the product of the hypothetical biosynthetic pathway might be the postulated structural component of complex I. As mentioned (see Section 3.1), loss of the 39-kDa subunit leads to an inactive complex I in *N. crassa*. Assuming a synthetic function of this subunit, the mutant enzyme would be inactive because an essential redox group is not properly made. The chromophore visible in redox difference

spectra of wild-type complex I, but not the mutant enzyme, might be the first
hint of that group.

5. ORIGIN AND EVOLUTION OF COMPLEX I

Parts of complex I are homologous to other plastidal and bacterial enzymes, namely, to a plastidal complex possibly functioning as ferredoxin–
plastoquinone oxidoreductase and to bacterial NAD^+-reducing hydrogenases,
formate hydrogenlyase, and cation transporters (Table III). These parts most
likely reflect functional modules for electron transfer and proton transport.
During evolution, fusion of preexisting protein assemblies constituting the
modules led to the formation of complex I and other electron transfer complexes (Friedrich and Weiss, 1997).

Gene encoding homologues to 11 of the "minimal" subunits of complex I
were found in the genome of chloroplasts and cyanobacteria (Shinozaki *et al.*,
1986; Ohyama *et al.*, 1986; Ogawa, 1991; Shimada and Sugiura, 1991; Ellersiek
and Steinmüller, 1992). The genes are organized in four transcriptional units in
the same order as the complex I genes in purple bacteria. The gene products are
subunits of a plastidal complex located on the stroma thylakoids in close connection to photosystem I (Berger *et al.*, 1993a,b). This complex lacks homologues to the three "minimal" subunits that make up the NADH dehydrogenase part of complex I, containing the FMN and four to five FeS clusters. The
electron input device of the complex is unknown, but recently it was found that

Table III
Homologous Subunits Found in Complex I
and Related Bacterial Enzymes[a]

Complex I	NAD^+-reducing hydrogenase	Formate hydrogenlyase	Glucose dehydrogenase	$K^+–H^+$ antiporter
75 kDa, N-term.	γ			
51 kDa	α, C-term.			
24 kDa	α, N-term.			
49 kDa	β (?)	HycE, C-term.		
30 kDa		HycE, N-term.		
PSST	δ (?)	HycG		
TYKY		HycF		
ND1		HycD	N-term.	
ND4				PhaD
ND5		HycC		PhaA

[a]Adapted from Friedrich and Weiss (1997).

the plastidal ferredoxin–NADPH reductase is associated with it (Guedeney *et al.*, 1996). The complex would then work as an NADPH–plastoquinone oxido-reductase participating in a cyclic electron flow passing the electrons from photosystem I back to the plastoquinone pool (Friedrich *et al.*, 1995; Bendall and Manasse, 1995; Mi *et al.*, 1992; Kubicki *et al.*, 1996).

The NADH dehydrogenase part of complex I, which is missing in the plastidal complex, is related to the diaphorase part of the NAD^+-reducing hydrogenase (Pilkington *et al.*, 1991). The soluble enzyme is found in chemo-lithotrophic purple bacteria (Tran-Betcke *et al.*, 1990), sulfate-reducing purple bacteria (Malki *et al.*, 1995), and cyanobacteria (Schmitz *et al.*, 1995). In *Alcaligenes eutrophus* this hydrogenase is composed of the four subunits α, β, γ, and δ. The β–δ dimer is engaged in the splitting of hydrogen and the α–γ dimer in NAD^+-reduction. The latter diaphorase dimer contains a structural unit of about 900 amino acids with a noncovalently bound FMN and several FeS clusters (Friedrich and Schwarz, 1993). The α subunit is a fusion product of the 51- and 24-kDa subunits of complex I, while the γ subunit is homologous to the N-terminal 200 amino acids of the 75-kDa subunit (Table III). The diaphorase genes are arranged in the so-called *hox*-loci in the same order as the corresponding complex I genes in bacteria.

The formate hydrogenlyase complex of *E. coli* is expressed only under strictly anaerobic conditions and couples the oxidation of formate by the formate dehydrogenase with proton reduction by the hydrogenase part. The genes of the formate hydrogenlyase are organized in the *hyc*-operon (Böhm *et al.*, 1990; Sauter *et al.*, 1992). *Hyc*E, the large subunit of the hydrogenase part, is a fusion of the 40- and the 49-kDa subunits of complex I (Videira and Azevedo, 1994). The conserved residues in *Hyc*E that coordinate the Ni-atom of the Ni-Fe hydrogenases (Volbeda *et al.*, 1995), however, are not found in the 49-kDa subunit. *Hyc*G, the small hydrogenase subunit, is related to PSST and coordinates one [4Fe-4S] cluster. *Hyc*F, a typical 2 [4Fe-4S] ferredoxin, is related to TYKY. It contains two similar moieties that most likely have arisen by gene duplication.

Homologues of the large and small hydrogenase subunits of the formate hydrogenlyase are present in most of the known hydrogenases (Przybyla *et al.*, 1992). It was therefore proposed that PSST and the 49-kDa subunit of complex I are related to hydrogenases in general (Weidner *et al.*, 1993; Albracht, 1993). This would imply that the hydrogenase part (the β–δ dimer) of the above-mentioned NAD^+-reducing hydrogenases is also related to complex I (Table III). In fact, the conserved sequence motifs of the hydrogenase subunits (Voor-douw, 1992) are found in the two corresponding complex I subunits; but apart from these motifs, there is no significant sequence similarity between the two complex I subunits and the two hydrogenase subunits.

Two of the membrane-intrinsic subunits of complex I have homologues among the *hyc* gene products as well. *Hyc*D is related to ND1 and *Hyc*C is homologous to ND5. As mentioned earlier, ND1 is also related to glucose–ubiquinone oxidoreductase (glucose dehydrogenase). The N-terminal region of the glucose dehydrogenase is predicted to fold to five membrane-spanning helices and to contain the ubiquinone reduction site (Yamada *et al.*, 1993). This region shows sequence similarity to a consensus sequence derived from all known homologues of NuoH, including *Hyc*D (Friedrich *et al.*, 1990).

Other homologues to membrane intrinsic complex I subunits have been identified in bacterial cation–H^+ antiporters (Friedrich and Weiss, 1997). PhaA a subunit of a K^+–H^+ antiporter of *R. meliloti* and a related protein (ORF-1 product) in *Bacillus* C-125 are homologous to ND5 of complex I and *Hyc*C of formate hydrogenlyase. PhaD is a homologue of ND4.

The origin of the other membrane-intrinsic subunits of complex I, ND2, ND3, ND4L, and ND6 is unknown. No homologous counterparts have so far been found in other proteins. ND2, -4, and -5 have very likely arisen from a common ancestor (Kikuno and Miyata, 1985).

An electron-transferring and proton-transporting complex probably was a common ancestor of complex I and the formate hydrogenlyase (Friedrich and Weiss, 1997). This ancestral enzyme might have worked already as a proton-translocating hydrogen–quinone oxidoreductase using the extended hydrogenase module made up of predecessors of PSST and the 30- and 49-kDa subunits, together with the ferredoxin-type TYKY predecessor. The ancestors of ND1 and ND4 constituted the ubiquinone reduction and proton transportation module. The membrane domain of complex I further evolved by gene duplication and acquisition of additional subunits, leading to subunits ND3, ND4L, and ND6. This complex then diverged to form the NADH–ubiquinone oxidoreductase of the purple bacterial respiratory chain by fusion with the diaphorase–NADH dehydrogenase module and the putative NADPH–plastoquinone oxidoreductase of the cyanobacterial photosynthetic electron transfer chain.

6. PERSPECTIVES

While the other respiratory enzymes are being studied on the basis of an at least partially solved high-resolution structure (Abrahams *et al.*, 1994; Iwata *et al.*, 1995; Tsukihara *et al.*, 1996; Yu *et al.*, 1996), complex I is left far behind. Neither the number of the prosthetic groups involved in electron transfer nor the number and location of substrate binding sites is known exactly. In view of the large number of different subunits that complex I contains, even in prokaryotes, the low redox potential of many of the prosthetic groups involved and the particular problems in the spectroscopic analysis of the complex, a considerable effort will be necessary to catch up in the understanding of this enzyme.

The membrane part certainly holds the secret of how complex I achieves the high H^+/e^- ratio. Current models of a Q-cycle mechanism attempt to explain the proton pumping by a sophisticated arrangement of different Q-binding sites. The experimental data from detailed EPR and inhibitor studies supporting these multiple binding sites do leave much of the details to speculation. A pure Q cycle would imply a function of the membrane-intrinsic subunits solely in the formation of a suitable stage for the Q chemistry.

Some points that have been dealt with in this chapter, however, indicate that complex I might hold an as yet unknown structural component. No convincing complex intrinsic target for the two biosynthetic subunits of the mitochondrial complex I have yet been found. Characterization of the respective mutants suggest that the ACP and the 39-kDa subunit have a particular essential function for complex I. Transfer of acyl groups and redox reactions of β-hydroxy–keto groups is common to a variety of biosynthetic pathways. A first glimpse of the product of this pathway could well be visible in the redox difference spectra. The absorbance increase around 295 nm, which is not readily attributable to FMN or FeS clusters, is seen in wild-type complex I but not in the complex lacking the 39-kDa subunit. Of course, the experimental data are still quite scarce and the proposed function of the biosynthetic subunits is highly speculative, but the *N. crassa* mutants available provide a good approach to put these speculations on a more solid basis.

ACKNOWLEDGMENTS. We thank T. Friedrich for helpful discussion and B. Brors for support in phylogenetic analysis. The work in the authors' laboratory was supported by grants from the Deutsche Forschungsgemeinschaft and the Fonds der Chemischen Industrie.

7. REFERENCES

Abrahams, J. P., Leslie, A. G. W., Lutter, R., and Walker, J. E., 1994, Structure at 2.8-Angstrom resolution of F-1-ATPase from bovine heart mitochondria, *Nature* **370**:621–628.

Adams, C. W., Fornwald, J. A., Schmidt, F. J., Rosenberg, M., and Brawner, M. E., 1988, Gene organization and structure of the *Streptomyces lividans* gal operon, *J. Bacteriol.* **170**:203–212.

Albracht, S. P. J., 1993, Intimate relationships of the large and the small subunits of all nickel hydrogenases with two nuclear-encoded subunits of mitochondrial NADH:ubiquinone oxidoreductase, *Biochim. Biophys. Acta* **1144**:221–224.

Alves, P. C., and Videira, A., 1994, Disruption of the gene coding for the 21.3-kDa subunit of the peripheral arm of complex I from *Neurospora crassa*, *J. Biol. Chem.* **269**:7777–7784.

Arizmendi, J. M., Runswick, M. J., Skehel, J. M., and Walker, J. F., 1992, NADH:ubiquinone oxidoreductase from bovine heart mitochondria: A fourth nuclear coded subunit with a homologue encoded in chloroplast genomes, *FEBS Lett.* **301**:237–242.

Azevedo, J. E., and Videira, A., 1994, Characterization of a membrane fragment of respiratory chain complex I from *Neurospora crassa*. Insights on the topology of the ubiquinone binding site, *Int. J. Biochem.* **26**:505–510.

Azevedo, J. E., Duarte, M., Belo, J. A., Werner, S., and Videira, A., 1994, Complementary DNA sequences of the 24 kDa and 21 kDa subunits of complex I from *Neurospora*, *Biochim. Biophys. Acta* **1188**:159–161.

Baens, M., Chaffanet, M., Cassiman, J. J.,van den Berghe, H., and Marynen, P., 1993, Construction and evaluation of a *hnc*DNA library of human 12p transcribed sequences derived from a somatic cell hybrid, *Genomics* **16**:214–218.

Beinert, H., and Albracht, S. P. J., 1982, New insights, ideas and unanswered questions concerning iron–sulphur clusters in mitochondria, *Biochim. Biophys. Acta* **683**:245–277.

Beld, M., Martin, C., Huits, H., Stuitje, A. R., and Gerats, A. G., 1989, Flavonoid synthesis in *Petunia hybrida*: Partial characterization of dihydroflavonol-4-reductase genes, *Plant Mol. Biol.* **13**:491–502.

Bendall, D. S., and Manasse, R. S., 1995, Cyclic photophosphorylation and electron transport, *Biochim. Biophys. Acta* **1229**:23–38.

Benecke, R., Strümper, P., and Weiss, H., 1992, Electron transfer complex I defect in idiopathic dystonia, *Ann. Neurol.* **32**:683–686.

Berger, S., Ellersiek, U., Kinzelt, D., and Steinmüller, K., 1993a, Immunopurification of a sub-complex of the NAD(P)H-plastoquinone oxidoreductase from the cyanobacterium *Synechocystis* sp. PCC6803, *FEBS Lett.* **326**:246–250.

Berger, S., Ellersiek, U., Westhoff, P., and Steinmüller, K., 1993b, Studies on the expression of NDH-H, a subunit of the NAD(P)H-plastoquinone oxidoreductase of higher-plant chloroplasts, *Planta* **190**:25–31.

Böhm, R., Sauter, M., and Böck, A., 1990, Nucleotide sequence and expression of an operon in *Escherichia coli* coding for formate hydrogenlyase, *Mol. Microbiol.* **4**:231–243.

Braun, M., Bungert, S., and Friedrich, T., 1998, Characterization of the overproduced NADH dehydrogenase fragment of the NADH:ubiquinone oxidoreductase (complex I) from *Escherichia coli*, *Biochemistry* **37**:1861–1867.

Brown, M. D., and Wallace, D. C., 1994, Molecular basis of mitochondrial DNA disease, *J. Bioenerg. Biomembr.* **26**:273–289.

Brody, S., and Mikolajczyk, S., 1988, *Neurospora* mitochondria contain an acyl-carrier protein, *Eur. J. Biochem.* **173**:353–359.

Brody, S., Oh, C., Hoja, U., and Schweizer, E., 1997, Mitochondrial acyl-carrier protein is involved in lipoic acid synthesis in *Saccharomyces cerevisiae*, *FEBS Lett.* **408**:217–220.

Buchanan, S., and Walker, J. E., 1996, Large-scale chromatographic purification of F_1F_0-ATPase and complex I from bovine heart mitochondria, *Biochem. J.* **318**:343–349.

Bult, C. J., White, O., Olsen, G. J., Zhou, L., Fleischmann, R. D., Sutton, G. G., Blake, J. A., Fitzgerald, L. M., Clayton, R. A., Gocayne, J. D., Kerlavage, A. R., Dougherty, B. A., Tomb, J.-F., Adams, M. D., Reich, C. I., Overbeek, R., Kirkness, E. F., Weinstock, K. G., Merrick, J. M., Glodek, A., Scott, J. L., Geoghagen, N. S. M., Weidman, J. F., Fuhrmann, J. L., Presley, E. A., Nguyen, D., Utterback, T. R., Kelley, J. M., Peterson, J. D., Sadow, P. W., Hanna, M. C., Cotton, M. D., Hurst, M. A., Roberts, K. M., Kaine, B. P., Borodovsky, M., Klenk, H.-P., Fraser, C. M., Smith, H. O., Woese, C. R., and Venter, J. C., 1996, Complete genome sequence of the methanogenic archeon, *Methanococcus jannaschii*, *Science* **273**:1058–1073.

Burbaev, D. S., Moroz, I. A., Kotlyar, A. B., Sled', V. D., and Vinogradov, A. D., 1989, Ubisemiquinone in the NADH:ubiquinone reductase region of the mitochondrial respiratory chain, *FEBS Lett.* **254**:47–51.

Büschges, R., Bahrenberg, G., Zimmermann, M., and Wolf, K., 1994, NADH:ubiquinone oxido-reductase in obligate aerobic yeasts, *Yeast* **10**:475–479.

Chen, S., and Guillory, R. J., 1981, Studies on the interaction of arylazido-β alanyl NAD$^+$ with the mitochondrial NADH dehydrogenase, *J. Biol. Chem.* **256**:8318–8323.

Chomyn, A., Mariottini, P., Cleeter, M. W. J., Ragan, C. I., Matsuno-Yagi, A., Hatefi, Y., Doolittle, R. F., and Attardi, G., 1985, Six unidentified reading frames of human mitochondrial DNA encode components of the respiratory-chain NADH dehydrogenase, *Nature* **314**:592–597.

Cooper, J. M., Mann, V. M., Krige, D., and Schapira, A. H. V., 1992, Human mitochondrial complex I dysfunction, *Biochim. Biophys. Acta* **1101**:198–203.

Daum, G., 1985, Lipids of mitochondria, *Biochim. Biophys. Acta* **822**:1–42.

Degli Esposti, M., Ghelli, A., Ratta, M., Cortes, D., and Estornell, E., 1994, Natural substances (acetogenins) from the family *Annonaceae* are powerful inhibitors of mitochondrial NADH dehydrogenase (complex I), *Biochem. J.* **301**:161–167.

Degli Esposti, M. D., Ngo, A., Mcmullen, G. L., Ghelli, A., Sparla, F., Benelli, B., Ratta, M., and Linnane, A. W., 1996, The specificity of mitochondrial complex I for ubiquinones, *Biochem. J.* **313**:327–334.

de Jong, A. M. P., and Albracht, S. P. J., 1994, Ubisemiquinone as obligatory intermediates in the electron transfer from NADH to ubiquinone, *Eur. J. Biochem.* **222**:975–982.

de Jong, A. M. P., Kotlyar, A. B., and Albracht, S. P. J., 1994, Energy-induced structural changes in NADH-Q oxidoreductase of the mitochondrial respiratory chain, *Biochim. Biophys. Acta* **1186**: 163–171.

De Jonge, P. C., and Westerhoff, H. V., 1982, The proton-per-electron stoichiometry of "site 1" of oxidative phosphorylation at high protonmotive force is close to 1.5, *Biochem. J.* **204**:515–523.

Deng, P. S. K., Hatefi, Y., and Chen, S., 1990, *N*-Arylazido-β-alanyl-NAD$^+$, a new NAD$^+$ photo-affinity analogue. Synthesis and labeling of mitochondrial NADH dehydrogenase, *Biochemistry* **29**:1094–1098.

de Vries, S., and Grivell, L. A., 1988, Purification and characterization of a rotenone-insensitive NADH:Q$_6$ oxidoreductase from mitochondria of *Saccharomyces cerevisiae*, *Eur. J. Biochem.* **176**:377–384.

Duarte, M., Sousa, R., and Videira, A., 1995, Inactivation of genes encoding subunits of the peripheral and membrane arms of *Neurospora* mitochondrial complex I and effects on enzyme assembly, *Genetics* **139**:1211–1221.

Duarte, M., Finel, M., and Videira, A., 1996, Primary structure of a ferredoxin-like iron–sulfur subunit of complex I from *Neurospora crassa*, *Biochim. Biophys. Acta* **1275**:151–153.

Duarte, M., Schulte, U., and Videira, A., 1997, Identification of the TYKY homologous subunit of complex I from *Neurospora crassa*, *Biochim. Biophys. Acta* **1322**:237–241.

Dupuis, A., 1992, Identification of two genes of *Rhodobacter capsulatus* coding for proteins homologous to the ND1 and 23 kDa subunits of the mitochondrial complex I, *FEBS Lett.* **301**:215–218.

Dupuis, A., Skehel, J. M., and Walker, J. E., 1991a, NADH:ubiquinone reductase from bovine mitochondria: Complementary DNA sequence of a 19 kDa cysteine rich subunit, *Biochem. J.* **277**:11–15.

Dupuis, A., Skehel, J. M., and Walker, J. E., 1991b, Plant chloroplast genomes encode a homologue of a nuclear coded iron–sulfur protein subunit of bovine mitochondrial complex I, *Biochemistry* **30**:2954–2960.

Dupuis, A., Peinnequin, A., Chevallet, M., Lunardi, J., Darrouzet, E., Pierrard, B., Procaccio, V., and Issartel, J. P., 1996a, Identification of five *Rhodobacter capsulatus* genes encoding the equivalent of ND subunits of the mitochondrial NADH-ubiquinone oxidoreductase, *Gene* **167**:99–104.

Dupuis, A., Chevallet, M., Claustre, P., Darrouzet, E., Duborjal, H., Issartel, J. P., Lunardi, J.,

Pierrard, B., Procaccio, V., van Belzen, R., and Albracht, S. P. J., 1996b, Mutagenesis study of the NADH-CoQ reductase of *Rhodobacter capsulatus*, *EBEC Short Rep.* **9**:23.

Earley, F. G. P., and Ragan, C. I., 1984, Photoaffinity labeling of mitochondrial NADH dehydrogenase with arylazidomorphigenin, an analogue of rotenone, *Biochem. J.* **224**:525–534.

Earley, F. G. P., Patel, S. D., Ragan, C. I., and Attardi, G., 1987, Photolabeling of a mitochondrially encoded subunit of NADH dehydrogenase with [³H]dihydrorotenone, *FEBS Lett.* **219**:108–113.

Ellersiek, U., and Steinmüller, K., 1992, Cloning and transcription analysis of the *ndh*(A-I-G-E) gene cluster and the *ndh*D gene of the cyanobacterium *Synechocystis* sp. PCC6803, *Plant Mol. Biol.* **20**:1097–1110.

Fato, R., Estornell, E., Dibernardo, S., Pallotti, F., Castelli, G. P., and Lenaz, G., 1996, Steady-state kinetics of the reduction of coenzyme Q analogs by complex I (NADH-ubiquinone oxidoreductase) in bovine mitochondria and submitochondrial particles, *Biochemistry* **35**:2705–2716.

Fearnley, I. M., and Walker, J. E., 1992, Conservation of sequences of subunits of mitochondrial complex I and their relationships with other proteins, *Biochim. Biophys. Acta* **1140**:105–134.

Fearnley, I. M., Finel, M., Skehel, J. M., and Walker, J. E., 1991, NADH:ubiquinone oxidoreductase from bovine heart mitochondria: cDNA sequences of the import precursors of the nuclear coded 39 kDa and 42 kDa subunits, *Biochem. J.* **278**:821–829.

Fecke, W., Sled, V. D., Ohnishi, T., and Weiss, H., 1994, Disruption of the gene encoding the NADH-binding subunit of NADH:ubiquinone oxidoreductase in *Neurospora crassa*. Formation of a partially assembled enzyme without FMN and the iron–sulphur cluster N3, *Eur. J. Biochem.* **220**:551–558.

Finel, M., 1993, The proton-translocating NADH:ubiquinone oxidoreductase: A discussion of selected topics, *J. Bioenerg. Biomembr.* **25**:357–367.

Finel, M., Skehel, J. M., Albracht, S. P. J., Fearnley, I. M., and Walker, J. E., 1992, Resolution of NADH:ubiquinone oxidoreductase from bovine heart mitochondria into two subcomplexes one of which contains the redox centers of the enzyme, *Biochemistry* **31**:11425–11434.

Finel, M., Majander, A. S., Tyynela, J., de Jong, A. M. P., Albracht, S. P. J., and Wikström, M., 1994, Isolation and characterisation of subcomplexes of the mitochondrial NADH-ubiquinone oxidoreductase (complex I), *Eur. J. Biochem.* **226**:237–242.

Friedrich, B., and Schwarz, E., 1993, Molecular biology of hydrogen utilization in aerobic chemolithotrophs, *Annu. Rev. Microbiol.* **47**:351–383.

Friedrich, T., and Weiss, H., 1997, Modular evolution of the respiratory NADH:ubiquinone oxidoreductase and the origin of its modules, *J. Theort. Biol.* **187**:529–541.

Friedrich, T., Hofhaus, G., Ise, W., Nehls, U., Schmitz, B., and Weiss, H., 1989, A small isoform of NADH:ubiquinone oxidoreductase (complex I) without mitochondrially synthesized subunits is made in chloramphenicol treated *Neurospora crassa*, *Eur. J. Biochem.* **180**:173–180.

Friedrich, T., Strohdeicher, M., Hofhaus, G., Preis, D., Sahm, H. H., and Weiss, H., 1990, The same domain motif for ubiquinone reduction in mitochondrial or chloroplast NADH dehydrogenase and bacterial glucose dehydrogenase, *FEBS Lett.* **265**:37–40.

Friedrich, T., van Heek, P., Leif, H., Ohnishi, T., Forche, E., Kunze, B., Janssen, R., Trowitzsch-Kienast, W., Höfle, G., Reichenbach, H., and Weiss, H., 1994, Two binding sites of inhibitors in NADH:ubiquinone oxidoreductase (complex I), *Eur. J. Biochem.* **219**:691–698.

Friedrich, T., Steinmüller, K., and Weiss, H., 1995, The proton-pumping respiratory complex I of bacteria and mitochondria and its homologue in chloroplasts, *FEBS Lett.* **367**:107–111.

Gabler, L., Herz, U., Liddell, A., Leaver, C. J., Schröder, W., Brennicke, A., and Grohmann, L., 1994, The 42.5 kDa subunit of the NADH-ubiquinone oxidoreductase (complex I) in higher plants is encoded by the mitochondrial *nad7* gene, *Mol. Gen. Genet.* **244**:33–40.

Galante, Y. M., and Hatefi, Y., 1979, Purification and molecular properties of mitochondrial NADH dehydrogenase, *Arch. Biochem. Biophys.* **192**:559–568.

Guedeney, G., Corneille, S., Cuine, S., and Peltier, G., 1996, Evidence for an association of *ndh* B, *ndh* J gene products and ferredoxin-NADP-reductase as components of a chloroplastic NAD(P)H dehydrogenase complex. *FEBS Lett.* **378**:277–280.

Guenebaut, V., Mills, D., Weiss, H., and Leonard, K. R., 1996, Three dimensional structure of NADH-dehydrogenase from *Neurospora crassa* by electron microscopy and conical tilt reconstruction, *J. Mol. Biol.* **265**:409–418.

Harington, A., Herbert, C. J., Tung, B., Getz, G. S., and Slonimski, P. P., 1993, Identification of a new nuclear gene (CEM1) encoding a protein homologous to a β-keto-acyl synthase which is essential for mitochondrial respiration in *Saccharomyces cerevisiae*, *Mol. Microbiol.* **9**: 545–555.

Harkness, T. A. A., Rothery, R. A., Weiner, J. H., Werner, S., Azevedo, J. E., Videira, A., and Nargang, F. E., 1994, Disruption of the gene encoding the 78-kilodalton subunit of the peripheral arm of complex I in *Neurospora crassa* by repeat induced point mutation (RIP), *Curr. Genet.* **27**:339–350.

Hatefi, Y., Haavik, A. G., and Jurtshuk, P., 1960, Studies on the electron transfer system XXXII. Reduction of coenzyme Q by DPNH, *Biochem. Biophys. Res. Commun.* **3**:281–284.

Hatefi, Y., Haavik, A. G., and Griffiths, D. E., 1962, Studies on the electron transfer system. XL. Preparation and properties of mitochondrial DPNH-coenzyme Q reductase, *J. Biol. Chem.* **237**:1676–1680.

Heinrich, H., and Werner, S., 1992, Identification of the ubiquinone-binding site of NADH:ubiquinone oxidoreductase (complex I) from *Neurospora crassa*, *Biochemistry* **31**:11413–11419.

Heller, W., and Fokmann, G., 1993, Biosynthesis of flavonoids, in: *The Flavonoids: Advances in Research since 1986* (J. B. Harborne, ed.), pp. 499–535, Chapman & Hall, London.

Herz, U., Schröder, W., Liddel, A., Leaver, C. J., Brennicke, A., and Grohmann, L., 1994, Purification of the NADH:ubiquinone oxidoreductase (complex I) of the respiratory chain from the inner mitochondrial membrane of *Solanum tuberosum*, *J. Biol. Chem.* **269**:2263–2269.

Hofhaus, G., Weiss, H., and Leonard, K., 1991, Electron microscopic analysis of the peripheral and membrane parts of mitochondrial NADH dehydrogenase (complex I), *J. Mol. Biol.* **221**:1027–1043.

Horinouchi, S., Ishizuka, H., and Beppu, T., 1991, Cloning, nucleotide sequence, and transcriptional analysis of the NAD(P)-dependent cholesterol dehydrogenase gene from a *Nocardia* sp. and its hyperexpression in *Streptomyces* spp., *Appl. Environ. Microbiol.* **57**:1386–1393.

Ingledew, W. J., and Ohnishi, T., 1980, An analysis of some thermodynamic properties of iron–sulphur centres in site I of mitochondria, *Biochem. J.* **186**:111–117.

Ise, W., Haiker, H., and Weiss, H., 1985, Mitochondrial translation of subunits of the rotenone-sensitive NADH:ubiquinone reductase in *Neurospora crassa*, *EMBO J.* **4**:2075–2080.

Iwata, S., Ostermeier, C., Ludwig, B., and Michel, H., 1995, Structure at 2.8 Å resolution of cytochrome *c* oxidase from *Paracoccus denitrificans*, *Nature* **376**:660–669.

Kaneko, T., Tanaka, A., Sato, S., Kotani, H., Sazuka, T., Miyajima, N., Sugiura, M., and Tabata, S., 1995, Sequence analysis of the genome of the unicellular cyanobacterium *Synechocystis* sp. strain PCC6803. I. Sequence features in the 1 Mb region from map positions 64% to 92% of the genome, *DNA Res.* **2**:153–166.

Kean, E. A., Gutman, M., and Singer, T. P., 1971, Studies on the respiratory chain-linked nicotinamide adenine dinucleotide dehydrogenase, XXII: Rhein, a competitive inhibitor of the dehydrogenase, *J. Biol. Chem.* **246**:2346–2353.

Kikuno, R., and Miyata, T., 1985, Sequence homologies among mitochondrial DNA-coded URF2, URF4 and URF5, *FEBS Lett.* **189**:85–88.

Kotlyar, A. B., and Vinogradov, A. D., 1990, Slow active/inactive transition of the mitochondrial NADH ubiquinone reductase, *Biochim. Biophys. Acta* **1019**:151–158.

Kotlyar, A. B., Sled, V. D., Burbaev, D. S., Moroz, I. A., and Vinogradov, A. D., 1990, Coupling site I and the rotenone-sensitive ubisemiquinone in tightly coupled submitochondrial particles, *FEBS Lett.* **264:**17–20.

Krishnamoorthy, G., and Hinkle, P. C., 1988, Studies on the electron transfer pathway, topography of iron–sulphur centres, and site of coupling in NADH-Q oxidoreductase, *J. Biol. Chem.* **263:** 17566–17575.

Kubicki, A., Funk, E., Westhoff, P., and Steinmüller, K., 1996, Differential expression of plastome-encoded *ndh* genes in mesophyll and bundle-sheath chloroplasts of the C-4 plant *Sorghum bicolor* indicates that the complex I-homologous NAD(P)H-plastoquinone oxidoreductase is involved in cyclic electron transport, *Planta* **199:**276–281.

Kubo, T., Mikami, T., and Kinoshita, T., 1993, The sugar beet mitochondrial genome contains an ORF sharing sequence homology with the gene for the 30 kDa subunit of bovine mito-chondrial complex I, *Mol. Gen. Genet.* **241:**479–482.

Laterme, S., and Boutry, M., 1993, Purification and preliminary characterization of mitochondrial complex I (NADH:ubiquinone reductase) from the broad bean (*Vicia faba* L.), *Plant Physiol.* **102:**435–443.

Leif, H., Sled, V. D., Ohnishi, T., Weiss, H., and Friedrich, T., 1995, Isolation and characterization of the proton-translocating NADH:ubiquinone oxidoreductase from *Escherichia coli, Eur. J. Biochem.* **230:**538–548.

Lemasters, J. J., Grunwald, R., and Emaus, R. K., 1984, Thermodynamic limits to the ATP/site stoichiometries of oxidative phosphorylation by rat liver mitochondria, *J. Biol. Chem.* **259:** 3058–3063.

Leonard, K., Haiker, H., and Weiss, H., 1987, Three-dimensional structure of NADH:ubiquinone reductase (complex I) from *Neurospora* mitochondria determined by electron microscopy of membrane crystals, *J. Mol. Biol.* **194:**277–286.

Majander, A., Finel, M., and Wikström, M., 1994, Diphenyleneiodonium inhibits reduction of iron–sulfur clusters in the mitochondrial NADH-ubiquinone oxidoreductase (complex I), *J. Biol. Chem.* **269:**21037–21042.

Malki, S., Saimmaime, I., de Luca, G., Rousset, M., Dermoun, Z., and Belaich, J. P., 1995, Characterization of an operon encoding an NADP-reducing hydrogenase in *Desulfovibrio fructosovorans, J. Bacteriol.* **177:**2628–2636.

Matsushita, K., Ohnishi, T., and Kaback, H. R., 1987, NADH-ubiquinone oxidoreductases of the *Escherichia coli* aerobic respiratory chain, *Biochemistry* **26:**7732–7737.

Matsushita, K., Shinagawa, E., Adachi, O., and Ameyama, M., 1989, Reactivity with ubiquinone of quinoprotein D-glucose dehydrogenase from *Gluconobacter suboxydans, J. Biochem.* **105:** 633–637.

Merson-Davies, L. A., and Cundliffe, E., 1994, Analysis of five tylosin biosynthetic genes from the tylIBA region of the *Streptomyces fradiae* genome, *Mol. Microbiol.* **13:**349–355.

Mi, H., Endo, T., Schreiber, U., and Asada, K., 1992, Donation of electrons from cytosolic components to the intersystem chain in the cyanobacterium *Synechococcus* sp. PCC7002 as determined by the reduction of P700$^+$, *Plant Cell Physiol.* **33:**1099–1105.

Mikolajczyk, S., and Brody, S., 1990, *De novo* fatty acid synthesis mediated by acyl-carrier protein in *Neurospora crassa* mitochondria, *Eur. J. Biochem.* **187:**431–437.

Mizuno, Y., Ohta, S., Tanaka, M., Takamiya, S., Suzuki, K., Sato, T., Oya, H., Ozawa, T., and Kagawa, Y., 1989, Deficiencies in complex I subunits of the respiratory chain in Parkinson's disease, *Biochem. Biophys. Res. Commun.* **163:**1450–1455.

Møller, I.M., and Palmer, J. M., 1982, Direct evidence for the presence of a rotenone resistant NADH dehydrogenase on the inner surface of the inner membrane of plant mitochondria, *Physiol. Plant* **54:**267–274.

Nehls, U., Friedrich, T., Schmiede, A., Ohnishi, T., and Weiss, H., 1992, Characterization of assembly intermediates of NADH:ubiquinone oxidoreductase (complex I) accumulated in *Neurospora* mitochondria by gene disruption, *J. Mol. Biol.* **227**:1032–1042.

Nosek, J., and Fukuhara, H., 1994, NADH dehydrogenase subunit genes in the mitochondrial DNA of yeast, *J. Bacteriol.* **176**:5622–5630.

Ogawa, T., 1991, A gene homologous to the subunit-2 gene of NADH dehydrogenase is essential to inorganic carbon transport of Synechocystis PCC6803, *Proc. Natl. Acad. Sci. USA* **88**: 4275–4279.

Ohnishi, T., 1979, Mitochondrial iron–sulphur flavodehydrogenases, in: *Membrane Proteins in Energy Transduction* (R. A. Capaldi, ed.), pp. 1–87, Marcel Dekker, New York.

Ohnishi, T., 1993, NADH-quinone oxidoreductase, the most complex complex, *J. Bioenerg. Biomembr.* **25**:325–329.

Ohnishi, T., and Salerno, J. C., 1982, Iron–sulphur clusters in the mitochondrial electron transport chain, in: *Iron–Sulphur Proteins* (T. G. Spiro, ed.), pp. 285–327, John-Wiley & Sons, New York.

Ohnishi, T., Ragan, C. I., and Hatefi, Y., 1985, EPR studies of iron–sulfur clusters in isolated subunits and subfractions of NADH-ubiquinone oxidoreductase, *J. Biol. Chem.* **260**:2782–2788.

Ohyama, K., Fukuzawa, H., Kohchi, T., Shirai, H., Sano, T., Sano, S., Umesono, K., Shiki, Y., Takeuchi, M., Chang, Z., Aota, S.-i., Inokuchi, H., and Ozaki, H., 1986, Chloroplast gene organization deduced from complete liverwort *Marchantia polymorpha* chloroplast DNA, *Nature* **322**:572–574.

Paiva, N. L., Sun, Y., Dixon, R. A., VanEtten, H. D., and Hrazdina, G., 1994, Molecular cloning of isoflavone reductase from pea (*Pisum sativum* L): Evidence for a 3R-isoflavanone intermediate in (+)-pisatin biosynthesis, *Arch. Biochem. Biophys.* **312**:501–510.

Pilkington, S. J., and Walker, J. E., 1989, Mitochondrial NADH-ubiquinone reductase: Complementary DNA sequences of import precursors of the bovine and human 24 kDa subunit, *Biochemistry* **28**:3257–3264.

Pilkington, S. J., Skehel, J. M., Gennis, R. B., and Walker, J. E., 1991, Relationship between mitochondrial NADH-ubiquinone reductase and a bacterial NAD-reducing hydrogenase, *Biochemistry* **30**:2166–2175.

Przybyla, A. E., Robbins, J., Menon, N., and Peck, H. D., 1992, Structure–function relationships among the nickel-containing hydrogenases, *FEMS Microbiol. Rev.* **88**:109–136.

Ragan, C. I., 1987, Structure of NADH-ubiquinone reductase (complex I), *Curr. Top. Bioenerg.* **15**:1–36.

Ragan, C. I., 1990, Structure and function of an archetypal respiratory chain complex: NADH-ubiquinone reductase, *Biochem. Soc. Trans.* **18**:515–516.

Ragan, C. I., Galante, Y. M., Hatefi, Y., and Ohnishi, T., 1982, Resolution of mitochondrial NADH dehydrogenase and the isolation of two iron–sulfur proteins, *Biochemistry* **21**:590–594.

Rasmussen, A. G., Mendel-Hartvig, J., Møller, I., and Wiskich, J. T., 1994, Isolation of the rotenone-sensitive NADH-ubiquinone reductase (complex I) from red beet mitochondria, *Physiol. Plant* **90**:607–615.

Reith, M. E., and Munholland, J., 1995, Complete nucleotide sequence of the *Porphyra purpurea* chloroplast genome, *Plant Mol. Biol. Rep.* **13**:333–335.

Rheaume, E., Lachance, Y., Zhao, H. F., Breton, N., Dumont, M., de Launoit, Y., Trudel, C., Luu-The, V., Simard, J., and Labrie, F., 1991, Structure and expression of a new complementary DNA encoding the almost exclusive 3 beta-hydroxysteroid dehydrogenase/delta 5-delta 4-isomerase in human adrenals and gonads, *Mol. Endocrinol.* **5**:1147–1157.

Röhlen, D. A., Hoffmann, J., van der Pas, J. C., Nehls, U., Preis, D., Sackmann, U., and Weiss, H., 1991, Relationship between a subunit of NADH dehydrogenase (complex I) and a protein family including subunits of cytochrome reductase and processing protease from mitochondria, *FEBS Lett.* **278**:75–78.

Runswick, M. J., Fearnley, I. M., Skehel, J. M., and Walker, J. E., 1991, Presence of an acyl-carrier protein in NADH:ubiquinone oxidoreductase from bovine heart mitochondria, *FEBS Lett.* **286:**121–124.

Rydström, J., Montelius, J., Bäckström, D., and Ernster, L., 1978, The mechanism of oxidation of reduced nicotinamide dinucleotide phosphate by submitochondrial particles from beef heart, *Biochim. Biophys. Acta* **501:**370–380.

Sackmann, U., Zensen, R., Röhlen, D., Jahnke, U., and Weiss, H., 1991, The acyl-carrier protein in *Neurospora crassa* mitochondria is a subunit of NADH:ubiquinone reductase (complex I), *Eur. J. Biochem.* **200:**463–469.

Sakamoto, K., Miyoshi, H., Matsushita, K., Nakagawa, M., Ikeda, J., Ohshima, M., Dachi, O., Akagi, T., and Iwamura, H., 1996, Comparison of the structural features of ubiquinone reduction sites between glucose dehydrogenase in *Escherichia coli* and bovine heart mitochondrial complex I, *Eur. J. Biochem.* **237:**128–135.

Salerno, J. C., Ohnishi, T., Lim, J., Widger, R., and King, T. E., 1977, Spin coupling between electron carriers in the dehydrogenase segments of the respiratory chain, *Biochem. Biophys. Res. Commun.* **75:**618–624.

Sauter, M., Böhm, R., and Böck, A., 1992, Mutational analysis of the operon (*hyc*) determining hydrogenase 3 formation in *Escherichia coli*, *Mol. Microbiol.* **6:**1523–1532.

Schapira, A. H. V., Cooper, J. M., Dexter, D., Clark, J. B., Jenner, P., and Marsden, C. D., 1990, Mitochondrial complex I deficiency in Parkinson's disease, *J. Neurochem.* **54:**823–827.

Schmidt, M., Friedrich, T., Wallrath, J., Ohnishi, T., and Weiss, H., 1992, Accumulation of the preassembled membrane arm of NADH:ubiquinone oxidoreductase in mitochondria of manganese-limited grown *Neurospora crassa*, *FEBS Lett.* **313:**8–11.

Schmitz, O., Boison, G., Hilscher, R., Hundshagen, B., Zimmer, W., Lottspeich, F., and Bothe, H., 1995, Molecular biological analysis of a bi-directional hydrogenase from cyanobacteria, *Eur. J. Biochem.* **233:**266–276.

Schneider, R., Massow, M., Lisowsky, T., and Weiss, H., 1995, Different respiratory-defective phenotypes of *Neurospora crassa* and *Saccharomyces cerevisiae* after inactivation of the gene encoding the mitochondrial acyl carrier protein, *Curr. Genet.* **29:**10–17.

Schulte, U., and Weiss, H., 1995, Generation and characterization of NADH:ubiquinone oxido-reductase mutants in *Neurospora crassa*, *Methods Enzymol.* **260:**3–14.

Schulte, U., Fecke, W., Krüll, C., Nehls, U., Schmiede, A., Schneider, R., Ohnishi, T., and Weiss, H., 1994, *In vivo* dissection of the mitochondrial respiratory NADH:ubiquinone oxidoreduc-tase (complex I), *Biochim. Biophys. Acta* **1187:**121–124.

Shimada, H., and Sugiura, M., 1991, Fine structural features of the chloroplast genome: Compari-son of the sequenced chloroplast genomes, *Nucleic Acids Res.* **19:**983–995.

Shinozaki, K., Ohme, M., Tanaka, M., Wakasugi, T., Hayashida, N., Matsubayashi, T., Zaita, N., Chunwongse, J., Obokata, J., Yamaguchi-Shinozaki, K., Ohto, C., Torazawa, K., Meng, B. Y., Sugita, M., Deno, H., Kamogashira, T., Yamada, K., Kusuda, J., Takaiwa, F., Kato, A., Tohdoh, N., Shimada, H., and Sugiura, M., 1986, The complete nucleotide sequence of the tobacco chloroplast genome: its organization and expression, *EMBO J.* **5:**2043–2049.

Shintani, D. K., and Ohlrogge, J. B., 1994, The characterization of a mitochondrial acyl-carrier protein isoform isolated from *Arabidopsis thaliana*, *Plant Physiol.* **104:**1221–1229.

Shoffner, J. M., Watts, R. L., Juncos, J. L., Torroni, A., and Wallace, D. C., 1991, Mitochondrial oxidative phosphorylation defects in Parkinson's disease, *Ann. Neurol.* **30:**332–339.

Singer, T. P., and Ramsay, R. R., 1992, NADH:ubiquinone oxidoreductase, in: *Molecular Mecha-nisms in Bioenergetics* (L. Ernster, ed.), pp. 145–162, Elsevier, Amsterdam.

Singer, T. P., and Ramsay, R. R., 1994, The reaction sites of rotenone and ubiquinone with mitochondrial NADH dehydrogenase, *Biochim. Biophys. Acta* **1187:**198–202.

Sled', V. D., Friedrich, T., Leif, H., Weiss, H., Meinhardt, S. W., Fukumori, Y., Calhoun, M. W.,

Gennis, R. B., and Ohnishi, T., 1993, Bacterial NADH:quinone oxidoreductase: Iron–sulfur clusters and related problems, *J. Bioenerg. Biomembr.* **25:**347–356.

Sled', V. D., Rudnitzky, N. I., Hatefi, Y., and Ohnishi, T., 1994, Thermodynamic analysis of flavin in mitochondrial NADH:ubiquinone oxidoreductase (complex I), *Biochemistry* **33:**10069–10075.

Suzuki, H., and King, T. E., 1983, Evidence of an ubisemiquinone radical(s) from the NADH-ubiquinone reductase of the mitochondrial respiratory chain, *J. Biol. Chem.* **258:**352–358.

Tran-Betcke, A., Warnecke, U., Böcker, C., Zaborosch, C., and Friedrich, B., 1990, Cloning and nucleotide sequences of the genes for the subunits of NAD-reducing hydrogenase of *Alcaligenes eutrophus* H16, *J. Bacteriol.* **172:**2920–2929.

Tsukihara, T., Aoyama, H., Yamashita, E., Tomizaki, T., Yamaguchi, H., Shinzawaitoh, K., Nakashima, R., Yaono, R., and Yoshikawa, S., 1996, The whole structure of the 13-subunit oxidized cytochrome c oxidase at 2.8 Angstrom, *Science* **272:**1136–1144.

Tuschen, G., Sackmann, U., Nehls, U., Haiker, H., Buse, G., and Weiss, H., 1990, Assembly of NADH:ubiquinone reductase (complex I) in *Neurospora* mitochondria: independent pathways of nuclear encoded and mitochondrially encoded subunits, *J. Mol. Biol.* **213:**845–857.

van Belzen, R., and Albracht, S. P. J., 1989, The pathway of electron transfer in NADH:Q oxidoreductase, *Biochim. Biophys. Acta* **974:**311–320.

van Belzen, R., van Gaalen, M. C. M., Cuypers, P. A., and Albracht, S. P. J., 1990, New evidence for the dimeric nature of NADH:Q oxidoreductase in bovine heart submitochondrial particles, *Biochim. Biophys. Acta* **1017:**152–159.

van Belzen, R., de Jong, A. M. P., and Albracht, S. J. P., 1992, On the stoichiometry of the iron–sulphur clusters in mitochondrial NADH:Q oxidoreductase, *Eur. J. Biochem.* **209:**1019–1022.

van Belzen, R., Kotlyar, A. B., Dunham, W. R., and Albracht, S. P. J., 1996, EPR-spectroscopic studies on the iron–sulfur clusters 2 of complex I in coupled submitochondrial particles, *EBEC Short Rep.* **9:**24.

Videira, A., and Azevedo, J. E., 1994, Two nuclear-coded subunits of mitochondrial complex I are similar to different domains of a bacterial formate hydrogenlyase subunit, *Int. J. Biochem.* **26:**1391–1393.

Vinogradov, A. D., 1993, Kinetics, control, and mechanism of ubiquinone reduction by the mammalian respiratory chain-linked NADH-ubiquinone reductase, *J. Bioenerg. Biomembr.* **25:**367–376.

Vinogradov, A. D., Sled', V. D., Burbaev, D. S., Grivennikova, V. G., Moroz, I. A., and Ohnishi, T., 1995, Energy-dependent complex I-associated ubisemiquinones in submitochondrial particles, *FEBS Lett.* **370:**83–87.

Voelker, D. R., 1991, Organelle biogenesis and intracellular lipid transport in eukaryotes, *Microbiol. Rev.* **55:**543–560.

Volbeda, A., Charon, M. H., Piras, C., Hatchikian, E. C., Frey, M., and Fontecilla-Camps, J. C., 1995, Crystal structure of the nickel–iron hydrogenase from *Desulfovibrio gigas*, *Nature* **373:**580–587.

Voordouw, G., 1992, Evolution of hydrogenase genes, *Adv. Inorg. Chem.* **48:**397–422.

Walker, J. E., 1992, The NADH:ubiquinone oxidoreductase (complex I) of respiratory chains, *Q. Rev. Biophys.* **25:**253–324.

Walker, J. E., Arizmendi, J. M., Dupuis, A., Fearnley, I. M., Finel, M., Medd, S. M., Pilkington, S. J., Runswick, M. J., and Skehel, J. M., 1992, Sequences of twenty subunits of NADH:ubiquinone oxidoreductase from bovine heart mitochondria. Application of a novel strategy for sequencing proteins using the polymerase chain reaction, *J. Mol. Biol.* **226:**1051–1072.

Wang, D. C., Meinhardt, S. W., Sackmann, U., Weiss, H., and Ohnishi, T., 1991, The iron–sulfur clusters of two related forms of mitochondrial NADH:ubiquinone oxidoreductase made by *Neurospora crassa*, *Eur. J. Biochem.* **197:**257–264.

Weidner, U., Geier, S., Ptock, A., Friedrich, T., Leif, H., and Weiss, H., 1993, The gene locus of the proton translocating NADH:ubiquinone oxidoreductase in *Escherichia coli*, *J. Mol. Biol.* **233:** 109–122.

Weiss, H., and Friedrich, T., 1991, Redox-linked proton translocation by NADH-ubiquinone reductase (complex I), *J. Bioenerg. Biomembr.* **23:**743–754.

Weiss, H., Friedrich, T., Hofhaus, G., and Preis, D., 1991, The respiratory chain NADH dehydrogenase (complex I) of mitochondria, *Eur. J. Biochem.* **197:**563–576.

Wierenga, R. K., Terpstra, P., and Hol, W. G. J., 1985, Prediction of the occurrence of the ADP-binding βαβ-fold in proteins, using an amino acid sequence fingerprint, *J. Mol. Biol.* **187:** 101–107.

Xu, X., Matsuno-Yagi, A., and Yagi, T., 1993, DNA sequencing of the seven remaining genes of the gene cluster encoding the energy-transducing NADH-quinone oxidoreductase of *Paracoccus denitrificans*, *Biochemistry* **23:**968–981.

Yagi, T., 1986, Purification and characterization of NADH dehydrogenase complex from *Paracoccus denitrificans*, *Arch. Biochem. Biophys.* **250:**302–311.

Yagi, T., and Dinh, T. M., 1990, Identification of the NADH-binding subunit of NADH-ubiquinone oxidoreductase of *Paracoccus denitrificans*, *Biochemistry* **29:**5515–5520.

Yagi, T., Hon-nami, K., and Ohnishi, T., 1988, Purification and characterization of two types of NADH-quinone reductase from *Thermus thermophilus* HB-8, *Biochemistry* **27:**2008–2013.

Yagi, T., Xu, X., and Matsuno-Yagi, A., 1992, The energy-transducing NADH-quinone oxidoreductase (NADH-1) of *Paracoccus denitrificans*, *Biochim. Biophys. Acta* **1101:**181–184.

Yamada, M., Sumi, K., Matsushita, K., Adachi, O., and Yamada, Y., 1993, Topological analysis of quinoprotein glucose dehydrogenase in *Escherichia coli* and its ubiquinone-binding site, *J. Biol. Chem.* **268:**12812–12817.

Yano, T., Sled', V. D., Ohnishi, T., and Yagi, T., 1994, Identification of amino acid residues associated with the [2Fe-2S] cluster of the 25 kDa (Nqo2) subunit of the proton-translocating NADH-ubiquinone oxidoreductase of *Paracoccus denitrificans*, *FEBS Lett.* **354:**160–164.

Yano, T., Sled', V. D., Ohnishi, T., and Yagi, T., 1996, Expression and characterization of the flavoprotein subcomplex composed of 50-kDa (NQO1) and 25-kDa (NQO2) subunits of the proton-translocating NADH-quinone oxidoreductase of *Paracoccus denitrificans*, *J. Biol. Chem.* **271:**5907–5913.

Young, I. G., Rogers, B. L., Campbell, H. D., Jaworowski, A., and Shaw, D. C., 1981, Nucleotide sequence coding for the respiratory NADH dehydrogenase of *Escherichia coli*, *Eur. J. Biochem.* **116:**165–170.

Yu, C.-A., Xia, J.-Z., Kachurin, A. M., Yu, L., Xia, D., Kim, H., and Deisenhofer, J., 1996, Crystallization and preliminary structure of beef heart mitochondrial cytochrome-bc_1 complex, *Biochim. Biophys. Acta* **1275:**47–53.

Zensen, R., Husmann, H., Schneider, R., Friedrich, T., Peine, T., and Weiss, H., 1992, *De novo* synthesis and de-saturation of fatty acids at the mitochondrial acyl-carrier protein, a subunit of NADH:ubiquinone oxidoreductase in *Neurospora crassa*, *FEBS Lett.* **310:**179–182.

CHAPTER 15

Structure of F_1-ATPase and the Mechanism of ATP Synthesis– Hydrolysis

Mario A. Bianchet, Peter L. Pedersen, and L. Mario Amzel

1. INTRODUCTION

ATP synthesis—the phosphorylation of ADP by P_i in mitochondria, chloroplast, and bacteria—is a fundamental and complex biochemical pathway. The reaction involves the utilization of the proton electrochemical gradient produced by either the oxidation of substrates or the utilization of light quanta, for the generation of ATP levels up to 10^8 times the concentration expected from the hydrolytic equilibrium ($ATP + H_2O \leftrightarrow ADP + P_i$). The reaction in which the utilization of a Ha^+ gradient is coupled to the phosphorylation of ADP is carried out by a large, complex enzyme system: the ATP synthase.

The membranes of bacteria, chloroplasts, and mitochondria contain ATP synthases that utilize H^+ gradients generated across their membranes for the formation of as much as 98% of the ATP required by the organisms. All ATP

Mario A. Bianchet, Peter L. Pedersen, and L. Mario Amzel • Department of Biophysics and Biophysical Chemistry, and Department of Biological Chemistry, Johns Hopkins School of Medicine, Baltimore, Maryland 21205.

Frontiers of Cellular Bioenergetics, edited by Papa *et al.* Kluwer Academic/Plenum Publishers, New York, 1999.

synthases are composed of two main sectors: the integral membrane portion F_0 and the membrane associated portion F_1; both F_0 and F_1 are multisubunit proteins. It has been demonstrated that F_0 forms a transmembrane H^+ channel that directs H^+ ions to the F_1 sector where their translocation is coupled to the synthesis of ATP (Ysern *et al.*, 1988; Cross, 1981; Amzel and Pedersen, 1983; Senior, 1988; Futai *et al.*, 1989; Fillingame, 1990; Weber and Senior, 1997; Pedersen and Amzel, 1993; Boyer, 1993, 1997; Cross and Duncan, 1996). The F_1 sector contains all the catalytic and noncatalytic nucleotide binding sites.

In this chapter, we describe the three-dimensional structure of the F_1 sector of the rat liver ATP synthase obtained by single crystal X-ray diffraction methods, a methodology that provides the structural atomic detail necessary to propose molecular mechanisms of ATP synthesis.

One of the most interesting aspects of ATP synthesis is the ability of F_1 to synthesize enzyme bound ATP from ADP and P_i (the equilibrium constant for the enzyme-bound reaction $ADP + P_i \leftrightarrow ATP + H_2O$ is approximately 1.0) (Cross, 1988; Grubmeyer *et al.*, 1982; Cross *et al.*, 1982). This observation has led to proposals in which H^+ ions from the H^+ gradient bring about the release of enzyme bound ATP; both direct and indirect effects of the H^+ ions have been proposed. Most of the proposals invoke a complex pattern of nucleotide binding and release involving conformational changes in the enzyme called the "binding change mechanism" (Cross, 1981; Boyer, 1993). According to this mechanism, catalytic sites on each of the three β subunits exhibit different affinities for nucleotide, depending on whether only one or more than one of the sites is occupied. When F_1 binds only 1 mole of ATP(Mg) such that only one catalytic β site is occupied per mole enzyme (unisite conditions), that mole of ATP(Mg) is very strongly bound ($K_d \approx 10^{-12}$ M). This bound ATP undergoes a reversible hydrolysis at that site with a $K_{eq} \approx 1$, but products are not readily released (Pedersen and Amzel, 1993; Boyer, 1993). To exhibit high steady-state ATPase activity ($k_{cat} \approx 600$ sec^{-1}), additional ATP(Mg) must fill at least one other catalytic site in order to promote product release (multisite conditions), and maximal activity is only achieved when three catalytic sites are occupied. The complex kinetics of cooperativity between the catalytic sites of F_1 suggest that the enzyme has at least two equivalent catalytic (exchangeable) sites that are assumed to function alternately. Thus, the binding change mechanism is characterized as an "alternating site-binding change" process. These mechanisms have very explicit structural implications that can be be evaluated using the new structural information described below.

The F_1 sector of the ATP synthase—the object of our structural studies— is an oligomeric protein of five different subunits with a stoichiometry of $\alpha_3\beta_3\gamma\delta\varepsilon$ (Amzel and Pedersen, 1983). The amino acid sequences of the subunits of F_1 are now known in many species. The total number of residues in each subunit depends on the species: α, 509–513; β, 459–480; γ, 272–286; δ, 177–190; and ε, 132–146. Sequence comparison between different species has shown that the α and β subunits are well conserved in all species, while the minor subunits show a lesser degree of homology. The fraction of identical residues between different species found for the α subunit is greater than 60%; for the β subunit, the homology is even higher. Sequence alignment of α with β shows similarities throughout the polypeptide chains, suggesting that the two major subunits are evolutionarily related and share the same polypeptide fold. In addition, local homologies with proteins of different activities have been found for several regions of the α and the β subunits (Walker *et al.*, 1982; Weber and Senior, 1997). For β, the first homology region, known as the "glycine-rich loop" or "P-loop," is centered around amino acid 175 and the second around amino acid 264.

Since the identification of the F_1 portion of the ATP synthase as a structure protruding from the inner mitochondrial membrane, numerous studies utilizing electron microscopy have been reported (Boekema *et al.*, 1986; Lünsdorf *et al.*, 1984; Gogol *et al.*, 1989a–c; Wakabayashi *et al.*, 1977; Yoshimura *et al.*, 1989; Tiedge *et al.*, 1983; Tsprun *et al.*, 1984). However, the detailed atomic information about the complex structure of the ATP synthase necessary for mechanistic understanding can only be obtained by X-ray crystallography. X-ray models to 9Å resolution (Amzel *et al.*, 1982), 6.5Å resolution (Abraham *et al.*, 1995), 3.6Å resolution (Bianchet *et al.*, 1991), and two structures (bovine heart: Abrahams *et al.*, 1994; rat liver: Bianchet *et al.*, 1998) at 2.8Å resolution have been reported. In this chapter, we describe the structure of the F_1-ATPase based principally on the results of the rat liver enzyme (Bianchet *et al.*, 1998). The structure of the rat liver enzyme, determined in the presence of physiological concentrations of nucleotides, has nucleotides in all three β subunits and will be referred to as the "three-nucleotide structure," whereas the bovine heart structure, determined in the presence of limiting amounts of nucleotides, will be referred to as the "two-nucleotide structure."

2. STRUCTURE OF F_1-ATPase

Crystals of rat liver F_1-ATPase (Amzel and Pederson, 1978) contain protein that is native and can carry out all the expected activities. For example, protein from crystals redissolved by the addition of ammonium sulfate-free crystallization buffer not only contained all five subunits but also was active in ATP hydrolysis and was able to restore oligomycin–sensitive ATPase activity to F_1-depleted inner mitochondrial membranes (Bianchet *et al.*, 1998). The physiological relevance of the crystallization conditions was ascertained by showing that F_1 in the crystallization buffer could hydrolyze ATP upon the addition of Mg^{2+}. Most importantly, the two ways of initiating the reaction— addition of Mg^{2+} to an F_1 and ATP solution, or addition of ATP to an F_1 and Mg^{2+} solution—yield indistinguishable kinetics, suggesting that under the crystallization conditions the F_1–ATP complex is poised for hydrolysis and that release of bound ATP is not a prerequisite to the formation of bound ATP-Mg^{2+}.

2.1. Overall Structure

The unit cell of the rat liver crystals contains one third of one F_1 molecule ($\alpha\beta\gamma_{1/3}\delta_{1/3}\varepsilon_{1/3}$) in the asymmetric unit of the crystal. For each molecule, positioned around a crystallographic threefold axis, the α, β, and parts of the γ subunits are built. No interpretable density was found for the two smallest subunits, δ and ε.

Rat liver F_1 is best described as an "inverted apple," 110 Å in diameter and 105 Å in length (125 Å including the γ protrusion) (Fig. 1) formed by three α and three β subunits alternating around a threefold axis of symmetry (Bianchet *et al.*, 1991). The molecule is narrowest at the "top," where a depression surrounding the threefold axis marks the beginning of a central cavity. This cavity, which extends for the complete length of the molecule, is surrounded by the six major subunits, leaving a large channel in the central region that surrounds the threefold axis. The γ subunit occupies part of this channel. The regions of the γ subunit that were built in the model extend well past the end of the major subunits, forming a long narrow protrusion (the "stem" of the

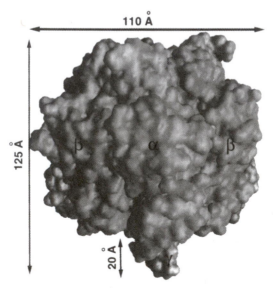

FIGURE 1. Space-filling model of the rat liver F_1-ATPase which is about 110 Å in diameter and 90 Å in length. The α subunits are in green and the β subunits in red. The γ subunit in purple is shown protruding about 20 Å from the bottom of the molecule. For color representation of this figure, see color tip facing page 366.

"apple"). There is ample evidence that this side, the "bottom" of the F_1, is closest to the membrane in the F_0F_1 complex.

2.1.1. α and β Subunits

The asymmetric unit of the rat liver F_1-ATPase R32 crystals contains one α subunit and one β subunit. Each of these subunits represents a superposition of the three copies present in an F_1 molecule. In the configuration of F_1 present in the crystal (i.e., all nucleotide binding sites occupied), the three αβ pairs that make up one F_1 molecule are in very similar conformations (Bianchet *et al.*, 1998). In a small number of places there were indications that the chain was present in more than one conformation. These alternative conformations represent minor differences among the three copies of the major subunits in the F_1 molecule that probably correspond to different functional states of the β subunit. In the structure of the bovine heart enzyme (Abrahams *et al.*, 1994), the structure of one β subunit, β_E (E, empty, i.e., no nucleotide), differs from that of the other two by a large hinge motion of the carboxy-terminal domain (some portions move more than 20 Å). No indication of such a conformation is found in the three-nucleotide structure. The main reason for this difference between the two structures appears to be the presence of nucleotides in all three β subunits in the rat liver crystals. This configuration of F_1 found in the rat liver crystals is probably an important intermediate in the reaction because, as shown by recent experiments (Weber and Senior, 1996; Lobau *et al.*, 1996, 1997), under physiological concentrations of nucleotides all three β subunits in one F_1 molecule have bound nucleotides.

The α and the β subunits have similar structures (Fig. 2). Both are composed of three domains: an N-terminal β-sheet domain, a nucleotide binding domain, and a C-terminal domain (Fig. 2). In the β subunit the N-terminal domain, a β barrel formed by six antiparallel strands and their connecting loops, extends from residue 9 to residue 81 (26 to 96 in α). The nucleotide binding domain, extending over residues 82 to 356 (97 to 373 in α), has a fold similar to the one found in other ATP-utilizing enzymes, such as elongation factor Tu, ras-P21, adenylate kinase, recA, and others. In the β subunit of F_1 it consists of a six-stranded parallel β-sheet structure with five connecting helices. Residues 82 to 135 form a long extended loop that allows the nucleotide fold to start at a position opposite to the end of the N-terminal domain. As in many other ATP-utilizing enzymes, the P-loop (GGAGVGKT; consensus sequence GX_4GKT/S) is at the beginning of the fold (residues 158–165), connecting the first helix and the first β-strand. Residues in the P-loop interact with the triphosphate moiety of ATP. The C-terminal domain extends from residue 357 to residue 477 (374 to 510 in α). It consists of six helices and contains several residues that also contribute to the nucleotide binding site. Residues 394 to 400 in β (sequence DELSEED), between the first and the second helix of the domain, occur in the region closest to the center of the molecule, facing the

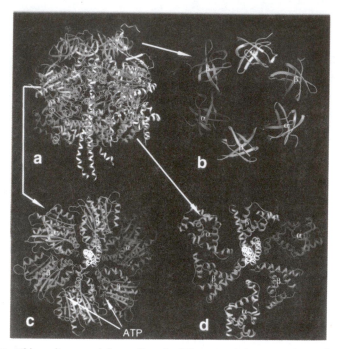

FIGURE 2. Ribbon diagram of the rat liver F_1-ATPase (a) "Side view" of complete model (viewed from a direction perpendicular to the threefold axis). (b) "Top view" of the NH_2-terminal domain. (c) "Top view" of the nucleotide binding domain. (d) "Top view" of the COOH-terminal domain. In the top views (c and d), the corresponding portion of the γ subunit in yellow is shown. For color representation of this figure, see color tip facing page 366. Figure was drawn with the program SETOR (Evans, 1993).

central channel that surrounds the threefold axis. The equivalent residues in the α subunit, 408 to 414 (SDLDAAT), are located further away from the threefold axis.

The structures of the α and β subunits in the three nucleotide structure are similar to those described as α_D and β_D (or the very similar α_T and β_T) in the two nucleotide structure (Abrahams *et al.*, 1994).

2.1.2. γ Subunit

Only three regions of the γ subunit were visible in the experimental electron density of the rat liver F_1 crystals. The sections of the γ subunit built consist of three stretches of α-helix: two of them (residues 1 to 41, and 206 to

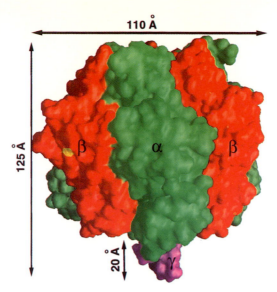

FIGURE 1. Space-filling model of the rat liver F_1-ATPase which is about 110 Å in diameter and 90 Å in length. The α subunits are in green and the β subunits in red. The γ subunit in purple is shown protruding about 20 Å from the bottom of the molecule.

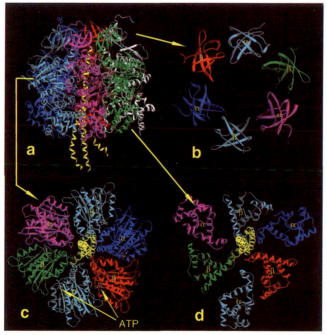

FIGURE 2. Ribbon diagram of the rat liver F_1-ATPase (a) "Side view" of complete model (viewed from a direction perpendicular to the threefold axis). (b) "Top view" of the NH_2-terminal domain. (c) "Top view" of the nucleotide binding domain. (d) "Top view" of the COOH-terminal domain. In the top views (c and d), the corresponding portion of the γ subunit in yellow is shown.

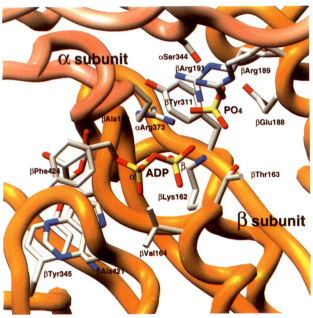

(a)

FIGURE 3. (a) Nucleotide binding site of the β subunit. The main chain is represented by a flexible cylinder (orange for the β subunit, pink for the α). The coloring of other atoms is the following: carbon, gray; oxygen, red; nitrogen, blue; phophorous, yellow. (b) Schematic diagram of nucleotide binding to the β subunit. Close interactions and hydrogen bonds (with distances in Å) and are shown. The ADP and P_i are shown in red. (c) Schematic diagram of nucleotide binding to the α subunit. Close interactions and hydrogen bonds (with distances in Å) and are shown. The ATP is shown in red and the Mg^{2+} in pink.

269) form an antiparallel coiled-coil that runs parallel to the threefold axis; the other (residues 74 to 87) forms a short rod that extends outward, making an angle of approximately 60° with the coiled-coil (Fig. 2). All three regions have interactions with the major subunits. In describing these interactions, we refer to the three copies of the α and β subunits as 1, 2, and 3. Residues γ 260 to 268 are in a region of the subunit that is a single α-helix very close to the threefold axis. It occurs at the top of the molecule surrounded by equivalent regions of the nucleotide binding domains of the three α (residues 286 to 291) and of the three β (residues 273 to 279) subunits. No other contacts of the γ subunit involve similar interactions with the three copies of the major subunits. Approximately 15 Å below the contact at the top of the subunits, residues γ 248 to 252 make contacts with residues of the second β subunit β_2 314 to 318. Residues γ 231 to 236 are in close proximity to residues β_1 389 to 391 (the beginning of the loop containing the DELSEED sequence). The region γ 74 to 79 has two sets of contacts: with residues γ 221 to 228 and with residues β_1 391 to 395. Asp 394 and Glu 395 are the first two residues of the DELSEED sequence and the nature of the contacts of this region with γ appears to depend critically on the conformation of the loop, as well as on the exact position of the two helices connected by the loop (residues 365 to 389 and residues 397 to 416). These contacts may be very important during cycling of the enzyme because they could couple changes in the position of the γ subunit to changes in the nucleotide binding sites of β (see Section 3). Subunit α_3 (residues α_3 402 to 410) makes contacts with residues γ 14 to 20; residues β_3 390 and 391 make contacts with residues γ 231 to 235. Subunit β_2 residues 384 to 396 have van der Waals clashes with residues γ 226 to 237. Some of the contacts between γ and the major subunits observed in the rat liver structure are also present in the two nucleotide structure (Abrahams *et al.*, 1994). However, alignment of the two structures shows that the positions and orientations of the γ subunits differ significantly. There is a small difference in the conformation of the γ subunit that results in a 4Å translation (the rat liver further from the center). This change affects contacts of γ with the major subunits, especially the interactions with the loop containing the DELSEED sequence. The differences in the contacts of the γ subunit between the two structures probably reflect changes that occur during catalytic cycling of the enzyme.

2.2. Nucleotide Binding Sites

The surfaces between the α and the β subunits involve large areas of contact and lie approximately in planes that contain the threefold axis of symmetry. The subunits make extensive contacts with each other, but neither $\alpha-\alpha$ nor $\beta-\beta$ contacts are present in the structure. In this arrangement there are two kinds of interfaces (called hereafter $\alpha-\beta$ and $\beta-\alpha$) that contain the nucleotide binding sites. Three of the nucleotide binding sites are formed mainly by residues of β with few contributions from α residues (β subunit sites at β–α interfaces) and three are formed mainly by α residues with few from β (α subunit sites at α–β interfaces).

2.2.1. Sites in β Subunits

The experimental results suggest that in the rat liver crystals one of the β subunits contains bound ADP and two contain ADP + P_i. This structure probably corresponds to the intermediate in the reaction that contains ADP, ADP + P_i, and ATP in equilibrium with ADP and P_i (see Section 3). Although the crystallization medium contained only ATP and no Mg^{2+}, low ATPase activity in the absence of Mg^{2+} ions or nonenzymatic hydrolysis of ATP

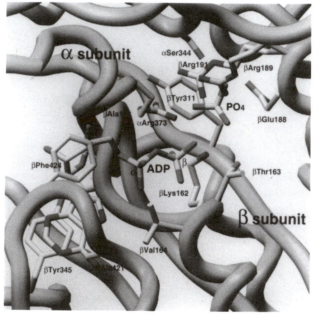

(a)

FIGURE 3. (a) Nucleotide binding site of the β subunit. The main chain is represented by a flexible cylinder (orange for the β subunit, pink for the α). The coloring of other atoms is the following: carbon, gray; oxygen, red; nitrogen, blue; phophorous, yellow. (b) Schematic diagram of nucleotide binding to the β subunit. Close interactions and hydrogen bonds (with distances in Å) and are shown. The ADP and P_i are shown in red. (c) Schematic diagram of nucleotide binding to the α subunit. Close interactions and hydrogen bonds (with distances in Å) are shown. The ATP is shown in red and the Mg^{2+} in pink. For color representation of this figure, see color tip facing page 367.

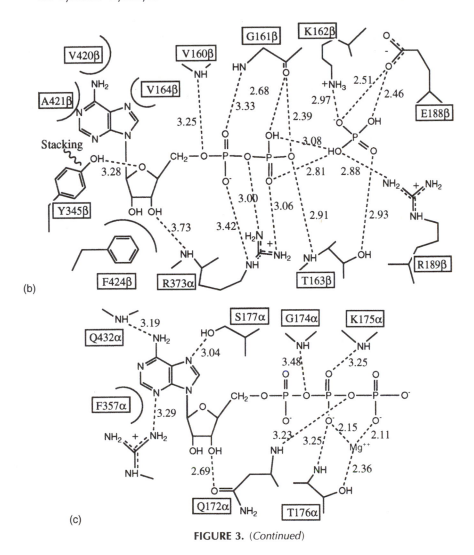

(b)

(c)

FIGURE 3. (*Continued*)

probably produced the ADP during the crystallization period (2 weeks to several months). Within experimental error, the conformations of the ADP molecules and that of the ADP + P_i are the same: The nucleotide is in the *anti* conformation, the sugar puckering is C3' *endo*, and the phosphates are in the same conformation up to the β phosphate. The adenine ring binds to a hydrophobic pocket defined by Tyr 345, Ala 421, Val 420, and Val 164 (Fig. 3a). No groups of the protein make hydrogen bonds to any of the adenine atoms;

therefore, although the pocket appears to be the right size to accommodate a purine, it lacks the residues necessary to be specific for adenine. This observation is consistent with the known lack of specificity of the enzyme in nucleotide triphosphate hydrolysis: F_1 hydrolyzes ATP, GTP, and ITP with similar kinetics (Cross *et al.*, 1982; Catterall and Pedersen, 1974). In contrast, main chain (from the P-loop) and side chain groups participate in hydrogen bonds to the ribose and the triphosphate (Fig. 3b). Other residues, which do not make direct contacts with nucleotide atoms, nevertheless appear to be important in defining the nucleotide binding site. Glu 188 is the most important among them. It is hydrogen bonded to one of the oxygens of the inorganic phosphate in a position in the site that makes it a strong candidate for donating a proton during ATP synthesis. It was also proposed that during ATP hydrolysis it functions as the catalytic base necessary for the hydrolysis of ATP (Abrahams *et al.*, 1994). Other residues present in the cavity provide multiple interactions with the nucleotide contacting residues.

Comparison of the nucleotide binding site of the three nucleotide structure (ADP + P_i or ADP and no Mg^{2+}) with the β_T subunit (AMP-PNP-Mg^{2+}) of the two-nucleotide structure shows that the absence of Mg^{2+} in the β subunit nucleotides has only a minor local effect on the conformation of the subunits. The only difference between the conformations of the nucleotides of the two enzymes is a shift in the positions of the β (and probably the γ) phosphate. The effect of Mg^{2+} is to bind between Thr 163 and the phosphates such that in the bovine heart β_T, the β and γ phosphates are farther away from the P-loop. The greater distance between the γ phosphate and Glu 188 in the bovine heart β_T allows a water molecule to bind between the two groups along the axial direction of the γ phosphate such that this water, deprotonated by Glu 188, can carry out a nucleophilic attack on the phosphate. In addition there are small movements (less than 0.8 Å) of some of the atoms in the P-loop. Thus, the presence of the "open" conformation in the bovine heart enzyme, which is the major difference between the structures of the F_1-ATPase in the bovine heart and the rat liver crystals, cannot be attributed to the presence or absence of Mg^{2+} (Boyer, 1997; Duncan *et al.*, 1995). Most likely, the presence of the open conformation in the bovine heart enzyme is due to the low ADP concentration in the crystallization medium (Lobau *et al.*, 1997).

2.2.2. Sites in α Subunits

All three α subunits contain bound ATP in the site at the α–β interface. Despite not having included Mg^{2+} in the crystallization medium, these bound ATPs appear to be in the form of ATP-Mg^{2+}. The adenine ring binds in a pocket in which it makes van der Waals contacts as well as hydrogen bonding interactions with several residues (Fig. 3c). This is in agreement with the high specificity of the α subunits for ATP in contrast to the β subunit. As in the case of the nucleotide binding site of the β subunit, several residues that do not contact the ATP directly interact with contacting residues and appear to be part of the binding cavity. Residue 208 of α—the residue equivalent to the catalytic Glu 188 of β—is a glutamine in the α subunit. This difference was suggested to explain a lack of catalytic activity of the α subunit (Abrahams *et al.*, 1994). This may represent an oversimplification because several other residues—Glu 328, Asp 269, and Asp 270—are in positions that could allow them to act as a catalytic base. In addition, in slow triphosphatases such as ras P21, Ef Tu, and others, which are 1000 times slower than F_1, the equivalent residue is a glutamine (Mildvan, 1997), suggesting that the site in the α subunit could be active but only at a rate about 1000 times slower than that of β.

3. MECHANISM OF ATP SYNTHESIS–HYDROLYSIS

In the "binding change mechanism" (Cross, 1981; Boyer, 1993, 1997) proposed for ATP synthesis (or H^+ translocation coupled ATP hydrolysis), the β subunits proceed sequentially through three conformational states: tight (T), loose (L), and open (O). The three different conformations of the β subunits observed in the bovine heart crystals*—β_T, β_D, and β_E—are considered to correspond to these states (Boyer, 1997; Abrahams *et al.*, 1994).† These sequential conformational changes are proposed to be coupled to movements of the γ subunit relative to the major subunits (Duncan *et al.*, 1995; Noji *et al.*, 1997; Sabbert *et al.*, 1996; Capaldi *et al.*, 1996).‡ Although the proposed general scheme does not provide a detailed sequence of all binding and debinding steps, it has some general characteristics that can be compared and contrasted with existing experimental evidence. In the described enzymatic cycle, F_1-ATPase alternates through species that have either one or two nucleotides bound at catalytic sites per F_1 molecule (Boyer, 1993; Cross *et al.*, 1984). This is at variance with existing evidence and in particular with recent observa-

---

*In the bovine heart crystals, nucleotide concentrations are 250 μM AMP–PNP, 5 μM ADP, with no phosphate present. Since the open site is unoccupied, the dissociation constant of this site for ADP is greater than 5 μM and for AMP–PNP (and ATP) it is greater than 250 μM. In addition, the amount of ADP is limited to one per F_1-ATPase.

†It should be noted that the binding change mechanism as proposed does not contain any species equivalent to the one observed in the bovine heart crystals. The species containing ATP (AMPPNP in the crystals), ADP, and an empty site has inorganic phosphate in the ADP site. Binding of P_i, as required for the proposed mechanism, will have an effect on the conformation of the subunit.

‡As described, the mechanism involves rotation of the major subunits around the γ subunit as part of the cycling of the enzyme. Although there is direct experimental evidence that γ interacts sequentially with each of the β subunits (Duncan *et al.*, 1995; Noji *et al.*, 1997; Sabbert *et al.*, 1996; Capaldi *et al.*, 1996), the data only show the existence of relative movements between γ and the major subunits (Noji *et al.*, 1997). Consensus is building that in the membrane-bound enzyme these interactions involve movements of the γ subunit and not rotations of α and β.

tions that indicate that at physiological concentrations of nucleotides, F_1 molecules contain three bound nucleotides in β subunits (Lobau *et al.*, 1996, 1997). Thus, the mechanism operating under physiological conditions must involve species containing two and three nucleotides bound at catalytic sites per F_1 molecule and no species containing only one bound nucleotide.* It is clear that the conformation of F_1 with three bound nucleotides at catalytic sites must differ significantly from the one observed in the two nucleotide structure, as the subunit $β_E$ in that structure does not have a well-formed nucleotide binding site (Abrahams *et al.*, 1994). The structure of the rat liver F_1-ATPase, determined in the presence of physiological concentrations of nucleotides, contains three bound nucleotides at catalytic sites consistent with a role as a cycling intermediate in ATP hydrolysis–synthesis.

Mechanisms of ATP synthesis–hydrolysis involving three F_1 bound nucleotides at catalytic sites recently have been proposed (Cross and Duncan, 1988; Lobau *et al.*, 1997). Due in part to the lack of structural information available at the time, however, these mechanisms still include some unlikely steps and intermediates. The most significant problems with the mechanisms are: (1) they do not explicitly exclude the possibility of releasing ADP + P_i during ATP synthesis, and (2) they require that nucleotides bind and remain bound to β subunits in the O conformation. Using information from the structure of the active form of the rat liver F_1-ATPase, we proposed a step-by-step mechanism for the reaction carried out by the enzyme that overcomes these and other difficulties.

In previous mechanisms, each of the three β subunits in the same F_1 molecule is in a different conformation, either T, L, or O. The reaction proceeds through concerted conformational changes that change all three subunits at once: T to O, L to T, and O to L. In trying to explain $P_i/H_2^{18}O$ exchange data (Cross, 1981) the T site formed immediately after the conformational change is said to contain tightly bound ADP + P_i that is in equilibrium with ATP (ADP + $P_i \rightleftharpoons$ ATP + H_2O, $K_{eq} \approx 1$). However, if this equilibrium can occur in the T site without any conformational changes, the change from T to O will occur part of the time with ADP + P_i in the T site, leading to the release of ADP + P_i instead of ATP when going in the synthetic direction. That is, there is no way proposed to ensure that the T site with ADP + P_i cannot change to the O configuration and release the ADP + P_i. Since this kind of uncoupling is incompatible with a

---

*The presence or absence of species containing one bound nucleotide per F_1 molecule will depend on the nucleotide concentrations. It will result from a balance between the rates of binding and release of nucleotides from the different binding sites. Since release is a unimolecular reaction and binding is bimolecular, only the latter will depend on the nucleotide concentration. The physiological concentrations of nucleotides favor full occupancy of sites: One site remains empty only for the time necessary to bind nucleotide.

tightly coupled synthase, more likely the T site has two possible conformations: T and T' (Fig. 4). T binds only ATP, and T' binds only ADP + P_i. The T and T' sites release their nucleotides very slowly (rates of unisite catalysis) (Grubmeyer *et al.*, 1982; Cross *et al.*, 1982). When one β subunit is in the T' conformation, the transition T to O cannot occur and therefore ADP + P_i cannot be released at cycling rates from T' sites. The form of the enzyme with T site and T' site are in fast equilibrium. This equilibrium not only explains the $P_i/H_2^{18}O$ exchange data, but also the concentration dependence of the exchange in both the hydrolytic and the synthetic directions (Boyer, 1993).*

In previous models, in the synthetic direction, ADP and P_i bind to the O site and remain bound until the energy input drives the concerted conformational change O to L, L to T, and T to O (Boyer, 1993, 1997; Abrahams *et al.*, 1994). This is an unlikely step because in the O conformation of the β subunit the nucleotide binding site is highly distorted and does not have the arrangement of residues that interact with nucleotides in the L or T conformations. More likely, after ADP and P_i binding, the β subunit in the O conformation goes through a local conformational change to a conformation, C (closed)†, that is more like the L and T conformations than the O. This change is local and does not involve the other β subunits, which remain in their original conformations (T and L). The resulting conformation has the subunits in the C, T, and L conformations (Fig. 4). This conformation of the F_1 molecule is the one that goes through the concerted conformational change C to L, L to T, and T to O.

Previous mechanisms had an additional difficulty. With the membrane-bound enzyme operating in the ATPase direction, the actual translocation of the protons occurred triggered only by the binding of ATP to the O site and before hydrolysis of ATP (Boyer, 1993; Cross *et al.*, 1984). The driving force was provided by the conformational change that transforms the ATP occupied O site to a T site. In view of what is said above about the nucleotide binding characteristics of the O site, binding of ATP to the O site appears to be inadequate to trigger the translocation of protons against their concentration gradient. In the model we propose, this step also involves the hydrolysis of ATP (Fig. 4), providing a more direct coupling between ATP hydrolysis and proton translocation. Finally, the equilibrium between T with bound ATP and T' with

---

*The extent of incorporation of ^{18}O into P_i under hydrolytic conditions increases markedly at low ATP concentrations. This is explained in the proposed model because the enzyme remains in the equilibrium between III and III' until it can bind ATP. In the synthetic direction, the enzyme remains in the same equilibrium until it binds ADP and P_i, in agreement with the experimental data that indicate that the exchange increases at low ADP and/or P_i concentrations. The increased exchange that is observed in the absence of a proton gradient is explained by the equilibrium between I and I'.

†The closed (C) conformation of the β subunit occurs right after the open (O) conformation binds nucleotide (hence, closes) and again just before the β subunit opens to release ATP.

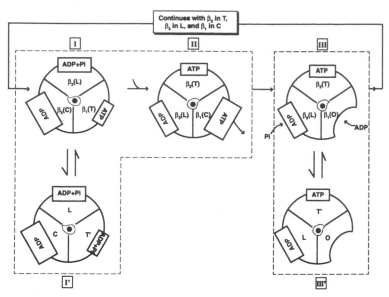

FIGURE 4. Proposed mechanism for ATP synthesis. The model involves two different configurations of F_1-ATPase: one with T, L, and C sites (I and II, boxed by the dashed line), the other
with T, L. and O sites (III, boxed by the dashed line). These two configurations are the only
ones necessary to describe all the species in the mechanism. Conformations T and T' of the β
subunit are highly similar, but T contains ATP and T' contains ADP + P_i. Species T' cannot
undergo the concerted molecular conformational changes. The invidual copies of the β
subunits in one F_1 molecule are labeled β_1, β_2, and β_3.

bound ADP and P_i can also take place with species I of Fig. 4. The mechanism
shown in Fig. 4 is a more detailed representation of the previously described
mechanisms (Bianchet *et al.*, 1998).

All these considerations suggest a mechanism for those events in ATP
synthesis that involve the F_1 portion of the synthase (Fig. 4). As in previous
proposals, changes in the nucleotide binding sites of the three β subunits of one
F_1 molecule are coupled to each other and to changes in the position of the γ
subunit. The position of the γ subunit also responds to changes in the F_0
portion, thus coupling proton translocation with events in the nucleotide binding sites. The contacts between the β and the γ subunits described above
(especially those involving residues 394 to 400 of β, DELSEED) mediate this
coupling.

In the structure of the rat liver enzyme determined under physiological
conditions, all three β subunits have bound nucleotides and adopt similar
conformations. This structure has the characteristics of the form of the enzyme
containing β subunits in the C, T (and T'), and L conformations (Fig. 4). The C

site has ADP, the L site has ADP and P_i, and the T site has ATP in equilibrium, with T′ containing ADP and P_i. Starting with this, the most stable species, ATP synthesis proceeds through a proton translocation-driven conformational change that results in the following concerted transformations of the binding sites: T to C, L to T, and C to L. Since the T site can only bind ATP, the transformation of the L site, which contains ADP and P_i, from L to T occurs with formation of ATP. The sites with ADP and P_i in the rat liver structure represent the species poised for the formation of the β–γ phosphate bond before the addition of Mg^{2+}. In the next step, ATP is released from the C site that changes to the O conformation. The resulting species has sites in the T, O, and L conformations with ATP in the T site and ADP in the L site. This is similar to the species found in the bovine heart crystals but with ATP instead of AMPPNP. This species, under physiological conditions, will immediately bind ADP to the O site, which will then change back to the C conformation. After P_i binding to the L site, the enzyme is ready for the next round of synthesis. During the proposed cycle, the γ subunit cycles through three different orientations following the conformational changes in the major subunits. In addition, the change of one β subunit from C to O is accompanied by a change in γ that results in its displacement of 4 Å away from the threefold axis. These movements are an ideal vehicle for coupling changes in the major subunits with changes in γ that can be transmitted to the F_0 region.

This mechanism of ATP synthesis–hydrolysis incorporates the two conformations for which known structures are available (bovine heart: Abrahams *et al.*, 1994; rat liver: Bianchet *et al.*, 1998) as the only intermediates in the reaction. In addition, it provides a rationale for most of the available experimental data in this complex system. Many important questions about this reaction remain unanswered. However, the two structures of the F_1 sector reported to date—the two-nucleotide and the three-nucleotide structures— have provided valuable information for understanding the mechanism of this important process.

ACKNOWLEDGMENTS. This work was supported by NIH Grants CA 10951 (to P.L.P.) and GM 25432 (to L.M.A.). Equipment used in this work was purchased and supported with funds from NSF, NIH, the Keck Foundation, and the Lucile P. Markey Charitable Trust.

4. REFERENCES

Abrahams, J. P., Leslie, A. G. W., Lutter, R., and Walker, J. E., 1994, *Nature* **370**:621–628.
Abrahams, J. P., Lutter, R., Todd, R. J., van Raaij, M. J., Leslie, A. G. W., and Walker, J. E., 1995, *Proc. Natl. Acad. Sci. USA* **92**:10964–10968.
Amzel, L. M., and Pedersen, P. L., 1978, *J. Biol. Chem.* **253**:2067–2069.

Amzel, L. M., and Pedersen, P. L., 1983, *Annu. Rev. Biochem.* **52**:801–824.

Amzel, L. M., McKinney, M., Narayanan, P., and Pederson, P. L., 1982, *Proc. Natl. Acad. Sci. (USA)* **79**:5852–5856.

Bianchet, M. A., Ysern, Q., Hullihen, J., Pedersen, P. L., and Amzel, L. M., 1991, *J. Biol. Chem.* **266**:21197–21201.

Bianchet, M. A., Hullihen, J., Pedersen, P. L., and Amzel, L. M., 1998, *Proc. Natl. Acad. Sci. (USA)* **95**:11065–11070.

Boekema, E. J., Berden, J. A., and Van Heel, M. G., 1986, *Biochim. Biophys. Acta* **851**:353–360.

Boyer, P. D., 1993, *Biochim. Biophys. Acta* **1140**:215–250.

Boyer, P. D., 1997, *Annu. Rev. Biochem.* **66**:714–749.

Capaldi, R. A., Aggeler, R., Wilkens, S., and Grüber, G., 1996, *Biochim. Biophys. Acta* **28**:397–402.

Catterall, W. A., Coty, W. A., and Pedersen, P. L., 1973, *J. Biol. Chem.* **248**:7427–7431.

Catterall, W. A., and Pedersen, P. L., 1974, in: *Membrane ATPases and Transport Process* (R. J. Bronk, ed.), pp. 63–68.

Cross, R. L., 1981, *Annu. Rev. Biochem.* **50**:681–714.

Cross, R. L., 1988, *J. Bioenerg. Biomembr.* **20**:395–405.

Cross, R. L., Grubmeyer, C., and Penefsky, H. S., 1982, *J. Biol. Chem.* **257**:12101–12105.

Cross, R. L., Cunningham, D., and Tamura, J. K., 1984, *Curr. Top. Cell. Regul.* **24**:365–378.

Cross, R. L., and Duncan, T. M., 1996, *J. Bioenerg. Biomembr.* **28**:403–408.

Duncan, T. M., Bulygin, V. V., Zhou, Y., Hutcheon, M. L., and Cross, R. L., 1995, *Proc. Natl. Acad. Sci. USA* **92**:10964–10968.

Evans, S. V., 1993, *J. Mol. Graphics* **11**:268–272.

Fillingame, R. H., 1990, in: *The Bacteria*, pp. 345–391.

Futai, M., Noumi, T., and Maeda, M., 1989, *Annu. Rev. Biochem.* **58**:111–136.

Gogol, E. P., Lücken, U., Bork, T., and Capaldi, R. A., 1989a, *Biochemistry* **28**:4709–4716.

Gogol, E. P., Aggeler, R., Sagermann, M., and Capaldi, R. A., 1989b, *Biochemistry* **28**:4717–4724.

Gogol, E. P., Lucken, U., Bork, T., and Capaldi, R. A., 1989c, *Biochemistry* **28**:4709–4716.

Grubmeyer, C., Cross, R. L., and Penefsky, H. S., 1982, *J. Biol. Chem.* **257**:12092–12100.

Lobau, S., Weber, J., Wilke-Mounts, S., and Senior, A. E., 1996, *J. Biol. Chem.* **272**:2648–2656.

Lobau, S., Weber, J., and Senior, A. E., 1997, *FEBS Lett.* **404**:15–18.

Lünsdorf, H., Ehrig, K., Friedl, P., and Schairer, H. U., 1984, *J. Mol. Biol.* **173**:131–136.

Mildvan, A., 1997, Proteins.

Noji, H., Yasuda, R., Yoshida, M., and Kinosita Jr., K., 1997, *Nature* **385**:299–302.

Pedersen, P. L., and Amzel, L. M., 1993, *J. Biol. Chem.* **269**:9937–9940.

Sabbert, D., Engelbrecht, S., and Junge, W., 1996, *Nature* **381**:623.

Senior, A. E., 1988, *Physiol. Rev.* **68**:177–231.

Tiedge, H., Schafer, G., and Mayer, F., 1983, *Eur. J. Biochem.* **132**:37–45.

Tsuprun, V. L., Mesyanzhinova, I. V., Koslov, I. A., and Orlova, E. V., 1984, *FEBS Lett.* **167**:285–289.

Wakabayashi, T., Kubota, M., Yoshida, M., and Kagawa, Y., 1977, *J. Mol. Biol,* **117**:515–519.

Walker, J. E., Saraste, M., Runswick, M. J., and Gay, N. J., 1982, *EMBO J.* **1**:945–951.

Weber, J., and Senior, A. E., 1996, *Biochim. Biophys. Acta* **1275**:101–104.

Weber, J., and Senior, A. E., 1997, *Biochim. Biophys. Acta* **1319**:19–58.

Yoshimura, H., Endo, S., Matsumoto, M., Nagayama, K., and Kagawa, Y., 1989, *J. Biol. Chem.* **106**:958–960.

Ysern, X., Amzel, L. M., and Pedersen, P. L., 1988, *J. Bioenerg. Biomembr.* **20**:423–450.

CHAPTER 16

Mechanism of ATP Synthesis by Mitochondrial ATP Synthase

Abdul-Kader Souid and Harvey S. Penefsky

1. INTRODUCTION

Adenosine triphosphate (ATP) is the major energy carrier in living cells, facilitating the flow of chemical potential energy from a variety of catabolic reactions to energy-requiring cellular processes (Lehninger, 1975). Oxidative phosphorylation is the major source of cellular ATP, and the enzymes and other molecules involved in the process are contained in the mitochondrial inner membrane, the plasma membrane of eubacteria, and chloroplast thylakoid membranes (Penefsky, 1979). In mitochondria, the respiratory chain catalyzes the oxidation of NADH or $FADH_2$ by oxygen. Energy released during the flow of a pair of reducing equivalents from NADH or $FADH_2$ to oxygen is captured by components of the respiratory chain and utilized to generate proton gradients across the mitochondrial inner membrane. The resulting chemiosmotic proton circuit drives the formation of ATP from ADP and Pi (Trumpower and Gennis, 1994; Skulachev, 1981; Mitchell, 1961).

Functionally, the mitochondrial oxidative phosphorylation system is composed of five complexes (Trumpower and Gennis, 1994; Hatefi, 1985): NADH–

Abdul-Kader Souid and Harvey S. Penefsky • State University of New York, Syracuse, New York 13210; and Public Health Research Institute, New York, New York 10028.

Frontiers of Cellular Bioenergetics, edited by Papa *et al.* Kluwer Academic/Plenum Publishers, New York, 1999.

ubiquinone oxidoreductase (complex I), succinate–ubiquinone oxidoreductase (complex II), ubiquinol–ferricytochrome c oxidoreductase (complex III), ferrocytochrome c–oxygen oxidoreductase (complex IV), and ATP synthase (F_oF_1, complex V).

The respiratory chain consists of complexes I to IV plus ubiquinone (Q) and cytochrome c. Complex V (ATP synthase, F_oF_1) catalyzes the formation of ATP from ADP and Pi at the expense of protonic energy generated by oxidations in the respiratory chain. F_oF_1 is also capable of ATP hydrolysis.

The mitochondrial F_oF_1 complex is composed of two major components: F_1 and F_o (Abrahams et al., 1994; Collinson et al., 1994; Walker and Collinson, 1994; Walker et al., 1991; Amzel and Pedersen, 1983; Amzel et al., 1982; Kagawa et al., 1979). F_1 is an extrinsic membrane enzyme. It appears in electron micrographs as a spherical projection (a knob 80 Å high and 100 Å across) into the matrix (Abrahams et al., 1994; Wilkens and Capaldi, 1994; Boekema et al., 1986). It has the catalytic properties of the enzyme and can be isolated as a water-soluble ATPase with a turnover number of about 600 sec^{-1} (Penefsky, 1985; Grubmeyer and Penefsky, 1981a,b). F_1 consists of five different subunits with the stoichiometry $3\alpha:3\beta:\gamma:\delta:\epsilon$. The catalytic sites are on the β subunits. Although each α subunit has a nucleotide-binding site, these sites do not participate directly in catalysis. The sequences of the α and β subunits are 20% identical; both are elongated masses and contain the P-loop nucleotide-binding motif (Walker et al., 1982). The α and β subunits are arranged alternately in a hexagonal array, like the segments of an orange. The central core of the molecule is occupied by the γ subunit, a 90-Å-long helical molecule (Abrahams et al., 1994; Gogol et al., 1989). The structure of the enzyme is asymmetric, a feature that may have mechanistic implications. The asymmetry was confirmed by electron microscopy (Wilkens and Capaldi, 1994; Boekema et al., 1986), low-resolution X-ray crystallography (Abrahams et al., 1993), and X-ray crystallography at 2.8 Å (Abrahams et al., 1994).

The F_o component of F_oF_1 is a membrane-embedded complex of protein and phospholipid that participates in proton transduction across the mitochondrial inner membrane. The number of subunits in mitochondrial F_o may vary with the source of the mitochondria (McEnery et al., 1984; Galante et al., 1979; Tzagoloff, 1979; Sebald and Wild, 1979; Serrano et al., 1976). Fourteen different subunits have been reported in mitochondrial F_o from bovine heart (Walker and Collinson, 1994; Walker et al., 1991; Galante et al., 1979).

In mitochondria, the ATPase is associated with a specific inhibitor protein, the oligomycin sensitivity-conferring protein (OSCP) and a coupling factor called F_6 (Hatefi, 1985; Penefsky, 1979). The ATPase-inhibiting protein binds to the β subunit of F_1 (1 mole per mole of F_1) and inhibits ATP hydrolysis (Heeke et al., 1993; Vazquez-Laslop and Dreyfus, 1990; Rouslin and Pullman, 1987; Schwerzmann and Pedersen, 1986; Hatefi, 1985; Gomez-Puyou et al.,

1982, 1983; Frangione *et al.*, 1981; Klein *et al.*, 1981; Pedersen *et al.*, 1981; Clintron and Pedersen, 1979; Gomez-Fernandez and Harris, 1978; Pullman and Monroy, 1963). The inhibitor peptide can be partially removed from its association with F_1 by washing submitochondrial particles with dilute solutions of KCl or by chromatography on Sephadex columns (Horstman and Racker, 1970). ATPase activity of the resulting enzyme is enhanced as much as 15-fold (Souid and Penefsky, 1995; Penefsky, 1985a). In *Escherichia coli* F_oF_1, the F_o component consists of three different subunits (a, b, and c) with the stoichiometry 1a:2b:9–12c (Schneider and Altendorf, 1987; Walker *et al.*, 1984; Foster and Fillingame, 1982; Dunn and Heppel, 1981).

Several published reviews should provide adequate discussions of the structural and functional aspects of F_oF_1 (Nakamoto, 1996; Capaldi *et al.*, 1994; Souid and Penefsky, 1994; Hatefi, 1993; Pedersen and Amzel, 1993; Issartel *et al.*, 1992; Penefsky and Cross, 1991; Fillingame, 1990; Senior, 1990). This chapter focuses on changes in the catalytic sites of F_oF_1 that account for net ATP synthesis. It begins with a review of the thermodynamic and kinetic properties of oxidative phosphorylation, since these aspects are fundamental to our current studies on the reaction mechanism of ATP synthesis (Souid and Penefsky, 1994, 1995; Penefsky and Cross, 1991).

2. THERMODYNAMIC RELATIONSHIPS IN OXIDATIVE PHOSPHORYLATION

The thermodynamics of oxidative phosphorylation describe relationships between the various processes involved in energy conversion by mitochondria. The conversion mechanism has attracted numerous investigators over the years (Murphy, 1989; Herweijer *et al.*, 1985; Wikstrom *et al.*, 1981; Zoratti *et al.*, 1981; Fillingame, 1980; Kagawa, 1978). According to the chemiosmotic hypothesis, respiration generates an electrochemical potential across the mitochondrial inner membrane, which drives ATP synthesis (Mitchell, 1961, 1979). The electrochemical processes can be described by three major categories of relationships. The first category describes the free energy change accompanying the electron transfer potential:

$$\Delta G^o = -nF\Delta E^o \tag{1}$$

in which ΔG^o, in kcal/mole, is the standard free energy available as a pair of reducing equivalents passes from NADH (or $FADH_2$) to oxygen; n is the number of equivalents transferred; F (Faraday constant) is the energy change as a mole of electrons falls through a potential of 1 V (23.062 kcal/volt per mole); E^o, in volts, is the standard oxidation–reduction potential [equivalent to the standard redox potential or the electromotive force, (EMF)]; ΔE^o is the difference between E^o of the electron donor systems (i.e., NAD^+–NADH or FAD^+–$FADH_2$ conjugate redox couples) and E^o of the electron acceptor system (i.e., $\frac{1}{2}O_2$–H_2O conjugate redox couple) (Stryer, 1988; Rottenberg, 1975, 1985; Ferguson and Sorgato, 1982; Lehninger, 1975, 1982). ΔE^o is the driving force of oxidative phosphorylation. The value of ΔE^o for NADH–oxygen couple (donor acceptor couple) is 1.14 V and for succinate–oxygen couple 0.79 V. Thus, ΔG^o for the oxidation of NADH by oxygen is -52.6 kcal/mole ($2 \times 23.062 \times 1.14$) and for succinate -39.4 kcal/mole.

The ΔG^o for the individual respiratory chain complexes is NADH–ubiquinone oxidoreductase (the span between NADH to coenzyme Q) -12.2 kcal/mole (0.27 V), ubiquinol–ferricytochrome c oxidoreductase (the span between cytochrome b and cytochrome c) -9.9 kcal/mole (0.22 V), and ferrocytochrome c–oxygen oxidoreductase (the span between cytochrome a and oxygen) -24 kcal/mole (0.53 V) (Stryer, 1988; Hatefi, 1985; Lehninger, 1975, 1982; Wikstrom et al., 1981). The respiratory chain is reversible except for the step cytochrome a_3 to oxygen. This irreversible step "pulls" the system toward substrate oxidation (Hatefi, 1985; Wikström et al., 1981).

The observed oxidation–reduction potential (E_h) is related to the standard oxidation–reduction potential of a given conjugate redox couple (E^o) by the Nernst equation:

$$E_h = E^o + 2.3RT/nF \log [\text{electron acceptor}]/[\text{electron donor}] \qquad (2)$$

where E_h, in volts, is the observed potential; E^o, in volts, is the standard oxidation–reduction potential; R is the gas constant (1.98×10^{-3} kcal/deg per mole); T is the absolute temperature (298°K); F (Faraday constant) is equal to 23.062 kcal/volts per mole; and n is the number of electrons transferred; $2.3RT/nF$ has the value 0.059 V when n is 1 and 0.03 when n is 2. Equation 2 can be simplified to:

$$E_h = E^o + 0.03 \log [\text{electron acceptor}]/[\text{electron donor}] \qquad (3)$$

Thus, the Nernst equation defines E^o as the midpoint potential at which 50% of the reductant is oxidized, the ratio [electron acceptor]/[electron donor] is 1 and E_h is equal E^o (Hatefi, 1985; Wikström et al., 1981; Lehninger, 1975).

The energy released during oxidations in the respiratory chain is captured in vectorial translocation of H^+ from the mitochondrial matrix to the intermembrane space. Consequently, the matrix becomes more alkaline and the inner membrane is negatively charged on the matrix side. This electrochemical potential ($\Delta\mu_{H^+}$) drives ATP synthesis. The limited permeability of the mitochondrial inner membrane to H^+ results in a slow rate of H^+ leakage and favors proton flow through F_oF_1.

The amount of ATP formed per pair of reducing equivalents transferred in the individual respiratory chain complexes is 1 ATP per pair of reducing equivalents at complex I and IV, and 0.5 at complex III (Hinkle *et al.*, 1991; Wikstrom *et al.*, 1981; Hinkle and Yu, 1979). Thus, the oxidation of NADH yields less than 3 ATP and that of succinate less than 2 (Hinkle *et al.*, 1991; Luvisetto and Azzone, 1989; Stoner, 1987; Beavis and Lehninger 1986; Hinkle and Yu, 1979; Van Dam *et al.*, 1980; Lemasters, 1984).

The second category of thermodynamic relationships describes the electrochemical H^+ potential:

$$\Delta\mu_{H+} = nF\Delta\varphi - 2.3RT\Delta pH \qquad (4)$$

in which $\Delta\mu_{H+}$, in kcal/mole, is the electrochemical H^+ gradient mitochondrial inner membrane; n, the number of H^+, is equal to 1; F is the Faraday constant (23.062 kcal/volt per mole); $\Delta\varphi$, in volts, is the electrical potential difference across the mitochondrial inner membrane (negative in the matrix); R is the gas constant (1.98×10^{-3} kcal/mole per deg); T is the absolute temperature (298°K); and ΔpH is the pH gradient across the inner membrane (pH in the matrix − pH in the intermembrane space) is 1.4 unit (Rottenberg, 1975, 1985; Berry and Hinkle, 1983; Lehninger, 1982; Ferguson and Sorgato, 1982). Since the matrix pH is increased, the term "$-2.3\,RT\Delta pH$" is always negative and $\Delta\varphi$ is negative, the two terms thus sum to a negative value for $\Delta\mu_{H+}$ (Rottenberg, 1985).

Peter Mitchell (1979) proposed a positive electrochemical force, the proton-motive force (Δp, in volts),

$$\Delta p = -\Delta\mu_{H+}/F \qquad (5)$$

Thus,

$$\Delta p = \Delta\varphi - 2.3RT/F\Delta pH \qquad (6)$$

in which $\Delta\varphi$, in volts, is the electrochemical H^+ gradient across the mitochondrial inner membrane (positive in the intermembrane space); ΔpH is the pH gradient across the inner membrane (pH in the intermembrane space − pH in the matrix), which is equal to −1.4 unit; and "2.3 RT/F" is equal to 59 mV (Stryer 1988; Rottenberg, 1985; Ferguson and Sorgato, 1982; Sorgato and Ferguson, 1979; Mitchell, 1966). Thus, both components of Δp are positive and sum to a positive value for Δp. This force reflects the spontaneous translocation of H^+ from the intermembrane space to the matrix.

In summary, $\Delta\mu_{H+}$ has units of free energy, kcal/mole. The electrochemical components add up to a negative value for $\Delta\mu_{H+}$. The Δp is expressed in electrical units—volts—and the electrochemical components add up to a positive value.

The Δp has been measured in mitochondria (Ferguson and Sorgato, 1982; Holian and Wilson, 1980; Nicholls, 1974; Padan and Rottenberg, 1973; Mitchell and Moyle, 1969) and submitochondrial particles (Berry and Hinkle,

1983; Ferguson and Sorgato, 1982; Branca et al., 1981; Sorgato and Ferguson, 1979; Sorgato et al., 1978; Rottenberg, 1975). The Δp during NADH oxidation and state 4 of respiration (i.e., during ADP depletion) in the mitochondria ranges from 140 to 240 mV (equivalent to 3.2 to 5.5 kcal/mole H^+) and in submitochondrial particles ranges from 200 to 240 mV (4.6 to 5.5 kcal/mole H^+, for conversion 1 kJ = 10.36 mV and 4.18 kJ = 1 kcal). The approximate values of the components of Δp in submitochondrial particles are the electric component ($\Delta\varphi$) 160 mV (equal to 3.7 kcal/mole H^+) and the chemical component (ΔpH) 88 mV (equal to 2 kcal/mole H^+) (Berry and Hinkle, 1983; Ferguson and Sorgato, 1982; Wikström et al., 1981).

The third category of thermodynamic relationships is described by the phosphorylation potential, ΔG_p:

$$\Delta G_p = \Delta G^o + RT \ln [ATP]/[ADP][Pi] \tag{7}$$

in which ΔG_p, in kcal/mole, is the free energy of ATP synthesis by mitochondria or submitochondrial particles; ΔG^o is equivalent to the absolute value of the standard free energy of ATP hydrolysis, 7 to 9 kcal/mole (Rosing and Slater, 1971; Phillips et al., 1969); R is the gas constant (1.98×10^{-3} kcal/mole per deg); and T is the absolute temperature (298°K) (Rottenberg, 1985; Lehninger, 1982; Lehninger et al., 1980; Klingenberg, 1975; Slater et al., 1973).

The maximum value of ΔG_p in mitochondria is 16 kcal/mole ATP, which is equivalent to 690 mV (for conversion, 1 kJ = 10.36 mV, and 4.18 kJ = 1 kcal) and in submitochondrial particles 10.6 kcal/mole ATP (equivalent to 460 mV) (Lehninger et al., 1980; Thayer et al., 1977; Ferguson and Sorgato, 1977). Thus, the oxidative phosphorylation system is capable of net ATP synthesis up to a very high ratio of ATP to ADP and Pi (10^2–10^4). At these ratios (or ΔG_p), ATP synthesis ceases (i.e., state 4 of respiration) (Ferguson and Sorgato, 1977, 1982).

Since $\Delta\mu_{H+}$ drives ΔG_p and the circuit is fully reversible, the steady-state ratio $\Delta G_p/\Delta\mu_{H+}$ should be related to the ratio H+/ATP, n:

$$n = \Delta G_p/\Delta\mu_{H+} \tag{8}$$

The reported values for n are between 3 and 4 (Pitard et al., 1996; Berry and Hinkle, 1983; Lehninger et al., 1980; Mitchell and Moyle, 1968).

Similarly, during steady-state respiration the relationship between ΔE_h and Δp is given by equation 9,

$$2\Delta E_h = n\Delta p \tag{9}$$

where n is the number of H^+ per pair of electrons over each span of the respiratory chain components that translocates nH^+ per pair of electrons (Ferguson and Sorgato, 1982; Wikström et al., 1981; Mitchell and Moyle, 1969).

3. RATE CONSTANTS, EQUILIBRIUM CONSTANTS, AND FREE ENERGY CHANGES DURING ATP SYNTHESIS

This section discusses methods used to determine the first-order and second-order rate constants describing ATP dissociation and binding at high-affinity catalytic sites during unisite catalysis. These numbers permit calculation of equilibrium constants and the free energy changes during ATP synthesis. The equilibrium constant for the dissociation of product ATP from high-affinity catalytic sites of the membrane-bound enzyme can be calculated from the ratio of the rate constants for the dissociation step (first-order rate constant, "off" rate, k_{-1}) and the step of ATP binding (second-order rate constant, "on" rate, bimolecular rate constant, k_{+1}).

The step of dissociation of product ATP can be written as:

$$F_1/ATP \overset{k_{-1}}{\rightarrow} F_1 + ATP \tag{10}$$

The forward rate (v_f) for this reaction can be expressed as:

$$v_f = \frac{d[ATP]}{dt} = -\frac{d[F_1/ATP]}{dt} = k_{-1}[F_1/ATP] \tag{11}$$

$[F_1/ATP]$ is equal to the original concentration of the enzyme-product complex, $[F_1/ATP]_0$, minus the concentration of product ATP (i.e., free ATP), $[ATP]$. Thus, Eq. 11 can be arranged to:

$$v_f = k_{-1}\{[F_1/ATP]_0 - [ATP]\} \tag{12}$$

and Eqs. 11 and 12 can be rearranged to:

$$\frac{d[ATP]}{[F_1/ATP]_0 - [ATP]} = k_{-1} dt \tag{13}$$

Integrating Eq. 13 gives,

$$\log \frac{[F_1/ATP]_0 - [ATP]}{[F_1/ATP]_0} = -\frac{k_{-1}t}{2.303} \tag{14}$$

In Eq. 14, $[ATP]$ (i.e., free ATP) is equal to the amount of glucose-6-phosphate formed in the presence of glucose and hexokinase (Souid and Penefsky, 1995). $[F_1/ATP]_0$ is equal to the total amount of added radioactive ATP minus small amounts of free radioactive Pi in the ATP solution (less than 2%) minus the free ATP (less than 3%). Thus, the term $\{[F_1/ATP]_0 - [ATP]\}$ in Eq. 14 represents bound ATP and $[F_1/ATP]_0 - [ATP]/[ATP]_0$ represents the fraction of total ATP bound in catalytic sites, the fraction remaining. A plot of log fraction remaining versus time (i.e., a first-order plot) gives a straight line, and k_{-1} can be calculated from the slope, slope $= -k_{-1}/2.303$ (Souid and Penefsky, 1995).

The step of binding of ATP in the catalytic sites of the enzyme can be written as:

$$F_1 + \text{ATP} \xrightarrow{k_{+1}} F_1/\text{ATP} \qquad (15)$$

The rate (v_b) of this reaction can be expressed as:

$$v_b = d[F_1/\text{ATP}]/dt = k_{+1}[F_1]/[\text{ATP}] \qquad (16)$$

$[F_1]$ is equal to the original concentration of the enzyme, $[F_1]_o$, minus the concentration of enzyme–product complex formed; $[F_1/\text{ATP}]$ $[\text{ATP}]$ is equal to the original concentration of ATP, $[\text{ATP}]_o$, minus $[F_1/\text{ATP}]$. Thus, Eq. 16 can be arranged to give:

$$v_b = k_{+1}\{[F_1]_o - [F_1/\text{ATP}]\}\{[\text{ATP}]_o - [F_1/\text{ATP}]\} \qquad (17)$$

Equations 16 and 17 can be rearranged to:

$$d[F_1/\text{ATP}]/\{[F_1]_o - [F_1/\text{ATP}]\}\{[\text{ATP}]_o - [F_1/\text{ATP}]\} = k_{+1}/dt \qquad (18)$$

Integrating Eq. 18 gives:

$$\ln \frac{[F_1]_o\{[\text{ATP}]_o - [F_1/\text{ATP}]\}}{[\text{ATP}]_o\{[F_1]o - [F_1/\text{ATP}]\}} = \{[\text{ATP}]_o - [F_1]_o\}k_{+1}t \qquad (19)$$

$[\text{ATP}]_o$ was determined spectrophotometrically, $[F_1]_o$ was calculated on the assumption that each milligram of submitochondrial particles contains 0.4 nmole of F_1 (Harris et al., 1977; Beltran et al., 1986; Matsuno-Yagi and Hatefi, 1988), and $[F_1/\text{ATP}]$ was set to equal the amount of radioactive Pi formed following addition of cold ATP in cold-chase experiments or the amount of radioactive ATP that is inaccessible to hexokinase in the hexokinase method (Souid and Penefsky, 1995). A plot of Eq. 19 (second-order plot) gives a straight line, and k_{+1} can be calculated from the slope:

$$\text{slope} = k_{+1}\{[\text{ATP}]_o - [F_1]_o\}.$$

The equilibrium constant, $K_{\text{eq ATP}}$ (dissociation constant, $K_{\text{d ATP}}$), the standard free energy, ΔG^o, and the observed free energy changes, ΔG_p, can be calculated as follows; the equilibrium reaction for the step of ATP release can be written as:

$$F_1/\text{ATP} \underset{k_{+1}}{\overset{k_{-1}}{\rightleftharpoons}} F_1 + \text{ATP} \qquad (20)$$

At equilibrium, v_f (Eq. 11) is equal to v_b (Eq. 16). Thus,

$$k_{-1}[F_1/\text{ATP}] = k_{+1}[F_1][\text{ATP}] \tag{21}$$

Equation 21 can be arranged to give:

$$[F_1][\text{ATP}]/[F_1/\text{ATP}] = k_{-1}/k_{+1} = K_{d\,\text{ATP}} = K_{eq\,\text{ATP}} \tag{22}$$

The standard free energy can be calculated from the equilibrium constant using the following expression:

$$\Delta G^o = -RT \ln K_{eq} \tag{23}$$

The observed free energy changes during transition from nonenergized to energized states are calculated using the following expression:

$$\Delta G_p = -RT \ln \Delta K_{eq} \tag{24}$$

where ΔK_{eq} is the difference in equilibrium constants during transition from nonenergized to energized states.

4. UTILIZATION OF INTRINSIC BINDING ENERGY AND THE BINDING CHANGE MECHANISM

Jencks (1975, 1989) has proposed that the intrinsic binding energy is a driving force in enzyme catalysis. High-affinity binding of ATP in the active site of an enzyme is thought to "pull" the equilibrium toward ATP formation (Jencks, 1989). In contrast, binding of ADP and Pi is relatively weak. The weak binding of ADP and Pi can be related to destabilization forces in the catalytic site. Release of these forces favors ATP formation. Thus, the energy difference between tight binding of ATP and loose binding of ADP and Pi (the interaction energy) is a critical factor in the formation of ATP. The enzyme–substrate complex represents a higher energized state, so that ADP and Pi can readily react to form ATP.

A representation of the reaction mechanism for ATP synthesis in accord with Jencks' suggestions is shown in Fig. 1 (Souid and Penefsky, 1994). The catalytic steps are shown in the upper part and the associated energy levels in the lower part of Fig. 1. The first catalytic step couples binding of ADP and Pi to proton translocations from the intermembrane space (I) to the mitochondrial matrix (M), Fig. 1, top. The captured energy is utilized to overcome the loss of entropy associated with binding of ADP and Pi (Jencks, 1989). This energy is stored in a higher energized state of the enzyme, E2 (Fig. 1, bottom: ATP synthesis). Destabilization forces are responsible for the energized E2–ADP–Pi complex. Release of these forces favors ATP formation. The reversible

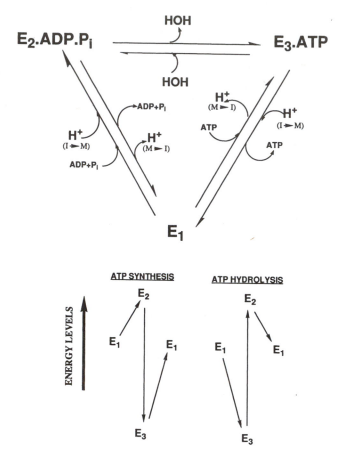

FIGURE 1. The reaction mechanism of ATP synthesis and hydrolysis according to the intrinsic binding energy concept (see text). The catalytic steps at a single site are summarized in the top and the energy levels are summarized in the bottom of the figure. Adapted from Souid and Penefsky (1994), with permission.

formation of ATP occurs independent of proton translocation, that is, the reaction is insensitive to uncouplers. ATP formation is driven by the free energy of binding of product ATP (Fig. 1, bottom: ATP synthesis). In the final step, the energy of proton translocation from I to M is utilized to lower the affinity of the enzyme for ATP, promoting its release.

According to Boyer (1993), the energy of proton translocations is utilized to alter conformational states of F_oF_1 that result in ATP formation. Based on the observation that the $Pi \rightleftharpoons HOH$ oxygen exchange reaction catalyzed by submitochondrial particles is less sensitive to uncouplers than ATP synthesis, Boyer and co-workers (1973) concluded that the reversible formation of ATP is independent of proton flux. In contrast, Pi $\rightleftharpoons$ ATP exchange, a reaction requiring dissociation and rebinding of ATP, is highly sensitive to uncouplers. Thus, the energy of proton transductions is required primarily for the release of bound ATP (Boyer, 1993; Boyer et al., 1973).

5. KINETIC ASPECTS OF UNISITE CATALYSIS

Unisite catalysis is defined as ATP hydrolysis in a single catalytic site of multisite ATPases (Souid and Penefsky, 1994; Penefsky and Cross, 1991; Penefsky, 1986). Catalysis at a single site of the enzyme is characterized by the following properties: (1) The affinity of the site for ATP is very high, $K_{a\,ATP} = 10^{12}$ M^{-1}; (2) negative cooperativity in binding; (3) the rate of hydrolysis of ATP in a single catalytic site is slow; the turnover number is 10^{-3} sec^{-1}; and (4) cooperativity between catalytic sites; ATP hydrolysis is accelerated to 600 sec^{-1} (as much as five orders of magnitude) when substrate is made available to additional sites on the enzyme (Souid and Penefsky, 1994; Penefsky and Cross, 1991).

The elementary steps for ATP hydrolysis in a single catalytic site of soluble beef heart F_1 (Cross et al., 1982; Grubmeyer et al., 1982) and membrane-bound F_oF_1 (Penefsky, 1985a) are determined under unisite conditions. These steps are summarized below:

$$
\begin{array}{cccc}
\text{(step 1)} & \text{(step 2)} & \text{(step 3)} & \text{(step 4)} \\
& & \text{Pi} & \text{ADP} \\
& & \nearrow & \nearrow \\
F_1 + \text{ATP} \rightleftharpoons F_1/\text{ATP} & \rightleftharpoons F_1/\text{ADP/Pi} & \rightleftharpoons F_1/\text{ADP} & \rightleftharpoons F_1
\end{array}
$$

Step 1 represents the formation of the enzyme–substrate complex (F_1/ATP). The bimolecular rate constant describing the binding of [γ-^{32}P]ATP by F_1 is fast ($k_{+1} = 1.6 \times 10^6$ M^{-1} sec^{-1}). However, the first-order rate constant describing [γ-^{32}P]ATP dissociation from the enzyme–substrate complex is slow ($k_{-1} = 7 \times 10^{-6}$ sec^{-1}). Thus, the dissociation constant ($K_{d\,ATP}$) for ATP bound in a high-affinity catalytic site is 10^{-12} M. That is, under unisite conditions, addition of [γ-^{32}P]ATP to F_1 is followed by rapid binding of ATP and virtually no dissociation.

Step 2 is the catalytic step. Products remain bound in the catalytic site. The equilibrium constant (K_2) is near unity, suggesting that this step takes place with little or no change in Gibbs free energy.

Steps 3 and 4 are product release steps. Pi and ADP dissociate from the enzyme at slow and apparently equal rates. The k_{-3} and k_{-4} are 4×10^{-3} sec^{-1}.

These latter rates define the net rate of hydrolysis under unisite conditions (Cunningham and Cross, 1988).

Thus, the properties of unisite catalysis support the binding change mechanism (Boyer, 1993; Rosing *et al.*, 1977; Kayalar *et al.*, 1977). According to Boyer and co-workers, reversible ATP formation takes place with little or no change in Gibbs free energy, is not coupled to proton translocation, and is insensitive to uncouplers. However, the release of product ATP is the energy-requiring step. In addition, Boyer (1993) has proposed that binding of ADP and Pi are energy-requiring reactions (Fig. 1).

The energy-dependent release of ATP bound in high-affinity catalytic sites of submitochondrial particles from beef heart was studied under unisite conditions during oxidation of reduced respiratory chain substrates (Penefsky, 1985b). Rapid rates of ATP dissociation were observed. The rates were commensurate with the rate of ATP synthesis catalyzed by the same particles (Penefsky, 1985b). The NADH-stimulated dissociation of $[\gamma\text{-}^{32}P]ATP$ was prevented by either FCCP or oligomycin (Penefsky, 1985b).

The properties of the high-affinity catalytic site of beef heart mitochondrial ATPase support the conclusion that this site is a competent catalytic site. It participates in ATP hydrolysis at V_{max} rates, is accessible to cooperative interactions between catalytic subunits, and can utilize energy released during proton flux (Souid and Penefsky, 1994; Penefsky and Cross, 1991; Penefsky, 1985b). Unisite catalysis closely similar to that of beef heart mitochondria has been described for soluble F_1 from *Escherichia coli* plasma membranes (Park *et al.*, 1994; Senior and Al Shawi, 1992; Senior, 1992; Al Shawi *et al.*, 1990; Al Shawi and Senior, 1988; Noumi *et al.*, 1986; Duncan and Senior, 1985) and from yeast mitochondria (Mueller, 1989). There were also similarities in chloroplast F_0F_1-type ATPase. In contrast, Yohda and Yoshida (1987) found little or no evidence for unisite catalysis by the ATPase of the thermophilic bacterium PS3 when ATP was used as a substrate. However, an acceleration in the hydrolysis of trinitrophenyl-ATP (TNP-ATP) was observed when the enzyme was switched from conditions of unisite to multisite catalysis (Hisabori *et al.*, 1992). Such an acceleration is diagnostic of unisite catalysis.

A study of the soluble F_1 from *E. coli* plasma membranes found that enzyme preparations lacking the δ subunit exhibited the rate accelerations characteristic of unisite hydrolysis of ATP, while enzyme preparations containing a full complement of the δ subunit did not show such accelerations (Xiao and Penefsky, 1994). Reconstitution of a five-subunit enzyme by incubating four-subunit preparations (lacking the δ subunit) with a purified preparation of subunit δ was accompanied by disappearance of the response to a cold chase. Because the δ subunit was required for binding of the soluble enzyme to F_1-depleted vesicles (Mendel-Hartvig and Capaldi, 1991; Engelbrecht and Junge, 1990; Sternweis and Smith, 1977; Futai *et al.*, 1974; Bragg *et al.*, 1973), it might be expected that membrane-bound *E. coli* F_oF_1 would not exhibit rate accelerations characteristic of unisite catalysis. An absence of cold-chase acceleration of ATP hydrolysis by *E. coli* vesicles was in fact noted. However, after extraction of the vesicles with KCl, cold-chase accelerations were observed; moreover, ATP bound in high-affinity catalytic sites of membrane-bound *E. coli* F_1 was subject to energy-dependent dissociation (Xiao and Penefsky, 1994).

6. ENERGY-DEPENDENT DISSOCIATION OF PRODUCT ATP FROM HIGH-AFFINITY CATALYTIC SITES

As discussed above, the affinity of nonenergized submitochondrial particles for ATP is high ($K_{d\ ATP} = 4.3 \times 10^{-12}$ M). However, in the presence of 1 mM NADH, 0.1 mM ADP, and 5 mM Pi, $K_{d\ ATP}$ is 6×10^{-5} M (60 μM) (Table I). Thus, the affinity for product ATP decreases more than seven orders of magnitude during respiration.

The above $K_{d\ ATP}$ of 60 μM is obtained with suboptimal concentrations of ADP because hexokinase is rate limiting in the measurement. But maximum rates of ATP synthesis (1 μmole ATP/min per mg) are observed with ETPH(Mg^{2+}) particles at 1 mM ADP. The turnover number calculated from the latter rate is 44 mole of ATP per mole of F_1 per second. If the turnover number is set equal to k_{-1}, the calculated $K_{d\ ATP}$ at 30°C in the presence of 1 mM ADP is in the millimolar range. Such a dissociation constant is consistent with the millimolar range of ATP concentration in the mitochondrial matrix (Souid and Penefsky, 1995; Klingenberg, 1975), and in fact is required if ATP is to be formed in the mitochondrial matrix.

The observed free energy change (ΔG) during the transition from non-energized membranes to membranes energized by NADH alone is 8 kcal/mole

Table I
Rate Constants, Equilibrium Constants, and Free Energy Changes during ATP Synthesis Catalyzed by ETPH (Mg^{2+})[a]

Additions	k_{+1} ($M^{-1}\ sec^{-1}$)	k_{-1} (sec^{-1})	K_d (M)	$\Delta G°$ (kcal/mole)	ΔG (kcal/mole)
None	2.3×10^5	10^{-6}	4.3×10^{-12}	15.4	—
ADP	1.6×10^5	10^{-6}	6.3×10^{-12}	15.2	—
NADH	1.0×10^5	0.32	3.2×10^{-6}	7.4	8
NADH + ADP	3.8×10^4	2.3	6×10^{-5}	5.7	9.7

[a]Adapted from Souid and Penefsky (1995). Where shown, the concentration of ADP was 0.1 mM and of NADH, 1 mM.

and by NADH and ADP is 9.7 kcal/mole (Table I). The ΔG value of 8 kcal/mole should be compared with the standard free energy of hydrolysis of ATP of -8.4 kcal/mole (Rosing and Slater, 1971).

The change in $K_{d\ ATP}$ accompanying energization is expressed almost entirely as an increase in the rate of ATP dissociation. The k_{-1} increases by five orders of magnitude over the rate observed with nonenergized particles. Only small changes are observed in k_{+1} (Table I).

The rate of energy-dependent dissociation of ATP is increased in the presence of ADP. The enhancement is concentration dependent and at 0.1 mM ADP is sevenfold (Table I). The enhancement also is sensitive to FCCP (Souid and Penefsky, 1995). Thus, proton flux brings about cooperative interactions between the subunits of the enzyme that further facilitate the release of product ATP. This conclusion is compatible with earlier findings of cooperativity with regard to both ATP hydrolysis (Grubmeyer and Penefsky, 1981a,b) and synthesis (Matsuno-Yagi and Hatefi, 1985). The observed free energy change associated with the cooperative effects of ADP is calculated to be 1.7 kcal/mole (Table I).

The contribution of subunit cooperativity to ATP synthesis is of considerable importance, since the rate of ATP synthesis at very low concentration of ADP is negligible (Perez and Ferguson, 1990a,b; Matsuno-Yagi and Hatefi,

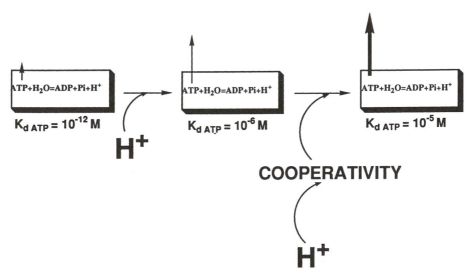

FIGURE 2. A schematic presentation of the changes in the affinity of a single catalytic site of the enzyme for product ATP during transition from the nonenergized to the energized state. As can be seen, vectorial protons bring about cooperative interactions between the catalytic subunits that facilitate product ATP release (see text).

ETPH(Mg^{++})

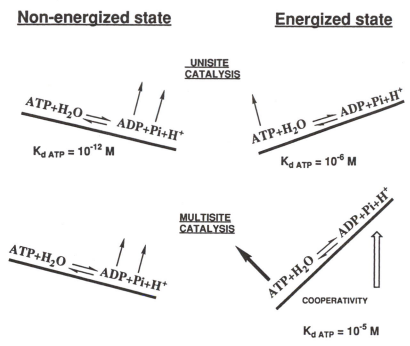

FIGURE 3. Changes in the kinetic aspects of a single catalytic site of the enzyme during transitions from the nonenergized to the energized state and from the unisite to the multisite catalysis. As can be seen, vectorial protons lower the affinity for product ATP by six orders of magnitude (unisite catalysis mode). In the presence of ADP (multisite catalysis mode), a further tenfold decrease in the affinity for ATP is observed in the energized state.

1985). Further support for the importance of subunit cooperativity is shown in Table I, where it can be seen that in the presence of NADH alone, $K_{d\ ATP}$ is about 3 μM. Thus, the affinity for product ATP in the absence of ADP is too high to permit efficient net synthesis of ATP (Souid and Penefsky, 1995). The value of 3 μM for $K_{d\ ATP}$ in the presence of NADH alone may be compared with earlier reports of $K_{d\ ATP}$ in light-energized spinach chloroplasts of 7 μM (Magnusson and McCarty, 1976) and of 16 μM in phosphorylating vesicles from *Paracoccus denitrificans* by Perez and Ferguson (1990 a,b).

The energy made available to the enzyme by proton translocations and the cooperative interactions between the catalytic subunits determine the affinity for product ATP. This in return determines the concentration of ATP in the mitochondrial matrix. A schematic illustration of the observed affinity changes for product ATP as a result of proton generation during respiration is presented in Figs. 2 and 3. Figure 2 shows that vectorial protons lower the affinity of the catalytic site for product ATP. In addition, in the presence of ADP, translocated protons bring about cooperative interactions between the subunits, which further facilitate product ATP release. Figure 3 summarizes the changes in $K_{d\ ATP}$ that occur as a result of transitions from nonenergized to energized states and from unisite to multisite catalysis. In the unisite mode, the catalytic site has high affinity for ATP ($K_{d\ ATP} = 10^{-12}$ M). The enzyme is capable of ATP hydrolysis. However, during NADH oxidation and in the absence of ADP, the affinity for product ATP decreases six orders of magnitude, facilitating ATP release. In the presence of NADH and ADP (energized, multisite catalysis mode), further enhancement in the rate of ATP release is observed ($K_{d\ ATP}$ is about 10-fold higher).

The question may be raised whether ATP is hydrolyzed during ATP synthesis? Since $K_{d\ ATP}$ is in the millimolar range during oxidative phosphorylation, one would expect binding of product ATP to catalytic sites if the matrix concentration of ATP is high enough. However, cooperative interactions between catalytic subunits on energized membranes or the presence of added ADP or ATP promote dissociation of bound product ATP rather than hydrolysis. Moreover, measurements of ATP hydrolysis by ETPH(Mg^{2+}) during oxidative phosphorylation indicate a low rate of hydrolysis that can be equated with the small amount of oligomycin-sensitive ATPase present (A.K. Souid and H.S. Penefsky, unpublished data).

ACKNOWLEDGMENT.   Supported in part by Research Grant GM21737 from the National Institutes of Health, United States Public Health Service.

7. REFERENCES

Abrahams, J. P., Luter, R., Todd, R. J., van Raaji, M. J., Leslie, A. G., and Walker, J. E., 1993, Inherent asymmetry of the structure of F1-ATPase from bovine heart mitochondria at 6.5 A resolution. *EMBO J.* **12:**1775–1780.

Abrahams, J. P., Leslie, A. G. W., Lutter, R., and Walker, J. E., 1994, Structure at 2.8 A resolution of F1-ATPase from bovine heart mitochondria. *Nature* **370:**621–628.

Al Shawi, M. K., and Senior, A. E., 1988, Complete kinetics and thermodynamic characterization of the unisite catalytic pathway of *Escherichia coli* F_1-ATPase. Comparison with mitochondrial F_1-ATPase and application to the study of mutant enzymes. *J. Biol. Chem.* **263:** 19640–19648.

Al Shawi, M. K., Parsonage, D., and Senior, A. E., 1990, Thermodynamic analyses of the catalytic pathway of Fl-ATPase from *Escherichia coli.* Implications regarding the nature of energy coupling by Fl-ATPases. *J. Biol. Chem.* **265:**4402–4410.

Amzel, L. M., McKinney, M., Narayanan, P., and Pedersen, P. L., 1982, Structure of the mito-chondrial F_1-ATPase at 9-A resolution. *Proc. Natl. Acad. Sci. USA* **79:**5852–5856.

Amzel, L. M., and Pedersen, P. L., 1983, Proton ATPases: Structure and mechanism. *Annu. Rev. Biochem.* **52:**801–824.

Beavis, A. D., and Lehninger, A. L., 1986, The upper and lower limits of the mechanistic stoi-chiometry of mitochondrial oxidative phosphorylation. Stoichiometry of oxidative phospho-rylation. *Eur. J. Biochem.* **158:**315–322.

Beltran, C., Tuena de Gomez-Puyou, M., Darszon, A., and Gomez-Puyou, A., 1986, Simultaneous synthesis and hydrolysis of ATP regulated by the inhibitor protein in submitochondrial particles. *Eur. J. Biochem.* **160:**163–168.

Berry, E. A., and Hinkle, P. C., 1983, Measurement of the electrochemical proton gradient in submitochondrial particles. *J. Biol. Chem.* **258:**1474–1486.

Boekema, E. J., Berden, J. A., and Van Heel, M. G., 1986, Structure of mitochondrial F_1-ATPase by electron microscopy and image processing. *Biochim. Biophys. Acta* **851:**353–360.

Boyer, P. D., 1993, The binding change mechanism for ATP synthase, some probabilities and possibilities. *Biochim. Biophys. Acta* **1140:**215–250.

Boyer, P. D., Cross, R. L., and Momsen W., 1973, A new concept for energy coupling in oxidative phosphorylation based on a molecular explanation of the oxygen exchange reactions. *Proc. Natl. Acad. Sci.* **70:**2837–2839.

Bragg, P. D., Davies, P. L., and Hou, C., 1973, Properties of nonheme iron in a cell envelope fraction from *Escherichia coli. Arch. Biochem. Biophys.* **159:**664–671.

Branca, D., Ferguson, S. J., and Sorgato, M. C., 1981, Clarification of factors influencing the nature and magnitude of the protonmotive force in bovine heart submitochondrial particles. *Eur. J. Biochem.* **116:**341–346 .

Capaldi, R. A., Aggeler, R., Turina, P., and Wilkens, S., 1994, Coupling between catalytic sites and the proton channel in F_1F_o-type ATPases. *Trends Biochem. Sci.* **19:**284–289 .

Clintron, N. M., and Pedersen, P. L., 1979, Purification of an ATPase inhibitor peptide fraction from rat liver mitochondria. *J. Biol. Chem.* **254:**3439–3443.

Collinson, I. R., Runswick, M. J., Buchanan, S. K., Fearnley, I. M., Skehel, J. M., van Raafij, M. J., Griffiths, D. E., and Walker, J. E., 1994, F_o membrane domain of ATP synthase from bovine heart mitochondria: Purification, subunit composition, and reconstitu-tion with F_1-ATPase. *Biochemistry* **33:**7971–7978.

Cross, R. L., Grubmeyer, C., and Penefsky, H. S., 1982, Mechanism of ATP hydrolysis by beef heart mitochondrial ATPase. Rate enhancements resulting from cooperative interactions be-tween multiple catalytic sites. *J. Biol. Chem.* **257:**12101-12105.

Cunningham, D., and Cross, R. L., 1988, Catalytic site occupancy during ATP hydrolysis by MF_1-ATPase. Evidence for alternating high affinity sites during steady-state turnover. *J. Biol. Chem.* **263:**18850–18856.

Dunn, S. D., and Heppel, L. A., 1981, Properties and functions of the subunits of *Escherichia coli* coupling factor ATPase. *Arch. Biochem. Biophys.* **210:**421–436.

Duncan, T. M., and Senior, A. E., 1985, The defective proton-ATPase of uncD mutants of *Esche-richia coli.* Two mutations which affect the catalytic mechanism. *J. Biol. Chem.* **260:**4901–4907.

Engelbrecht, S., and Junge, W., 1990, Subunit delta of $H^{(+)}$-ATPases: At the interface between proton flow and ATP synthesis. *Biochim. Biophys. Acta* **1015:**379–390.

Ferguson, S. J., and Sorgato, M. C., 1977, The phosphorylation potential generated by respiring bovine heart submitochondrial particles. *Biochem. J.* **168:**299–303.

Ferguson, S. J., and Sorgato, M. C., 1982, Proton electrochemical gradients and energy-transduction processes. *Annu. Rev. Biochem.* **51:**185–217.

Fillingame, R. H., 1980, The proton translocating pumps of oxidative phosphorylation. *Annu. Rev. Biochem.* **49:**1079–1113.

Fillingame, R. H., 1990, Molecular mechanics of ATP synthesis by F_1F_o-type H^+-transporting ATP synthases. In *The bacteria* (T. A. Krulwich, ed.), vol. XII pp. 345–391, Academic Press, New York.

Foster, D. L., and Fillingame, R. H., 1982, Stoichiometry of subunits in the H^+- ATPase complex of *Escherichia coli. J. Biol. Chem.* **257:**2009–2015.

Frangione, B., Rosenwasser, E., Penefsky, H. S., and Pullman, M. E., 1981, Amino acid sequence of the protein inhibitor of mitochondrial adenosine triphosphatase. *Proc. Natl. Acad. Sci. USA* **78:**7403–7407.

Futai, M., Sternweiss, P. C., and Heppel, L. A., 1974, Purification and properties of reconstitutively active and inactive adenosinetriphosphatase *Escherichia coli. Proc. Natl. Acad. Sci.* **71:**2725–2729.

Galante, Y. M., Wong, S. Y., and Hatefi, Y., 1979, Composition of complex V of the mitochondrial oxidative phosphorylation system. *J. Biol. Chem.* **254:**12372–12378.

Gogol, E. P., Lucken, U., Bork, T., and Capaldi, R. A., 1989, Molecular architecture of *Escherichia coli* F_1 adenosinetriphosphatase. *Biochem.* **28:**4709–4716.

Gomez-Fernandez, J. C., and Harris, D. A., 1978, A thermodynamic analysis of the interaction between the mitochondrial coupling adenosine triphosphatase and its naturally occurring inhibitor protein. *Biochem. J.* **176:**967–975.

Gomez-Puyou, M. T., Muller, U., Devars, S., Nava, A., and Dreyfus, G., 1982, Functional and immunological characterization of ATPase inhibitor proteins from heart, liver, and yeast mitochondria. *FEBS Lett.* **146:**168–172.

Gomez-Puyou, M. T., Muller, U., Dreyfus, G., Ayala, G., and Gomez-Puyou, A., 1983, Regulation of the synthesis and hydrolysis of ATP by mitochondrial ATPase. Role of the natural ATPase inhibitor protein. *J. Biol. Chem.* **258:**13680–13684.

Grubmeyer, C., and Penefsky, H. S., 1981a, The presence of two hydrolytic sites on beef heart mitochondrial adenosine triphosphatase. *J. Biol. Chem.* **256:**3718–3727.

Grubmeyer, C., and Penefsky, H. S., 1981b, Cooperativity between catalytic sites in the mechanism of action of beef heart mitochondrial adenosine triphosphatase. *J. Biol. Chem.* **256:**3728–3734.

Grubmeyer, C., Cross, R. L., and Penefsky, H. S., 1982, Mechanism of ATP hydrolysis by beef heart mitochondrial ATPase. Rate constants for elementary steps in catalysis at a single site. *J. Biol. Chem.* **257:**12092–12100.

Harris, D. A., Radda, G. K., and Slater, E. C., 1977, Tightly bound nucleotides of the energy-transducing ATPase, and their role in oxidative phosphorylation. *Biochim. Biophys. Acta* **459:** 560–572.

Hatefi, Y., 1985, The mitochondrial electron transport and oxidative phosphorylation system. *Annu. Rev Biochem.* **54:**1015–1069.

Hatefi, Y., 1993, ATP synthesis in mitochondria. *Eur. J. Biochem.* **218:**759–767.

Heeke, G. V., Deforce, L., Schnizer, R. A., Shaw, R., Couton, J. M., Shaw, G., Song, P., and Schuster, S. M., 1993, Recombinant bovine heart mitochondrial F1-ATPase inhibitor protein: Overproduction in *Escherichia coli*, purification, and structural studies. *Biochem.* **32:**10140–10149.

Herweijer, M. A., Berden, J. A., Kemp, A., and Slater, E. C., 1985, Inhibition of energy-transduction reactions by 8-nitreno-ATP covalently bound to bovine heart submitochondrial particles: direct interaction between ATPase and redox enzymes. *Biochim. Biophys. Acta* **809:** 81–89.

Hinkle, P. C., and Yu, M. L., 1979, The phosphorus/oxygen ratio of mitochondrial oxidative phosphorylation. *J. Biol. Chem.* **254:**2450–2455.

Hinkle, P. C., Kumar, M. A., Resetar, A., and Harris, D. L., 1991, Mechanistic stoichiometry of mitochondrial oxidative phosphorylation. *Biochem.* **30:**3576–3582.

Hisabori, T., Muneguki, E., Odaka, M., Yokoyama, K., Mochizuki, K., and Yoshida, M., 1992, Single site hydrolysis of 2′,3′-O-(2,4,6-trinitrophenyl)-ATP by the F_1-ATPase from thermophilic bacterium PS3 is accelerated by the chase-addition of excess ATP. *J. Biol. Chem.* **267:** 4551–4556.

Holian, A., and Wilson, D. F., 1980, Relationship of transmembrane pH and electrical gradients with respiration and adenosine 5′-triphosphate synthesis in mitochondria. *Biochem.* **19:**4213–4221.

Horstman, L. L., and Racker, E., 1970, Partial resolution of the enzyme catalyzing oxidative phosphorylation. XXII. Interaction between mitochondrial adenosine triphosphatase inhibitor and mitochondrial adenosine triphosphatase. *J. Biol. Chem.* **245:**1336–1344.

Issartel, J. P., Dupuis, A., Garin, J., Lunardi, J., Michel, L., and Vigaris, P. V., 1992, The ATP synthase (F_o-F_1) complex in oxidative phosphorylation. *Experientia* **48:**351–362.

Jencks, W. P., 1975, Binding energy, specificity, and enzymic catalysis: The circa effect. *Adv. Enzymol.* **43:**219–410.

Jencks, W. P., 1989, Utilization of binding energy and coupling rules for active transport and other coupled vectorial processes. *Methods in Enzymology* **171:**145–164.

Kagawa, Y., 1978, Proton translocating ATPase: Its pump, gate, and channel. *Adv. Biophys.* **10:** 209–247.

Kagawa, Y., Sone, N., Hirata, H., and Yoshida, M., 1979, Structure and function of H^+-ATPase. *J. Bioenerg.* **11:**39–78.

Kayalar, C., Rosing, J., and Boyer, P.D., 1977, An alternating site sequence for oxidative phosphorylation suggested by measurement of substrate binding patterns and exchange reaction. *J. Biol. Chem.* **252:**2486–2491.

Klein, G., Satre, M., Dianoux , A.C., and Vignais, P.V., 1981, Photoaffinity labeling of mitochondrial adenosine triphosphatase by an azido derivative of the natural adenosine triphosphate inhibitor. *Biochem.* **20:**1339–1344.

Klingenberg, M., 1975, Energetic aspects of transport of ADP and ATP through the mitochondrial membrane. *Ciba Found. Symp.* **31:**105–120.

Lehninger, A. L., 1975, Bioenergetic principles and the ATP cycle, in *Biochemistry* (A. L. Lehninger, ed.), pp. 387–416 and 477–542. Worth Publishers, Inc., New York.

Lehninger, A. L., Reynafarje, B., Alexandre, A., and Villalobo, A., 1980, Respiration-coupled H^+ ejection by mitochondria. *Annals New York Academy of Science* **341:**585–592.

Lehninger, A. L., 1982, Bioenergetic principles and the ATP cycle. In *Biochemistry* (A. L. Lehninger, ed.), pp. 361–396 and 467–510. Worth Publishers, Inc., New York.

Lemasters, J. J., 1984, The ATP-to-oxygen stoichiometries of oxidative phosphorylation by rat liver mitochondria. An analysis of ADP-induced oxygen jumps by linear nonequilibrium thermodynamics. *J. Biol. Chem.* **259:**13123–13130.

Luvisetto, S., and Azzone, G. F., 1989, Nature of proton cycling during gramicidin uncoupling of oxidative phosphorylation. *Biochem.* **28:**1100–1118.

Magnusson, R. P., and McCarty, R. E., 1976, Light-induced exchange of nucleotides into coupling factor in 1 spinach chloroplast thylakoids. *J. Biol. Chem.* **251:**7417–7422.

Matsuno-Yagi, A., and Hatefi, Y., 1985, Studies on the mechanism of oxidative phosphorylation. *J. Biol. Chem.* **260:**14424–14427.

Matsuno-Yagi, A., and Hatefi, Y., 1988, Estimation of the turnover number of bovine heart F_oF_1 complexes for ATP synthesis. *Biochem.* **27:**335–340.

McEnery, M. W., Buhle, E. L., Aebi, U., and Pedersen, P. L., 1984, Proton ATPase of rat liver mitochondria. Preparation and visualization of a functional complex using the novel zwitterionic detergent 3-[(3-cholamidopropyl)dimethylammonio]-1-propanesulfonate. *J. Biol. Chem.* **259:**4642–4651.

Mendel-Hartvig, J., and Capaldi, R. A., 1991, Structure-function relationships of domains of the delta subunit in *Escherichia coli* adenosine triphosphatase. *Biochem. Biophys. Acta* **1060**:115–124.

Mitchell, P., 1961, Coupling of phosphorylation to electron and hydrogen transfer by a chemiosmotic type of mechanism. *Nature* **191**:144–148.

Mitchell, P., 1966, Chemiosmotic coupling in oxidative and photosynthetic phosphorylation, Glynn Research, Ltd., Bodmin, U.K.

Mitchell, P., and Moyle, J., 1968, Proton translocation coupled to ATP hydrolysis in rat liver mitochondria. *Eur. J. Biochem.* **4**:530–539.

Mitchell, P., and Moyle, J., 1969, Estimation of membrane potential and pH difference across the cristae membrane of rat liver mitochondria. *Eur. J. Biochem.* **7**:471–484.

Mitchell, P., 1979, Keilin's respiratory chain concept and its chemiosmotic consequences. *Science* **206**:1148–1159.

Mueller, D. M., 1989, A mutation altering the kinetic responses of the yeast mitochondrial F_1-ATPase. *J. Biol. Chem.* **264**:16552–16556.

Murphy, M. P., 1989, Slip and leak in mitochondrial oxidative phosphorylation. *Biochim. Biophys. Acta* **977**:123–141.

Nakamoto, R. K., 1996, Mechanisms of active transport in the F_oF_1 ATP synthase. *J. Membr. Biol.* **151**:101–111.

Nicholls, D. G., 1974, The influence of respiration and ATP hydrolysis on the proton-electrochemical gradient across the inner membrane of rat-liver mitochondria as determined by ion distribution. *Eur. J. Biochem.* **50**:305–315.

Noumi, T., Taniai, M., Kanazawa, H., and Futai, M., 1986, Replacement of arginine 246 by histidine in the beta subunit of *Escherichia coli* H^+-ATPase resulted in loss of multi-site ATPase activity. *J. Biol. Chem.* **261**:9196–9201.

Padan, E., and Rottenberg, H., 1973, Respiratory control and the proton electrochemical gradient in mitochondria. *Eur. J. Biochem.* **40**:431–437.

Park, M. Y., Omote, H., Maeda M., and Futai, M., 1994, Conserved Glu-181 and Arg-182 residues of *Escherichia coli* $H^{(+)}$-ATPase (ATP synthase) beta subunit are essential for catalysis: Properties of 33 mutants between beta Glu-161 and beta Lys-201 residues. *J. Biochem.* **116**:1139–1145.

Pedersen, P. L., Schwerzmann, K., and Cintron, N., 1981, Regulation of the synthesis and hydrolysis of ATP in biological systems: Role of peptide inhibitors of H^+-ATPase. *Curr. Top. Bioenerg.* **11**:149–199.

Pedersen, P. L., and Amzel, L. M., 1993, ATP synthase. Structure, reaction center, mechanism, and regulation of one of nature's most unique machines. *J. Biol. Chem.* **268**:9937–9940.

Penefsky, H. S., 1979, Mitochondrial ATPase. *Adv. Enz.* **49**:223–280.

Penefsky, H. S., 1985a, Reaction mechanism of the membrane-bound ATPase of submitochondrial particles from beef heart. *J. Biol. Chem.* **260**:13728–13734.

Penefsky, H. S., 1985b, Energy-dependent dissociation of ATP from high affinity catalytic sites of beef heart mitochondrial adenosine triphosphatase. *J. Biol. Chem.* **260**:13735–13741.

Penefsky, H. S., 1986, Rate constants and equilibrium constants for the elementary steps of ATP hydrolysis by beef heart mitochondrial ATPase. *Methods Enzymol.* **126**:608–618.

Penefsky, H. S., and Cross, R. L., 1991, Structure and mechanism of F_oF_1-type ATP synthases and ATPases. *Adv. Enzym. Rel. Areas Mol. Biol.* **64**:173–214.

Perez, J. A., and Ferguson, S.J., 1990a, Kinetics of oxidative phosphorylation in Paracoccus denitrificans. 1. Mechanism of ATP synthesis at the active site(s) of F_oF_1-ATPase. *Biochemistry* **29**:10503–10518.

Perez, J. A., and Ferguson, S. J., 1990b, Kinetics of oxidative phosphorylation in Paracoccus denitrificans. 1. Evidence for a kinetic and thermodynamic modulation of F_oF_1-ATPase by the activity of the respiratory chain. *Biochemistry* **29**:10518–10526.

Phillips, R. C., George, P., and Rutman, R. J., 1969, Thermodynamic data for the hydrolysis of adenosine triphosphate as a function of pH, Mg^{2+} ion concentration, and ionic strength. *J. Biol. Chem.* **244**:3330–3342.

Pitard, B., Richard P., Dunach M., and Rigaud, J., 1996, ATP synthesis by F_oF_1 ATP synthase from thermophilic bacillus PS3 reconstituted into liposomes with bacteriorhodopsin. 2. Relationships between proton motive force and ATP synthesis. *Eur. J. Biochem.* **235**:779–788.

Pullman, M. E., and Monroy, G. C., 1963, A naturally occurring inhibitor of mitochondrial adenosine triphosphatase. *J. Biol. Chem.* **238**:3762–3769.

Rosing, L., and Slater, E. C., 1971, The value of ΔG^o for the hydrolysis of ATP. *Biochim. Biophys. Acta* **267**:275–290.

Rosing, J., Kayalar, C., and Boyer, P.D., 1977, Evidence for energy-dependent change in phosphate binding for mitochondrial oxidative phosphorylation based on measurements of medium and intermediate phosphate-water exchange. *J. Biol. Chem.* **252**:2478–2485.

Rottenberg, H., 1975, The measurement of transmembrane electrochemical proton gradients. *J. Bioenerg.* **7**:61–74.

Rottenberg, H., 1985, Proton-couple energy conversion: chemiosmotic and intramembrane coupling. *Modern Cell Biology* **4**:47–83.

Rouslin, W., and Pullman, M. E., 1987, Protonic inhibition of the mitochondrial adenosine 5′-triphosphatase in ischemic cardiac muscle. Reversible binding of the ATPase inhibitor protein to the mitochondrial ATPase during ischemia. *J. Mol. Cell Cardiol.* **19**:661–668.

Schneider, E., and Altendorf, K., 1987, Bacterial adenosine s′-triphosphate synthase (F_1F_o): Purification and reconstitution of F_o complexes and biochemical and functional characterization of their subunits. *Microbiol. Rev.* **51**:477–497.

Schwerzmann, K., and Pedersen, P., 1986, Regulation of the mitochondrial ATP synthase/ATPase complex. *Arch. Biochem. Biophys.* **250**:1–18.

Sebald, W., and Wild, G., 1979, Mitochondrial ATPase complex from *Neurospora crassa. Methods Enzymol.* **55**:344–351.

Senior, A. E., 1990, The proton-translocating ATPase of *Escherichia coli. Annu. Rev. Biophys. Bioyphs. Chem.* **19**:7–41.

Senior, A. E., 1992, Further examination of seventeen mutations in *Escherichia coli* F_1-ATPase beta-subunit. *J. Bioenerg. Biomembr.* **24**:479–484.

Senior, A. E., and Al Shawi M. K., 1992, Further examination of seventeen mutations in *Escherichia coli* F1-ATPase. *J. Biol. Chem.* **267**:21471–21478.

Serrano, R., Kanner, B. I., and Racker, E., 1976, Purification and properties of the proton-translocating adenosine triphosphatase complex of bovine heart mitochondria. *J. Biol. Chem.* **251**:2453–2461.

Skulachev, V. P., 1981a, Chemiosmotic proton circuits in biological membranes, in *Chemiosmotic proton circuits in biological membranes* (V. P. Skulachev and P. C. Hinkle, eds.), pp. 3–46, Addison-Wesley, Reading, Mass.

Slater, E. C., Rosing, J., and Mol A., 1973, The phosphorylation potential generated by respiring mitochondria. *Biochim. Biophys. Acta* **292**:534–553.

Sorgato, M. C., Ferguson, S. J., Kell, D. B., and John, P., 1978, The protonmotive force in bovine heart submitochondrial particles. Magnitude, sites of generation and comparison with the phosphorylation potential. *Biochem. J.* **174**:237–256.

Sorgato, M. C., and Ferguson, S. J., 1979, Variable proton conductance of submitochondrial particles. *Biochem.* **18**:5737–5742.

Souid, A. K., and Penefsky, H. S., 1994, Mechanism of ATP synthesis by mitochondrial ATP synthase from beef heart. *J. Bioenerg. Biomem.* **26**:627–630.

Souid, A. K., and Penefsky, H. S., 1995, Energetics of ATP dissociation from the mitochondrial ATPase during oxidative phosphorylation. *J. Biol. Chem.* **270**:9074–9082.

Sternweis, P. C., and Smith, J. B., 1977, Characterization of the purified membrane attachment (beta) subunit of the proton translocating adenosine triphosphatase from *Escherichia coli*. *Biochemistry* **16**:4020–4025.

Stoner, C. D., 1987, Determination of the P/2e$^-$ stoichiometries at the individual coupling sites in mitochondrial oxidative phosphorylation. Evidence for maximum values of 1.0, 0.5, and 1.0 at sites 1, 2, and 3. *J. Biol. Chem.* **262**:10445–10453.

Stryer, L., 1988, Oxidative phosphorylation, in *Biochemistry* (L. Stryer, ed.), pp. 396–426, W. H. Freeman and Company, New York.

Thayer, W. S., Tu, Y. L., and Hinkle, P. C., 1977, Thermodynamics of oxidative phosphorylation in bovine heart submitochondrial particles. *J. Biol. Chem.* **252**:8455–8457.

Trumpower, B. L., and Gennis, R. B., 1994, Energy transduction by cytochrome complexes in mitochondrial and bacterial respiration: The enzymology of coupling electron transfer reactions to transmembrane proton translocation. *Annu. Rev. Biochem.* **63**:675–716.

Tzagoloff, A., 1979, Oligomycin-sensitive ATPase of Saccharomyces cerevisiae. *Methods Enzymol.* **55**:351–358.

Walker, J. E., Saraste, M., Runswick, M., and Gay, N.J., 1982, Distantly related sequences in the alpha- and beta-subunits of ATP synthase, myosin, kinases and other ATP-requiring enzymes and a common nucleotide binding fold. *EMBO J.* **1**:945–951.

Walker, J. E., Saraste, M., and Gay, N.J., 1984, The unc operon. Nucleotide sequence, regulation and structure of ATP-synthase. *Biochim. Biophys. Acta* **768**:164–200.

Walker, J. E., Lutter, R., Dupuis, A., and Runswick, M. J., 1991, Identification of the subunits of F_1F_o-ATPase from bovine heart mitochondria. *Biochemistry* **30**:5369–5378.

Walker, J. E., and Collinson, I. R., 1994, The role of the stalk in the coupling mechanism of F_1F_o-ATPases. *FEBS Lett.* **346**:39–43.

Wikstrom, M., Krab, K., and Saraste, M., 1981, Proton-translocating cytochrome complexes. *Annu. Rev. Biochem.* **50**:623–655.

Wilkens, S., and Capaldi, R. A., 1994, Asymmetry and structural changes in ECF_1 examined by cryoelectronmicroscopy. *Biol. Chem. Hoppe-Seyler* **375**:43–51.

Van Dam, K., Westerhoff, H. V., Krab, K., Van der Meer, R., and Arents, J. C., 1980, Relationship between chemiosmotic flows and thermodynamic forces in oxidative phosphorylation. *Biochim. Biophys. Acta* **591**:240–250.

Vazquez-Laslop, N. and Dreyfus, G., 1990, The native mitochondrial F_1-inhibitor protein complex carries out uni- and multisite ATP hydrolysis. *J. Biol. Chem.* **265**:19002–19006.

Xiao, R., and Penefsky, H. S., 1994, Subunit catalysis and the delta subunit of F_1-ATPase in *Escherichia coli*. *J. Biol. Chem.* **269**:19232–19237.

Yohda, M., and Yoshida, M., 1987, Single-site catalysis of F_1-ATPase from thermophilic bacterium PS3 and its dominance in steady-state catalysis at low ATP concentration. *J. Bioch.* **102**: 875–883.

Zoratti, M., Pietrobon, D., Conover, T., and Azzone, G. F., 1981, On the role of $\Delta\mu_H$ as an intermediate in ATP synthesis, in *Vectorial reactions in electron and ion transport in mitochondria and bacteria* (F. Palieri *et al.*, eds.), pp. 331–338.

CHAPTER 17

Mutational Analysis of ATP Synthase

An Approach to Catalysis and Energy Coupling

Masamitsu Futai and Hiroshi Omote

1. INTRODUCTION

The ATP synthases (F_oF_1) of *Escherichia coli*, mitochondria, and chloroplasts (for reviews, see Futai and Omote, 1996; Fillingame, 1990; Senior, 1990; Futai *et al.*, 1989) are closely similar and have conserved basic subunit structures: a catalytic sector F_1—$\alpha_3\beta_3\gamma\delta\varepsilon$—and a transmembrane proton pathway F_o—a $b_2 c_{10-12}$. The ATP synthesis by F_oF_1 is driven by a transmembrane electrochemical proton gradient established by the electron transfer chain. The gradient is required for catalytic turnover: ATP release from and ADP plus Pi (phosphate) binding to the catalytic site in the β subunit. The cooperativity between the three catalytic sites has been established. However, elucidation of the mechanism of energy coupling between catalysis and the electrochemical proton gradient is still at an early stage.

The F_1 is an asymmetric molecule, as judged easily from its subunit stoichiometry (α:β:γ:δ:ε, 3:3:1:1:1). Bovine F_1 at 6 Å resolution clearly showed

Masamitsu Futai and Hiroshi Omote • Department of Biological Sciences, Institute of Scientific and Industrial Research, Osaka University, Ibaraki, Osaka 567, Japan.

Frontiers of Cellular Bioenergetics, edited by Papa *et al.* Kluwer Academic/Plenum Publishers, New York, 1999.

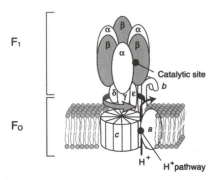

FIGURE 1. Schematic higher-ordered structure of *E. coli* F_oF_1. Flow of protons and subunit rotation are shown by arrows.

an asymmetric organization of subunits (Abrahams *et al.*, 1993). The higher-ordered structure of the bovine $\alpha_3\beta_3\gamma$ complex was solved at 2.8 Å resolution (Abrahams *et al.*, 1994); as schematically shown in Fig. 1, the α and β subunits have similar structures and are arranged alternately around the amino and carboxyl-terminal α-helices of the γ subunit. The α and β subunits have three distinct domains: the domain at the top of the complex, a six-stranded β barrel (an amino terminal region); the central domain, an α-helix–β sheet domain forming a catalytic site in the β subunit or a noncatalytic site in the α subunit; and the carboxyl-terminal domain, bundles of seven and six helices for the α and β subunits, respectively. Consistent with the mechanism of catalysis described below, the three β subunits (empty, ADP-bound, and ATP-bound forms) in the bovine structure have different conformations. The crystal structure of rat F_1 seems to be slightly different from the beef F_1 structure (Bianchet *et al.*, 1991), although higher-resolution results have not been reported yet. The differences between the two crystal structures, if any, will be interesting from the functional aspect of the enzyme; it may be possible that the states of the enzyme may be different, as the crystallization conditions are not identical. The X-ray structure of the $\alpha_3\beta_3$ complex of *Bacillus* PS3 was also solved recently (Shirakibara *et al.*, 1987).

E. coli F_oF_1 was highly purified by column chromatography (Schneider and Altendolf, 1982) or density gradient centrifugation (Foster and Fillingame, 1979). One-step purification from an overproducing strain was also reported (Moriyama *et al.*, 1991). The F_1 sector can be solubilized and purified as a soluble protein (Futai *et al.*, 1989). Extensive mutational analysis of the *E. coli* enzyme has been carried out taking advantage of the easy genetic manipulation of the organism (Futai and Omote, 1996; Fillingame, 1990; Senior, 1990; Futai *et al.*, 1989). The bovine and bacterial subunits exhibit high homologies: The amino acid identities of the α, β, and γ subunits from the two organisms are 55, 72, and 29%, respectively. Therefore, we can interpret mutational defects of the enzyme with reference to the higher-ordered structure of the bovine $\alpha_3\beta_3\gamma$ complex (Abrahams *et al.*, 1993, 1994), although it is still desirable to solve the higher-ordered structure of the *E. coli* enzyme. Comparison of the higher-ordered structures of the wild-type and mutant enzymes may be also very interesting. Furthermore, the holoenzyme (F_oF_1) of mitochondria or bacteria should be crystallized to obtain the entire higher-ordered structure.

In this chapter, we discuss the results of mutational analysis of ATP synthase, hoping to visualize catalysis and the coupling mechanism at the level of amino acid residues. We mainly discuss and focus on the results of our own studies. Excellent studies on the enzyme not cited in this chapter have been reviewed from different aspects (Futai and Omote, 1996; Boyer, 1993; Hatefi, 1993; Allison *et al.*, 1992; Amzel, 1992; Capaldi *et al.*, 1992; Duncan and Cross, 1992; Fillingame, 1990, 1992; Gromet-Elhanan, 1992; Kagawa *et al.*, 1992; Penefsky and Cross, 1991; Senior, 1988, 1990; Futai *et al.*, 1987, 1989; Walker *et al.*, 1984, 1987; Vignais and Lunardi, 1985; Futai and Kanazawa, 1983).

2. CATALYSIS AND CATALYTIC SITES IN THE F_1 SECTOR

2.1. ATP Synthesis and Hydrolysis by F_oF_1

ATP synthesis by *E. coli* F_oF_1 has been studied with membrane vesicles or F_oF_1 reconstituted into liposomes, and comparable turnover numbers have been obtained with the two systems (Fischer *et al.*, 1994). The purified bovine or *E. coli* F_1 sector showed two types of kinetics in ATP hydrolysis (Fischer *et al.*, 1994; Grubmeyer *et al.*, 1982; Duncan and Senior, 1985): unisite (single site) and multisite (steady state) catalysis, measured in the presence of a substoichiometric and an excess amount of ATP, respectively. The association constant in unisite catalysis, K_1 (k_{+1}/k_{-1}), for ATP binding–release is about 10^9 M^{-1} (*E. coli* value). The equilibrium constant (K_2) in unisite catalysis (F_1–ATP $\leftrightarrow$ F_1–ADP–Pi) is almost unity, indicating that F_1–ATP can be formed from F_1–ADP–Pi, with little free energy change (Noumi *et al.*, 1986; Duncan and Senior, 1985; Grubmeyer *et al.*, 1982). These results also suggest that F_1–ATP undergoes a hydrolysis–synthesis cycle a number of times, as demonstrated by the incorporation of oxygen of $[^{18}O]$ H_2O into Pi (O'Neal and Boyer, 1984). The formation of F_1–ATP was actually demonstrated upon the addition of ADP and Pi to F_1 (Sakamoto and Tonomura, 1983; Yoshida, 1983; Feldman and Sigman, 1982). In multisite catalysis, ATP binds to the three catalytic sites, and the rate of ATP hydrolysis is 10^5- to 10^6-fold higher than that of unisite catalysis. The higher multisite rate is due to cooperativity between the three catalytic sites in the β subunits. The requirement of Mg^{2+} for cooperative binding was clearly shown using the intrinsic fluorescence of the βTrp-331 residue (βTyr-331 → Trp mutant) (Weber *et al.*, 1994). Catalytic cooperativity in ATP synthesis was also shown (Matsuno-Yagi and Hatefi, 1985; Hatefi, 1993).

These observations support the "binding change mechanism" (Boyer, 1989), which predicts that an electrochemical proton gradient is not required for the synthesis of F_1–ATP from F_1–ADP–Pi, but is required for the dissociation of product ATP from and the binding of ADP and Pi to the catalytic sites. The rate of dissociation of ATP from a single catalytic site of mitochondrial F_oF_1 was actually five orders of magnitude higher in the presence of NADH and about six orders of magnitude in the presence of NADH and ADP (Souid and Penefsky, 1995).

Certain mutant enzymes are only defective in multisite catalysis, as discussed in Section 2.4. Similarly, inhibitors specific for multisite catalysis have been found: Azide (Noumi *et al.*, 1987c) or dicyclohexylcarbodiimide (DCCD) (Tommasino and Capaldi, 1985) binds to the β subunit and strongly inhibits multisite but not unisite catalysis, indicating that they inhibit the catalytic cooperativity required for multisite catalysis but not the chemical reaction itself. DCCD actually binds to βGlu-192, which is not in the catalytic site (Yoshida *et al.*, 1982). *N*-Ethylmaleimide alkylation to Cys-89 of the chloroplast γ subunit inhibited the multisite catalysis more than 90%, but had no effect on the unisite catalysis (Zhang and Jagendolf, 1995). Thus, unisite and multisite catalysis can be differentiated by mutations and inhibitors.

2.2. βLys-155 and βThr-156 Are Catalytic Residues

The importance of the glycine-rich sequence or phosphate-binding loop (P-loop) [Gly-X-X-X-X-Gly-Lys-Thr/Ser] in the nucleotide binding proteins has been suggested (Fillingame, 1990; Senior, 1990; Futai *et al.*, 1989). Mutation and affinity labeling results indicated that ßLys-155 (Omote *et al.*, 1992; Senior and Al-Shawi, 1992; Ida *et al.*, 1991; Tagaya *et al.*, 1988) and ßThr-156 (Omote *et al.*, 1992) are essential for catalysis. The P-loop of bovine F_1 is in the central nucleotide binding domain of the β subunit, and extends in the direction of the carboxyl- and amino-terminal α-helices of the γ subunit in the center of the F_1 sector (Abrahams *et al.*, 1993) (Fig. 2). Our real interest started when we found an *E. coli* mutant mapped to the P-loop [Gly-Gly-Ala-Gly-Val-Gly-Lys-Thr-Ala, *E. coli* β subunit positions 149–157]: The first P-loop mutant, βAla-151→Val, showed reduced multisite and altered unisite catalysis (Hsu *et al.*, 1987).

The covalent binding of one mole of an ATP affinity analogue, adenosine

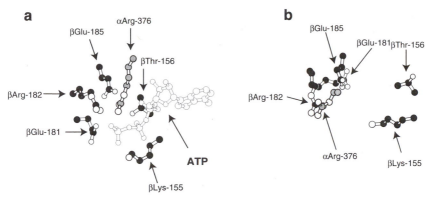

FIGURE 2. Catalytic site of *E. coli* F$_1$ ATPase. (a) The residues around ATP bound to the catalytic site in the β subunit. (b) The same region of the empty β subunit having no bound nucleotide. Comparison of (a) and (b) suggest conformational changes of a catalytic site after the release of ATP. βLys-155 and βThr-156 are residues in the P-loop and βGln-181, βArg-182, and βGlu185 are in the GERXXE sequence (see text and Fig. 3). The positions of amino acid residues were deduced from the bovine X-ray structure (Abrahams *et al.*, 1994). This figure and Figs. 4–6 were prepared by MOLSCRIPT (Kraulis, 1991).

triphosphopyridoxal, to βLys-155 per mole of F$_1$ completely inhibited the F$_1$-ATPase activity (Ida *et al.*, 1991; Noumi *et al.*, 1987b), consistent with the location of βLys-155 at the catalytic site near the β–γ phosphate moiety of ATP (Fig. 2a). Mutants of βLys-155 (βLys-155 → Ala, Ser, or Thr, or βThr-155/βLys-156) exhibited no oxidative phosphorylation or membrane ATPase activity (Omote *et al.*, 1992). Furthermore, the purified βAla-155 or βSer-155 mutant F$_1$ exhibited neither unisite nor multisite catalysis, although mutant enzymes bound ATP under both sets of catalytic conditions. Similar results were obtained with the βGln-155 mutant (Senior and Al-Shawi, 1992). These results suggest that the ε amino moiety of the βLys-155 side chain is essential for the binding of ADP or ATP at the catalytic site of the β subunit, as shown in Fig. 2. The difference of the direction of the βLys-155 side chain in the nucleotide-free and ATP-bound β subunit (Fig. 2) indicates the change of the catalytic site after the release of ATP.

The enzyme with a glycine insertion between βLys-155 and βThr-156 or replacement of the entire P-loop with that of adenylate kinase (<u>Gly</u>-Gly-Pro-Gly-Ser-<u>Gly</u>-<u>Lys</u>-Gly-<u>Thr</u>) showed no activity, suggesting that the hydroxyl moiety of the threonine residue should be at position 156 (Takeyama *et al.*, 1990). Consistent with this notion, replacement of the βThr-156 residue led to defective ATP synthesis and essentially no membrane ATPase activity, except that the βSer-156 mutant was active in ATP synthesis and exhibited ~1.5-fold higher membrane ATPase activity than the wild type (Omote *et al.*, 1992). Similarly, the yeast residue corresponding to βThr-156 can be replaced only by serine (Shen *et al.*, 1994). The purified βAla-156 and βCys-156 enzymes exhibited no multisite or unisite catalysis and showed no detectable ATP binding (k_{+1}) (Omote *et al.*, 1992). The mutant and wild-type β subunit bound aurovertin, but its fluorescence in the mutant and wild-type enzymes showed different responses to Mg^{2+}. Van Raaji *et al.* (1996) showed recently that aurovertin bound in a cleft between the nucleotide binding site and the carboxyl-terminal domain of bovine F_1. Thus, the fluorescence change of aurovertin is due to a conformation change transmitted from the catalytic site. These results suggest that βThr-156, possibly its hydroxyl moiety, is essential for catalysis and may contribute to Mg^{2+} binding. In a model of the ATP-bound catalytic site (Abrahams *et al.*, 1994), the hydroxyl moiety of the βThr-156 side chain can be placed close to the Mg (Fig. 2).

The α subunit also has a P-loop in the central domain similar to the β subunit. The bovine residues corresponding to αLys-175 and αThr-176 are near the β–γ phosphate moiety of ATP and Mg in the noncatalytic site (Abrahams *et al.*, 1994). However, mutational studies indicated that they are not absolutely essential for catalysis (Jounouchi *et al.*, 1993; Rao *et al.*, 1988). Amino acid replacements of these residues often caused defective assembly of the F_1 sector.

2.3. βGlu-181, βArg-182, and βGlu-185 in the Conserved GER XXE Sequence Are Catalytic Residues

Together with the P-loop, the GERXXE (Gly-Glu-Arg-X-X-Glu) sequence is conserved among the β subunits of ATP synthases, subunit A of vacuolar-type ATPases, and other ATP binding proteins (Fillingame *et al.*, 1990; Senior, 1990; Futai *et al.*, 1989; Futai and Omote, 1996; Amano *et al.*, 1994; Eichelberg *et al.*, 1994) (Fig. 3). Consistent with the high degree of conservation, mutations introduced at βGlu-181, βArg-182, or βGlu-185 gave seriously defective enzymes (Park *et al.*, 1994). Mutants at position 181 showed no ATP synthesis and only negligible membrane ATPase activity. Purified βAla-181 and βGln-181 F_1 showed no multisite and very slow unisite catalysis. These defects could be attributed to the decreased rates (k_{+2} and k_{-2}) of unisite catalysis (F_1–ATP ↔ F_1–ADP–Pi) (Park *et al.*, 1994; Senior and Al-Shawi, 1992). However, the βAsp-181 enzyme showed detectable multisite and unisite catalysis (27 and 21% of the wild-type levels, respectively) (Park *et al.*, 1994). These results suggest that the carboxyl side chain of βGlu-181 is essential for the catalysis. βGlu-181 could be located at the catalytic site in the higher-ordered structure, its carboxyl side chain being hydrogen bonded to the water molecule close to the γ–α phosphate moiety of ATP (Fig. 2). Therefore, this residue could activate the water molecule when the enzyme hydrolyzes ATP.

E. coli	GGAGVGKT VNMMEL IRN IA IEHSGYS

S. cerevisiae	GAFGCGKT VISQSLSKYSNSDA - - - I
Bovine	GAFGCGKT VISQSLSKYSNSDV - - - I
M. barkeri	GPFGSGKT VTQQSLAKWSDTE I - - - V
N. crassa	GAFGCGKT VISQSVSKFSNSDV IVYV
H. salinarium	GPFGSGKT VTQQSLAKFADAD IVVY I

P-loop

E. coli	VFAGV GERTRE GNDFYHEMTDSNV IDK

S. cerevisiae	IYVGC GERGNE MAEVLMEFPELYTEMS
Bovine	IYVGC GERGNE MSEVLRDFPELTMEVD
M. barkeri	VYIGC GERGNE MADVLSEFPELEDPQT
N. crassa	- - - GC GERGNE MAEVLKDFPELSIEVD
H. salinarium	- - - GC GERGNE MTEVIEDFPELPDPQT

GERXXE sequence

FIGURE 3. The P-loop (GXXXXGKT) and GERXXE sequences are conserved among the β subunits of ATP synthases from different origins, subunit A of vacuolar-type ATPase, and other ATP binding proteins. *E. coli* β subunit residues βLys-155, βThr-156, βGlu-181, βArg-182, and βGlu-185 are shown to be located in the catalytic site as catalytic residues (see also Fig. 2). Sequences are cited from database (Swiss Prot. Release34).

Mutant studies also suggested that βArg-182 is essential for substrate binding and catalysis (Park *et al.*, 1994). The purified βGln-182 enzyme showed essentially no multisite or unisite catalysis. The βLys-182 enzyme had low but significant activity, showing 1 and 85% of the multisite and unisite wild-type rates, respectively. Thus, the positive charge of the arginine side chain positioned at the catalytic site (Abrahams *et al.*, 1994) may be required for the catalysis (Fig. 2).

The βGlu-185 residue is the last residue of the conserved GER XXE sequence and is located at the catalytic site close to the Mg (Figs. 2, 3) (Abrahams *et al.*, 1994). All mutants at this position except βAsp-185 could not grow through oxidative phosphorylation and essentially had no membrane ATPase activities (Omote *et al.*, 1995; Noumi *et al.*, 1987a). The βAsp-185 mutant showed positive growth by oxidative phosphorylation, 20% of the wild-type membrane ATPase level, and similar unisite catalysis to the wild type. The purified F_1 of the βGln-185 and βCys-185 mutants exhibited no multisite catalysis, although they showed substantial unisite catalysis (Omote *et al.*, 1995). Interestingly, the βCys-185 enzyme modified with iodoacetate (carboxymethylated cysteine at position 185) of all three β subunits exhibited about one third of the wild-type multisite catalysis. The purified F_1 with βS-carboxylmethyl Cys-185 or βAsp-185 showed no Mg^{2+} inhibition. Mutant F_1 showed low sensitivity to azide, possibly because they exhibit altered dissociation constants for Mg ADP. Enzyme complexes with Mg–ADP are known to be

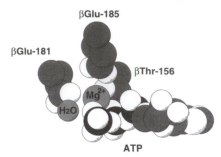

FIGURE 4. Location of Mg^{2+} and the βGlu-185 residue at the catalytic site of the ATP synthase. The higher-ordered structure around Mg^{2+} and βGlu-185 in the catalytic site is shown. The positions of amino acid residues are from the higher-ordered structure of bovine F_1 (Abrahams *et al.*, 1994).

inhibited by azide (Murataliev *et al.*, 1991). These results taken together suggest that the carboxyl moiety of βGlu-185 may form a Mg^{2+} binding site and that it is essential for the cooperativity of the enzyme. As shown in the model, βGlu-185 is located close to Mg in the ADP-bound or ATP-bound β subunit, but Mg is not found in the empty β subunit (Fig. 4), suggesting that Mg–ADP is released in the ATP hydrolysis reaction.

The bovine F_1 structure indicates that the adenine ring of ATP in the β subunit is surrounded by hydrophobic residues (Abrahams *et al.*, 1994). The residues forming the adenine-binding site were defined by affinity labeling of bovine F_1. The 2-azido ATP and 8-azido ATP bound to βTyr-331 (bovine Tyr-345) (Garin *et al.*, 1986) and βTyr-297 (bovine Tyr-311) (Hollemans *et al.*, 1983), respectively, in the hydrophobic region.

2.4. Interaction(s) between the α and β Subunits

Conformational transmission among catalytic sites in the β subunits is essential for cooperativity of the enzyme. Studies involving mutations and their suppression can be a possible approach for examining functional interaction(s) between the different domains within the enzyme. The defect due to a mutation in one domain may be suppressed by the second site-mutations in another domain, if the two domains functionally interact. This approach is possible if the residue of the first mutation affects catalysis seriously but is not absolutely essential, although the first and second mutations are not necessarily at or close to the catalytic site. This approach can be applied for studying the domain–domain interaction(s) within the same subunit or between different subunits.

βLys-155 and βThr-156 are catalytic residues, as discussed in Section 2.2. However, other residues in the P-loop, such as βGly-149 and βGly-150, are not absolutely essential and can be replaced by Ser or Ala without loss of activity (Iwamoto *et al.*, 1991). Mutations in these nonessential residues may cause defects, possibly by changing the conformations of essential residues in the P-loop. The βCys-149 mutant is one such example and is defective in ATP synthesis and multisite catalysis. The defect of the βCys-149 mutation was suppressed by the replacements of the βGly-172, βSer-174, βGlu-192, or βVal-198 residue (Iwamoto *et al.*, 1993). The single βSer-174 → Phe mutation was suppressed by βGly-149 → Ser (Iwamoto *et al.*, 1991), βGly-149 → Cys (Iwamoto *et al.*, 1993), or other mutations (Miki *et al.*, 1990). Our initial interpretation of these results was that βGly-149 and the residues with the second mutation are nearby and actually interacting. However, the corresponding bovine residues that suppressed the βCys-149 mutant are located near the outer surface of the β subunit and not actually near βGly-149, which faces the two central α-helices of the γ subunit (Abrahams *et al.*, 1994) (Fig. 5). Thus, the βCys-149 mutant may be defective in the conformational transmission from the P-loop to the outer domain(s), and such conformational transmission may be essential during multisite catalysis and ATP synthesis. It is of interest to solve the higher-ordered structures of the single and double mutant enzymes and compare them with that of the wild type.

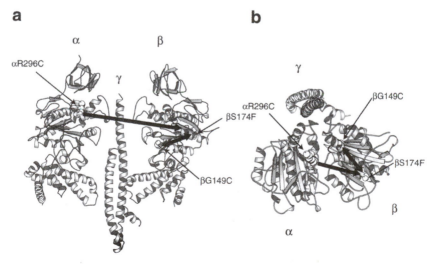

FIGURE 5. Long-range suppression of βGly-149 and βSer-174 mutations. Mutations are indicated by two single code amino acid abbreviations, such as βG149C (βGly-149 → Cys), and their suppression is indicated by arrows (⇒: βG149C ⇒ βS174F, the βGly149 → Cys mutation was suppressed by βSer-174 → Phe. Amino acid residues are mapped on the schematic higher-ordered structure of F$_1$. (a) Side view; (b) top view.

As described above, three β subunits synthesize or hydrolyze ATP in a cooperative manner. As each β subunit is located next to the α subunit in the hexagonal arrangement of the major subunits, conformational change(s) of one β subunit should be transmitted to another β subunit through the adjacent α subunit. The requirement of the functional interaction between the α and β subunits for catalytic cooperatively is supported by the early mutants mapped around position 370 of the α subunit such as αSer-373 → Phe (Noumi *et al.*, 1984; Maggio *et al.*, 1987) and α Arg-376 → Cys (Soga *et al.*, 1989). The mutant multisite catalysis was ~1/1000-fold lower than that of the wild type, but they exhibited normal or substantial unisite catalysis. In the higher-ordered structure (Fig. 2), αArg-376 is in the catalytic site close to the γ phosphate moiety of ATP, whereas αSer-373 at the α–β interface is at least several angstroms away from the γ phosphates.

The functional α–β subunit interaction(s) was suggested clearly by the results of mutation–suppression. Replacing βSer-174 with other residues reduced multisite catalysis, depending on the sizes of the residues: The larger the side chain volume of the residue introduced, the lower the enzyme activity became. Both βLeu-174 and βPhe-174 exhibited about 10% of the wild-type ATPase activity (Omote *et al.*, 1994). However, the βPhe-174 mutant was also defective in energy coupling, although the βLeu-174 mutant was only defective in catalysis. Interestingly, the defect of the βPhe-174 mutant in energy coupling was suppressed by the second mutation in the α subunit (αArg-296 → Cys) (Omote *et al.*, 1994). The αCys-296/βPhe-174 double mutant enzyme was coupled efficiently, although its membrane ATPase activity remained essentially the same as that of the βPhe-174 single mutant. The two bovine residues corresponding to αArg-296 and βSer-174 are not located nearby in the higher-ordered structure (Abrahams *et al.*, 1994) (Fig. 5). However, these two residues are located at the interface between the two subunits, suggesting the importance of the functional α–β interaction for energy coupling. As described above, mutations in βSer-174 or its nearby residues suppressed the βGly-149 → Cys mutation of the P-loop mutation (Iwamoto *et al.*, 1993). Thus, it may be possible to assume that the conformational change in the P-loop during the catalysis may be transmitted to the region around αArg-296 through the domain containing βSer-174, thus enabling the proper α–β subunit interaction.

3. THE γ SUBUNIT IN CATALYSIS AND ENERGY COUPLING

3.1. Mutations in the γ Subunit Cause Defective Catalysis and Assembly

The γ subunit is required for the formation of a minimal catalytic $\alpha_3\beta_3\gamma$ complex having ATPase activity (Dunn and Futai, 1980; Futai, 1977). The *E.*

coli (Miki *et al.*, 1986) or yeast (Paul *et al.*, 1994) mutant lacking this subunit had no stable F_1 attached to F_o. Furthermore, mutations in the carboxyl- and amino-terminal domains of the γ subunit severely affected enzyme assembly: γGln-269 → end had normal F_oF_1 assembly, whereas γGln-261 → end lost the assembly of the entire enzyme (Miki *et al.*, 1986). The F_1 sector was also lost on the eight amino acid deletion in the amino-terminus (between γLys-21 and γAla-28) (Kanazawa *et al.*, 1985). Consistent with the essential roles in enzyme assembly, the amino- and carboxyl-terminal α-helices of the γ subunit are positioned in the central space of the α–β subunit hexamer (Abrahams *et al.*, 1994) (Fig. 1).

The sequence conservation of the γ subunit is low: When the known γ subunit sequences are aligned to obtain maximal identity, 28 of the 286 total residues in the *E. coli* subunit are conserved (Nakamoto *et al.*, 1992). The conserved regions are the amino-terminal around the γMet-23 residue and the carboxyl terminal region between γGlu-238 and γAla-283, both located in the amino- and carboxyl-terminal α-helices of the γ subunit, respectively. The carboxyl-terminal domain is important for normal enzyme activity: Removal of the region between γLeu-278 and the carboxyl-terminus (γVal-286) lowered the activity (Iwamoto *et al.*, 1990). Furthermore, the γGln-269 → end mutant exhibited essentially no ATP synthesis or hydrolysis activity. The γGln-269 → Glu, Arg, or Leu mutant (Iwamoto *et al.*, 1990) showed reduced activity. In this regard, the bovine residues corresponding to γGln-269 and βAsp-305 form a hydrogen bond (Abrahams *et al.*, 1994). Other missense mutants, γThr-273 → Val or Gly and γGlu-275 → Lys or Gln, also had decreased activities (Iwamoto *et al.*, 1990). These results suggest that the residues in the carboxyl α-helix are required for normal catalysis, possibly for the interaction with the β subunit.

Of special interest was that three mutants (γGln-269 → Leu, γGlu-275 → Lys, and γThr-277 → end) exhibited the same reduced ATPase activity level (about 14% of the wild-type one) but formed ATP-dependent electrochemical proton gradients of different magnitudes (Iwamoto *et al.*, 1990). Thus, the same amount of ATP hydrolyzed formed a different electrochemical proton gradient, depending on the mutant. This finding was the initial indication that the γ subunit carboxyl-terminus participates in energy coupling.

3.2. Roles of the γ Subunit in Energy Coupling

Inspired by the amino-terminal deletion mutant of the γ subunit (Kanazawa *et al.*, 1985), conserved or homologous amino acid residues were replaced systematically in the amino-terminal region. Most replacements, except those of γMet-23, had little effect on ATP synthesis or hydrolysis (Shin *et al.*, 1992). The γMet-23 → Arg or Lys mutant grew very slowly through oxidative phosphorylation. Surprisingly, the membranes of γArg-23 and γLys-23 mutants had substantial ATPase activities (100 and 65% of the wild-type level, respectively), but formed much lower ATP-dependent electrochemical proton gradients than the wild type. Other mutants such as γAsp-23 and γLeu-23 were similar to the wild type, indicating that γMet-23 itself is not essential, although this residue is conserved among the γ subunits so far sequenced. Thus, introduction of the positively charged side chain at position 23 reduced efficiency of the energy coupling between catalysis and proton transport.

Interestingly, one of the eight amino acid substitutions (γArg-242 → Cys, γGln-269 → Arg, γAla-270 → Val, γIle-272 → Thr, γThr-273 → Ser, γGlu-278 → Gly, γIle-279 → Thr, and γVal-280 → Ala) suppressed the defect of the γLys-23 mutant (Nakamoto et al., 1993). The γGln-269 → Arg or γThr-273 → Ser mutant showed reduced oxidative phosphorylation and energy coupling as a single mutation, whereas these mutations in combination with γLys-23 caused substantially increased activity. Further studies defined the critical domains for both the catalytic function and energy coupling: Second-site mutations that suppressed the primary mutation, γGln-269 → Glu or γThr-273 → Val, were mapped at the amino-terminal region (residues 18, 34, and 35), and near the carboxyl-terminus (residues 236, 238, 242, and 246) (Nakamoto et al., 1995).

From these mutation–suppression studies, we concluded initially (before the bovine crystal structure) that γMet-23, γArg-242, and the region between γGln-269 and γVal-280 interact with each other and mediate energy coupling (Nakamoto et al., 1993, 1995). As expected, the amino- and carboxyl-terminal α-helices of the γ subunit closely interact and the residue corresponding to γMet-23 in the amino-terminal α-helix is actually near the residue corresponding to γArg-242 in the carboxyl-terminal α-helix (Fig. 6). However, the carboxyl-terminal α-helix between γGln-269 and γVal-280 does not directly interact with the γMet-23 residue or the domain nearby. These results suggest that the two regions functionally interact through long-range conformational transmission including movement such as rotation of the γ subunit. The three γ subunit segments defined by suppressor mutagenesis, γ21-35, γ236-246, and γ269-280, constitute a critical domain for such conformation transmission in energy coupling (Nakamoto et al., 1993, 1995).

3.3. Interactions between the β and γ Subunits

The γ subunit conformational changes have been indicated by cross-linking and tryptic digestion experiments. Cross-linking of the γ subunit (γSer-8 → Cys mutant) with the β subunit by a bifunctional reagent was shown (Aggeler et al., 1993). The β subunit site of cross-linking in the presence of Mg^{2+} ADP was localized in the βVal-145–βLys-155 segment (containing the P-loop). This cross-linking seems to be reasonable from the higher-ordered structure: The side chain of γSer-8 extends to the P-loop of the β subunit (ADP-bound form).

The γ subunit conformation during ATPase catalysis has been studied

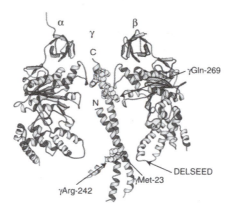

FIGURE 6. The three γ subunit segments defined by the suppressor mutations. The results of suppressor mutagenesis are mapped on the model of F_1 of the γ subunit. The carboxyl- and amino-terminal α helices are positioned following the higher-ordered structure of bovine F_1 (Abrahams *et al.*, 1994). The sequence homologies in the two regions between the bovine and *E. coli* are high: γ15–γ30, 75%; and γ237–γ285, 53%.

using a fluorescent maleimide bound to a Cys residue incorporated through mutagenesis. The fluorescence probe bound to γCys-106 responded to ATP hydrolysis in the catalytic site (Turina and Capaldi, 1994a). The position of the probe seems to extend from the γ subunit in the direction of the β subunit. Further studies indicated that ATP binding caused fluorescence enhancement and this change was reversed upon bond cleavage to yield an F_1–ADP–Pi complex (Turina and Capaldi, 1994). Thus, the γ subunit possibly changes its conformation upon ATP binding and changes to a different form after ATP hydrolysis.

A frameshift mutant having 16 unrelated carboxyl-terminal residues could not grow by oxidative phosphorylation and exhibited low membrane ATPase activity (Iwamoto *et al.*, 1990). Two methods predicted that the mutant carboxyl-terminus became much longer and formed a long β-strand extending about 60 Å from γThr-277. This β-strand may interact with the upper β-barrel or a part of the nucleotide binding domain, causing defective β–γ subunit conformational transmission.

The βArg-52 → Cys or βGly-150 → Asp mutation could suppress the deleterious effects of the γ frameshift mutation, indicating that the altered β–γ subunit interaction in the γ frameshift was restored by the β subunit mutation (Jeanteur-De Beukelar *et al.*, 1995). The βArg-52 → Cys mutation suppressed the γ frameshift, possibly by restoring the deleterious interactions between the putative carboxyl-terminal β-strand of the γ frameshift and the β-barrel domain. As expected, the bovine residue corresponding to βArg-52 locates in the β-barrel and its side chain extends toward the central space formed from the

$\alpha-\beta$ hexamer complex. This notion was further supported by the suppressing effect of βVal-77 $\rightarrow$ Ala. Bovine Ile-84, corresponding to βVal-77, is located in the β-sheet connecting the β-barrel with the nucleotide binding domain. The defective energy coupling of the γ frameshift was also suppressed by the βAsp-150 mutation of the P-loop, possibly through alteration of the mode of conformational transmission between the catalytic site in the β subunit and the frameshift γ subunit. The frameshift mutant may provide a further approach for elucidating the roles of the γ subunit. It may be of interest to introduce a fluorescent probe at the carboxyl-terminus of the frameshift γ subunit or to cross-link its carboxyl-terminus with the β-barrel.

The β subunit DELSEED domain (Asp-Glu-Leu-Ser-Glu-Glu-Asp, positions 380–386) (Fig. 6) has been shown to interact tightly with a segment of the γ subunit by Capaldi and co-workers (1992). Cysteine mutants of the region, βGlu-381 $\rightarrow$ Cys and βGlu-384 $\rightarrow$ Cys, could form a disulfide bond, possibly with γCys-87 (Aggeler *et al.*, 1995). The cross-linking was the highest with the β subunit having ATP at the catalytic sites and was much lower with ADP. This region is also the binding site for inhibitors such as quinacrine mustard (Bullough *et al.*, 1989) and dequalinium (Zhuo *et al.*, 1993). Aurovertin binds to βArg-398 near the DELSEED sequence (Lee *et al.*, 1989). The DELSEED domain forms a loop interacting with the γ subunit in the higher-ordered structure (Abrahams *et al.*, 1994). A conformational change of this domain of the chloroplast subunit was suggested by chemical modification studies (Komatsu-Takaki, 1995).

The DELSEED domain has also been shown to bind the ϵ subunit (Komatsu-Takaki, 1995; Mendel-Hartvig and Capaldi, 1991). Mutation studies (Aggeler *et al.*, 1995; Haughton and Capaldi, 1995) indicated that ϵCys-108 (wild type, Ser) forms a disulfide bond with mutant βCys-381 or βCys-383 of the β subunits that interact with the γ subunit. The $\beta-\epsilon$ disulfide formation was the highest with ADP and lower with ATP. The $\beta-\gamma$ or $\beta-\epsilon$ disulfide bond formation inhibits ATPase activity in proportion to the formation of the bond. The disulfide bond formation between the β and ϵ subunits is consistent with the earlier finding that the ϵ subunit and an inhibitory monoclonal antibody interact with the carboxyl-terminal region of the β subunit (Tozer and Dunn, 1987). Catalytically, the ϵ subunit inhibits product release from the catalytic site in unisite catalysis (Dunn *et al.*, 1987). These results suggest a model in which the β subunit movement is relative to the γ or ϵ subunit, and such movement is essential for catalytic cooperativity.

3.4. The Stalk Region of $F_o F_1$

The transport of protons through F_o causes a series of successive conformational changes and should drive ATP synthesis in the β subunit. The catalytic sector $\alpha_3\beta_3\gamma$ complex is connected to the F_o sector by a stalk domain (~45 Å long and 25–30 Å in diameter), as visualized by electron microscopy (Gogol *et al.*, 1983). Biochemical studies indicated that the δ and ε subunits of F_1, the cytoplasmic domain of F_o subunit *b* (Perlin *et al.*, 1983), and the amino-terminus of subunit *a* are required (Eya *et al.*, 1988) for binding F_1 ($\alpha_3\beta_3\gamma$). Thus, it is reasonable to assume that these subunits constitute the bacterial stalk structure. The mitochondrial stalk appears to be more complex and is formed from the γ, δ, ε, OSCP, F_6, *b*, and *d* subunits (Walker and Collinson, 1994). The higher-ordered structure of the stalk remains unknown, except that the structure of the ε subunit was solved recently by nuclear magnetic resonance (NMR) (Wilkens *et al.*, 1995).

The δ subunit (176 residues) in F_1 is sensitive to proteases, but is resistant in F_oF_1 (Gavilanes-Ruiz *et al.*, 1988), δCys-140 of this subunit forms a disulfide bond with an α subunit residue (Tozer and Dunn, 1986), suggesting functional interaction between the α and δ subunits. The isolated δ subunit has an elongated shape (Sternweis and Smith, 1977), and mutagenesis experiments suggested that its overall structure is necessary, but that individual residues may not have strict functional roles (Hazard and Senior, 1994a,b; Jounouchi *et al.*, 1992b).

In contrast to the δ subunit, the ε subunit (138 amino acid residues) has been suggested to have regulatory roles (Dunn *et al.*, 1987). The mutants with the ε subunit lacking 68–70 residues from the carboxyl-terminus (ε subunit with 68–70 residues) were active in ATP synthesis, suggesting that the carboxyl-terminal half of the subunit possibly forms different domains from the amino-terminal half and is not necessary for the binding of F_1 to F_o to form a functional complex (Kuki *et al.*, 1988). Consistent with these results, the ε subunit of *Chlorobium limicola* comprises 88 residues, being 50 residues shorter than the *E. coli* ε subunit (Xie *et al.*, 1993). The ε subunit lacking 15 amino-terminal residues is functionally competent, whereas the same subunit lacking 16 amino-terminal residues is defective in energy coupling (Jounouchi *et al.*, 1992a). It is of interest that the ε subunit lacking 15 residues from the amino-terminus and 4 residues from the carboxyl-terminus was defective. These results suggest that both terminal regions affect the conformation of the functional domain and are required for formation of the competent subunit. The higher-ordered structure determined on NMR clearly indicates that the carboxyl-terminal region between εGlu-91 and εMet-138 forms α helical domains and the amino-terminal half constitutes β-sheet domains (Wilkens *et al.*, 1995). The presence of the two distinct domains is consistent with the result of mutational analysis (Kuki *et al.*, 1988).

Binding of the ε subunit inhibits the off-rate of Pi from F_1 and also affects aurovertin D binding to the β subunit (Dunn *et al.*, 1987). The ε subunit binding to F_1 is controlled by the presence of Pi. The trypsin cleavage rate of the ε

subunit was fast in the presence of adenine nucleotide with or without Mg^{2+}, but became slower when Pi was additionally present. The half-maximal concentration of Pi (50 μM) is similar to that for high-affinity binding of Pi to F_1.

The isolated ε subunit binds to the γ subunit with an affinity of ~3 nM (Dunn, 1982). The subunit is also cross-linked with one β subunit, forming an ester linkage between βGlu-381 and εSer-108 (Dallmann et al., 1992). The $\beta-\varepsilon$ interaction was shown to be altered by a detergent, lauryldimethylamine oxide (Dunn et al., 1990). Mutation studies indicated that εHis-38 lies near the γ subunit. Furthermore, the cAsp-44 $\rightarrow$ Cys mutant subunit could form a cross-link with the γ (positions between 202 and 286) and ε subunits, indicating that the γ and ε subunits actually interact with the c subunits (Watts et al., 1995). Thus, the γ subunit spans the entire F_1 complex from the β-barrel of the β subunit to the c subunit. This structure is consistent with the fact that the γ subunit transduces conformation changes between the catalytic site and the proton pathway.

4. SUMMARY

ATP synthase is a proton pump coupling the electrochemical proton gradient and a chemical reaction, "ATP $\leftrightarrow$ ADP + Pi," and is conserved among eukaryotic and prokaryotic cells. The higher-ordered structure of the catalytic sector $\alpha_3\beta_3\gamma$ complex has been established by X-ray diffraction. The catalytic site and proton pathway have been studied extensively by means of mutational studies of the E. coli enzyme. It should be noted that most of the catalytic residues and/or essential residues were identified before the higher-ordered structure was determined, indicating that systematic mutagenesis on conserved amino acid residues has been a useful approach. However, it was not possible to interpret the results correctly without the higher-ordered structure. Although interpretations were made applying the bovine F_1 structure, it is still desirable to know the crystal structure of the E. coli enzyme. Furthermore, the higher-ordered structure of the entire F_oF_1 is essential to understand the mechanism of the enzyme.

The next obvious question is what is the molecular basis of the energy coupling between the transport and the chemical reaction. "Conformation transmission through subunit rotation" is a fascinating model for the coupling. Results supporting intersubunit rotation of the γ subunit in bacterial (Duncan et al., 1995) and chloroplast (Sabbert et al., 1996) F_1 were obtained recently. The rotation seems to be a logical mechanism for the γ subunit to interact successively with the three β subunits. The rotation is also consistent with the structure and function of the F_o sector, because Asp-61 of the c subunits (10–12 copies in one F_oF_1 complex) and residues in the a subunit (one copy) participate in proton transport. The cross-linking experiment (Watts *et al.*, 1995) suggested that the γ subunit may interact with one c subunit in the entire F_o complex. Thus, downhill movement of protons through F_o may rotate subunit c and the stalk region, including the γ subunit. It is imperative in this model that the $\alpha_3\beta_3$ assembly is fixed to the bilayer. The stationary part of the complex may be formed from the $\alpha_3\beta_3$ assembly, the a and the b subunits, because this subunit has a single transmembrane region and most of the carboxyl-terminal extends in the direction of the F_1 sector. We need more structural information on F_o to plan experiments showing the actual rotation of the enzyme subunits in ATP synthesis.

ACKNOWLEDGMENT. We are grateful to Drs. John Walker and J. P. Abrahams for providing us with the coordinates of the bovine $\alpha\beta\gamma$ complex. Without the coordinates, the detailed discussion in this chapter would not have been possible. The work in our laboratory described in this chapter was supported by grants from the Japanese Ministry of Education, Science, and Culture, and the Human Frontier Science Program. We also thank the co-workers whose names appeared in our publications.

5. REFERENCES

Abrahams, J. P., Lutter, R., Todd, R. J., van Raaij, M. J., Leslie, A. G. W., and Walker, J. E., 1993, Inherent asymmetry of the structure of F_1-ATPase from bovine heart mitochondria at 6.5 Å resolution, *EMBO J.* **12:**1775–1780.

Abrahams, J. P., Leslie, A. G. W., Lutter, R., and Walker, J. E., 1994, Structure at 2.8 Å resolution of F_1-ATPase from bovine heart mitochondria, *Nature* **370:**621–628.

Aggeler, R., Haughton, M. A., and Capaldi, R. A., 1995, Disulfide bond formation between the COOH-terminal domain of the β subunits and the γ and ε subunits of the *Escherichia coli* F_1-ATPase: Structural implication and functional consequences, *J. Biol. Chem.* **270:**9185–9191.

Aggeler, R., Cai, S. X., Keana, J. F. W., Koike, T., and Capaldi, R. A., 1993, The γ-subunit of the *Escherichia coli* F_1-ATPase can be cross-linked near the glycine-rich loop region of a β-subunit when $ADP+Mg^{2+}$ occupies catalytic sites but not when $ATP+Mg^{2+}$ is bound, *J. Biol. Chem.* **268:**20831–20837.

Allison, W. S., Jault, J.-M., Zhuo, S., and Paik, S. R., 1992, Functional sites in F_1-ATPases: Location and interactions, *J. Bioenerg. Biomembr.* **24:**469–477.

Amano, T., Yoshida, M., Matsumoto, Y., and Nishikawa, K., 1994, Structural model of the ATP-binding domain of the F_1-β subunit based on analogy to the RecA protein, *FEBS Lett.* **351:** 1–5.

Amzel, L. M., Bianchet, M. A., and Pedersen, P. L., 1992, Quaternary structure of ATP synthase: Symmetry and asymmetry in the F_1 moiety, *J. Bioenerg. Biomembr.* **24:**429–433.

Bianchet, M., Ysern, X., Hullihen, J., Pedersen, P. L., and Amzel, L. M., 1991, Mitochondrial ATP synthase: Quaternary structure of the F_1 moiety at 3.6 Å determined by X-ray diffraction analysis, *J. Biol. Chem.* **266:**21197–21201.

Boyer, P. D., 1989, A perspective of the binding change mechanism for ATP synthesis, *FASEB J.* 3:2164–2178.

Boyer, P. D., 1993, The binding change mechanism for ATP synthase—Some probabilities and possibilities, *Biochim. Biophys. Acta* 1140:215–250.

Bullough, D. A., Ceccarelli, E. A., Verburg, J. G., and Allison, W. S., 1989, Localization of sites modified during inactivation of the bovine heart mitochondrial F_1-ATPase by quinacrine mustard using [^{3}H]aniline as a probe, *J. Biol. Chem.* 264:9155–9163.

Capaldi, R. A., Aggeler, R., Gogol, E. P., and Wilkens, S., 1992, Structure of the *Escherichia coli* ATP synthase and role of the γ and ε subunits in coupling catalytic site and proton channeling functions, *J. Bioenerg. Biomembr.* 24:435–439.

Dallmann, H. G., Flynn, T. G., and Dunn, S. D., 1992, Determination of the 1-ethyl-3-[(3-dimethylamino)propyl]-carbodiimide-induced cross-link between the β and ε subunits of *Escherichia coli* F_1-ATPase, *J. Biol. Chem.* 267:18953–18960.

Duncan, T. M., and Cross, R. L., 1992, A model for the catalytic site of F_1-ATPase based on analogies to nucleotide-binding domains of known structure, *J. Bioenerg. Biomembr.* 24: 453–461.

Duncan, T. M., and Senior, A. E., 1985, The defective proton-ATPase of *uncD* mutants of *Escherichia coli*: Two mutations which affect the catalytic mechanism, *J. Biol. Chem.* 260: 4901–4907.

Duncan, T. M., Bulygin, V. V., Zhou, Y., Hutcheon, M. L., and Cross, R. L., 1995, Rotation of subunits during catalysis by *Escherichia coli* F_1-ATPase, *Proc. Natl. Acad. Sci. USA* 92: 10964–10968.

Dunn, S. D., 1982, The isolated γ subunit of *Escherichia coli* F_1-ATPase binds the ε subunit, *J. Biol. Chem.* 257:7354–7359.

Dunn, S. D., and Futai, M., 1980, Reconstitution of a functional coupling factor from the isolated subunits of *Escherichia coli* F_1-ATPase, *J. Biol. Chem.* 255:113–118.

Dunn, S. D., Zadorozny, V. D., Tozer, R. G., and Orr, L. E., 1987, ε subunit of *Escherichia coli* F_1-ATPase: Effects on affinity for aurovertin and inhibition of product release in unisite ATP hydrolysis, *Biochemistry* 26:4488–4493.

Dunn, S. D., Tozer, R. G., and Zadorozny, V. D., 1990, Activation of *Escherichia coli* F_1-ATPase by lauryldimethylamine oxide and ethylene glycol: Relationship of ATPase activity to the interaction of the ε and β subunits, *Biochemistry* 29:4335–4340.

Eichelberg, K., Ginocchio, C. C., and Galán, J. E., 1994, Molecular and functional characterization of the *Salmonella typhimurium* invasion genes *ilvB* and *ilvC*: Homology of InvC to the F_oF_1ATPase family of proteins, *J. Bacteriol.* 176:4501–4510.

Eya, S., Noumi, T., Maeda, M., and Futai, M., 1988, Intrinsic membrane sector (F_o) of H$^+$-ATPase (F_oF_1) from *Escherichia coli*: Mutations in the *a* subunit give F_o with impaired proton translocation and F_1 binding, *J. Biol. Chem.* 263:10056–10062.

Feldman, R. I., and Sigman, D. S., 1982, The synthesis of enzyme-bound ATP by soluble chloroplast coupling factor 1, *J. Biol. Chem.* 257:1676–1683.

Fillingame, R. H., 1990, Molecular mechanics of ATP synthesis by F_1F_o-type H$^+$-transporting ATP synthases, in *The Bacteria* (T. A. Krulwich, Ed.), vol. XII, pp. 345–391, Academic Press, New York.

Fillingame, R. H., 1992, H$^+$ transport and coupling by the F_o sector of the ATP synthase: Insights into the molecular mechanism of function, *J. Bioenerg. Biomembr.* 24:485–491.

Fischer, S., Etzold, C., Turina, P., Deckers-Hebestreit, G., Altendorf, K., and Gräber, P., 1994, ATP synthesis catalyzed by the ATP synthase of *Escherichia coli* reconstituted into liposomes, *Eur. J. Biochem.* 225:167–172.

Foster, D. L., and Fillingame, R. H., 1979, Energy-transducing H$^+$-ATPase of *Escherichia coli*, *J. Biol. Chem.* 254:8230–8236.

Futai, M., 1977, Reconstitution of ATPase activity from the isolated α, β, and γ subunits of the coupling factor, F_1, of *Escherichia coli*, *Biochem. Biophys. Res. Commun.* **79:**1231–1237.

Futai, M., and Kanazawa, H.,1983, Structure and function of proton-translocating adenosine triphosphatase (F_oF_1): Biochemical and molecular biological approaches, *Microbiol. Rev.* **47:**285–312.

Futai, M,. and Omote, H., 1996, Transport processes in eukaryotic and prokaryotic organisms, in *Handbook of Biological Physics* vol. 2 (W. N. Konings, H. R. Kaback, and J. S. Lollcema, eds.), pp. 48–74, Elsevier, Amsterdam.

Futai, M., Noumi, T., Miki, J., and Maeda, M., 1987, Molecular biological studies of the F_1 portion of H^+-ATPase of *Escherichia coli*, *Chem. Scripta* **27B:**89–96.

Futai, M., Noumi, T., and Maeda, M., 1989, ATP synthase (H^+-ATPase): Results by combined biochemical and molecular biological approaches, *Annu. Rev. Biochem.* **58:**111–136.

Garin, J., Boulay, F., Issartel, J. P., Lunardi, J., and Vignais, P. V., 1986, Identification of amino acid residues photolabeled with 2-azido[α-^{32}P] adenosine diphosphate in the β subunit of beef heart mitochondrial F_1-ATPase, *Biochemistry* **25:**4431–4437.

Gavilanes-Ruiz, M., Tommasino, M., and Capaldi, R. A., 1988, Structure–function relationships of the *Escherichia coli* ATP synthase probed by trypsin digestion, *Biochemistry* **27:**603–609.

Gogol, E. P., Lücken, U., and Capaldi, R. A., 1987, The stalk connecting the F_1 and F_o domains of ATP synthase visualized by electron microscopy of unstained specimens, *FEBS Lett.* **219:** 274–278.

Gromet-Elhanan, Z., 1992, Identification of subunits required for the catalytic activity of the F_1-ATPase, *J. Bioenerg. Biomembr.* **24:**447–452.

Grubmeyer, C., Cross, R. L., and Penefsky, H. S., 1982, Mechanism of ATP hydrolysis by beefheart mitochondrial ATPase: Rate constants for elementary steps in catalysis at a single site, *J. Biol. Chem.* **257:**12092–12100.

Hatefi, Y., 1993, ATP synthesis in mitochondria, *Eur. J. Biochem.* **218:**759–767.

Haughton, M. A., and Capaldi, R. A., 1995, Asymmetry of *Escherichia coli* F_1-ATPase as a function of the interaction of α–β subunit pairs with the γ and ε subunits, *J. Biol. Chem.* **270:** 20568–20574.

Hazard, A. L., and Senior, A. E., 1994a, Mutagenesis of subunit δ from *Escherichia coli* F_1F_o-ATP synthase, *J. Biol. Chem.* **269:**418–426.

Hazard, A. L., and Senior, A. E., 1994b, Defective energy coupling in δ-subunit mutants of *Escherichia coli* F_1Fo-ATPsynthase, *J. Biol. Chem.* **269:**427–432.

Hollemans, M., Runswick, M. J., Fearnley, I. M., and Walker, J. E., 1983, The sites of labeling of the β-subunit of bovine mitochondrial F_1-ATPase with 8-azido-ATP, *J. Biol. Chem.* **258:** 9307–9313.

Hsu, S. Y., Noumi, T., Takeyama, M., Maeda, M., Ishibashi, S., and Futai, M., 1987, β subunit of *Escherichia coli* F_1-ATPase: An amino acid replacement within a conserved sequence (G-X-X-X-X-G-K-T/S) of nucleotide-binding proteins, *FEBS Lett.* **218:**222–226.

Ida, K., Noumi, T., Maeda, M., Fukui, T., and Futai, M., 1991, Catalytic site of F_1-ATPase of *Escherichia coli*: Lys-155 and Lys-201 of the β subunit are located near the γ-phosphate group of ATP in the presence of Mg^{2+}, *J. Biol. Chem.* **266:**5424–5429.

Iwamoto, A., Miki, J., Maeda, M., and Futai, M., 1990, H^+-ATPase γ subunit from *Escherichia coli*: Role of the conserved carboxylic terminal region, *J. Biol. Chem.* **265:**5043–5048.

Iwamoto, A., Omote, H., Hanada, H., Tomioka, N., Itai, A., Maeda, M., and Futai, M., 1991, Mutations in Ser-174 and the glycine-rich sequence (Gly-149, Gly-150, and Thr-156) in the β subunit of *Escherichia coli* H^+-ATPase, *J. Biol. Chem.* **266:**16350–16355.

Iwamoto, A., Park, M. Y., Maeda, M., and Futai, M., 1993, Domains near ATP γ phosphate in the catalytic site of H^+-ATPase: Model proposed from mutagenesis and inhibitor studies, *J. Biol. Chem.* **268:**3156–3160.

Jeanteur-De Beukelar, C., Omote, H., Iwamoto-Kihara, A., Maeda, M., and Futai, M., 1995, β–γ subunit interaction is required for catalysis by H^+-ATPase (ATP synthase): β subunit amino acid replacements suppress a γ subunit mutation having a long unrelated carboxyl terminus, *J. Biol. Chem.* **270:**22850–22854.

Jounouchi, M., Takeyama, M., Noumi, T., Moriyama, Y., Maeda, M., and Futai, M., 1992a, Role of the amino terminal region of the ε subunit of *Escherichia coli* H^+-ATPase, *Arch. Biochem. Biophys.* **292:**87–94.

Jounouchi, M., Takeyama, M., Chaiprasert, P., Noumi, T., Moriyama, Y., Maeda, M., and Futai, M., 1992b, *Escherichia coli* H^+-ATPase: Role of the δ subunit in binding F_1 to the F_o sector, *Arch. Biochem. Biophys.* **292:**376–381.

Jounouchi, M., Maeda, M., and Futai, M., 1993, The α subunit of ATP synthase(F_oF_1): The Lys-175 and Thr-176 residues in the conserved sequence(Gly-X-X-X-X-Gly-Lys-Thr/Ser) are located in the domain required for stable subunit-subunit interaction, *J. Biochem.* **114:**171–176.

Kagawa, Y., Ohta, S., Harada, M., Kihara, H., Ito, Y., and Sato, M., 1992, The αβ complexes of ATP synthase: The $\alpha_3\beta_3$ oligomer and $\alpha_1\beta_1$ protomer, *J. Bioenerg. Biomembr.* **24:**441–445.

Kanazawa, H., Hama, H., Rosen, B. P., and Futai, M., 1985, Deletion of seven amino acid residues from the γ subunit of *Escherichia coli* H^+-ATPase causes total loss of F_1 assembly on membranes, *Arch. Biochem. Biophys.* **241:**364–370.

Komatsu-Takaki, M., 1995, Effects of energization and substrates on the reactivities of lysine residues of the chloroplast ATP synthase β subunit, *Eur. J. Biochem.* **228:**265–270.

Kraulis, P. J., 1991, MOLSCRIPT: A program to produce both detailed and schematic plots of protein structures. *J. Appl. Crystallogr.* **24:**946–950.

Kuki, M., Noumi, T., Maeda, M., Amemura, A., and Futai, M., 1988, Functional domains of ε subunit of *Escherichia coli* H^+-ATPase (F_oF_1), *J. Biol. Chem.* **263:**17437–17442.

Lee, R. S.-F., Pagan, J., Satre, M., Vignais, P. V., and Senior, A. E., 1989, Identification of a mutation in *Escherichia coli* F_1-ATPase β-subunit conferring resistance to aurovertin, *FEBS Lett.* **253:**269–272.

Matsuno-Yagi, A., and Hatefi, Y., 1985, Studies on the mechanism of oxidative phosphorylation, *J. Biol. Chem.* **260:**14424–14427.

Mendel-Hartvig, J., and Capaldi, R. A., 1991, Catalytic site nucleotide and inorganic phosphate dependence of the ε subunit in *Escherichia coli* adenosinetriphosphatase, *Biochemistry* **30:** 1278–1284.

Maggio, M. B., Pagan, J., Parsonage, D., Hatch, L., and Senior, A. E., 1987, The defective proton-ATPase of *uncA* mutants of *Escherichia coli*: Identification by DNA sequencing of residues in the α-subunit which are essential for catalysis or normal assembly, *J. Biol. Chem.* **262:**8981– 8984.

Miki, J., Takeyama, M., Noumi, T., Kanazawa, H., Maeda, M., and Futai, M., 1986, *Escherichia coli* H^+-ATPase: Loss of the carboxyl terminal region of the γ-subunit causes defective assembly of the F_1 portion, *Arch. Biochem. Biophys.* **251:**458–464.

Miki, J., Fujiwara, K., Tsuda, M., Tsuchiya, T., and Kanazawa, H., 1990, Suppression mutations in the defective β subunit of F_1-ATPase from *Escherichia coli*, *J. Biol. Chem.* **265:**21567–21572.

Moriyama, Y., Iwamoto, A., Hanada, H., Maeda, M., and Futai, M., 1991, One-step purification of *Escherichia coli* H^+-ATPase (F_oF_1) and its reconstitution into liposomes with neurotransmitter transporters, *J. Biol. Chem.* **266:**22141–22146.

Murataliev. M. B., Milgrom, Y. M., and Boyer, P. D., 1991, Characteristics of the combination of inhibitory Mg^{2+} and azide with the F_1 ATPase from chloroplasts, *Biochemistry* **30:**8305– 8310.

Nakamoto, R. K., Shin, K., Iwamoto, A., Omote, H., Maeda, M., and Futai, M., 1992, *Escherichia coli* F_oF_1-ATPase: Residues involved in catalysis and coupling, *Ann. NY Acad. Sci.* **671:** 335–344.

Nakamoto, R. K., Maeda, M., and Futai, M., 1993, The γ subunit of the *Escherichia coli* ATP synthase: Mutations in the carboxyl-terminal region restore energy coupling to the amino-terminal mutant, γMet-23→Lys, *J. Biol. Chem.* **268:**867–872.

Nakamoto, R. K., Al-Shawi, M. K., and Futai, M., 1995, The ATP synthase γ subunit: Suppressor mutagenesis reveals three helical regions involved in energy coupling, *J. Biol. Chem.* **270:**14042–14046.

Noumi, T., Futai, M., and Kanazawa, H., 1984a Replacement of serine-373 by phenylalanine in the α subunit of *Escherichia coli* F$_1$-ATPase results in loss of steady state catalysis by the enzyme, *J. Biol. Chem.* **259:**10076–10079.

Noumi, T., Taniai, M., Kanazawa, H., and Futai, M., 1986, Replacement of arginine 246 by histidine in the β subunit of *Escherichia coli* H$^+$-ATPase resulted in loss of multi-site ATPase activity, *J. Biol. Chem.* **261:**9196–9201.

Noumi, T., Azuma, M., Shimomura, S., Maeda, M., and Futai, M., 1987a, *Escherichia coli* H$^+$-ATPase: Glutamic acid 185 in β subunit is essential for its structure and assembly, *J. Biol. Chem.* **262:**14978–14982.

Noumi, T., Tagaya, M., Miki-Takeda, K., Maeda, M., Fukui, T., and Futai, M., 1987b, Loss of unisite and multisite catalysis by *Escherichia coli* F$_1$ through modification with adenosine tri- or tetraphosphopyridoxal, *J. Biol. Chem.* **262:**7686–7692.

Noumi, T., Maeda, M., and Futai, M., 1987c, Mode of inhibition of sodium azide on H$^+$-ATPase of *Escherichia coli*, *FEBS Lett.* **213:**381–384.

Omote, H., Maeda, M., and Futai, M., 1992, Effects of mutations of conserved Lys-155 and Thr-156 residues in the phosphate-binding glycine-rich sequence of the F$_1$-ATPase β subunit of *Escherichia coli*, *J. Biol. Chem.* **267:**20571–20576.

Omote, H., Park, M.-Y., Maeda, M., and Futai, M., 1994, The α/β subunit interaction in H$^+$-ATPase (ATP synthase): An *Escherichia coli* α subunit mutation (αArg-296 → Cys) restores coupling efficiency to the deleterious β subunit mutant (βSer-174→Phe), *J. Biol. Chem.* **269:**10265–10269.

Omote, H., Nga, P. L., Park, M.-Y., Maeda, M., and Futai, M., 1995, β subunit Glu-185 of *Escherichia coli* H$^+$ATPase (ATP synthase) is an essential residue for cooperative catalysis, *J. Biol. Chem.* **270:**25656–25660.

O'Neal, C. C., and Boyer, P. D., 1984, Assessment of the rate of bound substrate interconversion and of ATP acceleration of product release during catalysis by mitochondrial adenosine triphosphatase, *J. Biol. Chem.* **259:**5761–5767.

Park, M.-Y., Omote, H., Maeda, M., and Futai, M., 1994, Conserved Glu-181 and Arg-182 residues of *Escherichia coli* H$^+$-ATPase (ATP synthase) β subunit are essential for catalysis: Properties of 33 mutants between βGlu-161 and βLys-201 residues, *J. Biochem.* **116:**1139–1145.

Paul, M.-F., Ackermann, S., Yue, J., Arselin, G., Velours, J., and Tzagoloff, A. 1994, Cloning of the yeast ATP3 gene coding for the γ-subunit of F$_1$ and characterization of *atp3* mutants, *J. Biol. Chem.* **269:**26158–26164.

Penefsky, H. S., and Cross, R. L., 1991, Structure and mechanism of F$_o$F$_1$-type ATP synthases and ATPases, *Adv. Enzymol.* **64:**173–213.

Perlin, D. S., Cox, D. N., and Senior, A. D., 1983, Integration of F$_1$ and the membrane sector of the proton-ATPase of *Escherichia coli*: Role of subunit "b" (*uncF* protein), *J. Biol. Chem.* **258:**9793–9800.

Rao, R., Pagan, J., and Senior, A. E., 1988, Directed mutagenesis of the strongly conserved lysine 175 in the proposed nucleotide-binding domain of α-subunit from *Escherichia coli* F$_1$-ATPase, *J. Biol. Chem.* **263:**15957–15963.

Sakamoto, J., and Tonomura, Y., 1983, Synthesis of enzyme-bound ATP by mitochondrial soluble F$_1$-ATPase in the presence of dimethylsulfoxide, *J. Biochem.* **93:**1601–1614.

Schneider, E., and Altendorf, K., 1982, ATP synthetase (F$_1$F$_o$) of *Escherichia coli* K-12 high-yield preparation of functional F_o by hydrophobic affinity chromatography, *Eur. J. Biochemistry* **126:**149–153.

Senior, A. E., 1988, ATP synthesis by oxidative phosphorylation, *Physiol. Rev.* **68:**177–231.

Senior, A. E., 1990, The proton-translocating ATPase of *Escherichia coli*, *Annu. Rev. Biophys. Chem.* **19:**7–41.

Senior, A. E., and Al-Shawi, M. K., 1992, Further examination of seventeen mutations in *Escherichia coli* F_1-ATPase β-subunit, *J. Biol. Chem.* **267:**21471–21478.

Sabbert, D., Engelbrecht, S., and Junge, W., 1996, Intersubunit rotation in active F-ATPase, *Nature* **381:**623–625.

Shen, H., Yao, B.-Y., and Mueller, D. M., 1994, Primary structural constraints of P-loop of mitochondrial F_1-ATPase from yeast, *J. Biol. Chem.* **269:**9424–9428.

Shin, K., Nakamoto, R. K., Maeda, M., and Futai, M., 1992, F_oF_1-ATPase γ subunit mutations perturb the coupling between catalysis and transport, *J. Biol. Chem.* **267:**20835–20839.

Shirakara, Y., Leslie, A. G. N., Abraham, J. P., Walker, J. E., Veda, T., Sakimoto, Y., Kambara, M., Saika, K., Kagawa, Y., and Yoshida, M., 1997, The crystal structure of the nucleotide-free $\alpha_3\beta_3$ subcomplex of the F_1-ATPase from the thermophilic *Bacillus* P53 is a symmetric trimmer, *Structure* **5:**825–836.

Soga, S., Noumi, T., Takeyama, M., Maeda, M., and Futai, M., 1989, Mutational replacements of conserved amino acid residues in the α subunit change the catalytic properties of *Escherichia coli* F_1-ATPase, *Arch. Biochem. Biophys.* **268:**643–648.

Souid, A.-K., and Penefsky, H. S., 1995, Energetics of ATP dissociation from the mitochondrial ATPase during oxidative phosphorylation, *J. Biol. Chem.* **270:**9074–9082.

Sternweis, P. C., and Smith, J. B., 1977, Characterization of the purified membrane attachment (δ) subunit of the proton translocating adenosine triphosphatase from *Escherichia coli*, *Biochemistry* **16:**4020–4025.

Tagaya, M., Noumi, T., Nakano, K., Futai, M., and Fukui, T., 1988, Identification of α-subunit Lys-201 and β-subunit Lys-155 at the ATP-binding site in *Escherichia coli* F_1-ATPase, *FEBS Lett.* **233:**347–351.

Takeyama, M., Ihara, K., Moriyama, Y., Noumi, T., Ida, K., Tomioka, N., Itai, A., Maeda, M., and Futai, M., 1990, Identification of α-subunit Lys-210 and β-subunit Lys-155 at the ATP-binding sites in *Escherichia coli* F_1-ATPase, *J. Biol. Chem.* **265:**21279–21284.

Tommasino, M., and Capaldi, R. A., 1985, Effect of dicyclohexylcarbodiimide on unisite and multisite catalytic activities of the adenosinetriphosphatase of *Escherichia coli*, *Biochemistry* **24:**3972–3976.

Tozer, R. G., and Dunn, S. D., 1986, Column centrifugation generates an intersubunit disulfide bridge in *Escherichia coli* F_1-ATPase, *Eur. J. Biochem.* **161:**513–518.

Tozer, R. G., and Dunn, S., 1987, The ε subunit and inhibitory monoclonal antibodies interact with the carboxyl-terminal region of the β subunit of *Escherichia coli* F_1-ATPase, *J. Biol. Chem.* **262:**10706–10711.

Turina, P., and Capaldi, R. A., 1994a, ATP hydrolysis-driven structural changes in the γ-subunit of *Escherichia coli* ATPase monitored by fluorescence from probes bound at introduced cysteine residues, *J. Biol. Chem.* **269:**13465–13471.

Turina, P., and Capaldi, R., 1994b, ATP binding causes a conformational change in the γ subunit of the *Escherichia coli* F_1ATPase, which is reversed on bond cleavage, *Biochemistry* **33:**14275–14280.

Van Raaji, M. J., Abrahams, J. P., Leslie, A. G. W., and Walker, J. E., 1996, The structure of bovine F_1-ATPase complexed with the antibiotic inhibitor aurovertin B, *Proc. Natl. Acad. Sci. USA* **93:**6913–6917.

Vignais, P. V., and Lunardi, J., 1985, Chemical probes of the mitochondrial ATP synthesis and translocation, *Annu. Rev. Biochem.* **54:**977–1014.

Walker, J. E., and Collinson, I. R., 1994, The role of the stalk in the coupling mechanism of F_1F_o-ATPases, *FEBS Lett.* **346**:39–43.

Walker, J. E., Saraste, M., and Gay, N. J., 1984, The *unc* operon: Nucleotide sequence, regulation and structure of ATP-synthase, *Biochim. Biophys. Acta* **768**:164–200.

Walker, J. E., Cozens, A. L., Dyer, M. R., Fearnley, I. M., Powell, S. J., and Runswick, M. J., 1987, Studies of the genes for ATP synthases in eubacteria, chloroplasts and mitochondria: Implications for structure and function of the enzyme, *Chem. Scripta* **27B**:97–105.

Watts, S. D., Zhang, Y., Fillingame, R. H., and Capaldi, R. A., 1995, The γ subunit in the *Escherichia coli* ATP synthase complex (ECF_1F_o) extends through the stalk and contacts the c subunits of the F_o part, *FEBS Lett.* **368**:235–238.

Weber, J., Wilke-Mounts, S., and Senior, A. E., 1994, Cooperativity and stoichiometry of substrate binding to the catalytic sites of *Escherichia coli* F_1-ATPase: Effects of magnesium, inhibitors, and mutation, *J. Biol. Chem.* **269**:20462–20467.

Wilkens, S., Dahlquist, F. W., McIntosh, L. P., Donaldson, L. W., and Capaldi, R. A., 1995, Structural features of the ε subunit of the *Escherichia coli* ATP synthase determined by NHR spectroscopy, *Nature Struct. Biol.* **2**:961–967.

Xie, D.-L., Holger, L., Hauska, G., Maeda, M., Futai, M., and Nelson, N., 1993, The *atp2* operon of the green bacterium *Chlorobium limicola*, *Biochim. Biophys. Acta* **1172**:267–273.

Yoshida, M., 1983, The synthesis of enzyme-bound ATP by the F_1-ATPase from the thermophilic bacterium PS3 in 50% dimethylsulfoxide, *Biochem. Biophys. Res. Commun.* **114**:907–912.

Yoshida, M., Allison, W. S., Esch, F. S., and Futai, M., 1982, The specificity of carboxyl group modification during the inactivation of the *Escherichia coli* F_1-ATPase with dicyclohexyl [C^{14}]carbodiimide, *J. Biol. Chem.* **257**:10033–10037.

Zhang, S., and Jagendorf, A. T., 1995, Some unique characteristics of thylakoid unisite ATPase, *J. Biol. Chem.* **270**:6607–6614.

Zhuo, S., Paik, S. R., Register, J. A., and Allison, W. S., 1993, Photonactivation of the bovine heart mitochondrial F_1-ATPase by [^{14}C] dequalinium cross-links phenylalanine-403 or phenylalanine-406 of an α subunit to a site or sites contained within residues 440–459 of a β subunit, *Biochemistry* **32**:2219–2227.

CHAPTER 18

Analysis of the Nucleotide Binding Sites of ATP Synthase and Consequences for the Catalytic Mechanism

J. A. Berden and A. F. Hartog

1. INTRODUCTION

The complexity of the mechanism of ATP hydrolysis and ATP synthesis by the ATP synthase is illustrated by the large number of nucleotide binding sites. Many efforts have been made to characterize all six binding sites and to analyze the role of the nucleotides at each site. In the present chapter we will describe, starting from some well-established facts concerning the properties of F_1 (the hydrophilic part of the enzyme, containing all nucleotide binding sites), the results of studies on the nucleotide binding sites of the enzyme and draw some conclusions on the catalytic mechanism of ATP hydrolysis and synthesis. We will mainly describe studies with the mitochondrial system from bovine heart;

J. A. Berden and A. F. Hartog • E. C. Slater Institute, BioCentrum, University of Amsterdam, Plantage Muidergracht 12, 1018 TV Amsterdam, The Netherlands.

Frontiers of Cellular Bioenergetics, edited by Papa *et al.* Kluwer Academic/Plenum Publishers, New York, 1999.

however, where relevant, studies on the enzyme from other sources will be mentioned. At present, the subunit interactions and the conformational changes within the ATP synthase are studied in much detail in order to obtain a better insight into what really happens during catalysis. But conclusions on the precise mechanism of catalysis have to be in agreement with the properties of the nucleotide binding sites. Some features of the ATP synthase, considered to be well established, may serve as starting point for our analysis. We name the following:

1. F_1 contains six nucleotide binding sites, located at each of the three α and three β subunits of the enzyme. After the evidence for the presence of three α and three β subunits in mitochondrial F_1 (Todd *et al.*, 1980; Stutterheim *et al.*, 1981), the presence of six sites for adenine nucleotides was firmly established for F_1 from various sources (Wagenvoord *et al.*, 1980; Senior and Wise, 1983; Boulay *et al.*, 1985; Weber *et al.*, 1985). As far as the nomenclature is concerned, we will call them α or β sites, depending on the mainly contributing subunit, although it has been shown in many experiments (Wagenvoord *et al.*, 1977; Lübben *et al.*, 1984; Bar-Zvi *et al.*, 1983; Boulay *et al.*, 1985; Cross *et al.*, 1987) and confirmed by the crystal structure (Abrahams *et al.*, 1994) that both sites are located at interfaces between α and β subunits. This implies that modification of an amino acid belonging to a β subunit may be the result of occupation of an α site by a modifying ligand. Covalent modification of a certain subunit by an adenine nucleotide analogue is therefore no proof for the modification of either an α or a β site, unless additional data are available.

2. The catalytic sites are located on β subunits. This conclusion was already drawn from the initial studies on nucleotide binding (Grubmeyer and Penefsky, 1981a,b; Senior and Wise, 1983), further established by the formulation of a consensus sequence for ATPases (Walker *et al.*, 1982), and finally unequivocally confirmed by the reported crystallographic data (Abrahams *et al.*, 1994).

3. The catalytic mechanism of ATP synthase implies the cooperative involvement of more than one catalytic site, and these catalytic sites show negative cooperativity of binding and positive cooperativity of catalysis. The large number of arguments for these two assumptions have been well described by Boyer in a recent review (Boyer, 1993) and they may be considered as established facts.

On the basis of the formulated assumptions, we will try to describe the properties of each of the six binding sites and the role of the nucleotides at these sites in the process of ATP synthesis and hydrolysis. We will refer to the sites, if appropriate, with a number, as represented in Fig. 1.

1	2	3
4	5	6

Site 1: Tight non-exchangeable binding site
Site 2: High-affinity catalytic binding site
Site 3: Low-affinity catalytic binding site
Site 4: Tight non-catalytic non-exchangeable binding site
Site 5: Regulatory non-catalytic binding site
Site 6: Low-affinity non-catalytic binding site

FIGURE 1. Schematic representation of the nucleotide binding sites of F_1. The upper row represents the three β-sites (1 to 3) in order of decreasing affinity for nucleotides. The lower row represents the three α-sites (4 to 6) in order of decreasing affinity.

2. RELEVANT ELEMENTS IN THE CHARACTERIZATION OF NUCLEOTIDE BINDING SITES

2.1. The Different Aspects of Nucleotide Binding

When studying the nucleotide binding sites of F_1 it is important to differentiate between the various elements that are relevant for the characterization of a nucleotide binding site. In the introduction to several papers, the error is made that rapid exchangeability is identified with catalytic involvement, tight binding with a noncatalytic role, and nonexchangeability with localization on an α subunit. However, a catalytic site may contain a tightly bound nucleotide, a noncatalytic site may contain a rapidly exchangeable nucleotide, and a β subunit may contain a binding site that is not participating in rapid catalysis, although it is well established now that only β sites have catalytic potential. It is therefore relevant to define the property that is determined in a specific experiment and each has to be defined with respect to several aspects: binding affinity (tight or loose), exchangeability (nonexchangeable, slowly exchangeable, rapidly exchangeable), catalytic involvement, and localization.

The statement that the catalytic sites are cooperative may be a useful tool in the analysis of experiments in which several sites are modified at the same time with a covalently binding analogue: if the inhibition curve, that is, the curve relating activity with the number of modified sites, is linear, maximally one catalytic site is involved in the modification, independent of the total number of sites that has to be modified to obtain full inhibition. If more than one catalytic sites are involved, the inhibition curve cannot be linear.

It has been well established that all three β subunits are structurally different, both at the level of specificity of ligands (Bragg and Hou, 1990) and at the level of interaction with the small subunits (Haughton and Capaldi, 1995, 1996). This aspect will be treated in other chapters and we will not use it in our argumentation, since it is difficult to establish whether this nonequivalence remains constant during catalysis.

2.2. Different Types of F_1 Preparations

2.2.1. Different Forms of Mitochondrial F_1

For the correct analysis of experiments in which nucleotides have been bound to the enzyme, or exchanged or removed, it is important to have sufficient knowledge of the starting situation. The most widely used preparation of bovine heart F_1 is isolated according to the procedure of Knowles and Penefsky (1972). Before use, the preparation is freed of loosely bound nucleotides by ammonium sulfate precipitation and column centrifugation (Penefsky, 1977) in the presence of EDTA. This preparation contains three tightly bound nucleotides (Garrett and Penefsky, 1975; Rosing et al., 1975; Kironde and Cross, 1986; Wagenvoord et al., 1980; van Dongen et al., 1986; Edel et al., 1992), but when Mg^{2+} has been added before the removal of free and loosely bound nucleotides, about four nucleotides remain bound (Kironde and Cross, 1986; Edel et al., 1995). On the other hand, the preparations mostly used by Allison's group (Bullough et al., 1987, 1988) and the pig heart enzyme used by Gautheron's group (Penin et al., 1979) contain only two tightly bound nucleotides, and such enzyme preparations respond differently toward added nucleotides. The most well-known example is the so-called hysteretic inhibition after preincubation of the enzyme with ADP plus Mg^{2+}. Upon addition of ATP, the hydrolysis activity starts at a noninhibited rate but slows down to 15–20% after a few minutes. This behavior is not seen when the enzyme is prepared according to the procedure of Knowles and Penefsky (1972).

Many investigators have performed experiments with nucleotide-depleted enzyme. The procedure to remove tightly bound nucleotides has been established by Garrett and Penefsky (1975) and can be used to study the localization of these nucleotides and to determine how many of these sites are catalytic, since upon rebinding of nucleotides the three sites with the highest affinity (Weber et al., 1985; Jault and Allison, 1994) will still be the sites that originally contained tightly bound nucleotides. On the other hand, however, the enzyme has become a different enzyme as a consequence of the glycerol treatment and it is not allowed to assume that the properties of the tight binding sites have not changed at all. Upon addition of adenine nucleotides or suitable analogues, the originally nonexchangeable sites are no longer the sites with the highest affinity, but the tight catalytic site now has the highest affinity (Lunardi *et al.*, 1987; Milgrom and Boyer, 1990). It is evident that these sites are no longer nonexchangeable under the usual conditions (Weber *et al.*, 1985; Lunardi *et al.*, 1987).

2.2.2. F_1 Preparations from Different Sources

Although the homology between F_1 from various sources is large enough to assume that the mechanism of catalysis is the same, there may be differences in some specific aspects. As an example, it has been shown that with bovine heart F_1, both the V_{max} of ATP hydrolysis and the apparent cooperativity are influenced by anions (Ebel and Lardy, 1975). With yeast F_1, however, the presence of anions only affects the apparent cooperativity (linearization of the Lineweaver–Burk plot), but not the V_{max} (Recktenwald and Hess, 1977; Stutterheim *et al.*, 1980). This may be due to a difference in the rate of dissociation of product relative to the rate of catalysis, but it shows that care has to be taken when results with enzyme preparations from different sources are compared. Preparations of ATP synthase from bacteria, chloroplasts, and mammals also respond differently toward the activating anion sulfite, an inhibitor of ATP synthesis in some preparations but not in others (Bakels *et al.*, 1994). A well-known example is also the presence of tightly bound nucleotides. The presence of tightly bound nucleotides in bovine heart F_1, discovered in the 1970s (Harris *et al.*, 1973; Rosing *et al.*, 1975), was important for the development of the idea that catalysis occurs when a nucleotide is tightly bound at a catalytic site (Boyer *et al.*, 1973; Slater *et al.*, 1974), but F_1 preparations from the thermophilic bacterium PS3 do not contain any tightly bound nucleotide (Kagawa *et al.*, 1976) and the tightly bound nucleotides in the *Escherichia coli* enzyme preparations may be bound at sites that are partly different from the sites that contain the tightly bound nucleotides in the mitochondrial enzyme.

Isolated CF_1 is inactive and has to be activated, so that there may be differences in nucleotide binding between activated and nonactivated enzyme.

2.2.3. Differences in Nucleotide Binding between Isolated F_1 and F_1-F_0

Fewer data are available on the binding of nucleotides to isolated F_1-F_0 and to submitochondrial particles. For the CF_1-F_0, the data on the modification of all six binding sites with 2-azido-AT(D)P (Possmayer, 1995) give relevant information, and for MF_1-F_0, recent data are available from the work by Beharry and Bragg (1996). The latter data show some differences between the isolated F_1 and the complete ATP synthase; although these differences seem to be restricted to the exchange behavior of only one high-affinity site, care has to be taken when data from isolated F_1 and the complete enzyme are compared. The studies of van der Zwet-de Graaff *et al.* (1997) show that the noncatalytic α sites in submitochondrial particles (SMP) bind FSBA in a way similar to that in isolated F_1 and with the same consequence for the ATPase and ITPase activity. At this point it may be relevant to mention that F_1 as isolated in our laboratory (according to Knowles and Penefsky, 1972) and the F_1 from the Allison laboratory react differently with FSBA. This result may indicate that our preparation is a better model for the intact system than the Allison preparation, but this has to be verified for each aspect. In fact, we have argued (Hartog *et al.*, 1997) that the main reason for the difference in FSBA binding may not be the type of preparation as such, but the treatment with glycerol to obtain nucleotide-depleted enzyme as starting material. On the other hand, the absence of hysteretic inhibition in SMP also indicates that the enzyme preparation with two tight nucleotides is indeed not as good a model for the intact system as the enzyme with three or four tightly bound nucleotides. Thus, when differences between these two types of F_1 preparations are seen, the choice is for the behavior of the Knowles–Penefsky preparation as a model for the native system.

3. PROPERTIES OF THE NUCLEOTIDE BINDING SITES OF MF$_1$

3.1. Characterization of Nucleotide Binding Sites on Basis of Affinity and Exchangeability

3.1.1. Isolated F$_1$

From the earliest data on the binding of adenine nucleotides to isolated F_1, it has been clear that the enzyme contains a number of tightly bound nucleotides that are not removed upon gel filtration and ammonium sulfate precipitation. Most of the bound ligand finally can be removed upon frequent repetition of these steps (van Dongen and Berden, 1987), but the most common procedure is to pass the enzyme slowly through a Sephadex column in 50% glycerol (Garrett and Penefsky, 1975). After this step only 0.2–0.5 nucleotide/F_1 remains.

The literature reports on the number of tightly bound nucleotides are in good agreement. After column centrifugation according to Penefsky (1977) and ammonium sulfate precipitation in the absence of Mg^{2+} ions, three nucleotides per F_1 remain, either two ATP and one ADP (Rosing *et al.*, 1975) or two ADP and one ATP (Garrett and Penefsky, 1975). The conversion of one bound ATP to ADP is slow but enhanced by Mg^{2+} (Edel *et al.*, 1992), so we may suppose that the one ATP that is converted into ADP is bound at a catalytic site. This agrees with the finding that upon treatment of the enzyme with NAP_3-2-azido-ADP one nucleotide exchanges and only one ATP and one ADP are left (Edel *et al.*, 1995). Also after the binding of the maximal amount of 8-nitreno-adenine nucleotides (4 mole/mole F_1), just one ADP and one ATP remain bound (Wagenvoord *et al.*, 1980; van Dongen *et al.*, 1986).

When the enzyme with three bound nucleotides is treated with a suitable analogue like NAP_3-2-azido-ADP in the presence of Mg^{2+} (Edel *et al.*, 1995), one adenine nucleotide disappears and two molecules of the analogue are bound quite strongly. One is bound at the exchangeable site (exchange occurs with a K_D of about 30 μM) and the second occupies a noncatalytic site with a low K_D in the presence of Mg^{2+} (Edel *et al.*, 1995). The binding of ADP to this site has been studied by Kironde and Cross (1987), who measured both the rate of binding and of dissociation, resulting in a K_D of 50 nM. It should be kept in mind that the measured k_{off} is the same value as reported by Grubmeyer *et al.* (1982) for the k_{off} of product ADP during single-site catalysis (3.10^{-4} sec^{-1}); it is the opinion of the present authors that the low rate of dissociation measured by Grubmeyer and co-workers during single-site catalysis is not the rate of dissociation from the catalytic site but from a noncatalytic site. The rate of dissociation from the catalytic site itself is in the order of 0.05–0.1 sec^{-1} (Milgrom and Murataliev, 1987; Berden *et al.*, 1991). The low rate of dissociation of ADP from the noncatalytic site implies that the bound ADP can be exchanged only at a very long time scale. The experiments of Kironde and Cross (1986) also clearly show that at four sites nucleotides are so strongly bound in the presence of Mg^{2+} that they remain bound after column centrifugation. Of these four bound nucleotides only one is rapidly exchanged during catalysis.

Treatment of F_1 with 100 mM pyrophosphate results in loss of the rapid exchangeable (catalytic) nucleotide and the slowly exchangeable (noncatalytic) nucleotide, while the two nonexchangeable nucleotides remain bound (Hartog *et al.*, 1997). Preliminary data indicate that in this preparation one of the latter is exchangeable with ATP but not with GTP.

The two residual sites, not mentioned so far, show a much larger K_D. In the experiments of Edel *et al.* (1993) it is obvious that the nucleotide at the tight exchangeable site is much more strongly bound than suggested by the K_D of 30 μM for the exchange reaction (see also van Dongen and Berden, 1987). From the available data on cooperativity one may assume then that an added nucleotide binds with a K_D of 30 μM at an empty catalytic site, but that by cooperativity the nucleotide bound at the tight catalytic site then becomes more loosely bound and is also replaced with the added ligand. Upon column centrifugation, one bound ligand with a K_D of 30 μM dissociates, and upon removal of this nucleotide from one catalytic site, the nucleotide at the other site becomes tightly bound and does not dissociate any more. The conclusion then is that an additional catalytic site binds ADP or ATP with a K_D of 30 μM. A similar K_D is found upon binding of 8-azido-ATP in the presence of EDTA (van Dongen *et al.*, 1986).

Finally, the sixth site has a still lower affinity, and this should be either the third catalytic site or an additional noncatalytic site. The finding that 8-azido-AT(D)P can be covalently bound at a low-affinity noncatalytic site on an α subunit (Edel *et al.*, 1993) strongly indicates that the sixth binding site is a noncatalytic α site, implying that one of the three nucleotides bound at high-affinity noncatalytic sites is bound at a β subunit. This preliminary conclusion agrees with the report from Aloise *et al.* (1991) that in MF_1 one β site is not available for BzADP plus Mg^{2+}, while in TF_1, not containing tightly bound nucleotides, all three β-sites are available. This point will be discussed further.

Also, F_1 from *E. coli* usually contains three tightly bound nucleotides. It has been reported that all three bound nucleotides are nonexchangeable and from this it has been concluded that only noncatalytic α sites are involved (Senior, 1990). Weber *et al.* (1993) have shown, however, using an enzyme with a Trp at position 331 of the β-subunits, that more than 0.5 nucleotide is bound at a β subunit; therefore, we would like to propose that of the three tightly bound noncatalytic nucleotides in the *E. coli* F_1 one is bound at a β subunit. The binding of an adenine nucleotide at a single catalytic site is apparently less tight than in MF_1, such a nucleotide not being present after isolation of the enzyme.

3.1.2. Nucleotide-Depleted F_1

Schäfer's group has reported data on both binding of photoaffinity analogues to nucleotide-depleted F_1 and the resulting inhibition after covalent modification (Lübben *et al.*, 1984; Weber *et al.*, 1985). The authors differentiated three high-affinity sites and three low-affinity sites. With one analogue two sites were modified, resulting in complete inhibition (Lübben *et al.*, 1984), with another analogue, three sites were modified (Weber *et al.*, 1985). Since the relation between binding and inhibition was linear, only one of the sites was catalytic in both cases. If more sites would have been directly involved in catalysis, the relation covalent binding versus inhibition would not have been linear. These results agree with the data of Jault and Allison (1994), who showed that in nucleotide-depleted enzyme three sites can bind ADP with high affinity in the absence of Mg^{2+} and four in the presence of Mg^{2+}. Only one of the sites was catalytic; this site now is the site with the highest affinity (Lunardi *et al.*, 1987; Milgrom and Boyer, 1990) and the originally nonexchangeable sites are now exchangeable after the glycerol treatment.

Of the two noncatalytic sites with high affinity in the absence of Mg^{2+} one was the site responsible for hysteretic inhibition by bound ADP (Jault and Allison, 1994). Hysteretic inhibition means that addition of ATP plus Mg^{2+}

results in full activity of the enzyme, but with a half-time of about 1 min the enzyme becomes largely inhibited (Di Pietro *et al.*, 1980). Since the enzyme as studied by Kironde and Cross (1986) and by us does not show any hysteretic inhibition by ADP, none of the three sites that are occupied with ADP in that enzyme in the presence of Mg^{2+} is responsible for hysteretic inhibition. The only high-affinity site in enzyme preparations isolated according to the procedure of Knowles and Penefsky that is not occupied with ADP is the site that contains ATP. In addition, it is well established that in preparations isolated in the laboratories of Gautheron and Allison (which show hysteretic inhibition after incubation with ADP), only two high-affinity sites are occupied, both with ADP (Penin *et al.*, 1979; Bullough *et al.*, 1988). The conclusion then has to be that the hysteretic site is the site that contains ATP in the preparations isolated according to Knowles and Penefsky and is not occupied in the preparations from the laboratories of Gautheron and Allison. Since Allison has reported (Bullough *et al.*, 1988) that in his preparation the bound ADP is lost upon treatment with FSBA, a ligand that is specific for α sites, the two ADP per F_1 are bound at noncatalytic α sites. If now the hysteretic site should be an α site, it would imply that all three α sites bind ADP tightly and that there is no low-affinity noncatalytic α site. If, however, the hysteretic site is a β site, the one β site that contains bound ATP in the Knowles and Penefsky (1972) preparation is not catalytic and one α subunit may contain a low-affinity binding site, as is concluded by Jault and Allison (1993) and by our group (Edel *et al.*, 1993).

3.1.3. F_1-F_0

Affinity studies with the complete enzyme have been performed both with SMP and isolated F_1-F_0. In SMP, the number of tightly bound nucleotides has been reported to be four (Matsuno-Yagi and Hatefi, 1993), but these data are possibly not fully reliable because of the inaccuracy in the determination of the enzyme concentration and the possible binding of adenine nucleotides to other enzymes. The data on isolated F_1-F_0 are more reliable, but also these data confirm the presence of four tightly bound nucleotides (Beharry and Bragg, 1996), of which one can be removed by GTP, just like in SMP (Matsuno-Yagi and Hatefi, 1993). This one binding site certainly is catalytic. One of the three remaining bound nucleotides is bound at the high-affinity site at which in isolated F_1 a bound nucleotide exchanges very slowly (a noncatalytic site). As far as high-affinity binding is concerned, the data on F_1-F_0 are therefore in agreement with the data obtained with F_1 isolated according to Knowles and Penefsky (1972). Each molecule contains three high-affinity sites for non- or slowly exchangeable nucleotides and one high-affinity catalytic site, available to GTP. The one difference between isolated F_1 and F_1-F_0, as reported by Beharry and Bragg (1996), is the exchangeability in F_1-F_0 of one of the two

1 ATP	2	3
4 ADP	5 ADP	6

1	2	3
4 ATP	5 ADP	6 ADP

FIGURE 2. Localization of the three high-affinity noncatalytic nucleotide binding sites. (A) It is assumed that one low-affinity noncatalytic binding site exists, only occupied at high concentration of ligand (site 6). (B) It is assumed that all noncatalytic sites are α sites. The consequence is that all noncatalytic sites have a high affinity for ligands.

nucleotides that are nonexchangeable in isolated F_1. In Fig. 2, the two possible localizations of the high-affinity noncatalytic sites are depicted and in the model of our preference (A) one tightly bound nonexchangeable nucleotide is bound at a β site.

Considering the fact that in many preparations of F_1 (Bullough *et al.*, 1987) and in CF_1 (Fromme and Gräber, 1989) the first β site can perform (slow) hydrolysis, the exchangeable nucleotide in F_1-F_0 will be at site 1 and not at site 4. The nucleotide at site 4 is probably the nucleotide with the highest affinity, since even after dissociation of all β subunits from the enzyme one ADP remains bound to the residual $\alpha_3\gamma\delta\varepsilon$ moiety (Hartog *et al.*, 1992).

The rate of the exchange of the exchangeable high-affinity, not GTP-replaceable nucleotide has not been determined, and on the basis of the present data it remains theoretically possible that this site is involved in multisite catalysis in the intact system, but not when GTP is the substrate. In this interpretation only two of the three tightly bound nucleotides that cannot be replaced with GTP are located on α subunits. This leaves one α site available for low-affinity binding. In the alternative model (Fig. 2B), no low-affinity noncatalytic α site exists, in contrast with experimental data (Edel *et al.*, 1993, 1995; Jault and Allison, 1994). In SMP, the studies with 8-azido-ATP show that there are two available binding sites, one of them a catalytic β site and one a noncatalytic α site (Sloothaak *et al.*, 1985). In SMP, in contrast to isolated F_1, no additional sites are available for binding of 8-azido-ATP in the presence of Mg^{2+} (Sloothaak *et al.*, 1985). Since the enzyme in SMP contains four tightly bound nucleotides (Matsuno-Yagi and Hatefi, 1993), the data show that also in

1 AXP	2 tight catalytic	3 (8N-ATP)
4 AXP	5 AXP	6 (8N-ATP)

FIGURE 3. Occupation of nucleotide binding sites in SMP, treated with 8-azido-ATP. Phosphorylating SMP contain four bound nucleotides of which one is catalytic. Upon incubation with 8-azido-ATP plus EDTA, followed by illumination, two sites become occupied with 8-nitreno-ATP, one α site, and one β site. Upon further incubation with 8-azido-ADP plus Mg^{2+}, followed by illumination, no further site is modified.

SMP a low-affinity noncatalytic site is available for binding an added ligand (see Fig. 3).

The finding that in F_1-F_0 and in SMP three noncatalytic sites are occupied with bound nucleotides also agrees with the finding of van der Zwet-de Graaff (1996) that in SMP no high-affinity noncatalytic site is available for 2-azido-ATP, so that these sites are all occupied. The occupation in phosphorylating SMP of the high-affinity noncatalytic site that is slowly exchangeable in isolated F_1 (site 5 in Fig. 2A; site 6 in Fig. 2B) may be due to the fact that these particles are prepared in the presence of Mn^{2+} and Mg^{2+}. The high-affinity catalytic site, on the other hand, can be easily modified with 2-azido-ATP (van der Zwet-de Graaff, 1996; Martins and Penefsky, 1994).

After the discovery of the tightly bound nucleotides, Harris investigated whether they could be involved in ATP synthesis. Indeed, he measured some additional exchange under energized conditions, but this exchange is probably too slow to be compatible with direct participation of one of these sites in catalysis (Harris et al., 1977, 1978). In fact, one additional nucleotide exchanged (Harris et al., 1977). Recent experiments by van der Zwet-de Graaff with 2-azido-ATP in our laboratory (not shown) showed that only one (catalytic) site in SMP retained tightly bound 2-azido-AT(D)P, whether the SMP were just incubated with the ligand, incubated under conditions of hydrolysis of 2-azido-ATP, or incubated under conditions of synthesis of 2-azido-ATP, so no additional exchange of a tightly bound nucleotide with 2-azido-AT(D)P under energized conditions was found. Therefore, some data indicate no exchange of one of the two "nonexchangeable" nucleotides (I. van der Zwet-de Graaff, unpublished observation), some data indicate exchange of one of them under energized conditions (Harris et al., 1977), and some data (Beharry and Bragg, 1996) indicate a similar exchange even in the absence of energetization. The possible exchange of one tightly bound nucleotide at a site that is not involved in GTP hydrolysis suggests to us that the site involved may be able to perform slow catalysis with ATP or ADP as substrate under certain conditions or in certain preparations. A possible involvement of this site in ATP synthesis will be discussed later, but it seems not exchangeable enough to be involved in ATP hydrolysis. Also, CF_1 contains four bound nucleotides (Shapiro *et al.*, 1991), but isolated CF_1-F_0 contains three bound nucleotides (Possmayer, 1995). Two of them are bound at α sites and one at a β site. The latter site is involved in activation of the enzyme and can perform unisite catalysis (Possmayer, 1995).

3.2. Characterization of Binding Sites on Basis of Localization

The site of modification has been studied with various covalently binding ligands. In view of the present knowledge of the structure of F_1, only a few are still relevant. In all cases, 2-nitreno-AT(D)P has been shown to modify only amino acids belonging to β subunits, whether the azido-compounds were bound at catalytic or at noncatalytic sites. It has been shown by Garin *et al.* (1986) that amino acids around Tyr345 were modified by 2-azido-ATP when bound at the tight catalytic site. Cross *et al.* (1987) have further shown that principally only two amino acids were modified: one in the case of binding to catalytic sites (β Tyr345 of the mitochondrial enzyme) and the other in the case of binding to noncatalytic sites (β-Tyr368 of the mitochondrial enzyme). This result has been very useful in the analysis of functional studies, since now one could analyze which type of site was modified under special conditions. For the chloroplast enzyme (CF_1-F_0), Possmayer (1995) has shown that in the nonacti-vated enzyme three sites indeed can be specifically modified at the equivalent tyrosine of Tyr345 and three at the equivalent tyrosine of Tyr-368. For the mitochondrial enzyme, however, data for all six sites are not available. The structural data (Abrahams *et al.*, 1994) suggest that also in MF_1 the structure of all three β sites could be identical, as well as that of the three α sites; however, Lunardi *et al.* (1987) have reported that with nucleotide-depleted enzyme the first molecule of ligand did indeed modify β-Tyr345 (a catalytic β site), but subsequent binding resulted in modification of fragment Gly72–Arg83 of the β subunit. This result has not been verified by other groups and the published structure of the enzyme cannot easily accommodate modification of this pep-tide fragment by specific binding at a nucleotide binding site, whether it is an α or a β site. The authors used cyanogen bromide, however, to obtain their peptide fragments, and Hartog *et al.* (1992) have reported that this procedure never shows modification of β-Tyr368, not even in preparations in which modification of this amino acid can be shown using tryptic digestion.

In the enzyme from *E. coli* at least the three β sites have the same structure, since in the mutant β-Tyr331Trp the binding of ATP or ADP induces the same fluorescence change at all three sites (Weber *et al.*, 1993).

We have found (J. A. Berden and A. F. Hartog, unpublished observations) that in MF_1 2 mole of 2-nitreno-ADP can be bound at Tyr345 and two at

Tyr368. But the localization of 2-azido-AT(D)P when bound at one of the two nonexchangeable sites has never been demonstrated. Just this point is relevant, since a main issue of the still-existing controversies on the mechanism of F_1 is whether both nonexchangeable sites are α sites or whether one of them is a β site. Our finding strongly indicates that one of the two nonexchangeable nucleotides is bound at a β site, but direct evidence is not yet available. Hartog *et al.* (1992) have tried to modify specifically one of these two nonexchangeable sites by reconstituting LiCl-dissociated enzyme, containing one residual nucleotide at the $α_3γδε$ moiety of the enzyme, by reconstitution in the presence of 2-azido-ATP, followed by exchange of all ligand at catalytic and low-affinity sites with ATP. Their results were not conclusive, however, because of a low level of covalent binding.

Several ligands that inhibit catalysis by binding at a catalytic site do not bind in the pocket where the Tyr345 is located, although this pocket contains the ADP or AMPPNP in the crystals of Abrahams *et al.* (1994) and also binds FSBI, Tyr345 being modified (Bullough and Allison, 1986b). NbfCl and 8-azido-ATP bind at a different site, modifying β-Tyr311 (Andrews *et al.*, 1984; Hollemans *et al.*, 1983). This region is not identifiable in the known crystal structure as a nucleotide binding site (Abrahams *et al.*, 1994). NbfCl and 8-azido-ATP are not real analogues of AD(T)P (8-azido-ATP adapts the syn-configuration instead of the anticonfiguration), so they may not fit in the same binding pocket as AD(T)P. The solution to this problem can be derived from the studies with FSBA (Hartog *et al.*, 1997). Although at the suitable pH in the (usually nucleotide-depleted) preparations of Allison's group only modification of β-Tyr368 has been reported as the result of specific binding (Bullough and Allison, 1986a), we have found that all three α subunits contain two binding sites for FSBA, and the same result has also been obtained with SMP (van der Zwet-de Graaff *et al.*, 1997). The second site has about the same distance to the P-loop as the first one, but in a different direction, and contains at least α-Tyr244 as a modifiable amino acid. In the preparation of Allison, which is not depleted of bound nucleotides, etheno-FSBA also modifies this amino acid (Verburg and Allison, 1990). Binding of FSBA at either site induces a partial inhibition of catalysis. We may speculate then that the β subunits also contain such a second binding pocket in the analogous position and this is the site where 8-azido-ATP and NbfCl bind, modifying β-Tyr311 (Hollemans *et al.*, 1983; Andrews *et al.*, 1984). In this respect it is interesting that the aluminum—fluoride complex of 8-azido-ADP modified β-Tyr345 (Garin *et al.*, 1994), so the adenosine moiety of the 8-azido-adenine nucleotides has two possible positions, depending on the conformation of the molecule. Both binding pockets are at a suitable distance from the P-loop, so that in both positions the di- or triphosphate moiety binds at this loop. We have further speculated (Hartog *et al.*, 1997) that the absence of such a second binding pocket in the crystal structure (Abrahams *et al.*, 1994) may be a consequence of the pretreatment of the enzyme with glycerol to remove endogenous nucleotides.

Vogel and Cross (1991) have shown that F_1 and nucleotide-depleted F_1 have one high-affinity site for Ap_4A. Since the binding of 1 Ap_4A/F_1 inhibits only partially, the site is probably noncatalytic. The binding is stronger, however, than that of ADP, and the authors proposed that Ap_4A binds to both a catalytic and a noncatalytic site, suggestive for a structure of the binding sites similar to that of myokinase. The crystal structure, however, has shown that the distance between a catalytic and a noncatalytic site cannot be bridged by four phosphate groups and from our FSBA studies (Hartog *et al.*, 1997) the most likely explanation is that Ap_4A binds at the high-affinity, slowly exchangeable, noncatalytic α site, but that the two adenosine groups each occupy one of the binding pockets mentioned above, connected via the P-loop.

Many studies with covalently binding ligands have indicated that the nucleotide binding sites are located at or near interfaces between an α and β subunit. The modification of either α or β is usually not enough to determine whether the α or β site is modified, except when using 8-azido-adenine nucleotides. Van Dongen and Berden (1986) have shown that binding to a (catalytic) β site results in modification of the β subunit only, while binding at an α site results in equal modification of both subunits. The concomitant modification of one α and one β site with radioactive 8-nitreno-ATP results in a 1:3 distribution of the label over the α and β subunits. F_1 preparations containing 8-nitreno-ADP at both an α and a β site could be further modified with 2 mole 8 nitreno-ADP per mole F_1. Also, for this additional modification the distribution of the label over α and β subunits was 1:3, indicating that again both an α and a β site were modified. In total, then, two α and two β sites can be modified with 8-nitreno-AXP (see Fig. 4). The reliability of the label distribution for determining the type of modified site(s) was confirmed by Edel *et al.* (1993). These authors modified three nucleotide binding sites by incubating F_1

1 ATP	2 8N-ADP/Mg	3 8N-ATP
4 ADP	5 8N-ADP/Mg	6 8N-ATP

FIGURE 4. Occupation of nucleotide binding sites of MF_1 after treatment with 8-azido-AXP. Isolated MF_1 was twice treated with radioactive 8-azido-ATP plus EDTA and after that with radioactive 8-azido-ADP plus Mg^{2+}, as described by van Dongen and Berden (1986). Both treatments resulted in the modification of one α site and one β site, as determined from the distribution of label over α and β subunits. After the treatment, two bound adenine nucleotides were still present.

with 8-azido-ATP in the presence of EDTA, followed by illumination. The distribution of label between α and β subunits was 1:2, and this should indicate that one β site and two α sites were modified. The linear relation between level of modification and inhibition of activity indeed confirmed that only one catalytic β site was modified, so the other two were α sites.

3.3. Characterization of Binding Sites on Basis of Catalytic Involvement

In the literature one may find suggestions that only high-affinity sites are catalytic (Weber et al., 1985) or that all rapidly exchangeable sites are catalytic (Cross and Nalin, 1982; Kironde and Cross, 1986; Senior, 1990). Both statements may be wrong. Also, low-affinity sites may be catalytic and exchangeable sites may be noncatalytic. It is well known that one catalytic site has a very high affinity when the other catalytic sites are not occupied (Grubmeyer et al., 1982; Cross et al., 1982) and it is wrong to state that the three tightly bound nucleotides are bound at noncatalytic sites, unless it has been shown that they are not rapidly exchangeable.

In mitochondrial F_1, one of the tightly bound nucleotides is bound at a catalytic site: It is rapidly exchangeable and covalent modification causes complete inhibition of activity. The other two tightly bound nucleotides are noncatalytic, since they do not exchange during catalysis (Kironde and Cross, 1986; Edel et al., 1992). The same holds for the slowly exchangeable site with a high affinity for ADP in the presence of Mg^{2+}. ADP (or a suitable analogue) at this site remains bound during catalysis and does not exchange within 15 min. Binding of ADP at this site even causes a partial inhibition of the ATPase activity (Edel et al., 1992, 1995).

In F_1 from E. coli, the three tightly bound nucleotides are not easily exchangeable and therefore noncatalytic (Perlin et al., 1984; Senior, 1990). The affinity of the tight catalytic site for nucleotides in the E. coli enzyme is apparently too low to retain a nucleotide upon isolation. But the noncatalytic character of the tight binding sites does not imply that these sites are all located on α subunits, as is usually suggested (Senior, 1990). The preparation may be compared with the preparations of mitochondrial enzyme with four nucleotides [e.g., F_1 depleted of medium and loosely bound nucleotides in the presence of Mg^{2+} (Edel et al., 1995), F_1-F_0 (Beharry and Brag, 1996), or SMP (Matsuno-Yagi and Hatefi, 1993)], but then treated with GTP to exchange the nucleotide at the catalytic site with GTP, an analogue that is not retained after column chromatography. In all these preparations one bound nucleotide may be bound at a noncatalytic β site.

Starting with an enzyme with four nucleotides, one of which is catalytic, at least one of the two residual sites is catalytic (in E. coli F_1 with no catalytic site occupied, at least two of the three additional sites are catalytic), since it has been shown first by Grubmeyer *et al.* (1982) that upon binding of ATP to a second catalytic site ATP hydrolysis occurs at a high rate at both sites. The K_D of this second catalytic site has to be around 30 μM, the K_M for ATP hydrolysis (at concentrations of ATP in the order of 1–50 μM). Also, the K_D for the exchange of the nucleotide at the high-affinity catalytic site has this value (van Dongen and Berden, 1987; Edel *et al.*, 1995), since for exchange at the high-affinity catalytic site added nucleotide first has to bind at a low-affinity catalytic site.

The sixth site has to be catalytic as well according to the model for three catalytic sites, but it is noncatalytic when only two sites perform multisite catalysis. Murataliev and Boyer (1994) assume the presence of a catalytic site with a K_D of 2 mM, but such a K_M for ATP hydrolysis has never been observed, and it would therefore imply that this site is usually unoccupied. This does not fit with our own data nor with the data on F_1-F_0 (Beharry and Bragg, 1996) or the data on *E. coli* F_1 (Weber *et al.*, 1993). The latter authors show that all three β-sites have to be filled for a high rate of catalysis, so it can be excluded that a catalytic site is not occupied at 0.5 mM ATP, under which conditions the rate of ATP hydrolysis is nearly maximal. Allison has concluded on the basis of the complicated pre-steady-state kinetics (Jault and Allison, 1993) that a low-affinity site for ATP has to be present that has regulatory properties. A similar conclusion had been drawn already by Recktenwald and Hess (1977) and Stutterheim *et al.* (1980) in order to explain the kinetics of the yeast enzyme and the effect of anions. It is evident from the steady-state kinetics that a site with a relatively low affinity (but a K_D value of 2 mM seems much too high) affects catalysis, since the Lineweaver–Burk plot shows negative cooperativity. Either it is a third catalytic site, whose participation in catalysis causes the Michaelis constant of the catalytic reaction to increase because of negative binding cooperativity (the K_M increases to about 200–300 μM) or it is a regulatory site whose occupation with ligand influences the affinity of the two catalytic sites for substrate. Many experiments have been performed to obtain a direct answer to this question, especially in our own group.

We have shown that with 8-azido-ATP in the presence of EDTA two sites can be modified upon illumination, both with isolated F_1 (Wagenvoord *et al.*, 1977; van Dongen and Berden, 1986) and with submitochondrial particles (Sloothaak *et al.*, 1985). Originally, Wagenvoord *et al.* (1977) had concluded that both sites were catalytic and the linear inhibition curve obtained was interpreted by these authors as indicative of the presence of two independent catalytic sites. But this interpretation was in contradiction with the principle of cooperativity between catalytic sites (Grubmeyer *et al.*, 1982; Cross *et al.*, 1982). The solution was found by van Dongen and Berden (1986) who demonstrated that in fact two sites of different types were modified upon illumination in the presence of 8-azido-ATP plus EDTA: One site bound 8-azido-ATP with a relatively high affinity; its modification with 8-nitreno-ATP caused complete inhibition of ATPase activity and the modification was restricted to amino acids of β subunits. The second site showed a lower affinity for 8-azido-ATP; its modification by 8-nitreno-ATP did not contribute to the inhibition and the modification involved amino acids of both α and β subunits in about equal amounts. Subsequent modification of two additional sites in isolated F_1 with 8-azido-ADP in the presence of Mg^{2+} again resulted in the modification of both a β and an α site, at that time called an α–β site. Edel *et al.* (1993) modified with 8-azido-ADP in the presence of EDTA three sites, of which two were noncatalytic and one was catalytic. This was concluded on the basis of both a linear inhibition curve and an 1 : 2 distribution of label over α and β subunits. Edel *et al.* (1993) also succeeded in binding 8-nitreno-ATP to just the postulated low-affinity noncatalytic site by illuminating F_1, after preincubation with 8-azido-ADP plus EDTA, under conditions of turnover with ATP as substrate, so that no catalytic site was occupied with 8-azido-ATP during illumination. The exchange at the noncatalytic site with added ATP was apparently slow enough not to interfere significantly with the modification of this site, resulting in about equal modification of α and β subunits. The obtained modification caused a linearization of the Lineweaver–Burk plot to a single high K_M value for ATP, but the V_{max} of ATP hydrolysis was hardly affected. This site, therefore, was really a regulatory noncatalytic site.

To ascertain that the high-affinity slowly exchangeable noncatalytic site was not modified with this procedure, the preparation with nearly one noncatalytic site modified was further incubated with NAP_3-2-azido-ADP and illuminated. The partial (about 40%) inhibition of the ATPase activity that is a characteristic for the binding of ADP or a suitable analogue in the anticonfiguration to the high-affinity slowly exchangeable noncatalytic site, covalent or noncovalent (Edel *et al.*, 1992), was indeed obtained (see Fig. 5). Covalent modification amounted to 60% of the site.

The results described can lead to only one conclusion: Since in these experiments two different noncatalytic α sites are identified and can be covalently modified, while two adenine nucleotides still remain bound to the enzyme, apparently also at noncatalytic sites since they do not exchange, one β site contains a tightly bound nucleotide and the number of residual catalytic sites is just two and cannot be three.

Another relevant experiment has been done by Nieboer *et al.* (1987). They showed that after dissociation, induced by LiCl, of preparations containing 0–1 mole of covalently bound Nbf per mole F_1 at a catalytic site, reconstitution resulted in a partial activation without loss of ligand. These experiments were a follow-up of experiments by Wang (1985) and the data showed that after dissociation–reconstitution, assuming random incorporation of modified and

1 ATP	2	3
4 ADP	5 NAP₃-2N-ADP	6 8N-ATP

FIGURE 5. Presence in isolated MF_1 of two noncatalytic regulatory α sites available for modification. F_1, containing three tightly bound nucleotides, was incubated with 8-azido-ATP, followed by illumination in the presence of added ATP plus Mg^{2+}, resulting in the modification of one regulatory α site. After that, the enzyme was incubated with NAP₃-2-azido-ADP plus Mg^{2+}, resulting in occupation of an additional noncatalytic site with ligand, inducing a partial inhibition of enzyme activity (see Edel *et al.*, 1993). The nucleotides at sites 1 and 4 are not replaced during this treatment.

nonmodified β subunits in the reconstituted enzyme molecules, the molecules with one Nbf-modified β subunit can hydrolyze ATP at a rate equal to 33% of the uninhibited rate. This result indicates that in one of the three molecules the modification does not inhibit activity any more, since a partial activity of each molecule, induced by Nbf binding, is not possible. Molecules with two modified β subunits were not active. Although it had been shown (Wang, 1985) that upon reconstitution modified and nonmodified β subunits are incorporated randomly, the results were not accepted by, for example, Boyer (1993), the main argument being that modified β subunits might not be incorporated randomly and molecules with either nonmodified or modified subunits might be preferentially formed.

Miwa *et al.* (1989) later improved this type of experiment. These authors did not use modification by a ligand followed by dissociation–reconstitution; they used isolated genetically modified β subunits from TF_1 for the reconstitution of intact enzyme molecules, containing specific amounts of the modified subunit. They prepared enzyme preparations containing 0, 1, 2, or 3 modified β subunits and showed that enzyme with one modified β subunit had in one case (the mutant E190Q) 8% activity and 45% activity in the other case (E201Q).

Enzymes with two mutated β subunits were essentially inactive. In a recent paper Amano *et al.* (1996) reported that the 8% activity found in the enzyme containing one mutant β subunit (E190Q) was at least largely if not completely due to scrambling, and they concluded that three catalytically active β subunits are therefore required for multisite catalysis. This conclusion seems quite premature, since another explanation is also possible. Since the change of Glu190 into a Gln resulted in a low activity in the experiment of Miwa *et al.* when one β subunit was modified, this mutation apparently also affected activity when present in the noncatalytic β subunit, since according to a two-site model the activity should be 33% and not 8% when the structure of the third β site does not impair catalysis. When this 8% is now brought back to about 0, it could mean that no activity is possible when the structure of the binding site in the noncatalytic β-subunit is such that it cannot undergo the conformational change accompanying the hydrolysis of ATP. If a certain mutation inhibits all activity, the experiment has no discriminatory power, since there can be many reasons why a change in a noncatalytic site inhibits activity. For example, modification of any noncatalytic site in CF_1-F_0 with 2-nitreno-ATP inhibits the enzyme completely (Possmayer, 1995).

The discriminatory experiment is the other one, not repeated by Amano *et al.* (1996), that shows substantial activity with one modified β subunit. The mutation was such (E201Q) that the modified subunit could no longer bind adenine nucleotides. The originally reported activity (45%) was even higher than the statistical value of 33% (in 33% of the molecules, the modified subunit will have the position of the first noncatalytic β subunit). If now 8% activity is due to scrambling of the reconstituted TF_1, the real activity is 35%, quite close to the theoretical value of 33% and in full agreement with the activity measured by Nieboer *et al.* (1987) with reconstituted enzyme containing one Nbf-modified β subunit. One may ask the question, then, why in the one case no activity is present when one subunit is mutated and in the other case there is. A second question is whether 33% activity of the molecules with one nonfunctional β subunit contradicts any three-site rotational model for catalysis. The most simple answer to the first question is that in the one case the mutated binding region can accommodate an adenine nucleotide but cannot produce the required conformational change accompanying binding and hydrolysis of the substrate at that site (Al-Shawi and Senior, 1988) or at another site (Turina and Capalsi, 1994), just as when 2-nitreno-ATP at a noncatalytic site of CF_1-F_0 inhibits all activity. The other mutation, allowing activity when one mutated β subunit is present in a noncatalytic position, cannot accommodate an adenine nucleotide in the binding site, but the required conformational change accompanying binding and hydrolysis of substrate at a catalytic site is apparently still possible, similar to the case in which Nbf is bound to the subunit.

The answer to the second question must be positive, we think, despite Boyer's suggestion of slippage (Boyer, 1993). If slippage were possible, no ligand would be fully inhibitory when modifying one β site. In addition, all known inhibitors of a catalytic site cause full inhibition of enzyme activity: Partial inactivation when one catalytic site is not functional seems mechanistically impossible to us. If we take the Nbf modification, binding of Nbf at a catalytic β subunit induces full inhibition, so there is no slip through such a site. And if with another position of the same β subunit in the enzyme there is activity (even full activity), that same β subunit cannot be catalytic in that position. How does a rotational model work? The driving force for the conformational changes, including rotation over 120° in one direction, is the binding of ATP to an empty site that then adopts a so-called tight conformation, resulting in the relative movement of the γ subunit, such that the interaction between the two subunits changes (Turina and Capaldi, 1994). But if during the reaction cycle the β subunit occupying the position (relative to the γ subunit) specific for the empty subunit cannot perform the proper conformational change upon binding of the substrate and does not induce the movement of the γ subunit, the whole process of catalysis comes to a halt. Slippage is not an alternative explanation.

4. NUCLEOTIDE BINDING SITES AND CATALYSIS

4.1. A Model of the Nucleotide Binding Sites of MF_1 and Consequences for the Catalytic Mechanism

From the presented data on the properties of the nucleotide binding sites, a clear picture emerges of the six nucleotide binding sites of mitochondrial F_1. In the preparations of mitochondrial F_1, isolated according to Knowles and Penefsky and depleted of loosely bound nucleotides in the presence of EDTA, three tightly bound nucleotides are present (Edel *et al.*, 1995). In this preparation, two molecules of ligand can be bound at additional noncatalytic sites on α subunits. One site has a high affinity, in the presence of Mg^{2+}, for ADP and analogues in the anticonfiguration (like Ap_4A, 2-azido-ADP, NAP_3ADP) and the other one has a low affinity for ATP (Jault and Allison, 1994), but a reasonable affinity for 8-azido-AT(D)P (Edel *et al.*, 1993). ADP at the first site remains bound during catalysis and causes a partial inhibition of the enzyme activity, just like the analogues, whether the ligand is bound covalently or nonvocalently. This site is occupied with ADP when the enzyme is depleted of loosely bound nucleotides in the presence of Mg^{2+} (enzyme with four bound nucleotides) (Edel *et al.*, 1995) and it is also occupied in the preparations of *E. coli* F_1, SMP, and F_1-F_0. In the latter case, the ligand might be ATP and not ADP. The second site can be easily occupied with 8-azido-ATP in the presence of EDTA, and binding of 8-nitreno-ATP at this site induces an upward shift of the K_M value and linearizes the Lineweaver–Burk plot, but has no effect on the V_{max}. This result explains why at increasing concentrations of ATP the K_M value of the ATPase reaction increases and the Lineweaver–Burk plot shows negative cooperativity. This site is also important for the explanation of the effect of anions (Recktenwald and Hess, 1977; Stutterheim *et al.*, 1980; Hartog *et al.*, 1997).

Of the three tightly bound nucleotides in the F_1 preparation described above, one is rapidly exchanged by any nucleotide and this is bound at a catalytic site. The two others are nonexchangeable and therefore noncatalytic. Since the enzyme has only three sites for adenine nucleotides on α subunits and two can be modified with added ligands, only one of the two nonexchangeable nucleotides is bound at an α subunit and the other at a β subunit. This conclusion is crucial, since this leaves only two sites of β subunits for performing the catalytic process, contrary to the generally accepted ideas about the catalytic mechanism of F_1 (see also Muneyuki *et al.*, 1994).

This conclusion fits with all data on the catalytic sites of mitochondrial F_1, since all experiments intended to identify catalytic sites have demonstrated the presence of only two catalytic sites. The model of the binding sites can then be depicted as in Fig. 6. The main questions to be answered are the following: Why is it concluded in so many studies that the enzyme contains three catalytic sites? Are the properties of all sites the same in enzymes of different origin? Is there a difference between isolated F_1 and F_1-F_0?

A partial answer to the first question is that a three-site rotational model is attractive for reasons of symmetry, and differentiation between a three- and a two-site model is hardly possible in mechanistic, structural, and kinetic studies. Second, it is attractive to assign all high-affinity noncatalytic sites to one type of subunit and all exchangeable sites to the other type. Third, and now we also come to an answer to the second question, in many cases the first β site can indeed perform catalysis. This is the case in all forms of the chloroplast enzyme, the enzyme from PS3, and certain forms of the mitochondrial enzyme. But in two systems in which it was shown that all β sites are empty before ATP was added and in which the rate of catalysis at the first site was measured under

1 tightly bound ATP	2 tight catalytic site	3 loose catalytic site
4 tightly bound ADP	5 high-affinity regulatory site	6 low-affinity regulatory site

FIGURE 6. Schematic characterization of the six nucleotide binding sites of MF_1. Sites 1 and 4 contain a tightly bound, usually nonexchangeable, nucleotide. When ATP at site 1 is replaced by ADP, the enzyme shows hysteretic inhibition. The sites 2 and 3 are catalytically involved in multisite catalysis. Site 5 is a high-affinity site at which bound ADP induces a partial inhibition of ATP hydrolysis. The rate of dissociation of ADP from this site is very slow. Site 6 is a low-affinity site. Binding of ATP at this site decreases the affinity of the catalytic sites for substrate ATP, thereby inducing apparent negative cooperativity of the ATPase reaction.

conditions of full multisite catalysis (chloroplast enzyme and the preparation of Allison's group), the rate of catalysis by the first β site was very slow and not compatible with an equivalent participation in multisite catalysis. The data obtained with the chloroplast enzyme have been confirmed several times and the rate of catalysis at the first site is only 0.5 $\sec^{-1}$, while the overall rate of hydrolysis is about 80 $\sec^{-1}$ (Fromme and Gräber, 1989; Labahn, 1991). The other catalytic sites, therefore, must have made many turnovers before the ADP dissociates from the first site.

Similar data have been obtained with the MF_1 preparation from Allison's group. In contrast to Penefsky's group (Grubmeyer *et al.*, 1982), Bullough *et al.* (1987) showed that the dissociation of ADP from the first catalytic site is only slightly enhanced by addition of ATP, while the overall rate of catalysis was much faster. Penefsky (1988) has rejected these results, but the arguments are not convincing. Penefsky did not take into account that the preparation of Bullough *et al.* was different from his own. The enzyme used by Bullough *et al.* showed hysteretic inhibition after incubation with ADP, a difference from the Penefsky preparation, and contained only two bound ADP at noncatalytic sites. We have argued that the hysteretic site has to be the ATP-containing site in the Penefsky preparation, what appears to be the first β site. In the preparation of Bullough *et al.*, therefore, no β site is occupied. Added ATP binds at the first β site and can be slowly hydrolyzed, just like in activated CF_1. Upon addition of more ATP, both the second and third β site bind ATP and perform dual-site catalysis, while the first site still turns over very slowly, albeit a little bit faster than before. Penefsky's own preparation, however, contained three tightly bound nucleotides and these are located, as we have explained above, at sites 1, 2, and 4, with site 5 occupied only when the enzyme is treated with Mg^{2+} in the presence of nucleotides. When ATP is added to this enzyme, it binds at site 3 (with low affinity) and multisite catalysis is performed. To obtain high-affinity binding at one catalytic site and unisite catalysis, site 2 has to be emptied (at least in part of the enzyme molecules); this was performed with phosphate. [We have shown that with phosphate, and even better with pyrophosphate, site 2 can be emptied, so that only the two nonexchangeable nucleotides remain bound (Hartog *et al.*, 1997).] Site 2 now binds substrate with a high affinity and can perform slow uni-site catalysis. Upon addition of a high concentration of ATP also site 3 becomes occupied and multi-site catalysis occurs on sites 2 and 3. Penefsky, therefore, measured a large enhancement of the rate of dissociation of ADP from the site that was originally performing unisite catalysis.

Support for this model is also delivered by the radiation-inactivation studies of Ma *et al.* (1993) with *E. coli* F_1. Their data show that in multisite catalysis three α–β pairs are involved, in agreement with the conclusion of Weber *et al.* (1993) that all three β sites have to be occupied for multisite catalysis. For unisite catalysis, however, not one but two α–β pairs are required; this means that one pair is required that does not turn over, since only one nucleotide binding site is turning over under these conditions.

A special position is taken by the TF_1. This enzyme does not contain tightly bound nucleotides, although a high-affinity site is present when the γ subunit is part of the molecule (Kaibara *et al.*, 1996). At the first β site, therefore, catalysis can occur, and an increased rate of catalysis at this site upon addition of high concentrations of ATP has been observed. We are not sure whether this increased rate is compatible with participation in multisite catalysis. The data of Noji *et al.* (1997) show that in TF_1 indeed all three β sites can participate in (rotatory) catalysis, but even in this case rotation does not seem to be an essential requirement for catalysis.

The third question regards possible differences between F_1 and F_1-F_0. The study of Beharry and Bragg (1996) showed only one difference, and this is the finding that in F_1-F_0 only one bound nucleotide was not exchangeable instead of two in F_1. Considering the data on the first β site in several preparations and the finding that in F_1 the ADP at site 4 remains bound even after dissociation of the β subunits from the rest of the enzyme (Hartog *et al.*, 1992), the non-exchangeable nucleotide will be the nucleotide at site 4, while the nucleotide (ATP) at site 1 is exchangeable. As we have seen, exchangeability of the nucleotide at site 1 does not mean that the site is participating in multisite catalysis. To draw such a conclusion, proper kinetic data should be obtained on the rate of exchange.

4.2. Possible Role of the Nucleotide Binding Sites in Catalysis of ATP Synthesis

Despite some papers in which an opposite conclusion is drawn (Wang *et al.*, 1987; Syroeshkin *et al.*, 1995), there is no reason to assume that the role of the various nucleotide binding sites is different in synthesis and hydrolysis. The kinetics may be different because of differences in the relative rates of the backward reactions, but sites that are catalytic in the one direction will be catalytic as well in the other direction, and the effects of the noncatalytic sites on the catalytic sites will also be the same. We have seen, however, that the first β site is principally catalytic, although not involved in multisite catalysis of ATP hydrolysis. It might be possible, then, that during ATP synthesis, when the membranes are energized by a proton-motive force, the rate of catalysis by site 1 increases due to an increased rate of dissociation of ATP. In intact phosphorylating particles, however, the exchange of an additional tightly bound nucleotide (at site 1) is either not observed (van der Zwet-de Graaff, 1996) or quite slow (Harris *et al.*, 1977). Also, inhibitor studies do not reveal an essential difference between inhibition of ATP synthesis and of ATP hydrolysis.

In one system, however, differences in site involvement between hydrolysis and synthesis are consistently observed and that is in particles reconstituted from isolated F_1 and F_1-depleted particles. This type of system has been used by several authors (Kohlbrenner and Boyer, 1982; Matsuno-Yagi and Hatefi, 1984; Wang et al., 1987); in our laboratory, van der Zwet-de Graaff (1996) used this system. The conclusion that can be drawn from these studies is that with inhibitors that are bound at just one catalytic site, like DCCD, NbfCl, FSBI, and (when bound at the tight catalytic site only) 2-azido-ATP, synthesis of ATP is only partially inhibited when one site is occupied, while hydrolysis is fully inhibited. With FSBI, van der Zwet-de Graaff (1996) found no real difference between its effect on synthesis and on hydrolysis when it was used in intact submitochondrial particles, but in particles reconstituted with FSBI-inhibited F_1, synthesis was much less inhibited than hydrolysis, while hydrolysis was inhibited to the same extent as was measured with the F_1 before reconstitution. Matsuno-Yagi and Hatefi (1984) showed that upon reconstitution with DCCD-treated F_1, synthesis is only 60% inhibited when hydrolysis is nearly fully inhibited. Kohlbrener and Boyer (1982) showed the same for Nbf-treated F_1 and so did Wang et al. (1987). To inhibit ATP synthesis completely, Matsuno-Yagi and Hatefi needed 2 mole of DCCD per mole F_1. This difference cannot be due to overcapacity of the ATP synthesis system, since in these reconstituted particles the synthesis capacity is quite low because of leakiness of the membranes; in addition, with other inhibitors of the ATPsynthase, like various F_0 inhibitors, a proportional relation between inhibition of synthesis and inhibition of hydrolysis is found (Matsuno-Yagi and Hatefi, 1984). The only possible conclusion is that in this system the knockout of one catalytic β site only partially inhibits ATP synthesis activity, while ATP hydrolysis is fully inhibited.

The explanation that we like to propose is that in the reconstituted system (but not in intact SMP) the nucleotide at site 1 can take part in multisite catalysis. In intact particles the dissociation of nucleotide is too slow to have a significant effect, but apparently in this system it is much faster. This does not mean that a three-site mechanism operates, but that a two-site mechanism is operative all the time, shifting from one couple of sites to another. Probably site 1 is still the slowest, so that when only a couple consisting of sites 1 and 2, or sites 1 and 3, is operative, the rate of ATP synthesis is slower than when sites 2 and 3 are working. When two sites are inhibited, all activity is lost.

It is not clear what causes this specific behavior of the reconstituted system. It may be a matter of reconstitution (a different interaction between the F_1 and F_0), but it may also be that a change has occurred as a consequence of the isolation of F_1, and such a change is only revealed when the enzyme is reconstituted with particles to make it capable of ATP synthesis. The occurrences of ATP synthesis catalyzed by an enzyme of which one site is inhibited favors our view that in essence a two-site catalytic mechanism is operative in the ATPsynthase.

5. INTERPRETATION OF SOME DATA REPORTED TO BE INDICATIVE FOR A THREE-SITE MECHANISM OF CATALYSIS

Only a few experiments reported in the literature really claim to prove a three-site mechanism of catalysis, although such a mechanism has been widely accepted. We have already discussed the experiment reported by Amono *et al.* (1996), and we will discuss three other findings here.

The experiment of Duncan *et al.* (1995) is quite elegant and is the first functional study that in principle can directly discriminate between a three-site and a two-site mechanism of catalysis without any further assumptions. But the resulting data are not decisive. One β subunit of *E. coli* F_1 was covalently linked with the γ subunit (S–S bridge) and then the two other β subunits, radio-actively labeled, were reconstituted; after reconstitution, the covalent linkage was broken by reduction. After 10 sec of catalysis, the reaction was stopped; by oxidation, the γ subunit was again cross-linked with a β subunit and the radioactivity in the cross-linked product was determined. The question, then, was asked whether the radioactivity fitted with a full randomization of the β–γ interaction, or whether one β subunit did not see the γ subunit at the position of the cysteine that was involved in the cross-link. The result was that after 10 sec of turnover the label was in fact distributed over 2.5 subunits; that is, the result was just in between the expected distribution for a two- and a three-site mechanism.

Duncan *et al.* (1995) concludes that these data show participation of three sites, but in fact they show the opposite. Only two sites are really performing rapid catalysis, while in about 50% of the molecules the third site has made a turnover. That is fairly slow and is not in agreement with a three-site mecha-nism. This interpretation assumes the absence of any experimental error in the numbers, which is unlikely; therefore, a further test should be made. If our interpretation is correct, the distribution of label shifts from involvement of two sites during very short periods of catalysis to (apparent) involvement of three sites during long periods of catalysis, since slow turnover occurs at site 1. We hope that such an experiment can be done.

A recent experiment by Sabbert *et al.* (1996) seems quite straightforward. We suppose, however, that the formula used by Sabbert *et al.* is derived for the case where the original polarization is close to 0.3 and not when this value is only 0.08. If the decrease of the polarization also has an (small) absolute contribution, not significant when the starting value is 0.3 but significant when it is 0.08, the experiment is not discriminating. On the other hand, although a rotation over at least 200° does not fit with a two-site mechanism, we also have to take into account that the first site in CF_1 turns over at a low rate. The relatively very low rate of multisite catalysis in the reported experiment is therefore a weak point: After 100 msec, about 10% of the site not participating in multisite catalysis may have undergone one turnover, thereby contributing to the dislocation of the probe.

Finally, we would like to give our interpretation of the structural data of Abrahams *et al.* (1994). The authors explain the data from their MF_1 crystals in terms of a rotary three-site mechanism of catalysis. We have concluded above that the F_1, whether isolated or in conjunction with F_0, contains only two sites that directly participate in multisite hydrolysis or synthesis of ATP. Is it possible to interpret the data satisfactorily in terms of a two-site mechanism of catalysis?

The enzyme used for the crystallization was first depleted of endogenous nucleotides with 50% glycerol treatment and then incubated with the analogue AMPPNP. The concentration was probably high enough to fill all six nucleotide binding sites (Cross and Nalin, 1982). The crystals formed were not homogeneous. It was then found that upon addition of a low amount of ADP homogeneous crystals were formed. The resulting enzyme contained an AMPPNP molecule at each α subunit and at one of the β subunits, while ADP was present at one other β subunit. The third β subunit was empty. This occupation is to be expected, since ADP rapidly exchanges at the high-affinity catalytic site; but when this site is occupied with ADP and the low-affinity noncatalytic site is filled with an ATP analogue, the affinity of the second catalytic site is lowered so much that ADP does not bind at this site when the concentration is 50 μM, nor does AMPPNP. We may expect, then, from the binding properties of the various binding sites, that site 1 of our model contains AMPPNP, ADP is bound at the first catalytic site (site 2), and site 3 is empty. Abrahams *et al.* (1994) conclude from the structure of the sites that the ADP-containing site has the highest affinity. This is not in contradiction with the fact that site 1 is originally a nonexchangeable site, since after treatment with glycerol the nonexchangeable sites have a lowered affinity and the catalytic site has the highest affinity for ADP (see Section. 3.1.2.).

The observed occupation of the β sites is not a real proof against the possible involvement of the first site in multisite catalysis, but if the first site were really catalytic, one would have expected that both high-affinity catalytic sites were occupied by ADP. As far as the α sites is concerned, it is unfortunate that Abrahams *et al.* (1994) have assumed that the main interaction between α and β subunits occurs at the catalytic site and have named the α subunits accordingly. One may expect, however, that the potentially regulatory α sites are more relevant for intersubunit interactions [the mutation of α-Gln173 in the yeast enzyme destroys catalytic cooperativity (Jault *et al.*, 1991)], and if we like to think in terms of three α–β pairs, each pair should contain the two subunits around an α site.

The data reported by Abrahams *et al.* (1994) do indeed suggest this type of pair formation. The α site with the highest affinity seems to be the site on α_{DP}, so that this should be site 4 and the α_{DP} subunit should form a pair with β_T. The lowest affinity is ascribed to the site on subunit α_{TP}, so that this should be site 6 and the subunit should form a pair with β_E. The three α–β pairs in our model, 1–4, 2–5, and 3–6, then are not formed from the two subunits around the β sites but from the subunits around the α sites. Also, the configuration of the α and β sites indicates that at the α sites the subunits have more interactions than at the β sites (see scheme of Fig. 7).

Does the structure give any further indication about the catalytic mechanism? Abrahams *et al.* (1994) point out that the structure is very suitable for rotation of the γ subunit within the central cavity, since only a few interactions have to be broken for such a rotation. Since a partial movement is not excluded by the structure, the proposal for rotation is probably made on the basis of the expectation that, if there had been no rotation, a larger barrier would have been present for moving on in one direction, forcing the γ subunit to move back to the original position during the next turnover.

But let us look at the data. We know that the enzyme is inactive, probably not because of the presence of AMPPNP at the noncatalytic sites instead of AT(D)P (we do not know the consequences of this replacement), but certainly because of the occupation of the high-affinity catalytic site with the ADP in the absence of phosphate (Drobinskaya *et al.*, 1985). But this will not strongly affect, we suppose, the structure of the various sites and subunits. Binding of ATP at the empty site, followed by closing of the site into a tight binding site, induces a conformational change that finally results in release of ADP from the β_D. This effect will be conveyed via the γ subunit (the contribution of the α subunits is less clearly defined) that changes its conformation (or position) by the tightening of the empty site and then pushes the ADP-containing site outward, so that this site opens up and the ADP dissociates. In the next catalytic cycle, the reverse will happen unless there is a vectorial force favoring a movement of the γ subunit in the same direction. From the structure such a vectorial force is not evident. Therefore, we do not consider the reported

1		2		3	
β_T/AMP-PNP		β_D/ADP		β_E	
4		5		6	
α_{DP}/AMP-PNP		α_E/AMP-PNP		α_{TP}/AMP-PNP	

FIGURE 7. Occupation of the nucleotide binding sites in the crystals of Abrahams *et al.* (1994). The description of the different nucleotide binding sites by Abrahams *et al.* (1994) is used to compare their properties with the characteristics of each of the sites in our model. The arguments for the identification are given in the text.

structure of mitochondrial F_1 as favoring a rotational movement during catalysis, but in full agreement with an alternating dual-site catalytic mechanism.

6. CONCLUSIONS: A PROPOSAL FOR THE OVERALL MECHANISM OF THE ATP SYNTHASE

In his review, Boyer (1993) agrees that no proof has ever been given of the supposition that all three β sites pass sequentially through the same states. We have discussed the more recent data favoring such a mechanism.

In the present summary of data on the binding of nucleotides, we have seen that several experiments really show that only two β sites can be directly involved in catalysis. The most convincing experiments are the binding studies with the 8- and 2-azido-adenine nucleotides. When four sites can be specifically modified, two of them being catalytic β sites and two noncatalytic α sites, while at the same time the enzyme still contains two bound adenine nucleotides, no other conclusion is possible. In several forms of F_1, the first β site may be slowly hydrolyzing ATP, but it is not participating equally in multisite catalysis. The argument of Boyer (1993) in his review that we might not have been able to modify the third catalytic site because of its low affinity seems to lack any credibility, as it was shown that all six binding sites were actually occupied.

It is not yet clear why in our preparations of MF_1 the first β site does not turn over at all and contains a tightly bound, nonexchangeable ATP, while in other preparations of MF_1 or enzyme preparations from other sources this site does perform slow catalysis. Recent experiments have indicated that the manner of treatment of the enzyme may be relevant. After treatment with pyrophosphate to remove all bound nucleotides except the two nonexchangeable ones (Hartog et al., 1997), preliminary data suggest that under these conditions one of them (at site 1) can exchange with added ATP or 2-azido-ATP but not with GTP. This will be further investigated as well as our previous conclusion that of the two tightly bound nonexchangeable nucleotides the ADP is bound at an α site and the ATP at a β site (Hartog et al., 1992). This seems in contrast with the data of Possmayer (1995) on modification of CF_1-F_0 with 2-azido-AXP: The β sites always contained the di- or monophosphate, while the α sites were specific for the triphosphates.

The presented evidence for an alternating site mechanism of F_1 does not mean that no rotation is involved in the mechanism of ATP hydrolysis or synthesis. Considering the proposed structure of F_0 (Birkenhäger et al., 1995; Sing et al., 1996; Takeyasu et al., 1996), a rotational movement of the c subunits enabling them to deliver (or receive) protons at the right spot of interaction between the stalk region (including the γ subunit) and the F_0 seems very attractive and fits with the analogy between the ATP synthase and various ATP-driven systems that induce a rotational movement, like the flagellar motor and the gyrase. A rotational movement can be obtained by connecting a load to the c subunits, so that the energy of ATP hydrolysis is not just converted into a proton-motive force but directly into work. But when the cylinder of c subunits rotates during catalysis, the point or region of interaction with the stalk region of the enzyme will have to remain in position, most likely via a stator (Kagawa and Hamamoto, 1996), and the conformational changes in the stalk accompanying binding and release of protons will preferentially be part of a two-state system with just two conformations. An ATP-driven motor is just like the engine of a car: A back-and-forth movement, induced by the fuel combustion, is transformed into a rotational movement. The ATP synthase preferentially works in the opposite direction, synthesizing ATP, and it may have kept the rotation of the proton-delivering system because of its efficiency. Such a rotational system for the delivery of protons may also have made it possible to vary the proton–ATP ratio, being four in chloroplasts (van Walraven *et al.*, 1996), but certainly not more than three in mitochondria. Therefore, while the ATP hydrolysis will induce a rotational movement of the c subunits and in the reverse reaction the c subunits deliver their protons to the stalk region via a rotation, in the F_1 part of the enzyme there is no rotational movement of the γ subunit relative to the three α–β pairs.

ACKNOWLEDGMENTS. The part of the reviewed work conducted in this laboratory was supported in part by grants from the Netherlands Organization for the Advancement of Scientific Research (NWO) under the auspices of the Netherlands Foundation for Chemical Research (SON).

7. REFERENCES

Abrahams, J. P., Leslie, A. G. W., Lutter, R., and Walker, J. E., 1994, Structure at 2.8 Å resolution of F_1-ATPase from bovine heart mitochondria, *Nature* **370:**621–628.

Aloise, P., Kagawa, Y., and Coleman, P. S., 1991, Comparative Mg^{2+}-dependent sequential covalent binding stoichiometries of 3'-O-(4-benzoyl)benzoyl adenosine 5'-diphosphate of MF_1, TF_1 and the $\alpha_3\beta_3$ core complex of TF_1, *J. Biol. Chem.* **266:**10368–10376.

Al-Shawi, M. K., and Senior, A. E., 1988, Complete kinetic and thermodynamic characterization of the unisite catalytic pathway of *Escherichia coli* F_1-ATPase, *J. Biol. Chem.* **263:**19640–19648.

Amano, T., Hisabori, T., Muneyuki, E., and Yoshida, M., 1996, Catalytic activities of $\alpha_3\beta_3\gamma$ complexes of F_1-ATPase with 1, 2, or 3 incompetent catalytic sites, *J. Biol. Chem.* **271:**18128–18133.

Andrews, W. W., Hill, F. C., and Allison, W. S., 1984, Identification of essential tyrosine residue in the β subunit of bovine heart mitochondrial F_1-ATPase that is modified by 7-chloro-4-nitro[^{14}C]benzofurazan, *J. Biol. Chem.* **259:**8219–8225.

Bakels, R. H. A., Walraven, H. S. van, Wielink, J. E. van, Zwet-de Graaf, I. van der, Krenn, B. E., Krab, K., Berden, J. A., and Kraayenhof, R., 1994, The effect of sulfite on the ATP hydrolysis and syntheses activity of membrane-bound H^+-ATPsynthase from various species, *Biochem. Biophys. Res. Commun.* **201**:487–492.

Bar-Zvi, D., Tiefert, M. A., and Shavit, N., 1983, Interaction of the chloroplast ATPsynthase with the photoreactive nucleotide 3'-O-(4-benzoyl)benzoyl adenosine 5'-diphosphate, *FEBS Lett.* **160**:233–238.

Beharry, S., and Bragg, P. D., 1996, The bound adenine nucleotides of purified bovine mitochondrial ATPsynthase, *Eur. J. Biochem.* **240**:165–172.

Berden, J. A., Hartog, A. F., and Edel, C. M., 1991, Hydrolysis of ATP by F_1 can be described only on the basis of a dual-site mechanism, *Biochim. Biophys. Acta* **1057**:151–156.

Birkenhäger, R., Hoppert, M., Deckers-Hebestreit, G., Mayer, F., and Altendorf, K., 1995, The F_0 complex of the *Escherichia coli* ATPsynthase. Investigation by electron spectroscopic imaging and immunoelectron microscopy, *Eur. J. Biochem.* **230**:58–67.

Boulay, F., Dalbon, P., and Vignais, P. V., 1985, Photoaffinity labeling of mitochondrial adenosine-triphosphatase by 2-azidoadenosine 5'-[α-^{32}P]diphosphate, *Biochemistry* **24**:7372–7379.

Boyer, P. D., 1993, The binding change mechanism for ATPsynthase. Some probabilities and possibilities, *Biochim. Biophys. Acta* **1140**:215–250.

Boyer, P. D., Cross, R. L., and Momsen, W., 1973, A new concept for energy coupling in oxidative phosphorylation based on a molecular explanation of the oxygen exchange reaction, *Proc. Natl. Acad. Sci. USA* **70**:2837–2839.

Bragg, P. D., and Hou, C., 1990, Reaction of membrane-bound F_1-adenosine triphosphatase of *Escherichia coli* with chemical ligands and the asymmetry of β subunits, *Biochim. Biophys. Acta* **1015**:216–222.

Bullough, D. A., and Allison, W. S., 1986a, Three copies of the β subunit must be modified to achieve complete inactivation of the bovine mitochondrial F_1-ATPase by 5'-*p*-fluorosulfonyl-benzoyladenosine, *J. Biol. Chem.* **261**:5722–5730.

Bullough, D. A., and Allison, W. S., 1986b, Inactivation of the bovine heart mitochondrial F_1-ATPase by 5'-*p*-fluorosulfonylbenzoyl[^{3}H]inosine is accompanied by modification of tyrosine 345 in a single β subunit, *J. Biol. Chem.* **261**:14171–14177.

Bullough, D. A., Verburg, J. G., Yoshida, M., and Allison, W. A., 1987, Evidence for functional heterogeneity among the catalytic sites of the bovine heart mitochondrial F_1-ATPase, *J. Biol. Chem.* **262**:11675–11683.

Bullough, D. A., Brown, E. L., Saario, J. D., and Allison, W. S., 1988, On the location and function of the noncatalytic sites on the bovine heart mitochondrial F_1-ATpase, *J. Biol. Chem.* **263**: 14053–14060.

Cross, R. L., and Nalin, C. M., 1982, Adenine nucleotide binding sites on beef heart F_1-ATPase, *J. Biol. Chem.* **257**:2874–2881.

Cross, R. L., Grubmeyer, C., and Penefsky, H. S., 1982, Mechanism of ATP hydrolysis by beef heart mitochondrial ATPase, *J. Biol. Chem.* **257**:12101–12105.

Cross, R. L., Cunningham, D., Miller, C. G., Xue, Z., Zhou, J.-M., and Boyer, P. D., 1987, Adenine nucleotide binding sites on beef heart F_1-ATPase: Photoaffinity labeling of β-subunit Tyr-368 at a noncatalytic site and β Tyr-345 at a catalytic site, *Proc. Natl. Acad. Sci. USA* **84**:5715–5719.

Di Pietro, A., Penin, F., Godinot, C., and Gautheron, D. C., 1980, "Hysteretic" behavior and nucleotide binding sites of pig heart mitochondrial F_1 adenosine 5'-triphosphate, *Biochemistry* **19**:5671–5678.

Drobinskaya, I. Ye., Kozlov, I. A., Murataliev, M. B., and Vulfson, E. N., 1985, Tightly bound adenosine diphosphate, which inhibits the activity of mitochondrial F_1ATPase, is located at the catalytic site of the enzyme, *FEBS Lett.* **182**:419–424.

Duncan, T. M., Bulygin, V. V., Zhou, Y., Hutcheon, M. L., and Cross, R. L., 1995, Rotation of subunits during catalysis by *Escherichia coli* F_1-ATPase, *Proc. Natl. Acad. Sci. USA* **92:** 10964–10968.

Ebel, R. E., and Lardy, H. A., 1975, Stimulation of rat liver mitochondrial adenosinetriphosphatase by anions, *J. Biol. Chem.* **250:**191–196.

Edel, C. M., Hartog, A. F., and Berden, J. A., 1992, Inhibition of mitochondrial F_1-ATPase activity by binding of (2-azido-)ADP to a slowly exchangeable non-catalytic nucleotide binding site, *Biochim. Biophys. Acta* **1101:**329–338.

Edel, C. M., Hartog, A. F., and Berden, J. A., 1993, Identification of an exchangeable non-catalytic site on mitochondrial F_1-ATPase which is involved in the negative cooperativity of ATP hydrolysis, *Biochim. Biophys. Acta* **1142:**327–335.

Edel, C. M., Hartog, A. F., and Berden, J. A., 1995, Analysis of the inhibitory non-catalytic ADP binding site on mitochondrial F_1, using NAP_2-$2N_3$ADP as probe. Effects of the modification on ATPase and ITPase activity, *Biochim. Biophys. Acta* **1229:**103–114.

Fromme, P., and Gräber, P., 1989, Heterogeneity of ATP-hydrolyzing sites on reconstituted CF_0F_1, *FEBS Lett.* **259:**33–36.

Garin, J., Boulay, F., Issartel, J. P., Lunardi, J., and Vignais, P. V., 1986, Identification of amino acid residues photolabeled with 2-azido[α-^{32}P] adenosine diphosphate in the β subunit of beef heart mitochondrial F_1-ATPase, *Biochemistry* **25:**4431–4437.

Garin, J., Vinçon, M., Gagnon, J., and Vignais, P. V., 1994, Photolabeling of mitochondrial F_1-H^+ ATPase by 2-azido[^{3}H]ADP and 8-azido[^{3}H]ADP entrapped as flourometal complexes into the catalytic sites of the enzyme, *Biochemistry* **33:**3772–3777.

Garrett, N. E., and Penefsky, H. S., 1975, Interaction of adenine nucleotides with multiple binding sites on beef heart mitochondrial adenosine triphosphatase, *J. Biol. Chem.* **250:**6640–6647.

Grubmeyer, C., and Penefsky, H. S., 1981a, The presence of two hydrolytic sites on beef heart mitochondrial adenosine triphosphatase, *J. Biol. Chem.* **256:**3718–3727.

Grubmeyer, C., and Penefsky, H. S., 1981b, Cooperativity between catalytic sites in the mechanism of action of beef heart mitochondrial adenosine triphosphatase, *J. Biol. Chem.* **256:**3728–3734.

Grubmeyer, C., Cross, R. L., and Penefsky, H. S., 1982, Mechanism of ATP hydrolysis by beef-heart mitochondrial ATPase, *J. Biol. Chem.* **257:**12092–12100.

Harris, D. A., Rosing, J., van der Stadt, R. J., and Slater, E. C., 1973, Tight binding of adenine nucleotides to beef-heart mitochondrial ATPase, *Biochim. Biophys. Acta* **314:**149–153.

Harris, D. A., Radda, G. K., and Slater, E. C., 1977, Tightly bound nucleotides of the energy-transducing ATPase, and their role in oxidative phosphorylation, *Biochim. Biophys. Acta* **459:** 560–572.

Harris, D. A., Gomez-Fernandez, J. C., Klungsöyr, L., and Radda, G. K., 1978, Specificity of nucleotide binding and coupled reactions utilising the mitochondrial ATPase, *Biochim. Biophys. Acta* **504:**364–383.

Hartog, A. F., Edel, C. M., Lubbers, F. B., and Berden, J. A., 1992, Characteristics of the non-exchangeable nucleotide binding sites of mitochondrial F_1 revealed by dissociation and reconstitution with 2-azido-ATP, *Biochim. Biophys. Acta* **1100:**267–277.

Hartog, A. F., Edel, C. M., Braham, J., Muijsers, A. O., and Berden, J. A., 1997, FSBA modifies both α- and β-subunits of F_1 specifically and can be bound together with AXP at the same α-subunit, *Biochim. Biophys. Acta* **1318:**107–122.

Haughton, M. A., and Capaldi, R. A., 1995, Asymmetry of *Escherichia coli* F_1-ATPase as a function of the interaction of α-β subunit pairs with the γ and the ε subunits, *J. Biol. Chem.* **270:**20568–20574.

Haughton, M. A., and Capaldi, R. A., 1996, The *Escherichia coli* F_1-ATPase mutant βTyr-297-Cys: Functional studies and asymmetry of the enzyme under various nucleotide conditions based on reaction of the introduced Cys with *N*-ethylmaleimide and 7-chloro-4-nitrobenzofurazan, *Biochim. Biophys. Acta* **1276:**154–160.

Hollemans, M., Runswick, M. J., Fearnley, I. M., and Walker, J. E., 1983, The sites of labeling of the β-subunit of bovine mitochondrial F_1-ATPase with 8-azido-ATP, *J. Biol. Chem.* **258:** 9307–9313.

Jault, J.-M., and Allison, W. S., 1993, Slow binding of ATP to noncatalytic nucleotide binding sites which accelerates catalysis is responsible for apparent negative cooperativity exhibited by the bovine mitochondrial F_1-ATPase, *J. Biol. Chem.* **268:**1558–1566.

Jault, J.-M., and Allison, W. S., 1994, Hysteretic inhibition of the bovine heart mitochondrial F_1-ATPase is due to saturation of noncatalytic sites with ADP which blocks activation of the enzyme by ATP, *J. Biol. Chem.* **269:**319–325.

Jault, J.-M., Di Pietro, A., Falson, P., and Gautheron, D. C., 1991, Alteration of apparent negative cooperativity of ATPase activity by α-subunit glutamine 173 mutation in yeast mitochondrial F_1, *J. Biol. Chem.* **266:**8073–8078.

Kagawa, Y., and Hamamoto, T., 1996, The energy transmission in ATPsynthase: From the γ-c rotor to the α3β3 oligomer fixed by OSCP-b stator via the βDELSEED sequence, *J. Bioenerg. Biomembr.* **28:**421–431.

Kagawa, Y., Sone, N., Yoshida, M., Hirata, H., and Okamoto, H., 1976, Proton translocating ATPase of a thermophilic bacterium, *J. Biochem.* **80:**141–151.

Kaibara, C., Matsui, T., Hisabori, T., and Yoshida, M., 1996, Structural asymmetry of F_1-ATPase caused by the γ subunit generates a high affinity nucleotide binding site, *J. Biol. Chem.* **271:** 2433–2438.

Kironde, F. A. S., and Cross, R. L., 1986, Adenine nucleotide-binding sites on beef heart F_1-ATPase, *J. Biol. Chem.* **261:**12544–12549.

Kironde, F. A. S., and Cross, R. L., 1987, Adenine nucleotide binding sites on beef heart F_1-ATPase, *J. Biol. Chem.* **262:**3488–3495.

Knowles, A. F., and Penefsky, H. S., 1972, The subunit structure of beef heart mitochondrial adenosine triphosphatase, *J. Biol. Chem.* **247:**6617–6623.

Kohlbrenner, W. E., and Boyer, P. D., 1982, Catalytic properties of beef heart mitochondrial ATPase modified with 7-chloro-4-nitro-2-oxa-1,3-diazole, *J. Biol. Chem.* **257:**3441–3446.

Labahn, A., 1991, Die H⁺-ATPase aus Chloroplasten: Die kinetik der mit einem protonentransport gekoppelten ATP-synthese auf einem bindungsplatz, PhD Thesis, Physikalische und angewandte chemie der technischen universität Berlin.

Lübben, M., Lücken, U., Weber, J., and Schäfer, G., 1984, Azidonaphtoyl-ADP: A specific photolabel for the high-affinity nucleotide-binding sites of F_1-ATPase, *Eur. J. Biochem.* **143:** 483–490.

Lunardi, J., Garin, J., Issartel, J.-P., and Vignais, P. V., 1987, Mapping of nucleotide-depleted mitochondrial F_1-ATPase with 2-azido-[α³²P]adenosine diphosphate, *J. Biol. Chem.* **262:**15172– 15181.

Ma, J. T., Wu, J. J., Tzeng, C. M., and Pan, R. L., 1993, Functional size analysis of F-ATPase from *Escherichia coli* by radiation inactivation, *J. Biol. Chem.* **268:**10802–10807.

Martins, I. S., and Penefsky, H. S., 1994, Covalent modification of catalytic sites on membrane-bound beef heart mitochondrial ATPase by 2-azido-adenine nucleotides, *Eur. J. Biochem.* **224:** 1057–1065.

Matsuno-Yagi, A., and Hatefi, Y., 1984, Inhibitory chemical modifications of F_1-ATPase: Effects on the kinetics of adenosine 5′-triphosphate synthesis and hydrolysis in reconstituted systems, *Biochemistry* **23:**3508–3514.

Matsuno-Yagi, A., and Hatefi, Y., 1993, Studies on the mechanism of oxidative phosphorylation, *J. Biol. Chem.* **268:**1539–1545.

Milgrom, Y. M., and Boyer, P. D., 1990, The ADP that binds tightly to nucleotide-depleted mitochondrial F_1-ATPase and inhibits catalysis is bound at a catalytic site, *Biochim. Biophys. Acta* **1020**:43–48.

Milgrom, Y. M., and Murataliev, M. B., 1987, Steady-state rate of F_1-ATPase turnover during ATP hydrolysis by the single catalytic site, *FEBS Lett.* **212**:63–67.

Miwa, K., Ohtsubo, M., Denda, K., Hisabori, T., Date, T., and Yoshida, M., 1989, Reconstituted F_1-ATPase complexes containing one impaired β subunit are ATPase-active, *J. Biochem.* **106**:679–683.

Muneyuki, E., Hisabori, T., Allison, W. S., Jault, J.-M., Sasayama, T., and Yoshida, M., 1994, Catalytic cooperativity of beef heart mitochondrial F_1-ATPase revealed by using 2′,3′-*O*-(2,4,6-trinitrophenyl)-ATP as a substrate; an indication of mutually activating catalytic sites, *Biochim. Biophys. Acta* **1188**:108–116.

Murataliev. M. B., and Boyer, P. D., 1994, Interaction of mitochondrial F_1-ATPase with trinitrophenyl derivatives of ATP and ADP, *J. Biol. Chem.* **269**:15431–15439.

Nieboer, P., Hartog, A. F., and Berden, J. A., 1987, Dissociation-reconstitution experiments support the presence of two catalytic β-subunits in mitochondrial F_1, *Biochim. Biophys. Acta* **894**: 277–283.

Noji, H., Yasuda, R., Yoshida, M., and Kinosita, Jr., K., 1997, Direct observation of the rotation of F_1-ATPase, *Nature* **386**:299–302.

Penefsky, H. S., 1977, Reversible binding of P_i by beef heart mitochondrial adenosine triphosphatase, *J. Biol. Chem.* **252**:2891–2899.

Penefsky, H. S., 1988, Rate of chase-promoted hydrolysis of ATP in the high affinity catalytic site of beef heart mitochondrial ATPase, *J. Biol. Chem.* **263**:6020–6022.

Penin, F., Godinot, C., and Gautheron, D., 1979, Optimization of the purification of mitochondrial F_1-adenosine triphosphatase, *Biochim. Biophys. Acta* **548**:63–71.

Perlin, D. S., Latchney, L. R., Wise, J. G., and Senior, A. E., 1984, Specificity of the proton adenosinetriphosphatase of *Escherichia coli* for adenine, guanine, and inosine nucleotides in catalysis and binding, *Biochemistry* **23**:4998–5003.

Possmayer, F. E., 1995, Charakterisierung der nucleotidbindungsplätze der H^+-ATPase aus Chloroplasten, PhD Thesis, Biologisches institut der universität Stuttgart.

Recktenwald, D., and Hess, B., 1977, Allosteric influence of anions on mitochondrial ATPase of yeast, *FEBS Lett.* **76**:25–28.

Rosing, J., Harris, D. A., Slater, E. C., and Kemp, A., 1975, The possible role of tightly bound adenine nucleotide in oxidative and photosynthetic phosphorylation, *J. Supramol. Struct.* **3**: 284–296.

Sabbert, D., Engelbrecht, S., and Junge, W., 1996, Intersubunit rotation in active F-ATPase, *Nature* **381**:623–625.

Senior, A. E., 1990, The proton-translocating ATPase of *Escherichia coli*, *Annu. Rev. Biophys. Chem.* **19**:7–41.

Senior, A. E., and Wise, J. G., 1983, The proton–ATPase of bacteria and mitochondria, *J. Membr. Biol.* **73**:105–124.

Shapiro, A. B., Huber, A. H., and McCarty, R. E., 1991, Four tight nucleotide binding sites of chloroplast coupling factor 1, *J. Biol. Chem.* **266**:4194–4200.

Sing, S., Turina, P., Bustamante, C. J., Keller, D. J., and Capaldi, R., 1996, Topographical structure of membrane-bound *Escherichia coli* F_1F_0 ATPsynthase in aqueous buffer, *FEBS Lett.* **397**:30–34.

Slater, E. C., Rosing, J., Harris, D. A., van der Stadt, R. J., and Kemp, A., 1974, The identification of functional ATPase in energy-transducing membranes, in *Membrane Proteins in Transport and Phosphorylation* (G. F. Azzone, M. E., Klingenberg, E. Quagliariello, and N. Siliprandi, eds.), pp. 137–147, North-Holland Publishing Company, Amsterdam.

Sloothaak, J. B., Berden, J. A., Herweijer, M. A., and Kemp, A., 1985, The use of 8-azido-ATP and

8-azido-ADP as photoaffinity labels of the ATP synthase in submitochondrial particles: Evidence for a mechanism of ATP hydrolysis involving two independent catalytic sites? *Biochim. Biophys. Acta* **809**:27–38.

Stutterheim, E., Henneke, M. A. C., and Berden, J. A., 1980, Studies on the structure and conformation of yeast mitochondrial ATPase using aurovertin and methanol as probes, *Biochem. Biophys. Acta* **592**:415–430.

Stutterheim, E., Henneke, M. A. C., and Berden, J. A., 1981, Subunit composition of mitochondrial F_1-ATPase isolated from *Saccharomyces carlsbergensis, Biochim. Biophys. Acta* **634**: 271–278.

Syroeshkin, A. V., Vasilyeva, E. A., and Vinogradov, A. D., 1995, ATP synthesis catalyzed by the mitochondrial F_1F_0-ATPsynthase is not a reversal of its ATP activity, *FEBS Lett.* **366**:29–32.

Takeyasu, K., Omote, H., Nettikadan, S., Tokumasu, F., Iwamoto-Kihara, A., and Futai, M., 1996, Molecular imaging of *Escherichia coli* F_0F_1-ATPase in reconstituted membranes using atomic force microscopy, *FEBS Lett.* **392**:110–113.

Todd, R. D., Griesenbeck, T. A., and Douglas, M. G., 1980, The yeast mitochondrial adenosine triphosphatase complex, *J. Biol. Chem.* **255**:5461–5467.

Turina, P., and Capaldi, R., 1994, ATP binding causes a conformational change in the γ-subunit of the *Escherichia coli* F_1ATPase which is reversed on bond cleavage, *Biochemistry* **33**:14275–14280.

van der Zwet-de Graaff, I., 1996, ATP synthesis in submitochondrial particles: Effect of modification of nucleotide binding sites of F_1F_0-ATPsynthase, PhD Thesis, University of Amsterdam.

van der Zwet-de Graaff, I., Hartog, A. F., and Berden, J. A., 1997, Modification of membrane-bound F_1 by FSBA: Sites of binding and effect on activity, *Biochim. Biophys. Acta* **1318**: 123–132.

van Dongen, M. B. M., and Berden, J. A., 1986, Demonstration of two exchangeable non-catalytic and two cooperative catalytic sites in isolated bovine heart mitochondrial F_1, using the photoaffinity labels [2-^{3}H]8-azido-ATP and [2-^{3}H]8-azido-ADP, *Biochim. Biophys. Acta* **850**: 121–130.

van Dongen, M. B. M., and Berden, J. A., 1987, Exchange and hydrolysis of tightly bound nucleotides in normal and photolabelled bovine heart mitochondrial F_1, *Biochim. Biophys. Acta* **893**:22–32.

van Dongen, M. B. M., Geus, J. P. de, Korver, T., Hartog, A. F., and Berden, J. A., 1986, Binding and hydrolysis of 2-azido-ATP and 8-azido-ATP by isolated mitochondrial F_1: Characterisation of high-affinity binding sites, *Biochim. Biophys. Acta* **850**:359–368.

van Walraven, H. S., Strotmann, H., Schwarz, O., and Rumberg, B., 1996, The H^+/ATP coupling ratio of the ATP synthase from thiol-modulated chloroplasts and two cyanobacterial strains is four, *FEBS Lett* **379**:309–313.

Verburg, J. G., and Allison, W. S., 1990, Tyrosine α244 is derivatized when the bovine heart mitochondrial F_1-ATPase is inactivated with 5′-p-Fluorosulfonylbenzoylethenoadenosine, *J. Biol. Chem.* **265**:8065–8074.

Vogel, P. D., and Cross, R. L., 1991, Adenine nucleotide-binding sites on mitochondrial F_1-ATPase, *J. Biol. Chem.* **266**:6101–6105.

Wagenvoord, R. J., van der Kraan, I., and Kemp, A., 1977, Specific photolabeling of beef heart mitochondrial ATPase by 8-azido-ATP, *Biochim. Biophys. Acta* **460**:17–24.

Wagenvoord, R. J., Kemp, A., and Slater, E. C., 1980, The number and localisation of adenine nucleotide binding sites in beef-heart mitochondrial ATPase (F_1) determined by photolabelling with 8-azido-ATP and 8-azido-ADP, *Biochim. Biophys. Acta* **593**:204–211.

Walker, J. E., Saraste, M., Runswick, M. J., and Gay, N. J., 1982, Distantly related sequences in the α and β subunits of ATPsynthase, myosin, kinases and other ATP-requiring enzymes and a common nucleotide binding fold, *EMBO J.* **1**:945–951.

Wang, J. H., 1985, Functionally distinct β subunits in F_1-adenosinetriphosphatase, *J. Biol. Chem.* **260**:1374–1377.

Wang, J. H., Cesana, J., and Wu, J. C., 1987, Catalytic hydrolysis and synthesis of adenosine 5'-triphosphate by stereoisomers of covalently labeled F_1-adenosinetriphosphatase and reconstituted submitochondrial particles, *Biochemistry* **26**:5527–5533.

Weber, J., Lücken, U., and Schäfer, G., 1985, Total number and differentiation of nucleotide binding sites on mitochondrial F_1-ATPase, *Eur. J. Biochem.* **148**:41–47.

Weber, J., Wilke-Mounts, S., Lee, R. S.-F., Grell, E., and Senior, A. E., 1993, Specific placement of tryptophan in the catalytic sites of *Escherichia coli* F_1-ATPase provides a direct probe of nucleotide binding: Maximal ATP hydrolysis occurs with three sites occupied, *J. Biol. Chem.* **268**:20126–20133.

Coupling Structures and Mechanisms in the Stalk of the Bovine Mitochondrial F_oF_1-ATP Synthase

Sergio Papa, Ting Xu, Antonio Gaballo, and Franco Zanotti

1. INTRODUCTION

The ATP synthases of coupling membranes, the central enzymes in cellular energy metabolism, are made up of a globular catalytic moiety, F_1, which protrudes out of the inner side of the membrane, a membrane-integral proton-translocating moiety, F_o, and a stalk that provides structural and functional connection between F_1 and F_o. A definite breakthrough has been made in understanding, at the molecular level, the mechanism of catalysis in the F_oF_1-ATP synthase. The X-ray crystallographic analysis of the structure of the bovine mitochondrial F_1-ATPase (Abrahams *et al.*, 1994; Chapter 15, this volume) has revealed the atomic structure of this section (Fig. 1) and provided a structural basis for the binding change, rotatory mechanism of the catalytic process in F_1, in which the γ subunit rotates in the central cavity of the $\alpha 3\beta 3$

Sergio Papa, Ting Xu, Antonio Gaballo, and Franco Zanotti • Institute of Medical Biochemistry and Chemistry, University of Bari, I-70124 Bari, Italy. *Permanent address of T.X.:* Institute of Biophysics, Academia Sinica, Beijing 100101, China.

Frontiers of Cellular Bioenergetics, edited by Papa *et al.* Kluwer Academic/Plenum Publishers, New York, 1999.

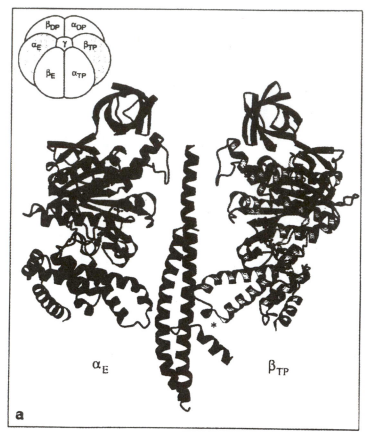

FIGURE 1. The three-dimensional structure of F_1-ATPase. (a) Subunits α_E, γ, and β_{TP}. The *
indicates a "catch" interaction of the DELSEED loop and γ subunit. (b) α_{TP}, γ, and β_E. The
arrow indicates the disruption of the β-sheet in the nucleotide binding domain. The *
indicates a "catch" interaction between the subunit β and the C-terminus of γ subunit. For
details, see text and Abrahams *et al.* (1994), from which the figure is reproduced with
permission. For color reproduction of this figure, see color tip facing this page.

hexamer (Boyer, 1997). Biochemical (Duncan *et al.*, 1995) and physical (Sabbert *et al.*, 1996) evidence for a rotatory movement of the γ subunit relative to
the $\alpha 3 \beta 3$ hexamer of F_1 has also been obtained. Very recently, physical
unidirectional rotation of γ subunit driven by ATP hydrolysis in the immobilized $\alpha 3 \beta 3$ subcomplex of F_1 has been demonstrated (Noji *et al.*, 1997). The
structural organization and the mechanism of the proton-translocating F_o sector
(Schneider and Altendorf, 1987; Fillingame, 1996; Howitt *et al.*, 1996), the
mechanism by which the energy provided by transmembrane proton transloca-

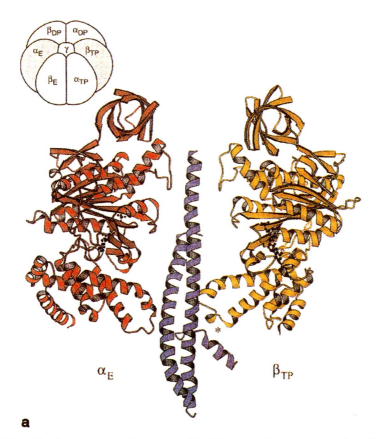

α_E β_{TP}

a

FIGURE 1. The three-dimensional structure of F_1-ATPase. (a) Subunits α_E, γ, and β_{TP}. The *
indicates a "catch" interaction of the DELSEED loop and γ subunit. (b) α_{TP}, γ, and β_E. The
arrow indicates the disruption of the β-sheet in the nucleotide binding domain. The *
indicates a "catch" interaction between the subunit β and the C-terminus of γ subunit. For
details, see text and Abrahams *et al.* (1994), from which the figure is reproduced with
permission.

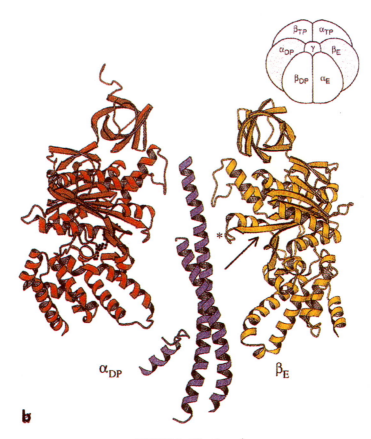

FIGURE 1. (*Continued*)

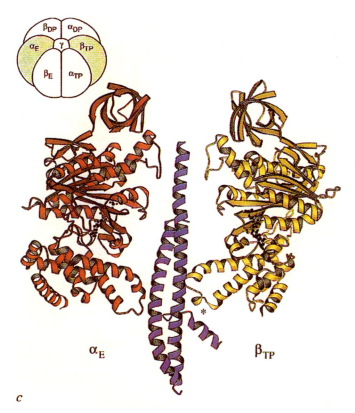

FIGURE 1. (*Continued*)

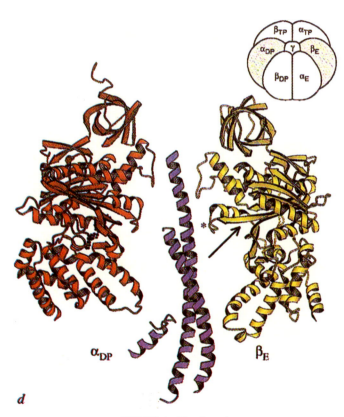

FIGURE 1. (*Continued*)

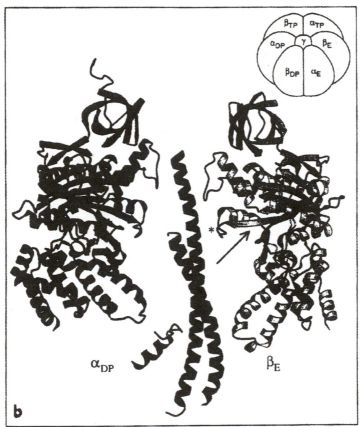

FIGURE 1. (*Continued*) For color reproduction of this figure, see color tip facing this page.

tion through the F$_o$ sector is transferred to F$_1$ to drive ATP synthesis, and how the reverse process, i.e., ATP-driven uphill proton translocation, occurs are less understood. Solutions to these problems and in particular that of energy coupling between proton transport and catalysis in the ATP synthase rest on the elucidation of the structural and functional organization of the components of the stalk. In this chapter we review recently acquired information on the F$_1$ and F$_o$ subunits constituting the stalk, their stoichiometry and near-neighbor relationships. The proton gate and energy-coupling function of the stalk as well as the molecular organization of the stalk subunits in relation to the "revolving" element and the "stator" in the proposed rotatory motor of ATP synthase are examined.

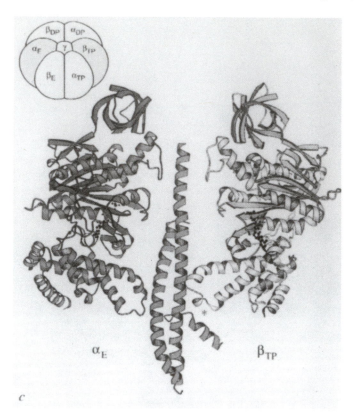

FIGURE 1. (*Continued*) For color reproduction of this figure, see color tip facing page 461.

2. THE SUBUNITS CONSTITUTING THE STALK AND THEIR STOICHIOMETRY

The stalk was first observed to connect the protruding F_1 particles to the matrix side of the negatively stained mitochondrial membrane (Fernandez-Moran, 1962; Racker and Horstmann, 1967). Later, it was clearly visualized by cryoelectronmicroscopy of the ATP synthase in *Escherichia coli* membrane, which showed its real existence in the native enzyme (Gogol *et al.*, 1987). The stalk is made up of interdigitated extensions of F_1 and F_o subunits, which assure transmission of the energy provided by a downhill flow of protons through the F_o domain in the membrane to the membrane-extrinsic F_1 domain where ATP is synthesized from ADP and Pi (Papa *et al.*, 1984a).

In eukaryotic and prokaryotic ATP synthases, the F_1 moiety has the

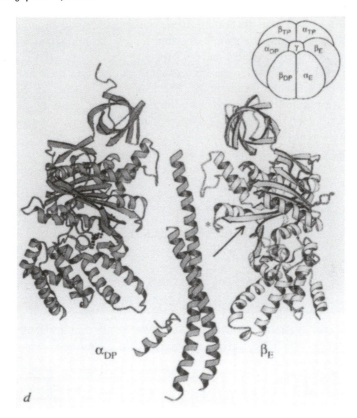

FIGURE 1. (*Continued*) For color reproduction of this figure, see color tip facing page 461.

universal 3α:3β:γ:δ:ε subunit composition (see Chapters 2 and 17, this volume). The crystal structure of bovine mitochondrial F$_1$-ATPase, analyzed by X-ray diffraction at 2.8 Å resolution, shows a stem protruding about 40 Å from the center of the bottom of the spherical body (Abrahams *et al.*, 1994; Chapter 15, this volume) (Fig. 1). The crystal structure shows that the stem, which is a vestige of the ≈45 Å-long stalk, consists of the extension of the C-terminal and N-terminal α-helices of the γ subunit which run in an antiparallel coiled-coil structure in the central cavity of F$_1$ formed by alternating 3α and 3β subunits. In the stem, a third short α-helix, residues 73–90, of the γ subunit is identified. The remaining central part of the γ subunit and the δ and ε subunits are not seen in the crystal structure. Cross-linking and mutational analysis indicate that in the *E. coli* F$_o$F$_1$ complex the ε subunit (equivalent to the δ subunit of the mitochondrial F$_1$) is located on the stalk (Capaldi *et al.*, 1994;

Wilkens *et al.*, 1995). At the bottom of the sphere, the crystal structure of the mitochondrial F_1 shows the stem to be surrounded by alternating bundles of C-terminal α-helices of the 3α and 3β subunits. The N-terminal six-stranded β-barrels of the 3α-3β subunits are located at the top and their nucleotide-binding domains at the center of F_1. The C-terminal bundles of the 3α-3β subunits are reached by F_0 subunits contributing to the stalk, in which the central part of the γ subunit and ε subunits (in the *E. coli* enzyme) penetrate for most of its length (Wilkens *et al.*, 1995).

In prokaryotic and eukaryotic ATP synthases, F_0 has three conserved membrane-intrinsic subunits: a and c, which are essential for proton transport (Schneider and Altendorf, 1985; Fillingame, 1996), and the b subunit in *E. coli*, which is the least conserved part of the F_0F_1-ATP synthase and has counterpart(s) in the eukaryotic enzymes (Papa *et al.*, 1989; Herrmann *et al.*, 1993). The bovine mitochondrial F_0 has seven additional constituents, i.e., subunits OSCP, d, e, f, g, F_6, and A_6L (Collinson *et al.*, 1994a; Belogrudov *et al.*, 1996) (Table I).

The F_0 subunits contributing to the stalk in the mitochondrial ATP synthase have been identified with different approaches: (1) limited proteolysis of F_0 and F_1 subunits (Houstek *et al.*, 1988; Zanotti *et al.*, 1988; Papa *et al.*, 1989; Heckman *et al.*, 1991; Collinson *et al.*, 1994b); (2) immunodetection by subunit-specific antibodies raised against purified F_0 subunits (Papa *et al.*, 1990; Heckman *et al.*, 1991; Belogrudov *et al.*, 1996); (3) cross-linking of F_0 and F_1 subunits (Dupuis *et al.*, 1985; Archinard *et al.*, 1986; Papa *et al.*, 1990; Zanotti *et al.*, 1992; Belogrudov *et al.*, 1995; Watts *et al.*, 1995); and (4) *in vitro* assembly of stalk complexes from F_0 subunits, produced by heterologous expression in *E. coli*, and isolated F_1 from bovine mitochondrial ATP synthase (Collinson *et al.*, 1994c, 1996).

The F_0 subunits extending in the stalk have to be covered by the F_1 subunits that contribute to this structure; thus, F_1 should shield the F_0 components of the stalk from digestion by proteolytic enzymes, which, on the other hand, will digest, after F_1 removal, the membrane-extrinsic parts of F_0 subunits in the stalk. By using this approach, Papa and co-workers (Houstek *et al.*, 1988; Zanotti *et al.*, 1988; Papa *et al.*, 1989) showed that, after F_1 removal, trypsin digested progressively with the incubation time a substantial part of the F_0I-PVP(b) subunit from its C-terminal end. Collinson *et al.* (1994b) confirmed this observation and showed that the digestion could reach the Lys120-Arg121 bond, the remaining N-terminal part apparently being shielded from trypsin digestion. After F_1 removal, trypsin was also found to completely digest OSCP and F_6 (Houstek *et al.*, 1988; Heckman *et al.*, 1991; Collinson *et al.*, 1994b). The trypsin digestion of F_0I-PVP(b) in F_1-depleted vesicles of the mitochondrial membrane resulted in inhibition of proton conduction in F_0, loss of oligomycin sensitivity of this process (Fig. 2), as well as of oligomycin sen-

Table I

The F$_o$ Subunits of Bovine Heart Mitochondrial and *E. coli* ATP Synthase[a]

	Mitochondria				*E. coli*				Structure homology	Sequence identity
Subunits	Genes[b]	Stoichiometry	M$_r$ (kDa)	Subunits	Genes	Stoichiometry	M$_r$ (kDa)			
FoI-PVP(b)	N	1	24.6	b	*uncF*	2	17.3		+	Weak
ATP6(a)	M	1	24.8	a	*uncB*	1	30.3		++	24%
OSCP	N	1	21.0							
d	N	1	18.5							
g	N	1	11.3							
f	N	1	10.2							
F$_6$	N	1	9.0							
IF$_1$	N	1	9.6							
e	N	1–2	8.2							
c	N	9–12	8.0	c	*uncE*	9–12	8.3		+++	21%
A6L	M	1	7.9							

[a]The information given in the table represents a summary of data from Walker *et al.* (1984), Papa *et al.* (1984a), Collinson *et al.* (1994a), Cox *et al.* (1992), Belogrudov *et al.* (1996), Fillingame (1996), and other publications quoted therein.
[b]N, Nuclear; M, Mitochondrial.

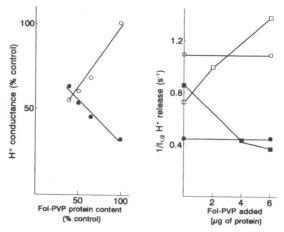

FIGURE 2. Relationship between trypsin digestion of the F_oI-PVP(b) and oligomycin-sensitive H^+ conduction in vesicles of bovine inner mitochondrial membrane deprived of F_1 (USMP) (left panel). Reconstitution by the addition of isolated F_oI-PVP(b) subunit of oligomycin-sensitive H^+ conduction in liposomes reconstituted with F_o extracted from trypsinized USMP (right panel). Left panel: H^+ conduction in trypsinized USMP in the absence (○) and presence of oligomycin (●). Right panel: Liposomes reconstituted with F_o extracted from USMP; F_o-liposomes (○); F_o-liposomes plus oligomycin (●); liposomes reconstituted with F_o extracted from trypsinized USMP (□); liposomes reconstituted with F_o extracted from trypsinized USMP +oligomycin (■). For details, see Zanotti *et al.* (1988), from which the figure is reproduced with permission.

sitivity of the ATPase activity of F_1 reconstituted with F_o-vesicles (Fig. 3). Proton conduction in F_o and its oligomycin sensitivity were fully restored by the addition of the isolated native F_oI-PVP(b) protein alone to digested F_o membrane vesicles; OSCP and F_6 were without effect in this respect (Zanotti *et al.*, 1988) (Fig. 2, Table II). Addition of OSCP and F_6 together with F_oI-PVP(b) were, on the other hand, necessary to fully restore oligomycin sensitivity of the reconstituted F_1-ATPase (Fig. 3) (Guerrieri *et al.*, 1991). It was also found that F_oI-PVP(b) promoted oligomycin inhibition of H^+ conduction when added to F_o membrane vesicles from *E. coli*, and that addition of this protein together with OSCP and F_6 promoted oligomycin inhibition of the F_1-ATPase from *E. coli* or bovine mitochondria when reconstituted with F_o vesicles from *E. coli* (Fig. 3) (Zanotti *et al.*, 1994). These observations provided important information: (1) The facility with which addition of the isolated F_oI-PVP(b) can reconstitute with mitochondrial F_o vesicles in a functional state in which the endogenous subunit has been truncated by trypsin or with *E. coli* F_o vesicles favors the structure of the F_o sector in which the F_oI-PVP(b) is associated with

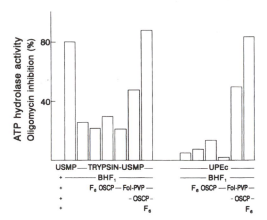

FIGURE 3. Reconstitution by the addition of isolated mitochondrial F_o subunits of oligomycin sensitivity of the ATP hydrolysis in trypsinized USMP and in *E. coli* F_o vesicles deprived of F_1 (UPEc), reconstituted with F_1 isolated from bovine heart mitochondria (BHF_1). The USMP experiments are reproduced from Guerrieri *et al.* (1991) and those on the *E. coli* vesicles from Zanotti *et al.* (1994).

its N-terminal hydrophobic region at the membrane periphery of the bundle of the α-helices of c subunits (see Schneider and Altendorf, 1987; Fillingame, 1996), rather than being located in the central pore of the bundle (Howitt *et al.*, 1996); (2) the membrane-extrinsic C-terminal part of F_oI-PVP has a critical role in the functional organization of the membrane-intrinsic components of the proton channel in F_o, as also seems to be the case of the analogous subunit of the yeast mitochondrial ATP synthase (Paul *et al.*, 1992) and subunit b of *E. coli* (Schneider and Altendorf, 1984, 1985) and is responsible for the oligomycin inhibition of proton conduction in F_o; and (3) the C-terminal part of F_oI-PVP(b), like that of subunit b of *E. coli*, represents an important component of the stalk, which together with OSCP and F_6 plays a central role in functional coupling of F_1 with F_6.

The above conclusions were substantiated by the results of quantitative immunoblotting detection, using subunit specific antibodies, of F_o subunits of the bovine ATP synthase in mitoplast and inside-out vesicles of the inner mitochondrial membrane retaining (ESMP) or deprived of the F_1 moiety (USMP). The results showed that the F_o subunits F_oI-PVP(b), d, F_6, and OSCP are exposed at the matrix, but not at the cytosolic side of the inner mitochondrial membrane, while subunits c and a were occluded to their antibodies on both sides. A6L appeared to be anchored to the membrane with its N-terminal region, the C-terminal part being exposed at the matrix side (Heckman *et al.*, 1991) (see Table III). More recently, it has been shown that subunit f and g are

Table II
Modulation of H$^+$ Conduction in Trypsinized F$_o$ Liposomes by Added γ and F$_o$ Subunits: Effect of Diamide[a]

| | H$^+$ release (nmole/min per mg protein) | | | |
| | $-$ γ subunit | | $+$ γ subunit | |
Liposomes	$-$ diamide	$+$ diamide	$-$ diamide	$+$ diamide
F$_o$	720	579	309	1413
Trypsinized F$_o$	283	184	250	504
+ FoI-PVP(b)	607	366	540	861
+ OSCP	295	302	289	302
+ F$_6$	308	285	257	566
+ OSCP, F$_6$	301	283	303	527
+ FoI-PVP(b), OSCP	584	428	517	900
+ FoI-PVP, F$_6$	683	462	475	1157
+ FoI-PVP, OSCP, F$_6$	757	637	334	1607

[a]Reproduced with permission from Zanotti *et al.* (1992).

both exposed with the N-terminus at the matrix side and the C-terminus at the cytosolic side of the membrane, while the e subunit is exposed only at the cytosolic side (Belogrudov *et al.*, 1996).

Guided by the results of limited proteolysis of F$_o$ subunits, Collinson *et al.* (1994c) overexpressed in *E. coli* the entire F$_o$I-PVP(b) subunit as well as its C-terminal domain (b'), F$_6$, OSCP, and the d subunit and studied their association *in vitro*. They obtained an assembly of a stoichiometric quaternary complex of OSCP:F$_6$:d and the entire or C-terminal part of subunit F$_o$I-PVP(b). This quaternary complex produced a stoichiometric pentameric complex with isolated F$_1$-ATPase. The isolated OSCP formed a binary complex with F$_1$, but in order to have a stoichiometric association of F$_o$I-PVP(b) with F$_1$, OSCP and F$_6$ also had to be present. Further addition of subunit d or d', lacking residue 1 to 14, give the pentameric complex F$_1$:OSCP:F$_o$I-PVP(b)(or b'):d(or d'):F$_6$. Collinson and co-workers concluded that the quaternary OSCP:F$_o$I-PVP(b):d:F$_6$ complex obtained *in vitro* constitutes the essential part of the stalk in the native F$_1$F$_o$ ATP synthase. This is a conclusion, however, that needs verification by independent methods such as cross-linking experiments and analysis of the catalytic and proton-motive activity of the pentameric F$_1$:OSCP:F$_o$I-PVP(b): F$_6$:d complex.

More recently, Collinson *et al.* (1996) have revisited the stoichiometric relationships of the components of the stalk in the bovine mitochondrial enzyme, which is an issue that had previously produced controversial results

Table III
The Membrane Topology and Cysteine Residues
of Bovine Heart Mitochondrial ATPase F_0 Subunits[a]

Subunits	Transmembrane helices	N-terminus[b]	C-terminus	Cysteines
FoI-PVP(b)	2	M •	M	C197
ATP6(a)	5–7	?	?	0
OSCP	0	M	M	C118
d	0	M	M	C100
g	1	M	C	0
f	1	M	C	C72
F_6	0	M	M	0
e	1	M	C	0
C	2	C	c	C64
A6L	1	C	M	0

[a]The information given in the table represents a summary of data from Walker *et al.* (1987), Houstek *et al.* (1988), Oda *et al.* (1989), Joshi and Burrows (1990), Guerrieri *et al.* (1991), Heckman *et al.* (1991), Belogrudov *et al.* (1996), and publications quoted therein.
[b]M, Matrix side; C, cytosolic side.

(cf. Penin *et al.*, 1985; Lippe *et al.*, 1988; Heckman *et al.*, 1991; Collinson *et al.*, 1994c). These authors, by careful quantitation of radioactive *S*-carboxymethylation, radioactive N^{ε}-acetimidation, and quantitative N-terminal sequence analysis of F_1 and F_0 subunits in the quaternary OSCP:F_0I-PVP(b):F_6:d complex, the pentameric F_1:OSCP:F_0I-PVP(b):F_6:d complex, and in a functionally intact isolated F_0F_1-ATP synthase, found that the three preparations consisted of equimolar quantities of their constituent proteins giving the stoichiometric ratios: 3α:3β:1γ:1δ:1ε:1OSCP:1F_0I-PVP(b):1d:1F_6 (see, however, Heckman *et al.*, 1991; Belogrudov *et al.*, 1996). The situation in mitochondria is different from that in the *E. coli* enzyme, which has two copies of the b subunit (Walker *et al.*, 1987; Fillingame, 1990; Dunn, 1992). It is possible that in mitochondria the function of the two b subunits is accomplished by a single copy of F_0I-PVP(b) and another or other components of the stalk. This would be similar to that of the ATP synthase of photosynthetic bacteria and chloroplasts, which have two F_0 subunits as possible counterparts of the b doublet in *E. coli* (Cozens and Walker, 1987; Herrmann *et al.*, 1993).

3. THE PROTON GATE IN THE STALK

The F_0 and F_1 components of the stalk in the ATP synthase have to be involved in the gating of proton translocation and its coupling to hydrodehydration catalysis in F_1, thus preventing dissipative proton flow in the absence of ATP synthesis, as well as ATP hydrolysis decoupled from uphill proton translocation. Disorganization of the F_1-F_0 connection, which can be effected by preparation of "inside-out" vesicles of the inner mitochondrial membrane in the presence of the Mg^{2+} chelator EDTA (ESMP), results in passive trans-membrane proton diffusion through F_0 in the absence of ATP synthesis. Removal of the F_1 moiety from F_0 (USMP) gives further stimulation of proton diffusion, which can be blocked by adding back F_1 (Pansini *et al.*, 1978; Xu *et al.*, 1996). Treatment of ESMP with butanedione, which modifies arginine residues (Guerrieri and Papa, 1981), diamide, which oxidizes vicinal dithiols to disulfides (Zanotti *et al.*, 1985), or ethoxyformic anhydride, which modifies histidine residues (Guerrieri *et al.*, 1987a), all resulted in a dramatic stimulation of proton diffusion in F_0, which was blocked by oligomycin (Fig. 4). This enhancement of oligomyciin-sensitive proton conduction is symptomatic of the opening of the gate in the H^+ channel. Using membrane vesicles with various degrees of the resolution of the F_0F_1 complex, it was shown that the stimulatory effect of ethoxyformic anhydride was due to modification of critical histidine residues in the ATP synthase inhibitor protein (Guerrieri *et al.*, 1987a). This protein associates reversibly and stoichiometrically with the ATP synthase complex at the connection between F_1 and F_0 and inhibits both ATP hydrolysis and H^+ conduction (Guerrieri *et al.*, 1987b; Papa *et al.*, 1996). The

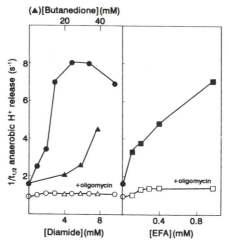

FIGURE 4. Titration of the effect of chemical modification of aminoacids on oligomycin-sensitive H^+ conduction in vesicles of bovine inner mitochondrial membrane with the F_0F_1 complex (ESMP). ESMP were preincubated with diamide (●) or butanedione (▲) at the concentrations reported in the figure, as described in Zanotti *et al.* (1992) and Guerrieri and Papa (1981), respectively; with ethoxyformic anhydride (EFA) (■) as described in Guerrieri *et al.* (1987a).

stimulatory effect caused by butanedione modification persisted after removal of the ATPase inhibitor and F$_1$ and F$_o$ in the vesicles, and thus could be ascribed to modification of arginine residues in F$_o$ subunits (Guerrieri and Papa, 1981). A good candidate for this effect was thought to be the conserved Arg45 in the polar loop connecting the two transmembrane α-helices of subunit c in F$_o$, which is apparently exposed at the matrix side of the membrane (Papa *et al.*, 1984b). Mutational analysis of the c subunit in the *E. coli* ATP synthase has in fact shown that conversion of this arginine to a lysine (R41K) resulted in a high proton permeability of membrane with bound F$_1$ and F$_o$ and suppression of ATP-driven proton translocation (Fraga *et al.*, 1994; Fillingame, 1996).

Identification of the subunits involved in the stimulatory effect of diamide on proton conduction revealed that diamide induced cross-linking in ESMP of the F$_o$I-PVP(b) subunit and the γ subunit of F$_1$ through a disulfide bridge between their single cysteines (Papa *et al.*, 1990; Zanotti *et al.*, 1992) (Fig. 5). An earlier claim of diamide-induced cross-linking of two F$_o$I-PVP(b) copies (Lippe *et al.*, 1988) is disproved by the demonstration of the existence in the

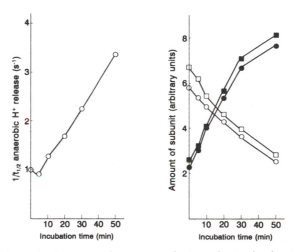

FIGURE 5. Relationship between enhancement of H$^+$ conduction by diamide and cross-linking of F$_o$I-PVP(b) and γ subunits in ESMP. ESMP were treated with 80 μM diamide. The content of F$_o$I-PVP(b) (○) and γ (□) subunits in ESMP treated with diamide was estimated by immunoblot densitometric analysis using specific polyclonal antibodies against the two subunits. For determination of the F$_o$I-PVP(b) and γ subunits in the cross-linking product, ESMP, treated with diamide, were subjected to sodium sodecyl sulfate–polyacrylamide gel electrophoresis (SDS-PAGE). The section of the gels above the γ subunit was cut and electroeluted. After treatment of the electroeluted protein with 20 mM DTT, the material was subjected to a second electrophoretic run, followed by immunodection of F$_o$I-PVP(b) (●) and γ (■) subunits released. For other experimental details, see Zanotti *et al.* (1992), from which the figure is reproduced with permission.

ATP synthase of only one copy of this subunit. The F_oI-PVP(b) cysteine is located at position 197 in the C-terminal region, already shown by limited proteolysis to be an essential component of the stalk. The γ cysteine is located next to the C-terminus of the small α-helical segment of the central part of this subunit, which in the crystal structure of F_1 appears to be situated in the stem vestige of the stalk (Abrahams *et al.*, 1994; Chapter 15, this volume). Thus, these domains of F_oI-PVP(b) and γ subunits play a critical role in the gate function of the stalk. Addition of the γ subunit but not of α and β subunits to F_o liposomes inhibited passive H^+ conduction, which was abolished when the reconstituted γ subunit was cross-linked to F_oI-PVP(b) by diamide (Tables II and IV, Fig. 6). It was concluded that the γ subunit, extending throughout the stalk from the bottom of F_1 to the polar loop of subunit c, closes the matrix mouth of the proton channel in F_o. The long central part of the γ subunit (likely from residue 46 to 208) is flanked on one side of the stalk by the C-terminal region of F_oI-PVP(b), which might keep the γ subunit in place. In the absence of catalysis, the γ subunit is in a state that results in the inhibition of proton diffusion in F_o. During ATP synthesis, as well as hydrolysis, the γ subunit changes to a state that allows proton flow through F_o, coupled to catalysis in F_1. Opening the gate also occurs artificially when the γ subunit is displaced from its native position by disulfide cross-linking to F_oI-PVP(b).

4. NEAR-NEIGHBOR RELATIONSHIPS OF STALK SUBUNITS

Further insight into the topography and near-neighbor relationships of F_1 and F_o subunits contributing to the stalk of the ATP synthase has been provided

Table IV
Effect of F_1 and F_1 Subunits
on H^+ Conduction in F_o Liposomes
in the Absence and Presence of Diamide[a]

Addition	H^+ release (nmole/min^{-1} per mg protein)	
	− diamide	+ diamide
None	720	579
F_1	432	1276
α/β	669	594
γ	309	1414
$\gamma + \alpha/\beta$	321	771

[a]The data presented are from Zanotti *et al.* (1992) and unpublished results of the authors.

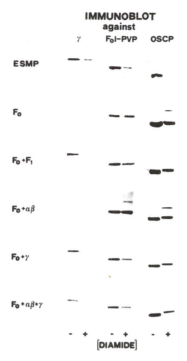

FIGURE 6. Effect of added purified F_1 and isolated F_1 subunits on the formation of cross-linking products induced by diamide in purified F_o. The α/β and γ subunits were purified by electroelution from SDS-PAGE. Where indicated, isolated peptides, at the concentration of 4 µg/mg F_o protein, were incubated with F_o for 10 min, before treatment with 2 mM diamide (incubation time, 2 min). After electrophoresis, proteins were electrotransferred to nitrocellulose and immunoblotted with anti-γ, anti-F_oI-PVP(b), or anti-OSCP sera. From unpublished experiments of A. Gaballo, F. Zanotti, and S. Papa.

by cross-linking experiments and mutational analysis. In mitochondria, the F_oI-PVP(b) subunit is anchored to the membrane by two α-helices and the C-terminal region extends at one side of the stalk, flanking the central part of the γ subunit. In the *E. coli* enzyme, this part of the γ subunit (residue 200–286) could be cross-linked with the polar loop of subunit c in which a cysteine had been site-directed at position 44, thus near the conserved essential Arg41 (Watts *et al.*, 1995). This substantiated the evidence that the central part of the γ subunit, emerging from the cavity in the 3α–3β hexamer of F_1, extends throughout the stalk and contacts the polar loop of the c subunit exposed at the inner side of F_o in the membrane. Cross-linking experiments in the *E. coli* enzyme showed that the ε subunit, homologous to subunit δ of the mitochondrial enzyme, is bound to α and β subunits with the C-terminal domain, its amino-terminal domain interacts with the γ subunit (Aggeler *et al.*, 1995; Capaldi *et al.*, 1994) and the region around residue 31 is physically close to the polar loop of subunit c (Zhang and Fillingame, 1995). Thus, in *E. coli*, subunit ϵ extends in the stalk from the α–β bottom of F_1 to the inner polar surface of subunit c in F_0. Mutagenesis experiments have indicated that in *E. coli* the C-terminal region of δ subunit, homologous to the mitochondrial OSCP (Ovchinnikov *et al.*, 1984), is involved in the functional binding of F_1 to F_0 (Jounouchi *et al.*, 1992; Hazard and Senior, 1994). *E. coli* ATP synthase defective in the δ subunit loses the capacity to translocate protons (Noumi and Kanazawa, 1983; Humbert *et al.*, 1983). The bovine ϵ and δ subunits, produced by overexpression in *E. coli*, form *in vitro* a 1:1 stoichiometric complex (Orriss *et al.*, 1996). This dimer, however, did not bind to the tetrameric OSCP:F_0I-PVP(b):F_6:d.

Direct accessibility to a monoclonal antibody and lack of reaction with hydrophobic probes show an external location of OSCP in the stalk (Archinard *et al.*, 1986). OSCP could be directly linked to α and β subunits by zero-length cross-linkers (Dupuis *et al.*, 1985; Archinard *et al.*, 1986). Furthermore, it has been found that F_1 protects OSCP and OSCP F_1 against mild trypsin proteolysis (Hundal *et al.*, 1983; Dupuis *et al.*, 1985), which results in the cleavage of the first 15 N-terminal residues of the α-subunit and the first 5–7 residues of the β subunit (Walker *et al.*, 1985; Xu *et al.*, 1996). Since the crystal structure of F_1 shows both the α and β N-termini to be located at the top of the F_1 sphere, it has been inferred, from these observations, the OSCP might extend from the stalk around the external surface of the F_1 to its top (Collinson *et al.*, 1996). Reconstitution experiments of OSCP-depleted F_0F_1 complexes with deletion mutants of OSCP expressed in *E. coli* provided evidence indicating that the N-terminus of OSCP, residues 16–28, is essential for the binding of OSCP to F_1; the C-terminus, residues 181–186, is important for interaction of OSCP with F_6, and $\Delta\mu H^+$-driven ATP synthesis and the central part forms helices contributing to the stalk (Joshi *et al.*, 1996).

The topography and near-neighbor relationships of subunits d, F_6, A6L, OSCP, and F_0I-PVP(b) with each other and the F_1 subunits in the stalk have been examined by Belogrudov *et al.* (1995) using cross-linkers of different chemical specificities and cross-linking lengths from 0 to 10 Å (see Table V). Subunit OSCP, d, and F_6 could be cross-linked to α and/or β but not to γ or δ. F_0I-PVP(b), when modified by the heterobifunctional photoactivable sulfhydryl reagent MBP, cross-linked with its Cys197 to F_6 but not to α and β. It should be noted that Cys197 is the same residue that is disulfide bridged under the action of diamide to Cys91 of γ (Papa *et al.*, 1990). Thus, the C-terminal domain around Cys197 of F_0I-PVP(b) is apparently distant from α and β but close to both the short α-helix (residue 73–90) of γ and F_6 (Tables IV and V).

OSCP cross-linked to F_0I-PVP(b), α, and β. A6L, which is anchored to

Table V
The Near-Neighbor Relationship of F_o and F_1 Subunits in the Stalk of Bovine ATP Synthase[a]

Subunits	Fo1-PVP(b)[b]	a	OSCP	d	g	f	F_6	e	c	A6L	γ	δ
Fo1-PVP(b)												
a	+											
OSCP*	+	−										
d	+	−	−									
g	?	−	−	−								
f	?	−	−	−	+							
F_6	+	−	−	−	−	−						
e	?	−	−	−	+	−	−					
c	?	+	−	−	−	−	−	−				
A6L	−	?	−	+	−	+	−	−	−			
γ	+	−	−	−	−	−	−	−	+	−		
δ	−	−	−	−	−	−	−	−	+	−	+	
α/β	+	−	+	+	−	−	+	−	−	−	+	+

[a]The information given in the table represents a summary of data from Dupuis *et al.* (1985), Joshi and Burrows (1990), Papa *et al.* (1990), Zanotti *et al.* (1992), Aggeler *et al.* (1995), Watts *et al.* (1995), Zhang and Fillingame (1995), Belogrudov *et al.* (1995, 1996), and publications quoted therein.
[b]The sign + indicates a contact; the sign − indicates the absence of contact.

the membrane by its N-terminus, cross-linked only to d. F_6 and d cross-linked to F_oI-PVP(b). OSCP, F_6, and d did not cross-link to one another. The single Cys in OSCP is thus distant from Cys91 in γ, but it is close to and can disulfide bridge with a Cys present in another yet to be identified component of F_o of molecular weight of $\approx$ 19 kDa (Fig. 6) (a possible candidate with a molecular weight of 18.6 kDa and a single Cys is subunit d). It should be noted here, however, that while a positive result obtained with a cross-linking reagent reveals a contact between the subunits involved, a negative result does not exclude contacts.

5. THE STALK, ENERGY COUPLING, AND ROTATORY MOTOR IN THE ATP SYNTHASE

The mechanism of ATP synthesis and hydrolysis is reviewed in Chapter 16, this volume. It is now generally accepted that the energy provided by transmembrane $\Delta\mu H^+$ is utilized for the release of ATP formed at the high-affinity catalytic site of F_1 from ADP and Pi in a reaction that per se involves a small ΔG (Penefsky, 1985). The endergonic release of ATP is thought to result from sequential and cooperative affinity-binding changes for ATP, ADP, and Pi at the catalytic site in the three β subunits induced by deformation caused by $\Delta\mu H^+$-driven rotation of the single copy of the γ subunit (alternating binding changes, rotatory catalysis) (Boyer, 1997).

Definite support in favor of the rotatory catalytic mechanism was provided by the crystal structure of mitochondrial F_1. In the structure of the bovine heart enzyme crystallized in the presence of limiting amount of AMP-PNP, an inhibitory analogue of ATP, and ADP, the three β subunits showed different conformations. two bind AMP-PNP and ADP at the catalytic sites, respectively; the third, which has the catalytic site empty, differs from the other two by a large hinge motion of the carboxy-terminal domain (Abrahams *et al.*, 1994). In the structure of the rat liver enzyme crystallized in the presence of a physiological concentration of nucleotides, the three β subunits show similar conformation with all the three nucleotide binding sites occupied (Chapter 15, this volume). Bianchet *et al.* (Chapter 15, this volume) have proposed a version of the rotatory mechanism in which they incorporate the two conformations found in the bovine heart and rat liver F_1, respectively. The crystal structures of both the bovine heart and rat liver show that the C-terminus of the γ subunit, forming an antiparallel coiled-coil structure with the N-terminus, runs in the central cavity of the α3β3 hexamer and comes in contact with the catalytic site in the central part of β. On the other hand, the small α-helix (residue 73–90) in the stalk domain of γ can contact the C-terminal domain of α and β (Abrahams *et al.*, 1994). Consistent with a central role of γ in energy coupling in the F_oF_1 complex is the demonstration using fluorescent and cross-linking reagents of conformational and position changes of subunits γ and ε induced by ATP hydrolysis in the *E. coli* enzyme (Capaldi *et al.*, 1994). Evidence for ATP-driven rotation (<100 msec) of the γ subunit relative to the α3β3 hexamer was obtained by monitoring polarized absorption relaxation of the eosin-labeled γ in F_1-ATPase from spinach chloroplasts (Sabbert *et al.*, 1996). A rotatory motion of the γ subunit relative to the β subunits was also demonstrated in *E. coli* F_1 (Duncan *et al.*, 1995) and F_oF_1 complex (Zhou *et al.*, 1996), using mutational insertion of a Cys (Asp 380 Cys) in the [380]DELSEED[386] sequence of β, which appears as a β–γ contact site in the crystal structure of F_1 and its cross-linking with Cys87 in the γ subunit. Very recently, physical unidirectional anticlockwise rotation (viewed from the membrane side) of the γ subunit driven by ATP hydrolysis in the α3β3γ subcomplex from a thermophilic bacterium coated on a glass plate has been observed (Noji *et al.*, 1997) (Fig. 7). This shows that the F_1-ATPase can function, at least in the isolated state, as a rotatory device during catalysis. So far, no direct evidence is available, however, showing rotation of the membrane intrinsic F_o subunit c (Cross and Duncan, 1996) or a (Howitt *et al.*, 1996), driven by $\Delta\mu H^+$ or ATP hydrolysis. Also, it is unknown how the rotation of F_o subunits is coupled with rotation of γ in the α3β3 hexamer of F_1 to drive ATP synthesis.

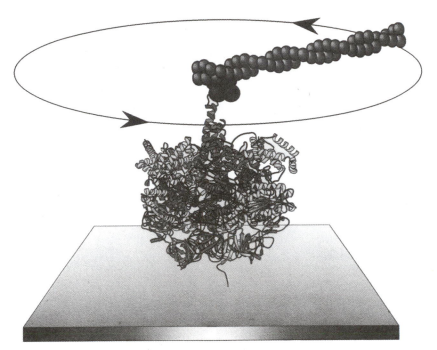

FIGURE 7. Schematic illustration of the system used by Noji *et al.* (1997) for the observation by epifluorescence microscope of the rotation of the γ subunit tagged by a fluorescent actin filament in the $\alpha_3\beta_3\gamma$ subcomplex from the thermophilic *Bacillus* PS3 fixed onto a glass surface through the β subunits. Courtesy of Dr. Masasuke Yoshida. For color reproduction of this figure, see color tip facing page 478.

Efficient conversion of osmotic into mechanical and the latter into chemical energy, and vice versa, by a rotatory mechanism requires the existence of a stable stator in the ATP synthase that keeps the $\alpha3\beta3$ hexamer immobilized relative to the immobile part of the F_o sector and at the same time allows rapid and efficient clockwise and anticlockwise rotation (during ATP synthesis and hydrolysis, respectively) of the γ subunit on an axis of rotation possibly coincident with the sixfold axis of pseudosymmetry of F_1 (see Figs. 8 and 9). The work on the *E. coli* enzyme reviewed above would suggest that the ε subunit (equivalent to the mitochondrial δ subunit) is associated with and rotates together with the central part of γ in the stalk (Aggeler *et al.*, 1997).

Genetic and proteolytic experiments have provided evidence of the N-terminal part of the α subunit requirement for the binding of the δ subunit to *E.*

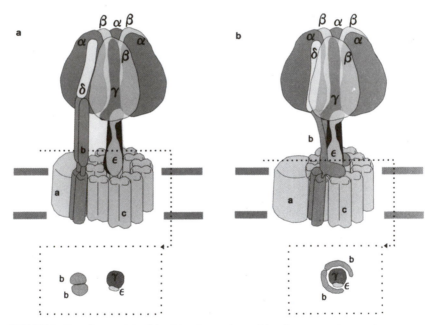

FIGURE 8. Tentative models of the F_oF_1 ATP synthase of the plasma membrane of *E. coli*. The F_1 structure is a sketch based on the crystal structure of bovine F_1 (Abrahams *et al.*, 1994; Chapter 15, this volume). The structure of the F_o sector and the stalk is drawn on the basis of the observations reviewed and their interpretation in this chapter. Model (a) reproduces a structure proposed by Ogilvie *et al.* (1997), which assumes the existence of two stalks in the F_oF_1 complex: A central one composed of the γ and ε subunits and a second lateral stalk made up by the b dimer. Model (b) envisages a single stalk made up by the elongated stem of γ and thin central parts of the two b subunits. For color reproduction of this figure, see color tip facing page 479.

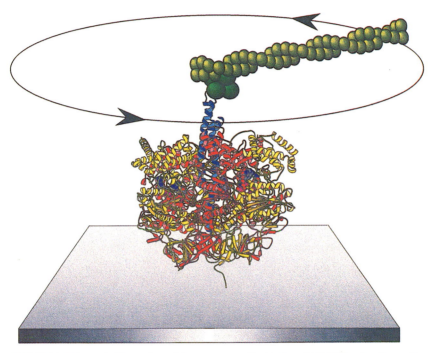

FIGURE 7. Schematic illustration of the system used by Noji *et al.* (1997) for the observation by epifluorescence microscope of the rotation of the γ subunit tagged by a fluorescent actin filament in the α₃β₃γ subcomplex from the thermophilic *Bacillus* PS3 fixed onto a glass surface through the β subunits. Courtesy of Dr. Masasuke Yoshida.

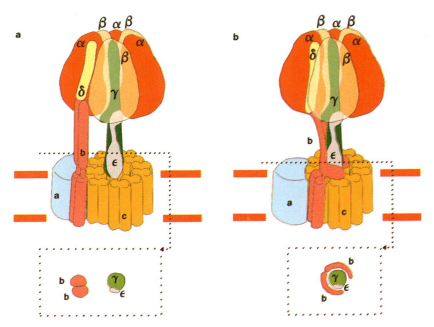

FIGURE 8. Tentative models of the F_oF_1 ATP synthase of the plasma membrane of *E. coli*. The F_1 structure is a sketch based on the crystal structure of bovine F_1 (Abrahams *et al.*, 1994; Chapter 15, this volume). The structure of the F_o sector and the stalk is drawn on the basis of the observations reviewed and their interpretation in this chapter. Model a reproduces a structure proposed by Ogilvie *et al.* (1997), which assumes the existence of two stalks in the F_oF_1 complex: A central one composed of the γ and ε subunits and a second lateral stalk made up by the b dimer. Model b envisages a single stalk made up by the elongated stem of γ and thin central parts of the two b subunits.

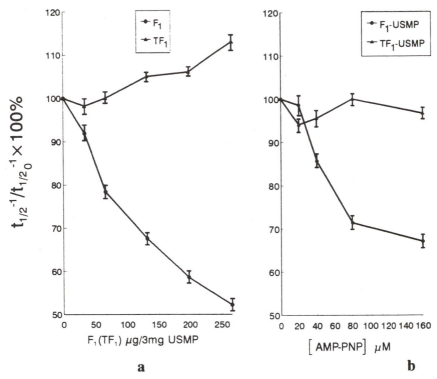

FIGURE 9. The effect of (a) purified F_1 and trypsinized F_1 and of (b) AMP-PNP on proton conduction in USMP. The results are expressed as percentages of inhibition. (a) F_1 or TF_1, in the amounts shown in the figure, were added to the assay medium containing USMP. After 10 min incubation, the proton conduction was measured. (b) USMP were reconstituted with F_1 or TF_1 and then preincubated with AMP-PNP before starting the measurement of H^+ conduction. For other details, see Xu *et al.* (1996), from which the figure is reproduced with permission.

coli F_1 (Dunn *et al.*, 1980; Meggio *et al.*, 1988). Recently, Ogilvie *et al.* (1997) have mutagenized glutamine-2 of the α subunit of *E. coli* to cysteine and showed this to be cross-linked by Cu^{2+} treatment to Cys64 and Cys140 of the δ subunit (cf. Ziegler *et al.*, 1994). These authors concluded that the δ subunit is located on the surface of F_1, on one α, or between one α–β pair, from their N-termini at the top of F_1, pointing toward the F_1 bottom with the C-terminal region, which binds the two b subunits of F_o (Mendel-Hartvig and Capaldi, 1991; Perlin *et al.*, 1983). Ogilvie *et al.* (1997) and Wilkens *et al.* (1997) have thus proposed that the dimer of the b subunits forms a second stalk, which is located asymmetrically at one side of the central stalk; the central stalk is composed of the protruding part of the γ and the ε subunits (see Fig. 8a). These authors suggest that the lateral stalk would have been missed in averaged microscopy images of the *E. coli* F_oF_1, which show only a single central stalk (Gogol *et al.*, 1987). The δ-b complex would constitute the stator of the ATP synthase motor, as indicated by the fact that cross-linking of δ with α does not affect the function of the ATP synthase (Ogilvie *et al.*, 1997). It also could be possible, however, that thin elongated central parts of the two b subunits, departing from their docking site on the δ subunit [b subunit, however, also has contacts with the α and β subunits (see Aris and Simoni, 1983)], run along the surface of the γ stem, thus constituting a single stalk (as seen, in fact, in cryoelectronmicroscopy images) with enlarged central parts close to the NH_2-terminal membrane anchor, thus forming a ring structure surrounding the γ–ε complex. This could in fact appear in electronmicroscopy as part of the membrane F_o sector (Fig. 8b). A ring structure formed by lobes of the b subunits surrounding γ–ε complex at its contact with the rotating bundle of F_o c subunits might prevent tumbling of the γ–ε complex when it is made to rotate by the proton-driven revolution of the c subunits.

The main component of the stator in mitochondrial F_o seems to be the F_oI-PVP(b), which, emerging from its N-terminal membrane anchor, extends in the stalk with the C-terminal part touching the C-terminal region of α and β at the bottom of F_1. In the bovine ATP synthase, the C-terminus of F_oI-PVP(b) around Cys197 is apparently bent away from the α–β contacts (Belogrudov *et al.*, 1995) and lies in a site, also occupied by F_6, close to the Cys91 at the C-terminus of the α-helix, residues 73–90, of γ. It seems likely that this α-helix can, during the energy-linked rotation of γ, touch the DELSEED segment in the β subunit when occupied by ATP, inducing a conformational wave that reaches the catalytic site and induces the release of bound ATP (Fig. 1). A second catch between γ and β in the empty state is formed by hydrogen bonds between β-Asp316, β-Thr318, and β-Asp319 and γ-Arg254 and γ-Gln255 (Abrahams *et al.*, 1994).

The central role of the α-helix (residues 73–90) of subunit γ for energy transfer in the F_oF_1 complex and its topological interaction with the F_oI-PVP(b) in the stator seems to be well illustrated by the diamide-induced disulfide cross-linking of these subunits. As reported, diamide treatment of the F_oF_1-ATP synthase in ESMP in the absence of a respiratory $\Delta\mu H^+$ results in disulfide cross-linking between the Cys197 in the carboxy-terminal region of the F_oI-PVP(b) subunit and Cys91 in the small α-helix of subunit F_1γ, both located in the stalk. The same disulfide cross-linking is produced by diamide treatment of F_o liposomes supplemented with isolated γ. The F_oI-PVP(b) and γ cross-linking results in both cases in a one-order enhancement of oligomycin-sensitive decay of $\Delta\mu H^+$. In ESMP and MgATP particles, the cross-linking is accompanied by decoupling of respiratory ATP synthesis. These effects are consistent with the view that F_oI-PVP(b) and $F_1\gamma$ are components of the stator and rotor of the proposed rotatory motor, respectively. The fact that Cys197 of F_oI-PVP(b) and Cys91 of γ can form a direct disulfide bridge shows that the carboxyl-terminal region of the first and the short central α-helix of the second subunits are, at least in the resting state of the enzyme, in direct contact.

When ESMP is treated with diamide in the presence of $\Delta\mu H^+$ generated by respiration, no cross-linking between F_oI-PVP(b) and γ subunits or the associated effects on proton conduction and ATP synthesis are observed. Cross-linking and functional effects are restored in respiring ESMP by $\Delta\mu H^+$-collapsing agents as well as by oligomycin or DCCD. These observations indicate that the torque generated by $\Delta\mu H^+$ decay through F_o induces a relative motion and/or a separation of the F_oI-PVP(b) subunit and γ that places the two cysteine residues at a distance at which they cannot be engaged in disulfide bridging. It is possible that the relative position of the stator and rotor, or simply of F_oI-PVP(b) and γ, change from a contiguous to a separate position, depending whether F_oF_1 complex is in the resting or active state, respectively.

A particular position in the mitochondrial stator is apparently held by OSCP. This subunit is considered to be homologous to the δ subunit of the *E. coli* enzyme (Ovchinnikov *et al.*, 1984). It can be noted, however, that when the F_1 and F_o moieties are split apart, OSCP remains stoichiometrically attached to F_o while δ copurifies with F_1 in the *E. coli* enzyme (Collinson *et al.*, 1994a). OSCP is covered by F_1, which protects it from proteolytic digestion (Houstek *et al.*, 1988; Heckman *et al.*, 1991; Collinson *et al.*, 1994b). OSCP appears to be close to the α and β subunit with which it cross-links (Belogrudov *et al.*, 1995) and to be superficially located with respect to other F_o components of the stalk (Archinard *et al.*, 1986). Together with F_oI-PVP(b) and F_6 OSCP is required for the conferral of oligomycin sensitivity to F_1 (Guerrieri *et al.*, 1991). It is possible that OSCP, which is present in a single copy, is associated with the C-terminal domain of F_oI-PVP(b) also present in a single copy; with this and F_6, OSCP constitutes one part of the stator in the stalk. Selective cleavage in purified mitochondrial F_1 of the N-terminus of subunits α and β by trypsin impaired gating of proton translocation in F_o by F_1 (Xu *et al.*, 1996) (Fig. 9). On the other hand, the N-termini of α and β were occluded toward trypsin digestion in native F_oF_1 complex in the membrane, with protection exerted by OSCP (T. Xu and S. Papa, manuscript in preparation). It is possible that OSCP extends from the stalk along the surface of the $\alpha 3\beta 3$ hexamer so as to cover the N-termini of α and β located at the top of F_1 (Collinson *et al.*, 1996). It is also possible, however, that the influence of the N-terminal segments of α and/or β on the coupling of ATP hydrolysis to proton pumping and the protective effect of OSCP on proteolytic cleavage of the N-termini of α and β involve long-range conformational changes from the top to the bottom of F_1 (Xu *et al.*, 1996, 1998).

Another component of the stalk could be represented by subunit d extending from contact site(s) in F_1 to contact site(s) in A6L, f, and/or g subunits anchored in the F_o sector. Definition of the topological organization of the components connecting the F_1 and F_o sector as well as the possibility that the stator and the rotor can form, at least transiently, two separate stalks (see the average electron microscopy images of a V-type ATPase reported by Boekema *et al.*, 1997) rests, however, on higher-resolution structure analysis of the F_oF_1 complex. It can be mentioned in this respect that intersubunit distances calculated from phosphorescence and fluorescence energy transfer in the mitochondrial F_oF_1 between the lone tryptophan residue in the ε subunit and a fluorescent probe attached to Cys118 of OSCP seem to indicate that these two subunits are in close proximity, if not in direct contact, and do not support location of OSCP in a separate structure at a considerable distance, larger than 37 Å, from the ε subunit (Gabellieri *et al.*, 1997). Clearly, the stage is set for further studies on the structural organization and dynamics of F_1 and F_o stalk components in energy coupling.

ACKNOWLEDGMENTS. This work has been supported by grants No. 96.03707. CT14 of Consiglio Nazionale delle Ricerche and No. 97.01167.PF49, Italy, and by a grant from the MURST National Project on Bioenergetics and Membrane Transport. T. Xu is supported by grants of the Italian Research Council, the UNESCO Global Network for Molecular and Cell Biology, and the University of Bari. The authors are grateful to Dr. Nazzareno Capitanio for drawing of Fig. 8.

6. REFERENCES

Abrahams, J. P., Leslie, A. G. W., Lutter, R., and Walker, J. E., 1994, Structure at 2.8 Å resolution of F_1-ATPase from bovine heart mitochondria, *Nature* **370:**621–628.

Aggeler, R., Haughton, M. A., and Capaldi, R. A., 1995, Disulfide bond formation between the COOH-terminal domain of the beta subunits and the gamma and epsilon subunits of the *Escherichia coli* F_1-ATPase: Structural implications and functional consequences, *J. Biol. Chem.* **270:**9185–9191.

Aggeler, R., Ogilvie, I., and Capaldi, R. A, 1997, Rotation of a γ-ε subunit domain in the *Escherichia coli* F_1F_o-ATP synthase complex, the γ-ε subunits are essentially randomly distributed relative to the α3β3delta domain in the intact complex, *J. Biol. Chem* **272:**19621–19624.

Archinard, P., Godinot, C., Comte, J., and Gautheron, D. C., 1986, Topography of oligomycin sensitivity conferring protein in the mitochondrial adenosinetriphosphatase-ATP synthase, *Biochemistry* **25:**3397–3404.

Aris, J. P., and Simoni, R. D., 1983, Crosslinking and labeling of the *Escherichia coli* F_1F_o-ATP synthase reveal a compact hydrophilic portion of F_o close to an F_1 catalytic subunit, *J. Biol. Chem.* **258:**14599–14609.

Belogrudov, G. I., Tomich, J. M., and Hatefi, Y., 1995, ATP synthase complex. Proximities of subunits in bovine submitochondrial particles, *J. Biol. Chem.* **270:**2053–2060.

Belogrudov, G. I., Tomich, J. M., and Hatefi, Y., 1996, Membrane topography and near-neighbor relationships of the mitochondrial ATP synthase subunits e,f and g, *J. Biol. Chem.* **271:**20340–20345.

Boekema, E. J., Ubbink-Kok, T., Lolkema, J. S., Brisson, A., and Konings, W. N., 1997, Visualization of a peripheral stalk in V-type ATPase: Evidence for the stator structure essential to rotational catalysis, *Proc. Natl. Acad. Sci. USA* **94:**14291–14293.

Boyer, P. D., 1997, The ATP synthase—A splendid molecular machine, *Annu. Rev. Biochem.* **66:** 717–749.

Capaldi, R. A., Aggeler, R., Turina, P., and Wilkens, S., 1994, Coupling between catalytic sites and the proton channel in F$_1$F$_o$-type ATPases, *Trends Biochem. Sci.* **19:**284–289.

Collinson, I. R., Runswick, M. J., Buchanan, S. K., Fearnley, I. M., Skehel, J. M., van Raaij, M. J., Griffiths, D. E., and Walker, J. E., 1994a, F$_o$ membrane domain of ATP synthase from bovine heart mitochondria: Purification, subunit composition, and reconstitution with F$_1$-ATPase, *Biochemistry* **33:**7971–7978.

Collinson, I. R., Fearnley, I. M., Skehel, J. M., Runswick, M. J., and Walker, J. E., 1994b, ATP synthase from bovine heart mitochondria: Identification by proteolysis of sites in F$_o$ exposed by removal of F$_1$ and oligomycin-sensitivity conferral protein, *Biochem. J.* **303:**639–645.

Collinson, I. R., van Raaij, M. J., Runswick, M. J., Fearnley, I. M., Skehel, J. M., Orriss, G. L., Miroux, B., and Walker, J. E., 1994c, ATP synthase from bovine heart mitochondria. *In vitro* assembly of a stalk complex in the presence of F$_1$-ATPase and in its absence, *J. Mol. Biol.* **242:** 408–421.

Collinson, I. R., Skehel, J. M., Fearnley, I. M., Runswick, M. J., and Walker, J. E., 1996, The F$_1$F$_o$-ATPase complex from bovine heart mitochondria: The molar ratio of the subunits in the stalk region linking the F$_1$ and F$_o$ domains, *Biochemistry* **35:**12640–12646.

Cox, G. B., Devenish, R. J., Gibson, F., Howitt, S. M., and Nagley, P., 1992, The structure and assembly of ATP synthase, in *Molecular Mechanism in Bioenergetics* (L. Ernster, ed.), pp. 283–315, Elsevier, Amsterdam.

Cozens, A. L., and Walker, J. E., 1987, The organization and sequence of the genes for ATP synthase subunits in the cyanobacterium *Synechococcus* 6301. Support for an endosymbiotic origin of chloroplasts, *J. Mol. Biol.* **194:**359–383.

Cross, R. L., and Duncan, T. M., 1996, Subunit rotation in F$_1$F$_o$-ATP synthases as a means of coupling proton transport through F$_o$ to the binding changes in F$_1$, *J. Bioener. Biomembr.* **28:** 403–408.

Duncan, T. M., Bulygin, V. V., Zhou, Y., Hutcheon, M. L., and Cross, R. L., 1995, Rotation of subunits during catalysis by *Escherichia coli* F$_1$-ATPase, *Proc. Natl. Acad. Sci. USA* **92:**10964–10968.

Dunn, S. D., 1992, The polar domains of the b subunit of *Escherichia coli* F$_1$F$_o$-ATPase forms an elongated dimer that interacts with the F$_1$ sector, *J. Biol. Chem.* **267:**7630–7636.

Dunn, S. D., Heppel, L. A., and Fullmer, C. S., 1980, the NH$_2$-terminal portion of the subunit α of *E. coli* F$_1$ ATPase is required for binding the δ subunit, *J. Biol. Chem.* **255:**6891–6896.

Dupuis, A., Issartel, J. P., Lunardi, J., Satre, M., and Vignais, P. V., 1985, Interactions between the oligomycin sensitivity conferring protein (OSCP) and beef heart mitochondrial F$_1$-ATPase. 1. Study of the binding parameters with a chemically radiolabeled OSCP, *Biochemistry* **24:** 728–733.

Fernandez-Moran, H., 1962, Cell membrane ultrastructure. Low temperature electron microscopy and x-ray diffraction studies of lipoprotein components in lamellar systems, *Circulation* **26:** 1039–1065.

Fillingame, R. H., 1990, Molecular mechanics of ATP synthesis for F$_1$F$_o$-type H$^+$-translocating

ATP synthases, in *The Bacteria* (T. A. Krulwich, Ed.), vol. XII, pp. 345–391, Academic Press, New York.

Fillingame, R. H., 1996, Membrane sectors of F and V-type H^+-transporting ATPases, *Curr. Opin. Struct. Biol.* **6:**491–498.

Fraga, D., Hermolin, J., Oldenburg, M., Miller, M. J., and Fillingame, R. H., 1994, Arginine 41 of subunit c of *Escherichia coli* H^+-ATP synthase is essential in binding and coupling of F_1 to F_0, *J. Biol. Chem.* **269:**7532–7537.

Gabellieri, E., Strambini, G. B., Baracca, A., and Solaini, G., 1997, Structure mapping of the ε-subunit of mitochondrial H^+-ATPase complex (F_1), *Biophys. J.* **72:**1818–1827.

Gogol, E. P., Lucken, U., and Capaldi, R. A., 1987, The stalk connecting the F_1 and F_0 domains of ATP synthase visualized by electron microscopy of unstained specimens, *FEBS Lett.* **219:** 274–278.

Guerrieri, F., and Papa, S., 1981, Effect of chemical modifiers on amino acid residues on proton conduction by the H^+-ATPase of mitochondria, *J. Bioenerg. Biomembr.* **13:**393–400.

Guerrieri, F., Zanotti, F., Che, Y. W., Scarfò, R., and Papa, S., 1987a, Inactivation of the mitochondrial ATPase inhibitor protein by chemical modification with diethylpyrocarbonate, *Biochim. Biophys. Acta* **892:**284–293.

Guerrieri, F., Scarfò, R., Zanotti, F., Che, Y. W., and Papa, S., 1987b, Regulatory role of the ATPase inhibitor protein on proton conduction by mitochondrial H^+-ATPase complex, *FEBS Lett.* **213:**67–72.

Guerrieri, F., Zanotti, F., Capozza, G., Colaianni, G., Ronchi, S., and Papa, S., 1991, Structural and functional characterization of subunits of the F_0 sector of the mitochondrial F_0F_1-ATP synthase, *Biochim. Biophys. Acta* **1059:**348–354.

Hazard, A. L., and Senior, A. E., 1994, Defective energy coupling in delta-subunit mutants of *Escherichia coli* F_1F_0-ATPsynthase, *J. Biol. Chem.* **269:**427–432.

Heckman, C., Tomich, J. M., and Hatefi, Y., 1991, Mitochondrial ATP synthase complex. Membrane topography and stoichiometry of the F_0 subunits, *J. Biol. Chem.* **266:**13564–13571.

Herrmann, R. G., Steppuhn, J., Herrmann, G. S., and Nelson, N., 1993, The nuclear-encoded polypeptide Cf_0-II from spinach is a real, ninth subunit of chloroplast ATP synthase, *FEBS Lett.* **326:**192–198.

Houstek, J., Kopecky, J., Zanotti, F., Guerrieri, F., Jirillo, E., Capozza, G., and Papa, S., 1988, Topological and functional characterization of the F_0I subunit of the membrane moiety of the mitochondrial H^+-ATP synthase, *Eur. J. Biochem.* **173:**1–8.

Howitt, S. M., Rodgers, A. J. W., Hatch, L. P., Gibson, F., and Cox, G. B., 1996, The coupling of the relative movement of the a and c subunits of the F_0 the conformational changes in the F_1-ATPase, *J. Bioenerg. Biomembr.* **28:**415–420.

Humbert, R., Brusilow, W. S. A., Gunsalus, R. P., Klionsky, D. J., and Simoni, R. D., 1983, *Escherichia coli* mutants defective in the uncH gene, *J. Bacteriol.* **153:**416–422.

Hundal, T., Norling, B., and Ernster, L., 1983, Lack of ability of trypsin–treated mitochondrial F_1-ATPase to bind the oligomycin-sensitivity conferring protein (OSCP), *FEBS Lett.* **162:**5–10.

Jounouchi, M., Takeyama, M., Chaiprasert, P., Noumi, T., Moriyama, Y., Maeda, M., and Futai, M., 1992, *Escherichia coli* H^+-ATPase: Role of the δ subunit in binding F_1 to the F_0 sector, *Arch. Biochem. Biophys.* **292:**376–381.

Joshi, S., and Burrows, R., 1990, ATP synthase complex from bovine heart mitochondria. Subunit arrangement as revealed by nearest neighbor analysis and susceptibility to trypsin, *J. Biol. Chem.* **265:**14518–14525.

Joshi, S., Cao, G. J., Nath, C., and Shah, J., 1996, Oligomycin sensitivity conferring protein of mitochondrial ATP synthase: Deletions in the N-terminal end cause defect in interaction with F_1, while deletion in the c-terminal end cause defects in interactions with F_0, *Biochemistry* **35:**12094–12103.

Lippe, G., Sala, F. D., and Sorgato, M. C., 1988, ATP synthase complex from beef heart mitochondria. Role of the thiol group of the 25-kDa subunit of F_o in the coupling mechanism between F_o and F_1, *J. Biol. Chem.* **263**:18627–18634.

Maggio, M. B., Parsonage, D., and Senior, A. E., 1988, A mutation in the α-subunit of F_1-ATPase from *Escherichia coli* affect the binding of F_1 to the membrane, *J. Biol. Chem.* **263**:4619–4623.

Mendel-Hartvig, J., and Capaldi, R. A., 1991, Structure–function relationship of domains of the δ subunit in *Escherichia coli* adenosine triphosphatase, *Biochim. Biophys. Acta* **1060**:115–124.

Noji, H., Yasuda, R., Yoshida, M., and Kinosita, K., 1997, Direct observation of the rotation of F_1-ATPase, *Nature* **386**:299–302.

Noumi, T., and Kanazawa, H., 1983, Mutants of *Escherichia coli* H^+-ATPase defective in the δ subunit of F_1 and the b subunit of F_o, *Biochem. Biophys. Res. Commun.* **111**:143–149.

Oda, T., Futaky, S., Kitagawa, K., Yoshihara, Y., Tani, I., and Higuti, T., 1989, Orientation of chargerin II (A6L) in the ATP synthase of rat liver mitochondria determined with antibodies against peptides of the protein, *Biochem. Biophys. Res. Commun.* **165**:449–456.

Ogilvie, I., Aggeler, R., and Capaldi, R. A, 1997, Crosslinking of the δ subunit to one of the three α subunit has no effect on functioning, as expected if δ is a part of the stator that links the F_1 and F_o parts of the *Escherichia coli* ATP synthase, *J. Biol. Chem.* **272**:16652–16656.

Orriss, G. L., Runswick, M. J., Collinson, I. R., Miroux, B., Fearnley, I. M., Skehel, J. M., and Walker, J. E., 1996, The δ and ε-subunits of bovine F_1-ATPase interact to form a heterodimeric subcomplex, *Biochem. J.* **314**:695–700.

Ovchinnikov, Y. A., Modyanov, N. N., Grinkevich, V. A., Aldanova, N. A., Trubetskaya, O. E., Nazimov, I. V., Hundal, T., and Ernster, L., 1984, Amino acid sequence of the oligomycin sensitivity-conferring protein (OSCP) of beef-heart mitochondria and its homology with the δ-subunit of the F_1-ATPase of *Escherichia coli*, *FEBS Lett.* **166**:19–22.

Pansini, A., Guerrieri, F., and Papa, S., 1978, Control of proton conduction by the H^+-ATPase in the inner mitochondrial membrane, *Eur. J. Biochem.* **92**:545–551.

Papa, S., Altendorf, K., Ernster, L., Packer, L., eds., 1984a, *H^+-ATPase (ATP Synthase): Structure, Function, Biogenesis. The F_oF_1 Complex of Coupling Membranes*, ICSU Press, Miami, FL.

Papa, S., Guerrieri, F., Zanotti, F., and Scarfò, R., 1984b, Flow and interactions of protons in the H^+-ATPase of mitochondria, in *Information and Energy Transduction in Biological Membranes* (C. L. Bolis, E. J. M. Helmreich, and H. Passow, eds.), pp. 187–197, Alan R. Liss, New York.

Papa, S., Guerrieri, F., Zanotti, F., Houstek, J., Capozza, G., and Ronchi, S., 1989, Role of the carboxyl-terminal region of the PVP protein (F_oI subunit) in the H^+ conduction of F_oF_1 H^+-ATP synthase of bovine heart mitochondria, *FEBS Lett.* **249**:62–66.

Papa, S., Guerrieri, F., Zanotti, F., Fiermonte, M., Capozza, G., and Jirillo, E., 1990, The γ subunit of F_1 and the PVP protein of F_o (F_oI) are components of the gate of the mitochondrial F_oF_1 H^+-ATPsynthase, *FEBS Lett.* **272**:117–120.

Papa, S., Zanotti, F., Cocco, T., Perrucci, C., Candita, C., and Minuto, M., 1996, Identification of functional domains and critical residues in the adenosintriphosphatase inhibitor protein of mitochondrial F_oF_1 ATP synthase, *Eur. J. Biochem.* **240**:461–467.

Paul, M. F., Guerin, B., and Velours, J., 1992, The C-terminal region of subunit 4 (subunit b) is essential for assembly of the F_o portion of yeast mitochondrial ATP synthase, *Eur. J. Biochem.* **205**:163–172.

Penefsky, H. S., 1985, Energy-dependent dissociation of ATP from high affinity catalytic sites of beef heart mitochondrial adenosine triphosphatase, *J. Biol. Chem.* **260**:13735–13741.

Penin, F., Archinard, P., Moradi-Améli, M., and Godinot, C., 1985, Stoichiometry of the oligomycin-sensitivity-conferring protein (OSCP) in the mitochondrial F_oF_1-ATPase determined by an immunoelectrotransfer blot technique, *Biochim. Biophys. Acta* **810**:346–353.

Perlin, D. S., Cox, D. N., and Senior, A. D., 1983, Integration of F_1 and the membrane-sector of the proton-ATPase of *Escherichia coli*, *J. Biol. Chem.* **258**:9793–9800.

Racker, E., and Horstmann, L. L., 1967, Partial resolution of the enzymes catalyzing oxidative phosphorylation, *J. Biol. Chem.* **242**:2547–2551.

Sabbert, D., Engelbrecht, S., and Junge, W., 1996, Intersubunit rotation in active F-ATPase, *Nature* **381**:623–625.

Schneider, E., and Altendorf, K., 1984, Subunit b of the membrane moiety (F_o) of ATP synthase (F_1F_o) from *Escherichia coli* is indispensable for H^+ translocation and binding of the water-soluble F_1 moiety, *Proc. Natl. Acad. Sci. USA* **81**:7279–7283.

Schneider, E., and Altendorf, K., 1985, Modification of subunit b of the F_o complex from *Escherichia coli* ATP synthase by a hydrophobic maleimide and its effects on F_o functions, *Eur. J. Biochem.* **153**:105–109.

Schneider, E., and Altendorf, K., 1987, Bacterial ATP synthase (F_1F_o): Purification and reconstitution of F_o complexes and biochemical and functional characterization of their subunits, *Microbiol. Rev.* **51**:477–497.

Walker, J. E., Saraste, M., and Gay, N. J., 1984, The *unc* operon nucleotide sequence, regulation and structure of ATP synthase, *Biochim. Biophys. Acta* **768**:164–200.

Walker, J. E., Fearnley, I. M., Gay, N. J., Gibson, B. W., Northrop, F. D., Powell, S. J., Runswick, M. J., Saraste, M., and Tybulewicz, W. L. J., 1985, Primary structure and subunit stoichiometry of F_1-ATPase from bovine mitochondria, *J. Mol. Biol.* **184**:677–701.

Walker, J. E., Runswick, M. J., and Poulter, L., 1987, ATP synthase from bovine mitochondria. The characterization and sequence analysis of two membrane-associated subunits and of the corresponding cDNAs, *J. Mol. Biol.* **197**:89–100.

Watts, S. D., Zhang, Y., Fillingame, R. H., and Capaldi, R. A., 1995, The γ subunit in the *Escherichia coli* ATP synthase complex (ECF_1F_o) extends through the stalk and contacts the c subunits of the F_o part, *FEBS Lett.* **368**:235–238.

Wilkens, S., Dahlquist, F. W., McIntosh, L. P., Donaldson, L. W., and Capaldi, R. A., 1995, Structural features of the ε subunit of the *Escherichia coli* ATP synthase determined by NMR spectroscopy, *Nature Struct. Biol.* **2**:961–967.

Wilkens, S., Dunn, S. D., Chandler, J., Dahlquist, F. W, and Capaldi, R. A, 1997, Solution structure of the N-terminal domain of the d subunit of the *E. coli* ATPsynthase, *Nature Struct. Biol.* **4**:198–201.

Xu, T., Candita, C., and Papa, S., 1996, The effect of mild trypsin digestion of F_1 on energy coupling in the mitochondrial ATP synthase, *FEBS Lett.* **397**:308–312.

Xu, T., Candita, C., Amoruso, G., and Papa, S., 1998, The N-termini of the α and β subunits at the top of F_1 stabilize the energy-transfer function in the mitochondrial F_1F_o ATP synthase, *Eur. J. Biochem.* **252**:155–161.

Zanotti, F., Guerrieri, F., Scarfò, R., Berden, J., and Papa, S., 1985, Effect of diamide on proton translocation by the mitochondrial H^+-ATPase, *Biochem. Biophys. Res. Commun.* **132**:985–990.

Zanotti, F., Guerrieri, F., Capozza, G., Houstek, J., Ronchi, S, and Papa, S., 1988, Identification of nucleus-encoded F_oI protein of bovine heart mitochondrial H^+-ATPase as a functional part of the F_o moiety, *FEBS Lett.* **237**:9–14.

Zanotti, F., Guerrieri, F., Capozza, G., Fiermonte, M., Berden, J., and Papa, S., 1992, Role of F_o and F_1 subunits in the gating and coupling function of mitochondrial H^+-ATP synthase. The effect of dithiol reagents, *Eur. J. Biochem.* **208**:9–16.

Zanotti, F., Guerrieri, F., Deckers-Hebestreit, G., Fiermonte, M., Altendorf, K., and Papa, S., 1994, Cross-reconstitution studies with polypeptides of *Escherichia coli* and bovine heart mitochondrial F_oF_1 ATP synthase, *Eur. J. Biochem.* **222**:733–741.

Zhang, S., and Fillingame, R. H., 1995, Subunits coupling H^+ transport and ATP synthesis in the

Escherichia coli ATP synthase. Cys–Cys crosslinking of F_1 subunit ε to the polar loop of F_o subunit c, *J. Biol. Chem.* **270**:24609–24614.

Zhou, S., Duncan, T. M., Bulygin, V. V., Hutcheon, M. L., and Cross, R. L., 1996, ATP hydrolysis by membrane-bound *Escherichia coli* F_oF_1 causes rotation of the gamma subunit relative to the beta subunits, *Biochim. Biophys. Acta* **1275**:96–100.

Ziegler, M., Xiao, R., and Penefsky, H. S., 1994, Close proximity of Cys[64] and Cys[140] in the δ subunit of *Escherichia coli* F_1-ATPase, *J. Biol. Chem.* **269**:4233–4239.

CHAPTER 20

The Mitochondrial Carrier Protein Family

Ferdinando Palmieri and Ben van Ommen

1. INTRODUCTION

Mitochondria and cytosol cooperate in a large number of metabolic cellular processes. Therefore, a continuous flux of metabolites, nucleotides, and cofactors into and out of the mitochondria is necessary. The outer mitochondrial membrane is permeable to small molecules, due to nonspecific pore proteins. The inner mitochondrial membrane, however, is impermeable to most solutes (and even to very small anions and cations such as Cl^- and H^+ ions), and therefore forms a barrier for the translocation of essential compounds. Specific carrier proteins inserted into the inner mitochondrial membrane have been discovered that facilitate the transport of a variety of metabolites. They may function as simple exchange proteins, but can also be involved in regulation and maintaining of physiological balances between cytosol and mitochondria necessary for proper cellular functioning. For example, the mitochondria supply ATP, generated by the oxidative phosphorylation pathway, to the complete cellular system. A transport protein is located in the inner mitochondrial membrane that exports ATP and imports ADP, the adenine nucleotide carrier.

Ferdinando Palmieri and Ben van Ommen • Department of Pharmaco-Biology, Laboratory of Biochemistry and Molecular Biology, University of Bari, I-70125 Bari, Italy.

Frontiers of Cellular Bioenergetics, edited by Papa *et al.* Kluwer Academic/Plenum Publishers, New York, 1999.

The inorganic phosphate necessary for the phosphorylation of ADP is transported into the mitochondria by the phosphate carrier, presumably acting as a phosphate–proton cotransporter. Other carriers have been identified for the transport of metabolites involved in the citric acid cycle, fatty acid oxidation, gluconeogenesis, lipogenesis, transfer of reducing equivalents, urea synthesis, amino acid degradation, and many other functions shared between cytosol and mitochondria.

The early work on mitochondrial carrier proteins was characterized by laborious purification of the transporters (reviewed by Krämer and Palmieri, 1989; Palmieri, 1994), which are present in minute quantities in the inner mitochondrial membrane, with the exception of the adenine nucleotide carrier and the uncoupling protein of brown adipose tissue. The purification was followed by functional reconstitution into artificial membranes and subsequent characterization. It became evident that mitochondrial metabolite carriers had a number of structural and functional features in common, and this resulted in the definition of the mitochondrial carrier family. The amino acid sequencing of some of these carriers and the elucidation of DNA sequences indicated a structural homology of the carriers, most likely evolved from a common ancestor. The first functional expression of a mitochondrial carrier (the 2-oxoglutarate transport protein) in *Escherichia coli* initiated a new era of carrier research due to the availability of large amounts of pure protein and the possibility to selectively modify the amino acid sequence. This chapter discusses recent developments in the mitochondrial metabolite transporter family research, with special attention to the structural aspects. For an overview of the earlier literature, the reader is referred to previous reviews (LaNoue and Schoolwerth, 1979; Schoolwerth and LaNoue, 1985; Walker, 1992; Krämer and Palmieri, 1992; Palmieri, 1994). Furthermore, a minireview series on specific mitochondrial anion transport systems has been published, edited by Pedersen (1993).

2. PRIMARY STRUCTURE AND EXTENSION OF THE FAMILY

At present, the primary structures of seven biochemically well-characterized mitochondrial carrier proteins have been reported from at least one species. These are the adenine nucleotide carrier (Aquila *et al.*, 1982), the uncoupling protein (Aquila *et al.*, 1985), the phosphate carrier (Runswick *et al.*, 1987), the 2-oxoglutarate carrier (Runswick *et al.*, 1990), the citrate carrier (Kaplan *et al.*, 1993), the carnitine carrier (Indiveri et al, 1997), and the dicarboxylate carrier (L. Palmieri *et al.*, 1996). The publications reporting the first sequence associated with a specific transport function are mentioned. The amino acid sequences of these proteins are presented in Fig. 1. The first six sequences

FIGURE 1. Primary structure and alignment of the seven sequenced mitochondrial carrier proteins. The sequences shown are the bovine T1 isoform of the adenine nucleotide carrier (Aquila *et al.*, 1982), the rat uncoupling protein (Aquila *et al.*, 1985), the bovine phosphate carrier (Runswick *et al.*, 1987), the bovine 2-oxoglutarate carrier (Runswick *et al.*, 1990), the rat citrate carrier (Kaplan et al, 1993), the rat carnitine carrier (Indiveri et al, 1997), and the yeast dicarboxylate carrier (L. Palmieri *et al.*, 1996). The asterisks indicate residues completely conserved or substituted in a conservative manner in at least six of the seven carriers. Abbreviations: AAC, bovine adenine nucleotide carrier; UCP, rat uncoupling protein; PiC, bovine phosphate carrier; OGC, bovine 2-oxoglutarate carrier; CIC, rat citrate carrier; CAC, rat carnitine carrier; DIC, yeast dicarboxylate carrier.

presented are derived from mammals, while the dicarboxylate carrier sequence is from *Saccharomyces cerevisiae*. The mammalian sequences were elucidated either by purification and sequencing of the full-length peptide chain or by molecular biological techniques initiating with partial amino acid sequence information. The last strategy was also adopted for cloning and sequencing of the carnitine carrier (Indiveri *et al.*, 1997), which is the most recently published structure of a mitochondrial carrier of known function. The dicarboxylate carrier sequence was established by overexpressing the *S. cerevisiae* DNA in *E. coli* and defining its function after reconstitution into liposomes (L. Palmieri *et al.*, 1996). All sequences encode a protein of approximately 30–34 kDa.

A characteristic feature of the sequences of the seven mitochondrial carriers of known function determined so far (Fig. 1) is the presence of three related internal domains. This striking pattern of homologous repeats was first identified (Saraste and Walker, 1982) in the published sequence of the adenine nucleotide carrier (Aquila *et al.*, 1982). Figure 2 is a dot-plot comparison of all seven sequences of known function demonstrating the three repetitive elements at about 100 amino acids distance. The repetitive domains of one carrier are related to those found in the other proteins. Each of the three related domains contains two hydrophobic regions, separated by hydrophilic parts. Comparison of the sequences of the seven mitochondrial transport proteins reveals major similarities. All carriers contain the characteristic sequence motif P-h-D/E-X-h-K/R-X-R/K-(20–30 residues)-D/E-G-(4 residues)-a-K/R-G, where h represents a hydrophobic and a represents an aromatic residue. This sequence motif is present in nearly all three related internal domains of all seven biochemically characterized carriers sequenced to date. The alignment of 21 sequences of the seven functionally characterized carriers reported in Fig. 1 shows that 68 residues out of a total of 314 (oxoglutarate carrier) are completely conserved or substituted in a conservative manner in at least six of the seven sequences. The sequence similarities, the tripartite organization, and the presence of six hydrophobic regions show that all seven mitochondrial transport proteins mentioned above belong to the same protein carrier family. Most likely, their structure originates from a common ancestor by two-tandem gene duplication.

Members of the mitochondrial carrier protein family by now have been identified in a large number of species. The adenine nucleotide carrier has been described in 24 species, yielding 39 different sequences including isoforms. These species include mammals, plants, yeast, protozoa, and insects. The 2-oxoglutarate carrier has been described in cow, man, rat, *Caenorhabditis elegans*, and millet (plant), with a total of seven sequences. The citrate carrier by now is known in five species (see this Section). The phosphate carrier protein has been described in man, cow, pig, rat, yeast, and *C. elegans*. The uncoupling protein is known in hamster, mouse, rat, sheep, rabbit, cow, and

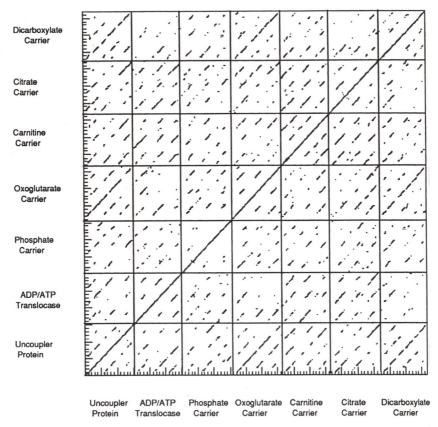

Uncoupler ADP/ATP Phosphate Oxoglutarate Carnitine Citrate Dicarboxylate
Protein Translocase Carrier Carrier Carrier Carrier Carrier

FIGURE 2. Pairwise dot–plot comparison of the seven mitochondrial carrier proteins of known function sequenced so far. The comparisons were made with the computer program DIAGON. The sequences used are cited in the legend of Fig. 1.

man. Most of the data on sequence information have been summarized by D. R. Nelson, and are available through the internet (http://drnelson.utmem.edu/homepage.html).

Interspecies comparison of a single carrier protein also reveals that these proteins are relatively highly conserved. For example, examination of the citrate carrier from five different species (human, bovine, rat, yeast, and the putative citrate carrier from *C. elegans*) indicates 83 amino acids that are totally conserved and 106 amino acids that are substituted in a highly conserved manner. Again, the identical amino acids are mostly located in the sequence motif that is characteristic for the mitochondrial transport protein family. The amino acid sequence of the bovine citrate carrier shows a 93.8% identity with rat, 94.5% with the human carrier, 63.0% with the putative *C. elegans* carrier, and 32.2% with the yeast carrier (Iacobazzi *et al.*, 1996a).

For at least two members of the carrier family, isoforms have been described. The adenine nucleotide carrier has three isoforms in man and yeast, encoded by different genes. The three human isoforms (T1, T2, and T3 or AAC1, AAC2, and AAC3), detected both at the level of DNA and expressed protein (reviewed by Walker and Runswick, 1993), are expressed differently in various tissues. The function of these isoforms is not yet clear. Also, for the phosphate carrier, two isoforms were found in man, cow, and rat (Dolce *et al.*, 1994). These two isoforms are not encoded by different genes, but are generated by alternative splicing of two exons, resulting in a change of 13 and 11 amino acids in man and cow, respectively. The two isoforms A and B of the phosphate carrier are tissue specific as demonstrated both at the transcript and protein level. Thus, by using two DNA probes that are specific for bovine exons IIIA and IIIB, respectively (Dolce *et al.*, 1996), and two antibodies that are specific for bovine isoforms A and B, respectively (Fiermonte *et al.*, 1997), it was shown that isoform A is expressed in large amounts only in bovine heart, skeletal muscle, and diaphragm, whereas isoform B is ubiquitously expressed in all bovine tissues examined, although at different levels. In order to investigate whether the two phosphate carrier isoforms show differences in the transport function, the two bovine isoforms were expressed in *E. coli* with a C-terminal His-tail, and subsequently purified and reconstituted into liposomes. Comparative studies showed that both isoforms act similarly in mediating the transmembrane phosphate–OH- and phosphate–phosphate exchanges. However, isoform B had a two- to threefold higher transport activity and a two- to threefold higher affinity toward phosphate as compared to isoform A (Fiermonte *et al.*, 1997).

Apart from the sequenced carrier proteins, three other proteins of the inner mitochondrial membrane have been purified and characterized as transport proteins after functional reconstitution into liposomes, but their sequences are as yet unknown. These are the aspartate–glutamate carrier (Bisaccia *et al.*, 1992), the pyruvate carrier (Bolli *et al.*, 1989), and the ornithine carrier (Indiveri *et al.*, 1992a). The molecular weight of these proteins is similar to those of the transporters with known sequence (32–34 kDa).

Furthermore, there is evidence for a number of other mitochondrial transport proteins, although no conclusive proof for their existence has been provided. A mitochondrial carrier that transports ATP-Mg^{2+} in exchange for phosphate has been described (Aprille, 1993), which is involved in charging the mitochondrial matrix with adenine nucleotides immediately after birth of mammals, i.e., during the adaptation to independent aerobic life. In contrast to the adenine nucleotide carrier, the ATP-Mg/Pi carrier does not exchange adenine nucleotides but increases the net content of adenine nucleotides of the mitochondria. This carrier has been kinetically characterized but not purified, and thus no structural data are available. Also, for the transport of branched-chain α-keto acids in and out of the mitochondria, kinetic data suggest the presence of a specific carrier protein differing from the pyruvate carrier already characterized (Patel *et al.*, 1980; Hutson and Rannels, 1985). Apart from the glutamate–aspartate exchange, as mediated by the purified carrier mentioned above, evidence also exists for a glutamate–H^+ symporter or glutamate–OH^- antiporter (Schoolwerth *et al.*, 1984). Glutamine is transported into the mitochondria in an electroneutral manner (Soboll *et al.*, 1991). Although glutamine itself is uncharged, the proton gradient across the mitochondrial membrane is the driving force for mitochondrial glutamine uptake, since its hydrolysis to glutamate in the matrix yields a proton. The transport of glutathione into the mitochondria also appears to be carrier-mediated (Kurosawa *et al.*, 1990). Microinjection of hepatic mRNA into oocytes resulted in augmented transport activity of glutathione (Fernandez-Checa *et al.*, 1996). Other carriers have been suggested for thiamine (Barile *et al.*, 1990), choline (Porter *et al.*, 1992), spermine (Toninello *et al.*, 1988), coenzyme A (Tahiliani and Neely, 1987), *N*-acetylglutamate (Meijer *et al.*, 1982), and FAD (Tzagoloff *et al.*, 1996).

Thus, the family of mitochondrial transport proteins consists of a number of biochemically characterized members, of which only ten have been purified and seven sequenced. The possible inclusion of the as yet not fully characterized carriers mentioned above into the mitochondrial carrier family awaits further structural information.

With the availability of DNA data banks, it has become possible to search for DNA sequence similarities in either completely or partially unraveled genomes. By now, the complete genome of *S. cerevisiae* is available. Comparison of the DNA sequence of the known members of the mitochondrial carrier family with the genomic DNA of *S. cerevisiae* resulted in the detection of 30 new putative members of unknown function and in the identification of the dicarboxylate carrier (L. Palmieri *et al.*, 1996) (see Section 5). Also, the genome of the nematode worm (*C. elegans*) is partially sequenced, and 11 sequences possibly encoding proteins related to the mitochondrial carrier family were encountered. The adenine nucleotide carrier, the phosphate carrier and the 2-oxoglutarare carrier appeared to be encoded by seven of these genes (Runswick *et al.*, 1994). Other putative mitochondrial carriers are a protein associated with Grave's disease (Zarrilli *et al.*, 1989; Fiermonte *et al.*, 1992), and a protein from the ciliated protozoa *Oxytricia fallax* (Williams and Herrick, 1991). It is clear that for proper functioning of cellular metabolism, a large number of compounds needs to be transported in and out of the mitochondria; consequently, the family of mitochondrial carrier proteins in the near future will undergo substantial extension. It is of interest that two other members of the mitochondrial carrier family, identified on the basis of the primary structure analogy, are not located in the inner mitochondrial membrane. These are the maize brittle 1 protein (Sullivan *et al.*, 1991), located in amyloplasts, and the PMP47 protein from *Candida boidinii*, located in peroxisomes (Jank *et al.*, 1993; Sakai *et al.*, 1996). These findings would indicate that metabolite carrier proteins of this family are not restricted to mitochondria, but represent a more extended family in eukaryotes.

3. SECONDARY STRUCTURE AND TRANSMEMBRANE TOPOLOGY

Based on the tripartite structure and the hydrophobic profile of the adenine nucleotide carrier, Saraste and Walker (1982) proposed that each repetitive domain consists of two transmembrane α-helices, with the complete protein thus forming a structure of six helices, referred to as helices I–VI. The two helices of the individual repetitive domains are linked by an extensive polar segment, named A, B, and C. The three repeats are connected by shorter peptide hydrophilic chains (a' and b'). Inherent to this model is the location of the C-terminus and the N-terminus on the same side of the inner mitochondrial membrane. Figure 3 demonstrates the proposed model and the nomenclature. Later, this model was adopted also for the other sequenced members of this carrier family, i.e., the phosphate carrier (Capobianco *et al.*, 1991), the 2-oxo-glutarate carrier (Runswick *et al.*, 1990; Bisaccia *et al.*, 1994), the carnitine carrier (Indiveri *et al.*, 1997), the uncoupling protein (Miroux *et al.*, 1993), and the citrate carrier (Kaplan *et al.*, 1993). Carrier research has focused on the confirmation of this hypothetical model, applying a number of biochemical techniques, among which are covalent labeling with substrate analogues or inhibitors (Dalbon *et al.*, 1988; Mayinger *et al.*, 1989), chemical modification (Bogner *et al.*, 1986), and use of impermeable reagents such as proteolytic enzymes and peptide-specific antibodies (Brandolin *et al.*, 1989; Capobianco *et al.*, 1991; Bisaccia *et al.*, 1994).

Indeed, recent developments in this area confirm the proposed structure. For example, Bisaccia et al. (1994), using a specific antibody raised against the C-terminus (residues 303–314) of the bovine heart 2-oxoglutarate carrier, demonstrated that the C-terminal region of the oxoglutarate carrier is exposed exclusively to the cytoplasmic side of the inner mitochondrial membrane, because the anti-C-terminal antiserum reacted with the oxoglutarate carrier in intact mitoplasts and its reactivity was not enhanced by permeabilization of the mitoplasts. Later, the same strategy was employed to identify the N-terminus of the oxoglutarate carrier on the same side as the C-terminus, i.e., the cytosolic side, by using antibodies directed to the N-terminal region of the bovine heart oxoglutarate carrier and intact mitoplasts (Bisaccia *et al.*, 1995). Furthermore, limited proteolysis using proteinase K was performed on the oxoglutarate carrier

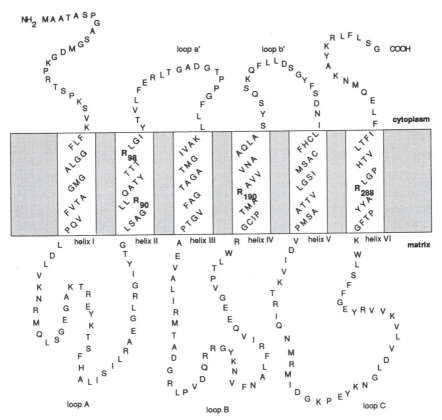

FIGURE 3. Proposed two-dimensional model of the 2-oxoglutarate carrier protein. This model is based on the hydrophobic profile and on immunochemical and enzymatic experimental results (Bisaccia *et al.*, 1994). The intramembrane segments are indicated as I–VI and are linked by extensive hydrophilic regions (A, B, and C) on the matrix side and by shorter loops (a' and b') on the cytosolic side.

incorporated in the membrane of proteoliposomes (Bisaccia *et al.*, 1994). In these vesicles the oxoglutarate carrier is oriented inside out compared to mitochondria (Indiveri *et al.*, 1991; Bisaccia *et al.*, 1994). Two types of proteoliposomes were used: "normal" and partially inverted proteoliposomes. In normal proteoliposomes, cleavage occurred only at loops A and B (after residues V39, Y61, F169, and R182) (see Fig. 3), while in inverted proteoliposomes, additional cleavages at the C-terminus, after residues A5 and S22 and at loop a' after T103, were found (Bisaccia *et al.*, 1994). Thus, it was concluded that loops A and B are located at the matrix side of the mitochondrial inner membrane, while the C-terminus and loop a' are exposed to the cytosolic side. For the phosphate carrier, it was shown that both the C- and N-terminus, as well as loops a' and b', are exposed to the cytosolic side, while loop B was found to be oriented toward the matrix (Capobianco *et al.*, 1991; Palmieri *et al.*, 1992a, 1993a). The same approach was chosen for the citrate carrier, again proving the positioning of the C- and N-terminus on the cytosolic side (Capobianco *et al.*, 1995).

The secondary structure of the uncoupling protein was investigated using ten antibodies raised against small peptide fragments obtained by expression of small subsequences of the uncoupling protein as fusion proteins (Miroux *et al.*, 1993). It was demonstrated that the first and third helices are oriented with their N-terminal extremities into the cytosol, while the second and fourth helices are oriented with their N-terminal extremities into the matrix. The C-terminal region of the uncoupling protein (Eckerskorn and Klingenberg, 1987) and the N-terminal region of the adenine nucleotide carrier (Brandolin *et al.*, 1989) are also exposed on the cytosolic side of the mitochondrial membrane. After stabilizing the conformation of the latter carrier by the selective inhibitor bongkrekic acid, the loops A, B, and C were accessible to proteases from the matrix side of the inner mitochondrial membrane (Marty *et al.*, 1991). Using the membrane-impermeable fluorescent sulfhydryl reagent eosin 5-maleimide (EMA) and the membrane-permeable *N*-ethylmaleimide (NEM), Majima et al. (1993) were able to discriminate between the four cysteines present in the adenine nucleotide carrier. Three of the four cysteines (C56, C159, and C256) are located in loops A, B, and C respectively, while C128 is located in the third putative transmembrane helix. Indeed, using submitochondrial particles, in which the membrane is predominantly inverted as compared to mitochondria, C128 did not react either with EMA or NEM, whereas C159 reacted very rapidly and C56 and C256 more slowly with EMA. It was concluded that C128 is buried in the membrane and all three EMA-reacting cysteines are located on the matrix side of the mitochondrial membrane, thus confirming the six-helical structure of this carrier, with the termini located on the cytosolic side of the membrane.

It is interesting that ADP prevented the labeling of the three "loop" cysteines (Majima *et al.*, 1994), and the specific transport inhibitor, bongkrekic acid, which is thought to bind to the carrier only at the matrix side, inhibited the labeling of C159 and C256 but not that of C56, when added on the matrix side but not when added on the cytosolic side. The other well-known inhibitor of the adenine nucleotide carrier, carboxyactractyloside, inhibited EMA binding only if added from the cytosolic side. Therefore, it was suggested that the ADP transport involves regulatory action of the loops containing C159 and C256, and that these loops may intrude into the membrane as hairpin structures during the conformational change occurring in the transport process.

Indeed, this hypothesis had been proposed before, since the segments of the polypeptide chain of the adenine nucleotide carrier spanning residues F153–M200 and Y250–M281 (belonging to the matrix loops B and C in the model) are covalently labeled by two nonpermeant specific ligands, azido atractyloside and azido (α-^{32}P)ADP, respectively, when these are added from the cytoplasmic side of the membrane (Brandolin *et al.*, 1993). Similarly, loop C of the uncoupling protein was shown to be photolabeled by 2-azido ADP (Winkler and Klingenberg, 1992) and loop A of the phosphate carrier to react, at the level of C41 (rat liver), with the impermeable eosin maleimide (Houstek and Pedersen, 1985). Further studies on the ADP–ATP carrier, applying a cross-linking technique using the copper-o-phenanthroline complex, also indicated the importance in the transport activity of the first matrix-oriented loop, containing C56, since only this cysteine appeared to be involved in dimerization. Of the two inhibitors mentioned above, only carboxyactractoside prevented the cross-linking (Majima *et al.*, 1995), suggesting that also this loop, although located on the matrix side, may protrude into the membrane during transport activity.

4. EXPRESSION

The research on mitochondrial carrier proteins, being integral membrane proteins, until recently could not profit from the molecular biological techniques of cloning and expression in *E. coli* that have been so helpful in the structural research of many globular soluble proteins. The first mitochondrial carrier protein expressed in *E. coli* was the bovine 2-oxoglutarate carrier (Fiermonte *et al.*, 1993). Besides the high level of protein expression, a major breakthrough was made by solubilizing the protein, which accumulated in inclusion bodies in the bacteria, using the detergent sarcosyl, and reconstituting the solubilized protein in its functionally active form. Actually, this was the first time that a eukaryotic membrane protein had been overexpressed in *E. coli* and renatured. Similar protocols have subsequently been used for the expression of the phosphate carrier (Wohlrab and Briggs, 1994; Fiermonte *et al.*, 1997) and the citrate carrier (Kaplan *et al.*, 1995) in *E. coli*. Apart from the expression of the mitochondrial carrier proteins in *E. coli*, the yeast system also has been used for expression and site-directed mutagenesis studies (for example, see Phelps and Wohlrab, 1991; Nelson and Douglas, 1993). By now, the expression technique is being widely applied in the carrier research area and allows for the application of specific mutations and the use of large amounts of protein to be used in structural and functional studies.

5. THE YEAST GENOME: 35 CARRIERLIKE SEQUENCES

The yeast genome by now has been fully sequenced, thus allowing for sequence comparison of this genome with the typical mitochondrial carrier protein internal sequence repeat, as mentioned in Section 2. Thus, 35 genes encoding for "putative" mitochondrial carrier family members could be identified and have been summarized in Table I, in collaboration with Dr. J. E. Walker. Only a few of these genes encoded for known yeast mitochondrial carrier proteins. Among these is the adenine nucleotide carrier, which is represented by three genes, referred to as AAC1, AAC2, and AAC3, of which only the AAC1 is normally expressed. Also, the phosphate carrier had already been identified to be encoded by the gene MIR1. This sequence has 38% homology in amino acid sequence with YER053c (V-C1). However, this latter protein seems not to be expressed or is expressed in a nonfunctional form, since yeast deleted of MIR1 is not viable (Murakami *et al.*, 1990). Last, the citrate carrier gene was already known (Kaplan *et al.*, 1995). The other 29 sequences had no known function; they might code for the other transport proteins that have already been characterized biochemically, as well as for carriers of yet as unknown function. It should be noted, however, that not all of these 29 sequences necessarily code for unique carrier proteins. First, it is possible that a number of genes express isoforms of a single carrier. Apart from the example presented above, this has been noted for the two genes MRS3 and MRS4 (Wiesenberger *et al.*, 1991), which are 75% identical and therefore are likely to be isoforms. Second, it is possible that not all the sequences are expressed into functionally active proteins.

The sequence II-C5 (Table I), encoding the yeast citrate carrier, has a 38% homology with the citrate carrier sequence from rat (Kaplan *et al.*, 1995). In general, however, this homology approach is not sufficient to identify the yeast putative carriers of unknown function, since for most of the 29 sequences with unknown function the percentage of homology is too low (lower than 30%) to allow for accurate conclusions. Therefore, the strategy to be adopted for the identification is to express each gene product in *E. coli* and define its function after reconstitution into liposomes. Using this strategy, the yeast dicarboxylate carrier (XII-C1 in Table I), which had already been purified and biochemically characterized from rat liver (Bisaccia *et al.*, 1988) and also from yeast (Lancar-Benba *et al.*, 1996), but not yet sequenced, has now been identified in our laboratory in collaboration with Dr. J. E. Walker (L. Palmieri *et al.*, 1996). Further extension of this approach will assign a function to some of the carrierlike genes and either identify them as known members of the family or reveal them to be new members, thus extending the carrier protein family. Subsequently, the information obtained in yeast will help to identify analogous genes in eukaryotic species including man.

6. GENOMIC STRUCTURE AND EVOLUTION

The gene sequence and organization of the adenine nucleotide carrier in *Neurospora crassa*, man, maize, *Chlamydomonas*, and wheat (Arrends and

Table I

Location in the Genomic Sequence of *S. cerevisiae* of the Coding Sequences
for Members of the Family of Mitochondrial Carrier Proteins[a]

Chromosome	Strand	Position	Identity or acronym[b]	Standard nomenclature
II	−	163,956–163,000	II-C1 (AAC2)	YBLO30 c
II	+	415,937–416,860	II-C2 (AAC3)	YBRO85 w
II	+	449,622–450,611	II-C3 (YMC2)	YBR104 w
II	+	607,606–608,739	II-C4 (RIM2)	YBR192 w
II	−	784,528–783,629	II-C5 (citrate carrier)	YBR291 c
IV	−	104,552–103,650	IV-C1 (YMH1)	YDL198 c
IV	−	247,611–246,688	IV-C2	YDL119 c
IV	−	1,401,248–1,399,740	IV-C3	YDR470 c
V	+	144,326–145,333	V-C1	YEL006 w
V	−	259,638–258,736	V-C2	YER053 c
VI	+	242,450–242,986	VI-C1	YFR045 w
VII	+	676,615–677,559	VII-C1	YGR096 w
VII	−	1,007,301–1,006,201	VII-C2	YGR257 c
VIII	+	108,805–109,878	VIII-C1	YHR002 w
IX	+	97,395–98,330	IX-C1 (FLX1)	YIL134 w
IX	+	344,059–345,180	IX-C2 (SC4 554)	YIL006 w
X	+	160,317–161,261	X-C1 (MRS3)	YJL133 w
X	−	577,878–576,943	X-C2 (phosphate carrier)	YJR077 c
X	+	609,464–610,432	X-C3 (ACR1)	YJR095 w
XI	+	216,990–217,964	XI-C1 (PMT1)	YKL120 w
XI	−	533,104–532,190	XI-C2 (MRS4)	YKR052 c
XII	−	827,869–926,973	XII-C1 (dicarboxylate carrier)	YLR348 c
XIII	−	388,243–387,314	XIII-C1 (AAC1)	YMR056 c
XIII	−	594,472–593,366	XIII-C2	YMR166 c
XIII	+	751,960–752,904	XIII-C3	YMR241 w
XIV	+	471,376–472,860	XIV-C1	YNL083 w
XIV	−	625,828–624,971	XIV-C2 (PET8)	YNL003 c
XV	−	514,276–513,291	XV-C1	YOR100 c
XV	−	570,805–569,927	XV-C2 (ARG 11)	YOR130 c
XV	+	758,327–759,250	XV-C3	YOR222 w
XVI	−	299,502–298,570	XVI-C1 (LPI11)	YPL134 c
XVI	−	584,036–583,056	XIV-C2	YPR011 c
XVI	−	603,352–600,644	XVI-C3	YPR021 c
XVI	+	673,765–674,888	XVI-C4 (YMC1)	YPR058 w
XVI	−	792,199–791,213	XVI-C5	YPR128 c

[a]The data were obtained from the Saccharomyces Genome Database (SGD) located in the Department of
Genetics, School of Medicine, Stanford University.
[b]AAC, ADP–ATP carrier.

Sebald, 1984; Cozens *et al.*, 1989; Bathgate *et al.*, 1989; Sharpe and Day, 1993; Iacobazzi *et al.*, 1996b), the uncoupling protein in mouse and man (Kozak *et al.*, 1988; Cassard *et al.*, 1990), the oxoglutarate (Iacobazzi *et al.*, 1992) and phosphate (Dolce *et al.*, 1994) carriers in man and cow, and the citrate carrier in man (Iacobazzi *et al.*, 1997) have been determined. Their genes are relatively small, ranging from about 3.0 kb (ADP–ATP carrier, oxoglutarate carrier, and citrate carrier) to about 13 kb (uncoupling protein), with the phosphate carrier having an intermediate length of about 9.0 kb. Despite their relatively small size, the genes for the mitochondrial carriers are characterized by a relatively high number of introns (ranging from 5 to 8 in the human genes, with the exception of those for the adenine nucleotide carrier isoforms).

Another feature of the genes coding for mitochondrial carrier proteins is the fact that the introns are found in equivalent positions, indicating that the structures of these genes are closely related. For example, they all contain an intron after the first and an intron before the sixth transmembrane segment. Furthermore, there is a tendency for the introns to interrupt the coding sequence in or near to the extramembranous loops (Cozens *et al.*, 1989; Kozak *et al.*, 1988; Iacobazzi *et al.*, 1992; Dolce *et al.*, 1994). In the human citrate carrier gene (Iacobazzi *et al.*, 1997), the first intron is located in the middle of the loop between the first and the second transmembrane segment, the second intron is at the beginning of the second loop, the third and the fourth introns are at the beginning and the end of the third hydrophilic loop, the fifth and the sixth introns are before and after the fifth transmembrane segment, and the seventh intron is just at the beginning of the sixth transmembrane segment (Fig. 4).

The tendency of the introns to interrupt the sequences coding for the extramembrane loops rather than those encoding the transmembrane α-helices presumably reflects the course of evolution of the carrier genes. This course includes two tandem gene duplications of an ancestral gene encoding the 100 amino acid repeat. The 100 amino acid repeat itself may originate from an earlier duplication of a single hydrophobic α-helix, although there is no evident sequence similarity to support the latter proposal. The introns in the coding sequences for the extramembrane parts may be relics of these earlier fusion events.

The chromosomal mapping of some of the mitochondrial carrier genes in humans has been reported (Table II).

7. STRUCTURE AND FUNCTION STUDIES: THE USE OF MUTANT PROTEINS

Site-directed mutagenesis of specific amino acids has proved to be a valuable tool in the research into the functioning and the three-dimensional

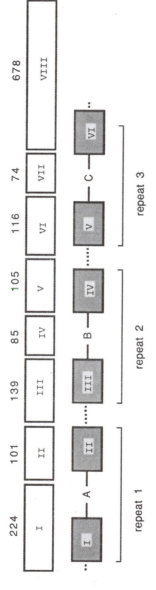

FIGURE 4. Alignment of the exons encoding the human citrate carrier gene with a secondary structural model of the mitochondrial carrier protein family. The exons are represented by white boxes containing Roman numerals, and their sizes are presented above. The gaps between the boxes indicate the positions of the introns in the human citrate carrier gene. The shaded boxes I–VI represent hydrophobic segments that may be folded into transmembrane helices. The repeats 1, 2, and 3 represent the three related domains of the mitochondrial carriers.

Table II
Chromosomal Localization of Mitochondrial
Carrier Genes in Humans

Gene	Chromosome	Reference
Uncoupling protein	4q31	Cassard *et al.* (1990)
Adenine nucleotide carrier (T1)	4	Li *et al.* (1989)
Phosphate carrier	12q23	Marsh *et al.* (1995)
Citrate carrier	22q11	Heisterkamp *et al.* (1995)
Carnitine carrier	3p21.31	Viggiano *et al.* (1997)

structure of proteins and enzymes. As soon as the mitochondrial carriers could be functionally expressed, this technique was embraced by a number of groups to study the various carriers.

7.1. Role of Cysteines

Cysteine was one of the first amino acids on which attention was focused. Previous studies on a number of mitochondrial carriers had indicated that cysteines might be involved in the transport activity, since chemical modification of sulfhydryl groups often resulted in inhibition of transport. In general, the mitochondrial carrier protein family appeared to contain two types of cysteines, the modification of the first group resulting in inhibition of the transport and the modification of the second group resulting in a reversible change of transport function from coupled antiport to uniport, characterized by a more or less complete loss of substrate specificity (Dierks *et al.*, 1990; Indiveri *et al.*, 1992b; Palmieri *et al.*, 1992b, 1993b). For the 2-oxoglutarate carrier, chemical modifications had indicated a possible role in transport of cysteine residues due to the inhibitory action of mercurial agents (Quagliariello and Palmieri, 1972), whereas maleimides did not inhibit the transport activity (Bisaccia *et al.*, 1985, 1988). Subsequently, each of the three cysteines (C221 and C224, supposed to be located in the fifth transmembrane helix, and C184, located in the fourth transmembrane segment close to the matrix side in mitochondria and to the external surface in proteoliposomes) were substituted for a serine residue either individually or all three together. None of the four mutants showed significantly different transport activity as compared to the wild-type carrier (F. Palmieri *et al.*, 1996, and unpublished results). Thus, the inhibition obtained after chemical modification of each of these three cysteines either results from sterical hindrance or from conformational change. Anyway, the sulfhydryl groups of the oxoglutarate carrier play no role in the mechanism of transport. Other studies have shown that the reactivity and/or the accessibility of one of the cysteines of the oxoglutarate carrier (C184) for SH reagent is increased by incubation with the substrates oxoglutarate or malate (Capobianco *et al.*, 1996).

The presence of the substrates increased the binding of pyrenylmaleimide to the reconstituted carrier as well as the degree of inhibition of the reconstituted transport activity caused by pyrenylmaleimides, other maleimides, and mercurials. This result is consistent with the assumption that substrate binding causes a change in the tertiary structure of the carrier protein which involves the region of C184. Furthermore, the two buried cysteines C221 and C224 form a disulfide bridge in the native protein, as was shown by chemical modification with pyrenylmaleimide and phenylmaleimide under reducing and nonreducing conditions (Bisaccia *et al.*, 1996a). Another interesting aspect of the cysteine residues of the 2-oxoglutarate carrier is that, using the disulfide-forming reagent Cu^{2+}-phenanthroline, a dimer of the carrier is formed due to cross-linking of the two cysteines 184. Extending this observation, it was observed that the rate of cross-linking is independent of the protein concentration, implying that the dimeric configuration is already present before the reaction (Bisaccia *et al.*, 1996b). Thus, the 2-oxoglutarate carrier exists as a dimeric structure.

The same arguments were used by Klingenberg and Appel (1989) as evidence for the dimeric state of the uncoupling protein in the solubilized state. It remains to be established with certainty that this configuration is obligatory for the transport function for these carriers and perhaps for the other members of the family. Also, for the adenine nucleotide carrier, cysteine residues essential for transport activity have been suggested (Hoffmann *et al.*, 1994). In this case, site-directed mutagenesis indicated that three out of the four cysteines could be replaced by serine residues without a significant change in transport activity. With the substitution of C73 by serine, full restoration of activity in proteoliposomes was achieved only after the addition of cardiolipin to phospholipids during reconstitution (Klingenberg and Nelson, 1994).

Last, in the case of the uncoupling protein, sequential replacement of each cysteine by a serine did not affect either its uncoupling activity or the regulation of proton transport by nucleotides or fatty acids (Arechaga *et al.*, 1993). It therefore was established that none of the seven cysteines present in the uncoupling protein was essential for its proper functioning, and that the results obtained by chemical modification were presumably due to steric effects.

7.2. Role of Charged Residues

Since most mitochondrial carrier proteins transport charged compounds through the membrane, it is obvious that a number of charged residues will be of extreme importance in forming a hydrophilic channel. All seven sequenced carriers contain a number of positively charged residues (arginine or lysine) in their transmembrane helices. These positively charged residues have been the target of some mutation studies. In the adenine nucleotide carrier, which contains an arginine in each second helix of the threefold repeated structure, each of these three arginines has been substituted (R96H, R204L, and R294A). Also, each of the arginines of the arginine triplet R252–R253–R254 present in the matrix loop C has been substituted for an isoleucine. The yeast cells expressing the mutated adenine nucleotide carrier presented a very poor functioning in various respects (glycerol-negative growth, very low ATP synthesis rate in mitochondria, and decreased respiratory capacity). The reconstituted mutant proteins had highly reduced ADP–ADP and ADP–ATP exchanges, indicating involvement of these arginines in the exchange mechanism. The mutant R294A, however, still showed significant ADP–ADP exchange, whereas the ATP–ADP exchange was extremely low. It was concluded that this arginine residue is essential for the transport involving an electrical imbalance, but not for the electronically neutral ADP–ADP exchange (Nelson et al., 1993; Klingenberg and Nelson, 1994; Heidkämper et al., 1996; Müller et al., 1996). Interestingly, of all mutations tested for the adenine nucleotide carrier (some 50 in total), the common feature of the mutations that resulted in the inability to grow on glycerol plates was a loss of charge. All other mutations resulted in viable cells (Nelson, 1996). These glycerol-negative mutants were used to select second-site mutants. Spontaneous second-site mutations in the arginine 294-less mutant (R294A) involved the loss of a negative charge at position 45, or the change of a neutral amino acid into another neutral residue. Therefore, it was concluded that the negative charge at position 45 served as a complementary charge for the positively charged R294. Mutation of R96 into six different amino acids did not result in the loss of a negative charge, although the structural model constructed for the adenine nucleotide carrier might suggest D149 to be the complementary negative charge for R96 (Nelson, 1996).

Four arginine residues also are located in the intramembrane region of the 2-oxoglutarate carrier (R90 and R98 in the second transmembrane helix, R190 in the fourth transmembrane helix, and R288 in the sixth transmembrane helix). Using the expression system in E. coli, all these positively charged residues were individually substituted for a leucine residue, with subsequent loss of the oxoglutarate–oxoglutarate exchange, while only the R288L mutant retained a very low transport activity toward malate. Mutating these arginine residues into lysine residues resulted in loss of activity in the case of R190 and R98, indicating that a positive charge either from lysine or arginine at the positions 90 and 288 suffices for transport activity, but that the arginines at position 190 and 98 are essential (F. Palmieri et al., 1996, and unpublished results). Unlike the adenine nucleotide transporter, the 2-oxoglutarate carrier does not contain any negatively charged residue in the transmembrane helices, excluding the possibility of interhelical stabilization due to charge pairs. Therefore, in order to neutralize the positively charged residues, either negatively charged residues located on one of the loops need to penetrate the core of the protein or the negatively charged substrates are necessary to complement the positively charged amino acid residues. Interestingly, arginine-specific chemical modification of the 2-oxoglutarate carrier also results in loss of activity (Stipani *et al.*, 1996). In total, the 2-oxoglutarate carrier was mutated at 56 positions in our laboratory. Of these, only seven resulted in significant loss of transport activity, all of which were charged residues.

The phosphate carrier has been the target of a number of mutations involving negatively charged amino acid residues (Phelps *et al.*, 1996). Five charged residues (H32 in helix 1, E126 and E137 in helix 3, and D39 and D236 at the matrix ends of the first and fifth helices) were found to be indispensable for normal growth in yeast and transport function. A model was proposed in which two phosphate carrier protein molecules form a homodimeric protein in which a channel is formed, walled by helices 1 and 3 of both subunits. The six charged residues (H32, E126, and E137 of both subunits) are proposed to form the proton cotransport channel. The earlier results that indicated an inhibitory disulfide formation under oxidizing conditions between the two C28 residues (also located in the first helix) of two phosphate carrier molecules are in agreement with this model (Phelps and Wohlrab, 1993).

7.3. Role of Tryptophan Residues

Recently, the tryptophan residues of the yeast adenine nucleotide carrier have been the subject of mutational studies. One of the classical approaches in the biophysical research into the structure of proteins has been the study of the intrinsic tryptophan fluorescence. Since most proteins contain more than one tryptophan, however, decisive conclusions were difficult. The three tryptophan residues (W87, W126, and W235) of the adenine nucleotide carrier were individually mutated to their tyrosine counterparts. After determining their proper functioning (Le Saux *et al.*, 1996), these mutants were applied in fluorescence studies involving binding of substrates and inhibitors. Conformational changes, known to occur after binding of either ATP, bongkrekic acid, or carboxyatractyloside, were reflected in changes of tryptophan fluorescence, indicating the tryptophans to be suitable intrinsic reporters for conformational changes (Roux *et al.*, 1996). It is clear that the combination of mutant proteins and these biophysical techniques can be exploited to a great extent in the structural research into the mitochondrial carrier protein family.

7.4. Mutations that Affect Carrier Regulation

The uncoupling protein uncouples the mitochondrial respiration from ATP production in brown adipose tissue by introducing a proton pathway through the inner mitochondrial membrane, bypassing the ATP synthase, and thus explaining the high thermogenic activity of this tissue. This carrier is regulated both by free fatty acids, which increase transport activity, and by purine nucleotides, which inhibit the proton transport (Rial *et al.*, 1983). The uncoupling protein also displays chloride channel properties, with the chloride transport inhibited by nucleotides (Jezek *et al.*, 1994; Huang and Klingenberg, 1996). The nucleotide binding site has been postulated to be located in helix six (Mayinger and Klingenberg, 1992). Bouillaud *et al.* (1994) have identified three amino acids (F267–K268–G269), the deletion of which resulted in loss of regulatory capacity of nucleotides. Interestingly, this sequence was determined by comparison of the uncoupling protein sequence with the adenine nucleotide carrier and the estrogen receptor, of which a DNA (nucleotide) binding region was already identified. Meanwhile, it was established that mutation of R279 into L or Q also abolished the inhibitory effect of nucleotide binding to the uncoupling protein (Murdza-Inglis *et al.*, 1994).

The adenine nucleotide carrier appears not only to act as a highly selective transport protein, but can be switched into a channel-like pore protein. The channel opening is proposed to be caused by calcium binding to cardiolipin, tightly bound to the adenine nucleotide carrier, thus releasing positive charges within the carrier that open the gate (Brustovetsky and Klingeberg, 1996). This channel protein may be the previously described mitochondrial permeability transition pore involved in the ischemia–reperfusion damage and cytosolic calcium oscillations.

8. BIOGENESIS OF MITOCHONDRIAL METABOLITE CARRIERS

The carrier proteins of the mitochondrial inner membrane are encoded by nuclear genes and are typically synthesized as mature-sized proteins on cytosolic polysomes. In only a few cases, such as the phosphate carrier of mammalian mitochondria (Zara *et al.*, 1991), the carrier proteins are synthesized with cleavable N-terminal extensions (presequences). Studies on the role of the presequence of the mitochondrial phosphate carrier have shown that the major import information must reside in the mature part of the phosphate carrier, as in the case of the other presequence-deficient members of the mitochondrial carrier family (Zara *et al.*, 1992).

The mechanisms of targeting to and translocation into mitochondria have mainly been studied with fungal mitochondria, from *S. cerevisiae* and *Neurospora crassa*. The carrier proteins are recognized by the receptor Tom70 (for "translocase of the outer mitochondrial membrane 70-kDa subunit") on the mitochondrial surface and are subsequently transferred to a second receptor Tom22 (Söllner *et al.*, 1990; Hines *et al.*, 1990; Dietmeier *et al.*, 1993; Kiebler *et al.*, 1993). Then they are transported across the outer membrane by a general import pore (GIP) (Kiebler *et al.*, 1990) and interact with a *trans* site at the intermembrane space side (Pfanner *et al.*, 1987). Translocation into the inner membrane requires a membrane potential, but is not mediated by the general import machinery that translocates presequence-carrying preproteins (Pfanner *et al.*, 1994).

Sirrenberg *et al.* (1996) identified a protein of the inner mitochondrial membrane, termed Tim22, that is required for insertion of the precursors of the ADP–ATP carrier and phosphate carrier into the inner membrane. Tim22 is part of a larger complex that seems to mediate insertion of carrier proteins. The import pathway of inner membrane carrier proteins is thus distinct from the general import pathway of presequence-carrying preproteins in a number of aspects. Common elements for both pathways are the central receptor Tom22, the general import pore of the outer membrane, and the requirement for a membrane potential across the inner membrane. In the few cases where carrier proteins are synthesized with a presequence, the exact mechanisms of targeting and translocation are not yet clarified.

The yeast phosphate carrier/p32 has been proposed to function as an import receptor on the outer mitochondrial membrane (Pain *et al.*, 1990). However, this is unlikely in view of the following three observations: (1) the submitochondrial localization and the mechanism of import (Dietmeier *et al.*, 1993), (2) the characterization of the yeast mutant lacking the phosphate carrier/p32 for transport properties and preproteins import (Zara *et al.*, 1996), and (3) the functional expression of the phosphate carrier protein in yeast (Phelps and Wohlrab, 1991).

9. THE CARNITINE CARRIER: A FATAL GENETIC DISORDER

The carnitine carrier transports acylcarnitine esters into the mitochondria, in exchange for free carnitine, an essential step in the process of long-chain fatty acid β oxidation. A deficiency of this carrier has been described (reviewed by Pande and Murthy, 1994). So far, several case reports have been published describing neonatal death with the main features being hypoketotic hypoglycemia, mild hyperammonemia, variable dicarboxylic aciduria, and hepatomegaly with abnormal liver functions. In all cases the carnitine carrier transport activity, as measured in cultured skin fibroblasts, was very low. However, on the molecular level the deficiency had not been characterized until recently.

In fact, the rat carnitine carrier cDNA was cloned and sequenced (Indiveri *et al.*, 1997). Using this sequence and a human cDNA as template, the human cDNA of the carnitine carrier was constructed, coding for a 301-amino acid protein with a 90% sequence identity as compared to the rat protein (Huizing *et al.*, 1997a). Using Northern blot analysis, it was observed that the carnitine carrier was expressed with different expression levels in various tissues (high levels in heart, skeletal muscle, and liver). Sequencing of the cDNA of a patient with a carnitine carrier deficiency revealed a homozygous cytosine nucleotide insertion. This insertion provoked a frameshift and an extension of the open reading frame with 23 novel codons (Huizing *et al.*, 1997a). In a second patient we found two heterozygous deletions in the carnitine carrier cDNA, one in each allele, resulting in a drastically mutated carrier protein. In both mutations, a premature stop codon was detected, as a result of a frameshift. This resulted in the absence of the C-terminal part of the protein: only 139 or 211 of the 301 amino acids were translated in the two mutations observed, respectively (Huizing *et al.*, 1997b). Consequently, a nonfunctioning carrier was expressed. So far, the frequency of these inborn errors is not established, but the availability of the probes and techniques allows for a reliable (and prenatal) diagnosis of patients at risk.

For the adenine nucleotide carrier, circumstantial evidence has been presented for a possible deficiency, related to cardiomyopathy (Schultheiss, 1992; Bakker *et al.*, 1993). In addition, patients with a deficiency in the aspartate–malate shuttle (Hayes *et al.*, 1987; Brivet *et al.*, 1997), the ornithine carrier (Inoue *et al.*, 1988), the pyruvate carrier (Selak *et al.*, 1997), and the protein import machinery (Schapira *et al.*, 1990) have been reported. However, it must be stressed that in all these cases the molecular defect at the genomic level has not yet been identified.

10. CONCLUSIONS AND PERSPECTIVES

As discussed in this chapter, the mitochondrial carrier protein research in the last few years has made substantial progress in a number of areas. The overexpression in yeast and more particularly in *E. coli* and subsequent functional incorporation into liposomes have been a major breakthrough. As a result, mutated proteins could be studied in detail, revealing mechanistic information. Also, a shift has taken place from classical purification to the application of molecular biological techniques, which made grateful use of the availability of genome data banks. This has resulted in the rapid expression and identification of some mitochondrial carrier proteins, and it is to be expected that in the near future, following the strategy described in Section 5, a fairly complete range of carrier proteins will be described in yeast. The progress in the human genome data bank will result in a logical extension of this work. Also, the work on the carnitine carrier has demonstrated the importance of the molecular biological techniques in the application of the fundamental knowledge into the medical area. The genetic disorders or possible polymorphisms need to be charted.

The research area that is still awaiting its breakthrough is the structural biochemistry of the mitochondrial carrier proteins, closely related to the understanding of the transport mechanism. Being integral membrane proteins, until now it has been impossible to obtain either crystals or a sufficiently high protein density in artificial membranes to allow for electron microscopic analysis of its structure (two-dimensional crystallization), as has been successfully applied for a small number of other integral membrane proteins. Until this milestone has been achieved, the structural work will depend for the major part on the application of biophysical and biochemical techniques, making use of selective probes incorporated in the proteins at strategic locations by means of specific mutations. This type of work has already been applied in the structural research of some other carrier proteins, gradually giving insight into their three-dimensional structure and functioning.

ACKNOWLEDGMENTS. The research in the authors' laboratory was supported by grants from the Ministero dell'Università e della Ricerca Scientifica e Tecnologica (MURST) and from the Italian Consiglio Nazionale delle Ricerche (CNR). The financial support of the CNR Target Project on "Biotechnology" in particular is gratefully acknowledged. F. P. is indebted to Dr. J. E. Walker and Dr. N. Pfanner for their continuous support and collaboration.

11. REFERENCES

Adrian, G. S., McCammon, M. T., Montgomery, D. L., and Douglas, M. G., 1986, Sequences required for delivery and localization of the ADP/ATP translocator to the mitochondrial inner membrane, *Mol. Cell. Biol.* **6**:626–634.

Aprille, J. R., 1993, Mechanism and regulation of the mitochondrial ATP-Mg/Pi carrier, *J. Bioenerg. Biomembr.* **25**:473–481.

Aquila, H., Misra, D., Eulitz, M., and Klingenberg, M., 1982, Complete amino acid sequence of the ADP/ATP carrier from beef heart mitochondria, *Hoppe-Seyler's Physiol. Chem.* **363**:345–349.

Aquila, H., Link, T. A., and Klingenberg, M., 1985, The uncoupling protein from brown fat mitochondria is related to the mitochondrial ADP/ATP carrier. Analysis of sequence homologies and of folding of the protein in the membrane, *EMBO J.* **4**:2369–2376.

Arechaga, I., Raimbault, S., Prieto, S., Levi-Meyrueis, C., Zaragoza, P., Miroux, B., Ricquier, D., Bouillaud, F., and Rial, E., 1993, Cysteine residues are not essential for uncoupling protein function, *Biochem. J.* **296**:693–700.

Arends, H., and Sebald, W., 1984, The nucleotide sequence of the cloned mRNA and gene of the ADP/ATP carrier from *Neurospora crassa*, *EMBO J.* **3**:377–382.

Bakker, H. D., Scholte, H. R., Van der Bogert, C., Ruitenbeek, W., Jeneson, J. A. L., Wanders, R. J.

A., Abeling, N. G. G. M., Dorland, B., Sengers, R. C. A., and Van Gennip, A. H., 1993, Deficiency of the adenine nucleotide translocator in muscle of a patient with myopathy and lactic acidosis: A new mitochondrial defect, *Pediatr. Res.* **33**:412–417.

Barile, M., Passarella, S., and Quagliarello, E., 1990, Thiamine pyrophosphate uptake into isolated rat liver mitochondria, *Arch. Biochem. Biophys.* **280**:352–357.

Bathgate, B., Baker, A., and Leaver, C. J., 1989, Two genes encode the adenine nucleotide translocator of maize mitochondria. Isolation, characterization and expression of the structural genes, *Eur. J. Biochem.* **183**:303–310.

Bisaccia, F., Indiveri, C., and Palmieri, F., 1985, Purification of reconstitutively active a-oxoglutarate carrier from pig heart mitochondria, *Biochim. Biophys. Acta* **810**:362–369.

Bisaccia, F., Indiveri, C., and Palmieri F., 1988, Purification and reconstitution of two carriers from rat liver mitochondria: The dicarboxylate and the 2-oxoglutarate carrier, *Biochim. Biophys. Acta* **933**:229–240.

Bisaccia, F., De Palma, A., and Palmieri, F., 1992, Identification and purification of the aspartate/glutamate carrier from bovine heart mitochondria, *Biochim. Biophys. Acta* **1106**:291–296.

Bisaccia, F., Capobianco, L., Brandolin, G., and Palmieri, F., 1994, Transmembrane topography of the mitochondrial oxoglutarate carrier assessed by peptide-specific antibodies and enzymatic cleavage, *Biochemistry* **33**:3705–3713.

Bisaccia, F., Capobianco, L., Mazzeo, M., De Palma, A., and Palmieri, F., 1995, Further insight into the structural properties of the mitochondrial oxoglutarate carrier, in *Progress in Cell Research*, vol. 5 (F. Palmieri *et al.*, eds.), pp. 95–100, Elsevier, Amsterdam.

Bisaccia, F., Capobianco, L., Mazzeo, M., and Palmieri, F., 1996a, The mitochondrial oxoglutarate carrier protein contains a disulfide bridge between intramembranous cysteines 221 and 224, *FEBS Lett.* **392**:54–58.

Bisaccia, F., Zara, V., Capobianco, L., Iacobazzi, V., Mazzeo, M., and Palmieri, F., 1996b, The formation of a disulfide cross-link between the two subunits demonstrates the dimeric structure of the mitochondrial oxoglutarate carrier, *Biochim. Biophys. Acta* **1292**:281–288.

Bogner, W., Aquila, H., and Klingenberg, M., 1986, The transmembrane arrangement of the ADP/ATP carrier as elucidated by the lysine reagent pyridoxal 5-phosphate, *Eur. J. Biochem.* **161**:611–620.

Bolli, R., Nalecz, K. A., and Azzi, A., 1989, Monocarboxylate and a-ketoglutarate carriers from bovine heart mitochondria. Purification by affinity chromatography as immobilized 2-cyano-4-hydroxycinamate, *J. Biol. Chem.* **264**:18024–18030.

Bouillaud, F., Arechaga, I., Petit, P. X., Raimbault, S., Levi-Meyrueis, C., Casteilla, L., Laurent, M., Rial, E., and Ricquier, D., 1994, A sequence related to a DNA recognition element is essential for the inhibition by nucleotides of proton transport through the mitochondrial uncoupling protein, *EMBO J.* **13**:1990–1997.

Brandolin, G., Boulay, F., Dalbon, P., and Vignais, P. V., 1989, Orientation of the N-terminal region of the membrane-bound ADP/ATP carrier protein explored by antibodies and an arginine-specific endoprotease: Evidence that the accessibility of the N-terminal residues depends on the conformational state of the carrier, *Biochemistry* **28**:1093–1100.

Brandolin, G., Le Saux, A., Trezeguet, V., Lauquin, G. J. M., and Vignais, P. V., 1993, Chemical, immunological, enzymatic, and genetic approaches to studying the arrangement of the peptide chain of the ADP/ATP carrier in mitochondrial membrane, *J. Bioenerg. Biomembr.* **25**:459–472.

Brivet, M., Slama, A., Rustin, P., Poggi, F., Boutron, A., Rabier, D., Munnich, A., Saudubray, J. M., and Legrand, A., 1997, A mitochondrial encephalomyopathy with a presumptive defect at the level of aspartate/malate shuttle, abstract P69 in *Abstract Book 7th International Congress of Inborn Errors of Metabolism*, Vienna.

Brustovetsky, N., and Klingenberg, M., 1996, Mitochondrial ADP/ATP carrier can be reversibly converted into a large channel by Ca^{2+}, *Biochemistry* **35**:8483–8488.

Capobianco, L., Brandolin, G., and Palmieri, F., 1991, Transmembrane topography of the mitochondrial phosphate carrier explored by peptide-specific antibodies and enzymatic digestion, *Biochemistry* **30**:4963–4969.

Capobianco, L., Bisaccia, F., Michel, A., Sluse, F. E., and Palmieri, F., 1995, The N- and C-temini of the tricarboxylate carrier are exposed to the cytoplasmic side of the inner mitochondrial membrane, *FEBS Lett.* **357**:297–300.

Capobianco, L., Bisaccia, F., Mazzeo, M., and Palmieri, F., 1996, The mitochondrial oxoglutarate carrier: Sulphydryl reagents bind to cysteine-184, and this interaction is enhanced by substrate binding, *Biochemistry* **35**:8974–8980.

Cassard, A.M., Bouillard, F., Mattei, M.G., Hentz, E., Raimbault, S., Thomas, M., and Ricquier, D., 1990, Human uncoupling protein gene: Structure, comparison with rat gene, and assignment to the long arm of chromosome 4, *J. Cell. Biochem.* **43**:255–264.

Cozens, A. L., Runswick, M. J., and Walker, J. E., 1989, DNA sequence of two expressed nuclear genes for human mitochondrial ADP/ATP translocase, *J. Mol. Biol.* **206**:261–280.

Dalbon, P., Brandolin, G., Boulay, F., Hoppe, J., and Vignais, P. V., 1988, Mapping of the nucleotide binding sites in the ADP/ATP carrier of beef heart by photolabeling with 2-azido-(^{32}P)adenosine diphosphate, *Biochemistry* **27**:5141–5149.

Dierks, T., Salentin, A., and Krämer, R., 1990, Pore-like and carrier-like properties of the mitochondrial aspartate/glutamate carrier after modification by SH-reagents: Evidence for a preformed channel as a structural requirement of carrier-mediated transport, *Biochim. Biophys. Acta* **1028**:281–288.

Dietmeier, K., Zara, V., Palmisano, A., Palmieri, F., Voos, W., Schlossmann, J., Moczko, M., Kispal, G., and Pfanner, N., 1993, Targeting and translocation of the phosphate carrier/p32 to the inner membrane of yeast mitochondria, *J. Biol. Chem.* **268**:25958–25964.

Dolce, V., Iacobazzi, V., Palmieri, F., and Walker, J. E., 1994, The sequences of human and bovine genes of the phosphate carrier from mitochondria contain evidence of alternatively spliced forms, *J. Biol. Chem.* **269**:10451–10460.

Dolce, V., Fiermonte, G., and Palmieri, F., 1996, Tissue-specific expression of the two isoforms of the mitochondrial phosphate carrier in bovine tissues, *FEBS Lett.* **399**:95–98.

Eckerskorn, C., and Klingenberg, M., 1987, In the uncoupling protein from brown adipose tissue the C-terminus protrudes to the c-side of the membrane, as shown by tryptic cleavage, *FEBS Lett.* **226**:166–170.

Fernandez-Checa, J. C., Yi, J.-R., Ruiz, C. G., Ookhtens, M., and Kaplowitz, N., 1996, Plasma Membrane and Mitochondrial transport of hepatic reduced glutathione, *Semin. Liver Dis.* **16**:147–157.

Fiermonte, G., Runswick, M. J., Walker, J. E., and Palmieri, F., 1992, Sequence and pattern of expression of a bovine homologue of a human mitochondrial transport protein associated with Grave's disease, *DNA Sequence* **3**:71–78.

Fiermonte, G., Walker, J. E., and Palmieri, F., 1993, Abundant bacterial expression and reconstitution of an intrinsic membrane-transport protein from bovine mitochondria, *Biochem. J.* **294**: 293–299.

Fiermonte, G., Dolce, V., and Palmieri, F., 1997, Expression in *E. coli*, functional characterization and tissue distribution of isoforms A and B of the phosphate carrier from bovine mitochondrial, *J. Biol Chem.* **273**:22782–22787.

Hayes, D. J., Taylor, D. J., Bore, P. J., Hilton-Jones, D., Arnold, D. L., Squier, M. V., Gent, A. E., and Radda, G. K., 1987, An unusual metabolic myopathy: A malate-aspartate shuttle defect, *J. Neurol. Sci.* **82**:27–39.

Heidkämper, D., Müller, V., Nelson, D. R., and Klingenberg, M., 1996, Probing the role of positive residues in the ADP/ATP carrier from yeast. The effect of six arginine mutations on transport and the four ATP versus ADP exchange modes. *Biochemistry* **35**:16144–16152.

Heisterkanp, N., Mulder, M. P., Langeveld, A., Ten Hoeve, J., Wang, Z., Rof, B. A., and Groffen J., 1995, Localization of human mitochondrial citrate transporter protein gene to chromosome 22Q11 in the DiGeorge syndrome critical region, *Genomics* **29**:451–456.

Hines, V., Brandt, A., Griffiths, G., Horstmann, H., Brütsch, H., and Schatz, G., 1990, Protein import into yeast mitochondria is accelerated by the outer membrane protein MAS70, *EMBO J.* **9**:3191–3200.

Hoffmann, B., Stöckl, A., Schlame, B., Beyer, K., and Klingenberg, M., 1994, The reconstituted ADP/ATP carrier has an absolute requirement for cardiolipin as shown in cysteine mutants, *J. Biol. Chem.* **269**:1940–1944.

Houstek, J., and Pedersen, P. L., 1985, Adenine nucleotide and phosphate transport systems of mitochondria, *J. Biol. Chem.* **260**:6288–6295.

Huang, S.-G., and Klingenberg, M., 1996, Chloride channel properties of the uncoupling protein from brown adipose tissue mitochondria: A patch clamp study, *Biochemistry* **35**:16806–16814.

Huizing, M., Iacobazzi, V., IJlst, L., Savelkoul, P., Ruitenbeek, W., van den Heuvel, L., Indiveri, C., Smeitink, J., Trijbels, F., Wanders, R., and Palmieri, F., 1997a, Cloning of the human carnitine–acylcarnitine carrier cDNA, and identification of the molecular defect in a patient, *Am. J. Hum. Genet.* **61**:1239–1245.

Huizing, M., Wendel, U., Ruitenbeek, W., Iacobazzi, V., IJlst, L., Veenhuizen, P., Savelkoul, P., van den Heuvel, L. P., Smeitink, J. A. M., Wanders, R., Trijbels, F., and Palmieri, F., 1997b, Carnitine–acylcarnitine carrier deficiency: Identification of the molecular defect in a patient, *J. Inherit. Metab. Dis.* **21**:262–267.

Hutson, S. M., and Rannels, S. L., 1985, Characterization of a mitochondrial transport system for branched chain α-keto acids, *J. Biol. Chem.* **260**:14189–14193.

Iacobazzi, V., Palmieri, F., Runswick, M. J., and Walker, J. E., 1992, Sequences of the human and bovine genes for the mitochondrial 2-oxoglutarate carrier, *DNA Sequence* **3**:79–88.

Iacobazzi, V., De Palma, A., and Palmieri, F., 1996a, Cloning and sequencing of the bovine cDNA encoding the mitochondrial tricarboxylate carrier protein, *Biochim. Biophys. Acta* **1284**:9–12.

Iacobazzi, V., Poli, A., Blanco, A., and Palmieri, F., 1996b, Nucleotide sequences of two genes (Accession Nos. X95863 for ANT-G1 and X95864 for ANT-G2) encoding the adenine nucleotide translocator of wheat mitochondria (PGR 96-016), *Plant Physiol.* **110**:1435–1436.

Iacobazzi, V., Lauria, G., and Palmieri, F., 1997, Organization and sequence of the human gene for the mitochondrial citrate transport protein, *DNA Sequence* **7**:127–139.

Indiveri, C., Dierks, T., Krämer, R., and Palmieri, F., 1991, Reaction mechanism of the reconstituted oxoglutarate carrier from bovine heart mitochondria, *Eur. J. Biochem.* **198**:339–347.

Indiveri, C., Tonazzi, A., and Palmieri, F., 1992a, Identification and purification of the ornithine/citrulline carrier from rat liver mitochondria, *Eur. J. Biochem.* **207**:449–454.

Indiveri, C., Tonazzi, A., Dierks, T. Krämer, R., and Palmieri, F., 1992b, The mitochondrial carnitine carrier: Characterization of SH-groups relevant for its transport function, *Biochim. Biophys. Acta* **1140**:53–58.

Indiveri, C., Iacobazzi, V., Giangregorio, N., and Palmieri, F., 1997, The mitochondrial carnitine carrier protein: cDNA cloning, primary structure, and comparison with other mitochondrial transport proteins, *Biochem. J.* **321**:713–719.

Inoue, I., Saheki, T., Kayanuma, K., Uono, M., Nakajima, M., Takeshita, K., Koike, R., Yuasa, T., Miyatake, T., and Sakoda, K., 1988, Biochemical analysis of decreased ornithine transport activity in the liver mitochondria from patients with hyperornithinemia, hyperammonemia and homocitrullinuria, *Biochim. Biophys. Acta* **964**:90–95.

Jank, B., Habermann, B., Schweyen, R. J., and Link, T. A., 1993, PMP47, a peroxisomal homologue of mitochondrial solute carrier proteins, *Trends Biochem. Sci.* **18:**427–428.

Jezek, P., Orosz, D. E., Modriansky, M., and Garlid, K. D., 1994, Transport of anions and protons by the mitochondrial uncoupling protein and its regulation by nucleotides and fatty acids, *J. Biol. Chem.* **269:**26184–26190.

Kaplan, R. S., Mayor, J. A., and Wood, D. O., 1993, The mitochondrial tricarboxylate transport protein: cDNA cloning, primary structure, and comparison with other mitochondrial transport protein, *J. Biol. Chem.* **268:**13682–13690.

Kaplan, R. S., Mayor, J. A., Gremse, D. A., and Wood, D. O., 1995, High level expression and characterization of the mitochondrial citrate transport protein from the yeast *Saccharomyces cerevisiae*, *J. Biol. Chem.* **270:**4108–4114.

Kiebler, M., Pfaller, R., Söllner, T., Griffiths, G., Horstmann, H., Pfanner, N., and Neupert, W., 1990, Identification of a mitochondrial receptor complex required for recognition and membrane insertion of precursor proteins, *Nature* **348:**610–616.

Kiebler, M., Kell, P., Schneider, H., van der Klei, I. J., Pfanner, N., and Neupert, W., 1993, The mitochondrial receptor complex: A central role of MOM22 in mediating preprotein transfer from receptors to the general insertion pore, *Cell* **74:**483–492.

Klingenberg, M., and Appel, M., 1989, The uncoupling protein dimer can form a disulfide crosslink between the mobile SH-groups, *Eur. J. Biochem.* **180:**123–131.

Klingenberg, M., and Nelson, D. R., 1994, Structure–function relationships of the ADP/ATP carrier, *Biochim. Biophys. Acta* **1187:**241–244.

Kozak, L. P., Britton, J. H., Kozak, U. C., and Wells, J. M., 1988, The mitochondrial uncoupling protein gene. Correlation of exon structure to transmembrane domains, *J. Biol. Chem.* **263:** 12274–12277.

Krämer, R., and Palmieri, F., 1989, Molecular aspects of isolated and reconstituted carrier proteins from animal mitochondria, *Biochim. Biophys. Acta* **974:**1–23.

Krämer, R., and Palmieri, F., 1992, Metabolite carriers in mitochondria, in *Molecular Mechanisms in Bioenergetics* (L. Ernster, ed.), pp. 359–384, Elsevier, Amsterdam.

Kurosawa, K., Hayashi, N., Sato, N., Kamada, T., and Tagawa, K., 1990, Transport of glutathione across the mitochondrial membranes, *Biochem. Biophys. Res. Commun.* **167:**367–372.

Lancar-Benba, J., Foucher, B., and Saint-Macary, M., 1996, Characterization, purification and properties of the yeast mitochondrial diacrboylate carrier (*Saccharomyces cerevisiae*), *Biochemie* **78:**195–200.

LaNoue, K. F., and Schoolwerth, A. C., 1979, Metabolite transport in mitochondria, *Annu. Rev. Biochem.* **48:**871–922.

Le Saux, A., Roux, P., Trézéguet, V., Fiore, C., Schwimmer, C., Dianoux, A.-C., Vignais, P. V., Brandolin, G., and Lauquin, G. J.-M., 1996, Conformational changes of the yeast mitochondrial adenosine diphosphate/adenosine triphosphate carrier studies through its intrinsic fluorescence. 1. Tryptophanyl residues of the carrier can be mutated without impairing protein activity, *Biochemistry* **35:**16116–16124.

Li, K., Warner, C. K., Hodge, J. A., Minoshima, S., Kudoh J., Fukuyama, R., Maekawa, M., Shimizu, Y., Shimizu, N., and Wallace, D. C., 1989, A human muscle adenine nucleotide translocator gene has four exons. Is located on chromosome 4, and is differentially expressed, *J. Biol. Chem.* **264:**13998–14004.

Majima, E., Koike, H., Hong, Y., Shinohara, Y., and Terada, H., 1993, Characterization of cysteine residues of mitochondrial ADP/ATP carrier with the SH-reagents eosin 5-maleimide and *N*-ethylmaleimide, *J. Biol. Chem.* **268:**22181–22187.

Majima, E., Shinohara, Y., Yamaguchi, N., Hong, Y., and Terada, H., 1994, Importance of loops of mitochondrial ADP/ATP carrier for its transport activity deduced from reactivities of its cysteine residues with the sulphydryl reagent eosin 5 maleimide, *Biochemistry* **33:**9530–9536.

Majima, E., Ikawa, K., Takeda, M., Hashimoto, M., Shinohara, Y., and Terada, H., 1995, Translocation of loops regulates transport activity of mitochondrial ADP/ATP carrier deduced from
formation of a specific intermolecular disulfide bridge catalyzed by copper-o-phenanthroline,
J. Biol. Chem. **270:**29548–29554.

Marsh, S., Carter, N. P., Dolce, V., Iacobazzi, V., and Palmieri, F., 1995, Chromosomal localization
of the mitochondrial phosphate carrier gene PHC to 12q23, *Genomics* **29:**814–815.

Marty, I., Brandolin, G., and Vignais, P. V., 1991, Topography of the membrane bound ADP/ATP
carrier assessed by enzymatic proteolysis, *Biochemistry* **31:**4058–4065.

Mayinger, P., and Klingenberg, M., 1992, Labeling of two different regions of the nucleotide
binding site of the uncoupling protein from brow adipose tissue mitochondria with two ATP
analogs, *Biochemistry* **31:**10536–10543.

Mayinger, P., Winkler, E., and Klingenberg, M., 1989, The ADP/ATP carrier from yeast 4AAC27 is
uniquely suited for the assignment of the binding center by photoaffinity labeling, *FEBS Lett.*
244:421–426.

Meijer, A. J., Van Woerkom, G. M., Wanders, R. J. A., and Lof, C., 1982, Transport of
N-acetylglutamate in rat-liver mitochondria, *Eur. J. Biochem.* **124:**325–330.

Miroux, B., Froscard, V., Raimbault, S., Ricquier, D., and Bouillaud, F., 1993, The topology of the
brown adipose tissue mitochondrial uncoupling protein determined with antibodies against its
antigenic sites revealed by a library of fusion proteins, *EMBO J.* **12:**3739–3745.

Müller,V., Basset, G., Nelson, D. R., and Klingenberg M., 1996, Probing the role of positive
residues in the ADP/ATP carrier from yeast. The effect of six arginine mutations on oxidative
phosphorylation and AAC expression. *Biochemistry* **35:**16132–16143.

Murakami, H., Blobel, G., and Pain, D., 1990, Isolation and characterization of a gene for a yeast
mitochondrial import receptor, *Nature* **347:**488–491.

Murdza-Inglis, D., Modriansky, M., Patel, H. V., Woldegriorgis, G., Freeman, K. B., and Garlid,
K.D., 1994, A single mutation in uncoupling protein of rat brown adipose tissue mitochondria
abolishes GDP sensitivity of H$^+$ transport, *J. Biol. Chem.* **269:**7435–7438.

Nelson, D. R., 1996, The yeast ADP/ATP carrier. Mutagenesis and second-site revertants, *Biochem.
Biophys. Acta* **1275:**133–137.

Nelson, D. R., and Douglas, G., 1993, Function-based mapping of the yeast mitochondrial ADP/
ATP translocator by selection for second site revertants, *J. Mol. Biol.* **230:**1171–1182.

Nelson, D. R., Lawson, J. E., Klingenberg, M., and Douglas, M. G., 1993, Site-directed mutagenesis of the yeast mitochondrial ADP/ATP translocator. Six arginines and one lysine are
essential. *J. Mol. Biol.* **230:**1159–1170.

Pain, D., Murakami, H., and Blobel, G., 1990, Identification of a receptor for protein import into
mitochondria, *Nature* **347:**444–449.

Palmieri, F., 1994, Mitochondrial carrier proteins, *FEBS Lett.* **346:**48–54.

Palmieri, F., Bisaccia, F., Capobianco, L., Dolce, V., Iacobazzi, V., Indiveri, C., and Zara, V., 1992a,
Structural and functional properties of two mitochondrial transport proteins: The phosphate
carrier and the oxoglutarate carrier, in *Molecular Mechanisms of Transport* (E. Quagliariello
and F. Palmieri, eds.), pp. 151–158, Elsevier, Amsterdam.

Palmieri, F., Bisaccia, F., Iacobazzi, V., Indiveri, C., and Zara, V., 1992b, Mitochondrial substrate
carriers, *Biochim. Biophys. Acta* **1101:**223–227.

Palmieri, F., Bisaccia, F., Capobianco, L., Dolce, V., Fiermonte, G., Iacobazzi, V., and Zara, V.,
1993a, Transmembrane topology, genes and biogenesis of the mitochondrial phosphate and
oxoglutarate carriers, *J. Bioenerg. Biomembr.* **25:**493–501.

Palmieri, F., Indiveri, C., Bisaccia, F., and Krämer, R., 1993b, Functional properties of purified and
reconstituted mitochondrial metabolite carriers, *J. Bioenerg. Biomembr.* **25:**525–535.

Palmieri, F., Bisaccia, F., Capobianco, L., Dolce, V., Fiermonte, G., Iacobazzi, V., Indiveri, C., and

Palmieri, L., 1996, Mitochondrial metabolite transporters, *Biochim. Biophys. Acta* **1275:** 127–132.

Palmieri, L., Palmieri, F., Runswick, M. J., and Walker, J. E., 1996, Identification by bacterial expression and functional reconstitution of the yeast genomic sequence encoding the mitochondrial dicarboxylate carrier protein, *FEBS Lett.* **399:**299–302.

Pande, S. V., and Murthy, M. S. R., 1994, Carnitine-acylcarnitine translocase deficiency: Implications in human pathology, *Biochim. Biophys. Acta* **1226:**269–276.

Patel, T. B., Waymack, P. P., and Olson, M. S., 1980, The effect of the monocarboxylate translocator inhibitor, α-cyanocinnamate, on the oxidation of the branched chain α-keto acids in rat liver, *Arch. Biochem. Biophys.* **201:**629–635.

Pedersen, P. L., series ed., 1993, Mitochondrial anion transport systems, *J. Bioenerg. Biomembr.* **25:**431–545.

Pfanner, N., Tropschug, M., and Neupert, W., 1987, Mitochondrial protein import: Nucleoside triphosphates are involved in conferring import-competence to precursors, *Cell* **49:**815–823.

Pfanner, N., Craig, E. A., and Meijer, M., 1994, The protein import machinery of the mitochondrial inner membrane, *Trends Biochem. Sci.* **19:**368–372.

Phelps, A., and Wohlrab, H., 1993, Cys28 of the mitochondrial phosphate transport protein is responsible for the inhibition of transport by oxygen, *FASEB J.* **7:**321.

Phelps, A., Briggs, C., Mincone, L., and Wohlrab, H., 1996, Mitochondrial phosphate transport protein. Replacements of glutamic, aspartic, and histidine residues affect transport and protein conformation and point to a coupled proton transport path, *Biochemistry* **35:**10757–10762.

Porter, R. K., Scott, J. M., and Brand, M. D., 1992, Choline transport into rat liver mitochondria. Characterization and kinetics of a specific transporter, *J. Biol. Chem.* **267:**14637–14646.

Quagliariello, E., and Palmieri, F., 1972, Kinetics of substrate uptake by mitochondria. Identification of carrier sites for substrates and inhibitors, in *Biochemistry and Biophysics of Mitochondrial Membranes* (G. F. Azzone *et al.*, eds.), pp. 659–680, Academic Press, New York.

Rial, E., Poustie, A., and Niccholls, D. G., 1983, Brown-adipose-tissue mitochondria: The regulation of the 32000 Mr uncoupling protein by fatty acids and purine nucleotides, *Eur. J. Biochem* **137:**197–203.

Roux, P., Le Saux, A., Trézéguet, V., Schwimmer, C., Dianoux, A.-C., Vignais, P.V., Lauquin, G. J.-M., and Brandolin, G., 1996, Conformational changes of the yeast mitochondrial adenosine diphosphate/adenosine triphosphate carrier studies through its intrinsic fluorescence. 2. Assignment of tryptophanyl residues of the carrier to the responses to specific ligands, *Biochemistry* **35:**16125–16131.

Runswick, M. J., Powell, S. J., Nyren, P., and Walker, J. E., 1987, Sequence of the bovine mitochondrial phosphate carrier protein: Structural relationship to ADP/ATP translocase and the brown fat mitochondria uncoupling protein, *EMBO J.* **6:**1367–1373.

Runswick, M. J., Walker, J. E., Bisaccia, F., Iacobazzi, V., and Palmieri, F., 1990, Sequence of the bovine 2-oxoglutarate/malate carrier protein: Structural relationship to other mitochondrial transport proteins, *Biochemistry* **29:**11033–11040.

Runswick, M. J., Philippides, A., Lauria, G., and Walker, J. E., 1994, Extension of the mitochondrial transporter super-family: Sequences of five members from the nematode worm, *Caernorhabditis elegans*, *DNA Sequence* **4:**281–291.

Sakai, Y., Saiganji, A., Yurimoto, H., Takabe, K., Saiki, H., and Kato, N., 1996, The absence of Pmp47, a putative yeast peroxisomal transporter, causes a defect in transport and folding of a specific matrix enzyme, *J. Cell Biol.* **134:**37–51.

Saraste, M., and Walker, J. E., 1982, Internal sequence repeats and the path of polypeptide in mitochondrial ADP/ATP translocase, *FEBS Lett.* **144:**250–254.

Schapira, A. H. V., Cooper, J. M., Morgan-Hughes, J. A., Landon, D. N., and Clark, J. B., 1990, Mitochondrial myopathy with a defect of mitochondrial protein transport, *N. Engl. J. Med.* **323**:37–42.

Schoolwerth, A. C., and LaNoue, K. F., 1985, Transport of metabolic substrates in renal mitochondria, *Annu. Rev. Physiol.* **47**:143–171.

Schoolwerth, A. C., LaNoue, K. F., and Hoover, W. J., 1984, Effect of pH on glutamate efflux from rat kidney mitochondria, *Am. J. Physiol.* **246**:F266–271.

Schultheiss, H. P., 1992, Dysfunction of the ADP/ATP carrier as causative factor for the disturbance of the myocardial energy metabolism in dilated cardiomyopathy, *Basic Res. Cardiol.* **87**(Suppl. 1):311–320.

Selak, M. A., Grover, W. D., Foley, C. M., Miles, D. K., and Salganicoff, L. 1997, Possible defect in pyruvate transport in skeletal muscle mitochondria from four children with encephalopathies and myopathies. Abstract No. 59, International Conference of Mitochondrial Diseases: "Challenges in the Study of Encephalomyopathies of Mitochondrial Origin," Philadelphia.

Sharpe, J. A., and Day, A., 1993, Structure, evolution and expression of the mitochondrial ADP/ATP translocator gene from *Chlamylomonas reinhardii*, *Mol. Gen. Genet.* **237**:134–144.

Sirrenberg, C., Bauer, M. F., Guiard, B., Neupert, W., and Brunner, M., 1996, Import of carrier proteins into the mitochondrial inner membrane mediated by Tim22, *Nature* **384**:582–588.

Soboll, S., Lenzen, C., Rettich, D., Gründel, S., and Ziegler, B., 1991, Characterization of glutamate uptake in rat liver mitochondria, *Eur. J. Biochem.* **197**:113–117.

Söllner, T., Pfaller, R., Griffiths, G., Pfanner, N., and Neupert, W., 1990, A mitochondrial import receptor for the ADP/ATP carrier, *Cell* **62**:107–115.

Stipani, I., Mangiullo, G., Stipani, V., Daddabbo, L., Natuzzi, D., and Palmieri, F., 1996, Inhibition of the reconstituted mitochondrial oxoglutarate carrier by arginine-specific reagents, *Arch. Biochem. Biophys.* **331**:48–54.

Sullivan, T. D., Strelow, L. I., Illingworth, C. A., Phillips, R. L., and Nelson Jr., O. E., 1991, Analysis of maize brittle 1 alleles and a defective suppressor-mutator-induced mutable allele, *Plant Cell* **3**:1337–1348.

Tahiliani, A. G., and Neely, J. R., 1987, A transport system for coenzyme A in isolated rat heart mitochondria, *J. Biol. Chem.* **262**:11607–11610.

Toninello, A., Miotto, G., Siliprandi, D., Siliprandi, N., and Garlid, K., 1988, On the mechanism of spermine transport in liver mitochondria, *J. Biol. Chem.* **263**:19407–19411.

Tzagoloff, A., Jang, J., Glerum, D. M., and Wu, M., 1996, FLX1 codes for a carrier protein involved in maintaining a proper balance of flavin nucleotides in yeast mitochondria, *J. Biol. Chem.* **271**:7392–7397.

Viggiano, L., Iacobazzi, V., Marzella, R., Cassano, C., Rocchi, M., and Palmieri, F., 1997, Assignment of the carnitine/acylcarnitine translocase gene (CACT) to human chromosome band 3p21.31 by *in situ* hybridization, *Cytogenet. Cell. Genet.* **79**:62–63.

Walker, J. E., 1992, The mitochondrial transporter family, *Curr. Opin. Struct. Biol.* **2**:519–526.

Walker, J. E., and Runswick, M. J., 1993, The mitochondrial transport protein superfamily, *J. Bioenerg. Biomembr.* **25**:435–446.

Wiesenberger, G., Link, T. A., von Ahsen, U., Waldherr, M., and Schweyer, R. J., 1991, MRS3 and MRS4, two suppressors of mtRNA splicing defects in yeast, are new members of the mitochondrial carrier family, *J. Mol. Biol.* **217**:23–27.

Williams, K. R., and Herrick, G., 1991, Expression of the gene encoded by a family of macronuclear chromosomes generated by alternative DNA processing in *Oxytrichia fallax*, *Nucleic Acids Res.* **19**:4717–4724.

Winkler, E., and Klingenberg, M., 1992, Photoaffinity labeling of the nucleotide binding site of the uncoupling protein from hamster brown adipose tissue, *Eur. J. Biochem.* **203**:295–304.

Wohlrab, H., and Briggs, C., 1994, Yeast mitochondrial phosphate transport protein expressed in

Escherichia coli. Site-directed mutations at threonine-43 and at a similar location in the second tandem repeat (Isoleucine-141), *Biochemistry* **33:**9371–9375.

Zara, V., Rassow, J., Wachter, E., Tropschug, M., Palmieri, F., Neupert, W., and Pfanner, N., 1991, Biogenesis of the mitochondrial phosphate carrier, *Eur. J. Biochem.* **198:**405–410.

Zara, V., Palmieri, F., Mahlke, K., and Pfanner, N., 1992, The cleavable presequence is not essential for import and assembly of the phosphate carrier of mammalian mitochondria, but enhances the specificity and efficiency of import, *J. Biol. Chem.* **267:**12077–12081.

Zara, V., Dietmeier, K., Palmisano, A., Vozza, A., Rassow, J., Palmieri, F., and Pfanner, N., 1996, Yeast mitochondria lacking the phosphate carrier/p32 are blocked in phosphate transport but can import preproteins after regeneration of a membrane potential, *Mol. Cell. Biol.* **16:**6524–6531.

Zarrilli, R., Oates, E. L., McBride, O. W., Lerman, M. L., Chan, J. Y., Santisteban, P., Ursini, M. V., Notkins, A. L., and Kohn, L. D., 1989, Sequence and chromosomal assignment of a novel cDNA identified by immunoscreening of a thyroid expression library: Similarity to a family of mitochondrial solute carrier proteins, *Mol. Endocrinol.* **3:**1498–1508.

CHAPTER 21

Structure and Evolution of Metazoan Mitochondrial Genome

Cecilia Saccone

Over the last decade, the development of new genetic and molecular biology technologies has greatly contributed to the knowledge of the organization and expression of the mitochondrial (mt) genome. Human mitochondrial DNA was the first eukaryotic genome to be completely sequenced in 1980, added another dimension to our knowledge and opened new and otherwise impracticable approaches in both theoretical and experimental studies.

Studies on the mt genome produced a spectacular advance in the field of molecular evolution. Mitochondrial DNA has became the most popular molecule for evolutionary studies and also has had a strong impact in other disciplines such as paleontology, taxonomy, and many branches of applied biology.

In this chapter, we shall focus on the organization and evolution of mtDNA in animals. In particular, after a short introduction on the general features of the mt genome in various organisms, we shall briefly treat three aspects of mtDNA evolution: (1) the evolution of metazoan mtDNA (variation both between and within the phyla); (2) The evolution of the mammalian genomes (interspecies variation); and (3) the evolution of the human genome (intraspecies variation).

Cecilia Saccone • Department of Biochemistry and Molecular Biology, C.S.M.M.E. CNR, University of Bari, I-70125 Bari, Italy.

Frontiers of Cellular Bioenergetics, edited by Papa *et al.* Kluwer Academic/Plenum Publishers, New York, 1999.

1. GENERAL FEATURES OF THE MITOCHONDRIAL GENOME IN VARIOUS ORGANISMS

The mt genome shows a great variability in terms of structure, gene content, organization and mode of expression in the different organisms. This extraordinary diversity probably reflects the different evolutionary pathways that generated the segregation of genetic information in the eukaryotic cell in different cellular compartments such as the nuclei and the mitochondria. Several features, however, are common to the majority of mt genomes.

1.1. Genome Shape and Size

The circular double-stranded structure appears to be almost a constant feature of mtDNA, which exhibits an extraordinary variability in length, particularly in the lower eukaryotes and in plants. The size ranges from 19 kilobase-pairs (kbp) in fungi to 2500 kbp in plants.

In Ciliata, e.g., *Tetrahymena pyryformis* and *Paramecium aurelia*, the genome is linear with a double helix molecule 46 and 40 kbp long, respectively. In the algae *Chlamydomonas reinhardtii*, besides the linear (major species) of mt genome (16 pb long), a circular (minor species) mtDNA might also be present (Ma *et al.*, 1992). In the kinetoplasts of the Tripanosomatidae, *Crithidia fasciculata*, *Leishmania tarentolae*, and *Trypanosoma brucei*, the mt genome has a very peculiar structure, being composed of an intricate network containing two types of molecules: many thousands of minicircles (1–3 kbp, depending on the species), for 90–95% of the DNA, and 50–100 maxicircles (20–40 kbp). Minicircles lack long amino acid coding frames and maxicircles are the equivalent of mtDNA in other organisms (Benne, 1985).

In higher plants, the mt genome ranges from 200 to 2500 kbp, and it has been demonstrated that its size can vary even sevenfold within the same family. This peculiar feature is not related to the number of repeated sequences and detectable translation products. The entire genetic complexity is contained in a circular chromosome (master) that may be resolved into subgenomic circles by recombination through directly repeated sequences. This mechanism is also involved in the rapid and extensive rearrangements that characterize the evolution of plant mtDNA. Moreover, frequent acquisitions of genes from nuclei and chloroplasts take place, resulting in the formation of mosaic genomes (Newton, 1988; Levings and Brown, 1989; Gray *et al.*, 1989; Palmer, 1990, for reviews). Table I reports the size and shape of some mt genomes completely sequenced in lower eukaryotes and in plants.

In contrast to such a great variability, the size of the mt genome in all Metazoa is extremely small compared to that of other eukaryotes, with a roughly constant length (14–19 kbp). Table II lists the mt genomes completely

Table I
Size and Shape of mtDNA in Various Organisms

Organism	Accession no.	Size (bp)	Shape
Plantae			
Marchantia polymorpha	M68929	186,609	Circular
Prototheca wickerhamii	U02970	55,328	Circular
Chlamydomonas reinhardtii	U03843	15,758	Linear
Fungi			
Saccharomyces cerevisiae	M62622	78,520	Circular
Schizosaccharomyces pombe	X54421	19,431	Circular
Podospora anserina	X55026	100,314	Circular
Protozoa			
Paramecium aurelia	X15917	40,469	Linear
Acanthamoeba castellani	U12386	41,591	Circular

sequenced in Metazoa. More recently, several cases in the literature have been described where the length of mtDNA exceeds the normal, rather constant average size found in animal cells. Such a difference in length can range from twofold the average size (14–28 kbp) to even more (39 kbp, as in sea scallop). In such cases, however, a duplication of some genomic regions seems to have occurred. Length variations are more frequent in Invertebrata and in poikilo-thermic Vertebrata, and it appears to generate very rapidly and distribute both within (heteroplasmy) and between individuals. In general, length differences are confined to the control region, but in some cases they are dispersed and/or also include structural genes (Saccone and Sbisà, 1994; Saccone, 1994; Wolstenholme, 1992, for review).

Several hypotheses have been offered to explain why the metazoan mt genome has attained such a small but constant size. A small genome has several advantages, such as a faster replication and a constitutive type of transcription. However, the "race for replication" hypothesis of A. C. Wilson (personal communication), according to which those genomes that replicate the fastest will win, has not found experimental support so far. Recently, Moraes and Schon (1995) have found no difference in the replication rate of a hetero-plasmic population of normal and partially deleted human mt DNA genomes in fibroblasts.

1.2. Genome Organization

Despite such a variety of structures and size, the information content of all mt genomes is not dramatically different in the various organisms. Since the information content of the mtDNA is sufficient to code for ribosomal RNA

Table II
Completely Sequenced Metazoan mtDNAs

Organism	Accession no.	Genome (bp)	Main noncoding region (bp)
Anellida			
Lumbricus terrestris	U24570	14998	384
Arthropoda			
Artemia franciscana	X69067	15822	1822
Anopheles gambiae	L20934	15363	519
Anopheles quadrimaculatus	L04272	15455	625
Apis mellifera	L06178	16343	827
Drosophila melanogaster	U37541	19517	4601
Drosophila yakuba	X03240	16019	1077
Locusta migratoria	X80245	15722	875
Echinodermata			
Arbacia lixula	X80396	15719	136
Asterina pectinifera	D16387	16260	445
Paracentrotus lividus	J04815	15696	132
Strongylocentrotus purpuratus	X12631	15650	121
Mollusca			
Albinaria coerulea	X83390	14130	nd[a]
Cepaea nemoralis	U23045	14100	nd
Nematoda			
Ascaris suum	X54253	14284	886
Caenorhabditis elegans	X54252	13794	466
Agnatha			
Petromyzon marinus	U11880	16201	491
Osteichthyes			
Crossostoma lacustre	M91245	16558	896
Cyprinus carpio	X61010	16575	928
Oncorhynchus mykiss	L29771	16642	1003
Protopterus dolloi	L42813	16646	1184
Aves			
Gallus gallus	X52392	16775	1227
Amphibia			
Xenopus laevis	M10217	17553	2134
Mammalia			
Balaenoptera musculus	X72204	16402	936
Balaenoptera physalus	X61145	16398	929
Bos taurus	J01394	16338	910
Cavia porcellus[b]	—	16801	1357
Didelphis virginiana	Z29573	17084	1615
Equus caballus	X79547	16660	1192
Erinaceus europaeus	X88898	17447	1988
Felis catus	U20753	17009	1560
Gorilla gorilla	X93347	16412	964
	D38114	16364	918
Halichoerus grypus	X72004	16797	1360

Table II
(*Continued*)

Organism	Accession no.	Genome (bp)	Main noncoding region (bp)
Homo sapiens	V00662	16569	1122
	X93334	16570	1123
Mus domesticus	L07905	16303	nd
Mus musculus	V00711	16295	879
Ornithorhynchus anatinus	X83427	17019	1559
Pan paniscus	D38116	16563	1121
Pan troglodytes	X93335	16561	1113
	D38113	16554	1113
Phoca vitulina	X63726	16826	1391
Pongo pygmaeus	D38115	16389	917
Rattus norvegicus	X14848	16300	894

[a]nd, Not defined.
[b]Sequence not available in databases (D'Erchia *et al.*, 1996).

species (two of three) for a reduced but complete set of transfer RNAs (with some exceptions) and a small set of proteins (13 in Metazoa), in general (with some exceptions like plants) the differences concern mainly the noncoding and the regulatory regions. Indeed, the mt genome is a good example of several different strategies that the eukaryotic cell can use to express the same information content. Among these, certainly the most peculiar is RNA editing, which is particularly active in Protozoa and in plants (Gray *et al.*, 1992).

Despite the relative constant gene content, gene order and organization vary strikingly in the various organisms. The mt genome organization in lower eukaryotes, like *Saccharomyces cerevisiae*, is very loose (the two thirds contain A-T rich, noncoding sequences), and several genes, in particular the apocytochrome *b*, cytochrome oxidase subunit I, and the 21S rRNA, are discontinuous. The number of introns and G-C- and A-T-rich mini-inserts within the genes are strain dependent. Some introns of these split genes code for proteins involved in RNA processing or intron transposition (de Zamaroczy and Bernardi, 1985; Grivell, 1987).

In plants, where, as reported previously, the mt genome is much larger than the fungal or animal counterparts, gene organization is highly dispersed. Extensive noncoding sequences separating the coding regions and introns have been detected in several genes. The mtDNAs show a tri- or multipartite structural organization, and due to frequent recombinations, they are continuously rearranged. A consequence of this peculiar evolutionary pattern is that the genome organization varies greatly in linear gene order. Thus, highly conserved coding sequences are often flanked by completely different sequences in the mtDNAs of plant species, even closely related between them. Moreover, chloroplast DNA sequences and plasmidelike sequences are present in the mtDNA (Gray *et al.*, 1992).

The gene structure and organization of Metazoa markedly differ from that of yeast and plant mtDNAs. The most distinctive feature of the metazoan mt genome is its extremely compact gene organization. Apart from the replication origin regions, the genome is saturated with discrete gene products. Genes coding for transfer, ribosomal, and messenger RNAs lack intronic sequences and flanking untranslated regions; they are often contiguous and sometimes slightly overlapped or separated by only few nucleotides.

In conclusion, the features of the mt genomes are of both prokaryotic (e.g., naked DNA, absence of introns in Metazoa) and eukaryotic (e.g., presence of introns in lower eukaryotes and plants, presence of 5'-end polyadenylated mRNAs in Metazoa) nature. This indicates that the endosymbiont from which mtDNA originated has followed different evolutionary pathways in the various organisms.

For many years, our studies (Slater *et al.*, 1968) have focused on the mt genome of Metazoa. The following sections are devoted to their structure and evolution.

2. THE EVOLUTION OF THE MITOCHONDRIAL DNA IN METAZOA

As stressed above the size of the mtDNA is roughly constant in all Metazoa. The gene order and organization, however, vary from phylum to phylum due to gene rearrangements that consist of a different distribution of genes between the two strands (polarity inversions), gene transpositions, and gene losses.

2.1. Completely Sequenced Genomes

Owing to the reduced size of the molecule, the sequencing of the metazoan mt genome has become very popular, providing much information about its variations both between and within phyla. Table II lists the metazoan mt genomes that have been completely sequenced. The smallest is that of the nematode *Caenorhabditis elegans*, with 13,794 bp; the largest is *Drosophila melanogaster*, with 19,517 bp. In Mammalia, 19 species belonging to 9 orders are represented.

2.2. Variation between Phyla

The gene organization of some animal mtDNAs are shown in Figs. 1 and 2. When comparing the mt gene organization in Nematoda, Insecta, Echinoder-

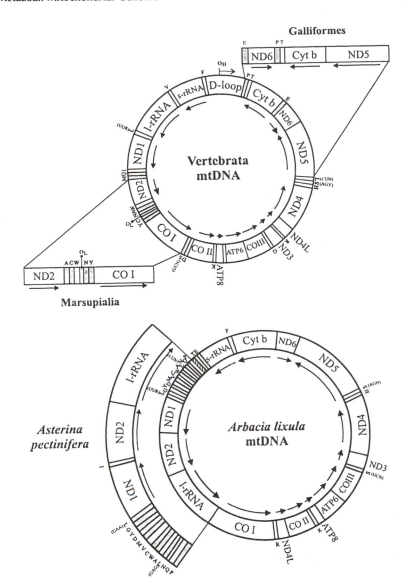

FIGURE 1. Comparative gene organization in Vertebrata and Echinodermata mtDNAs. ATP6, ATP8, ATPase subunits; CO I, II, III, cytochrome oxidase subunits; Cytb, cytochrome *b*; ND 1, 2, 3, 4, 4L, 5, 6, NADH dehydrogenase subunits; s-rRNA, 12S ribosomal RNA; l-rRNA, 16S ribosomal RNAs. Genes for tRNAs are indicated by the one-letter code for their corresponding amino acid. Shaded tRNAs are coded by the complementary strand. O$_H$ and O$_L$ indicate the origin of replication of the H and L strand. Arrows indicate the polarity of transcription. Rearrangements of Marsupialia and Galliformes with respect to Vertebrata and *Asterina pectinifere* with respect to *Arbacia lixula* are indicated. For further details, see Table III.

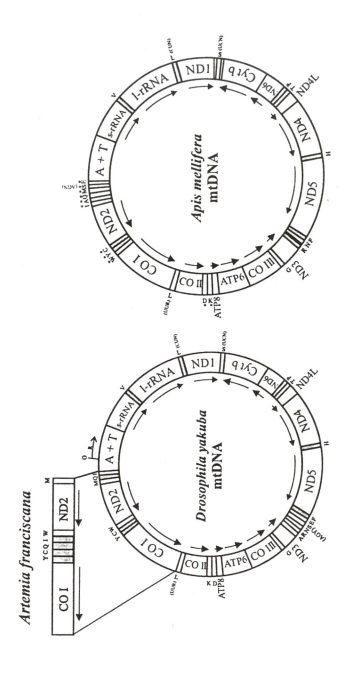

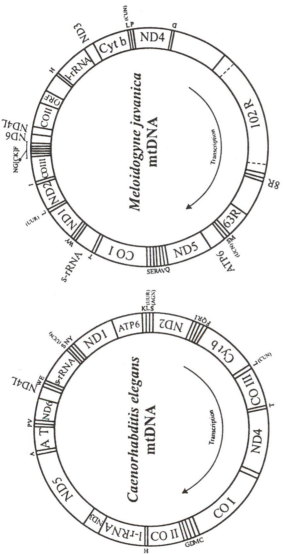

FIGURE 2. Comparative gene organization in Arthropoda and Nematoda mtDNAs. In *Apis mellifera*, tRNAs marked with a star are in a different relative position to their counterparts in *Drosophila yakuba*. In *Meloidogyne javanica*, 8R, 63R, and 102R correspond to directly repeated sequences. A+T, main noncoding region. For further details, see legend to Fig. 1 and Table III.

mata, and Vertebrata, the most remarkable feature is the lack of correlation of tRNA distribution in the four phyla. In Vertebrata, the tRNA genes are scattered along the molecule and make up a sort of punctuation that functions as signals for the RNA processing. In Echinodermata, 15 out of the 22 tRNA genes are in a cluster and in Nematoda all genes are coded for by the same strand.

Cantatore et al. (1987), on the basis of comparative analyses, suggested events of duplication and remolding of tRNA genes during the evolutionary rearrangement of mt genomes in the sea urchin *Paracentrotus lividus*. In particular, a high homology between the Leu^{CUN} and the Leu^{UUR} tRNA genes was observed, together with an altered location of the Leu^{CUN} gene in the mtDNA of sea urchin compared to Vertebrata. In addition, a sequence (72 bp long) containing a trace of the old Leu^{CUN} gene at its original location was observed in the sea urchin, where it coded for an extra amino acid sequence at the amino-termini of the ND5 gene. These data were interpreted by assuming that during evolution a tRNA gene lost its function and became part of a protein-coding gene. This loss was accompanied by the gain of a new tRNA through duplication and divergence from a tRNA gene specific to a different family of codons, namely, Leu^{UUR} tRNA gene. On the basis of these assumptions and of other observations indicating that tRNAs are present at the end of duplications and deletions, it has been suggested that tRNAs should be considered to be mobile elements involved in gene rearrangement (Moritz et al., 1987; Saccone et al., 1990a).

It is well known that gene rearrangement events may be used to reconstruct the ancestral gene order and then the phylogenetic interrelationships between organisms. In this case, by assuming that tRNAs move independently from other genes, the possible phylogenetic interrelationships between Metazoa can be established because of the transposition events of the ribosomal and messenger genes. The unrooted tree constructed by Cantatore et al. (1987) using this criterion, with the genomes of Vertebrata, Nematoda, Echinodermata, and Insecta, places Echinodermata and Vertebrata more closely in relation, since they are separated by four events only.

2.3. Variation within the Phylum

In contrast to the great variability of gene organization between phyla, it seems that the gene order is kept constant within each phylum also between species that separated over a large time span. Only a few transposition events seem to be tolerated. In Echinodermata, for example, between sea urchins and the sea stars, which diverged about 500 million years ago, an inversion of only a 4.6-kbp fragment took place.

In Table III and Figs. 1 and 2, the major variations in gene organization in

Table III
Variations of Gene Organization in Various Metazoan mtDNAs

Organism	Variation	Reference
Vertebrata		
Homo sapiens		Anderson *et al.* (1981)
Gallus gallus	Transposition of ND6 and tRNA^Glu	Desjardins and Morais (1991)
Marsupialia (3 species)	3 species: rearrangement of tRNAs	Pääbo *et al.* (1991)
Arthropoda		
Drosophila yakuba		Wolstenholme (1992)
Apis mellifera	11 tRNAs moved to different locations	Crozier and Crozier (1993)
Artemia franciscana	Rearrangement of tRNAs	Valverde *et al.* (1994)
Echinodermata		
Arbacia lixula		De Giorgi *et al.* (1996)
Asterina pectinifera	Inversion of 4.6 kb-long fragment	Asakawa *et al.* (1995)
Nematoda		
Caenorhabditis elegans		Okimoto *et al.* (1992)
Ascaris suum	Different location of A+T region	Okimoto *et al.* (1992)
Meloidogyne javanica	Extensive rearragement	Okimoto *et al.* (1991)

the various metazoan mt genomes are reported. Within Vertebrata, Galliformes display a transposition of the DNA fragment containing the tRNA^{GLU} and the ND6 genes (Desjardins and Morais, 1991) and Marsupialia show rearrangements of tRNAs located in the proximity of the light-strand replication origin (Pääbo *et al.*, 1991).

Within Insecta, 11 tRNAs move to a different position in *Apis mellifera* (Crozier and Crozier, 1993) compared to *Drosophila yakuba* (Wolstenholme and Clary, 1985); in *Artemia* and *Drosophila* the location of two tRNAs is different and in one of them their orientation also has changed (Valverde *et al.*, 1994).

The conservation of gene organization seems to be less stringent in Nematoda. As a matter of fact, *Ascaris suum* differs from *Caenorhabditis elegans* only in the location of the A+T region, while extensive rearrangements have occurred in the mtDNA of *Meloidogyne javanica* (Okimoto *et al.*, 1991, 1992).

2.4. The Noncoding Regions

A peculiar feature of metazoan mt genomes is the presence of a main noncoding region that contains the regulatory elements for the replication and expression of the mtDNA. Regardless of the high degree of conservation of the coding genes, this region shows great variability in length and base composition. It ranges from 121 in sea urchin to 1988 in hedgehog, 2134 in frog, and 4601 in *Drosophila*.

Differently from Vertebrata, in Invertebrata the structure and evolution of the regulatory region has not yet been fully characterized. In *Ascaris* and *Drosophila* this region is called AT-rich region for its extremely high A+T content. In *Ascaris* the main AT region is 886 bp long and a smaller noncoding sequence of 117 bp, which can be folded into a stem and loop structure, is also present. In *Drosophila* the AT region is extremely polymorphic; it varies in sequence and length both in different species (1077 bp in *D. yacuba* and 4601 in *D. melanogaster*) and within individuals of the same species, or in different mtDNA molecules of a single fly. The putative promoters and the replication origin are contained in two conserved regions, one of which can form a hairpin structure.

In Echinodermata the main noncoding region (121–445 bp) is located in the tRNA cluster. It appears as a condensate version of the vertebrate replication origin and the nascent strand coincides with a very stable stem–loop structure.

In Vertebrata the main noncoding region is called the D-loop-containing region, because starting from the strand replication origin (O_H) the heavy strand displaces (D) the parental one, creating a triple-strand structure. This region, ranging from 879 to 2134 bp, also contains the promoters for both the heavy (HSP) and light strands (LSP) (see Section 3.2). The two strands are called heavy (H) and light (L) according to their isopycnic sedimentation in the cesium–chloride gradient. The other noncoding region contains the origin of the light strand replication (O_L); it is only 30 bp long and is flanked by five tRNA genes. This region can be folded in a stable stem and loop structure that is very conserved in all Vertebrata except birds. Indeed, the sequence equivalent to the O_L has not been found at the same position in chicken and quail mt genomes.

In rat and humans the two origins of replication show an intrinsic DNA curvature, correlated with periodic distribution of dinucleotides in the sequence and involved in protein interactions (Pepe *et al.*, 1989, Gadaleta *et al.*, 1996).

2.5. Base Composition

Table IV reports the nucleotide composition percentage of the complete genome of several Metazoa and of the third codon positions of protein coding genes. It can be observed that the A-T/G-C content is highly variable, and in *Drosophila* the genome is extremely A-T rich. A peculiar feature is the asymmetric distribution of the complementary nucleotides between the two strands, which is particularly relevant for C-G distribution. The nucleotide composition percentage at the third codon positions of protein-coding genes shows that G-ending codons are avoided. Thus, the bias against the use of G is a general peculiarity of the metazoan mtDNA (Saccone *et al.*, 1981; Pepe *et al.*, 1983). The only possible exception to this rule are Nematoda whose complete genome base composition displays a relatively high content of G, while the C content is surprisingly low. Such anomalous behavior is also reflected in the nucleotide composition percentage at the third codon positions. From the data reported in Table IV, it is possible to see that in Nematoda and partially in Mollusca the bias against the use of G as the third codon position is obscured.

The asymmetric base composition of the metazoan mtDNA is also illustrated in Fig. 3, which reports the degree of asymmetry of the third codon positions in the quartets (family codons) of mt mRNA genes in different mammalian species. C/G asymmetry is always much higher than A/T asymmetry. In humans (results not shown), the asymmetry index calculated for each individual gene shows that only in the ND4L is A/T asymmetry slightly higher than C/G asymmetry (Saccone *et al.*, 1993).

Figure 4 reports the asymmetry index of sea urchin and vertebrate genes. Also, in the case of sea urchin the C/G asymmetry is higher than A/T. Furthermore, in all sea urchins the asymmetry index for C/G is almost always about 0.4, while the asymmetry index for A/T is close to zero in the case of *A. lixula*. These data strongly indicate that in this organism there is no difference in the use of A or T at the third codon positions of the quartet families. Thus, the nucleotide composition of the sense strand of the entire mtDNA of *A. lixula* shows the tendency to decrease the content of C and increase that of T, which is not common to other Echinodermata. This directionality toward a lower C and a higher T content has been taken as an indication that *A. lixula* evolved under a different directional mutation pressure than *P. lividus* and *S. purpuratus* (De Giorgi *et al.*, 1996).

We have suggested that the strong compositional bias, in particular the trend to avoid G at the third codon position, might also explain several deviations of the mt code with respect to the universal one. In animal mitochondria, the initiation codon most often used is AUA instead of AUG, which is a G-ending codon. One of the three nonsense codons, UGA, becomes an additional tryptophan codon, because of the use of two contiguous G (as in the canonical UGG tryptophan triplet), which is very rare in the sense strand of metazoan mt DNA. In other words, the compositional bias was so strong as to influence the genetic code itself. Different models can be suggested to explain the peculiar compositional bias of the mammalian genome. Metabolic discrimination between nucleotide bases and/or replication errors followed by biased repair could account for this property. Other explanations may be based on the mechanism of mtDNA replication, which, being asymmetric, leaves one DNA strand as a single, unprotected filament for two thirds of the replication cycle.

Table IV
Nucleotide Composition of the Complete Genome
and of the Third Codon Position of Protein Coding Genes
of the Completely Sequenced Metazoan mtDNAs

Organism	Nucleotide composition percentage of the complete genomes				Nucleotide composition percentage at the third codon position of protein coding genes			
	T	C	A	G	T	C	A	G
Anellida								
Lumbricus terrestris	31.8	22.5	29.8	15.8	29.9	24.3	35.5	10.3
Arthropoda								
Artemia franciscana	33.5	17.9	31.0	17.7	39.6	17.0	29.9	13.5
Anopheles gambiae	37.5	12.9	40.0	9.5	47.0	4.3	45.3	3.4
Anopheles quadrimaculatus	37.1	13.4	40.2	9.3	46.1	5.0	44.3	4.6
Apis mellifera	41.6	9.6	43.2	5.5	47.6	2.8	47.6	2.0
Drosophila melanogaster	40.4	10.3	41.8	7.6	49.0	3.0	45.4	2.6
Drosophila yakuba	39.1	12.2	39.5	9.2	48.5	3.3	45.3	2.9
Locusta migratoria	30.8	14.6	44.5	10.1	42.3	8.4	44.8	4.5
Echinodermata								
Arbacia lixula	33.0	20.4	29.5	17.0	33.7	18.0	36.3	11.9
Asterina pectinifera	28.9	24.6	32.4	14.1	27.5	24.3	35.7	12.5
Paracentrotus lividus	29.5	22.5	30.8	17.2	26.2	22.2	39.1	12.5
Stronglyocentrotus purpuratus	30.2	22.7	28.7	18.4	27.1	24.6	33.8	14.5
Mollusca								
Albinaria coerulea	37.9	13.8	32.7	15.5	42.0	9.6	38.5	9.9
Cepaea nemoralis	33.6	18.9	26.2	21.3	36.3	20.1	26.0	17.6
Nematoda								
Ascaris suum	49.8	7.7	22.2	20.4	62.5	2.5	11.6	23.4
Caenorhabditis elegans	44.8	8.9	31.4	14.9	49.7	4.7	36.6	9.0
Agnatha								
Petromyzon marinus	30.4	23.8	32.3	13.5	33.5	21.6	41.1	3.8
Osteichthyes								
Crossostoma lacustre	25.0	28.6	31.9	16.9	20.1	34.8	35.7	9.5
Cyprinus carpio	24.9	27.4	31.6	15.8	18.9	31.2	44.1	5.8
Oncorhynchus mykiss	26.2	28.9	27.9	17.0	23.8	34.0	33.4	8.8
Protopterus dolloi	29.0	26.5	28.8	15.7	28.5	28.5	34.7	8.4
Aves								
Gallus gallus	23.8	32.4	30.2	13.5	13.8	42.8	37.7	5.7
Amphibia								
Xenopus laevis	30.0	23.5	33.0	13.5	30.0	22.4	42.7	4.9
Mammalia								
Balaenoptera musculus	26.6	27.6	32.8	13.0	19.7	34.3	42.0	4.0
Balaenoptera physalus	26.7	27.3	32.7	13.3	19.9	33.5	41.5	5.1
Bos taurus	27.2	25.9	33.4	13.4	20.9	30.8	42.7	5.6
Cavia porcellus	28.6	24.8	32.0	14.6	24.4	28.0	40.1	7.5

Table IV
(*Continued*)

Organism	Nucleotide composition percentage of the complete genomes				Nucleotide composition percentage at the third codon position of protein coding genes			
	T	C	A	G	T	C	A	G
Mammalia (*cont.*)								
Didelphis virginiana	31.5	21.1	35.3	12.0	31.9	19.1	46.1	2.9
Equus caballus	25.9	28.5	32.2	13.4	17.7	36.2	39.7	6.4
Erinaceus europaeus	33.4	20.1	34.0	12.5	37.5	17.5	40.4	4.6
Felis catus	27.1	26.2	32.6	14.1	21.5	30.7	40.5	7.3
Gorilla gorilla[a]	25.2	30.7	30.9	13.2	18.5	39.4	35.8	6.3
Halichoerus grypus	25.3	27.5	33.0	14.3	17.8	33.8	40.5	7.9
Homo sapiens[b]	24.7	31.2	30.9	13.1	16.5	41.4	35.7	6.4
Mus musculus	28.7	24.4	34.5	12.3	24.1	26.5	45.3	4.1
Ornithorhynchus anatinus	31.4	23.5	31.4	13.6	31.4	25.0	38.0	5.6
Pan paniscus	25.3	30.7	31.3	12.7	18.2	39.6	37.1	5.1
Pan troglodytes[c]	25.2	30.8	31.1	12.9	18.3	39.6	36.3	5.8
Phoca vitulina	25.3	27.4	33.0	14.3	17.8	33.8	40.5	7.9
Pongo pygmaeus	23.8	32.4	30.5	13.2	14.7	43.7	35.4	6.2
Rattus norvegicus	27.3	26.3	34.0	12.4	21.0	31.9	42.9	4.2

[a]Accession No. D38114.
[b]Accession No. V00662.
[c]Accession No. D38113.

The possibility, too, of a damage directly at level of the RNAs cannot be excluded (Saccone and Sbisà, 1994).

3. THE EVOLUTION OF MITOCHONDRIAL DNA IN MAMMALS

It has been widely reported that mtDNA evolves faster than nuclear DNA. This refers particularly to nucleotide substitutions that represent the major if not unique mechanism of evolution in Mammalia. Insertions, deletions, or other mechanisms are indeed negligible and very rare except in the D-loop region (see this Section).

A correct determination of the nucleotide substitution rate strictly depends on the mathematical model used in the measurement. Based on the observation that the nucleotide substitution pattern of mtDNA in Mammalia is rather peculiar (because of the asymmetry in base composition), a new model, the stationary Markov clock, has been devised in Saccone's group (Saccone *et al.*,

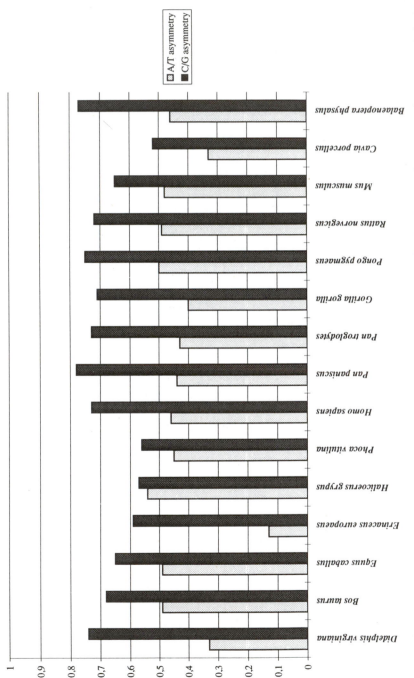

FIGURE 3. Asymmetry index of the third codon position of the quartet family codons of mt-mRNA for 15 mammalian species. The asymmetry index is calculated as $|A-T|/(A+T)$ (for A/T asymmetry) and $|C-G|/(C+G)$ (for C/G asymmetry), where A, T, C, and G are the occurrences of the four nucleotides at the third codon positions of the quartet family codons.

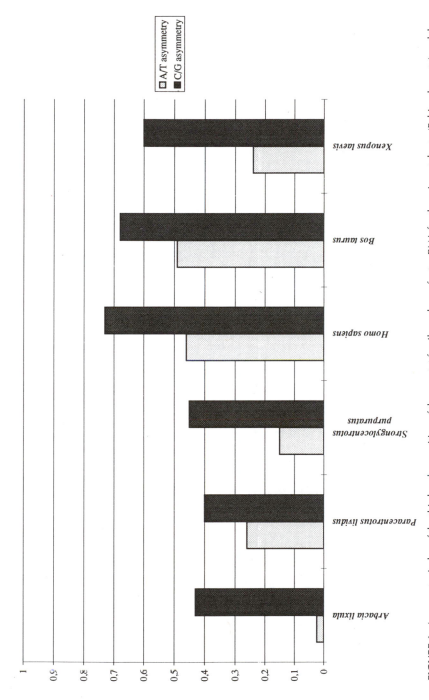

FIGURE 4. Asymmetry index of the third codon position of the quartet family codons of mt-mRNA for three Invertebrata (Echinodermata) and three Vertebrata (Mammalia, Amphibia) species. For further details, see legend to Fig. 3.

1990b). This method does not follow the a priori assumptions that are at the basis of other stochastic models available in the literature.

The stationary Markov clock model used in the quantitative analyses of mtDNA allows one to (1) determine the four-nucleotide substitution rate matrix, R_{ij}, which describes the propensity of nucleotide j to be substituted by nucleotide i independently of time T; (2) calculate the effective average evolutionary rate of sequence sites (i.e., the rate of silent codon position in mRNA genes); and (3) construct phylogenetic trees. By using this method the genetic distances between Primates (Holmes *et al.*, 1989), several orders of mammals (Pesole *et al.*, 1991; Saccone *et al.*, 1991a) and, very recently, the cladistic relationship between human populations, and thus the coalescence time of mtDNA of our species, have also been determined (Pesole *et al.*, 1992) (see this Section).

As far as the average evolutionary rate of the mt genome in Mammalia is concerned, we would like to point out that the rate of silent substitutions is definitely higher in mitochondrial than in nuclear genes, whereas the replacement rate varies from gene to gene and in several cases can be lower in mitochondria than in nuclei. This applies to the subunits of the cytochrome–oxidase complex, where the faster evolution of some nuclear-coded subunits compared to that of mt-coded subunits has been demonstrated (Saccone *et al.*, 1991b).

With respect to the regulatory regions, comparisons between mitochondrial and nuclear elements are meaningless due to the peculiar mechanisms of replication and expression in mitochondria and to the diverse structure of the two genomes.

3.1. Phylogenetic Reconstructions

The mitochondrial genome, because of its maternal inheritance, lack of recombination, and especially the presence of orthologous genes, has become the most often used molecule in the field of molecular phylogeny. Conversely, nuclear genes can be paralogous or can evolve under different evolutionary pressures, which thus could lead to controversial or unreliable results.

The mitochondrial genome revealed a powerful tool for dealing with the complex problem of the higher-level classification of the eutherian mammals. In particular, two issues can be addressed: (1) the reconstruction of the phylogenetic interordinal relationships; and (2) the testing of the monophyly or polyphyly of each order.

Both single mt genes and the entire mtDNA have been used to partially reconstruct the mammalian tree. With regard to single genes, most evolutionary studies have focused on cytochrome *b* and cytochrome oxidase subunit II genes. Analyses performed on cytochrome *b* for a great number of mammalian species (Irwin *et al.*, 1991) have provided indications for a close association between Cetacea and Artiodactyla, in agreement with paleontological and morphological data. The cytochrome oxidase subunit II gene has been used to investigate the relationships among the orders Primates, Dermoptera, Scandentia, and Chiroptera that morphological studies included in the superordinal Archonta. This study has demonstrated the common origin of Primates, Dermoptera, and Scandentia, but has placed the Chiroptera outside this clade (Adkins and Honeycutt, 1991).

Over the last years, the improvement of the DNA-sequencing techniques has increased the number of complete mt genomes available for evolutionary studies, allowing the possibility of obtaining more reliable phylogenies than those from single genes. Evolutionary studies performed recently on the mt genome of human, cow, whale, seal, mouse, rat, and opossum have provided stronger support to the Artiodactyla-Cetacea clade and have shown a close relationship between this clade and the order Carnivora (Arnason *et al.*, 1991; Arnason and Johnsson, 1992). Moreover, the studies have revealed that Rodentia diverged before the Primates, Carnivora, and Artiodactyla-Cetacea clade, with living Marsupialia being the earliest branch of the mammalian tree (Janke *et al.*, 1994).

Regarding the intraordinal relationships, the analyses of mt genomes have allowed the investigation of the position of the guinea pig, *Cavia porcellus*, within the complex order of Rodentia. The traditional subdivision of the order Rodentia in three suborders—Myomorpha, Hystricomorpha, and Sciuromorpha—had not been questioned until Graur *et al.* (1991) proposed the polyphyly of the order based on the analyses of nuclear proteins and suggested the elevation of the suborder Hystricomorpha, to which the guinea pig belongs, to the status of order. Several findings have since been reported, both for and against this phylogeny, without solving the debate. We have presented findings based on the analysis of the mt genome of the guinea pig (both the 13 protein-coding genes and the rRNA genes) that strongly support the hypothesis of rodent polyphyly (D'Erchia *et al.*, 1996). The study has included all the complete mt genomes available in the literature and representative of eight mammalian orders and has been performed by using three completely different methods: the maximum parsimony [a deterministic method applied to the derived protein sequences (Swofford, 1993)], the stationary Markov model [a stochastic method applied to nucleotide sequences (Saccone *et al.*, 1990b)], and the protein maximum-likelihood method (Adachi and Hasegawa, 1994).

In Fig. 5, the consensus tree obtained by applying the three methods is shown. The different analytical approaches and the large sets of data used in the studies have yielded congruent results, providing strong evidence for the inclusion of the guinea pig in a new mammalian order, different from Rodentia. Indeed, the guinea pig does not cluster with Myomorpha rodents (mouse and

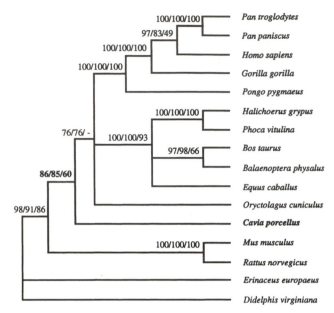

FIGURE 5. Consensus tree describing the phylogenetic relationship among 16 mammalian species, representative of 10 different taxa (Primates, Artiodactyla, Myomorpha, Hystrico-morpha, Lagomorpha, Cetacea, Carnivora, Perissodactyla, Insectivora, Metatheria) (D'Erchia *et al.*, 1996). The three values reported on each node represent the bootstrap values obtained from the analyses of (1) the 13 mitochondrial concatenated protein sequences by using maximum parsimony method; (2) the asynonymous codon positions of 13 concatenated protein coding genes by using the stationary Markov model; and (3) the concatenated 12S and 16S rRNA genes by using the stationary Markov model.

rat), since it appears as a sister group of the clade including Primates, Carnivora, Perissodactyla, Artiodactyla-Cetacea, and Lagomorpha. Moreover, analyses strongly confirm the phylogenetic relationships mentioned above and reject the validity of the cohort Glires (the Rodentia and Lagomorpha clade) supported by morphological data.

Molecular approaches have already yielded significant insights in the field of mammalian phylogeny, but because of the extraordinary complexity of the mammalian class more mitochondrial data and wider taxonomical samples are needed to clarify the entire mammalian scenario.

3.2. The D-Loop Region

As previously reported, the mammalian mtDNA possesses a unique main noncoding region, called the D-loop, which contains the regulatory elements for the replication and expression of the genome. Regardless of its functional importance, this is the most rapidly evolving part of the mtDNA. However, detailed comparative analyses of this region have revealed several peculiar well-conserved features in the evolution of vertebrates (Brown *et al.*, 1986; Saccone *et al.*, 1991a; Sbisà *et al.*, 1997). In particular, the 5' and 3' ends contain thermodynamically stable secondary cloverleaflike structures and short conserved sequences, namely conserved sequence box (CSBs) and termination-associated sequences (TASs), which are associated with the start and stop sites of the nascent H strand. How the function in this region is preserved in spite of such a high primary structure diversity remains to be clarified.

The D-loop regions show great variability in length and base composition among mammals, making their alignment very difficult. Figure 6 shows the structural organization of the D-loop region of nine mammalian orders. The organization is similar in all the organisms considered, and on the basis of both the degree of conservation and the base content, the D-loop has been divided into three domains. It is possible to distinguish a highly conserved central domain flanked by two hypervariable regions, extended TAS (ETAS) and CSB, whose rapid evolution is responsible for heterogeneity in both length and base composition. They are prone to insertion and deletion of elements and to the generation of repeats by replication slippage. The repeated sequences seem to be peculiar to each order and are inserted in the two peripheral domains generally at the same locations.

Table V reports the lengths and base compositions of the three domains. In all organisms but Primates, the D-loop is characterized by a high A+T content in all domains with A+T $\gg$ G+C in the peripheral ones. In Primates, A+T is greater than G+C in the ETAS and CSB domains, whereas the reverse base composition in the central domain is observed.

The central domain shows a constant length of 330–350 bp and the highest G content. The percentage of G increases to 22% in Primates, a value higher than that found in the other genomic regions coding for rRNA, tRNA, and mRNA. (GG)n repeats (n ranging from 3 to 6) are found almost exclusively here. In this domain, the degree of conservation is very high; conserved stretches are spaced by short sequences, which are peculiar to each order.

The 3' TAS domain, where synthesis of the heavy strand pauses, ranges from 230 in the rat (*Rattus Norvegicus*) to 365 in rabbit (*Oryctolagus cuniculus*). In addition to the TAS sequences previously identified, our alignments highlight two conserved blocks of about 60 bp, which we have called ETAS1 and ETAS2 (extended TAS), with variable distance between them and from the 3' end of the D-loop (Sbisà *et al.*, 1997). In opossum (*Didelphis virginiana*), repeated sequences between these two blocks are present.

The 5' CSB domain contains the principal regulatory elements of the mitochondrial genome: the two promoters (HSP and LSP) and the origin of

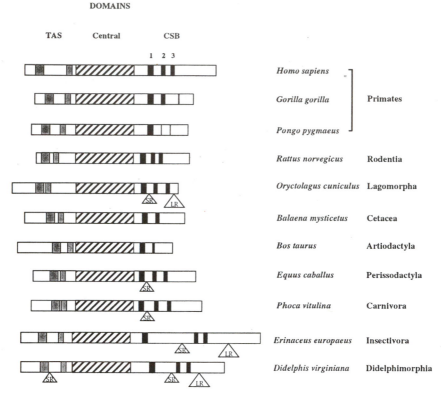

FIGURE 6. General scheme of the organization of the D-loop-containing region in various mammalian orders. ETAS, central and CSB domains are defined. In the ETAS domain, the shaded boxes represent the conserved sequences. In the CSB domain, the shaded boxes correspond to CSB1, CSB2, and CSB3 and the dotted boxes correspond to partial CSB sequences. SR, short repeats; LR, long repeats.

replication of the H strand (OH). This is the most variable region of the genome and its length ranges from 227 bp in the cow (*Bos taurus*) to 763 bp in the hedgehog (*Erinaceus europaeus*). The domain length increases greatly if the repeats, frequently found in this region, are considered. A peculiar feature of this domain is the small conserved sequence blocks (CSB1, CSB2, and CSB3) suggested to be involved in the synthesis of the H strand (Tullo *et al.*, 1995). The degree of similarity between the CSBs is variable. CSB1 is present in all the considered organisms, though it shows a different degree of homology. CSB2 and CSB3 are well conserved but are not present in all the organisms. CSB2 is only partially present and CSB3 is completely absent in the whale and

Table V
Length and Base Composition of the D-Loop Domains

Organism	D-loop length (bp)	TAS domain		Central domain		CSB domain	
		Length (bp)	(C + G)%	Length (bp)	(C + G)%	Length (bp)	(C + G)%
Homo sapiens	1122	347	45	333	52	442	44
Gorilla gorilla	918	240	42	328	52	350	48
Pongo pygmaeus	917	263	46	333	54	321	46
Rattus norvegicus	894	231	26	331	47	332	33
Oryctolagus cuniculus	1838 (966)[a]	365	32	339	48	262	38
Balaena mysticetus	917	295	32	345	49	277	39
Bos taurus	910	345	29	338	48	227	37
Equus caballus	1192 (960)[a]	245	37	341	48	374	41
Phoca vitulina	1391 (995)[a]	251	43	349	45	395	38
Erinaceus europaeus	1988 (1403)[a]	302	22	338	45	763	27
Didelphis virginiana	1615 (1193)[a]	313	23	324	45	556	24

[a]D-loop length excluding repeats.

cow. In Primates, it is noteworthy that these blocks are highly conserved in the human and chimp, but the CSB2 is only partially present in the orangutan and CSB3 is only partially present in the orangutan and gorilla. This finding opens the debate on the functional role of CSB2 and CSB3.

The region spanning between CSB1 and CSB2 is a site for insertions of repeated sequences in the rabbit (*Oryctolagus cuniculus*), horse (*Equus caballus*), seal (*Phoca vitulina*), hedgehog (*E. europaeus*), and opossum (*D. virginiana*). The region spanning between CSB3 and tRNA-Phe shows other repeated motives in the rabbit, opossum, and hedgehog.

Due to their peculiar evolution and different length, ETAS and CSB domains cannot be used to measure the genetic distance between species. Conversely, the central domain might be used as a molecular clock to allow calculation of divergence times between mammals and construct phylogenetic trees. However, the small dimension of the domain generates data with high statistic fluctuations (Saccone *et al.*, 1991a).

4. THE EVOLUTION OF THE HUMAN MITOCHONDRIAL GENOME

The variability of human DNA is a topic that piques the interest of researchers both for its evolutionary implications and for its impact in mitochondrial pathologies (Di Mauro and Wallace, 1993). Indeed, due to its high intraspecies variability and its many peculiar genetic features, mtDNA has proved to be an invaluable source of information in the reconstruction of the evolutionary history of modern humans (*Homo sapiens*). Human mitochondrial variability studies have been carried out using various methods applied to data derived either from the restriction analysis of the entire genome or from sequence data of the most variable region of mtDNA, the D-loop.

Such studies on the variability of mtDNA added to the knowledge of human evolution at different levels. Some studies considered mitochondrial variability of populations from different geographic areas worldwide to try to reconstruct their phylogenetic relationships, and thus to track back the origin and spread of modern man (Cann *et al.*, 1987; Vigilant *et al.*, 1991). Other studies focused on the characterization of populations in single continental areas in order to reconstruct their migration patterns (Torroni *et al.*, 1993a,b; Chen *et al.*, 1995). Still others centered on the microdifferentiation and demographic history of single populations (Ward *et al.*, 1991; Bertranpetit *et al.*, 1995, Bertorelle *et al.*, 1996).

Studies on the polymorphic sites have singled out characteristic patterns for many populations. Such patterns identify different DNA haplotypes and can be used as population markers (Torroni *et al.*, 1993a,b). Indeed, by tracking the distribution of DNA haplotypes in different geographic areas, population phylogenetic relationships, migrations, and colonizations can be reconstructed. These studies are generally based on data derived from the restriction analysis of the entire mitochondrial genome; however, recent comparison analysis based on sequence data showed that the polymorphic sites in the D-loop can be correlated with DNA haplotypes.

In 1987, Cann *et al.* studied mtDNA variability in individuals from different geographic areas and offered the hypothesis that modern humans originated in Africa no more than 300,000 years ago and spread from there to other continental areas, replacing the local populations of *Homo erectus* (the recent "out of Africa" hypothesis). Most of the evolutionary studies carried out following those of Wilson's group support the African origin of modern humans, but provide different dating (Table VI). These time discrepancies on the origin of modern man's origin are probably due to the different methods used to calibrate time and to the different mitochondrial DNA regions analyzed. Thus, the data in Table VI, are only partially comparable. Nucleotide substitution rate has not yet been assessed with adequate accuracy, despite it being a prerequisite in the estimate of a common ancestor of human mitochondrial DNA and in population genetic studies at the molecular level. Among the factors causing such poor accuracy in variability estimates is the assumption of all sites as equally variable. Since obviously the sites are not equally variable, a homogeneous nucleotide substitution pattern leads to underestimating evolutionary rate. Among the studies reported in Table VI only Pesole *et al.* (1992) and Hasegawa *et al.* (1991) consider a correction factor (f), corresponding to a fraction of variable sites, [$f = 0.28$ in Pesole *et al.* (1992)] to estimate the evolutionary rate.

In Table VII, the impact of the f factor on the substitution rate estimate, carried out with a modified version of the stationary Markov model (Saccone *et al.*, 1990b; Pesole *et al.*, 1992), is highlighted. The nucleotide substitution rates of the various mtDNA regions are reported both as a comparison between human mt sequences (H-H) and between human and chimpanzee sequences (H-C). Both for H-H and H-C comparison, the assumption of all sites as equally variable ($f = 1$) leads to an underestimation of the evolutionary rate. For some regions of mtDNA, a homogeneous rate model can be considered reliable, as it is the case of the silent positions (synonymous position) in coding regions; yet, for the D-loop region a correction factor is imperative. The high evolutionary rate causes multiple nucleotide substitutions, which are underestimated in the H-C comparison (saturation effect). Thus, in the D-loop for $f = 1$, the evolutionary rate estimated for the comparison H-C is paradoxically smaller than for the comparison H-H.

In order to reconstruct the evolutionary history of our species, studies also have been carried out on the variability of chromosome Y. Chromosome Y does not recombine for most of its length and is paternally inherited. Based on these

Table VI

Inferred Times of the Last Common Ancestor (LCA) from Human Mitochondrial DNA

Reference	LCA (yr)	Data[a]	Calibration time (Myr)	Method
Horai et al. (1995)	143,000 ± 18,000	Whole D-loop (S)	Human–chimp (4.9)	Not reported
Chen et al. (1995)	133,000–101,000	Coding regions (D, 3rd codon position) Whole mtDNA (R)	Central America colonization by Chibchan-speaking population	Maximum likelihood (Nei and Tajima, 1983)
Ruvolo et al. (1993)	536,000–129,000	COII + ND4/ND5 (S, 3rd codon position)	Human–chimp (5–7)	Maximum-likelihood (Felsenstein, 1990)
Pesole et al. (1992)	800,000–300,000	Whole D-loop (S, T)	Human–chimp (5–7)	Markov model improved version (Pesole et al., 1992)
Kocher and Wilson (1991)	157,000	ND4-ND5 (S, 3rd codon position)	Human–chimp (5)	Brown et al. (1982)
Hasegawa et al. (1991)	280,000–50,000	Whole D-loop (S, f) D-loop (HVS I + HVS II)(S, f) D-loop (HVS I + part HVS II) (S, f)	Human–chimp (4)	Hasegawa et al. (1985)
Vigilant et al. (1991)	249,000–166,000	D-loop (HVS I + HVS II) (S)	Human–chimp (4–6)	Wilson's method
Cann et al. (1987)	290,000–140,000	Whole mtDNA (R)	New Guinea colonization (0.040)	Wilson's method

[a]S, Sequence analysis data; R, restriction analysis data; f, correction factor; HVS, hypervariable sequence.

Table VII
Nucleotide Substitution Rate for Various Mitochondrial Genome Regions Obtained by Comparing Human–Human (H-H) and Human–Chimp (H-C) Sequences

Human deepest root (H-H)[a]		Human–chimp (H-C)[b]	
Sites	Sub/site yr × 10⁻⁹	Sites	Sub/site yr × 10⁻⁹
D-loop		D-loop	
$f = 1$	27.5	$f = 1$	13.2
$f = 0.28$	65.5	$f = 0.28$	94.2
Synonymous		Synonymous	
$f = 1$	13.7	$f = 1$	28.6
Asynonymous		Asynonymous	
$f = 1$	3.7	$f = 1$	2.1
12S		12S	
$f = 1$	8.7	$f = 1$	3.6
16S		16S	
$f = 1$	3.7	$f = 1$	6.0
tRNA		tRNA	
$f = 1$	5.0	$f = 1$	3.5

[a]Nucleotide substitution rate fixing the human deepest root at 400,000 years.
[b]Nucleotide substitution rate is calculated with a time of divergence human–chimp at 6 Myr.

features, it should provide information from the male side concerning the evolution of the species, since it can be regarded as the counterpart of mtDNA. Sequence analysis of different Y chromosome loci located in the sex-specific region or immediately adjacent to it have been carried out. In Table VIII, the proposed calculations from data obtained with chromosome Y are reported,

Table VIII
Inferred Times of the Last Common Ancestor (LCA) from Human Chromosome Y

Reference	LCA (yr)	Data	Calibration time (Myr)	Method
Dorit et al. (1995)	270,000 (0–800,000)	Locus ZFY (0.7 Kbp)	Human–chimp (5)	Not reported
Hammer (1995)	188,000 (51,000–411,000)	Locus YAP (2.6 Kbp)	Human–chimp (5)	Nei (1987)
Whitfield et al. (1995)	37,000–49,000	Locus SRY (18.3 Kbp)	Human–chimp (6)	Tamura and Nei (1993)

which have been brought to date for the origin of modern humans. The data emphasize that intraspecies variability is very low in some loci and even absent in others. Thus, the Y chromosome does not seem to contribute much information from an evolutionary viewpoint.

5. REFERENCES

Adachi, J., and Hasegawa, M., 1994, MOLPHY: Programs for Molecular Phylogenetics, Institute of Statistical Mathematics, Tokyo.

Adkins, R. M., and Honeycutt R. L., 1991, Molecular phylogeny of the superorder Archonta, Proc. Natl. Acad. Sci. USA 88:10317–10321.

Anderson, S., Bankier, A. T., Barrel, B. G., de Bruijen, M. H. L., Coulson, A. R., Drouin, J., Eperon, I. C., Nierlich, D. P., Roe, B. A., Sanger, F., Schreier, P. H., Smith, A. J. H., Staden, R., and Young, I. G., 1981, Sequence and organization of the human mitochondrial genome, Nature 290:457–465.

Arnason, U., and Johnsson, E., 1992, The complete mitochondrial DNA sequence of the harbor seal, Phoca vitulina, J. Mol. Evol. 34:493–505.

Arnason, U., Gullberg, A., and Widegren, B., 1991, The complete nucleotide sequence of the mitochondrial DNA of the fin whale Baleanoptera physalus, J. Mol. Evol. 33:556–568.

Asakawa, S., Himeno, H., Miura, K., and Watanabe, K., 1995, Nucleotide sequence and gene organization of the starfish Asterina pectinifera mitochondrial genome, Genetics 140:1047–1060.

Benne, R., 1985, Mitochondrial genes in trypanosomes, Trends Genet. 1:117–121.

Bertorelle, G., Calafell, F., Francalacci, P., Bertranpetit, J., and Barbujani, G., 1996, Geographic homogeneity and non-equilibrium patterns of mtDNA sequences in Tuscany, Italy, Hum. Genet. 98:145–150.

Bertranpetit, J., Sala, J., Calafell, F., Underhill, P. A., Moral, P., and Comas, D., 1995, Human mitochondrial DNA variation and the origin of Basques, Ann. Hum. Genet. 5955:63–81.

Brown, G. G., Gadaleta, G., Pepe, G., Saccone, C., and Sbisà, E., 1986, Structural conservation and variation in the D-loop containing region of vertebrate mitochondrial DNA, J. Mol. Biol. 92:503–511.

Brown, W. M., Prager, E. M., Wang, A., and Wilson, A. C., 1982, Mitochondrial DNA sequences of primates: Tempo and mode of evolution, J. Mol. Evol. 18:225–239.

Cann, R. L., Stoneking, M., and Wilson, A. C., 1987, Mitochondrial DNA and human evolution, Nature 325:31–36.

Cantatore, P., Gadaleta, M. N., Roberti, M., Saccone, C., and Wilson A. C., 1987, Duplication and remoulding of tRNA genes during the evolutionary rearrangement of mitochondrial genomes, Nature 329:853–855.

Chen, Y. S., Torroni, A., Excoffier, L., Santachiara-Benerecetti, A. S., and Wallace, D. C., 1995, Analysis of mtDNA variation in African populations reveals the most ancient of all human continent-specific haplogroups, Am. J. Hum. Genet. 57:133–149.

Crozier, R. H., and Crozier, Y. C., 1993, The mitochondrial genome of the honeybee Apis mellifera: Complete sequence and genome organization, Genetics 133:97–117.

De Giorgi, C., Martiradonna, A., Lanave, C., and Saccone, C., 1996, Complete sequence of the mitochondrial DNA in the sea urchin Arbacia lixula: Conserved features of the echinoid mitochondrial genome, Mol. Phyl. Evol. 5:323–332.

D'Erchia, A. M., Gissi, C., Pesole, G., Saccone, C., and Arnason, U., 1996, The guinea pig is not a rodent, Nature 381:597–600.

Desjardins, P., and Morais, R., 1991, Nucleotide sequence and evolution of coding and noncoding regions of a quail mitochondrial genome, *J. Mol. Evol.* **32**:153–161.

de Zamaroczy, M., and Bernardi, G., 1985, Sequence organization of the mitochondrial genome of yeast. A review, *Gene* **37**:1–17.

Di Mauro, S., and Wallace, D. C., eds., 1993, *Mitochondrial DNA in human pathology*, Raven Press, New York.

Dorit, R. L., Akashi, H., and Gilbert, W., 1995, Absence of polymorphism at the ZFY locus on the human Y chromosome, *Science* **268**:1183–1184.

Felsenstein, J., 1990, *PHYLIP: Phylogenetic inference package, version 3.3*, Department of Genetics, University of Washington, Seattle.

Gadaleta, G., D'Elia, D., Capaccio, L., Saccone, C., and Pepe, G., 1996, Isolation of a 25-kDa protein binding to a curved DNA upstream the origin of the L strand replication in the rat mitochondrial genome, *J. Biol. Chem.* **271**:13537–13541.

Graur, D., Hide, W. A., and Li, W.-H. 1991, Is the guinea pig a rodent? *Nature* **351**:649–652.

Gray, M. W., Cedergren, R., Abel, Y., and Sankoff, D., 1989, On the evolutionary origin of the plant mitochondrion and its genome, *Proc. Natl. Acad. Sci. USA* **86**:2267–2271.

Gray, M. W., Hanic-Joyce, P. J., and Covello, P. S., 1992, Transcription, processing and editing in plant mitochondria, *Annu. Rev. Plant Physiol. Plant Mol. Biol.* **43**:145–157.

Grivell, L. A., 1987, in: *Genetic Maps*, vol. 4 (S. J. O'Brien, ed.), pp. 290–297, Cold Spring Harbor Laboratory, Cold Spring Harbor, New York.

Hammer, M. F., 1995, A recent common ancestry for human Y chromosomes, *Nature* **378**: 376–378.

Hasegawa, M., Kishino, H., and Yano, T., 1985, Dating of human–ape splitting by a molecular clock of mitochondrial DNA, *Mol. Evol.* **22**:160–174.

Hasegawa, M., Kishino, H., and Saitou, N., 1991, On the maximum likelihood method in molecular phylogenetics, *J. Mol. Evol.* **32**:443–445.

Holmes, E. C., Pesole, G., and Saccone, C., 1989, Stochastic models of molecular evolution and the estimation of phylogeny and rates of nucleotide substitutions in the hominoid primates, *J. Hum. Evol.* **18**:775–794.

Horai, S., Hayasaka, K., Kondo, R., Tsugane, K., and Takahata, N., 1995, Recent African origin of modern humans revealed by complete sequences of hominoid mitochondrial DNAs, *Proc. Natl. Acad. Sci. USA* **92**:532–536.

Irwin, D. M., Kocher, T. D., and Wilson, A. C., 1991, Evolution of the cytochrome *b* gene of mammals, *J. Mol. Evol.* **32**:128–144.

Janke, A., Feldmaier-Fuchs, G., Thomas, W. K., von Haeseler, A. and Pääbo, S., 1994, The marsupial mitochondrial genome and the evolution of placental mammals, *Genetics* **137**:243–256.

Kocher, T. D., and Wilson, A. C., 1991, *Evolution of Life: Fossils, Molecules, and Culture* (S. Osawa and T. Honjio, eds.), pp. 391–413, Springer-Verlag, Tokyo.

Levings III, C. S., and Brown, G. G., 1989, Molecular biology of plant mitochondria, *Cell* **56**: 171–179.

Ma, D. P., King, Y.-T., Kim, Y., Luckett Jr., W. S., Boyle, J. A., and Chang, Y.-F., 1992, Amplification and characterization of an inverted repeat from the *Chlamidomonas reinhardtii* mitochondrial genome, *Gene* **119**:253–257.

Moraes, C. T., and Schon, E. A., 1995, Replication of a heteroplasmic population of normal and partially deleted human mitochondrial genomes, in *Progress in Cell Research*, vol. 5 (F. Palmieri, S. Papa, C. Saccone, and M. N. Gadaleta eds.), pp. 209–215, Elsevier Science, Amsterdam.

Moritz, C., Dowling, T. E., and Brown, W. M., 1987, Evolution of animal mitochondrial DNA: Relevance for population biology and systematics, *Annu. Rev. Ecol. Syst.* **18**:269–292.

Nei, M., 1987, *Molecular Evolutionary Genetics*, Columbia University Press, New York.

Nei, M., and Tajima, F., 1983, Maximum likelihood estimation of the number of nucleotide substitutions from restriction sites data, *Genetics* **105:**207–217.

Newton, J. K., 1988, Plant mitochondrial genomes: organization, expression and variation, *Annu. Rev. Plant Physiol. Plant Mol. Biol.* **39:**503–532.

Okimoto, R., Chamberlin, H. M., Macfarlane, J. L., and Wolstenholme, D. R., 1991, Repeated sequence sets in mitochondrial DNA molecules of rootknot nematodes (*Meloidogyne*): Nucleotide sequences, genome location and potential for host-race identification, *Nucleic Acids Res.* **19:**1619–1626.

Okimoto, R., Macfarlane, J. L., Clary, D. O., and Wolstenholme, D. R., 1992, The mitochondrial genome of two nematodes, *Caenorhabditis elegans* and *Ascaris suum*, *Genetics* **130:**471–498.

Pääbo, S., Thomas, W. K., Whitfield, K. M., Kumazawa Y., and Wilson, A. C., 1991, Rearrangements of mitochondrial transfer RNA genes in marsupials, *J. Mol. Evol.* **33:**426–430.

Palmer, J. D., 1990, Contrasting modes and temps of genome evolution in land plant organelles, *Trends Genet.* **6:**115–120.

Pepe, G., Holtrop, M., Gadaleta, G., Kroon, A. M., Cantatore, P., Gallerani, R., De Benedetto, C., Quagliariello, C., Sbisà, E., and Saccone, C., 1983, Nonrandom patterns of nucleotide substitutions and codon strategy in the mammalian mitochondriAl genes coding for identified and unidentified reading frames, *Biochem. Int.* **6:**553–563.

Pepe, G., Gadaleta, G., Palazzo, G., and Saccone, C., 1989, Sequence-dependent DNA curvature: Conformational signal present in the main regulatory region of the rat mitochondrial genome, *Nucleic Acids Res.* **17:**8803–8819.

Pesole, G., Bozzetti, M. P., Lanave, C., Preparata, G., and Saccone, C., 1991, Glutamine synthetase gene evolution: A good molecular clock, *Proc. Natl. Acad. Sci. USA* **88:**522–526.

Pesole, G., Sbisà, E., Preparata, G., and Saccone, C., 1992, The evolution of the mitochondrial D-loop region and the origin of modern man, *Mol. Biol. Evol.* **9:**587–598.

Ruvolo, M., Zehr, S., von Dornum, M., Pan, D., Chang, B., and Lin, J., 1993, Mitochondrial COII sequences and modern human origins, *Mol. Biol. Evol.* **10**(6):1115–1135.

Saccone, C., 1994, The evolution of mitochondrial DNA, *Curr. Opin. Genet. Dev.* **4:**875–881.

Saccone, C., and Sbisà, E., 1994, The evolution of mitochondrial genome, in *Principles of Medical Biology*, vol. 1B (E. E. Bittar, ed.), pp. 39–72, JAI Press, Greenwich, CT.

Saccone, C., Cantatore, P., Gadaleta, G., Gallerani, R., Lanave, C., Pepe, G., and Kroon, A. M., 1981, The nucleotide sequence of the large ribosomal RNA gene and the adjacent tRNA genes from rat mitochondria, *Nucleic Acids Res.* **9:**4139–4148.

Saccone, C., Attimonelli, M., De Giorgi, C., Lanave, C., and Sbisà, E., 1990a, The role of tRNA genes in the evolution of animal mitochondrial DNA, in *Structure, Function and Biogenesis of Energy Transfer System* (E. Quagliarello, S. Papa, F. Palmieri, and C. Saccone, eds.), pp. 93–96, Elsevier Science, Amsterdam.

Saccone, C., Lanave, C., Pesole, G., and Preparata, G. 1990b, The influence of base composition on quantitative estimates of gene evolution, in *Methods in Enzymology*, vol. 183 (R. F. Doolittle, ed.), pp. 570–583, Academic Press.

Saccone, C., Pesole, G., and Sbisà, E., 1991a, The main regulatory region of mammalian mitochondrial DNA: Structure-function model and evolutionary pattern, *J. Mol. Evol.* **33:**83–91.

Saccone, C., Pesole, G., and Kadenbach, B., 1991b, Evolutionary analysis of the nucleus-encoded subunits of mammalian cytochrome *c* oxidase, *Eur. J. Biochem.* **195:**151–156.

Saccone, C., Lanave, C., Pesole, G., and Sbisà, E., 1993, Peculiar features and evolution of mitochondrial genome of mammals, in *Mitochondrial DNA in Human Pathology* (S. Di Mauro, and D. C. Wallace, eds.), pp. 27–37, Raven Press, New York.

Sbisà, E., Tanzariello, F., Reyes, A., Pesole, G., and Saccone, C., 1997, Mammalian mitochondrial D-loop structural analysis: Identification of new conserved sequences and their functional and evolutionary implications, *Gene* **205:**125–160.

Slater, E. C., Tager, J. M., Papa, S., and Quagliarello, E., eds., 1968, *Biochemical Aspects of the Biogenesis of Mitochondria*, Adriatica Editrice, Bari, Italy.

Swofford, D. L., 1993, *PAUP: Phylogenetic Analysis Using Parsimony. Version 3.1.1*, Illinois Natural History Survey, Champaign, IL.

Tamura, K., and Nei, M., 1993, Estimation of the number of nucleotide substitutions in the control region of mitochondrial DNA in humans and chimpanzees, *Mol. Biol. Evol.* **10**:512–526.

Torroni, A., Sukernik, R., Schurr, I. T. G., Starikorskaya, Y., Cabell, B. M. F., Crawford, M. H., Comuzzie, A. G., and Wallace, D. C., 1993a, mtDNA variation of aboriginal Siberians reveals distinct genetic affinities with Native Americans, *Am. J. Hum. Genet.* **53**:591–608.

Torroni, A., Schurr, T. G., Cabell, M. F., Brown, M. D., Neel, J. V., Larsen, M., Smith, D. G., Vullo, C. M., and Wallace, D. C., 1993b, Asian affinities and continental radiation of the four founding Native American mtDNAs, *Am. J. Hum. Genet.* **53**:563–590.

Tullo, A., Rossmanith, W., Imre, E. M., Sbisà, E., Saccone, C., and Karwan, R. M., 1995, RNase mitochondrial RNA processing cleaves RNA from the rat mitochondrial displacement loop at the origin of heavy-strand DNA replication, *Eur. J. Biochem.* **227**:657–662.

Valverde, J. R., Batuecas, B., Moratilla, C., Marco, R., and Garesse, R., 1994, The complete mitochondrial DNA sequence of the crustacean *Artemia franciscana*, *J. Mol. Evol.* **39**:400–408.

Vigilant, L., Stoneking, M., Harpending, H., Hawkes, K., and Wilson A. C., 1991, African populations and the evolution of human mitochondrial DNA, *Science* **253**:1503–1507.

Ward, R. H., Frazier, B. L., Dew-Jager, K., and Pääbo, S., 1991, Extensive mitochondrial diversity within a single Amerindian tribe, *Proc. Natl. Acad. Sci. USA* **88**:8720–8724.

Whitfield, L. S., Sulston, J. E., and Goodfellow, P. N., 1995, Sequence variation of the human Y chromosome, *Nature* **378**:379–380.

Wolstenholme, D. R., and Clary, D. O., 1985, Sequence evolution of *Drosophila* mitochondrial DNA, *Genetics* **109**:725–744.

Wolstenholme, D. R., 1992, Animal mitochondrial DNA: Structure and evolution, *Int. Rev. Cytol.* **141**:173–216.

CHAPTER 22

Nuclear Transcription Factors in Cytochrome *c* and Cytochrome Oxidase Expression

Richard C. Scarpulla

1. NUCLEAR AND MITOCHONDRIAL CONTRIBUTIONS TO RESPIRATORY CHAIN EXPRESSION

A distinctive feature of eukaryotes is the compartmentalization of genetic systems for the replication and expression of nuclear, mitochondrial, and chloroplast genes. In vertebrate mitochondria, the compact circular genome is dedicated to the synthesis of 13 polypeptide subunits of the electron transport and oxidative phosphorylation systems (Attardi and Schatz, 1988). The genes for these subunits are interspersed with those encoding the rRNAs and tRNAs required for translation within the mitochondrial matrix. The mammalian mitochondrial genome, with its absence of introns, the minimization of intergenic and untranslated regions, and the expression of polygenic transcripts, represents a striking departure from nuclear genome organization and expression. Divergent promoters within a regulatory region, termed the D-loop, direct the synthesis of polygenic transcripts from both DNA strands. These tran-

Richard C. Scarpulla   •   Department of Cell and Molecular Biology, Northwestern Medical School, Chicago, Illinois 60611.

Frontiers of Cellular Bioenergetics, edited by Papa *et al.* Kluwer Academic/Plenum Publishers, New York, 1999.

scripts are subsequently cleaved to yield the individual functional RNAs. DNA replication is bidirectional, but initiation from each strand proceeds asymmetrically from separate origins located about two thirds of the genome apart (Clayton, 1991). By contrast, plants and unicellular eukaryotes, such as *Saccharomyces cerevisiae*, have larger genomes as well as more complex gene structures and arrangements (Tracy and Stern, 1995). Nevertheless, the polypeptide contribution of mitochondrial genomes to the respiratory apparatus is very similar among eukaryotes (Wolstenholme, 1992).

The limited coding capacity of the mitochondrial DNA necessitates that nuclear genes contribute the majority of respiratory subunits and all of the proteins necessary for mitochondrial transcription, translation, and DNA replication. In addition, the essential cofactors and substrates for oxidative phosphorylation are all produced through the action of nuclear gene products. Therefore, understanding the molecular mechanisms governing the interplay between nuclear and mitochondrial genetic systems in meeting cellular energy demands depends on a thorough analysis of the regulated expression of nuclear genes. An impressive body of work with the yeast, *S. cerevisiae*, has established that transcriptional mechanisms play an important role in governing the interactions between nucleus and mitochondria. This work has been the subject of several recent reviews (Poyton and McEwen, 1996; Grivell, 1995; Shyjan and Butow, 1993; Zitomer and Lowry, 1992; Poyton and Burke, 1992; Costanzo and Fox, 1990; Forsburg and Guarente, 1989a). By contrast, detailed molecular studies are less advanced in vertebrate systems. However, in recent years, a number of transcriptional regulatory proteins that are likely to play a role in nucleomitochondrial interactions in mammalian cells have been identified (Scarpulla, 1996, 1997; Nagley, 1991).

The major focus of this chapter will be on the recent progress made in mammalian systems in identifying nuclear transcription factors involved in the expression of cytochrome *c* and cytochrome oxidase genes. Regulatory models developed in *S. cerevisiae* provide a useful framework for comparison and will also be discussed.

2. REGULATED EXPRESSION OF CYTOCHROME *c* AND CYTOCHROME OXIDASE

Nucleomitochondrial interactions rely upon the action of both extra- and intracellular signals on the expression of nuclear genes. In yeast, the availability of oxygen and carbon sources is communicated to the transcriptional apparatus through site-specific regulatory proteins that act on the nuclear genes required for respiratory function (Zitomer and Lowry, 1992). In mammalian cells, nuclear respiratory genes appear to respond to a variety of signals including contractile activity, hormones, and second messengers (Luciakova and Nelson, 1992; Ku *et al.*, 1991; Williams *et al.*, 1987). However, in the majority of cases the specific regulatory proteins that mediate these responses have yet to be identified.

2.1. Cytochrome *c* Gene Expression

The primary structure of cytochrome *c* was among the first to be determined and the protein has been the subject of pioneering work in the areas of molecular evolution and protein structure and function (Dickerson, 1972). Detailed genetic analysis led to the deduction of a partial nucleotide sequence long before direct nucleic acid sequencing was feasible (Crow and Dove, 1990). For these reasons, the cytochrome *c* genes were among the first genes to be isolated from both yeast and mammals (Scarpulla *et al.*, 1981; Montgomery *et al.*, 1978).

In yeast, two isoforms of cytochrome *c* are expressed from unlinked genes (Sherman and Stewart, 1971). Iso-1-cytochrome *c*, the product of the *CYC1* locus, represents about 95% of the total, whereas iso-2-cytochrome *c*, the product of the *CYC7* locus, comprises the remainder. The rate of cytochrome *c* synthesis in yeast is glucose-repressed, and rapid derepression involves a step in mRNA synthesis (Zitomer and Nichols, 1978; Zitomer and Hall, 1976). A transcriptional mechanism was implicated by the finding that the mRNA half-life is the same under both conditions (Zitomer *et al.*, 1979). This conclusion is supported by experiments with a *CYC1–lacZ* fusion gene introduced into yeast cells on a plasmid vector (Guarente and Ptashne, 1981). In this fusion, 1100 nucleotides of the *CYC1* 5′-untranslated and flanking region directs the expression of β-galactosidase in yeast and confers glucose repression on enzyme synthesis. Several heme proteins, including iso-1-cytochrome *c*, are also repressed by anaerobic and heme-deficient conditions (Hortner *et al.*, 1982). As with glucose, regulated expression by heme and oxygen availability occurs at the level of mRNA synthesis (Laz *et al.*, 1984; Guarente and Mason, 1983).

Cytochrome *c* synthesis is also under the control of extracellular signals in mammalian cells. The best known example is the induction of cytochrome *c* by thyroid hormones in the tissues of whole animals. Thyroid hormones have long been known to regulate the rate of oxygen consumption and ATP utilization in responsive tissues (Sestoft, 1980; Tata *et al.*, 1963). Increased demands for ATP under hyperthyroid conditions are accompanied by increased mitochondrial mass and respiratory units within the mitochondrial inner membrane. Concomitant with these changes are elevated expression of key metabolic enzymes and respiratory cytochromes (Horrum *et al.*, 1985). The cytochrome *c* content of tissues has been correlated with the rate of oxidative metabolism and thyroidal status (Drabkin, 1950a,b). Thyroid hormone administration results in a net increase in the rate of cytochrome c synthesis in rat liver (Booth and Holloszy, 1975). Cytochrome c mRNA levels are also markedly elevated in the livers and kidneys of thyroidectomized rats treated with thyroid hormone, and this induction is accompanied by an increase in gene transcription (Scarpulla et al., 1986a). However, the slow kinetics of induction suggest that the cytochrome c gene is not a direct target for activation by thyroid hormone receptor. This interpretation is consistent with the observation that the cytochrome c promotor is not trans-activated by thyroid hormone receptor in the presence of thyroid hormone in transfected cells (L. Gopalakrishnan and R. C. Scarpulla, unpublished data).

More recently, cytochrome c mRNA levels were found to increase upon treatment of cultured cells with dibutyryl cyclic AMP or agents that increase intracellular levels of cAMP (Gopalakrishnan and Scarpulla, 1994). In this case, mRNA induction is rapid and insensitive to cycloheximide, indicating that direct gene activation is involved. Moreover, the relative gene transcription rate is also increased, with similar kinetics. The control of cytochrome c synthesis by cAMP may provide a mechanism for coordinating the regulation of cytochrome c gene expression with carbon source availability. Glucagon and epinephrine, which regulate the release of glucose from glycogen and the synthesis of glucose from amino acids, both act through G protein-linked receptors to stimulate adenylate cyclase (Linder and Gilman, 1992). In addition, epinephrine triggers the degradation of triacylglycerols, the storage form of fatty acids. Thus, transcriptional activation of cytochrome c expression under conditions of elevated cAMP may signal the availability of metabolic fuels to the respiratory apparatus. The molecular basis for the response to cAMP is discussed in Section 3.2.3.

In several mammalian species, a second isoform of cytochrome c is expressed during spermatogenic differentiation. The testicular isoform was first discovered in mouse testis and was found to differ from the somatic form in 13 of 104 amino acid positions (Hennig, 1975). Immunolocalization experiments indicated that the testicular protein is first expressed at the mid-pachytene stage of spermatogenic differentiation and persists in cells at later stages of differentiation (Goldberg et al., 1977; Wheat et al., 1977). Only the somatic form is present in Sertoli cells, interstitial cells, and spermatogonia demonstrating that expression of the testicular protein is restricted to germinal epithelium (Goldberg et al., 1977). The existence of a second cytochrome c gene was confirmed by the isolation and characterization of mouse and rat cDNA clones that encode the testis-specific variant and the subsequent isolation and sequencing of the rat gene (Virbasius and Scarpulla, 1988). The testis-specific gene differs markedly from the somatic form in its size and intron–exon structure. Moreover, the two genes are dissimilar in their untranslated and flanking regions including the flanking DNA upstream from the transcription initiation site. The mouse testis promoter binds tissue-specific proteins that may account for germ cell expression (Yiu *et al.*, 1997). Thus, developmental signals bring about the differential transcriptional expression of cytochrome *c* from distinct genetic loci in mammalian systems.

2.2. Cytochrome Oxidase Gene Expression

The cytochrome oxidase complex is composed of subunits encoded by both nuclear and mitochondrial genomes, making it a good model for investigating nucleomitochondrial interactions (Kadenbach *et al.*, 1995; Poyton and Burke, 1992; Lomax and Grossman, 1989). The three mitochondrial subunits have similarities to prokaryotic cytochrome oxidases in their primary structures. This sequence similarity and the fact that the prokaryotic enzyme is catalytically active in the absence of nuclear subunits is consistent with the notion that the mitochondrial oxidase subunits make up the catalytic core of the enzyme. The nuclear subunits (10 in mammals, 6 in yeast) contribute essential regulatory, catalytic, or assembly functions (Poyton and McEwen, 1996; Poyton *et al.*, 1988; Kadenbach and Merle, 1981).

Regulated expression of cytochrome oxidase genes is best understood in yeast for cytochrome oxidase subunits V (COX5) and VI (COX6). This work has been the subject of recent comprehensive reviews (Poyton and McEwen, 1996; Poyton and Burke, 1992). Like cytochrome *c*, COX5 is the product of two differentially regulated genes, *COX5a* and *COX5b* (Cumsky *et al.*, 1987). Disruption of both genes demonstrated that COX5 is essential to oxidase function. Strains whose COX5 was provided by a single copy of *COX5a* respire normally under conditions where multiple copies of *COX5b* are required (Trueblood and Poyton, 1987). Gene fusion experiments demonstrated that this results from a higher efficiency of expression of *COX5a* rather than from more efficient import into mitochondria or utilization of COX5a by the respiratory chain. High-efficiency expression of COX5a under oxidative growth conditions results in it being the predominant form present in yeast mitochondria. This is consistent with a differential transcriptional response of the *COX5a* and *COX6* genes to the presence of heme and oxygen (Hodge *et al.*, 1989). The presence of differentially regulated *COX5* genes is reminiscent of cytochrome *c* where differential expression of *CYC1* and *CYC7* genes is also dictated by the availability of oxygen.

By contrast to COX5, COX6 is the product of a single genetic locus (Wright *et al.*, 1984). Expression of the *COX6* gene is subject to catabolite repression by glucose and also to positive transcriptional regulation by heme. In this way, its pattern of expression resembles that of *CYC1* and *COX5a* and differs from *CYC7* and *COX5b*. Thus, in yeast, the genes encoding cytochrome *c* and cytochrome oxidase subunits fall into two classes that are defined by their transcriptional responses to oxygen and carbon source (Poyton and Burke, 1992). In the class represented by *CYC1*, *COX5a*, and *COX6*, high levels of expression are dictated by the availability of oxygen, heme, and nonfermentable carbon sources. These genes are thus most active during respiratory growth and contribute the predominant forms of their respective protein products to the respiratory chain. Expression of a second class of nuclear genes, represented by *CYC7*, *COX5b*, and others, are maximally expressed at the transcriptional level under anaerobic conditions. The protein isoforms encoded by certain of these hypoxic genes alter the kinetic properties of cytochrome oxidase to enhance activity under conditions of low oxygen tension (Allen *et al.*, 1995; Waterland *et al.*, 1991).

The rules governing the transcriptional expression of cytochrome oxidase in mammalian systems are not well understood, although a number of interesting phenomena have been described in recent years. As in yeast, the mammalian enzyme is composed of three mitochondria-encoded subunits (COX1, COX2, and COX3) that constitute the catalytic core. Cloned cDNAs for all ten of the mammalian oxidase nuclear subunits have been isolated and characterized in recent years (Lomax and Grossman, 1989). Three of the ten (COX6a, COX7a, and COX8) exist as tissue-specific isoforms that were initially designated either as liver or heart. The heart isoform is the major protein expressed in heart and skeletal muscle and the liver isoform is expressed in all other tissues (Seelan and Grossman, 1992). Several exceptions to this pattern exist. For example, both isoforms of COX6a (Anthony *et al.*, 1990) and COX7a (Van Beeumen *et al.*, 1990) have been found in human heart where there is an absence of the heart form of COX8 (Rizzuto *et al.*, 1989; Van Kuilenburg *et al.*, 1988). The nuclear genes encoding COX4 (Carter and Avadhani, 1991; Yamada *et al.*, 1990; Virbasius and Scarpulla, 1990; Bachman *et al.*, 1987), -5b (Basu and Avadhani, 1991), -6aL (Mell *et al.*, 1994), -6aH (Wan and Moreadith, 1995), -6c (Suske *et al.*, 1988), -7aL (Seelan and Grossman, 1993), -7aH (Seelan and Grossman, 1992), -7c (Seelan and Grossman, 1997), and -8H (Lenka *et al.*, 1996) have been isolated and characterized.

Many isolated examples of regulated expression of cytochrome oxidase subunits have been reported, but as of now no unifying principles have emerged from these studies. It has been known for many years that cytochrome oxidase enzyme activity is correlated with the levels of tissue-specific respiration (Stotz, 1939). In cases where it has been examined, mRNA content for both nuclear and mitochondrial subunits has also correlated with enzyme activity, suggesting that tissue-specific oxidase expression is controlled by mRNA synthesis or degradation (Gagnon *et al.*, 1991). Within a given tissue in rat, the level of cytochrome oxidase activity has been associated with the coordinate steady-state expression of mRNAs for COX3 and COX6c, although some differences are observed between mRNA and protein levels between muscle and nonmuscle tissues (Hood, 1990). These studies suggest that coordinate gene expression contributes to maintaining tissue-specific levels of cytochrome oxidase.

A number of *in vivo* model systems have been used to investigate the expression of nuclear and mitochondrial genes under changing physiological conditions. Chronic electrical stimulation of skeletal muscle was used to induce mitochondrial biogenesis and increase respiratory enzyme activities (Williams *et al.*, 1987). Stimulation of rabbit tibialis anterior muscles over a period of 10 to 21 days brought about a marked increase in mitochondrial citrate synthase and cytochrome oxidase activities (Williams *et al.*, 1987). This was accompanied by modest but specific increases in mRNAs for the F_1-ATPase β-subunit and COX6c and larger increases in mitochondrial cytochrome *b* transcripts. In a separate study, mRNAs for both mitochondrial COX3 and nuclear COX6c were coordinately elevated along with cytochrome oxidase activity upon extended electrical stimulation of the rat tibialis anterior muscle (Hood *et al.*, 1989). In endurance-trained human athletes, increases in mitochondrial volume in working muscles are accompanied by increased nucleus- and mitochondria-encoded mRNAs for several respiratory chain components, including cytochrome oxidase subunits (Puntschart *et al.*, 1995). Thus, it is likely that the induction of cytochrome oxidase gene expression in muscle plays a role in meeting increased demands for respiratory energy brought about by elevated contractile activity. These findings are consistent with the interpretation that terminally differentiated cells in complex organisms can adapt to changing energy demands by modulating the expression of respiratory chain genes.

Mammalian cytochrome oxidase expression also may be subject to hormonal and developmental controls. Differential screening of cDNA libraries constructed from Sertoli cells from hypophysectomized rats treated with follicle-stimulating hormone (FHS) yielded an FSH-induced COX1 cDNA clone (Ku *et al.*, 1991). In addition to FSH, forskolin and dibutyryl cAMP induce mRNAs for COX1 and -2 as well as mitochondrial 16S rRNA in these cells, suggesting that a cAMP-dependent signaling pathway operates on mitochondrial gene expression. The nucleus-encoded mRNA for COX5a is also markedly induced under the same conditions. Thus, an FSH-induced cAMP-dependent pathway may help coordinate the expression of respiratory genes encoded by both nucleus and mitochondria in Sertoli cells.

Many studies have also implicated thyroid hormones in the control of cytochrome oxidase expression. Several studies point to an increase in mRNAs for mitochondrial-encoded COX1, -2, and -3 in various tissues from thyroid hormone-treated rats (Wiesner *et al.*, 1992; Van Itallie, 1990) with no increase in nucleus-encoded COX4, -5a, and -6c (Wiesner *et al.*, 1992; Luciakova and Nelson, 1992; Van Itallie, 1990; Virbasius and Scarpulla, 1990; Joste *et al.*,

1989) except in one case where COX4 was induced (Wiesner *et al.*, 1992). In several of these studies, mRNAs encoding the subunits from other respiratory chain complexes are also induced by hormone (Wiesner *et al.*, 1992; Luciakova and Nelson, 1992; Virbasius and Scarpulla, 1990; Joste *et al.*, 1989). The physiological significance of these differential effects on mRNA expression are not understood, but they may reflect genuine differences in gene regulation. It is important to note that thyroid hormones have been implicated in numerous cellular responses, but relatively few genes have been shown to be the direct targets of receptor-dependent gene activation. Most of the observed changes in respiratory chain expression are likely to be indirect effects of thyroid hormone action.

Both mitochondria- and nucleus-encoded cytochrome oxidase mRNAs are markedly increased during mouse postnatal development, with several of the tissue-specific subunits exhibiting isoform switching (Kim *et al.*, 1995). These increases are accompanied by a proportionate increase in mitochondrial DNA. Although the largest postnatal differences occurred in ventricle and skeletal muscle, no COX mRNA induction is observed in the differentiation of myotubules from myoblasts in cell culture. A developmental program of COX gene expression is also supported by the finding that the mRNA for COX6aH is increased relative to that for COX6aL postnatally, resulting in the predominance of the H isoform in skeletal and cardiac muscle (Parsons *et al.*, 1996).

3. ACTIVATORS AND REPRESSORS OF CYTOCHROME *c* GENE TRANSCRIPTION

3.1. Positive and Negative Regulators of Cytochrome *c* in Yeast

An important objective is to associate the physiological changes in respiratory chain gene expression discussed in the previous sections with the specific regulatory molecules and pathways mediating these changes. This is most readily accomplished in microorganisms, such as yeast, where a combined genetic and biochemical approach is feasible. The ability to isolate or construct mutations in candidate regulatory genes is a valuable tool in establishing the *in vivo* regulatory function of a given transcriptional activator or repressor. In addition, as a facultative anaerobe, *S. cerevisiae* is genetically programmed to survive in both aerobic and anaerobic growth conditions. Studies with yeast cytochrome *c* thus have led to the identification of key regulatory molecules that govern the activation and repression of cytochrome *c* gene transcription in response to physiological signals.

Early work demonstrated that transcriptional regulation of the *CYC1* gene by heme is mediated by an upstream activation site (UAS1) located upstream from the proximal promoter (Guarente and Mason, 1983). This site directs the majority of transcription under conditions of glucose repression and is dependent on a *trans*-acting factor encoded by a gene designated *HAP1* (Guarente *et al.*, 1984). *HAP1* was subsequently found to be identical fo *CYP1*, a gene associated with the enhanced expression of *CYC7* (Vestweber and Schatz, 1988; Clavilier *et al.*, 1969). The *HAP1* (*CYP1*) gene encodes a transcriptional activator that affects *CYC7* expression upon binding a UAS (Iborra *et al.*, 1985). Heme is required for the binding of the HAP1 protein to the *CYC1* UAS1 *in vitro*, suggesting that the heme-dependence of USA1 *in vivo* is mediated by a heme–HAP1 interaction (Pfeifer *et al.*, 1987a). A second protein, RC2, also binds the HAP1 site in UAS1 and has the potential to competitively displace HAP1 binding, although its physiological function has not been established. Interestingly, although UASs from *CYC1* and *CYC7* differ in nucleotide sequence, HAP1 binds both sites with high affinity (Pfeifer *et al.*, 1987b). The protein also directs higher-level expression from *CYC1* UAS1 than from *CYC7* UAS and an internal deletion in HAP1 enhances this differential (Kim *et al.*, 1990). It has been proposed that the *CYC7* UAS induces a masking of the activation domain of HAP1, which is enhanced in the deletion by bringing DNA binding and activation domains into closer proximity.

Analysis of the HAP1 protein revealed a zinc finger DNA binding motif near its amino-terminus and a carboxy-terminal acidic transcriptional activation domain (Pfeifer *et al.*, 1989; Verdière *et al.*, 1985). A large internal region consisting of six conserved repeats of a short amino acid sequence was proposed to interfere with DNA binding in the absence of heme (Pfeifer *et al.*, 1989). The sequence resembles the heme binding sites found in several enzymes and its removal results in heme-independent, constitutive activation *in vivo* (Zhang and Guarente, 1995; Pfeifer *et al.*, 1989). Recent evidence indicates that this domain functions by blocking HAP1 dimerization in the absence of heme (Zhang *et al.*, 1993). The heme binding domain also may help sequester HAP1 in a high-molecular-weight cellular complex and may prevent it from binding its recognition site (Fytlovich *et al.*, 1993). In this model, a partially functional HAP1 dissociates from the complex when intracellular heme concentrations are high. A second region of HAP1, near the activation domain, also has been implicated in heme regulation, suggesting that multiple heme-responsive domains may modulate HAP1 function (Haldi and Guarente, 1995). Thus, the heme-dependence of HAP1 function *in vivo* may serve to modulate the expression of respiratory genes in response to oxygen availability through several interrelated mechanisms.

A second regulatory enhancer in the cytochrome *c* gene, designated UAS2, differs from UAS1 by its sensitivity to the availability of nonfermentable carbon sources and its inability to be derepressed by heme analogues (Guarente *et al.*, 1984). In addition, UAS2 is unaffected by mutations in *HAP1* but is sensitive to mutations in *HAP2*, a distinct regulatory locus that does not affect UAS1. HAP2 was subsequently found to encode a subunit of a hetero-trimeric complex that binds UAS2 (Forsburg and Guarente, 1989a; Hahn and Guarente, 1988; Olesen *et al.*, 1987). This complex is composed of subunits encoded by three distinct loci—*HAP2*, *-3*, and *-4*—which are all required for transcriptional activation through UAS2. The HAP2–HAP3 complex binds a sequence motif in UAS1 that resembles the CCAAT-box, the binding site for a family of sequence-specific activators in eukaryotic cells (Olesen *et al.*, 1987). One family member, CP1, is a CCAAT-box transcription factor that is thought to act as a global activator of many unrelated genes and resembles the HAP2–HAP3 complex in that it binds its recognition site as a heterodimer. Inter-estingly, HAP2 and HAP3 are functionally interchangeable with the human CP1 subunits (Chodosh *et al.*, 1988), suggesting that HAP2–HAP3 may have a much broader function in yeast as well. The HAP2 and -3 subunits have distinct regions that constitute a composite DNA binding domain for the complex (Xing *et al.*, 1993).

An additional subunit that was not observed *in vitro* was defined geneti-cally as the product of *HAP4*, a gene whose transcription is under catabolite repression and is induced by growth in lactate (Forsburg and Guarente, 1989b). The HAP4 protein is an inducible regulatory subunit that provides an acidic transcriptional activation domain to the complex but does not bind DNA by itself (Olesen and Guarente, 1990; Forsburg and Guarente, 1989b). As dis-cussed in Section 4.2.1., the mammalian nuclear respiratory factor 2 (NRF-2), which acts on cytochrome oxidase genes, also has its DNA binding and activation functions on different subunits (Gugneja *et al.*, 1995, 1996; J. V. Vir-basius *et al.*, 1993). A fourth subunit of the complex, HAP5 was recently isolated using the two-hybrid method and was found to be required for sequence-specific DNA binding (McNabb *et al.*, 1995). Thus, HAP2–HAP3–HAP5 represents an unusual example of a heterotrimeric DNA binding com-plex. It will be of interest to determine whether this DNA binding complex can associate with additional regulatory subunits to mediate transcriptional re-sponses to other physiological signals.

The proteins discussed thus far exert a positive effect on the expression of cytochrome *c* and other respiratory genes. In several instances, however, these genes are subject to negative regulation through the action of repressors of gene transcription. Transcription of the *CYC7* gene was observed to be under both positive and negative control mediated by distinct *cis*-acting elements (Zitomer *et al.*, 1987). In studies with a heme-repressed gene, *ANB1*, mutations in a second gene, *ROX1*, resulted in the constitutive expression of *ANB1* and other oxygen-regulated genes including *CYC7* (Lowry and Zitomer, 1984, 1988). A null mutation in *ROX1* enhanced expression of *CYC7*, suggesting that *ROX1* is a repressor of *CYC7* expression. The finding that ROX1 is also induced by heme indicated that *ROX1* encodes a repressor that inhibits expression of heme-repressed genes under aerobic growth conditions. Maximal heme induction of ROX1 is thought to require HAP1 (Deckert *et al.*, 1995a). ROX1 has been described as a high mobility group box protein that is capable of binding a hypoxic consensus sequence present in *ANB1*, *CYC7*, and *COX5b* genes among others (Balasubramanian *et al.*, 1993). Like other HMG proteins, ROX1 can bend DNA upon binding (Deckert *et al.*, 1995b). A transcriptional repression domain was localized toward the carboxy-terminal end of the molecule, but the mechanism of repression has yet to be determined. Thus, the regulated expression of cytochrome *c* in yeast results from the action of transcriptional activators and repressors that communicate the availability of metabolites to the transcriptional machinery.

3.2. Transcriptional Activators of Cytochrome *c* Expression in Mammalian Cells

3.2.1. Experimental Strategies for the Identification of Regulatory Proteins

The absence of a tractable genetic approach in mammalian cells necessitates that the identification of nuclear transcription factors involved in respiratory chain expression rely on molecular and biochemical methods. These entail the identification of *cis*-acting elements by transfectional analysis of mutated promoters and the isolation and characterization of the *trans*-acting factors that direct the activation of transcription through these elements. This reliance on the specificity of DNA–protein interactions as an indicator of biological specificity requires that several experimental criteria be met. First, the *cis*-acting elements identified by mutation should function within the proper promoter context. Second, nuclear protein interactions with these elements should be specific according to several independent criteria. The DNA–protein complexes formed should be competitively displaced by an excess of specific binding sites but not by nonspecific sequences. The protein should make nucleotide contacts within the mutationally defined element and point mutations of these contacts should abolish both factor binding and transcriptional activation. Third, the site-specific DNA binding and transcriptional activities should co-purify. Transcriptional activation can be monitored by *in vitro* transcription assays using a template containing the promoter binding site. Fourth, once the protein has been purified and a cDNA isolated, the recombinant protein expressed from the cDNA should display the same DNA binding and transcriptional specificities as the protein isolated from crude cellular extracts. Differences in the properties of cellular and recombinant proteins may indicate that a related but nonidentical protein has been cloned. Alternatively, the observed differences may result from posttranslational modifications that are not made in bacterial expression systems. Finally, specific antiserum against the recombinant protein should recognize the same DNA–protein complexes formed with the homogeneously pure cellular protein and with crude nuclear extracts.

The final proof of a regulatory function will require the manipulation of the gene encoding the transcription factor in an *in vivo* context and evaluating the expression of target genes. It is important to note that there are limitations to the approach outlined. For example, regulatory enhancers that are the targets of developmental or metabolic signaling pathways *in vivo* may not be detected by transient transfection of cultured cells. Thus, the *cis*-acting elements and *trans*-acting factors identified may be biased toward those that function in stable cell lines. Despite these caveats, several candidate transcription factors that act on subsets of respiratory chain and related genes have been isolated.

3.2.2. Characterization of Human and Rodent Somatic Cytochrome *c* Genes

The identification of nuclear transcription factors involved in respiratory chain expression was initiated in mammalian systems with the isolation and characterization of the rat somatic cytochrome *c* gene (Scarpulla *et al.*, 1981). The gene was isolated from a rat genomic library using the yeast *CYC1* gene as a probe. A large number of cytochrome *c* sequences were detected in mammalian genomes using the initial isolate, suggesting the presence of a large multigene family for cytochrome *c* (Scarpulla *et al.*, 1982). This was surprising because only a single somatic isoform of the protein had been described. Subsequent isolation and sequencing of many of these family members revealed that they are intronless pseudogenes that arise through mRNA retrotransposition (Scarpulla, 1984; Scarpulla and Wu, 1983). Similar cytochrome *c* sequence complexity was encountered in the human genome (Evans and Scarpulla, 1988a). As in rat, a single intron-containing functional human gene was isolated that accounted for the human protein expressed in somatic cells. The remainder of the family was composed of intronless pseudogenes. Although yeast and mammalian cytochromes *c* are highly divergent, expression of rat cytochrome *c* in yeast was able to complement a mutation in the *CYC1* gene (Scarpulla and Nye, 1986). Thus, cytochrome *c* protein function is highly conserved between yeast and mammals.

To determine whether a similar conservation existed in transcriptional control elements, an extensive deletion analysis of the cytochrome *c* promoter region was undertaken (Evans and Scarpulla, 1988b). As shown in Fig. 1, three major subregions of the promoter are required for maximal gene transcription in COS and CV-1 cells. Deletion of the most upstream region I had a large effect

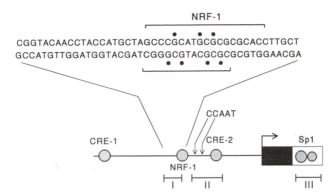

FIGURE 1. The rat cytochrome *c* promoter. The positions of recognition sites for CREB (CRE-1 and 2), NRF-1, and Sp1 (shaded spheres) are indicated within the 5'-flanking (solid line) and first intron (open box) of the rat cytochrome *c* promoter. Promoter regions I, II, and III are indicated below. The sequence of the region I-NRF-1 element is shown above with the NRF-1 footprint (brackets) and guanine contacts (filled spheres) on each DNA strand as indicated. The arrows denote putative binding sites for CCAAT transcription factors and the transcription initiation site.

on promoter activity but had no recognizable binding sites for known regulatory proteins. Region II is more proximal to the transcription start site and has tandem CCAAT-box similarities, whereas region III has consensus Sp1 binding sites and is localized to the 5'-end of the first intron. With the exception of the CCAAT-boxes in region II, none of the promoter regions has obvious similarities to the binding sites for the known yeast regulatory proteins.

3.2.3. Activation of the Cytochrome *c* Promoter by Sp1 and cAMP Response Element Binding Protein

DNAse I footprinting experiments were carried out to determine whether the *cis*-acting elements identified by mutagenesis were binding sites for nuclear DNA binding proteins (Evans and Scarpulla, 1989). Three distinct DNAse I footprints were found using a crude nuclear extract, and two of these coincided with promoter regions I and III. The DNAse I footprint in region III, within the first intron, coincides with a tandem direct repeat of Sp1 consensus binding sites (Fig. 1). The upstream site was a stronger match to the Sp1 consensus, and an intact upstream site is required for Sp1 binding to the downstream site. The strong correlation between promoter activity and the binding of Sp1 makes it likely that Sp1 or an Sp1-related protein is a major *trans*-activator of the intron promoter element. A third footprint is upstream from the previously mapped elements but coincides with a perfect consensus sequence for the activating transcription factor (ATF)–CREB family of transcriptional activators (Montminy *et al.*, 1986; Short *et al.*, 1986). Linker insertion mutagenesis of this site results in a modest reduction in basal promoter activity. A synthetic oligonucleotide containing this sequence directs cAMP-dependent expression from a heterologous promoter in transfected COS cells.

Additional DNAse I footprinting of an extended promoter fragment using BALB/3T3 nuclear extracts demonstrated the existence of a previously undetected footprint that was downstream from the CCAAT boxes (Gopalakrishnan and Scarpulla, 1994). This footprint is centered over a match to the CRE consensus (ACGTCA). The CRE sequences 5′ and 3′ relative to the nuclear respiratory factor 1 (NRF-1) site are designated CRE-1 and CRE-2, respectively (Fig. 1). Methylation interference and UV-induced DNA protein cross-linking assays demonstrate that both CREs make similar guanine nucleotide contacts upon specific binding of a 42-kDa protein from crude nuclear extracts. The mass of this protein is consistent with the binding of the cAMP response element binding protein CREB, and pure recombinant CREB binds both CREs *in vitro*. Both CREs also confer a response to cAMP on a heterologous viral promoter when cloned in *cis*. In addition, the response of transfected cytochrome *c* promoters to elevated cAMP levels requires both CREs and both are *trans*-activated specifically by a combination of CREB and protein kinase A. Thus, the transcriptional induction of cytochrome *c* expression by cAMP is mediated through CREB-dependent signaling.

3.2.4. NRF-1

Fine deletion mapping of region I demonstrated that it is a composite of two functionally distinct *cis*-acting elements (Evans and Scarpulla, 1989) (Fig. 1). One of these coincides with a strong DNAse I footprint over a GC-rich palindrome that had no obvious similarity to known transcription factor binding sites. The protein interacting at this site was designated as nuclear respiratory factor 1, because similarities to this sequence were present in the promoters of other newly isolated genes whose products contribute to mitochondrial respiratory function (Table I). These included genes encoding cytochrome oxidase and reductase subunits and the mitochondrial RNA processing RNA (Evans and Scarpulla, 1990). The latter encodes an RNA subunit of a ribonucleoprotein endonuclease, the MRP RNAse, that can cleave nascent mitochondrial light strand transcripts to generate primers for mitochondrial heavy strand DNA replication (Topper *et al.*, 1992; Chang and Clayton, 1989). This finding suggested that NRF-1 may provide a link between the expression of the respiratory chain and the replication of mitochondrial DNA.

Several criteria were applied to establish the functionality of the putative

Table I
NRF-1- and NRF-2-Dependent Genes[a]

NRF-1		NRF-2	NRF-1 + NRF-2	Neither
Cyt c	ATPSyn γ	COX4	COX5b	Cyc c$_1$
hQP	mtTFA	COX5b	COX7aL	ANT
COX5b	MRP RNA	COX7aL	mtTFA	COX6aH
COX6c	5-ALAS	COX7c		COX8H
COX7aL		ATPSyn β		ATPSyn α
		mtTFA		

[a]Nuclear genes that contribute to the synthesis or function of the respiratory apparatus in mammalian cells are categorized by the presence of functional promoter recognition sites for NRF-1, NRF-2, both NRF-1 and NRF-2, and neither. The gene names are abbreviated as follows: Cyt c, somatic cytochrome c (Evans and Scarpulla, 1988b, 1989); hQP, ubiquinone binding protein (Suzuki et al., 1989); COX cytochrome oxidase subunits 4 (Virbasius and Scarpulla, 1991), 5b (Basu and Avadhani, 1991), 6c (Suske et al., 1988), 6aH (Wan and Moreadith, 1995), 7aL (Seelan et al., 1996), 7c (Seelan and Grossman, 1997), 8H (Lenka et al., 1996); ATPSyn, ATP synthase subunits α (Breen et al., 1995), β (Villena et al., 1994), γ (Dyer et al., 1989); Cyt c$_1$, cytochrome c$_1$ (Li et al, 1996); ANT, adenine nucleotide transporter (Li et al., 1990) mtTFA, mitochondrial transcription factor A (Virbasius and Scarpulla, 1994; Tominaga et al., 1992); MRP RNA, mitochondrial RNA processing RNA (Topper and Clayton, 19990; Chang and Clayton, 1989); 5-ALAS, 5-aminolevulinate synthase (Braidotti et al., 1993).

NRF-1 binding sites from the various genes. Synthetic oligomers of each site compete for the binding of NRF-1 to the cytochrome c promoter. Each forms an NRF-1–DNA complex of identical size and specificity as measured by methylation interference footprinting and competition mobility shift and footprinting assays. All of the NRF-1 binding site sequences also direct transcription from a truncated cytochrome c promoter when cloned in cis. Thus, NRF-1 appears to be a novel transcription factor that contributes to expression of the respiratory apparatus in mammalian cells.

To explore this possibility further, NRF-1 was purified to near homogeneity using conventional chromatographic methods combined with DNA affinity chromatography (Chau et al., 1992). The protein was purified over 33,000-fold from a crude nuclear extract and all of the specific binding was assigned to a single 68-kDa a protein that was excised and renatured from sodium dodecyl sulfate–polyacrylamide gel electrophoresis (SDS-PAGE) gels. The purified protein displayed the same DNA binding specificity as the protein in crude nuclear extracts and UV-induced DNA–protein cross-linking demonstrated that the 68-kDa protein co-purified with the specific binding activity. Biologically active NRF-1 binding sites also are found in genes encoding eukaryotic initiation factor 2 α-subunit (eIF-2α) and tyrosine aminotransferase (Chau et al., 1992). These proteins participate in the rate-limiting steps of protein synthesis and tyrosine catabolism, respectively, and are not related directly to respiratory chain function. In addition, NRF-1 was identified independently as a protein factor that recognizes a GC-rich palindromic element in the eIF-2α promoter region (Jacob *et al.*, 1989). These results suggest that NRF-1 may help integrate the expression of other metabolic functions with respiratory chain expression.

The sequencing of tryptic peptides from the purified NRF-1 protein led to the isolation of a HeLa cell cDNA clone (C. A. Virbasius *et al.*, 1993). Although the cDNA reading frame specifies a protein of 54 kDa instead of the expected 68 kDa, the tryptic peptides are present in the sequence and the recombinant protein expressed from this clone migrates at 68 kDa on denaturing SDS gels. In addition, antiserum raised against the recombinant protein recognizes specifically the NRF-1–DNA complex formed with crude nuclear extracts, the purified HeLa protein, or recombinant NRF-1. The recombinant NRF-1 has the same binding site specificity as that of the purified HeLa cell protein and also directs the synthesis of properly initiated, NRF-1-dependent transcripts from specific promoter templates *in vitro*. These results established that the cDNA clone encodes NRF-1.

Deletion mapping demonstrated that the NRF-1–DNA binding domain (Fig. 2) does not have the common structural motifs associated with prokaryotic or eukaryotic regulatory proteins. A search of sequence databases revealed extensive sequence similarity with P3A2 (Calzone *et al.*, 1991) and *Drosophila* erect wing gene (EWG) (Desimone and White, 1993), two newly described regulatory factors. This similarity is restricted to the DNA binding domains of the three proteins, indicating that NRF-1, P3A2, and EWG represent a new family of sequence-specific regulatory factors that is defined by a common DNA binding motif (C. A. Virbasius *et al.*, 1993). The sequence divergence outside of the DNA binding domains is consistent with the disparate functions of the three proteins. P3A2 is a negative transcriptional regulator of developmentally expressed genes in sea urchins (Hoog *et al.*, 1991). It is thought to act

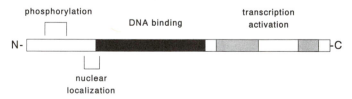

FIGURE 2. NRF-1 functional domains. The positions of the DNA binding (filled box) and transcription activation (shaded boxes) domains are indicated within the 503 amino acid NRF-1 protein sequences (open box). The locations of phosphorylation sites and the nuclear localization signal are indicated by brackets above and below, respectively.

by displacing a zinc finger activator that binds the same sequence. P3A2 is one of several transcription factors that act on a cytoskeletal actin gene that is expressed only in embryonic and larval ectoderm. The EWG protein is required for the viability of *Drosophila* embryos and for proper nervous system development (Desimone and White, 1993). The protein has been localized to the neuronal nuclei of 6- to 12-hr embryos, but no target genes have been identified.

Recently, the discovery of recognition sites for NRF-1 in two additional genes provided further evidence for an integrative role in respiratory chain expression (Table I). First, transcription of the gene encoding the ubiquitously expressed form of 5-aminolevulinate synthase (5-ALAS) has been found to be highly dependent on two tandem NRF-1 recognition sites (Braidotti *et al.*, 1993). These sequences have the correct DNA binding specificity and point mutations that disrupt NRF-1 binding and dramatically reduce promoter function in transfected cells. 5-ALAS, a mitochondrial matrix enzyme, catalyzes the first and rate-limiting step in the synthesis of heme for respiratory cytochromes. Thus, NRF-1 control over 5-ALAS expression may affect the availability of heme and consequently the function of cytochromes encoded by both nuclear and mitochondrial genomes. So far, there is no evidence that this model operates *in vivo*.

The human gene encoding mitochondrial transcription factor A (mtTFA) (Tominaga *et al.*, 1992) also has a functional NRF-1 recognition site within its promoter (Virbasius and Scarpulla, 1994). This site binds cellular and recombinant NRF-1 through characteristic guanine nucleotide contacts. Point mutations that eliminate NRF-1 binding to this site have a dramatic down effect on promoter activity and also severely dampen the effects of *cis*-acting mutations in Sp1 and NRF-2 recognition sites within the promoter. This dependence on an intact NRF-1 site is also observed in an *in vitro* transcription assay, where properly initiated mtTFA transcripts are not formed using templates with point mutations in the NRF-1 site. As a stimulatory factor for mitochondrial gene transcription, mtTFA promotes both the transcription and replication of mtDNA (Shadel and Clayton, 1993; Clayton, 1991). Therefore, its transcriptional requirement for NRF-1 provides a potential regulatory mechanism for nuclear control of mtDNA maintenance and gene expression.

To test models based on the control of nuclear–mitochondrial interactions by NRF-1 and related transcription factors, it ultimately will be necessary to manipulate the expression and function of these proteins in cultured cells and transgenic animals. This will require the isolation of the genes encoding these factors and the understanding of structure-to-function relationships in the encoded proteins. The human NRF-1 gene recently has been isolated and found to span over 65 kb on q31 of chromosome 7 (Gopalakrishnan and Scarpulla, 1995). A 5′-untranslated region composed of two noncoding exons was cloned via rapid amplification of cDNA ends–polymerase chain reaction (RACE-PCR) using a primer complementary to the first coding exon. The 5'-end of the RACE-PCR product coincides with a major transcription start site detected by S1 nuclease mapping of total RNA and the genomic sequence upstream from this position contains an active promoter. This promoter directs the transcription of properly initiated transcripts in transfected cells.

3.2.4.1. Structure-to-Function Relationships in NRF-1.

Several functional domains in NRF-1, shown schematically in Fig. 2, were identified by testing the ability of mutated NRF-1 molecules to *trans*-activate NRF-1-dependent promoters in transfected cells (Gugneja *et al.*, 1996). An amino-terminal domain just upstream from the DNA binding domain is highly conserved among NRF-1 family members and is required to direct NRF-1 to its site of action within the nucleus. This nuclear localization signal (NLS) is composed of three repeated clusters of basic residues. The first two clusters make up a composite NLS that combines with the third cluster to mediate efficient nuclear targeting.

A second region of NRF-1, which when mutated has a dramatic effect on the *trans*-activation of the reporter promoters, maps to the carboxy-terminal half of the molecule. This region appears to comprise a bipartite transcriptional activation domain (Gugneja *et al.*, 1996). This domain is unusual in that the sequence of the active regions do not fall into the known classes of activation domains defined by a preponderance of a particular amino acid residue. Instead, tandemly arranged repeated clusters of glutamine-containing hydrophobic residues are the major determinants of function. Alanine substitution mutagenesis demonstrated that the hydrophobic residues are essential to each of the active clusters, whereas the glutamines make no contribution to activity. Although much of the activation domain coincides with acidic regions of the molecule, the most active subdomain is devoid of acidic residues. These results demonstrate that DNA binding and activation functions are distinct and functionally independent.

Most recently, a particularly interesting functional domain in NRF-1 has been localized near the amino-terminus, just upstream from the NLS (Fig. 2). Extensive mutagenesis of this region has revealed that a cluster of serine residues serve as the targets of phosphorylation *in vivo* (Gugneja and Scarpulla, 1997). Phosphorylation of these serines results in a marked enhancement of NRF-1 DNA binding. Although NRF-1 homodimerization is required for DNA binding, phosphorylation stimulates binding without affecting the monomer–dimer equilibrium in the absence of DNA. These results raise the interesting possibility that NRF-1 may be modulated *in vivo* by kinase signaling pathways. As discussed in Section 4.1, ABF1, an activator of COX6 expression in yeast, appears to exist in multiple phosphorylated states (Weintraub *et al.*, 1992). It is of interest in this context that oxygen and heme-dependent phosphorylation of transcription factors have been described in bacterial oxygen sensing pathways (Gilles-Gonzalez *et al.*, 1991; David *et al.*, 1988).

3.2.4.2. Other NRF-1 Functions in Cell Growth and Development.

The discovery of functional NRF-1 sites in genes encoding eIF-2α and tyrosine aminotransferase prompted a search for other potential NRF-1 target genes. Published gene sequences were searched for matches to the NRF-1 consensus binding site. Matches were restricted to those sequences with only a single mismatch to the consensus of 14 functionally defined binding sites. In each case, the mismatched nucleotide was restricted to those present at the same position in one of the known binding sites. With these stringent criteria, 48 additional potential target genes were identified in computer databases (C. A. Virbasius *et al.*, 1993). In several cases the NRF-1 site is present in a similar position in the same gene from different species, indicating that it is functionally conserved. The most interesting of these genes encode rate-limiting enzymes of metabolic pathways and proteins involved in cell growth and cell cycle regulation. Although intriguing, a rigorous analysis of NRF-1 function within these genes will require an evaluation of each site within the proper promoter context and the identification of the cognate activator protein. Nevertheless, these observations make it likely that NRF-1 functions to integrate respiratory chain expression with other cellular functions.

In one well-documented case, NRF-1 has been implicated in the expression of the chicken histone *H5* gene during erythrocyte development (Gomez-Cuadrado *et al.*, 1992, 1995). A 75-kDa glycoprotein originally called chicken initiation binding repressor (cIBR) was isolated on the basis of its binding to sequences spanning the transcription start site of the *H5* gene. This factor represses transcription *in vitro* and mutations in its recognition site overcome this repression. cIBR was expressed in mature erythrocytes but not in early erythroid cells that express the *H5* gene. Instead, a nonglycosylated protein— chicken initiation binding factor (cIBF)—recognizes the same site and is present in transcriptionally active cells (Gomez-Cuadrado *et al.*, 1992). Subsequent purification and molecular cloning of cIBR revealed that it is the chicken homologue of NRF-1 and that cIBR and cIBF are most likely the products of the same gene (Gomez-Cuadrado *et al.*, 1995). Like NRF-1, cIBR is a phosphoprotein and binds its recognition site as a homodimer. The functional consequences of phosphorylation or glycosylation have not yet been established. Since NRF-1 has a well-defined transcriptional activation domain and clearly functions as an activator of respiratory gene transcription (Gugneja *et al.*, 1996), it will be of interest to determine whether glycosylation can modify its ability to activate transcription during development.

4. TRANSCRIPTIONAL ACTIVATORS AND REPRESSORS OF CYTOCHROME OXIDASE GENE EXPRESSION

4.1. Activators and Repressors of Yeast Cytochrome Oxidase Genes

Many of the same transcription factors regulating yeast cytochrome *c* expression also regulate the yeast *COX* genes. The *REO1* gene, a negative regulator of *COX5b*, was identified in a selection for increased *COX5b* expression under aerobic growth conditions (Trueblood *et al.*, 1988). This gene appears to be identical to *ROX1* (Zitomer and Lowry, 1992; Poyton and Burke, 1992). Both *HAP2* and *REO1* have been implicated in the differential expression of *COX5a* and *COX5b* (Trueblood *et al.*, 1988). A *HAP2* mutation reduces *COX5a* promoter activity without affecting that of *COX5b*. A mutation in the *REO1* gene allows growth in a nonfermentable carbon source of a strain with a null mutation in *COX5a* by increasing the expression of *COX5b*. An intact *REO1* gene is required for repression of *COX5b* transcription in the presence of heme, suggesting that heme acts through *REO1*. Multiple *cis*-acting elements that mediate transcriptional control of *COX5b* by oxygen (heme) and carbon source have been identified and some resemble those present in the *CYC* genes (Hodge *et al.*, 1990). Thus, the positive heme-dependent activation of *COX5a* transcription through *HAP2* and the negative heme-dependent repression of *COX5b* through *REO1* is similar to the differential regulation observed with *CYC1* and *CYC7*. The major difference is that, in contrast to *COX5b*, *CYC7* transcription is also inducible by heme through the action of *HAP1*. Other transcription factors involved in the aerobic repression of *COX5b* have also been identified (Lambert *et al.*, 1994).

The yeast *COX6* gene is also controlled by multiple transcription factors. Like *CYC1* and *COX5a*, heme-dependent transcriptional activation and glucose repression of *COX6* is promoted by *HAP2* (Trawick *et al.*, 1989a,b). At least one enhancer element, *COX6* UAS_6, mediates basal expression as well as the response to heme, glucose, and *HAP2* (Trawick *et al.*, 1989b). Glucose repression of *CYC1* and *COX6* appears to be controlled by a glucose-dependent signaling pathway that acts on many glucose-responsive genes, including those involved in sucrose fermentation (Carlson and Laurent, 1994; Wright and Poyton, 1900). At least two genes, *SSN6* and *SNF1* (the latter encoding a protein kinase) have been implicated in this pathway (Wright and Poyton, 1990). However, the precise link between these genes and the HAP2–HAP3–HAP4 complex has not yet been established.

Further dissection of UAS_6 suggested that HAP2 is required for glucose repression but does not bind to its consensus binding site *in vitro*, suggesting that HAP2 either acts indirectly or may require auxiliary factors for binding *in vitro*. A closely adjacent subdomain of UAS_6 binds transcription factor ABF1, which is required for transcription under repressing conditions (Trawick *et al.*, 1992). ABF1 has been found in association with the expression of many other genes that are not related to mitochondrial function (Dorsman *et al.*, 1990). Multiple electrophoretic forms of the ABF1–DNA complex have been ascribed to differences in phosphorylation (Silve *et al.*, 1992). Phosphorylation is enhanced by growth of cells on nonfermentable carbon sources and is diminished by growth in glucose. Increased phosphorylation of the ABF1–DNA complex is associated with increased *COX6* transcription. Although phosphorylation of ABF1 is affected by mutations in the *SNF1–SSN6* pathway of glucose repression, the SNF1 kinase did not phosphorylate ABF1. Nevertheless, the phosphorylated state of ABF1 may contribute to the carbon source control of *COX6* transcription.

4.2. Transcriptional Activators of Cytochrome Oxidase Expression in Mammalian Cells

4.2.1. Nuclear Respiratory Factor 2 in the Expression of *COX4* and *COX5b* Genes

At the time of the initial characterization of the rat cytochrome *c* promoter region, no other nuclear genes encoding respiratory subunits had been isolated from multicellular organisms. Therefore, the isolation of the rat gene encoding COX4 was undertaken in order to compare its *cis*-acting promoter elements with those found in the cytochrome *c* gene. Mammalian COX4 is homologous to the yeast COX5 subunits and a human cDNA clone had been isolated (Zeviani *et al.*, 1987). Using the human cDNA as a probe, the functional rat gene encoding COX4 and two processed pseudogenes were isolated from a genomic library (Virbasius and Scarpulla, 1990). The functional gene has five exons, whereas the pseudogenes are intronless and have defects in their coding regions that preclude COX4 function. In contrast to the single initiation site observed for cytochrome *c* (Scarpulla, 1984), transcription of COX4 mRNA is initiated at multiple sites spanning over 50 nucleotides. These sites were detected by S1 nuclease protection and confirmed by primer extension analysis (Virbasius and Scarpulla, 1990). The 5'-flanking region is GC-rich and devoid of CCAAT and TATA sequences. It was subsequently found that many of these features are conserved in the mouse gene (Carter and Avadhani, 1991). Interestingly, several differences with cytochrome *c* in the pattern of COX4 mRNA expression were found. Although the tissue-specific, steady-state levels of COX4 mRNA generally parallel that for cytochrome *c*, the overall range of expression among tissues is smaller compared to cytochrome *c*. Also, as noted, COX4 mRNA is not induced significantly by T3 in the tissues of thyroidectomized rats under conditions of marked induction of cytochrome *c* mRNA.

These results suggest that the transcription of respiratory chain genes is not subject to strict coordinate control in mammalian cells.

A number of deletion and site-directed point mutations within the promoter were analyzed for their effects on function in transfected COS cells (Virbasius and Scarpulla, 1991). In contrast to the cytochrome *c* promoter, the major determinants of *COX4* promoter function are localized to about 80 nucleotides partially overlapping the transcription start sites (Fig. 3). A single Sp1 site is present near the upstream boundary of this region and its removal resulted in a severalfold reduction in activity. Surprisingly, with the exception of this Sp1 site, none of the other *cis*-acting elements observed in the cytochrome *c* promoter are present in the *COX4* promoter. However, two identical, tandemly repeated, 11-nucleotide motifs are major contributors to promoter function. Each contained the GGAA core sequence that is characteristic of the binding site for the ETS-domain family of transcription factors (Karim *et al.*, 1990). Site-directed point mutations in the GGAA sequence reduce promoter activity and eliminate the specific binding of nuclear proteins to the tandem repeats. A specific binding protein made contact with the GGAA guanine nucleotides and appeared to be a single molecular species with a mass of 55 kDa. This contrasted with mobility shift assays that revealed multiple DNA–protein complexes. This protein was designated as NRF-2 because a functional

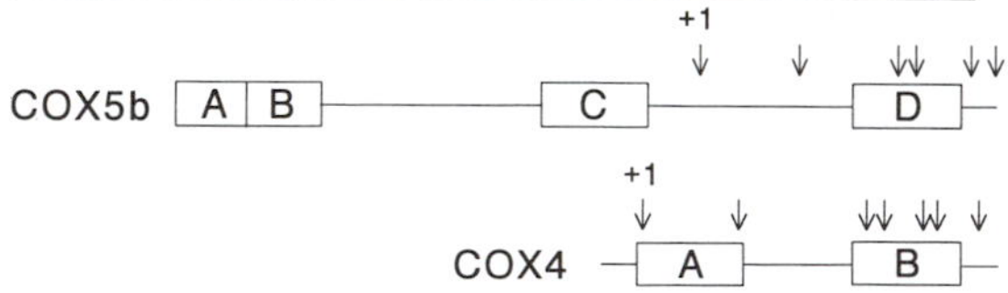

FIGURE 3. NRF-2 recognition sites in cytochrome oxidase promoters. The NRF-2 recognition sites in COX4 and COX5b promoters are aligned. The guanine nucleotides contacted upon NRF-2 binding are indicated above each sequence (closed spheres). Binding sites are labeled consecutively from the 5′-end of each sequence as depicted in the diagram below. The transcription initiation sites are marked by arrows with the most upstream position denoted as +1.

recognition site is also observed in the gene encoding the β-subunit of the ATP synthase.

Although the expression of many of the ETS-domain transcription factors is restricted to specific tissues, some have a broad tissue distribution (Rao *et al.*, 1989; Bhat *et al.*, 1987). The purification of NRF-2 was undertaken to examine the potential relationship between NRF-2 and the known ETS-domain family members. Five distinct polypeptides were co-purified with the NRF-2 binding activity to near homogeneity by using a combination of conventional and DNA affinity chromatography (Virbasius *et al.*, 1993a). The purified preparation displays the correct binding specificity and also stimulates NRF-2-site-dependent transcription *in vitro*. The final preparation yielded the same heterogeneous DNA–protein complexes detected in crude nuclear extracts, with the *COX4* NRF-2 binding sites indicating that the purified polypeptides are subunits of the same binding protein. Specific complexes are associated with presence of particular polypeptides in chromatographic fractions. Excision and renaturation of the various polypeptides from denaturing gel slices revealed that only one—designated the α-subunit—is capable of binding DNA (Fig. 4).

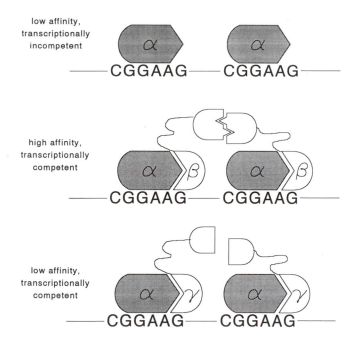

FIGURE 4. NRF-2 binding complexes. NRF-2α and -β subunit interactions with each other and with the tandem binding sites in *COX* promoters is depicted. The binding and transcriptional properties of each complex is summarized at the left.

The 56-kDa mass of this subunit is consistent with the mass of the DNA binding species detected by UV-induced DNA–protein cross-linking (Virbasius and Scarpulla, 1991). The others—designated β_1, β_2, γ_1, and γ_2—are unable to bind DNA by themselves but can associate with α to form the various DNA binding complexes (Fig. 4). Furthermore, the complexes differ in their affinity of binding to the tandem COX4 NRF-1 sites. The $\alpha\beta$ complexes bind the sites with high affinity, whereas the α and $\alpha\gamma$ complexes bind with about an order of magnitude lower affinity. The high-affinity binding by $\alpha\beta$ requires the presence of tandem sites because mutation in one of the *COX4* GGAA repeats results in lower affinity binding. These results are consistent with the high degree of synergism in transcriptional stimulation observed between the NRF-2 sites in the *COX4* promoter. Only the five NRF-2 polypeptides are detected in the pure preparation, indicating that they constitute the predominant protein responsible for the NRF-2 binding activity.

Two other ETS-domain transcription factors that are expressed ubiquitously have multiple interacting subunits. These include mouse (GA binding protein (GABP), a three-subunit activator of herpesvirus genes (Thompson *et al.*, 1991; LaMarco *et al.*, 1991; LaMarco and McKnight, 1989), and human E4TF1, which was initially identified as a two-subunit activator of adenovirus genes (Watanabe *et al.*, 1990). The binding properties of the NRF-2 subunits are reminiscent of the behavior of GABP, which has an α-subunit that binds DNA and two β subunits (β_1 and β_2) that associate with α in DNA binding complexes (Thompson *et al.*, 1991; LaMarco *et al.*, 1991). The GABP β_1 subunit forms a dimer in solution through a unique carboxy-terminal domain. This subunit, when complexed with α, also confers high-affinity binding to tandem ETS sites in a herpesvirus immediate early gene. Partial peptide sequencing of the NRF-2 subunits revealed near perfect matches to the mouse GABP subunits, suggesting that NRF-2 is the human homologue of GABP (J. V. Virbasius *et al.*, 1993). In addition, antiserum against GABP recognized the DNA–protein complexes formed with the *COX5b* ETS sites (Carter *et al.*, 1992).

Definitive proof of the precise relationship between NRF-2 and GABP came with the molecular cloning of all five NRF-2 subunits (Gugneja *et al.*, 1995). The results established that NRF-2 is the human homologue of mouse GABP. The differences in the number of subunits between GABP and NRF-2 result from the expression of human variants of the mouse GABP β_1 and β_2 subunits. The two additional human subunits (NRF-2β_1 and γ_1) have an insertion of 12 amino acids that is presumed to arise from the splicing of an alternative human exon. The function of this insertion is unknown. Three subunits of E4TF1 have now been identified and cloned (Watanabe *et al.*, 1993). These subunits are identical to NRF-2α, β_2, and γ_2. The activation of *COX4* and *COX5b* transcription represented the first known cellular function for these ubiquitous ETS-domain transcription factors.

Detailed analysis of GABP has elucidated functional domains required for DNA binding and subunit associations that are highly conserved in NRF-2 (Thompson *et al.*, 1991; LaMarco *et al.*, 1991). The DNA binding ETS domain is located near the carboxy-terminal end of the α subunit. This region and the ankyrin repeats at the amino-terminal ends of the β subunits (NRF-2β and γ) are required for the formation of heterodimers with α (Fig. 4). As mentioned above, GABP β$_1$ differs from β$_2$ by the presence of a homodimerization domain in its carboxy-terminus that facilitates high-affinity binding to tandem ETS sites. Recently, the transcriptional activation domain of NRF-2 has been localized to a common sequence upstream from the carboxy-terminal domains of the β and γ subunits (Gugneja *et al.*, 1995). Thus, all four of the NRF-2β and γ subunits can confer transcriptional competence to heteromeric complexes formed with α. No activation function is associated with α alone. The NRF-2 activation domain, like that in NRF-1, is not rich in a predominant amino acid residue and has a similar arrangement of glutamine-containing hydrophobic clusters. Alanine substitution mutagenesis demonstrated that the hydrophobic residues and not the glutamines within these clusters are required for activity (Gugneja *et al.*, 1996). Thus, NRF-1 and NRF-2 are related by common structural determinants required for transcriptional activation.

While the characterization of NRF-2 was in progress, the mouse gene encoding COX5b was isolated (Basu and Avadhani, 1991). Inspection of the transcription start site region revealed a tandem array of four potential NRF-2 recognition sites (J. V. Virbasius *et al.*, 1993). The fragment containing these sites behaves as a functional promoter in transfected cells. Purified NRF-2 forms specific complexes with all of the *COX5b* sites and contacts guanine nucleotides within the GGAA motifs (Fig. 3). Point mutation of the guanines within each site eliminates NRF-2 binding and reduces the activity of the proximal *COX5b* promoter. As with the *COX4* NRF-2 sites, the *COX5b* sites display a high degree of synergism in maximizing promoter activity and elimination of all of the sites by point mutation reduces the promoter activity to near background levels.

In addition to the ETS sites, an initiator element has been localized to the transcription start site region of *COX5b* (Basu *et al.*, 1993). This initiator overlaps an Sp1 consensus site, but binds transcription factor NF-E1 (YY-1) *in vitro* and can direct transcription through one of the multiple *COX5b* initiation sites. YY-1 has also been implicated in the negative control of the *COX5b* promoter (Basu *et al.*, 1997) and, along with NRF-2, in the expression of the *COX7c* gene (Seelan and Grossman, 1997). In contrast to NRF-1 and NRF-2 sites, however, the YY-1 initiator is not conserved among rodent and primate *COX5b* genes (Bachman *et al.*, 1996). Initiator function also has been ascribed to GABP (NRF-2) in mouse *COX4* and *-5b* expression (Sucharov *et al.*, 1995). When NRF-1 recognition sites from the *COX4* promoter are cloned into a truncated cytochrome *c* promoter in *cis*, however, they activate transcription through the cytochrome c initiation site, indicating that they can stimulate transcription without specifying the site of initiation (J. V. Virbasius *et al.*, 1993).

4.2.2. Transcription Factors in the Expression of Tissue-Specific *COX* Genes

As discussed, several of the mammalian COX subunits (-6a, -7a, and -8) are expressed from distinct genetic loci as tissue-specific isoforms (Lomax and Grossman, 1989). A heart isoform of these subunits is found predominantly in heart and skeletal muscle, whereas the liver isoform is present in most tissues. The heart and liver cytochrome oxidases have been distinguished by their kinetic properties and by their utilization of adenine nucleotides, leading to speculation that the two forms provide tissue-specific regulatory functions in meeting energy demands (Taanman *et al.*, 1994; Anthony *et al.*, 1993; Taanman and Capaldi, 1993; Rohdich and Kadenbach, 1993; Kadenbach *et al.*, 1991). Genes encoding several of these tissue-specific isoforms have been isolated in recent years, opening the way to understanding the transcriptional mechanisms governing their expression (Lenka *et al.*, 1996; Mell *et al.*, 1994; Smith and Lomax, 1993; Seelan and Grossman, 1993; Rizzuto *et al.*, 1989).

Recently, the mouse *COX6aH* gene has been characterized and the regions of the promoter responsible for tissue-specific transcription identified (Wan and Moreadith, 1995). The gene is present in single copy and, like other known *COX* genes, has heterogeneous transcription initiation sites. Interestingly, the muscle-specific expression pattern of the gene is maintained in cultured skeletal muscle cells. The COX6aH mRNA is expressed in differentiated Sol8 and C_2C_{12} myotubes but not in their respective myoblast precursors prior to differentiation. A transfected promoter region from this gene coupled to a luciferase reporter yields a higher level of expression in myotubes compared to myoblasts or fibroblasts. Analysis of a series of deletions within the 5'-flanking DNA was used to identify a 300-nucleotide segment that is sufficient for myotube-specific expression in Sol8 cells. Mutation of an E-box consensus sequence, the recognition site for the MyoD family of transcription factors (Olson, 1993), within this segment reduces expression in myotubes, myoblasts, and fibroblasts. However, mutation of a MEF2 consensus site eliminates expression in Sol8 myotubes without affecting expression in myoblasts or fibroblasts. The MEF2 proteins comprise a family of muscle-specific transcription factors that have been implicated in the tissue-specific expression of contractile proteins (Yu *et al.*, 1992; Gossett *et al.*, 1989).

In a similar study, the rat COX8H mRNA was found to be expressed at a higher level in cultured skeletal and cardiac myotubes relative to their respective undifferentiated myoblasts (Lenka *et al.*, 1996). A series of *COX8H* promoter deletions was analyzed in transfected myoblasts and myotubules and a stimulatory effect in myotubules was localized to a region of the promoter containing tandem E-box consensus binding sites. Point mutations in these sites reduced the elevated promoter activity in transfected myotubes and nuclear extracts from these cells yielded a complex pattern of DNA–protein interaction with the E-box elements. At least some of these complexes were recognized by an antibody specific for hetero- and homodimers containing transcription factor E2A, suggesting that E-box factors or related proteins are present in the complexes. The results with both the *COX6aH* and *COX8H* genes suggest that the expression of muscle-specific *COX* genes may be linked to the expression of the contractile apparatus through common regulatory factors. No obvious NRF-1 and NRF-2 recognition sites were found in either the *COX6aH* or *COX8H* promoters (Lenka *et al.*, 1996; Wan and Moreadith, 1995). This is consistent with the general observation that in gene pairs encoding ubiquitous and tissue-specific isoforms of a given protein (e.g., somatic and testicular cytochrome *c*), the NRF-1 site is associated with the ubiquitously expressed gene (C. A. Virbasius *et al.*, 1993).

Recent studies with the bovine *COX7aL* (liver isoform) gene have localized its promoter to 92 nucleotides of 5′-flanking DNA (Seelan *et al.*, 1996). The major determinants of promoter function within this region are attributed to specific recognition sites for NRF-1, NRF-2, and Sp1. Point mutations in the NRF-1 and the Sp1 sites result in diminished activity and certain pairwise combinations of these mutations result in a dramatic loss of activity. Purified recombinant NRF-1 and NRF-2 bind specifically to their proposed recognition sites through characteristic guanine contacts. In addition, anti-NRF-1 serum recognizes the specific DNA–protein complexes formed using either crude HeLa nuclear extracts or recombinant NRF-1. The data also suggest the possibility of interactions between the NRF-1 site and upstream Sp1 sites in maximizing promoter activity, although a direct interaction was not demonstrated. Thus, the *COX7aL* promoter resembles *COX4*, *COX5b*, and *mtTFA* promoters in both its compact structure and its reliance on some combination of NRF-1, NRF-2 and Sp1 transcription factors for activity.

5. PERSPECTIVE AND SUMMARY

Much progress has been made in understanding the transcriptional expression of cytochrome *c* and cytochrome oxidase genes, but major challenges remain. In yeast, specific transcriptional activators clearly have been identified as regulatory targets in the pathways utilized for sensing the availability of oxygen and carbon sources in the environment. Future work should clarify the specific connections between the constituents of these pathways and the transcriptional apparatus. For example, the means by which oxygen and heme activate transcription through the HAP2–HAP3–HAP4 complex and the potential links between this complex and the pathways of catabolite repression are not well understood. Future investigations of the various models proposed for oxygen sensing by HAP1 and the mechanisms of transcriptional regulation by ABF1 phosphorylation will also be of particular interest.

Substantial progress also has been made in describing changes in respiratory gene expression in vertebrate systems in response to physiologically important signals. However, little is known about the potential connections between these changes in gene expression and the nuclear transcription factors that contribute to promoter function. The one exception is the cAMP induction of the rat cytochrome c gene. In this case, the cAMP-dependent activation of the cytochrome c promoter by CREB and protein kinase A can account for the *in vivo* response of the endogenous gene to cAMP. An important future challenge will be to establish the regulatory roles of other transcription factors. Many questions remain to be answered. Do NRF-1, NRF-2, and other factors respond to metabolic or developmental signals in regulating respiratory chain expression? If so, what are the mechanisms that control the activities of these factors? Are they induced in response to upstream regulatory events? Are subunit interactions important in regulating the *in vivo* function of multisubunit activators? Do posttranslational modifications, such as the phosphorylation and glycosylation of NRF-1, regulate DNA binding, protein–protein interactions, or activation functions in response to physiological effectors? Future progress in these areas will depend on the development of cell culture models that mimic the regulation of respiratory gene expression *in vivo* and the application of transgenic technology in the evaluation of physiological functions.

It is apparent from the investigations conducted thus far that there is no universal mechanism for coordinating the transcription of all nuclear genes required for respiratory chain expression. For example, the differences observed between cytochrome c and COX4 expression in mammalian cells is reflected in differences in their promoter structures and the utilization of *trans*-acting factors. Nevertheless, a significant subset of nuclear respiratory genes can be grouped according to their utilization of NRF-1 and/or NRF-2 (Table I). Approximately 12 known genes in this category have functional NRF-1 and/or NRF-2 sites in their promoters. Of the 12, 9 are NRF-1 dependent, 6 are NRF-2 dependent, and 3 are dependent on both factors for maximal expression. These observations are consistent with a model, depicted in Fig. 5, where NRF-1 and NRF-2 help coordinate the expression of genes encoding respiratory subunits, components of the mitochondrial DNA transcription and replication machinery, and the rate-limiting enzyme in heme biosynthesis. This simplistic model, however, cannot account for the subset of genes in Table I whose promoters appear to be devoid of recognition sites for either of these factors. The

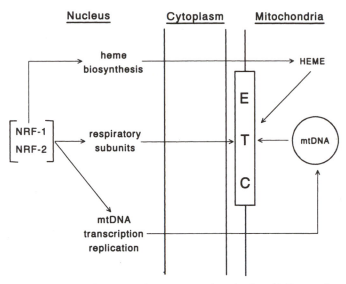

FIGURE 5. Potential roles of NRF-1 and NRF-2 in nucleomitochondrial interactions. Arrows denote the relationship between classes of NRF-1 and NRF-2 target genes and their contributions to the electron transport chain and oxidative phosphorylation system (ETC) of the inner mitochondrial membrane.

COX6aH and *ANT* genes are subject to muscle-specific transcription factors (Wan and Moreadith, 1995; Li *et al.*, 1990), whereas *cytochrome* c_1 and α-*ATPS* appear to be the targets of more ubiquitous activators (Li *et al.*, 1996; Breen *et al.*, 1995; Suzuki *et al.*, 1990). Thus, it is unlikely that all respiratory genes are regulated coordinately by common factors. Perhaps only those components that are rate-limiting for the assembly of respiratory complexes, for electron transfer, and for catalytic functions *in vivo* are subject to coordinate transcriptional control.

Finally, it is unlikely that the transcription factors used in respiratory chain expression are exclusive to this function. Combinatorial action of transcription factors within a given promoter context is often a major determinant of physiological function. The energy-generating systems of the mitochondria are intimately connected to a variety of cellular activities. Therefore, the biogenesis of the respiratory apparatus may well be integrated, at the transcriptional level, with that of other biological systems. For example, several of the transcription factors discussed in the context of cytochrome *c* and cytochrome oxidase gene expression also act on genes, specifying the cystosolic translational machinery. NRF-1 activates the expression of eIF-2α (Chau *et al.*, 1992; Jacob *et al.*, 1989) and NRF-2–GABP has been implicated in the expression of ribosomal protein genes (Genuario *et al.*, 1993). Likewise, as noted previously (Zitomer and Lowry, 1992), ROX1 and ABF1 participate in the expression of a translation initiation factor and ribosomal protein genes in yeast. Since protein synthesis is energy expensive, these observations may reflect a regulatory connection between major energy-producing and -expending cellular systems. Future studies with the transcriptional regulators of cytochrome *c* and cytochrome oxidase genes should continue to yield new insights into the mechanisms that integrate diverse cellular functions during cell growth and differentiation.

ACKNOWLEDGMENT. Work in the author's laboratory is supported by United States Public Health Service Grant GM32525-15 from the National Institutes of Health.

6. REFERENCES

Allen, L. A., Zhao, X.-J., Caughey, W., and Poyton, R. O., 1995, Isoforms of yeast cytochrome *c* oxidase subunit V affect the binuclear reaction center and alter the kinetics of interaction with the isoforms of yeast cytochrome *c*, *J. Biol. Chem.* **270**:110–118.

Anthony, G., Stroh, A., Lottspeich, F., and Kadenbach, B., 1990, Different isozymes of cytochrome *c* oxidase are expressed in bovine smooth muscle and skeletal or heart muscle, *FEBS Lett.* **277**:97–100.

Anthony, G., Reimann, A., and Kadenbach, B., 1993, Tissue-specific regulation of bovine heart cytochrome-*c* oxidase activity by ADP via interaction with subunit VIa, *Proc. Natl. Acad. Sci. USA* **90**:1652–1656.

Attardi, G., and Schatz, G., 1988, Biogenesis of mitochondria, *Annu. Rev. Cell. Biol.* **4**:289–333.

Bachman, N. J., Lomax, M. I., and Grossman, L. I., 1987, Two bovine genes for cytochrome *c* oxidase subunit IV:a processed pseudogene and an expressed gene, *Gene* **55**:219–229.

Bachman, N. J., Yang, T. L., Dasen, J. S., Ernst, R. E., and Lomax, M. I., 1996, Phylogenetic footprinting of the human cytochrome *c* oxidase subunit Vb promoter, *Arch. Biochem. Biophys.* **333**:152–162.

Balasubramanian, B., Lowry, C. V., and Zitomer, R. S., 1993, The Rox1 repressor of the *Saccharomyces cerevisiae* hypoxic genes is a specific DNA-binding protein with a high-mobility-group motif, *Mol. Cell. Biol.* **13**:6071–6078.

Basu, A., and Avadhani, N. G., 1991, Structural organization of nuclear gene for subunit Vb of mouse mitochondrial cytochrome *c* oxidase, *J. Biol. Chem.* **266**:15450–15456.

Basu, A., Lenka, N., Mullick, J., and Avadhani, N. G., 1997, Regulation of murine cytochrome oxidase Vb gene expression in different tissues and during myogenesis—Role of a YY-1 factor-binding negative enhancer, *J. Biol. Chem.* **272**:5899–5908.

Basu, A., Park, K., Atchison, M. L., Carter, R. S., and Avadhani, N. G., 1993, Identification of a transcriptional initiator element in the cytochrome *c* oxidase subunit Vb promoter which binds to transcription factors NF-E1 (YY-1, δ) and Sp1, *J. Biol. Chem.* **268**:4188–4196.

Bhat, N. K., Fisher, R. J., Fujiwara, S., Ascione, R., and Papas, T. S., 1987, Temporal and tissue-specific expression of mouse *ets* genes, *Proc. Natl. Acad. Sci. USA* **84**:3161–3165.

Booth, F. W., and Holloszy, J. O., 1975, Effect of thyroid hormone administration on synthesis and degradation of cytochrome c in rat liver, *Arch. Biochem. Biophys.* **167:**674–677.

Braidotti, G., Borthwick, I. A., and May, B. K., 1993, Identification of regulatory sequences in the gene for 5-aminolevulinate synthase from rat, *J. Biol. Chem.* **268:**1109–1117.

Breen, G. A. M., Jordan, E. M., and Vander Zee, C. A., 1995, ATPF1 binding site, a positive *cis*-acting regulatory element of the mammalian ATP synthase α-subunit gene, *J. Biol. Chem.* **269:**6972–6977.

Calzone, F. J., Hoog, C., Teplow, D. B., Cutting, A. E., Zeller, R. W., Britten, R. J., and Davidson, E. H., 1991, Gene regulatory factors of the sea urchin embryo. I. Purification by affinity chromatography and cloning of P3A2, a novel DNA binding protein, *Development* **112:** 335–350.

Carlson, M., and Laurent, B. C., 1994, The SNF/SWI family of global transcriptional activators, *Curr. Opin. Cell Biol.* **6:**396–402.

Carter, R. S., and Avadhani, N. G., 1991, Cloning and characterization of the mouse cytochrome c oxidase subunit IV gene, *Arch. Biochem. Biophys.* **288:**97–106.

Carter, R. S., Bhat, N. K., Basu, A., and Avadhani, N. G., 1992, The basal promoter elements of murine cytochrome c oxidase subunit IV gene consist of tandemly duplicated ets motifs that bind to GABP-related transcription factors, *J. Biol. Chem.* **267:**23418–23426.

Chang, D. D., and Clayton, D. A., 1989, Mouse RNase MRP RNA is encoded by a nuclear gene and contains a decamer sequence complementary to a conserved region of mitochondrial RNA substrate, *Cell* **56:**131–139.

Chau, C. A., Evans, M. J., and Scarpulla, R. C., 1992, Nuclear respiratory factor 1 activation sites in genes encoding the gamma-subunit of ATP synthase, eukaryotic initiation factor 2α, and tyrosine aminotransferase. Specific interaction of purified NRF-1 with multiple target genes, *J. Biol. Chem.* **267:**6999–7006.

Chodosh, L. A, Olesen, J., Hahn, S., Baldwin, A. S., Guarente, L., and Sharp, P. A., 1988, A yeast and a human CCAAT-binding protein have heterologous subunits that are functionally interchangeable, *Cell* **53:**25–35.

Clavilier, L., Pere, G., and Slonimski, P. P., 1969, Mise en evidence de plusiers loci independants impliques dans la synthese de l'iso-2-cytochrome c chez la levure, *Mol. Gen. Genet.* **104:** 195–218.

Clayton, D. A., 1991, Replication and transcription of vertebrate mitochondrial DNA, *Annu. Rev. Cell Biol.* **7:**453–478.

Costanzo, M. C., and Fox, T. D., 1990, Control of mitochondrial gene expression in *Saccharomyces cerevisiae, Annu. Rev. Genet.* **24:**91–113.

Crow, J. F., and Dove, W. F., 1990, Studies of yeast cytochrome c: How and why they started and why they continued, *Genetics* **125:**9–12.

Cumsky, M. G., Trueblood, C. E., Ko, C., and Poyton, R. O., 1987, Structural analysis of two genes encoding divergent forms of yeast cytochrome c oxidase subunit V, *Mol. Cell. Biol.* **7:**3511–3519.

David, M., Daveran, M., Batut, J., Dedieu, A., Domergue, O., Ghai, J., Hertig, C., Boistard, P., and Kahn, D., 1988, Cascade regulation of *nif* gene expression in *Rhizobium meliloti, Cell* **54:** 671–683.

Deckert, J., Perini, R., Balasubramanian, B., and Zitomer, R. S., 1995a, Multiple elements and auto-repression regulate Rox1, a repressor of hypoxic genes in *Saccharomyces cerevisiae, Genetics* **139:**1149–1158.

Deckert, J., Torres, A. M. R., Simon, J. T., and Zitomer, R. S., 1995b, Mutational analysis of Rox1, a DNA-bending, repressor of hypoxic genes in *Saccharomyces cerevisiae, Mol. Cell. Biol.* **15:** 6109–6117.

Desimone, S. M., and White, K., 1993, The *Drosophila* erect wing gene, which is important for both neuronal and muscle development, encodes a protein which is similar to the sea urchin P3A2 DNA binding protein, *Mol. Cell. Biol.* **13**(6):3641–3949.

Dickerson, R. E, 1972, The structure and history of an ancient protein, *Sci. Am.* **266**:58–72.

Dorsman, J. C., Van Heeswijk, W. C., and Grivell, L. A., 1990, Yeast general transcription factor GFI: Sequence requirements for binding to DNA and evolutionary conservation, *Nucleic Acids Res.* **18**:2769–2776.

Drabkin, D. L., 1950a, Cytochrome *c* metabolism and liver regeneration. Influence of thyroid gland and thyroxine, *J. Biol. Chem.* **182**:335–349.

Drabkin, D. L., 1950b, The distribution of the chromoproteins, hemoglobin, myoglobin, and cytochrome *c*, in the tissues of different species, and the relationship of the total content of each chromoprotein to body mass, *J. Biol. Chem.* **182**:317–333.

Dyer, M. R., Gay, N. J., Powell, S. J., and Walker, J. E., 1989, ATP synthase from bovine mitochondria: Complementary DNA sequence of the mitochondrial import precursor of the gamma-subunit and the genomic sequence of the mature protein, *Biochemistry* **28**:3670–3680.

Evans, M. J., and Scarpulla, R. C., 1988a, The human somatic cytochrome *c* gene: Two classes of processed pseudogenes demarcate a period of rapid molecular evolution, *Proc. Natl. Acad. Sci. USA* **85**:9625–9629.

Evans, M. J., and Scarpulla, R. C., 1988b, Both upstream and intron sequence elements are required for elevated expression of the rat somatic cytochrome *c* gene in COS-1 cells, *Mol. Cell. Biol.* **8**:35–41.

Evans, M. J., and Scarpulla, R. C., 1989, Interaction of nuclear factors with multiple sites in the somatic cytochrome *c* promoter. Characterization of upstream NRF-1, ATF and intron Spl recognition sites, *J. Biol. Chem.* **264**:14361–14368.

Evans, M. J., and Scarpulla, R. C., 1990, NRF-1: A *trans*-activator of nuclear-encoded respiratory genes in animal cells, *Genes Dev.* **4**:1023–1034.

Forsburg, S. L., and Guarente, L., 1989a, Communication between mitochondria and the nucleus in regulation of cytochrome genes in the yeast *Saccharomyces cerevisiae*, *Annu. Rev. Cell Biol.* **5**:153–180.

Forsburg, S. L., and Guarente, L., 1989b, Identification and characterization of HAP4: A third component of the CCAAT-bound HAP2/HAP3 heteromer, *Genes Dev.* **3**:1166–1178.

Fytlovich, S., Gervais, M., Agrimonti, C., and Guiard, B., 1993, Evidence for an interaction between the CYP1(HAP1) activator and a cellular factor during heme-dependent transcriptional regulation in the yeast *Saccharomyces cerevisiae*, *EMBO J.* **12**:1209–1218.

Gagnon, J., Kurowski, T. T., Wiesner, R. J., and Zak, R., 1991, Correlations between a nuclear and a mitochondrial mRNA of cytochrome *c* oxidase subunits, enzymatic activity and total mRNA content, in rat tissues, *Mol. Cell. Biochem.* **107**:21–29.

Genuario, R. R., Kelley, D. E., and Perry, R. P., 1993, Comparative utilization of transcription factor GABP by the promoters of ribosomal protein genes rpL30 and rpL32, *Gene Expr.* **3**:279–288.

Gilles-Gonzalez, M. A., Ditta, G. S., and Helinski, D. R., 1991, A haemoprotein with kinase activity encoded by the oxygen sensor of *Rhizobium meliloti*, *Nature* **350**:170–172.

Goldberg, E., Sberna, D., Wheat, T. E., Urbanski, G. J., and Margoliash, E., 1977, Cytochrome *c*: Immunofluorescent localization of the testis-specific form, *Science* **196**:1010–1012.

Gomez-Cuadrado, A., Rousseau, S., Renaud, J., and Ruiz-Carrillo, A., 1992, Repression of the H5 histone gene by a factor from erythrocytes that binds to the region of transcription initiation, *EMBO J.* **11**:1857–1866.

Gomez-Cuadrado, A., Martin, M., Noel, M., and Ruiz-Carrillo, A., 1995, Initiation binding receptor, a factor that binds to the transcription initiation site of the histone *h5* gene, is a glycosylated member of a family of cell growth regulators, *Mol. Cell. Biol.* **15**:6670–6685.

Gopalakrishnan, L., and Scarpulla, R. C., 1994, Differential regulation of respiratory chain subunits by a CREB-dependent signal transduction pathway. Role of cyclic AMP in cytochrome c and COXIV gene expression, *J. Biol. Chem.* **269**:105–113.

Gopalakrishnan, L., and Scarpulla, R. C., 1995, Structure, expression, and chromosomal assignment of the human gene encoding nuclear respiratory factor 1, *J. Biol. Chem.* **270**:18019–18025.

Gossett, L. A., Kelvin, D. J., Sternberg, E. A., and Olson, E. N., 1989, A new myocyte-specific enhancer-binding factor that recognizes a conserved element associated with multiple muscle-specific genes, *Mol. Cell. Biol.* **9**:5022–5033.

Grivell, L. A., 1995, Nucleo-mitochondrial interactions in mitochondrial gene expression, *Crit. Rev. Biochem. Mol. Biol.* **30**:121–164.

Guarente, L., and Mason, T., 1983, Heme regulates transcription of the *CYC1* gene of *S. cerevisiae* via an upstream activation site, *Cell* **32**:1279–1286.

Guarente, L., and Ptashne, M., 1981, Fusion of the *Escherichia coli lacZ* to the cytochrome c gene of *Saccharomyces cerevisiae*, *Proc. Natl. Acad. Sci. USA* **78**:2199–2203.

Guarente, L., LaLonde, B., Gifford, P., and Alani, E., 1984, Distinctly regulated tandem upstream activation sites mediate catabolite repression of the *CYC1* gene of *S. cerevisiae*, *Cell* **36**:503–511.

Gugneja, S., and Scarpulla, R. C., 1997, Serine phosphorylation within a concise amino-terminal domain in nuclear respiratory factor 1 enhances DNA binding, *J. Biol. Chem.* **272**:18732–18739.

Gugneja, S., Virbasius, J. V., and Scarpulla, R. C., 1995, Four structurally distinct, non-DNA-binding subunits of human nuclear respiratory factor 2 share a conserved transcriptional activation domain, *Mol. Cell. Biol.* **15**:102–111.

Gugneja, S., Virbasius, C. A., and Scarpulla, R. C., 1996, Nuclear respiratory factors 1 and 2 utilize similar glutamine-containing clusters of hydrophobic residues to activate transcription, *Mol. Cell. Biol.* **16**:5708–5716.

Hahn, S., and Guarente, L., 1988, Yeast HAP2 and HAP3: Transcriptional activators in a heteromeric complex, *Science* **240**:317–321.

Haldi, M. L., and Guarente, L., 1995, Multiple domains mediate heme control of the yeast activator HAP1, *Mol. Gen. Genet.* **248**:229–235.

Hennig, B., 1975, Change of cytochrome c structure during development of the mouse, *Eur. J. Biochem.* **55**:167–183.

Hodge, M. R., Kim, G., Singh, K., and Cumsky, M. G., 1989, Inverse regulation of the yeast COX5 genes by oxygen and heme, *Mol. Cell. Biol.* **9**:1958–1964.

Hodge, M. R., Singh, K., and Cumsky, M. G., 1990, Upstream activation and repression elements control transcription of the yeast *COX5b* gene, *Mol. Cell. Biol.* **10**:5510–5520.

Hood, D. A., 1990, Co-ordinate expression of cytochrome c oxidase subunit III and VIc mRNAs in rat tissue, *Biochem. J.* **269**:503–506.

Hood, D. A., Zak, R., and Pette, D., 1989, Chronic stimulation of rat skeletal muscle induces coordinate increases in mitochondrial and nuclear mRNAs of cytochrome-c-oxidase subunits, *Eur. J. Biochem.* **179**:275–280.

Hoog, C., Calzone, F. J., Cutting, A. E., Britten, R. J., and Davidson, E. H., 1991, Gene regulatory factors of the sea urchin embryo. II. Two dissimilar proteins, P3A1 and P3A2, bind to the same target sites that are required for early territorial gene expression, *Development* **112**:351–364.

Horrum, M. A., Tobin, R. B., and Ecklund, R. E., 1985, Thyroxine-induced changes in rat liver mitochondrial cytochromes, *Mol. Cell. Endocrinol.* **41**:163–169.

Hortner, H., Ammerer, E., Hartter, B., Hamilton, B., Rytka, J., Bilinski, T., and Ruis, H., 1982, Regulation of synthesis of catalase and iso-1-cytochrome c in *Saccharomyces cerevisiae* by glucose, oxygen and heme, *Eur. J. Biochem.* **128**:179–184.

Iborra, F., Francingues, M., and Guerineau, M., 1985, Localization of the upstream regulatory sites of yeast iso-2-cytochrome c gene, *Mol. Gen. Genet.* **199**:117–122.

Jacob, W. F., Silverman, T. A., Cohen, R. B., and Safer, B., 1989, Identification and characterization of a novel transcription factor participating in the expression of eukaryotic initiation factor alpha, *J. Biol. Chem.* **264**:20372–20384.

Joste, V., Goitom, Z., and Nelson, B. D., 1989, Thyroid hormone regulation of nuclear-encoded mitochondrial inner membrane polypeptides of the liver, *Eur. J. Biochem.* **184**:255–260.

Kadenbach, B., and Merle, P., 1981, On the function of multiple subunits of cytochrome c oxidase from higher eukaryotes, *FEBS Lett.* **135**:1–11.

Kadenbach, B., Stroh, A., Huther, F. J., Reimann, A., and Steverding, D., 1991, Evolutionary aspects of cytochrome c oxidase, *J. Bioenerg. Biomembr.* **23**:321–334.

Kadenbach, B., Barth, J., Akgün, R., Freund, R., Linder, D., and Possekel, S., 1995, Regulation of mitochondrial energy generation in health and disease, *Biochim. Biophys. Acta Mol. Basis Dis.* **1271**:103–109.

Karim, F. D., Urness, L. D., Thummel, C. S., Klemsz, M. J., McKercher, S. R., Celada, A., Van Beveren, C., Maki, R. A., Gunther, C. V., Nye, J. A., and Graves, B. J., 1990, The ETS-domain: A new DNA-binding motif that recognizes a purine-rich core DNA sequence, *Genes Dev.* **4**:1451–1453.

Kim, K., Lecordier, A., and Bowman, L. H., 1995, Both nuclear and mitochondrial cytochrome c oxidase mRNA levels increase dramatically during mouse postnatal development, *Biochem. J.* **306**:353–358.

Kim, K. S., Pfeifer, K., Powell, L., and Guarente, L., 1990, Internal deletions in the yeast transcriptional activator HAP1 have opposite effects at two sequence elements, *Proc. Natl. Acad. Sci. USA* **87**:4524–4528.

Ku, C. Y., Lu, Q., Ussuf, K. K., Weinstock, G. M., and Sanborn, B. M., 1991, Hormonal regulation of cytochrome oxidase subunit messenger RNAs in rat Sertoli cells, *Mol. Endocrinol.* **5**:1669–1676.

LaMarco, K., Thompson, C. C., Byers, B. P., Walton, E. M., and McKnight, S. L., 1991, Identification of Ets- and notch-related subunits in GA binding protein, *Science* **253**:789–792.

LaMarco, K. L., and McKnight, S. L., 1989, Purification of a set of cellular polypeptides that bind to the purine-rich *cis*-regulatory element of herpes simplex virus immediate early genes, *Genes Dev.* **3**:1372–1383.

Lambert, J. R., Bilanchone, V. W., and Cumsky, M. G., 1994, The *ORD1* gene encodes a transcription factor involved in oxygen regulation and is identical to *IXR1*, a gene that confers cisplatin sensitivity to *Saccharomyces cerevisiae*, *Proc. Natl. Acad. Sci. USA* **91**:7345–7349.

Laz, T. M., Peitras, D. F., and Sherman, F., 1984, Differential regulation of the duplicated isocytochrome c genes in yeast, *Proc. Natl. Acad. Sci. USA* **81**:4475–4479.

Lenka, N., Basu, A., Mullick, J., and Avadhani, N. G., 1996, The role of an E box binding basic helix loop helix protein in the cardiac muscle-specific expression of the rat cytochrome oxidase subunit VIII gene, *J. Biol. Chem.* **271**:30281–30289.

Li, K., Hodge, J. A., and Wallace, D. C., 1990, OXBOX, a positive transcriptional element of the heart-skeletal muscle ADP/ATP translocator gene, *J. Biol. Chem.* **265**:20585–20588.

Li, R., Luciakova, K., and Nelson, B. D., 1996, Expression of the human cytochrome c_1 gene is controlled through multiple Sp1-binding sites and an initiator region, *Eur. J. Biochem.* **241**:649–656.

Linder, M. E., and Gilman, A. G., 1992, G proteins, *Sci. Am.* **267**:56–65.

Lomax, M. I., and Grossman, L. I., 1989, Tissue-specific genes for respiratory proteins, *Trends Biochem. Sci.* **14**:501–504.

Lowry, C. V., and Zitomer, R. S., 1984, Oxygen regulation of anaerobic and aerobic genes mediated by a common factor in yeast, *Proc. Natl. Acad. Sci. USA* **81**:6129–6133.

Lowry, C. V., and Zitomer, R. S., 1988, *ROX1* encodes a heme-induced repression factor regulating *ANB1* and *CYC7* of *Saccharomyces cerevisiae*, *Mol. Cell. Biol.* **8:**4651–4658.

Luciakova, K., and Nelson, B. D., 1992, Transcript levels for nuclear-encoded mammalian mitochondrial respiratory-chain components are regulated by thyroid hormone in an uncoordinated fashion, *Eur. J. Biochem.* **207:**247–251.

McNabb, D. S., Xing, Y., and Guarente, L., 1995, Cloning of yeast *HAP5*: A novel subunit of a heterotrimeric complex required for CCAAT binding, *Genes Dev.* **9:**47–58.

Mell, O. C., Seibel, P., and Kadenbach, B., 1994, Structural organisation of the rat genes encoding liver- and heart-type of cytochrome *c* oxidase subunit VIa and a pseudogene related to the *COXVIa-L* cDNA, *Gene* **140:**179–186.

Montgomery, D. L., Hall, B. D., Gillam, S., and Smith, M., 1978, Identification and isolation of the yeast cytochrome *c* gene, *Cell* **14:**673–680.

Montminy, M. R., Sevarino, K. A., Wagner, J. A., Mandel, G., and Goodman, R. H., 1986, Identification of a cyclic-AMP-responsive element within the rat somatostatin gene, *Proc. Natl. Acad. Sci. USA* **83:**6682–6686.

Nagley, P., 1991, Coordination of gene expression in the formation of mammalian mitochondria, *Trends Genet.* **7:**1–4.

Olesen, J., Hahn, S., and Guarente, L., 1987, Yeast HAP2 and HAP3 activators both bind to the *CYC1* upstream activation site, UAS2, in an interdependent manner, *Cell* **51:**953–961.

Olesen, J., T. and Guarente, L., 1990, The HAP2 subunit of yeast CCAAT transcriptional activator contains adjacent domains for subunit association and DNA recognition: Model for the HAP2/3/4 complex, *Genes Dev.* **4:**1714–1729.

Olson, E. N., 1993, Regulation of muscle transcription by the MyoD family. The heart of the matter, *Circ. Res.* **72:**1–6.

Parsons, W. J., Williams, R. S., Shelton, J. M., Luo, Y. A., Kessler, D. J., and Richardson, J. A. 1996, Developmental regulation of cytochrome oxidase subunit VIa isoforms in cardiac and skeletal muscle, *Am. J. Physiol. Heart Circ. Physiol.* **270:**H567–H574.

Pfeifer, K., Arcangioli, B., and Guarente, L., 1987a, Yeast *HAP1* activator competes with the factor RC2 for binding to the upstream activation site UAS1 of the *CYC1* gene, *Cell* **49:**9–18.

Pfeifer, K., Prezant, T., and Guarente, L., 1987b, Yeast *HAP1* activator binds to two upstream activation sites of different sequence, *Cell* **49:**19–27.

Pfeifer, K., Kim, K.-S., Kogan, S., and Guarente, L., 1989, Functional dissection and sequence of yeast HAP1 activator, *Cell* **56:**291–301.

Poyton, R. O., and Burke, P. V., 1992, Oxygen regulated transcription of cytochrome *c* and cytochrome *c* oxidase genes in yeast, *Biochim Biophys. Acta Bio-Energ.* **1101:**252–256.

Poyton, R. O., and McEwen, J. E., 1996, Crosstalk between nuclear and mitochondrial genomes, *Annu. Rev. Biochem.* **65:**563–607.

Poyton, R. O., Trueblood, C. E., Wright, R. M., and Farrell, L. E., 1988, Expression and function of cytochrome *c* oxidase subunit isologues: Modulators of cellular energy production, *Ann. NY Acad. Sci.* **550:**289–307.

Puntschart, A., Claassen, H., Jostarndt, K., Hoppeler, H., and Billeter, R., 1995, mRNAs of enzymes involved in energy metabolism and mtDNA are increased in endurance-trained athletes, *Am. J. Physiol. Cell Physiol.* **269:**C619–C625.

Rao, V. N., Huebner, K., Isobe, M., ar-Rushdi, A., Croce, C. M., and Reddy, E. S. P., 1989, *elk*, tissue-specific *ets*-related genes on chromosomes X and 14 near translocation breakpoints, *Science* **244:**66–70.

Rizzuto, R., Nakase, H., Darras, B., Francke, U., Fabrizi, G. M., Mengel, T., Walsh, F., Kadenbach, B., DiMauro, S., and Schon, E. A., 1989, A gene specifying subunit VIII of human cytochrome *c* oxidase is localized to chromosome 11 and is expressed in both muscle and nonmuscle tissues, *J. Biol. Chem.* **264:**10595–10600.

Rohdich, F., and Kadenbach, B., 1993, Tissue-specific regulation of cytochrome *c* oxidase efficiency by nucleotides, *Biochemistry* **32:**8499–8503.

Scarpulla, R. C., 1984, Processed pseudogenes for rat cytochrome *c* are preferentially derived from
one of three alternate mRNAs, *Mol. Cell. Biol.* **4:**2279–2288.

Scarpulla, R. C., 1996, Nuclear respiratory factors and the pathways of nuclear–mitochondrial
interaction, *Trends Cardiovasc. Med.* **6:**39–45.

Scarpulla, R. C., 1997, Nuclear control of respiratory chain expression in mammalian cells, *J.
Bioenerg. Biomembr.* **29:**109–119.

Scarpulla, R. C., and Nye, S. H., 1986, Functional expression of rat cytochrome *c* in *Saccharomyces
cerevisiae, Proc. Natl. Acad. Sci. USA* **83:**6352–6356.

Scarpulla, R. C., and Wu, R., 1983, Nonallelic members of the cytochrome *c* multigene family of
the rat may arise through different messenger RNAs, *Cell* **32:**473–482.

Scarpulla, R. C., Agne, K. M., and Wu, R., 1981, Isolation and structure of a rat cytochrome *c* gene,
J. Biol. Chem. **256:**6480–6486.

Scarpulla, R. C., Agne, K. M., and Wu, R., 1982, Cytochrome *c* gene-related sequences in
mammalian genomes, *Proc. Natl. Acad. Sci. USA* **79:**739–743.

Scarpulla, R. C., Kilar, M. C., and Scarpulla, K. M., 1986, Coordinate induction of multiple
cytochrome *c* mRNAs in response to thyroid hormone, *J. Biol. Chem.* **261:**4660–4662.

Seelan, R. S., and Grossman, L. I., 1992, Structure and organization of the heart isoform gene for
bovine cytochrome *c* oxidase subunit VIIa, *Biochemistry* **31:**4696–4704.

Seelan, R. S., and Grossman, L. I., 1993, Structural organization and evolution of the liver isoform
gene for bovine cytochrome *c* oxidase subunit VIIa, *Genomics* **18:**527–536.

Seelan, R. S., and Grossman, L. I., 1997, Structural organization and promoter analysis of the
bovine cytochrome *c* oxidase subunit VIIc gene—A functional role for YY1, *J. Biol. Chem.*
272:10175–10181.

Seelan, R. S., Gopalakrishnan, L., Scarpulla, R. C., and Grossman, L. I., 1996, Cytochrome *c*
oxidase subunit VIIa liver isoform: Characterization and identification of promoter elements
in the bovine gene, *J. Biol. Chem.* **271:**2112–2120.

Sestoft, L., 1980, Metabolic aspects of the calorigenic effect of thyroid hormone in mammals, *Clin.
Endocrinol.* **13:**489–506.

Shadel, G. S., and Clayton, D. A., 1993, Mitochondrial transcription initiation: Variation and
conservation, *J. Biol. Chem.* **268:**16083–16086.

Sherman, F., and Stewart, J. W., 1971, Genetics and biosynthesis of cytochrome *c, Annu. Rev.
Genet.* **5:**257–296.

Short, J. M., Wynshaw-Boris, A., Short, H. P., and Hanson, R. W., 1986, Characterization of the
phosphoenolpyruvate carboxykinase (GTP) promoter-regulatory region, *J. Biol. Chem.* **261:**
9721–9726.

Shyjan, A. W., and Butow, R. A., 1993, Intracellular dialogue, *Curr. Biol.* **3:**398–400.

Silve, S., Rhode, P. R., Coll, B., Campbell, J., and Poyton, R. O., 1992, ABF1 is a phosphoprotein
and plays a role in carbon source control of *COX6* transcription in *Saccharomyces cerevisiae,
Mol. Cell. Biol.* **12:**4197–4208.

Smith, E. O., and Lomax, M. I., 1993, Structural organization of the bovine gene for the heart/
muscle isoform of cytochrome *c* oxidase subunit VIa, *Biochim. Biophys. Acta* **1174:**63–71.

Stotz, E., 1939, The estimation and distribution of cytochrome oxidase and cytochrome *c* in rat
tissues, *J. Biol. Chem.* **131:**555–565.

Sucharov, C., Basu, A., Carter, R. S., and Avadhani, N. G., 1995, A novel transcriptional initiator
activity of the GABP factor binding ets sequence repeat from the murine cytochrome *c*
oxidase Vb gene, *Gene Expr.* **5:**93–111.

Suske, G., Enders, C., Schlerf, A., and Kadenbach, B., 1988, Organization and nucleotide sequence of two chromosomal genes for rat cytochrome *c* oxidase subunit VIc: A structural and a processed gene, *DNA* **7**:163–171.

Suzuki, H., Hosokawa, Y., Toda, H., Nishikimi, M., and Ozawa, T., 1989, Isolation of a single nuclear gene encoding human ubiquinone-binding protein in complex III of mitochondrial respiratory chain, *Biochem. Biophys. Res. Commun.* **161**:371–378.

Suzuki, H., Hosokawa, Y., Toda, H., Nishikimi, M., and Ozawa, T., 1990, Common protein-binding sites in the 5′-flanking regions of human genes for cytochrome c_1 and ubiquinone-binding protein, *J. Biol. Chem.* **265**:8159–8163.

Taanman, J.-W., and Capaldi, R. A., 1993, Subunit VIa of yeast cytochrome *c* oxidase is not necessary for assembly of the enzyme complex but modulates the enzyme activity. Isolation and characterization of the nuclear-coded gene, *J. Biol. Chem.* **268**:18754–18761.

Taanman, J.-W., Turina, P., and Capaldi, R. A., 1994, Regulation of cytochrome *c* oxidase by interaction of ATP at two binding sites, one on subunit VIa, *Biochemistry* **33**:11833–11841.

Tata, J. R., Ernster, L., Lindberg, O., Arrhenius, E., Pederson, S., and Hedman, R., 1963, The action of thyroid hormones at the cell level, *Biochem. J.* **86**:408–428.

Thompson, C. C., Brown, T. A., and McKnight, S. L., 1991, Convergence of Ets- and notch-related structural motifs in a heteromeric DNA binding complex, *Science* **253**:762–768.

Tominaga, K., Akiyama, S., Kagawa, Y., and Ohta, S., 1992, Upstream region of a genomic gene for human mitochondrial transcription factor 1, *Biochim. Biophys. Acta Gene Struct. Expr.* **1131**:217–219.

Topper, J. N., Bennett, J. L., and Clayton, D. A., 1992, A role for RNAse MRP in mitochondrial RNA processing, *Cell* **70**:16–20.

Topper, J. N., and Clayton, D. A., 1990, Characterization of human MRP/Th RNA and its nuclear gene: Full-length MRP/Th RNA is an active endoribonuclease when assembled as an RNP, *Nucleic Acids Res.* **18**:793–799.

Tracy, R. L., and Stern, D. B., 1995, Mitochondrial transcription initiation: Promoter structures and RNA polymerases, *Curr. Genet.* **28**:205–216.

Trawick, J. D., Rogness, C., and Poyton, R. O., 1989a, Identification of an upstream activation sequence and other *cis*-acting elements required for transcription of *COX6* from *Saccharomyces cerevisiae*, *Mol. Cell. Biol.* **9**:5350–5358.

Trawick, J. D., Wright, R. M., and Poyton, R. O., 1989b, Transcription of yeast *COX6*, the gene for cytochrome *c* oxidase subunit VI, is dependent on heme and on the *HAP2* gene, *J. Biol. Chem.* **264**:7005–7008.

Trawick, J. D., Kraut, N., Simon, F. R., and Poyton, R. O., 1992, Regulation of yeast *COX6* by the general transcription factor ABF1 and separate HAP2- and heme-responsive elements, *Mol. Cell. Biol.* **12**:2302–2314.

Trueblood, C. E., and Poyton, R. O., 1987, Differential effectiveness of yeast cytochrome *c* oxidase subunit V genes results from differences in expression not function, *Mol. Cell. Biol.* **7**:3520–3526.

Trueblood, C. E., Wright, R. M., and Poyton, R. O., 1988, Differential regulation of the two genes encoding *Saccharomyces cerevisiae* cytochrome oxidase subunit V by heme and the *HAP2* and *REO1* genes, *Mol. Cell. Biol.* **8**:4537–4540.

Van Beeumen, J. J., Van Kuilenburg, A. B. P., Van Bun, S., Van den Bogert, C., Tager, J. M., and Muijsers, A. O., 1990, Demonstration of two isoforms of subunit VIIa of cytochrome *c* oxidase from human skeletal muscle: Implications for mitochondrial myopathies, *FEBS Lett.* **263**:213–216.

Van Itallie, C. M., 1990, Thyroid hormone and dexamethasone increase the levels of a messenger ribonucleic acid for a mitochondrially encoded subunit but not for a nuclear-encoded subunit of cytochrome *c* oxidase, *Endocrinology* **127**:55–62.

Van Kuilenburg, A. B. P., Muijsers, A. O., Demol, H., Dekker, H. L., and Van Beeumen, J. J., 1988, Human heart cytochrome c oxidase subunit VIII: Purification and determination of the complete amino acid sequence, *FEBS Lett.* **240:**127–132.

Verdière, J., Creusot, F., and Guerineau, M., 1985, Regulation of the expression of iso-2-cytochrome c gene in *S. cerevisiae:* Cloning of the positive regulatory gene *CYP1* and identification of the region of its target sequence on the structural gene *CYP3, Mol. Gen. Genet.* **199:**524–533.

Vestweber, D., and Schatz, G., 1988, Mitochondria can import artificial precursor proteins containing a branched polypeptide chain or a carboxy-terminal stilbene disulfonate, *J. Biol. Chem.* **107:**2045–2049.

Villena, J. A., Martin, I., Viñas, O., Cormand, B., Iglesias, R., Mampel, T., Giralt, M., and Villarroya, F., 1994, ETS transcription factors regulate the expression of the gene for the human mitochondrial ATP synthase β-subunit, *J. Biol. Chem.* **269:**32649–32654.

Virbasius, C. A., Virbasius, J. V., and Scarpulla, R. C., 1993, NRF-1, an activator involved in nuclear–mitochondrial interactions, utilizes a new DNA-binding domain conserved in a family of developmental regulators, *Genes Dev.* **7:**2431–2445.

Virbasius, J. V., and Scarpulla, R. C., 1988, Structure and expression of rodent genes encoding the testis-specific cytochrome c. Differences in gene structure and evolution between somatic and testicular variants, *J. Biol. Chem.* **263:**6791–6796.

Virbasius, J. V., and Scarpulla, R. C., 1990, The rat cytochrome c oxidase subunit IV gene family: Tissue-specific and hormonal differences in subunit IV and cytochrome c mRNA expression, *Nucleic Acids Res.* **18:**6581–6586.

Virbasius, J. V., and Scarpulla, R. C., 1991, Transcriptional activation through ETS domain binding sites in the cytochrome c oxidase subunit IV gene, *Mol. Cell. Biol.* **11:**5631–5638.

Virbasius, J. V., and Scarpulla, R. C., 1994, Activation of the human mitochondrial transcription factor A gene by nuclear respiratory factors: A potential regulatory link between nuclear and mitochondrial gene expression in organelle biogenesis, *Proc. Natl. Acad. Sci. USA* **91:**1309–1313.

Virbasius, J. V., Virbasius, C. A., and Scarpulla, R. C., 1993, Identity of GABP with NRF-2, a multisubunit activator of cytochrome oxidase expression, reveals a cellular role for an ETS domain activator of viral promoters, *Genes Dev.* **7:**380–392.

Wan, B., and Moreadith, R. W., 1995, Structural characterization and regulatory element analysis of the heart isoform of cytochrome c oxidase VIa, *J. Biol. Chem.* **270:**26433–26440.

Watanabe, H., Wada, T., and Handa, H., 1990, Transcription factor E4TF1 contains two subunits with different functions, *EMBO J.* **9:**841–847.

Watanabe, H., Sawada, J.-I., Yano, K.-I., Yamaguchi, K., Goto, M., and Handa H., 1993, cDNA cloning of transcription factor E4TF1 subunits with Ets and notch motifs, *Mol. Cell. Biol.* **13:**1385–1391.

Waterland, R. A, Basu, A., Chance, B., and Poyton, R. O., 1991, The isoforms of yeast cytochrome c oxidase subunit V alter the *in vivo* kinetic properties of the holoenzyme, *J. Biol. Chem.* **266:**4180–4186.

Weintraub, S. J., Prater, C. A., and Dean, D. C., 1992, Retinoblastoma protein switches the E2F site from positive to negative element, *Nature* **358:**259–261.

Wheat, T. E., Hintz, M., Goldberg, E., and Margoliash, E., 1977, Analyses of stage-specific multiple forms of lactate dehydrogenase and of cytochrome c during spermatogenesis in mouse, *Differentiation* **9:**37–41.

Wiesner, R. J., Kurowski, T. T., and Zak, R., 1992, Regulation by thyroid hormone of nuclear and mitochondrial genes encoding subunits of cytochrome-c-oxidase in rat liver and skeletal muscle, *Mol. Endocrinol.* **6:**1458–1467.

Williams, R. S., Garcia-Moll, M., Mellor, J., Salmons, S., and Harlan, W., 1987, Adaptation of skeletal muscle to increased contractile activity, *J. Biol. Chem.* **262:**2764–2767.

Wolstenholme, D. R., 1992, Animal mitochondrial DNA: Structure and evolution, *Int. Rev. Cytol.* **141:**173–216.

Wright, R. M., and Poyton, R. O., 1990, Release of two *Saccharomyces cerevisiae* cytochrome genes, *COX6* and *CYC1*, from glucose repression requires the *SNF1* and *SSN6* gene products, *Mol. Cell. Biol.* **10:**1297–1300.

Wright, R. M., Ko, C., Cumsky, M. G., and Poyton, R. O., 1984, Isolation and sequence of the structural gene for cytochrome *c* oxidase subunit VI from *Saccharomyces cerevisiae, J. Biol. Chem.* **259:**15401–15407.

Xing, Y., Fikes, J. D., and Guarente, L., 1993, Mutations in yeast HAP2/HAP3 define a hybrid CCAAT box binding domain, *EMBO J.* **12:**4647–4655.

Yamada, M., Amuro, N., Goto, Y., and Okazaki, T., 1990, Structural organization of the rat cytochrome *c* oxidase subunit IV gene, *J. Biol. Chem.* **265:**7687–7692.

Yiu, G. K., Murray, M. T., and Hecht, N. B., 1997, Deoxyribonucleic acid–protein interactions associated with transcriptional initiation of the mouse testis-specific cytochrome *c* gene, *Biol. Reprod.* **56:**1439–1449.

Yu, C. Y., Breitbart, R. E., Smoot, L. B., Lee, Y., Mahdavi, V., and Nadal-Ginard, B., 1992, Human myocyte-specific enhancer factor 2 comprises a group of tissue-restricted MADS box transcription factors, *Genes Dev.* **6:**1783–1798.

Zeviani, M., Nakagawa, M., Herbert, J., Lomax, M. I., Grossman, L. I., Sherbany, A. A., Miranda, A. F., DiMauro, S., and Schon, E. A., 1987, Isolation of a cDNA clone encoding subunit IV of human cytochrome *c* oxidase, *Gene* **55:**205–217.

Zhang, L., and Guarente, L., 1995, Heme binds to a short sequence that serves a regulatory function in diverse proteins, *EMBO J.* **14:**313–320.

Zhang, L., Bermingham-McDonogh, O., Turcotte, B., and Guarente, L., 1993, Antibody-promoted dimerization bypasses the regulation of DNA binding by the heme domain of the yeast transcriptional activator HAP1, *Proc. Natl. Acad. Sci. USA* **90:**2851–2855.

Zitomer, R. S., and Hall, B. D., 1976, Yeast cytochrome *c* messenger RNA. *In vitro* translation and specific immunoprecipitation of the *CYC1* gene product, *J. Biol. Chem.* **251:**6320–6326.

Zitomer, R. S., and Lowry, C. V., 1992, Regulation of gene expression by oxygen in *Saccharomyces cerevisiae, Microbiol. Rev.* **56:**1–11.

Zitomer, R. S., and Nichols, D. L., 1978, Kinetics of glucose repression of yeast cytochrome *c, J. Bacteriol.* **135:**39–44.

Zitomer, R. S., Montgomery, D. L., Nichols, D. L., and Hall, B. D., 1979, Transcriptional regulation of the yeast cytochrome *c* gene, *Proc. Natl. Acad. Sci. USA* **76:**3627–3631.

Zitomer, R. S., Sellers, J. W., McCarter, D. W., Hastings, G. A., Wick, P., and Lowry, C. V., 1987, Elements involved in oxygen regulation of the *Saccharomyces cerevisiae* CYC7 gene, *Mol. Cell. Biol.* **7:**2212–2220.

CHAPTER 23

Suppressor Genetics of the Mitochondrial Energy Transducing System

The Cytochrome bc_1 Complex

Jean-Paul di Rago and Piotr P. Slonimski

1. INTRODUCTION

Genetics is currently a popular science. In most cases it is reduced to a series of simple operations like the isolation of a gene, establishment of its nucleotide sequence, and conceptual translation into the protein sequence, sometimes followed by site-directed mutagenesis in order to test preconceived ideas about the structure–function relationships. However, what can be referred to as "suppressor genetics" is much less popular, especially in the study of mitochondrial energy transduction. The term *suppression* originated from classical genetics some 80 years ago (Sturtevant, 1920). It designated the reversal of a mutant phenotype via mutation at a locus distinct from that of the original mutation. Operationally, any *reversion*, whether partial or complete, can be classified as a suppressor mutation, whenever, by genetic or molecular experi-

Jean-Paul di Rago and Piotr P. Slonimski • Centre de Génétique Moléculaire, CNRS, 91190 Gif-sur-Yvette, France.

Frontiers of Cellular Bioenergetics, edited by Papa *et al.* Kluwer Academic/Plenum Publishers, New York, 1999.

ments, the nonidentity of the original wild type as well as the original and the "reverse" mutations can be demonstrated. In molecular terms, the DNA sequence of the suppressor mutation must be different from that of the wild type, and that of the mutant.

This suppressor mutation can be located (1) within the same gene as the primary (also called target) mutation, at the same or at a different nucleotide position; it is then referred to as intragenic suppression; (2) in a different gene of the same genome, i.e., intergenic suppression; and (3) in the case of the mitochondrial energy transduction system, it can be also intergenomic, since two different genomes, the mitochondrial and the nuclear one, participate in the biogenesis of the organelle.

Genetic approaches to the study of mitochondrial bioenergetics began in the late 1940s in the laboratory of Boris Ephrussi (see references in Ephrussi and Slonimski, 1955). The first suppressor mutations of respiratory deficiencies were reported in the 1960s (see references in Gorini and Beckwith, 1966; Clavilier *et al.*, 1969). Since that early period, suppressor genetics has been applied to the study of mitochondrial energy production on several occasions. However, it has not been applied in a systematic manner. Its main advantage to unravel unexpected and unpredicted relations between the various components of the system in our opinion has not been exploited sufficiently. In this chapter, we shall illustrate the genetic suppression approach to the study of yeast mitochondria by several examples focusing mainly on the cytochrome bc_1 complex (complex III) of the energy transduction apparatus. We shall begin analyzing various classes of primary mutations, which either modify or abolish the activity of the complex, followed by the analysis of intragenic and intergenic suppressions, which restore or alleviate the effects of primary mutations. Finally, we shall describe the first cases of intergenomic suppressions, which bring new insights into both the functional mechanism and evolutionary origin of the bc_1 complex.

2. THE CYTOCHROME bc_1 COMPLEX

The cytochrome bc_1 complex transfers electrons from ubiquinol to cytochrome c and couples this electron transfer to a translocation of protons across the inner mitochondrial membrane in which the complex resides (for a review, see Trumpower, 1990). This enzymatic activity is essential for the viability of most eukaryotic cells that rely on mitochondrial respiration to satisfy their ATP requirements. Cytochrome bc_1 complexes purified from several mitochondrial sources contain 10 or 11 polypeptide subunits, three of which bind metal redox centers (cytochrome b, cytochrome c_1, and the Rieske iron–sulfur protein). The cytochrome b is encoded by the mitochondrial DNA, while all the other subunits are specified by nuclear genes.

The cytochrome b is a transmembrane protein containing two hemes, called b_L and b_H, which are both located in the phospholipid bilayer. The former is close to the positive side of the membrane, while the latter is more deeply embedded in the middle of the membrane. The heme of cytochrome c_1 and the FeS cluster of the Rieske protein are on the positive side of the membrane (Ohnishi *et al.*, 1989; Yu *et al.*, 1996).

According to the Q cycle model (for a review, see Brandt and Trumpower, 1994), the cytochrome bc_1 complex contains two quinone-processing sites referred to as center P and center N. The center P catalyzes the oxidation of ubiquinol (QH_2) molecules on the electropositive side of the membrane, while center N catalyzes the reduction of ubiquinone (Q) molecules on the other side of the membrane. In the first half of the Q cycle, one molecule of ubiquinol is oxidized at center P. The first electron is transferred to the Rieske FeS cluster, and the resulting intermediate semiquinone reduces heme b_L. The two protons liberated by the oxidation of ubiquinol are deposited in the intermembrane space. The reduced FeS cluster is oxidized by cytochrome c_1, which then reduces cytochrome c, while reduced heme b_L is oxidized by heme b_H, which then reduces ubiquinone at center N to form a stable semiquinone intermediate (SQ_n). In the second part of the cycle, a novel molecule of ubiquinol is oxidized at center P, with the recycling of a second electron through the membrane by cytochrome b, which permits the full reduction of the semiquinone SQ_n produced during the first half of the cycle. The two protons required for the reduction of one molecule of ubiquinone at center N are taken from the matrix space. Thus, for each pair of electrons transferred from ubiquinol to cytochrome c, two protons are transported vectorially, from the matrix to the intermembrane space. This proton-motive activity results from the alternate oxidation and reduction of quinones on each side of the membrane, the protons being shuttled by means of the passive circulation of oxidized and reduced quinones through the phospholipid bilayer between the centers P and N.

The electron transfers catalyzed by the cytochrome bc_1 complex are inhibited by several compounds (for a review, see von Jagow and Link, 1986]. Myxothiazol (von Jagow *et al.*, 1984; Thierbach and Reichenbach, 1981), mucidin (identical to strobilurin A) (see von Jagow *et al.*, 1986; Becker *et al.*, 1981), and stigmatellin (von Jagow and Ohnishi, 1985; Thierbach *et al.*, 1984) are thought to impair the oxidation of ubiquinol at center P. Antimycin (Robertson *et al.*, 1984; Ohnishi and Trumpower, 1980; Slater, 1973), diuron (Convent *et al.*, 1978), and funiculosin (Nelson *et al.*, 1977) would act at center N by blocking the electron transfer between heme b_H and ubiquinone. Myxothiazol and stigmatellin both shift the optical spectrum of heme b_L, while antimycin shifts that of heme b_H and abolishes the stable electron paramagnetic resonance (EPR) signal of the intermediate semiquinone SQ_n. Thus, it is believed that the former are bound close to the positive side of the membrane, between the heme b_L and the FeS cluster, while the latter binds to a site close to heme b_H on the negative side of the membrane.

How is cytochrome b inserted in the membrane? Which domains bind the quinone substrates and the hemes? Which residues are important for catalysis and interaction of cytochrome b with other subunits of the cytochrome bc_1 complex? We have addressed these questions by means of genetic approaches in the budding yeast *Saccharomyces cerevisiae*. This eukaryote has the property of being able to satisfy its energy requirements by fermentation or respiration. Therefore, yeast cells can still grow in the absence of a functional cytochrome bc_1 complex provided that a fermentable carbon source is available. This offers the possibility to isolate and propagate viable mutant cells in which the function of the cytochrome bc_1 complex is altered or even completely absent. Furthermore, rules governing transmission and recombination of mitochondrial DNA in *S. cerevisiae* have been established, and methods were developed to localize point mutations within yeast individual mitochondrial genes (Dujon *et al.*, 1974; Slonimski and Tzagoloff, 1976; Kotylak and Slonimski, 1977; for review, see Dujon, 1981). Using different genetic screens, we collected more than 100 different cytochrome b mutants displaying a large variety of phenotypes. How these mutants were obtained and what information they provide on the structure and function of cytochrome b are the subject of the present chapter.

3. CYTOCHROME b INHIBITOR-RESISTANT MUTANTS

Some inhibitors of the cytochrome bc_1 complex are structurally related to ubiquinone and therefore are thought to displace or block the access of the quinone substrates to centers P and N (von Jagow and Link, 1986). To identify protein domains forming the centers P and N, an obvious genetic screen thus was to select for mutations giving rise to resistance to these inhibitors. In yeast, such mutations can be identified easily by searching for cells able to grow on a nonfermentable carbon source (such as glycerol or ethanol) in the presence of inhibitor concentrations that normally impair mitochondrial respiration.

Inhibitor-resistant mutations of two different types were obtained. The first are located in the nucleus and are responsible for pleiotropic nonspecific drug-resistant *in vivo* phenotypes (including increased resistance to cycloheximide, which inhibits cytoplasmic protein synthesis), indicating modifications of the cell permeability (Thierbach and Michaelis, 1982; Subik *et al.*, 1977). The second type of mutations result in specific inhibitor resistance at the level of the cytochrome bc_1 complex itself, as shown *in vitro* with mitochondrial membrane preparations. All mutants of this second type were, without a single exception among hundreds of independent isolates, located in the mitochondrial cytochrome *b* gene (di Rago *et al.*, 1990a, 1989; Thierbach and Michaelis, 1982; Colson and Slonimski, 1979; Colson *et al.*, 1977, 1979; Pratje and Michaelis, 1977; Subik *et al.*, 1977; Michaelis, 1976; Subik, 1975). It must be recalled that in yeast, the cytochrome *b* gene is split, i.e., the nucleotide sequence coding for the apo-cytochrome *b* is interrupted by intronic sequences that are removed posttranscriptionally to give the messenger RNA (Lazowska *et al.*, 1980, 1989; Jacq *et al.*, 1980). Most of the *S. cerevisiae* strains used for selection of cytochrome *b* mutants contain a "long" cytochrome *b* gene containing six exons (B1 to B6) distributed over a mitochondrial DNA (mtDNA) segment of 7172 base pairs (bp). Mutations leading to inhibitor-resistance were mapped in five exons. Center N inhibitor (antimycin, diuron, funiculosin) resistant mutations were located in exons B1 and B4, while center P inhibitor (myxothiazol, mucidin, stigmatellin) resistant mutations were located in B1, B3, B5, and B6. DNA sequence analysis revealed that most of the mutants resulted from one base pair substitution leading to a single amino acid replacement (di Rago *et al.*, 1986, 1989, 1990a; di Rago and Colson, 1988) (see Table I). Inhibitor-resistant mutants also were isolated from other organisms, including mouse (Howell and Gilbert, 1988; Howell *et al.*, 1987), bacteria (Daldal *et al.*, 1989), fungi (Weber and Wolf, 1988; Brunner *et al.*, 1987), and algae (Bennoun *et al.*, 1992). In most cases, the mutations were found at equivalent positions or in close vicinity of those identified in *S. cerevisiae* (Table I).

The inhibitor-resistance mutations affect distinct domains of cytochrome *b* (see Fig. 1). In yeast, those that give rise to resistance to center N inhibitors are clustered within two regions (residues 17–37 and 198–228), while those resulting in resistance to center P inhibitors are within two other regions (residues 129–147 and 256–275). There are cases where a mutation renders the cytochrome bc_1 complex less sensitive to more than one inhibitor. For example, all the mutants selected for myxothiazol resistance exhibit increased resistance to mucidine and vice versa (not surprisingly, the mutants selected for mucidin resistance and those selected for myxothizol resistance have identical lesions in their cytochrome *b*) (see Table I) (di Rago *et al.*, 1989). This may account for the competitive binding of myxothiazol and mucidin and the common presence in the structures of these inhibitors of a methoxyacrylate segment (Mansfield and Wiggins, 1990; Brandt *et al.*, 1988). Also, some diuron-resistant mutants are less sensitive to antimycin and 2-n-heptyl-4-hydroxyquinoline-N-oxide (HQNO) (Briquet and Goffeau, 1981). Importantly, there is no known case where a mutation selected for resistance to a center N inhibitor gives rise to resistance to a center P inhibitor and vice versa. These data corroborate

Table I

Mutations in the Cytochrome b of Mitochondrial Cytochrome bc_1 Complexes

Amino acid change	Growth phenotype	Species	References
I17F	diu^R	*S. cerevisiae*	di Rago *et al.* (1986)
W30C	sup[S206L]	*S. cerevisiae*	Coppée *et al.* (1994a)
N31K	diu^R	*S. cerevisiae*	di Rago *et al.* (1986)
G33D	deficient	*S. cerevisiae*	Coppée *et al.* (1994b)
G33A	sup[G33D]	*S. cerevisiae*	Coppée *et al.* (1994b)
G37V	ana^R	*S. cerevisiae*	di Rago and COlson (1988)
G37V	ana^R	*M. musculus*	Howell and Gilbert (1988)
A37V	ana^R	*S. pombe*	Weber and Wolf (1988)
A37G	diu^R	*S. pombe*	Weber and Wolf (1988)
I125T	sup[G137E]	*S. cerevisiae*	di Rago *et al.* (1990b)
A126T	sup[C133Y]	*S. cerevisiae*	di Rago *et al.* (1990b)
F129L	myx^R	*S. cerevisiae*	di Rago *et al.* (1989)
F129L	myx^R	*C. reinhardtii*	Bennoun *et al.* (1992)
L130M	sup[C133Y]	*S. cerevisiae*	di Rago *et al.* (1990a)
G131S	deficient	*S. cerevisiae*	Brivet-Chevillotte and diRago (1989)
Y132^c	myx^R	*C. reinhardtii*	Bennoun *et al.* (1992)
C133Y	deficient	*S. cerevisiae*	Lemesle-Meunier *et al.* (1993)
C133S	sup[C133E]	*S. cerevisiae*	di Rago *et al.* (1990a)
C133D	sup[C133Y]	*S. cerevisiae*	di Rago *et al.* (1990a)
C133N	sup[C133Y]	*S. cerevisiae*	di Rago *et al.* (1990a)
C133F	sup[C133Y]	*S. cerevisiae*	di Rago *et al.* (1990a)
G137R	mys^R	*S. cerevisiae*	di Rago *et al.* (1989)
G137R	muc^R	*S. cerevisiae*	di Rago *et al.* (1989)
G137E^d	deficient	*S. cerevisiae*	Tron and Lemesle-Meunier (1990)
G137V	deficient	*S. cerevisiae*	Tron and Lemsele-Meunier (1990)
G137A	sup[G137V]	*S. cerevisiae*	di Rago *et al.* (1990a)
G137S	myx^R	*C. reinhardtii*	Bennoun *et al.* (1992)
Q138Y	sup[Stop138]	*S. cerevisiae*	di Rago *et al.* (1990a)
W142R	deficient	*S. cerevisiae*	Bruel *et al.* (1995)
W142S	sup[W142R]	*S. cerevisiae*	Bruel *et al.* (1995)
W142T	sup[142R]	*S. cerevisiae*	Bruel *et al.* (1995)
W142K	sup[W142R]	*S. cerevisiae*	Bruel *et al.* (1995)
G142A	myx^R	*M. musculus*	Howell and Gilbert (1988)
H141Y	sup[G137E]	*S. cerevisiae*	di Rago *et al.* (1990b)
I147F	sti^R	*S. cerevisiae*	di Rago *et al.* (1989)
I147F	sup[G137E]	*S. cerevisiae*	di Rago *et al.* (1990b)
T147M	sti^R	*M. musculus*	Howell and Gilbert (1988)
F151C	deficient	*S. cerevisiae*	Lazowksa *et al.* (1989)
F151L	sup[G137E]	*S. cerevisiae*	di Rago *et al.* (1990b)
D171N	myopathy^b	*H. sapiens*	John and Neufeld (1991); Brown *et al.* (1992)
I176S	sup[C133Y]	*S. cerevisiae*	di Rago *et al.* (1990b)
L198F	fun^R	*S. cerevisiae*	di Rago *et al.* (1990c)
S206L	sup[S206L]	*S. cerevisiae*	Coppée *et al.* (1994a)

Table I
(*Continued*)

Amino acid change	Growth phenotype	Species	References
S206V	sup[S206L]	*S. cerevisiae*	Coppée *et al.* (1994a)
N208K	sup[S206L]	*S. cerevisiae*	Coppée *et al.* (1994a)
M208Y	sup[S206L]	*S. cerevisiae*	Coppée *et al.* (1994a)
M221K	deficient	*S. cerevisiae*	Lemesle-Meunier *et al.* (1993)
M221E	sup[M221K]	*S. cerevisiae*	Coppée *et al.* (1994b)
M221Q	sup[M221K]	*S. cerevisiae*	Coppée *et al.* (1994b)
S223P	sup[cor2][c]	*S. cerevisiae*	di Rago *et al.* (1997)
F225S	diu[R]	*S. cerevisiae*	di Rago *et al.* (1986)
F225L	diu[R]	*S. cerevisiae*	di Rago and Colson (1988)
K228M	diu[R]	*S. cerevisiae*	di Rago and Colson (1988)
K228M	ana[R]	*K. lactis*	Brunner *et al.* (1987)
G231D	HQNO[R]	*M. musculus*	Howell *et al.* (1987)
N256Y	myx[R]	*S. cerevisiae*	di Rago *et al.* (1989)
N256Y	muc[R]	*S. cerevisiae*	di Rago *et al.* (1989)
N256K	sup[G137E]	*S. cerevisiae*	di Rago *et al.* (1990b)
N256I	sup[L282F]	*S. cerevisiae*	di Rago *et al.* (1995)
G260A	sup[G131S]	*S. cerevisiae*	di Rago *et al.* (1990b)
L275F	myx[R]	*S. cerevisiae*	di Rago *et al.* (1989)
L275S	myx[R]	*S. cerevisiae*	di Rago *et al.* (1989)
L275S	muc[R]	*S. cerevisiae*	di Rago *et al.* (1989)
L275T	myx[R]	*S. cerevisiae*	di Rago *et al.* (1989)
L282F	deficient	*S. cerevisiae*	Lemesle-Meunier *et al.* (1993)
L282I	sup[L282F]	*S. cerevisiae*	di Rago *et al.* (1995)
L282C	sup[L282F]	*S. cerevisiae*	di Rago *et al.* (1995)
L282T	sup[L282F]	*S. cerevisiae*	di Rago *et al.* (1995)
D287H	sup[L282F]	*S. cerevisiae*	di Rago *et al.* (1995)
K288N	sup[G340E]	*S. cerevisiae*	di Rago *et al.* (1995)
G290D	myopathy[b]	*H. sapiens*	Dumoulin *et al.* (1996)
L294F	sti[R]	*M. musculus*	Howell and Gilbert (1988)
L355Stop	deficient	*S. cerevisiae*	di Rago *et al.* (1993)
G340E	deficient	*S. cerevisiae*	Lemesle-Meunier *et al.* (1993)
G340A	sup[G340E]	*S. cerevisiae*	di Rago *et al.* (1995)
G340V	sup[G340E]	*S. cerevisiae*	di Rago *et al.* (1995)
ΔG340	sup[G340E]	*S. cerevisiae*	di Rago *et al.* (1995)
C342G	sup[G340E]	*S. cerevisiae*	di Rago *et al.* (1995)
H343T	sup[C133Y]	*S. cerevisiae*	di Rago *et al.* (1995)
E345G	sup[G340E]	*S. cerevisiae*	di Rago *et al.* (1995)
V356M	myopathy[b]	*H. sapiens*	John and Neufeld (1991); Brown *et al.* (1992)

(*continued*)

Table I
(Continued)

Frameshift *S. cerevisiae* mutant in C-terminal tail and its revertants	
Frameshift: +T at V349/Stop and codon 382 (deficient)	di Rago *et al.* (1993)
Suppressors:	
−T at codon 345: V346Y + P347T + Y348I	di Rago *et al.* (1993)
−T at codon 347: Y348I	di Rago *et al.* (1993)
−T at codon 350: V349C	di Rago *et al.* (1993)
−G at codon 352: V349C + L350F + M351N	di Rago *et al.* (1993)
−A at codon 354: V349C + L350F + M351N + Q353T	di Rago *et al.* (1993)

[a]The structural alterations of mitochondrial cytochrome *b* listed in this table were characterized as (1) respiratory sufficient inhibitor-resistance mutations (*diu*, diuron; *ana*, antimycin; *fun*, funiculosin; *HQNO*, 2-heptyl-4-hydroxyquinoline N-oxide; *myx*, myxothiazol; *muc*, mucidin; and *sti*, stigmatellin); (2) respiratory deficient (loss-of-function); and (3) suppressors of cytochrome *b* deficiency mutations (*sup* followed by *brackets* by the original target mutation which is suppressed).

[b]The three mutations in human cytochrome *b* were detected in patients with Leber hereditary optic neuropathy (D171N, V356M) or progressive exercise muscle intolerance (G290D).

[c]The S223P mutation in *S. cerevisiae* cytochrome *b* was obtained as a functional suppressor of a cytochrome bc_1 complex deficiency due to alterations in the nuclear encoded core protein 2.

[d]The G173E mutation in *S. cerevisiae* cytochrome *b* is suppressed by a single amino acid replacement (P197T) in the Rieske iron–sulfur protein.

biochemical data indicating that center P and center N inhibitors bind to different sites in the cytochrome bc_1 complex.

The inhibitor-resistant mutants provided crucial information for building a folding model for cytochrome *b* with eight transmembrane segments (Crofts *et al.*, 1987; Brasseur, 1988). In this model, the center N inhibitor-resistant mutations affect residues that are located on the same side of the membrane, while center P inhibitor-resistant mutations all modify residues on the opposite side (Fig. 1). It is believed that residues affected by inhibitor-resistant mutations are at or in the immediate vicinity of the substrate binding sites (centers P and N). This interpretation is based on studies on the photosynthetic reaction center of bacteria and the photosystem II of chloroplasts. These complexes catalyze the reduction of ubiquinone at a site (called QB) that can be blocked selectively by several inhibitors including diuron and stigmatellin. The X-ray

$\longrightarrow$

FIGURE 1. Folding of the yeast cytochrome *b* in the inner mitochondrial membrane and locations of mutations that modify the activity of the protein. The histidine residues in helices B and D surrounded by lozenges are the ligands of hemes b_L (H82, H183) and b_H (H96, H197). Black circles indicate residues modified in respiratory growth deficient mutants. Residues changed in inhibitor-resistant mutants are surrounded by black (center P inhibitor-resistance) or open (center N inhibitor-resistance) squares. The corresponding amino-acid changes are listed in Table 1. STI, stigmatellin; MYX, myxothiazol; ANA, antimycin; DIU, diuron; FUN, funiculosin.

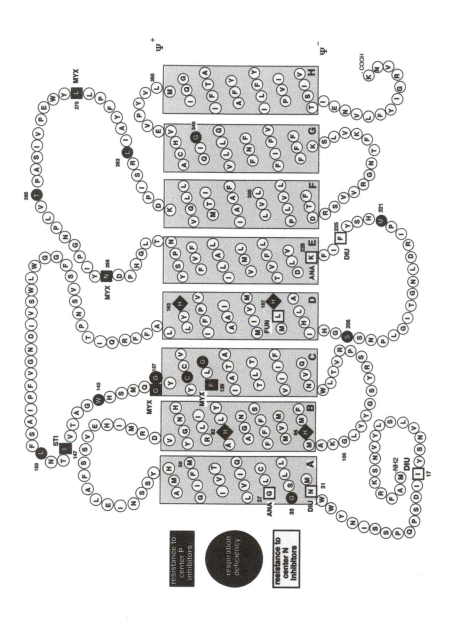

crystallographic structure of the reaction center has revealed that inhibitor-resistant mutations are proximal to the QB site (Deisenhofer *et al.*, 1984; Allen *et al.*, 1988). Thus, the membrane sidedness of the inhibitor-resistant sites in the eight-helix cytochrome *b* model is satisfactory, if we assume that center N inhibitor-resistant mutations belong to the ubiquinone reducing site on one side of the membrane, and if center P inhibitor-resistant mutations map to the ubiquinol oxidation site on the opposite side of the membrane.

Spectroscopic studies indicate that hemes b_L and b_H are bound to histidines, the heme planes being perpendicular to the plane of the membrane (Simpkin *et al.*, 1989; Erecinska *et al.*, 1978). In the eight-helix cytochrome *b* model, helices 2 and 4 contain four histidines (H82, H96, H183, and H198 of the yeast cytochrome *b*) that are conserved in all the known cytochrome *b* sequences (Degli Esposti *et al.*, 1993; Widger *et al.*, 1984; Saraste, 1984). Site-directed mutagenesis experiments in *Rhobacter sphaeroides* have provided strong evidence that H82 and H183 are the axial ligands of heme b_L, while H96 and H198 seem to be the ligands of heme b_H (Yun *et al.*, 1991). The former are close to the presumed positive side of the membrane (i.e., the side on which center P inhibitor-resistant mutations are located), while the latter are close to the negative side of the membrane (Fig. 1). This topology of the heme axial ligands relative to that of the inhibitor-resistant mutation sites is consistent with the separation of two quinone reactions sites, the centers P and N, connected via the hemes b_L and b_H.

4. CYTOCHROME *b* NONFUNCTIONAL MISSENSE MUTANTS

Only a limited number of structure modifications in cytochrome *b* were found by the analysis of inhibitor-resistant mutants. Although they are less sensitive to specific inhibitors, the electron transfer and proton translocation activities of the cytochrome bc_1 complex are not affected in an important manner by these mutations. This is to be expected, since the selection of mutants requires a good growth on respiratory substrates. To better define structure–function relations in cytochrome *b*, loss-of-function mutations that disrupt the activity of cytochrome bc_1 complex were searched for. In yeast, such alterations can be obtained easily by selecting cells that fail to grow on nonfermentable carbon sources. Dozens of yeast respiratory-deficient cytochrome *b* mutants were isolated and mapped on the mtDNA (Kotylak and Slonimski, 1977; Slonimski and Tzagoloff, 1976). Not surprisingly, numerous mutations affect the splicing of the cytochrome *b* pre-mRNA (introns represent about 85% of the nucleotide sequence of the cytochrome *b* gene and several of them code for proteins that are necessary for removing introns from the precursor transcript). Nevertheless, a dozen of missense exon mutations that affect the cytochrome *b* primary sequence itself (by replacing an amino acid present in the wild type by a different one in the mutant) could be identified (Coppée *et al.*, 1994; Lemesle-Meunier *et al.*, 1993; Tron and Lemesle-Meunier, 1990b; Brivet-Chevillotte and di Rago, 1989; Lazowska *et al.*, 1989) (see Table I).

Some mutants fail to assemble the cytochrome bc_1 complex as indicated by immunologic and spectral analysis. This was seen notably with different mutations at the level of highly conserved glycines (G33D, G131S, and G340E), which are therefore presumed to play some important structural role (see Section 5). In other mutants, the cytochrome bc_1 complex was found to be assembled, but its electron transfer activity was strongly affected, indicating more specific mutation effects. This was seen notably when the cysteine at position 133 was replaced by tyrosine. Since myxothiazol- and mucidine-resistant mutations were found at surrounding positions (129 and 137) (di Rago *et al.*, 1989), it may be that the mutated cysteine plays some role at center P; this was supported by showing a strong decrease in the amount of heme b_L in the mutant, which in the Q cycle is involved in the oxidation of QH_2 (Lemesle-Meunier *et al.*, 1993). A 2-nm shift in the optical spectrum of heme b_H was detected when the serine residue at position 206 was replaced by leucine, indicating that the center N is touched by this mutation (Lemesle-Meunier *et al.*, 1993). Consistent with this, alterations at surrounding positions (198, 225, and 228) were shown to confer an increased resistance to diuron, antimycin, and funiculosin (di Rago *et al.*, 1990a), three inhibitors that are presumed to block the reoxidation of heme b_H by ubiquinone at center N.

Importantly, in the eight-helix cytochrome *b* model, mutations that disrupt selectively the center N are all located on the presumed negative side of the membrane (i.e., the side where the center N inhibitor-resistant mutations are located), whereas mutations that affect the center P are on the opposite side of the membrane. This further supports the eight-helix model of cytochrome *b* and the necessity of two separate quinone processing catalytic sites.

5. INTRAGENIC SECOND-SITE SUPPRESSORS OF CYTOCHROME *b* NONFUNCTIONAL MUTANTS

When the function of a protein is abolished due to the presence of a missense mutation in its gene, it can be concluded that the mutated amino acid is involved in some critical constraint that cannot be satisfied in the mutant. However, discrete alterations may affect a protein in many different ways, and it is generally difficult to understand why the function is lost. Intragenic suppressor mutations may help to better define the roles of specific residues. If the original mutation alters an essential residue, i.e., a ligand of a redox metal center, only true back mutations restoring the wild-type protein sequence will be found. However, when the disrupted constraint can be met by several residues, novel functional variants of the protein may be found. Importantly, second-site reversion mutations may be obtained, and this can be particularly useful in revealing specific interactions between domains.

The reversion properties of cytochrome *b* nonfunctional respiratory deficient mutants can be analyzed conveniently in yeast. Cells issued from the original mutant, with or without a treatment by a specific mutagen, are simply plated on a nonfermentable carbon source in order to select the rare revertants clones, which are able to multiply under conditions where the original mutant is unable to do so (billions of cells can be tested on a single Petri dish, which makes feasible the selection of very rare, down to 10^{-9}, reversion events). More than 1000 genetically independent isolates of such revertants issued from different cytochrome *b* mutants were characterized using fast molecular screening procedures (di Rago *et al.*, 1990b,c). About 60 different intragenic pseudo-wild-type reversion mutations were identified (di Rago *et al.*, 1995, 1990b,c; Bruel *et al.*, 1995a; Coppée *et al.*, 1994a,b; Tron *et al.*, 1991). We will give a few examples of results to illustrate the type of information provided by these mutants.

Mutations at the level of three conserved glycines (G33D, G131S, and G34OE) were shown to have dramatic effects on the cytochrome bc_1 complex, suggesting that these glycines play some important structural role (Coppée *et al.*, 1994b; Lemesle-Meunier *et al.*, 1993; Brivet-Chevillotte and di Rago, 1989). Only true back mutations restoring the initial glycine 33 and a pseudoreversion that replaces the deleterious aspartic acid 33 by alanine were found in revertants issued from the mutant G33D (Coppée *et al.*, 1994b). It should be noted that several other amino acids (asparagine, histidine, tyrosine, and valine) could have been obtained by a single monosubstitution from the aspartic acid codon 33. Since none of these four amino acids has been found despite a large number (85) of analyzed isolates, it seems likely that they would have, like aspartic acid, some deleterious consequences when introduced at position 33. The restoration of cytochrome *b* function is only partial with alanine, showing that the presence a single $-CH3$ instead of $-H$ radical side chain at position 33 suffices to provoke some already important perturbations. Such data demonstrate that only very small-sized side chains can be tolerated at position 33. Similar conclusions were drawn from the analysis of revertants issued from the mutants G131S and G340E (di Rago *et al.*, 1995, 1990c).

In the eight-helix cytochrome *b* model, the glycines at positions 33, 131, and 340 are each located in the membrane. It was noted that helices A and D each contain a pair of conserved glycines (G33-G47 and G117-G131) that are spaced 13 residues apart as are the histidine ligands in helices B (H82–H96) and D (H183–H197) (Tron *et al.*, 1991). It was therefore concluded that the four glycines in helices A and C may be needed to accommodate the hemes in the phospholipid bilayer within a four-helix bundle structure. Similarly, the G340 residue could be important also for the packing of cytochrome b in the membrane (di Rago *et al.*, 1995).

A mutation at the level of another highly conserved glycine (G290D, in helix F) was detected in a 29-year-old man with progressive exercise muscle intolerance associated with a marked deficiency of complex bc_1 activity and a decreased amount of cytochrome b (Dumoulin *et al.*, 1996). Since severe respiratory deficiency is lethal in humans (but not in yeast), only heteroplasmic individuals can survive. This was the case for this patient who had still 20% of his mtDNA molecules of the G290 type and 80% of D290 type. This latter change has not been found in 150 control individuals. As in the *S. cerevisiae* mutants G33D, G131S, and G340E, it could be that the G290D mutation also affects the stability of cytochrome b, which could explain the observed mitochondrial disorders in this patient.

Replacement of cysteine 133 by tyrosine in yeast cytochrome b results in a total absence of growth on nonfermentable substrates (Lemesle-Meunier *et al.*, 1993). Mitochondrial respiration was recovered when the deleterious tyrosine was in turn replaced by either asparagine, aspartic acid, serine, or phenylalanine (di Rago *et al.*, 1990b). Thus, the size, rather than the nature, of amino acid side chain seems to be important at position 133 (the five residues that give an active cytochrome bc_1 complex all have a smaller side chain than tyrosine). Not surprisingly, only small-sized residues (cysteine, serine, valine, glycine) are found at the equivalent position in cytochromes b from other species. Second-site reversion mutations were found also at position 176 (isoleucine to serine) or 343 (histidine to threonine) (di Rago *et al.*, 1990c). These also lead to side chain size reduction, suggesting that Y133, I176, and H343 residues may be proximal to each other in the folded cytochrome b. However, it must be noted that size increases and decreases at interacting residues are not necessarily related in a simple complementary fashion and long-range mutation effects cannot be excluded.

Second-site reversion mutations may reveal interactions between charged amino acids; as such, spatial proximity of the mutated sites is strongly suggested, as illustrated by the following example. The cytochrome b function is lost when glycine 137 is replaced by negatively charged glutamic acid (Tron and Lemesle-Meunier, 1990). Function is regained upon replacement of polar uncharged asparagine 256 residue by positively charged lysine (di Rago *et al.*, 1990c). This suppressor mutation was detected in 16 independent revertants with two different codon changes. Since six other amino acids could have been derived *a priori* by monosubstitution from asparagine codon 256 (serine, threonine, isoleucine, aspartic acid, histidine, and tyrosine), and since none of these is positively charged at physiological pH, it seems that charge compensation between glutamic acid 137 and lysine 256 residues is responsible for the

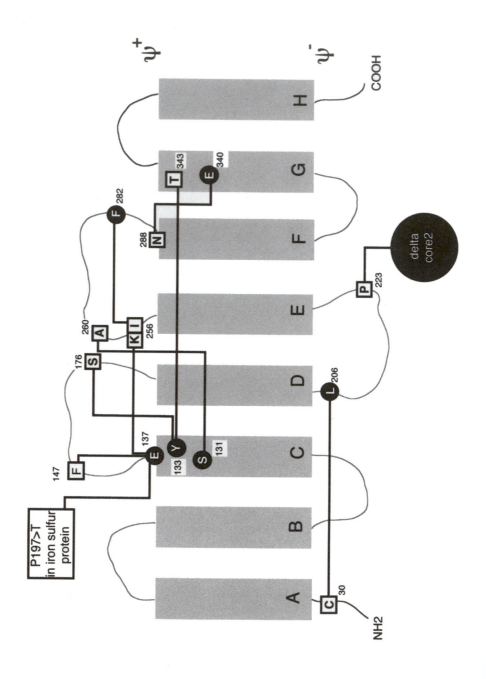

revertant phenotype; therefore, it was proposed that these two residues are proximal to each other in the three-dimensional structure of cytochrome *b*. Remarkably, other mutations at positions 137 (G137R) and 256 (N256Y) were identified in mutants selected for increased resistance to mucidin and myxothiazol (two inhibitors of center P) (di Rago *et al.*, 1989). This further supports the view (discussed in Section 3) that inhibitor-resistant mutations are proximal to the quinone substrate binding sites.

In total, seven "long-distance" interactions were characterized from yeast cytochrome *b* suppressor mutations of which some suggest contacts between transmembrane segments (C-G, F-G, and A-D) (di Rago *et al.*, 1995). According to the eight-helix cytochrome *b* model, there is never compensation across the membrane, i.e., the original mutations and their suppressor mutations are always on the same side of the membrane (see Fig. 2). By contrast, in the case of the yeast mitochondrial ATP–ADP translocase, most of the second-site suppressor mutations (14 of 15 isolated) are on the opposite side of the membrane relative to their parental mutations (Nelson and Douglas, 1993). Given the high number (more than 1000) of revertants analyzed, absence of compensation across the membrane in cytochrome *b* is not fortuitous. Rather, it strengthens the need for spatial separation of two catalytic quinone reaction sites.

6. CYTOCHROME *b* FRAMESHIFT MUTANTS AND THEIR REVERTANTS

No missense point mutation modifying the function of cytochrome *b* has been found in the C-terminal region (helix H). This may account for the high variability of this region among species (the homologous protein of chloroplasts has only seven transmembrane segments) (Degli Esposti *et al.*, 1993). However, frameshift mutations were used to probe the importance of this cytochrome *b* region in yeast. Usually such mutations result in the in-frame

←——

FIGURE 2. Topological locations of second-site suppressor mutations in yeast cytochrome *b*. Black-circled amino acid residues in cytochrome *b* lead to a respiratory growth-deficient phenotype. "Long-distance" intragenic second-site suppressor mutations are indicated: the amino acids introduced by the suppressor mutations are surrounded by squares and connected to the corresponding target residues by lines (di Rago *et al.*, 1995). For example, the growth on nonfermentable substrate is abolished upon replacement of C133 by Y133, and the respiratory growth is restored when I176 is replaced by S176 or when H343 is replaced by T343. The G137>E mutation in cytochrome *b* is suppressed by a single replacement P197>T in the nuclear-encoded iron–sulfur protein of the cytochrome bc_1 complex (di Rago, Slonimski, and Trumpower, in preparation). The S223>P mutation in cytochrome *b* suppresses a cytochrome bc_1 complex deficiency due to alterations in the nuclear-encoded core protein 2 gene (di Rago *et al.*, 1997).

introduction of a premature stop codon leading to the synthesis of truncated proteins, especially when they occur in a region remote from the normal C-terminus. Yet, we isolated a frameshift mutant that synthesizes a cytochrome *b* of nearly normal length but with a novel hydrophilic C-terminus (of 33 amino acids) due to a single base pair addition just before helix G (di Rago *et al.*, 1993). This mutant failed to assemble the cytochrome bc_1 complex as indicated by spectral and immunologic analysis. The assembly of an active cytochrome bc_1 complex was restored upon different base pair subtractions within a window of 24 bp before or after the original base pair addition, leading to single or multiple (up to five) consecutive amino acid replacements. Despite such important structure modifications the revertants were not different from the wild type in their respiratory capacity. These mutants provide evidence that helix G plays some role, but only its conserved hydrophobic character seems to be important, perhaps for an interaction with another polypeptide subunit of the cytochrome bc_1 complex.

7. EXTRAGENIC AND INTERGENOMIC SUPPRESSOR MUTATIONS AS A TOOL FOR THE STUDY OF POLYPEPTIDE SUBUNIT INTERACTIONS WITHIN THE CYTOCHROME bc_1 COMPLEX

The mitochondrial cytochrome bc_1 complex is a multisubunit enzyme containing 10 or 11 different polypeptide subunits that interact with each other as the pieces of a puzzle. Genetic approaches may be useful to probe such protein–protein interactions in revealing how a loss-of-function mutation in one of the enzyme subunits can be suppressed by a second mutation in another subunit. As already mentioned, the cytochrome *b* subunit of the cytochrome bc_1 complex is coded by the mitochondrial DNA, while all other subunits of the complex are specified by nuclear genes. This genetic compartmentalization is especially useful to analyze the interactions of cytochrome *b* with the rest of the complex. We offer two examples of such revertants. The first is a missense mutation of the Rieske iron–sulfur protein that restores the energy transduction functions of the cytochrome bc_1 complex in a cytochrome *b* mutant. The second is a point mutation in cytochrome *b* that suppresses a cytochrome bc_1 complex deficiency due to alterations in the nuclear-encoded core protein 2 gene.

7.1. A Missense Point Mutation in the Nuclear Rieske Protein Gene Compensates for Functional Defects due to a Mutation in the Mitochondrial Cytochrome *b* Gene

A practical means to search for mutations in a defined nuclear gene that compensate for a mutation in mtDNA is to work with the nuclear gene cloned in an episomal vector. The recombinant plasmid is mutagenized randomly, for example, by *in vitro* treatment with hydroxylamine, and yeast cells carrying the mtDNA mutation are transformed with the mutagenized plasmids. Rescuing plasmids are then extracted and the suppressor genes sequenced.

Using this procedure, we found that a single replacement (P197T) in iron–sulfur protein suppresses the effects of a point mutation (G137E) in cytochrome *b* (J.-P. di Rago, P. P. Slonimski, and B. L. Trumpower, in preparation). The original mutation in cytochrome *b* has some surprising properties (Tron and Lemesle-Meunier, 1990; Bruel *et al.*, 1995a). Mitochondrial membranes from this mutant retain a quite good ubiquinol-cytochrome *c* reductase activity (30%), which could be expected, by comparison with other mutants, to be sufficient for respiratory growth. In spite of this, the mutant does not grow on nonfermentable carbon sources, which raises the possibility of a decoupling of proton translocation from electron transfer. Consistent with this, defects in transmembrane potential and a decreased yield in ADP phosphorylation were found, suggesting a proton slip at the level of the mutated cytochrome bc_1 complex (i.e., protons go back to the matrix through the complex without any reverse electron transfer or ATP synthesis). Since the suppressor mutation P197T in iron–sulfur protein does not markedly improve the electron transfer activity but does efficiently improve the respiratory growth, it demonstrates that the energy coupling functions of the bc_1 complex have been restored by this intergenomic interaction.

In the crystal structure of a water-soluble fragment of bovine iron–sulfur protein determined recently (Iwata *et al.*, 1996), P197 is on the protein surface in the middle of a "Pro loop" at only 7.4 Å from L162, which has been presumed to be involved in QH_2 binding (Liebl *et al.*, 1995). In the eight-helix cytochrome *b* model, G137 is on the positive side of the membrane at the extremity of helix C. Thus, it may be that both residues are close to the QH_2 binding pocket.

As already described, G137E is suppressed also by N256K in cytochrome *b*, presumably due to a charge compensation (di Rago *et al.*, 1990c). Taken together, these data suggest that protons liberated by the oxidation of QH_2 are "trapped" by the acidic side chain of E137 and then conducted through the cytochrome bc_1 complex to the matrix, instead of being evacuated to the inter-membrane space. Thus, the introduction in the vicinity of E137 of either a positive charge (K instead of N at position 256 in cytochrome *b*) or a protonic hydroxylated side chain (T instead of P at position 197 in iron–sulfur protein) may prevent protons from being diverted by the mutant glutamate residue from their normal conduction pathways. The interaction between cytochrome *b*–G137E and iron–sulfur protein–P197T mutations illustrates well the intimate cooperation of cytochrome *b* with the iron–sulfur protein in the energy-transducing mechanism of the cytochrome bc_1.

7.2. A Point Mutation in the Mitochondrial Cytochrome *b* Gene Obviates the Requirement of the Nuclear-Encoded Core Protein 2 Subunit in the Cytochrome bc_1 Complex in *S. cerevisiae*

Cytochrome bc_1 complexes purified from various mitochondrial sources contain in addition to cytochrome *b*, cytochrome c_1, and iron–sulfur protein, seven to eight polypeptides that lack redox metal centers (two large proteins of 40–50 kDa referred to as core 1 and core 2, and 5 to 6 small polypeptides of less than 15 kDa). These are referred to as "supernumerary" or "extra" polypeptides because they have no counterparts in cytochrome bc_1 complexes purified from diverse bacteria, such as *Paracoccus denitrificans*, which contain cytochrome *b*, cytochrome c_1, and iron–sulfur protein only (Yang and Trumpower, 1986). Despite this, bacterial and mitochondrial cytochrome bc_1 complexes exhibit essentially identical electron transfer and proton-translocating activities, which raises the question of the biological functions of the extra polypeptides in mitochondrial cytochrome bc_1 complex (Yang and Trumpower, 1988).

Mutant studies in yeast have shown that the extra subunits of mitochondrial cytochrome bc_1 complex are all important, at least to some degree (Brandt *et al.*, 1994; Yang and Trumpower, 1994; Philips *et al.*, 1993; Schmitt and Trumpower, 1990; Schoppink *et al.*, 1989; Maarse *et al.*, 1988; Crivellone *et al.*, 1988; Oudshoorn *et al.*, 1987; Tzagoloff *et al.*, 1986). For example, when the gene of core protein 1 is inactivated, mitochondrial membranes lose their capacity to catalyze the oxidation of ubiquinol by cytochrome *c* (Tzagoloff *et al.*, 1986). In addition, as shown by immunologic analysis, several of the remaining polypeptides of the cytochrome bc_1 complex, among which is cytochrome *b*, are present in strongly decreased amounts. Since mRNA levels and synthesis of the missing polypeptides are not affected in pulse-labeled experiments, it is thought that in the absence of core protein 1, the rest of the complex is not properly assembled and that the unassembled polypeptides are rapidly degraded by mitochondrial proteases (Crivellone *et al.*, 1988). Similar defects were seen with mutations in other supernumerary subunits of the cytochrome bc_1 complex.

Why does the cytochrome bc_1 complex require so many additional factors to insert and function in the inner mitochondrial membrane but not in bacterial plasma membranes? It could be that similar factors do exist in bacteria but that they are bound less strongly to the cytochrome bc_1 complex and are lost during purification procedures. However, some studies have revealed that in some eukaryotes the cytochrome bc_1 complex has some additional activities that have no *raison d'être* in bacteria.

Indeed, it was found, quite unexpectedly, that cytochrome bc_1 complexes purified from plant sources and *N. crassa* cleave the N-terminus targeting sequences of precursors of mitochondrial proteins specified by nuclear genes

(Jänsch *et al.*, 1995; Eriksson *et al.*, 1994; Emmermann *et al.*, 1993; Braun *et al.*, 1992; Schulte *et al.*, 1989; for reviews, see Glaser *et al.*, 1996; Braun and Schmitz, 1995; Kalousek *et al.*, 1993). In yeast and mammals, this proteolytic activity is catalyzed by a couple of water-soluble proteins located in the matrix, which are referred to as β-MPP and α-MPP (mitochondrial processing peptidon) (Jensen and Yaffe, 1988; Yang *et al.*, 1988; Witte *et al.*, 1988). β-MPP is the catalytic subunit that cuts the substrate; α-MPP is an associated factor that increases (50 times) the activity of β-MPP.

The core proteins from plant mitochondria show high levels of sequence similarities with MPPs of yeast and mammals, and it thus appears that they are MPP. Interestingly, although the core proteins from yeast and mammals are proteolytically inactive, they exhibit significant sequence similarities with MPP, which indicates that the core proteins and MPP have a common phylogenetic origin. On this basis, it was hypothesized (Braun and Schmitz, 1995) that MPP has been derived from an ancestral protease that was already present in the prokaryotic progenitor of mitochondria (related endoproteases are found in some prokaryotes such as petrilysin of *Escherichia coli*). At the early stages of endosymbiosis, when MPP was still inefficient, it might have been advantageous to attach it to the inner mitochondrial membrane, i.e., close to the precursor entry sites. This could have been facilitated by interactions with a preexisting membrane protein such as the cytochrome bc_1 complex. Later in evolution, after development of a performing two-component protease system, it could have been beneficial to remove the protease from the cytochrome bc_1 complex for a separate regulation of respiration and protein processing activities (when yeast is growing under anaerobic conditions, proteins still have to be imported into the organelle, whereas the synthesis of respiratory enzymes is repressed). "Return" of the protease in a free form in the matrix might have been obtained by gene duplication of the protease genes, which could have allowed the progressive loss of the proteolytic activity of the membrane-bound protease, giving rise to the actual proteolytically inactive core proteins.

This elegant hypothesis raises several questions. Are there some reasons for the initial integration of MPP into the cytochrome bc_1 complex and not into another complex of the inner mitochondrial membrane? Also, what are the reasons for the maintenance of the core proteins in the cytochrome bc_1 complex after return of a free MPP in the matrix?

We have addressed these questions by searching for yeast mutants able to assemble an active cytochrome bc_1 complex despite a core protein deficiency (di Rago *et al.*, 1997). We started from a frameshift mutation in the nuclear gene of core protein 2 by which the last 204 residues of the wild-type protein are replaced by a novel sequence of 32 amino acids. Mitochondrial membranes from this mutant do not catalyze at all the oxidation of ubiquinol by cytochrome *c*. In addition, the rest of the complex is apparently degraded, as indicated by immunologic analysis. Interestingly, a cytochrome c reductase activity is still detectable (5–10%) and not all of the remaining subunits are degraded when the core protein 2 gene is completely deleted (Oudshoorn *et al.*, 1987), suggesting deleterious effects of the aberrant core protein 2.

Remarkably, assembly and activity of the cytochrome bc_1 are restored efficiently in this mutant upon a single mutation leading to replacement of the serine 223 residue by proline in cytochrome b. Surprisingly, this mutation not only corrects the phenotype resulting from the frameshift in the core protein 2 gene, it also obviates the need for core protein 2 in cytochrome bc_1 complex since it alleviates the respiratory deficiency resulting from the complete dele-tion of its gene. This is the only known homoplasmic point mutation of the mtDNA acting as a functional suppressor of a mutation located in a nuclear gene.

In the eight-helix cytochrome b model, the S223 residue is within the DE matrix-exposed connecting loop. The core proteins 1 and 2, which contribute about half of the protein mass of the cytochrome bc_1 complex, are held at the matrix side of the complex (Yu *et al.*, 1996; Gonzalez-Halphen *et al.*, 1988; Leonard *et al.*, 1981). Thus, the suppressor mutation S223P affects a subdomain of cytochrome b, which may be in contact with the core proteins. It is therefore believed that core protein 2 protects the DE connecting loop from proteolytic degradation (the S223P mutation may render it intrinsically resistant to mito-chondrial proteases, thus obviating the requirement for core protein 2) (di Rago *et al.*, 1997).

Strangely, the counterpart of cytochrome b in chloroplastic cytochrome $b_6 f$ complex is split into two subunits (cytochrome b_6 + subunit IV) at the level of the DE connecting loop of mitochondrial cytochrome b. No counterparts of the mitochondrial core proteins 1 and 2 have been found in the cytochrome $b_6 f$ complex (Takahashi *et al.*, 1996, and references therein), and therefore it may be speculated that the split nature of the chloroplastic homologue of mito-chondrial cytochrome b is a reminder of an ancient proteolytic site.

In view of these data, a possible scenario that merits consideration is that MPP itself participates in the proteolytic degradation of cytochrome b when the core proteins are absent. Indeed, the matrix-exposed connecting loops of cyto-chrome b are, like the substrate of MPP, enriched in basic residues that are thought to be critical for the proper insertion of cytochrome b into the inner mitochondrial membrane (Gavel and von Heijne, 1992; Hartl *et al.*, 1989; Hendrick *et al.*, 1989; von Heijne, 1986). This may be the reason, during evolution of mitochondria, for the initial integration of MPP in the cytochrome bc_1 complex and for the necessity of protecting the matrix-exposed connecting loops of cytochrome b after return of MPP as a free protease in the matrix in yeast and mammals.

8. REFERENCES

Allen, J. P., Feher, G., Yeates, T. O., Komiya, H., and Rees, D. C., 1988, Structure of the reaction center from *Rhodobacter sphaeroides* R-26: Protein–cofactor (quinones and Fe_2^+) interactions, *Proc. Natl. Acad. Sci. USA* **85**:8487–8491.

Becker, W. F., von Jagow, G., Anke, T., and Steglich, W., 1981, Oudemansin, strobilurin A, strobilurin B and myxothiazol: New inhibitors of the bc_1 segment of the respiratory chain with an E-β-methoxyacrylate system as common structural element, *FEBS Lett.* **132**:329–333.

Bennoun, P., Delosme, M., Godehardt, I., and Kück, U., 1992, New tools for mitochondrial genetic of *Chlamydomonas reinhardtii*—manganese mutagenesis and cytoduction, *Mol. Gen. Genet.* **234**:147–154.

Brandt, U., and Trumpower, B. L., 1994, The protonmotive Q cycle in mitochondria and bacteria, *Crit. Rev. Biochem. Mol. Biol.* **29**:165–197.

Brandt, U., Schagger, H., and von Jagow G., 1988, Characterisation of binding of the methoxyacrylate inhibitors to mitochondrial cytochrome c reductase, *Eur. J. Biochem.* **173**:499–506.

Brandt, U., Uribe, S., Schägger, H., and Trumpower B. L., 1994, Isolation and characterization of QCR10, the nuclear gene encoding the 8.5-kDa subunit 10 of the *Saccharomyces cerevisiae* cytochrome bc_1 complex, *J. Biol. Chem.* **269**:12947–12953.

Brasseur, R., 1988, Calculation of the three-dimensional structure of *Saccharomyces cerevisiae* cytochrome b inserted in a lipid matrix, *J. Biol. Chem.* **263**:12571–12575.

Braun, H.-P., and Schmitz, U. K., 1995, Are the "core" proteins of the mitochondrial bc_1 complex evolutionary relics of a processing protease? *Trends Biochem. Sci.* **20**:171–174.

Braun, H.-P., Emmermann, M., Kruft, V., and Schmitz, U. K., 1992, The general mitochondrial processing peptidase from potato is an integral part of cytochrome c reductase of the respiratory chain, *EMBO J.* **11**:3219–3227.

Briquet, M., and Goffeau A., 1981, Modification of the spectral properties of cytochrome b in mutants of *Saccharomyces cerevisiae* resistant to 3-(3,4-dichlorophenyl)-1,1-dimethylurea. Mapping at two distinct genetic loci of the split mitochondrial gene of cytochrome b, *Eur. J. Biochem.* **117**:333–339 .

Brivet-Chevillotte, P., and di Rago, J.-P., 1989, Electron-transfer restoration by vitamin K_3 in a complex III-deficient mutant of *S. cerevisiae* and sequence of the corresponding cytochrome b mutation, *FEBS Lett.* **255**:5–9.

Brown, M. D., Voljavec, A. S., Lott, M. T., Torroni, A., Yang, C.-C., and Wallace, D., 1992, Mitochondrial DNA complex I and III mutations associated with Leber's hereditary optic neuropathy, *Genetics* **130**:163–173.

Bruel, C., di Rago, J.-P., Slonimski, P. P., and Lemesle-Meunier, D., 1995a, Role of the evolutionary conserved cytochrome b tryptophan 142 residue in the ubiquinol oxidation catalyzed by the bc_1 complex in yeast *Saccharomyces cerevisiae*, *J. Biol. Chem.* **270**:22321–22328.

Bruel, C., Manon, S., Guerin, M., and Lemesle-Meunier, D., 1995b, Decoupling of the bc_1 complex in *S. cerevisiae*; point mutations affecting the cytochrome b gene bring new information about the structural aspect of the proton translocation, *J. Bioenerg. Biomembr.* **27**:527–539.

Brunner, L. A., Mendoza, R. V., and Tuena de Cobos, A., 1987, Extrachromosomal genetics in the yeast *Kluyveromyces lactis*. Isolation and characterization of antimycin-resistant mutants, *Curr. Genet.* **11**:475–482.

Clavilier, L., Pere, G., and Slonimski, P. P., 1969, Mise en évidence de plusieurs loci indépendants impliqués dans la synthèse de l'iso-2 cytochrome c chez la levure, *Mol. Gen. Genet.* **104**: 195–218.

Colson, A.-M., and Slonimski, P. P., 1979, Genetic localization of diuron- and mucidin-resistant mutants relative to a group of loci of the mitochondrial DNA controlling coenzyme QH_2-cytochrome c reductase in *Saccharomyces cerevisiae, Mol. Gen. Genet.* **167**:287–298.

Colson, A.-M., The van, L., Convent, B., Briquet, M., and Goffeau A., 1977, Mitochondrial heredity of resistance to 3-(3,4-dichlorophenyl)-1,1-dimethylurea, an inhibitor of cytochrome *b* oxidation, in *Saccharomyces cerevisiae, Eur. J. Biochem.* **74**:521–526.

Colson, A.-M., Michaelis, G., Pratje, E., and Slonimski P. P., 1979, Allelism relationships between diuron-resistant, antimycin-resistant, and funiculosin-resistant loci of the mitochondrial map in *Saccharomyces cerevisiae, Mol. Gen. Genet.* **167**:299–300.

Convent, B., Briquet, M., and Goffeau, A., 1978, Kinetic evidence for two sites in the inhibition by diuron of the electron transport in the bc_1 segment of the respiratory chain in *Saccharomyces cerevisiae, Eur. J. Biochem.* **92**:137–145.

Coppée, J. Y., Brasseur, G., Brivet-Chevillotte P., and Colson A. M., 1994a, Non-native intragenic reversions selected from *Saccharomyces cerevisiae* cytochrome *b*-deficient mutants. Structural and functional features of the catalytic center N domain, *J. Biol. Chem.* **269**:4221–4226.

Coppée, J.-Y., Tokutake, N., Marc, D., di Rago, J.-P., Miyoshi, H., and Colson, A.-M., 1994b, Analysis of revertants from respiratory deficient mutants within the center N of cytochrome *b* in *Saccharomyces cerevisiae, FEBS Lett.* **339**:1–6.

Crivellone, M. D., Wu, M., and Tzagoloff A., 1988, Assembly of the mitochondrial membrane system. Analysis of structural mutants of the yeast coenzyme QH_2-cytochrome c reductase complex, *J. Biol. Chem.* **263**:14323–14333.

Crofts, A. R., Robinson, H., Andrews, K., Van Doren, S., and Berry, E., 1987, Catalytic sites for reduction and oxidation of quinones, in *Cytochrome Systems* (S. Papa, B. Chance, and L. Ernster, eds.), pp. 617–624, Plenum Press, New York.

Daldal, F., Tokito, M. K., Davidson, E., and Faham, M., 1989, Mutations conferring resistance to quinol oxidation (Qz) inhibitors of the cyt bc_1 complex of *Rhodobacter capsulatus, EMBO J.* **8**:3951–3961.

Degli Esposti, M., De Vries, S., Crimi, M., Ghelli, A., Patarnello, T., and Meyer, A., 1993, Mitochondrial cytochrome *b*: Evolution and structure of the protein, *Biochim. Biophys. Acta* **1143**:243–271.

Deisenhofer, J., Epp, O., Miki, K., Huber, R., and Michel, H., 1984, X-ray structure analysis of a membrane protein complex. Electron density map at 3 Å resolution and a model of the chromophores of the photosynthetic reaction center from *Rhodopseudomonas viridis, J. Mol. Biol.* **180**:385–398.

di Rago, J.-P., and Colson, A.-M., 1988, Molecular basis for resistance to antimycin and diuron, Q-cycle inhibitors acting at the Q_i site in the mitochondrial ubiquinol–cytochrome c reductase in *Saccharomyces cerevisiae, J. Biol. Chem.* **263**:12564–12570.

di Rago, J.-P., Perea, J., and Colson A.-M., 1986, DNA sequence analysis of diuron-resistant mutations in the mitochondrial cytochrome *b* gene of *Saccharomyces cerevisiae, FEBS Lett.* **208**:208–210.

di Rago, J.-P., Coppée, J.-Y., and Colson, A.-M., 1989, Molecular basis for resistance to myxothiazol, mucidin, strobilurin A) and stigmatellin. Cytochrome *b* inhibitors acting at the center of the mitochondrial ubiquinol-cytochrome c reductase in *Saccharomyces cerevisiae, J. Biol. Chem.* **264**:14543–14548.

di Rago, J.-P., Netter P., and Slonimski P. P., 1990a, Pseudo-wild-type revertants from inactive-apocytochrome *b* mutants as a tool for the analysis of the structure/function relationships of the mitochondrial ubiquinol-cytochrome c reductase of *Saccharomyces cerevisiae, J. Biol. Chem.* **265**:3332–3339.

di Rago, J.-P., Netter P., and Slonimski P. P., 1990b, Intragenic suppressors reveal long distance interactions between inactivating and reactivating amino acid replacements generating three-dimensional constraints in the structure of mitochondrial cytochrome *b*, *J. Biol. Chem.* **265:** 15750–15757.

di Rago, J.-P., Perea, J., and Colson A.-M., 1990c, Isolation and RNA sequence analysis of cytochrome *b* mutants resistant to funiculosin, a center inhibitor of the mitochondrial ubiquinol–cytochrome *c* reductase in *Saccharomyces cerevisiae*, *FEBS Lett.* **263:**93–98.

di Rago, J.-P., Macadre, C., Lazowska, J., and Slonimski P.P., 1993, The C-terminal domain of yeast cytochrome *b* is essential for a correct assembly of the mitochondrial cytochrome *bc*₁ complex, *FEBS Lett.* **328:**153–158.

di Rago, J.-P., Hermann-Le Denmat, S., Pâques, F., Risler, J.-L., Netter, P., and Slonimski P. P., 1995, Genetic analysis of the folded structure of yeast mitochondrial cytochrome *b* by selection of intragenic second-site revertants, *J. Mol. Biol.* **248:**804–811.

di Rago, J.-P., Sohm, F., Boccia, C., Dujardin, G., Trumpower, B. L., and Slonimski P. P., 1997, A point mutation in the mitochondrial cytochrome *b* gene obviates the requirement for the nuclear encoded core protein 2 subunit in the cytochrome *bc*₁ complex in *Saccharomyces cerevisiae*, *J. Biol. Chem* **272:**4699–4704.

Dujon, B., 1981, Mitochondrial genetics and functions, in *The Molecular Biology of the Yeast Saccharomyces* (J. Strathern, E. Jones, and J. Broach, eds.), pp. 505–635, Cold Spring Harbor Laboratory, Cold Spring Harbor, New York.

Dujon, B., Slonimski, P. P., Weill, L., 1974, Mitochondrial genetics. IX. A model for recombination and segregation of mitochondrial genomes in *Saccharomyces cerevisiae*, *Genetics* **78:** 415–437.

Dumoulin, R., Sagnol, I., Ferlin, T., Bozon, D., Stepien, G., and Mousson, B., 1996, A novel gly290asp mitochondrial cytochrome *b* mutation linked to a complex III deficiency in progressive exercise intolerance, *Mol. Cell. Probes* **10:**389–391.

Emmermann, M., Braun, H.-P., Arretz, M., and Schmitz, U. K., 1993, Characterization of the bifunctional cytochrome *c* reductase-processing peptidase complex from potato mitochondria, *J. Biol. Chem.* **268:**18936–18942.

Ephrussi, B., and Slonimski, P. P., 1955, Yeast mitochondria: Subcellular units involved in the synthesis of respiratory enzymes in yeast, *Nature* **176:**1207–1209.

Erecinska, M., Wilson, D. F., and Blasie, J. K., 1978, Studies on the orientations of the mitochondrial redox carriers. I. Orientation of the hemes of cytochrome *c* oxidase with respect to the plane of a cytochrome oxidase–lipid model membrane, *Biochim. Biophys. Acta* **501:** 53–62.

Eriksson, A. C., Sjöling, S., and Glaser, E., 1994, The ubiquinol cytochrome *c* oxidoreductase complex of spinach leaf mitochondria is involved in both respiration and protein processing, *Biochim. Biophys. Acta* **1186:**221–231.

Gavel, Y., and von Heijne, G., 1992, The distribution of charged amino acids in mitochondrial inner-membrane proteins suggests different modes of membrane integration for nuclearly and mitochondrially encoded proteins, *Eur. J. Biochem.* **205:**1207–1215.

Glaser, E., Sjöling, S., Szigyarto, C., and Eriksson, A. C., 1996, Plant mitochondrial protein import: Precursor processing is catalysed by the integrated mitochondrial processing peptidase (MPP)/*bc*₁ complex and degradation by the ATP-dependent proteinase, *Biochim. Biophys. Acta* **1275:**33–37.

Gonzalez-Halphen, D., Lindorfer, M. A., and Capaldi R. A., 1988, Subunit arrangement in beef heart complex III, *Biochemistry* **27:**7021–7031.

Gorini, L., and Beckwith, J.R., 1966, Suppression, *Annu. Rev. Microbiol.* **20:**401–422.

Hartl, F.-U., Pfanner, N., Nicholson, D. W., and Neupert, W., 1989, Mitochondrial protein import, *Biochim. Biophys. Acta.* **988:**1–45.

Hendrick, J. P., Hodges, P. E., and Rosenberg, L. E., 1989, Survey of amino-terminal proteolytic cleavage sites in mitochondrial precursor proteins: Leader peptides cleaved by two matrix proteases share a three-amino acid motif, *Proc. Natl. Acad. Sci. USA* **86:**4056–4060.

Howell, N., and Gilbert, K., 1988, Mutational analysis of the mouse mitochondrial cytochrome *b* gene, *J. Mol. Biol.* **203:**607–618.

Howell N., Appel J., Cook, J. P., Howell, B., and Hauswirth, W. W., 1987, The molecular basis of inhibitor resistance in a mammalian cytochrome *b* mutant, *J. Biol. Chem.* **262:**2411–2414.

Iwata, S., Saynovits, M., Link, T. A., and Michel H., 1996, Structure of a water soluble fragment of the "Rieske" iron–sulfur protein of the bovine heart mitochondrial cytochrome bc_1 complex determined by MAD phasing at 1.5 Å resolution, *Structure* **4:**567–579.

Jacq, C., Lazowska J., and Slonimski P. P., 1980, Cytochrome B messenger RNA maturase encoded in an I intram regulated the expression of the split gene: I. Physician location and box sequence of intram mutations, in *Organisation and Expression of the Mitochondrial Genome* (A. M. Kroon and C. Saccone, eds.), pp. 139–152, North-Holland Publishing Co., Amsterdam.

Jänsch, L., Kruft, V., Schmitz, U. K., and Braun, H.-P., 1995, Cytochrome *c* reductase from potato does not comprise three core proteins but contains an additional low-molecular-mass subunit, *Eur. J. Biochem.* **228:**878–885.

Jensen, R. E., and Yaffe, M. P., 1988, Import of proteins into yeast mitochondria: The nuclear *MAS2* gene encodes a component of the processing protease that is homologous to the *MAS1*-encoded subunit, *EMBO J.* **7:**3863–3871.

John, D. R., and Neufeld, M. J., 1991, Cytochrome *b* mutations in Leber hereditary optic neuropathy, *Biochem. Biophys. Res. Commun.* **181:**1358–1364.

Kalousek, F., Neupert, W., Omura, T., Schatz, G., and Schmitz, U. K., 1993, Uniform nomenclature for the mitochondrial peptidases cleaving precursors of mitochondrial proteins, *Trends Biol. Sci.* **18:**249.

Kotylak, Z., and Slonimski, P. P., 1977, Mitochondrial mutants isolated by a new screening method based upon the use of the nuclear mutation op1, in *Mitochondria* (W. Bandlow, R. J. Schweyen, K. Wolf, and F. Kaudewitz, eds.), pp. 161–172, de Gruyter, Berlin.

Lazowska, J., Jacq, C., and Slonimski, P. P., 1980, Sequence of introns and flanking exons in wild-type and box3 mutants of cytochrome *b* reveals an interlaced splicing protein coded by an intron, *Cell* **22:**333–348.

Lazowska, J., Claisse, M., Gargouri, A., Kotylak, Z., Spyridakis, A., and Slonimski P. P., 1989, Protein encoded by the third intron of cytochrome *b* gene in *Saccharomyces cerevisiae* is an mRNA maturase. Analysis of mitochondrial mutants, RNA transcripts and evolutionary relationships, *J. Mol. Biol.* **205:**275–289.

Lemesle-Meunier, D., Brivet-Chevillote P., di Rago, J.-P., Slonimski, P. P., Bruel, C., Tron T., and Forget, N., 1993, Cytochrome *b* deficient mutants of the ubiquinol–cytochrome *c* oxidoreductase in *Saccharomyces cerevisiae*, *J. Biol. Chem.* **268:**15626–15632.

Leonard, K., Wingfield, P., Arad, T., and Weiss, H., 1981, Three-dimensional structure of ubiquinol: cytochrome *c* reductase from *Neurospora* mitochondria determined by electron microscopy of membrane crystals, *J. Mol. Biol.* **149:**259–274.

Liebl, U., Sled, V., Ohnishi, T., and Daldal F., 1995, Conserved non liganding residues of *Rhodobacter capsulatus* "Rieske" protein are essential for [2Fe-2S] cluster properties and communication with the quinone pool, in *Photosynthesis: From Light to Biosphere*, vol. II (R. Mathis, ed.), pp. 749–752, Kluwer Academic.

Maarse, A. C., De Haan, M., Schoppink, P. J., Berden, J. A., and Grivell, L. A., 1988, Inactivation of the gene encoding the 11-kDa subunit VIII of the ubiquinol–cytochrome *c* oxidoreductase in *Saccharomyces cerevisiae*, *Eur. J. Biochem.* **172:**179–184.

Mansfield, R. W., and Wiggins, T. E., 1990, Photoaffinity labelling of the beta-methoxyacrylate binding site in bovine heart mitochondrial cytochrome bc_1 complex, *Biochim. Biophys. Acta* **1015:**109–115.

Michaelis, G., 1976, Cytoplasmic inheritance of antimycin A resistance in *Saccharomyces cerevisiae*, *Mol. Gen. Genet.* **146**:133–137.

Nelson, B. D., Walter, P., and Ernster, L., 1977, Funiculosin: An antibiotic with antimycin-like inhibitory properties, *Biochim. Biophys. Acta* **460**:157–162.

Nelson, D. R., and Douglas, M. G., 1993, Function-based mapping of the yeast mitochondrial ADP/ATP translocator by selection for second-site revertants, *J. Mol. Biol.* **230**:1171–1182.

Ohnishi, T., and Trumpower B.L., 1980, Differential effects of antimycin on ubisemiquinone bound in different environments in isolated succinate–cytochrome *c* reductase complex, *J. Biol. Chem.* **255**:3278–3284.

Ohnishi, T., Schagger, H., Meinhardt, S. W., LoBrutto, R., Link, T. A., and von Jagow, G., 1989, Spatial organization of redox active centers in the bovine heart ubiquinol–cytochrome *c* oxidoreductase, *J. Biol. Chem.* **264**:735–744.

Oudshoorn, P., Van Steeg, H., Swinkels, B. W., Schoppink, P., and Grivell L. A., 1987, Subunit II of yeast QH_2:cytochrome-c oxidoreductase, *Eur. J. Biochem.* **163**:97–103.

Philips, J. D., Graham, L. A., and Trumpower, B. L., 1993, Subunit 9 of the *Saccharomyces cerevisiae* cytochrome bc_1 complex is required for insertion of EPR-detectable iron–sulfur cluster into the Rieske iron–sulfur protein, *J. Biol. Chem.* **268**:11727–11736.

Pratje, E., and Michaelis, G., 1977, Allelism studies of mitochondrial mutants resistant to antimycin A or funiculosin in *Saccharomyces cerevisiae*, *Mol. Gen. Genet.* **152**:167–174.

Robertson, D. E., Giangiacomo, K. M., de Vires, S., Moser, C. C., and Dutton, P. L., 1984, Two distinct quinone-modulated modes of antimycin-sensitive cytochrome *b* reduction in the cytochrome bc_1 complex, *FEBS Lett.* **178**:343–350.

Saraste, M., 1984, Location of haem-binding sites in the mitochondrial cytochrome *b*. *FEBS Lett.* **166**:367–372.

Schmitt, M. E., and Trumpower, B. L., 1990, Subunit 6 regulates half-of-the-sites reactivity of the dimeric cytochrome bc_1 complex in *Saccharomyces cerevisiae*, *J. Biol. Chem.* **265**:17005–17011.

Schoppink, P. J., Berden, J. A., and Grivell, L. A., 1989, Inactivation of the gene encoding the 14-kDa subunit VII of yeast ubiquinol. Cytochrome *c* oxidoreductase and analysis of the resulting mutant, *Eur. J. Biochem.* **181**:475–483.

Schulte, U., Arretz, M., Schneider, H., Tropschug, M., Wachter, E., Neupert, W., and Weiss H., 1989, A family of mitochondrial proteins involved in bioenergetics and biogenesis, *Nature* **339**:147–149.

Simpkin, D., Palmer, G., Devlin, F. J., McKenna, M. C., Jensen, G. M., and Stephens P. J., 1989, The axial ligands of heme in cytochromes: A near-infrared magnetic circular dichroism study of yeast cytochromes c, c_1, and *b* and spinach cytochrome *f*, *Biochemistry* **28**:8033–8039.

Slater, E. C., 1973, The mechanism of action of the respiratory inhibitor antimycin, *Biochim. Biophys. Acta* **301**:129–154.

Slonimski, P. P., and Tzagoloff A., 1976, Localization in yeast mitochondrial DNA of mutations expressing a deficiency of cytochrome oxidase and/or coenzyme QH_2-cytochrome *c* reductase, *Eur. J. Biochem.* **61**:24–41.

Sturtevant, A. H., 1920, Vermilion gene and gynandromorphism, *Proc. Soc. Exp. Biol. Med.* **17**:70–71.

Subik, J., 1975, Mucidin-resistant antimycin A-sensitive mitochondrial mutant of *Saccharomyces cerevisiae*, *FEBS Lett.* **59**:273–276.

Subik, J., Kovacova, V., and Takacsova, G., 1977, Mucidin resistance in yeast: Isolation, characterization and genetic analysis of nuclear and mitochondrial mucidin-resistant mutants of *Saccharomyces cerevisiae*, *Eur. J. Biochem.* **73**:275–286.

Takahashi, Y., Rahire, M., Breyton, C., Popot, J. L., Joliot, P., and Rochaix, J. D., 1996, The chloroplast ycf7 (petL) open reading frame of *Chlamydomonas reinhardtii* encodes a small functionally important subunit of the cytochrome b_6f complex, *EMBO J.* **15**:3498–3506.

Thierbach, G., and Michaelis, G., 1982, Mitochondrial and nuclear myxothiazol resistance in *Saccharomyces cerevisiae, Mol. Gen. Genet.* **186**:501–506.

Thierbach, G., and Reichenbach, H., 1981, Myxothiazol, a new inhibitor of the cytochrome bc_1 segment of the respiratory chain, *Biochim. Biophys. Acta* **638**:282–289.

Thierbach, G., Kunze, B., Reichenbach, H., and Höfle, G., 1984, The mode of action of stigmatellin, a new inhibitor of the cytochrome bc_1 segment of the respiratory chain, *Biochim. Biophys. Acta* **765**:227–235.

Tron, T., Infossi, P., Coppee, J. Y., and Colson, A. M., 1991, Molecular analysis of revertants from a respiratory-deficient mutant affecting the center o domain of cytochrome *b* in *Saccharomyces cerevisiae, FEBS Lett.* **278**:26–30.

Tron, T., and Lemesle-Meunier, D., 1990, Two substitutions at the same position in the mitochondrial cytochrome *b* gene of *Saccharomyces cerevisiae* induce a mitochondrial myxothiazol resistance and impair the respiratory growth of the mutated strains albeit maintaining a good electron transfer activity, *Curr. Genet* **18**:413–419.

Tron, T., Crimi, M., Colson, A. M., and Degli Esposti, M., 1991, Structure/function relationships in mitochondrial cytochrome *b* revealed by the kinetic and circular dichroic properties of two yeast inhibitor-resistant mutants, *Eur. J. Biochem.* **199**:753–760.

Trumpower, B. L., 1990, Cytochrome bc_1 complexes of microorganisms, *Microbiol. Rev.* **54**:101–129.

Tzagoloff, A., Wu, M., and Crivellone, M., 1986, Assembly of the mitochondrial membrane system. Characterization of *COR1*, the structural gene for the 44-kilodalton core protein of yeast coenzyme QH_2-cytochrome *c* reductase, *J. Biol. Chem.* **261**:17163–17169.

von Heijne, G., 1986, The distribution of positively charged residues in bacterial inner membrane proteins correlates with the *trans*-membrane topology, *EMBO J.* **5**:3021–3027.

von Jagow, G., and Link, T., 1986, Use of specific inhibitors on the mitochondrial bc_1 complex, *Methods Enzymol.* **126**:253–271.

von Jagow, G., and Ohnishi, T., 1985, The chromone inhibitor stigmatellin-binding to the ubiquinol oxidation center at the C-side of the mitochondrial membrane, *FEBS Lett.* **185**:311–315.

von Jagow, G., Ljungdahl, P. O., Graf, P., Ohnishi, T., and Trumpower, B. L., 1984, An inhibitor of mitochondrial respiration which binds to cytochrome *b* and displaces quinone from the iron–sulfur protein of the cytochrome bc_1 complex, *J. Biol. Chem.* **259**:6318–6326.

von Jagow, G., Gribble, G.W., and Trumpower B.L., 1986, Mucidin and strobilurin A are identical and inhibit electron transfer in the cytochrome bc_1 complex of the mitochondrial respiratory chain at the same site as myxothiazol, *Biochemistry* **25**:775–780.

Weber, S., and Wolf, K., 1988, Two changes of the same nucleotide confer resistance to diuron and antimycin in the mitochondrial cytochrome *b* gene of *Schizosaccharomyces pombe, FEBS Lett.* **237**:31–34.

Widger, W. R., Cramer, W. A., Hermann, R. G., and Trebst, A., 1984, Sequence homology and structural similarity between cytochrome *b* of mitochondrial complex III and the chloroplast b_6-f complex: Position of the cytochrome *b* hemes in the membrane, *Proc. Natl. Acad. Sci. USA* **81**:674–678.

Witte, C., Jensen, R. E., Yaffe, M. P., and Schatz, G., 1988, MAS1, a gene essential for yeast mitochondrial assembly, encodes a subunit of the mitochondrial processing protease, *EMBO J.* **7**:1439–1447.

Yang, M., and Trumpower, B. L., 1994, Deletion of QCR6, the gene encoding subunit six of the mitochondrial cytochrome bc_1 complex, blocks maturation of cytochrome c_1, and causes temperature-sensitive petite growth in *Saccharomyces cerevisiae, J. Biol. Chem.* **269**:1270–1275.

Yang, M., Jensen, R. E., Yaffe, M. P., Opliger, W., and Schatz, G., 1988, Import of proteins into yeast mitochondria: The purified matrix processing protease contains two subunits which are encoded by the nuclear *MAS1* and *MAS2* genes, *EMBO J.* **7**:3857–3862.

Yang, X. H., and Trumpower, B. L., 1986, Purification of a three-subunit ubiquinol-cytochrome c oxidoreductase complex from *Paracoccus denitrificans*, *J. Biol. Chem.* **261:**12282–12289.

Yang, X., and Trumpower, B. L., 1988, Protonmotive Q cycle pathway of electron transfer and energy transduction in the three-subunit ubiquinol–cytochrome c oxidoreductase complex of *Paracoccus denitrificans*, *J. Biol. Chem.* **263:**11962–11970.

Yu, C. A., Xia, J. Z., Kachurin, A. M., Yu, L., Xia, D., Kim, H., and Deisenhofer, J., 1996, Crystallization and preliminary structure of beef heart mitochondrial cytochrome bc_1 complex, *Biochim. Biophys. Acta* **1275:**47–53.

Yun, C. H., Crofts, A. R., and Gennis, R. B., 1991, Assignment of the histidine axial ligands to the cytochrome b_H and cytochrome b_L components of the bc_1 complex from *Rhodobacter sphaeroides* by site-directed mutagenesis, *Biochemistry* **30:**6747–6754.

CHAPTER 24

Tissue-Specific Expression of Cytochrome *c* Oxidase Isoforms and Role in Nonshivering Thermogenesis

Bernhard Kadenbach, Viola Frank, Dietmar Linder,
Susanne Arnold, Stefan Exner, and Maik Hüttemann

1. INTRODUCTION

Regulation of energy metabolism in mitochondria is essential for survival of animals, since variations of work load, speed of reactivity, amount of thermogenesis, and adaptation to starvation require specific adjustments of the rate of oxygen consumption and efficiency of energy transduction in the respiratory chain. The various aspects of regulation of oxygen consumption in animals have been reviewed by Skulachev (1997). For warm-blooded animals, regulation of thermogenesis is of particular importance. Apart from thermogenesis in brown adipose tissue of rodents via the uncoupling protein (Nedergaard and Cannon, 1984; Nicholls and Locke, 1984), the mechanism of nonshivering thermogenesis in mammals and birds is largely unknown (Block, 1994).

Bernhard Kadenbach, Viola Frank, Susanne Arnold, Stefan Exner, and Maik Hüttemann • Fachbereich Chemie, Philipps-Universität, D-35032 Marburg, Germany. **Dietmar Linder** • Biochemisches Institut am Klinikum, Justus Liebig Universität, D-35385 Giessen, Germany.

Frontiers of Cellular Bioenergetics, edited by Papa *et al.* Kluwer Academic/Plenum Publishers, New York, 1999.

The amount of heat produced in mitochondria by oxidative phosphorylation at a given rate of oxygen consumption would increase with decreasing efficiency of energy transduction (H^+/e^--stoichiometry). For complex I (NADH dehydrogenase) and III (cytochrome c reductase), a constant H^+/e^- stoichiometry is generally assumed (Papa *et al.*, 1991). In contrast, a decrease of H^+/e^- stoichiometry with increasing turnover rates has been described for cytochrome c oxidase (Papa *et al.*, 1991; Capitanio *et al.*, 1991, 1996). The authors postulated that the normal flow of electrons from Cu_A to the binuclear center heme a_3–Cu_B via heme a is bypassed at high rates of electron transport (Capitanio *et al.*, 1996). The involvement of heme a in proton translocation would imply that bypassing of heme a in the electron flow decreases the H^+/e^- stoichiometry. Decrease of coupling efficiency in cytochrome c oxidase should also increase the rate of electron flow through the enzyme.

Here, we describe the occurrence of isoforms for cytochrome c oxidase subunit VIa in mammals, birds, and fishes. The new mechanism for regulation of the efficiency of energy transduction in cytochrome c oxidase by halving the H^+/e^- stoichiometry via binding of ATP to subunit VIaH (heart type) at high intramitochondrial ATP/ADP-ratios is further investigated. This mechanism is postulated to participate in mammalian nonshivering thermogenesis.

2. SUBUNIT STRUCTURE AND ISOFORMS OF CYTOCHROME c OXIDASE

In mammals, cytochrome c oxidase, complex IV of the mitochondrial respiratory chain, consists of three mitochondrial-coded and ten nuclear-coded subunits that occur partly in tissue-specific isoforms (Capaldi, 1990; Kadenbach and Reimann, 1992). In contrast, no isoforms could be found for cytochrome c reductase (complex III), composed of one mitochondrial-coded and ten nuclear-coded subunits (Vázquez-Acevedo *et al.*, 1993). The NADH dehydrogenase (complex I) consists of seven mitochondrial-coded and more than 40 nuclear-coded subunits, some of which may occur in tissue-specific isoforms (Walker, 1992).

The physiological meaning of a variable number of subunits in cytochrome c oxidase from prokaryotes (2–4 subunits) and eukaryotes (7–13) is largely unknown. One remarkable property of "supernumerary" subunits in the eukaryotic enzyme is the occurrence of tissue-specific isoforms, demonstrated for subunit VII in *Dictyostelium discoideum* (Bisson and Schiavo, 1986), subunit V (corresponding to mammalian IV) in yeast (Cumsky *et al.*, 1987), and subunits VIa, VIIa, and VIII in mammals (Capaldi, 1990; Kadenbach and Reimann, 1992). In rat only one isoform for subunit VIIa and in human, sheep, and rabbit only one isoform for subunit VIII could be found

(Linder *et al.*, 1995). For subunit VIa tissue-specific isoforms are observed in all mammals. The heart type (VIaH) is expressed only in heart and skeletal muscle, not in smooth muscle (Anthony *et al.*, 1990). The heart-type genes for subunits VIa (Ewart *et al.*, 1991; Taanman *et al.*, 1992) and VIIa are also developmentally dependently expressed. In humans their expression switches from the liver type in fetal heart and skeletal muscle to the heart types after birth (Bonne *et al.*, 1993). The genes for the liver-type subunits (VIaL and VIIaL) are expressed in all other tissues (Kennaway *et al.*, 1990), and are suggested to represent "housekeeping" genes (Smith and Lomax, 1993).

The first supernumerary subunit in mammals for which a specific function could be found is the heart type of subunit VIa (VIaH), which binds ADP (or ATP) (Anthony *et al.*, 1993) and decreases the H^+/e^- stoichiometry of reconstituted cytochrome *c* oxidase from bovine heart (Frank and Kadenbach, 1996). It was postulated that this mechanism participates in mammalian thermogenesis at rest (i.e., during sleep) when the ATP/ADP-ratio is high (Rohdich and Kadenbach, 1993; Kadenbach *et al.*, 1995).

3. CYTOCHROME *c* OXIDASE FROM TURKEY

In order to investigate the general validity of the postulated mechanism of nonshivering thermogenesis, as described for mammals, we analyzed cytochrome *c* oxidase from a bird. The polypeptide composition of the isolated enzymes from turkey heart, liver, and leg muscle is shown in Fig. 1. Cytochrome *c* oxidase from turkey contains 13 subunits, as found in all mammals (Kadenbach *et al.*, 1986; Linder *et al.*, 1995). The N-terminal amino acid sequences of subunits VIa from turkey heart, liver, and skeletal muscle, and of subunits VIc (internal sequence), VIIa, and VIII from turkey heart and liver are compared with those from rat in Fig. 2. In contrast to all mammals, where different isoforms of subunit VIa are expressed in heart and liver (Linder *et al.*, 1995), in turkey the N-terminal amino acid sequence of subunit VIa is identical in both tissues, beginning with a sequence of five alanines. In cytochrome *c* oxidase from turkey leg muscle, however, subunit VIa is clearly different. For subunit VIIa, sequencing of 13 N-terminal amino acids gave no indication for tissue-specific isoforms in the heart and liver enzyme. Subunit VIII is clearly different in cytochrome *c* oxidase of turkey heart and liver, as found in all mammals except human (Linder *et al.*, 1995).

We also investigated the influence of high intraliposomal ATP/ADP-ratios on the H^+/e^- stoichiometry of the reconstituted enzymes from turkey heart and liver. No difference of H^+/e^- stoichiometries between proteoliposomes containing either 5 mM intraliposomal ATP or ADP was found for both enzymes (S. Arnold and B. Kadenbach, unpublished results). It remains to be demon-

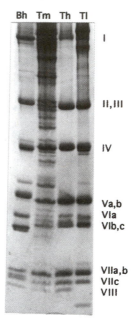

FIGURE 1. Comparison of the polypeptide composition of isolated COX from turkey heart (Th), liver (Tl), and leg muscle (Tm) with that of bovine heart (Bh). SDS-PAGE was performed as described by Kadenbach *et al.* (1986). COX from bovine heart and liver, and from turkey heart, liver, and leg muscle was isolated using nonionic detergents as previously described (Kadenbach *et al.*, 1986). SDS-PAGE was performed as described by Kadenbach *et al.* (1986).

strated whether the H^+/e^- stoichiometry of cytochrome c oxidase from turkey leg muscle is decreased by high intraliposomal ATP/ADP-ratios.

4. cDNAs OF SUBUNIT VIa OF TROUT LIVER AND CARP HEART

The above-described mechanism of thermogenesis in mammalian heart and skeletal muscle is not expected to occur in cold-blooded fishes. Therefore, we investigated the cDNAs for cytochrome c oxidase subunit VIa from rainbow trout liver and carp heart (Hüttemann *et al.*, 1997). The deduced amino acid sequences of the two cDNAs are compared with the corresponding amino acid sequences of the liver and heart type of subunit VIa from rat at optimal alignment in Fig. 3. The amino acid sequences of the mature proteins of the two fishes are 82% identical. Both cDNAs show a short precursor sequence similar to the heart type of mammalian subunit VIa. However, the two subunits from

Subunit VIa

```
                     10                    20
Rat-L        S S G A H G E E G S A R I W K A L T Y F . . .
Turkey-L     A A A A A H E G G G A R L W K T L ? F V . . .
                     10                    20
Rat-H        A S A S K G D H G G A G A N T W R L L T . . .
Turkey-H     A A A A A H E G G G A R L W K T L ? F V . . .
                     10
Turkey-M     A x R P H V E G M P . . .
```

Subunit VIc

```
                     10                    20                    30
Rat-L+H      M S S G A L L P K P Q M H D P L S K R L W V H I V G A F I D
Turkey-H         x x A L L P K P Q M R r L L A r x M K F . . . . . . . .
Turkey-L         x x A L L P K P Q M . . .
                     40                    50
Rat-L+H      D L G V A A A H K F G A A K P P K K A Y A D F Y R . . .
Turkey-H     . . . . . . . . K F G V A Q P R K R A Y A Q F Y K . . .
```

Subunit VIIa

```
                     10                    20
Rat-L+H      F E N K V P E K Q K L F Q E D N G M P V . . .
Turkey-L     x x N E V G E x Q K x F x E
Turkey-H     I h N e V g E G Q x L F f
```

Subunit VIII

```
                     10                    20
Rat-L        V H S K P P R E Q L G V L D I T I G L T . . .
Turkey-L     I V S H P P E H P V G T A E S . . .
                     10                    20
Rat-H        I S S K P A K S P T S A M D Q A V G M S . . .
Turkey-H     V T S K P P E H P V S P A E Q A I A N . . .
```

FIGURE 2. Comparison of N-terminal amino acid sequences of cytochrome *c* oxidase subunits from turkey (Tk) heart, liver, and leg muscle with those from rat. The liver-types of subunits are indicated by L, and the heart types by H. Subunit VIa from muscle is indicated by M. The underlined sequences indicate tissue-specific differences of amino acids in subunits of the enzymes from turkey. The subunits, separated by SDS-PAGE, were blotted onto PVDF membranes (ProBlott PVDF membrane, Applied Biosystems, Weiterstadt) by the semi-dry method described by Eckerskorn *et al.* (1988). The blotted polypeptides were stained with Coomassie blue and excised, and the N-terminal amino acid sequences were determined in a protein sequencer model 477A (Applied Biosystems, Weiterstadt).

fishes appear to be evolutionary equally distant from the mammalian heart type (VIaH) and liver type of subunit VIa (VIaL), as demonstrated in Fig. 4. This figure presents the percentage of identical amino acids between the sequences for subunit VIa of the two fishes and of the liver and heart type of human, bovine and rat. The three mammalian liver types of subunit VIa are more than 90% identical, and the heart types are 80% identical. Between the heart and liver types of subunit VIa the identity is only 55–61%. The sequences from the fishes are 56–60% identical to mammalian liver type and 46–60% identical to

```
            -26        -20                  -10              -1
Rat-L       M A S A V L S A S R V S G L L G R A L P R V G R P M
Trout (L)                       M A S P A S M A A R R V L
Carp  (H)                       M A M S P A A T V R R R A L
Rat-H                           M A L P L K V L S R S M

                       10                  20              30
Rat-L       S S G A H G - - E E G S A R I W K A L T Y F V A L P G V G V
Trout (L)   S A A S H A G H E G G S A R T W K I L S F V L A L P G V A V
Carp  (H)   A A A S Q G S H E G G - A R T W K I L S F V L A L P G V G V
Rat-H       A S A S K G D H G G A G A N T W R L L T F V L A L P S V A L

                       40                  50              60
Rat-L       S M L N V F L K S R H E E H E R P E F V A Y P H L R I R T K
Trout (L)   C I A N A Y M K M Q Q H S H E P P E F V A Y S H L R I R T K
Carp  (H)   C M A N A Y M K M Q A H S H D P P E F V P Y P H L R I R T K
Rat-H       C S L N C W M - - H A G H H E R P E F I P Y H H L R I R T K

                       70                  80
Rat-L       P F P W G D G N H T L F H N P H V N P L P T G Y E D E
Trout (L)   K W P W G D G N H S L F H N P H E N A L P E G Y E G P R H
Carp  (H)   P W P W G D G N H S L F H N A H I N A L P I G Y E G P H H
Rat-H       P F S W G D G N H T L F H N P H V N P L P T G Y E Q P
```

FIGURE 3. Comparison of deduced amino acid sequences of cytochrome *c* oxidase subunits VIa from carp heart and rainbow trout liver with those from rat. The sequences from fish enzymes are taken from Hüttemann *et al.* (1997) and the rat (Ra) sequences from Kadenbach and Reimann (1992). Amino acids different in carp and trout are underlined. Identical amino acids are bold-faced. L and H indicate the liver and heart type of subunit VIa; (L) and (H) indicate the origin of the cDNA from liver or heart, respectively.

mammalian heart type of subunit VIa. The same amino acid difference between subunit VIa from fish and from the mammalian liver as well as heart type suggests that only one isoform of subunit VIa occurs in fish. This is further supported by the fact that the trout cDNA was prepared from liver mRNA, and the carp cDNA from heart mRNA.

Also in cytochrome *c* oxidase from heart and liver of the partially endotherm tuna fish only one isoform of subunit VIa was found, in contrast to partially tissue-specific isoforms identified for subunits Va, VIc, VIIb, and VIII (Arnold *et al.*, 1997). Isoforms for subunits Va, VIc, and VIIb have not yet been described.

5. BINDING OF ATP AND ADP TO CYTOCHROME *c* OXIDASE

By equilibrium dialysis with [^{35}S]-ATPαS seven high-affinity binding sites for ATP, with a dissociation constant $K_d = 7.5$, were identified in isolated cytochrome *c* oxidase from bovine heart, six in the enzyme from bovine liver

	HuL	BoL	RaL	Trout	Carp	HuH	BoH	RaH
HuL	100	91	92	60	58	61	61	61
BoL		100	90	60	56	55	55	58
RaL			100	60	56	55	55	55
Trout				100	82	57	46	56
Carp					100	58	53	60
HuH						100	80	80
BoH							100	80
RaH								100

FIGURE 4. Relationship between amino acid sequences of subunits VIa from mammals and fishes. The numbers indicate the percentage of identical amino acids of the mature proteins as aligned in Fig. 2. The sequences from human, bovine, and rat liver type (HuL, BoL, RaL) and heart type (HuH, BoH, RaH) and of trout and carp (Tr, Ca), respectively, are compared.

($K_d = 12$), and one in the two subunit enzyme from *Paracoccus denitrificans* ($K_d = 20$) (Rieger *et al.*, 1995). Recently, the crystal structures of cytochrome *c* oxidase of *P. denitrificans* (Iwata *et al.*, 1995) and of bovine heart were determined at 2.8 Å resolution (Tsukihara *et al.*, 1995, 1996). The two structures are very similar in the "catalytic" subunits I–III, but differ in the total number of 4 subunits in the bacterial and 13 subunits in the mammalian enzyme (Kadenbach, 1995). In the crystal structure of the bovine heart enzyme two bound cholate molecules, which are structurally very similar to ADP, could be identified (Tsukihara *et al.*, 1996). One cholate molecule was located at the N-terminal (matrix-oriented) domain of the tissue-specific subunit VIaH, the other between subunits I and III. The binding site for ADP at the matrix-oriented N-terminal domain of subunit VIaH was previously postulated from kinetic studies (Anthony *et al.*, 1993). This binding site for ADP (and ATP) is different from another binding site for ATP at the cytosolic domain of subunit VIa of cytochrome *c* oxidase from bovine heart, liver and from yeast (Taanman *et al.*, 1994).

6. DECREASE OF H⁺/e⁻ STOICHIOMETRY IN CYTOCHROME *c* OXIDASE AT HIGH INTRALIPOSOMAL ATP/ADP-RATIOS

We have found that high intraliposomal (intramitochondrial) ATP/ADP-ratios decrease the respiratory control ratio (Rohdich and Kadenbach, 1993; Kadenbach *et al.*, 1995) and halve the H⁺/e⁻ stoichiometry of reconstituted cytochrome *c* oxidase from bovine heart (Frank and Kadenbach, 1996). The decrease of H⁺/e⁻ stoichiometry is tissue specific for heart and not observed with the enzyme from bovine liver as shown in Fig. 5. The H⁺/e⁻ stoichiometry decreases at high ATP/ADP-ratios to half of the value measured in the

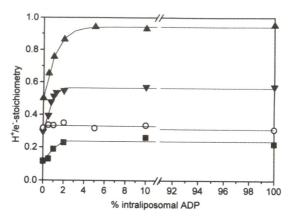

FIGURE 5. Influence of decreasing intraliposomal ATP/ADP-ratios on the H^+/e^- stoichiometry of reconstituted cytochrome c oxidase from bovine heart (■, ▲) and liver (○). The enzymes were reconstituted in liposomes by the hydrophobic adsorption method in the presence of the indicated percentage of ADP (ATP+ADP = 5 mM) followed by dialysis, and the H^+/e^- stoichiometry was measured by the reductant-pulse method as described previously (Frank and Kadenbach, 1996). Proteoliposomes (0.2 μM heme aa_3) were incubated in 1 mM K-Hepes, pH 7.0, 100 mM choline chloride, 5 mM KCl, and 1μg/μl valinomycin, and the pH was measured with a microcombination pH electrode. The H^+/e^- stoichiometry was determined from the initial pH decrease after addition of 7.7 nmole ferrocytochrome c (8 enzyme turnover). Squares and upward and downward triangles represent reconstitutions with different preparations of cytochrome c oxidase from bovine heart.

presence of ADP with the heart enzyme, but remains unchanged with cytochrome c oxidase from bovine liver (Fig. 5, open circles). The lack of an effect of high ATP/ADP-ratios on the H^+/e^- stoichiometry of the liver enzyme is not due to the lower maximal values, because preparations of the enzyme from bovine heart with H^+/e^- stoichiometries of 0.3 also showed 50% reduction at high intraliposomal ATP/ADP-ratios (Fig. 5, squares).

The decrease of H^+/e^- stoichiometry at high ATP/ADP-ratios is independent of magnesium ions or EDTA, as shown in Fig. 6. This result indicates that ATP (and ADP) are bound to the enzyme as free molecules, not complexed with magnesium ions. The data of Fig. 6 also exclude the higher complexing capacity of ATP, compared to ADP, as the cause of the decreased H^+/e^- stoichiometry.

The partial "uncoupling" of cytochrome c oxidase by high ATP/ADP-ratios is specific for adenine nucleotides. High intraliposomal GTP/GDP-ratios do not decrease the H^+/e^- stoichiometry of the heart enzyme (not shown). In order to exclude the possibility that GTP could have been partially hydrolyzed

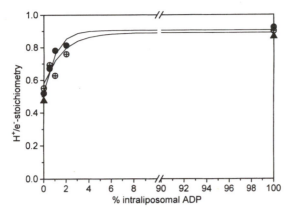

FIGURE 6. Effect of magnesium ions and EDTA on the decrease of H$^+$/e$^-$ stoichiometry of reconstituted cytochrome *c* oxidase from bovine heart at high ATP/ADP-ratios. The enzyme was reconstituted in liposomes in the presence of 5 mM ATP+ADP with the indicated percentage of ATP (▲). When indicated, the reconstitution medium contained in addition 2 mM MgCl$_2$ (●) or 2 mM EDTA (⊕).

to GDP and phosphate, 5 mM intraliposomal PNP–GMP, instead of GTP, was used, but no decrease of H$^+$/e$^-$ stoichiometry was obtained (not shown).

The molecular mechanism of partial uncoupling of energy transduction (decrease of H$^+$/e$^-$ stoichiometry) in cytochrome *c* oxidase is based on the exchange of bound ADP by ATP at the N-terminal, matrix-oriented domain of subunit VIaH (heart type), because a monoclonal antibody to subunit VIaH, which does not react with the liver type of subunit VIa (VIaL) (Anthony *et al.*, 1993), prevented the decrease of H$^+$/e$^-$ stoichiometry at high intraliposomal ATP/ADP-ratios (Fig. 7).

7. VARIATION OF H$^+$/e$^-$ STOICHIOMETRY AND NONSHIVERING THERMOGENESIS

The decrease of H$^+$/e$^-$ stoichiometry in cytochrome *c* oxidase at high rates of respiration, as described by Papa and co-workers (Papa *et al.*, 1991; Capitanio *et al.*, 1991, 1996), could have a physiological role by increasing oxidative metabolism at high substrate pressure, e.g., after excess uptake of foodstuff. The decrease of H$^+$/e$^-$ stoichiometry in cytochrome *c* oxidase at high matrix ATP/ADP-ratios, however, is postulated to participate in thermogenesis in heart and skeletal muscle at low work load (e.g., during sleep), when

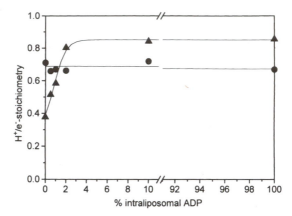

FIGURE 7. Influence of preincubation of cytochrome *c* oxidase from bovine heart with a monoclonal antibody to subunit VIa-H on the inhibition of H$^+$/e$^-$ stoichiometry by high-intraliposomal ATP/ADP-ratios after reconstitution. ●, Proteoliposomes reconstituted with untreated cytochrome *c* oxidase in the presence of the indicated amounts of nucleotides (ATP+ADP = 5 mM); ○, proteoliposomes reconstituted with cytochrome *c* oxidase, pre-treated with the monoclonal antibody to subunit VIa-H.

the ATP/ADP-ratio is elevated (Kadenbach *et al.*, 1995; Rohdich and Kaden-bach, 1993). The heart type of subunit VIa (VIaH) is only expressed in heart and skeletal muscle, not in liver, kidney, and smooth muscle (Anthony *et al.*, 1990). Heart and skeletal muscle vary their rate of respiration, and thus the amount of metabolic heat, up to a factor of five to ten.

The decrease of H$^+$/e$^-$ stoichiometry of reconstituted COX from bovine heart is only obtained at very high intraliposomal (matrix) ATP/ADP-ratios. ATP and ADP apparently bind to the same site—the N-terminal domain of subunit VIaH (Anthony *et al.*, 1993; Frank and Kadenbach, 1996)—because the same number of seven binding sites found for ATP at the bovine heart enzyme (Rieger *et al.*, 1995) also have been determined for ADP. However, the binding affinity for ADP is higher (lower K_d) than for ATP (J. Napiwotzki and B. Kadenbach, unpublished results).

As follows from Fig. 5, half-maximal decrease of H$^+$/e$^-$ stoichiometry is obtained at 99% ATP. This value is much higher than the measured ATP/ADP-ratios in rat liver *in vivo* of 86.5% ATP in the cytosol and of 50% ATP in the mitochondrial matrix (Schwenke *et al.*, 1981). However, measurement of total ADP levels from whole tissue extracts overestimates the free ADP concentrations by approximately 20-fold due to the large number of intracellular sites that sequester ADP (Veech *et al.*, 1979). The free intracellular ADP level is low and was determined to 0.05 μmole/g wet weight for mouse liver (Brosnan *et al.*, 1990) and 30–90 μm for rat heart (Wan *et al.*, 1993). The total concentrations of ATP and ADP in the mitochondrial matrix of rat liver *in vivo* was determined to 7.9 mM and 7.6 mM, respectively (Schwenke *et al.*, 1981). If only 1/20 of the ADP is free (0.38 mM), the ATP/ADP-ratio of free nucleotides would be 20.8 (95.4% ATP) in the mitochondrial matrix. This calculation could explain why the decrease of H^+/e^- stoichiometry of the reconstituted enzyme (Frank and Kadenbach, 1996), and the observed spectral changes of cytochrome *c* oxidase (Napiwotzki *et al.*, 1997) are obtained at very high ATP/ADP-ratios.

ACKNOWLEDGMENTS. We thank Prof. Dr. F. Lottspeich, MPI München-Martinsried, for internal amino acid sequencing of subunit VIc of the turkey enzymes. This chapter was supported by the Deutsche Forschungsgemeinschaft (Ka 192/28-3, Ka 192/32-1) and Fonds der Chemischen Industrie.

8. REFERENCES

Anthony, G., Stroh, A., Lottspeich, F., and Kadenbach, B., 1990, Different isozymes of cytochrome *c* oxidase are expressed in bovine smooth muscle and skeletal or heart muscle, *FEBS Lett.* **277**:97–100.

Anthony, G., Reimann, A., and Kadenbach, B., 1993, Tissue-specific regulation of bovine heart cytochrome *c* oxidase by ADP via interaction with subunit VIa, *Proc. Natl. Acad. Sci. USA* **90**:1652–1656.

Arnold, S., Lee, J., Kim, M. J., Song, E., Linder, D., Lottspeich, F., and Kadenbach, B., 1997, The subunit structure of cytochrome *c* oxidase from tuna heart and liver, *Eur. J. Biochem.* **248:** 99–103.

Bisson, R., and Schiavo, G., 1986, Two different forms of cytochrome *c* oxidase can be purified from the slime mold *Dictyostelium discoideum*, *J. Biol. Chem.* **261**:4373–4376.

Block, B. A., 1994, Thermogenesis in muscle, *Annu. Rev. Physiol.* **56**:535–577.

Bonne, G., Seibel, P., Possekel, S., Marsac, C., and Kadenbach, B., 1993, Expression of human cytochrome *c* oxidase subunits during fetal development, *Eur. J. Biochem.* **217**:1099–1107.

Brosnan, M. J., Chen, L., Van Dyke, T. A., and Koretsky, A. P., 1990, Free ADP levels in transgenic mouse liver expressing creatine kinase, *J. Biol. Chem.* **265**:20849–20855.

Capaldi, R. A., 1990, Structure and function of cytochrome *c* oxidase, *Annu. Rev. Biochem.* **59:** 569–596.

Capitanio, N., Capitanio, G., De Nitto, E., Billani, G., and Papa, S., 1991, H^+/e^- stoichiometry of mitochondrial cytochrome complexes reconstituted in liposomes. Rate-dependent changes of the stoichiometry in the cytochrome *c* oxidase vesicles, *FEBS Lett.* **288**:179–182.

Capitanio, N., Capitanio, G., Demarinis, D. A., De Nitto, E., Massari, S., and Papa, S., 1996, Factors affecting the H^+/e^- stoichiometry in mitochondrial cytochrome *c* oxidase: Influence of the rate of electron flow and transmembrane ΔpH^+, *Biochemistry* **35**:10800–10806.

Cumsky, M. G., Trueblood, C. E., Ko, C., and Poyton, R. O., 1987, Structural analysis of two genes encoding divergent forms of yeast cytochrome *c* oxidase subunit V, *Mol. Cell. Biol.* **7**:3511–3519.

Eckerskorn, C. W., Mewes, H. W., Goretzki, H., and Lottspeich, F., 1988, A new siliconized-glass fiber as support for protein-chemical analysis of electroblotted proteins, *Eur. J. Biochem.* **176:**509–519.

Ewart, D. G., Zhang, Y.-Z., and Capaldi, R. A., 1991, Switching of bovine cytochrome *c* oxidase subunit VIa isoforms in skeletal muscle during development, *FEBS Lett.* **292:**79–84.

Frank, V., and Kadenbach, B., 1996, Regulation of the H^+/e^--stoichiometry of cytochrome *c* oxidase from bovine heart by intraliposomal ATP/ADP ratios, *FEBS Lett.* **382:**121–124.

Iwata, S., Ostermeier, C., Ludwig, B., and Michel, H., 1995, Structure at 2.8 Å resolution of cytochrome *c* oxidase from *Paracoccus denitrificans*, *Nature* **376:**660–669.

Hüttemann, M., Exner, S., Arnold, S., Lottspeich, F., and Kadenbach, B., 1997, The cDNA sequences of cytochrome *c* oxidase subunit VIa from carp and rainbow trout suggest the absence of isoforms in fishes, *Biochim. Biophys. Acta* **1319:**14–18.

Kadenbach, B., 1995, The X-ray crystal structures of cytochrome *c* oxidases from *Paracoccus denitrificans* and bovine heart provide molecular mechanism of cell respiration, *Angew. Chem. Int. Ed. Engl.* **34:**2635–2637.

Kadenbach, B., and Reimann, A., 1992, Cytochrome *c* oxidase: Tissue-specific expression of isoforms and regulation of activity, in *New Comprehensive Biochemistry* (L. Ernster, ed.), pp. 241–263, Elsevier, Amsterdam.

Kadenbach, B., Jarausch, J., Hartmann, R., and Merle, P., 1983, Separation of mammalian cyto-chrome *c* oxidase into 13 poly-peptides by a sodium dodecyl sulfate-gel electrophoretic procedure, *Anal. Biochem.* **129:**517–521.

Kadenbach, B., Stroh, A., Ungibauer, M., Kuhn-Nentwig, L., Büge, U., and Jarausch, J., 1986, Isozymes of cytochrome *c* oxidase: Characterization and isolation from different tissues, *Methods Enzymol.* **126:**32–45.

Kadenbach, B., Barth, J., Akgün, R., Freund, R., Linder, D., and Possekel, S., 1995, Regulation of mitochondrial energy generation in health and disease, *Biochim. Biophys. Acta* **1271:**103–109.

Kennaway, N. G., Carrero-Valenzuela, R. D., Ewart, G., Balan, V. K., Lightowlers, R., Zhang, Y.-Z., Powell, B. R., Capaldi, R. A., and Buist, N. R. M., 1990, Isoforms of mammalian cytochrome *c* oxidase: Correlation with human cytochrome *c* oxidase deficiency, *Pediatric Res.* **28:**529–535.

Linder, D., Freund, R., and Kadenbach, B., 1995, Species-specific expression of cytochrome *c* oxidase isozymes, *Comp. Biochem. Physiol.* **112B:**461–469.

Napiwotzki, J., Shinzawa-Itoh, K., Yoshikawa, S., and Kadenbach, B., 1997, ATP and ADP bind to cytochrome *c* oxidase and regulate its activity, *Biol. Chem.* **378:**1013–1021.

Nedergaard, J., and Cannon, B., 1984, Thermogenic mitochondria, in *Bioenergetics* (L. Ernster, ed.), pp. 291–314, Elsevier, Amsterdam.

Nicholls, D. G., and Locke, R. M., 1984, Thermogenic mechanisms in brown fat, *Physiol. Rev.* **64:**1–64.

Papa, S., Capitanio, N., Capitanio, G., De Nitto, E., and Minuto, M., 1991, The cytochrome chain of mitochondria exhibits variable H^+/e^- stoichiometry, *FEBS Lett.* **288:**183–186.

Rieger, T., Napiwotzki, J., and Kadenbach, B., 1995, On the number of nucleotide binding sites in cytochrome *c* oxidase, *Biochem. Biophys. Res. Commun.* **217:**34–40.

Rohdich, F., and Kadenbach, B., 1993, Tissue-specific regulation of cytochrome *c* oxidase efficiency by nucleotides, *Biochemistry* **32:**8499–8503.

Schägger, H., and von Jagow, G., 1987, Tricine-sodium dodecyl sulfate-polyacrylamide gel electrophoresis for the separation of proteins in the range from 1 to 100 kDa, *Anal. Biochem.* **166:**368–379.

Schwenke, W. B., Soboll, S., Seitz, H. J., and Sies, H., 1981, Mitochondrial and cytosolic ATP/ADP ratios in rat liver *in vivo*, *Biochem. J.* **200:**405–408.

Segade, F., Hurlé, B., Claudio, E., Ramos, S., and Lazo, P., 1996, Identification of an additional member of the cytochrome *c* oxidase subunit VIIa family of proteins, *J. Biol. Chem.* **271:** 12343–12349.

Skulachev, V. P., 1996, Role of uncoupled and non-coupled oxidations in maintenance of safely low levels of oxygen and its one-electron reductants, *Q. Rev. Biophys.* **29:**169–202.

Smith, E. O., and Lomax, M. I., 1993, Structural organization of the bovine gene for the heart/ muscle isoform of cytochrome *c* oxidase subunit VIa, *Biochim. Biophys. Acta* **1174:**63–71.

Steverding, D., and Kadenbach, B., 1991, Influence of *N*-ethoxycarbonyl-2-ethoxy-1,2-dihydro-quinolin modification on proton translocation and membrane potential of reconstituted cyto-chrome *c* oxidase support "proton slippage," *J. Biol. Chem.* **266:**8097–8101.

Taanman, J.-W., Herberg, N. H., De Vries, H., Bolhuis, P. A., and Van den Bogert, C., 1992, Steady-state transcript levels of cytochrome *c* oxidase genes during human myogenesis indicate subunit switching of subunit VIa and co-expression of subunit VIIa isoforms, *Biochim. Bio-phys. Acta* **1139:**155–162.

Taanman, J.-W., Turina, P., and Capaldi, R. A., 1994, Regulation of cytochrome *c* oxidase by interaction of ATP at two binding sites, one on subunit VIa, *Biochemistry* **33:**11833–11841.

Tsukihara, T., Aoyama, H., Yamashita, E., Tomizaki, T., Yamaguchi, H., Shinzawa-Itoh, K., Nakashima, R., Yaono, R., and Yoshikawa, S., 1995, Structures of metal sites of oxidized bovine heart cytochrome *c* oxidase at 2.8 å, *Science* **269:**1069–1074.

Tsukihara, T., Aoyama, H., Yamashita, E., Tomizaki, T., Yamaguchi, H., Shinzawa-Itoh, K., Nakashima, R., Yaono, R., and Yoshikawa, S., 1996, The whole structure of the 13-subunit oxidized cytochrome *c* oxidase at 2.8 Å, Science **272:**1136–1144.

Vázquez-Acevedo, M., Antaramian, A., Corona, N., and González-Halphen, D., 1993, Subunit structures of purified beef mitochondrial cytochrome bc_1 complex from liver and heart, *J. Bioenerg. Biomembr.* **25:**401–410.

Veech, R. L., Lawson, J. W. R., Cornell, N. W., and Krebs, H. A., 1979, Cytosolic phosphorylation potential, *J. Biol. Chem.* **254:**6538–6547.

Walker, J. E., 1992, The NADH:ubiquinone oxidoreductase (complex I) of respiratory chains, *Q. Rev. Biophys.* **25:**253–324.

Wan, B., Doumen, C., Duszynski, J., Salama, G., Vary, T. C., and LaNoue, K. F., 1993, Effects of cardiac work on electrical potential gradient across mitochondrial membrane in perfused rat hearts, *Am. J. Physiol.* **265:**H453–H460.

CHAPTER 25

Mitochondrial DNA Mutations and Nuclear Mitochondrial Interactions in Human Disease

H. Cock, J.-W. Taanman, and A. H. V. Schapira

1. INTRODUCTION

The mitochondrial genome is dependent upon nuclear DNA for its replication, repair, transcription, and translation. In addition, both genomes encode proteins that are part of the multimeric complexes of the respiratory chain and oxidative phosphorylation system. It is hardly surprising therefore that one would expect there to be a significant degree of interaction between the genomes at the biochemical level. This interaction therefore is also a site for the development of pathology and it is this area on which this chapter will focus.

Leber's hereditary optic neuropathy (LHON) is an archetypal mitochondrial disease in terms of its inheritance pattern and its association with mitochondrial DNA mutations. However, there are numerous features of this disorder that indicate that additional factors must be involved in etiology and

H. Cock and J.-W. Taanman • Department of Clinical Neurosciences, Royal Free Hospital School of Medicine, London NW3 2PF, England.   **A. H. V. Schapira** • Department of Clinical Neurosciences, Royal Free Hospital School of Medicine, and Department of Clinical Neurology, Institute of Neurology, London NW3 2PF, England.

Frontiers of Cellular Bioenergetics, edited by Papa *et al.* Kluwer Academic/Plenum Publishers, New York, 1999.

pathogenesis. The first part of this chapter will therefore cover the emerging evidence for nuclear genomic involvement in LHON.

Deletions of mitochondrial DNA (mtDNA) have been identified in tissues from normal individuals (Coctopassi and Arnheim, 1990). The proportion of deleted mtDNA increases with age, and this has formed the basis for the mitochondrial theory of aging. This suggests that the increasing oxidative stress and damage that occurs with cellular aging results in damage to mtDNA. mtDNA is considered to be particularly susceptible to this type of injury because of its lack of a histone coat and the full complement of repair enzymes associated with the nuclear genome. The theory predicts that accumulating mtDNA mutations will affect the activity of the mitochondrial respiratory chain, and there is some evidence to support a decline in respiratory chain function with age. This has generated interest in progeria syndromes as a model of certain aspects of aging. Some recent work from our laboratory suggests that, at least in cultured cells from one individual with Hutchinson–Gilford progeria syndrome, there may be evidence of a primary mtDNA mutation.

Only approximately 60% of patients with mitochondrial encephalo-myopathy have identifiable mtDNA mutations. Statistically, defects of nuclear genes encoding respiratory chain subunits would be expected to constitute the highest proportion of molecular genetic abnormalities in mitochondrial disease. However, there has been only one report to date of a mitochondrial defect and this was in a gene encoding a subunit of succinate dehydrogenase (Bourgeron *et al.*, 1990). Several families have been described, however, with autosomal dominant inheritance of mitochondrial disease and this is associated with multiple deletions of mtDNA (Zeviani *et al.*, 1989, 1990). Identifying the molecular genetic basis of these families will provide valuable insight not only into the pathogenesis of these diseases but also into nuclear control of mtDNA replication.

Finally, mtDNA depletion syndrome is a disorder in which there is quantitative abnormality of mtDNA (Tritschler *et al.*, 1992). This appears to be the result of a failure to maintain mtDNA levels (Bodnar *et al.*, 1993, 1995). Again, an understanding of the molecular mechanisms that result in this disorder will provide information on the normal control of mtDNA levels by the nuclear genome.

2. NUCLEAR GENES IN LEBER'S HEREDITARY OPTIC NEUROPATHY

The association between LHON and mtDNA mutations is well documented. In keeping with a mitochondrial etiology, LHON displays strict maternal inheritance, and three mutations at positions 11,778 (Wallace *et al.*, 1988), 3,460 (Huoponen *et al.*, 1991), and 14,484 (Johns *et al.*, 1992) are generally accepted as having a primary role in disease pathogenesis. Each of these mutations has been documented only in LHON families and not in controls from a variety of phylogenetic backgrounds. The mutations do not require the presence of additional "secondary" mtDNA mutations. They affect a moderately or highly conserved amino acid, and they have been documented to exist in a heteroplasmic state in at least some families. A further mutation at 14,459 (Hely et al., 1994) also has been described in families with LHON and additional basal ganglia disease, and it too should probably be classed as primary. Despite the abundance of information regarding the association of mtDNA mutations with LHON, however, several clinical features of the disease imply that nuclear genes also might be involved in its expression.

Most striking is that males carrying a primary mtDNA LHON mutation are more likely to become clinically affected than females. Early reports probably overestimated the male bias (Seedorf, 1985; Van Senus, 1963; Imai and Moriwaki, 1936), since we now know the criteria set for penetrance used age limits that were inappropriate. Nonetheless, even the most recent and rigorous of pedigree studies in genetically confirmed pedigrees confirm an affected male-to-female ratio of between 2:1 and 4:1 in 11,778 and 3,460 LHON (Black et al., 1995; Harding et al., 1995; Riordan-Eva et al., 1995), and between 5:1 and 8:1 in 14,484 LHON (Harding et al., 1995; Riordan-Eva et al., 1995). Thus, an excess of male sufferers certainly exists and this cannot be explained on the basis of mitochondrial genetics alone. Even within the sexes, disease penetrance is variable: A homoplasmic male 11,778 patient has only ~50% chance of going blind, falling to ~15% if female (Riordan-Eva and Harding, 1995), and the age at onset in females is on average 2 to 15 years later than in males (Seedorf, 1985; Nikoskelainen et al., 1983). Although in a few cases this may relate to the mutation load (Smith et al., 1993; Zhu et al., 1992; Holt et al., 1989), most LHON families are homoplasmic for the pathogenic mutation, although variable penetrance is still apparent, indicating that additional factors are operating. Finally, despite the presence of near-homoplasmic pathogenic mtDNA mutations in all the tissues so far examined in LHON (Howell et al., 1994; Smith et al., 1994; Cornelissen et al., 1993; Larsson et al., 1991; Majander et al., 1991), the clinical symptoms are largely tissue specific, affecting only the optic nerves. Thus, some factor other than the presence of a pathogenic mtDNA mutation is influential in the final clinical picture. The suggestion that the clinical features of LHON, in particular the peculiar inheritance pattern, might result from the combination of a nuclear genetic factor and an unidentified "cytoplasmic" factor (Wallace, 1970) in fact predates the characterization of mtDNA (Anderson et al., 1981), but this notion not only has persisted but has gained increasing support in recent years. The favored current model (Wallace, 1987) is one in which both an mtDNA mutation and a nuclear gene (or genes), perhaps on the X chromosome, are required for disease penetrance. A nuclear gene(s) might also account for the tissue specificity of LHON, for instance, if it encoded for an optic nerve isoenzyme (Wallace, 1987).

This model has been studied in a number of ways including pedigree segregation analysis and chromosomal linkage studies. The role of nuclear genes in LHON can also be approached at a cellular level by examining the effect of different nuclear backgrounds on the expression of LHON mtDNA mutations *in vitro* using mitochondrial transfer techniques. The available information in each of these areas will be reviewed, including results from mitochondrial transfer studies in LHON undertaken within the Department of Clinical Neurosciences at the Royal Free Hospital School of Medicine.

2.1. Segregation Analysis in LHON Families

To test the hypothesis that clinical disease in LHON families resulted from simultaneous involvement of two loci, one mitochondrial and one X-linked, Bu and Rotter (1991) analyzed a total of 31 LHON pedigrees totaling more than 1200 individuals. Only maternal line family members whose age at examination or death was beyond the mean age of onset of the disease were included, thus minimizing the possibility of currently asymptomatic individuals who might later become affected being considered as clinically unaffected in the analysis. The pedigree data were shown to be most consistent with a two-locus disorder such as had been proposed by Wallace (1970). A similar study by Harding et al. (1995), who analyzed disease penetrance in 85 genetically confirmed LHON families, has since supported the suggestion of the presence of an X-linked visual loss susceptibility locus (VLSL) to account for disease penetrance in families with LHON mtDNA mutations. However, this explanation is still insufficient. If the X-linked VLSL were recessive, all male offspring of a clinically affected mother should be affected, whereas if the nuclear gene were dominant, there should be more affected females than males. Neither prediction agrees with actual observations (Mackey, 1993). However, Bu and Rotter (1992) further proposed that X-inactivation in a small number of embryonic precursor cells for optic tissue could account for the observed pattern of inheritance. Females could then be affected either through homozygosity for the X-linked VLSL or through disadvantageous X-inactivation. Both of the large pedigree studies have supported this model incorporating an X-linked VLSL, with a gene frequency of 0.08, a penetrance of 0.11 in heterozygous females, and with 40% of affected females being homozygous (Harding *et al.*, 1995; Bu and Rotter, 1991).

The factors controlling X inactivation are only recently being unraveled (Lyon, 1996), but it is well recognized in other X-linked recessive disorders that nearly all manifesting carriers show "skewed X inactivation": The normal X chromosome is preferentially inactivated, leading to expression of the disease as a heterozygote (Pegoraro *et al.*, 1995; Puck *et al.*, 1987). Thus, if an X-linked

VLSL exists in LHON, many, if not most, clinically affected females should show patterns of X inactivation that are more "skewed" than in unaffected females. In 35 females from 10 LHON pedigrees, skewed inactivation was observed in 2 of 10 (20%) affected LHON females and in 3 of 25 (12%) unaffected, thus not supporting the existence of a strong X-linked determinant (Pegoraro *et al.*, 1996). However, only 36% of their unaffected subjects were homoplasmic for a LHON mtDNA mutation, whereas 90% of the affected subjects were homoplasmic, which also may have influenced disease penetrance. In addition, those considered "unaffected" included subjects aged 19 to 59 years, and some may yet in fact become affected. These factors, in the context of a relatively small study, restrict the confidence with which conclusions can be drawn. Chalmers *et al.* (1996) also looked at X inactivation in heterozygous affected and unaffected females, and like Pegoraro *et al.* (1996) did not observe the skewed inactivation in the affected group predicted by an X-linked recessive VLSL despite applying more appropriate age criteria, although again the absolute numbers were relatively small (affected females $n = 23$; unaffected $n = 13$). Further studies addressing this point are awaited but are also likely to be hampered by small numbers given the relative sparseness of affected LHON females. Furthermore, both groups studied DNA from leukocytes as the most readily available source, and the absence of unbalanced X inactivation in these cells does not exclude its occurrence in the optic nerve. There are only few data concerning the tissue distribution of X inactivation. Gale *et al.* (1994) looked at X inactivation in blood, skin, muscle, and colon and found no case in which skewed inactivation was observed in other tissues but not in blood. There are few published data concerning X-inactivation patterns in tissues of neuroectodermal origin relative to blood. Thus, the existence of an X-linked VLSL, in combination with mtDNA mutations(s), remains the best of available models to account for the unusual pattern of disease penetrance in LHON.

2.2. Linkage Analysis in LHON

Initial attempts to identify an X-linked VLSL by linkage analysis were not successful (Chen *et al.*, 1989). However, this preliminary study predated the two-loci model of Bu and Rotter (1991) and set arbitrary values for gene frequency and penetrance in retrospect below those predicted by the results of the larger pedigree studies. Vilkki *et al.* (1991) studied genetic linkage to 15 X chromosome polymorphic markers in 115 individuals from six LHON pedigrees and proposed linkage to the locus DXS7 on the proximal Xp, with no evidence of heterogeneity among different families. Although arbitrary gene parameters were also used in this study, this linkage seemed valid over a broad range of options, including those supported by pedigree analysis (Harding *et al.*, 1995; Bu and Rotter, 1991). Linkage to DXS7 has not been demonstrated, however, in other families (Carvalho *et al.*, 1992; Sweeney *et al.*, 1992). A reevaluation of linkage of a VLSL gene to the X chromosome in the original families (Juvonen *et al.*, 1993) was subsequently carried out, using stricter age criteria for assigning "unaffected" status to individuals. Families were also studied individually according to their mtDNA mutation to avoid the effect of possible heterogeneity. This demonstrated that the earlier close linkage to DSX7 (Vilkki *et al.*, 1991) was implausible. These linkage data, while now in agreement with one another, have not excluded a VLSL from the entire X chromosome. Thus, more recently, Chalmers *et al.* (1996) performed linkage analysis in British and Italian families, using DNA polymorphisms that covered the entire X chromosome map, including the pseudoautosomal region, and applying appropriate age-related penetrance values and a penetrance value of 0.11 in females subject to X inactivation as predicted by the pedigree models. No evidence for a VLSL was provided by this study. The authors concede, however, that exclusion mapping can be inaccurate if the model applied is incorrect and genetic heterogeneity may mask linkage in some families. In order to address this problem, a subset of families with the 11,778 mtDNA mutation containing only affected males was analyzed further, but only 65% of the X chromosome was excluded in this way. Furthermore, linkage to the pseudoautosomal region could not be excluded because of the high rate of recombination in this area. Finally, the pitfalls of linkage analysis where additional variables, such as the mtDNA mutation load, may be operating and where arbitrary assumptions are made about the frequency and penetrance of that gene, as well as about the expression of the gene in heterozygous females, must be recognized (Juvonen *et al.*, 1993; Chen and Denton, 1991).

Thus, the reasons for the male bias in LHON remain undefined. Some areas of the X chromosome have not yet been excluded by linkage analysis, which itself has limitations in this condition. In addition, although there may indeed be nongenetic or hormonal influences, as supported by nonconcordance for at least age at onset in a pair of identical male twins (Johns *et al.*, 1994), the male preponderance may also represent a multifactorial genetic model, encompassing several X chromosome genes or a combination of autosomal and X-linked genes predisposing to visual loss in the presence of mtDNA mutations. Linkage studies under these circumstances would be even more difficult, and a different approach to the question of nuclear involvement in LHON would then be necessary. This has led to the development of mitochondrial transfer studies in LHON.

2.3. Mitochondrial Transfer Studies in LHON

The technique of fusing enucleated cells containing mutated mtDNA with a cell line devoid of its own mtDNA (rho^0), thus placing the mutant mitochondria in a new nuclear environment, was first developed by King and Attardi (1989). This technique is now well established within several groups, using a variety of different host rho^0 cell types into which potentially pathogenic mtDNA can be placed. Any transformed host cell line can theoretically be used, and rendered rho^0 by prolonged growth in ethidium bromide to deplete the host mtDNA (Desjardins *et al.*, 1985). Rho^0 cells derived from human osteosarcoma (143B) and HeLa (206) cell lines most commonly have been used (King and Attardi, 1989), but rho^0 cells derived from a human lung carcinoma (Dunbar *et al.*, 1995) and from lymphoblasts have also been reported (Jun *et al.*, 1996). In the majority of studies, experiments were performed in order to demonstrate that various biochemical defects, found in association with various disease-related mtDNA mutations, were indeed attributable to the mutant mtDNA, as they persisted in the presence of the new nuclear environment, thus supporting the pathogenicity of the mutant mtDNA (Chomyn *et al.*, 1991, 1994; Mariotti *et al.*, 1994; Hayashi *et al.*, 1993). This methodology, however, can also be used to demonstrate the reverse, i.e., that a new nuclear environment can correct mitochondrial functional defects that may be associated with disease, thus demonstrating that nuclear genes are involved in the pathogenesis of the mitochondrial dysfunction. Nuclear gene involvement using this approach so far has been used in mtDNA depletion syndromes (Bodnar *et al.*, 1993) and in some cases of complex IV deficiency associated with Leigh's syndrome (Tiranti *et al.*, 1995; Miranda *et al.*, 1989). Thus, functional analysis of cell lines containing LHON mtDNA mutations but a different nuclear environment might shed light on the role of nuclear genes in LHON.

All the primary LHON mtDNA mutations are in genes encoding subunits of complex I of the mitochondrial respiratory chain. A severe (~60%) defect in mitochondrial (rotenone sensitive) NADH–ubiquinone oxidoreductase (complex I) activity has been observed consistently in LHON mitochondrial fractions from cells containing the 3,460 mtDNA mutation, including lymphocytes (Majander *et al.*, 1991), platelets (Smith *et al.*, 1994; Howell *et al.*, 1991), and fibroblasts (Cock *et al.*, 1995a). The 3,460 mutation does not affect the proximal NADH dehydrogenase activity, only that of ubiquinone-dependent rotenone-sensitive electron transfer, in accordance with the proposed role of ND1 in rotenone and ubiquinone binding (Majander *et al.*, 1991). Data on complex I activity in intact mitochondria possessing the 3,460 mutation have not yet been published. In 11,778 LHON subjects, magnetic resonance spectroscopy (MRS) with ^{31}P as a noninvasive means of assessing energy metabolism *in vivo* has demonstrated alterations in muscle metabolism both during and after exercise (Cortelli *et al.*, 1991). In addition, polarography using both muscle cells (Larsson *et al.*, 1991) and lymphocytes (Majander *et al.*, 1991) from 11,778 subjects has demonstrated reduced rates of oxygen consumption in the presence of complex-I-linked substrates. Although spectrophotometric assays in these same studies have not consistently shown a complex I defect in mitochondrial preparations, a functional defect in intact 11,778 mitochondria is apparent. Complex I activity in mitochondrial fractions from 14,484 LHON fibroblasts has been reported as normal (Cock *et al.*, 1995b), but no data from intact mitochondria are as yet available. Thus, at least for 3,460 and 11,778 LHON, functional defects associated with the mtDNA mutations are available and can as such be used as biochemical markers for mitochondrial transfer studies. In addition, a 60% complex I defect has been observed in muscle mitochondria from subjects with the 14,459 LHON plus dystonia mtDNA mutation (Shoffner *et al.*, 1995).

Work undertaken the Royal Free Hospital School of Medicine has concentrated on the 3,460 mutation, as this has been associated with the most severe and easily detectable biochemical defect. An ~60% complex I defect was first confirmed in mitochondrial preparations from cultured fibroblasts from three members of a 3,460 LHON family compared to controls. The link between the presence of the 3,460 mutation and the complex I defect was further confirmed by demonstrating a clear relationship between mutant load and the severity of the biochemical defect in clonal cell lines derived from a heteroplasmic family member. In particular, complex I activity was observed to be comparable to that of control fibroblasts in a clonal line with 0% mutant mtDNA but a "LHON nucleus." Mitochondrial transfer cell lines were then established by fusion of enucleated 3,460 cytoplasts with human lung carcinoma (A549) rho^0 cells. Analysis of complex I activity in these fusion cybrids demonstrated that complex I defect previously identified in the 3,460 fibroblasts was no longer expressed in the presence of the new "non-LHON" A549 nuclear environment, despite persistence of homoplasmic 3,460 mutant mtDNA. Thus, these experiments have demonstrated that both the 3,460 mutant mtDNA and an LHON nucleus are required for expression of the complex I defect identified in these LHON families. This would fit with the two-loci model proposed to account for the clinical expression of LHON.

Several groups are addressing this same question with regard to other LHON mutations and have published conflicting data. The 60% complex I defect that has been observed in cells carrying the 14,459 LHON and dystonia mutation (Shoffner *et al.*, 1995) has been shown to persist in transmitochondrial cybrids generated using a lymphoblast rho^0 cell line, demonstrating that the biochemical defect in this case can be attributed to mtDNA. Furthermore, complex I-linked respiration in permeabilized cells containing 11,778 mutant mtDNA is reduced compared to controls whether studied in patients' fibroblasts or in transmitochondrial cybrids generated from rho^0 206 cells (Hofhaus *et al.*, 1996; Vergani *et al.*, 1995), thus indicating that this change of nuclear background has not influenced expression of the biochemical defect associated with 11,778 LHON.

In order to interpret these apparently conflicting results, several factors require further clarification. First, the possibility that the loss of the biochemical defect in 3,460-A549 rho^0 cybrids is due to a nonspecific loss of expression in transformed cells is excluded by the observation that a biochemical defect persists in transformed lymphoblasts from 3,460 subjects (Majander *et al.*, 1991). Second, as complex I defects have been shown to persist in A549 rho^0 cybrids containing high levels of the MELAS 3243 mutation (H. Cock, J.-W. Taanman, and A. H. V. Schapira, unpublished data), the restoration of complex I activities to control levels in 3,460-A549 rho^0 cybrids is not a nonspecific finding peculiar to the A549 rho^0 nuclear background. However, the range of data on complex I in transmitochondrial cybrids from all the reported studies is much greater than that observed in the original cell lines, indicating that interpretation of results from only a few fusion clones in each case may not be reliable. Vergani *et al.* (1995) reported on a total of eight 11,778 cybrid clones and Hofhaus *et al.* (1996) on only four. Jun *et al.* (1996) reported data from six 14,459 cybrid cell lines, and the Royal Free study to date has analyzed seven 3,460 cybrid clones. Certainly, studies other than our own have all demonstrated persisting functional defects associated with mtDNA mutations in new non-LHON nuclear environments, but as discussed by Hofhaus *et al.* (1996) in their paper, it is not known whether the 206 rho^0 nuclear background does carry the putative change in gene content or activity that contributes to the LHON phenotype in affected LHON individuals. The gene frequency of the proposed X-linked allele has been estimated to be 0.08 (Harding *et al.*, 1995; Bu and Rotter, 1991), and it is also possible that the 143B rho^0 cells used by Vergani *et al.* (1995) represents a "LHON-permissive" nuclear background, although this might be considered to be unlikely. The 14,459 study used yet another rho^0 cell type derived from lymphoblasts, but data from 14,459 families may not be representative of most LHON families given their additional neurological features.

However the transmitochondrial cybrid studies are interpreted in all the studies discussed, complex I activity was in some way influenced by the presence of a new nuclear environment, as evidenced by the increased spread in complex I activities compared to those of the original cell lines. In addition, the 3,460 data at least show that a different nuclear background can influence the expression of the biochemical defect associated with the presence of this mtDNA mutation. Further investigations using different nuclear backgrounds and analyzing more cybrid lines from both clinically affected and unaffected LHON individuals should be able to distinguish the effects of the mtDNA mutation from those of putative LHON nuclear factors and may subsequently allow identification of these factors.

Finally, however, it should be noted that an association between complex I dysfunction and the clinical phenotype in LHON families is by no means established, equal complex I defects being apparent in both clinically affected and unaffected individuals in all the published studies so far discussed. Some of these individuals may yet go on to become affected; but until affected and unaffected individuals can reliably be distinguished by a functional approach, the possibility that complex I dysfunction is not related to disease pathogenesis must remain open. An alternative autoimmune reaction generated through molecular mimicry between a defective complex I subunit and a structural optic nerve protein has been postulated and supported by preliminary data demonstrating circulating autoantibodies to human optic nerve protein in LHON subjects (Smith *et al.*, 1995). The results of mitochondrial transfer studies in this scenario might be less relevant to our understanding of disease pathogenesis in LHON, but would still provide information regarding the relative contribution of nuclear and mitochondrial genes to the normal function of complex I, which may have relevance to other conditions in which complex I function is altered.

2.4. Nuclear Genes and Mitochondrial Dysfunction in Hutchinson–Gilford Progeria Syndrome

Hutchinson–Gilford progeria syndrome (HGPS) is a rare condition (estimated incidence, 1 in 8 million births) often cited as a clinical model of accelerated aging (Mills and Weiss, 1990; DeBusk, 1972). Despite the clinical similarities with biological aging (Brown, 1992; Mills and Weiss, 1990) and the increasing evidence that abnormalities of mitochondrial function may play a role in the latter, there is little information regarding mitochondrial function in HGPS. The mitochondrial theory of aging (Ozawa, 1995; Wallace, 1992) is based on observations of declining mitochondrial respiratory chain (MRC) activity, in particular, complex IV activity (Hayashi *et al.*, 1994; Hsieh *et al.*, 1994), and the parallel accumulation of mtDNA mutations with increasing age (Ozawa *et al.*, 1995; Baumer *et al.*, 1994). Given the complexity of biological aging, the study of diseases such as HGPS might not only help to elucidate the pathogenetic mechanisms operating in this rare condition, but also shed light on aspects of the biological aging process.

HGPS was first described by Hutchinson, in 1886, and the early years of this century by Gilford (1904). Some 75 cases have since been reported in the literature (60 of which are reviewed by DeBusk, 1972). Typically, patients appear normal at birth but present within the first year of life with failure to thrive or with skin–hair abnormalities. There is a striking loss of subcutaneous fat, and limbs thus appear thin, with prominent joints. There also may be skeletal abnormalities such as short dystrophic clavicles, a pyriform thorax, and coxa valga. Dentition is abnormal and delayed, and sexual maturation does not occur. The skin has an "aged" appearance, being warm, thin, and dry, with little hair, prominent veins, and loose, wrinkled areas even by the age of 5 years. Later this is accompanied by patchy pigmentation and dystrophic nail changes. Arthritis and thinning bones are common, although these changes differ radiologically and pathologically from the osteoarthritis and osteoporosis that accompany normal aging. Intellect and cognitive function are typically normal, unless affected through cerebrovascular disease in later years. Insulin resistance may be a feature of HGPS (Villee et al., 1969), but no other endocrine abnormalities have been reported. Most importantly, there is widespread premature atherosclerosis, and vascular complications are the most common cause of death. Cardiac vessels are most severely affected, often causing symptomatic angina before the age of 10 and myocardial infarcts and cardiac failure in the teens. The mean age at death in reported series is 13.4 years, with a range of 7 to 27 years.

Certain features common to normal aging are not found in HGPS. These include senile cataracts, presbyacusis, arcus senilis, malignancies, or neurodegenerative changes such as senile plaques–neurofibrillary tangles. Despite these limitations, HGPS is generally considered to be the best disease model of accelerated aging of those available (Brown, 1992; Mills and Weiss, 1990).

The cause of HGPS remains unknown. The very consistent clinical phenotype suggests a genetic basis, and some reports have supported autosomal recessive inheritance (Khalifa, 1989; DeBusk, 1972), although there are too few familial cases to substantiate this. A defect in a helicase involved in nuclear DNA repair has recently been identified in another model of accelerated aging—Werner's syndrome (Yu et al., 1996)—but has not been found in HGPS subjects (Oshima et al., 1996). Defects in DNA repair mechanisms have been reported in five cases from two studies (Epstein et al., 1973; Wang et al., 1991), though others have not found this (Brown et al., 1980). Abnormalities of matrix proteins, notably increased hyaluronic acid excretion, have been reported in both HGPS and in normal aging (Sweeney and Weiss, 1992), and indeed most of the clinical features might be attributable to a connective tissue defect in which hyaluronic acid plays an important role. Other matrix protein abnormalities also have been reported in isolated cases (Clark and Weiss, 1995; Beavan et al., 1993; Giro and Davidson, 1993), but a primary pathogenic role has not been established for any. Of particular interest, in keeping with the suggestion that the telomere, which is a section of simple tandem repeats found at the end of nuclear chromosomes (Greider, 1990), may function as a biomarker of somatic cell aging (Allsopp et al., 1992), telomere length has been shown to be reduced in HGPS subjects compared to age-matched controls (Allsopp et al., 1992). Few studies regarding oxidative metabolism in HGPS exist, however, despite current interest in this area in biological aging. Early reports demonstrated normal oxygen utilization with complex-I-linked substrates in muscle mitochondria ($n = 2$) (Villee et al., 1969) and cultured fibroblasts ($n = 1$) (Goldstein et al., 1982) from patients with HGPS. However,

Goldstein *et al.* (1982) did observe increased lactate production in HGPS fibroblasts, suggesting a defect in oxidative metabolism might be present but not detectable by the methods then applied.

There are no recent reports concerning mitochondrial enzyme activities in HGPS, but work at the Royal Free Hospital School of Medicine has demonstrated a 60% defect in complex IV activity in cultured fibroblasts from two patients with HGPS, compared to nine age-matched controls, and a 50% decrease in complex II–III activity, implying that mitochondrial dysfunction might have a role in the pathogenesis of the disease, as has been proposed in biological aging. Preliminary mtDNA analysis has not demonstrated any excess of mtDNA with the common deletion in these patients, and mitochondrial proteins, studied by *in vitro* translation analysis and Western blots using monoclonal antibodies to complex IV proteins, were normal. Furthermore, fusion of enucleated HGPS fibroblasts with A549 rho^0 cells was able to restore the activities of all MRC enzymes to control levels, suggesting a role for nuclear and not mitochondrial genes in mitochondrial dysfunction in HGPS. However, although complex IV defects associated with the 3243 tRNA$^{Leu(UUR)}$ mtDNA mutation have been shown to persist in the A549 nuclear environment (Dunbar *et al.*, 1995), implying that the tissue type change alone is insufficient to correct the defect, an mtDNA mutation has not been excluded in the HGPS patients. An mtDNA mutation may well be present in a heteroplasmic state, and if this is the case, the apparent nuclear complementation might in fact be due either to selective fusion of low-mutant-containing cells or to selection of wild-type mtDNA in the fusion clones, rather than reflecting the presence of a new nuclear environment. A tendency to drift toward wild-type mtDNA was observed in A549 rho^0 fusion lines containing the 3243 tRNA$^{Leu(UUR)}$ mtDNA mutation in the study by Dunbar *et al.* (1995). Analysis of further fusion clones and sequencing of mitochondrial genes in HGPS is underway in order to clarify this.

Alternatively, the nature of the functional defects identified in HGPS might be considered more in keeping with nuclear rather than mitochondrial gene defects. No single mitochondrial or nuclear coding gene mutation would be expected to produce defects in both complex II–III and complex IV. Mutations affecting mtDNA tRNA genes preferentially affect complexes I and/or IV or all enzyme activities (Schapira, 1994). Cytochrome *b* (complex III) and the mitochondrially encoded cytochrome oxidase (COX) subunits are physically separated within the mitochondrial genome by complex I and V subunits, making it difficult to envisage any mtDNA rearrangement that could selectively affect complexes II–III and IV while sparing complex I. Furthermore, nuclear genes have been implicated by one study (Hayashi *et al.*, 1994) in the complex IV defect seen in biological aging: Placing fibroblast mitochondria from aging donors, expressing a complex IV defect, in a new nuclear environment restored complex IV activity in this study to levels comparable with those of younger patients (Hayashi *et al.*, 1994). However, this report studied only a small number of clones and the results should be regarded as provisional. The hypothesis that nuclear genes might be responsible for mitochondrial dysfunction in HGPS could be confirmed by a mirror fusion experiment in which mitochondria from control fibroblasts are introduced into an HGPS cell that itself has been rendered rho^0. However, this would require transformation of the HGPS cells in order to extend their replicative capacity. As senescence is a characteristic of major interest in HGPS, transformation might eliminate expression of genes important to the HGPS phenotype, as has been observed in other studies where preexisting differences in DNA repair were no longer expressed in transformed HGPS fibroblasts (Saito and Moses, 1991).

It is tempting to hypothesize a model in which telomere shortening (Allsopp *et al.*, 1992) might lead to damage to nuclear genes encoding mitochondrial proteins and consequent mitochondrial dysfunction. In turn, through free radical production (Zhang *et al.*, 1990), this might cause damage to mitochondrial proteins, lipids, and mtDNA itself, resulting in an escalation of mitochondrial dysfunction and ultimately in disruption of normal cellular function and cell death. Whether mitochondrial dysfunction in HGPS is of primary pathogenetic importance or a secondary common pathway, the further characterization of biochemical dysfunction in HGPS and detailed analysis of the relative contributions of mitochondrial and nuclear genes to this dysfunction might not reveal only information regarding nuclear mitochondrial interactions, but also perhaps information pertinent to the biochemical changes in biological aging.

3. NUCLEAR–MITOCHONDRIAL INTERACTIONS

Most of the human diseases associated with defects of mtDNA genes are either sporadic or maternally inherited. Nevertheless, some mitochondrial disorders accompanied with mtDNA abnormalities are transmitted as mendelian traits. These abnormalities can be either qualitative or quantitative.

3.1. Autosomal Dominant Progressive External Ophthalmoplegia

Mendelian-inherited, qualitative modifications of mtDNA are typically represented by large-scale, multiple deletions of the mitochondrial genome found in patients with autosomal dominant progressive external ophthalmoplegia (adPEO) (Kawai *et al.*, 1995; Suomalainen *et al.*, 1992a; Zeviani *et al.*, 1989, 1990; Ciafanoli *et al.*, 1991). This disorder is characterized by an adult onset of external ophthalmoplegia, ptosis, and general muscle weakness. In

addition, cataract, ataxia, deafness, severe retarded depression, and early death
have been reported (Kawai *et al.*, 1995; Suomalainen *et al.*, 1992a; Servidei *et
al.*, 1991; Zeviani *et al.*, 1990). Histochemical examination of skeletal muscle
biopsies from patients have revealed ragged-red fibers and decreased cytochrome *c* oxidase staining, and biochemical analyses have demonstrated partial
defects of complexes I and IV. Multiple deletions of mtDNA are present in
skeletal muscle from patients but not from unaffected family members (Suomalainen *et al.*, 1992a; Servidei *et al.*, 1991; Zeviani *et al.*, 1990). In one study
(Suomalainen *et al.*, 1992a), where additional tissues were available for molecular analyses, multiple deletions were also found in kidney, liver, heart, and
brain from a patient. No mtDNA deletions have been demonstrated in peripheral blood cells or cultured fibroblasts, myoblasts, or myotubes or innervated
muscle cells (Suomalainen *et al.*, 1992a; Zeviani, 1992; Servidei *et al.*, 1991).

Deleted mtDNA species coexist with wild-type mtDNA species (pleioplasmy) and may be as much as 60% of the total amount of mtDNA present
(Suomalainen *et al.*, 1992a, 1995). Deletions do not affect the region between
the D-loop and the origin of L-strand replication containing the ribosomal
RNA genes. Sequence analyses of deleted mtDNA molecules after polymerase
chain reaction (PCR) amplification have revealed that the deletions occur
across flanking direct repeats of variable length (Kawai *et al.*, 1995; Zeviani *et
al.*, 1989, 1990). Deletions seem to be cumulative with age and to correlate with
the severity of the disease (Suomalainen *et al.*, 1995; Servidei *et al.*, 1991). No
mtDNA deletions have been detected in healthy maternal progeny (Zeviani *et
al.*, 1990), and the patterns of inheritance in the different families clearly
indicate an autosomal dominant trait predisposing to large-scale deletions of
mtDNA (Kaukonen *et al.*, 1996; Kawai *et al.*, 1995; Suomalainen *et al.*, 1992a;
Servidei *et al.*, 1991; Zeviani *et al.*, 1990).

3.2. Other Mitochondrial Disorders Associated with Multiple Deletions of mtDNA

Large-scale, multiple deletions of mtDNA have also been reported in a
number of other mitochondrial diseases, some of which may be clinical variants of adPEO. Yuzaki and colleagues (1989) have demonstrated multiple deletions of mtDNA in skeletal muscle from two siblings with atrophy and mitochondrial myopathy associated with peripheral neuropathy. Like adPEO, the
breakpoints of the deletions were flanked by direct repeats, but unlike adPEO,
the family history was suggestive for autosomal recessive transmission.

Similar to the adPEO cases discussed, multiple deletions of mtDNA have
been reported in skeletal muscle tissue of a patient with progressive external
ophthalmoplegia and with a family history consistent with autosomal dominant
inheritance of mitochondrial myopathy (Otsuka *et al.*, 1990). The authors claim

that in addition to multiple deletions, the total amount of mtDNA was markedly reduced (15% of controls) in the affected tissue of their patient, in contrast to the other adPEO cases. However, no internal nuclear control was applied in the Southern blot experiment to ensure that equal amounts of total genomic DNA were present in the lanes.

Cormier and colleagues (1991) have described a case of progressive encephalomyopathy associated with multiple deletions of mtDNA in skeletal muscle tissue and leukocytes. Histopathologic examination of skeletal muscle showed typical mitochondrial myopathy features with ragged-red fibers. Deleted mtDNA species of a different size than in the proband were present in the clinically unaffected mother and a maternal aunt but not in other maternal relatives.

Two siblings with recurrent myoglobinuria and alcohol intolerance associated with multiple deletions of mtDNA in skeletal muscle have been reported by Ohno and colleagues (1991). The siblings appeared to have some deletions in common, while other deletions were patient specific. Histochemical and electron microscopic examination of muscle tissue disclosed patchy cytochrome c oxidase staining, ragged-red fibers, and morphologically abnormal mitochondria with paracrystaline inclusions. The family was nonconsanguineous and other relatives were healthy.

Suomalainen and colleagues (1992b) have described a woman and son both of whom died of idiopathic dilated cardiomyopathy. The maternal grandparents and two elder sisters of the son were clinically normal. Southern blot hybridization and PCR analysis revealed a high proportion of mtDNA with multiple deletions in the son's cardiac (50% of total mtDNA) and skeletal (65% of total mtDNA) muscle. Deleted mtDNA species were also demonstrated in paraffin embedded sections of the mother's cardiac muscle.

Two brothers with sideroblastic anemia, mild pancreatic insufficiency, and progressive muscle weakness have been reported with multiple deletions of mtDNA in skeletal muscle (29 and 22% of total mtDNA, respectively) and leukocytes (3 and 4% of total mtDNA, respectively) by Casademont and colleagues (1994). Analysis of muscle biopsies from both siblings revealed ragged-red fibers and reduced activity of respiratory enzyme complexes containing mtDNA-encoded subunits. Lower levels of multiple mtDNA deletions were present in skeletal muscle tissue and leukocytes from the asymptomatic mother but not the father. Deleted mtDNA species present in different family members were similar in size but not identical.

Recently, four patients in one generation of a multiple consanguineous pedigree were described who died with cardiomyopathy, cataracts, and lactic acidosis (Pitkänen *et al.*, 1996). Southern blot analysis revealed multiple deletions of mtDNA in cardiac but not in skeletal muscle. Biochemical studies showed a decrease in complex 1 activity in cardiac muscle (12% of controls) and fibroblasts (complex I + III, 55% of controls), while normal activity was found in skeletal muscle. Western blots and enzyme activity assays demonstrated grossly elevated levels of mitochondrial manganese–superoxide dismutase in cultured fibroblasts from one of the patients.

Barrientos and colleagues (1996) recently reported two families with Wolfram's syndrome who were harboring multiple deletions of mtDNA, as high as 85–90% in affected tissues (brain) of one patient, whereas other tissues from this patient and from other family members had deletion levels below 10% of the total amount of mtDNA. Wolfram's syndrome is a progressive neurodegenerative disease transmitted in an autosomal recessive mode. The disease locus has been assigned to chromosome 4p16 (Polymeropoulos *et al.*, 1994). Linkage to this region was confirmed in both families studied by Barrientos and colleagues.

Furthermore, large-scale multiple deletions of mtDNA have been reported in a number of sporadic cases. Recently, multiple deletions of mtDNA were demonstrated in various tissues of an infant with characteristics of Brachman–de Lange's syndrome (77% of total mtDNA in skeletal muscle) (Melegh *et al.*, 1996). Multiple deletions were found in skeletal muscle of a patient with inclusion body myositis (Oldfors *et al.*, 1993), a patient with multiple symmetric lipomatosis (Klopstock *et al.*, 1994), four patients with mitochondrial neurogastrointestinal encephalomyopathy (Hirano *et al.*, 1994), and an infant with characteristics of Brachmann–de Lange's syndrome (Melegh *et al.*, 1996). However, the levels of deleted mtDNA in these cases were well below the threshold of 60% of total mtDNA necessary to cause mitochondrial dysfunction in HeLa cells (Hayashi *et al.*, 1991) and therefore may not be of pathogenic significance. Multiple deletions affecting over 40% of the total amount of mtDNA have been found in a case of a 37-year-old woman with progressive external ophthalmoplegia, peripheral neuropathy, chronic intractable diarrhea, lacic acidosis, and ragged-red muscle fibers lacking cytochrome *c* oxidase activity (Uncini *et al.*, 1994), and in a case of a 25-year-old man with progressive external ophthalmoplegia, ptosis and ragged-red muscle fibers lacking cytochrome *c* oxidase activity (Campos *et al.*, 1996).

3.3. The Etiology of Multiple mtDNA Deletions

Patients with progressive external ophthalmoplegia, Kearns–Sayre's syndrome, or Pearson's syndrome associated with a single deletion of mtDNA are nearly all sporadic cases. This indicates that these single deletions are probably the result of single, clonally amplified mtDNA mutations of somatic origin. In contrast, mitochondrial disorders associated with multiple deletions of mtDNA are predominantly found in patients with a family history compatible with either autosomal dominant or autosomal recessive transmission of the disease.

Maternal inheritance of these multiple deletions are unlikely, as deleted mtDNA species vary in number, size, and breakpoints between different family members (Barrientos *et al.*, 1996; Casademont *et al.*, 1994; Suomalainen *et al.*, 1992a; Cormier *et al.*, 1991; Ohno *et al.*, 1991; Zeviani *et al.*, 1989, 1990). Thus, multiple deletions are most likely the result of multiple, clonally amplified mtDNA mutations due to a nuclear mutation present in the germ line.

Similar to the sporadic, single deletions of mtDNA, the breakpoints of multiple mtDNA deletions are usually sequence repeats, but in contrast with the single deletions, the multiple deletion breakpoints are often short (5–12 base pairs) and sometimes nonperfect or reverted (Kawai *et al.*, 1995; Casademont *et al.*, 1994; Cormier *et al.*, 1991; Zeviani *et al.*, 1989, 1990; Yuzaki *et al.*, 1989; Barrientos *et al.*, 1996). Such structures may facilitate deletion formation via homologous recombination (Mita *et al.*, 1990; Schon *et al.*, 1989). Although both single and multiple deletions are apparently generated by related mechanisms, the mechanism giving rise to multiple deletions seems to have less strict sequence requirements and seems to occur at a higher frequency.

The occurrence of adPEO in large, informative families has prompted linkage studies as a first step to identify the defective nuclear gene(s). Interestingly, distinct loci appear to be involved in different families. Suomalainen and colleagues (1995) assigned the disease locus to chromosome 10q23.3-24.3 in a Finnish family, but they excluded linkage to this locus in two Italian families despite a closely matching phenotype. A year later, the same research groups assigned the second locus to chromosome 3q14.1-21.2 in three Italian families, but they also showed that linkage to this chromosomal region was absent in three additional Italian families (Kaukonen *et al.*, 1996). Thus, at least three different nuclear genes appear to trigger multiple deletions of mtDNA.

Nuclear mutations causing multiple deletions of mtDNA could affect a gene product directly involved in replication of mtDNA, for example, DNA polymerase gamma or mitochondrial single-stranded binding protein (Zeviani *et al.*, 1995). However, if the defective gene product is directly involved in replication of mtDNA, an early onset of symptoms seems likely which is in disagreement with most cases (Kawai *et al.*, 1995). Alternatively, the defective gene product may not be involved in mtDNA replication but multiple deletions could be a nonspecific response to a pathological process in the affected tissue. A mild, tissue-specific secondary reaction, ultimately leading to damage of mtDNA, would not only explain better the late onset of the disease but also the differential tissue involvement.

Even in normal individuals, low levels (<2%) of deleted mtDNA species accumulate with age in tissues with a relatively high oxidative metabolism such as heart, skeletal muscle, and brain, a phenomenon that has been related to cumulative oxidative damage (Cortopassi and Arnheim, 1990; Corral-Debrinski *et al.*, 1992; Hayakawa *et al.*, 1992; Cortopassi *et al.*, 1992). Detailed molecular and biochemical analyses of one of the patients from the multiple consanguineous family with multiple deletions of mtDNA studied by Pitkänen and colleagues (1996) revealed a marked complex I deficiency in heart associated with an apparently induced expression of mitochondrial manganese–superoxide dismutase. The authors postulate that a nuclear-encoded defect in complex I induces the expression of manganese–superoxide dismutase through excessive free oxygen radical formation, as a side product of the impaired mitochondrial respiratory chain function. The increased production of free oxygen radicals may not be fully compensated by the increased levels of manganese–superoxide dismutase and cause oxidative damage leading to multiple deletions of mtDNA in the affected tissue (heart) of this patient, in a mechanism similar to that in aging individuals. Likewise, multiple deletions of mtDNA in, for example, adPEO could be the result of oxidative damage due to a defective respiratory chain. In this respect, it would be interesting to determine whether the expression of manganese–superoxide dismutase is induced in these patients. Tissue-specific accumulation of multiple mtDNA deletions could be a result of either the relatively high oxidative metabolism of the affected tissues (heart, muscle, brain) or a defect in a tissue-specific subunit of one of the respiratory chain enzymes. Nuclear-encoded, tissue-specific subunits of cytochrome *c* oxidase have been implicated in regulation of enzyme activity and may decouple electron transfer from proton translocation (Taanman *et al.*, 1994) leading to the production of free oxygen radicals. Over 65 subunits of the mitochondrial respiratory chain enzymes are encoded by nuclear DNA, and mutations in these polypeptides may in principle all lead to excessive free oxygen radical production. This could explain the clinical and genetic heterogeneity of diseases associated with multiple deletions of mtDNA.

3.4. mtDNA Depletion Syndrome

Mendelian-inherited quantitative modifications of mtDNA are typically represented by infants suffering mtDNA depletion syndrome. This disorder, first characterized by Moraes and colleagues in 1991, now has been diagnosed in over 30 patients and is emerging as an important cause of mitochondrial dysfunction in neonates and infants (Bakker *et al.*, 1996; Maaswinkel-Mooi *et al.*, 1996; Macmillan and Shoubridge, 1996; Parrot *et al.*, 1996; Mariotti *et al.*, 1995; Paquis-Flucklinger *et al.*, 1995; Poulton *et al.*, 1995; Bodnar *et al.*, 1993; Figarella-Branger *et al.*, 1992; Telerman-Toppet *et al.*, 1992; Mazziotta *et al.*, 1992; Tritschler *et al.*, 1992; Moraes *et al.*, 1991). Patients present in the early weeks of life (early onset) with muscle weakness and hepatic failure or renal tubulopathy, or at 12–14 months (late onset) with a slowly progressive encephalomyopathy. Patients often have lactic acidosis and show a deficiency of cytochrome *c* oxidase frequently associated with ragged-red muscle fibers. Early-onset patients usually die before 9 months of age. Southern blot analyses have demonstrated variable levels of mtDNA depletion in affected tissues (1– 12% of normal levels in early onset and 14–34% in late-onset patients), while unaffected tissues have relatively normal levels of mtDNA.

We have identified mtDNA depletion syndrome in a further three families, with five affected children dying of liver failure (Morris *et al.*, 1996). One of the patients was the first child of nonconsanguineous parents. Southern blot analysis of biopsy and postmortem samples from this child demonstrated a severe and progressive loss of mtDNA in liver as well as skeletal muscle tissue. On the other hand, analysis of increasing cell passages of the patient's cultured fibroblasts revealed only a transient decline in mtDNA levels. Another patient was the third child of first-cousin parents, the other siblings being healthy. This patient showed a severe depletion of mtDNA in a skeletal muscle biopsy specimen but not in cultured myoblasts. Finally, in the third pedigree, the first child was clinically normal but the next three children were all similarly affected. The third sibling of this nonconsanguineous family has been studied by us in detail. He exhibited a progressive loss of mtDNA in skeletal muscle tissue and had severely depleted levels of mtDNA in liver upon postmortem examination. Initially, mtDNA levels were normal in cultured myoblast from this patient but mtDNA levels declined over successive generations in culture.

3.5. The Etiology of mtDNA Depletion Syndrome

All patients reported with mtDNA depletion syndrome were born to clinically normal parents and siblings from both sexes were affected. Twenty-one of the reported cases were sporadic, and one was born to consanguineous parents (Morris *et al.*, 1996). The nine family histories of the remaining 14 reported cases were all positive for early-onset, fatal neuromuscular disorders. Two of these pedigrees were consanguineous (Bakker *et al.*, 1996; Mariotti *et al.*, 1995). All family histories were compatible with an autosomal recessive inheritance of the trait, with the possible exception of one family where the patient had two paternal half-siblings who both died of an undiagnosed myopathy at the age of 3 months (Mazziotta *et al.*, 1992). The pattern of transmission in this single family suggests either autosomal recessive inheritance with gonadal mosaicism or autosomal dominant inheritance with incomplete penetration, since an apparently asymptomatic father transmitted the disease and a third paternal half-sibling of the patient was clinically normal. Although mtDNA depletion syndrome may be genetically heterogeneous, the autosomal inheritance indicated by the nine pedigrees supports the involvement of a nuclear-encoded factor in the depletion. On the other hand, differential tissue involvement, even in related patients (Moraes *et al.*, 1991), and the family history in one case of neuromuscular disorders in the maternal lineage (Mariotti *et al.*, 1995) are suggestive for a heteroplasmic mtDNA mutation. Nevertheless, sequence analyses of mtDNA replication origins failed to demonstrate any potentially pathogenic mutations (Mariotti *et al.*, 1995; Moraes *et al.*, 1991) and two families showed an apparent paternal transmission of the disorder (Mazziotta *et al.*, 1992; Moraes *et al.*, 1991). Still, it cannot be excluded that in some pedigrees an mtDNA mutation outside the origins of replication is capable of interfering with DNA replication or that a seemingly neutral mtDNA polymorphism interacts with specific nuclear alleles.

We have studied cultured skin fibroblasts from two siblings with mtDNA depletion syndrome (Bodnar *et al.*, 1993, 1995). The cell cultures showed marked reductions in the enzyme activities of respiratory chain complexes with mtDNA-encoded subunits and a severe depletion of mtDNA. The most severely depleted cells were dependent on uridine and pyruvate for growth, a requirement well characterized in experimental cells devoid of mtDNA (rho^0 cells) (King and Attardi, 1989). Transfer of patient mitochondria, with residual levels of mtDNA, to experimental rho^0 cells led to restoration of mtDNA levels and enzyme activities, and hence demonstrates a nuclear involvement in the depletion (Bodnar *et al.*, 1993). Further studies have shown that the levels of mtDNA were unstable in the fibroblast cultures and declined with increasing cell passage. The remnant mtDNA, however, was both transcribed and translated, indicating that the primary defect is one of mtDNA replication or maintenance and not likely to be related to defects in mitochondrial transcription (Bodnar *et al.*, 1995).

It has been suggested that mtDNA depletion may be due to an error in the resumption of mtDNA replication after the replication arrest during oogenesis, fertilization, and early embryogenesis, causing reduced numbers of mtDNA in individual stem cell populations (Moraes *et al.*, 1991). This is apparently not so, at least not in some of the families we studied, because we observed a progressive loss of mtDNA in tissues and demonstrated initially normal levels of mtDNA in cultured myoblasts that became depleted over successive generations in culture. This progressive loss of mtDNA is in line with the progressive nature of the clinical features of the disease.

The early onset of symptoms and the apparent autosomal inheritance suggest that the depletion of mtDNA is caused by a mutation in a nuclear gene coding for a *trans*-acting factor directly involved in replication or maintenance of mtDNA. This *trans*-acting factor may be developmentally specific and only be expressed in the postnatal period when energy demands in many tissues will increase compared to the prenatal period. Although the basic mechanism of mammalian mtDNA replication has been elucidated, knowledge of regulatory factors is still poor and only a few protein factors directly involved in replication have been purified (Clayton, 1992). Furthermore, knowledge of maintenance of mammalian mtDNA is scarce, making it difficult to suggest candidate genes that could be defective in these patients. Nevertheless, mitochondrial transcription factor A (mtTFA), one of the few well-characterized factors involved in mtDNA replication, has been implicated in the disease. MtTFA is a component of the mitochondrial transcription machinery and is also likely to be involved in regulation of mtDNA copy number, since RNA primers synthesized by the mitochondrial transcription machinery are required for initiation of mtDNA replication (Clayton, 1992). Moreover, the proposed homologue of mtTFA in yeast (the ABF2 gene product) is essential for maintenance of mtDNA when yeast is grown on a fermentable carbon source such as glucose (Shadel and Clayton, 1993). Low levels of mtTFA polypeptide have been observed in tissues from three patients with mtDNA depletion and mtTFA mRNA levels were reduced (63% of controls) in the one patient investigated (Poulton *et al.*, 1994). However, sequencing of the mtTFA promoter and cDNA from one of the patients did not reveal any mutations (Morten *et al.*, 1996), and mitochondrial transcription of the residual mtDNA in cell cultures from patients studied by our group appears not to be impaired (Bodnar *et al.*, 1995; J.-W. Taanman *et al.*, unpublished observations). In addition, a defect in mtTFA is likely to be lethal early in embryonic development, whereas all patients with mtDNA depletion syndrome described so far were born after an uncomplicated, full-term pregnancy.

We are using myoblast cell lines derived for a patient that have normal mtDNA levels at early cell passages but become depleted with increasing passage number, as an *in vitro* model of the disease. As myoblasts are dormant most of the time, these cells probably reflect the situation that exists in the myoblast precursor population during embryological development. We plan to transfect these depleting myoblast cell lines with mammalian cDNA expression libraries to identify cDNA species that restore mtDNA levels and respiratory chain function. This approach is facilitated by the fact that, because of the impaired respiratory chain, the patient's cells are auxotrophic for uridine and pyruvate, thus providing a selectable marker to identify complementing cDNAs. Cloning of the nuclear gene(s) responsible for mtDNA depletion will not only assist genetic counseling and in the future the establishment of rational therapeutic approaches, but may also provide new insights in the communication between the nuclear and mitochondrial genomes in the cell.

ACKNOWLEDGMENT. The work was supported by a Medical Research Council, the Muscular Dystrophy Group of Great Britain, and the Royal National Institute for the Blind.

4. REFERENCES

Allsopp, R. C., Vaziri, H., Patterson, C., Goldstein, S., Younglai, E. V., Futcher, A. B., Greider, C. W., and Harley, C. B., 1992, Telomere length predicts replicative capacity of human fibroblasts, *Proc. Natl. Acad. Sci. USA* **89**:10114–10118.

Anderson, S., Bankier, A. T., de Bruijn, M. H. L., Coulson, A. R., Drouin, J., Eperon, I. C., Nierlich, D. P., Roe, B. A., Sanger, F., Schreier, P. H., Smith, A. J. H., Staden, R., and Young, I. G., 1981, Sequence and organization of the human mitochondrial genome, *Nature* **290**: 457–465.

Bakker, H. D., Scholte, H. R., Dingemans, K. P., Spelbrink, J. N., Wijburg, F. A., and Van den Bogert, C., 1996, Depletion of mitochondrial deoxyribonucleic acid in a family with fatal neonatal liver disease, *J. Pediatr.* **128**:683–687.

Barrientos, A., Volpini, V., Casademont, J., Genís, D., Manzanares, J.-M., Ferrer, I., Corral, J., Cardellach, F., Urbano-Márquez, Estivill, X., and Nunes, V., 1996, A nuclear defect in the 4p16 region predisposes to multiple mitochondrial DNA deletions in families with Wolfram syndrome, *J. Clin. Invest.* **97**:1570–1576.

Baumer, A., Zhang, C., Linnane, A. W., and Nagley, P., 1994, Age-related human mtDNA deletions: A heterogeneous set of deletions arising at a single pair of directly repeated sequences, *Am. J. Hum. Genet.* **54**:618–630.

Beavan, L. A., Quentin-Hoffmann, E., Schönherr, E., Snigula, F., Leroy, J. G., and Kresse, H., 1993, Deficient expression of decorin in infantile progeroid patients, *J. Biol. Chem.* **268**: 9856–9862.

Black, G. C. M., Criag, I. W., Oostra, R., Norby, S., Morten, K., Laborde, A., and Poulton, J., 1995, Leber's hereditary optic neuropathy: Implications of the sex ratio for linkage studies in families with the 3,460 ND1 mutation, *Eye* **9**:513–516.

Bodnar, A. G., Cooper, J. M., Holt, I. J., Leonard, J. V., and Schapira, A. H. V., 1993, Nuclear complementation restores mtDNA levels in cultured cells from a patient with mtDNA depletion, *Am. J. Hum. Genet.* **53**:663–669.

Bourgeron, I., Rasun, P., Cinenon, D., Birch-Machin, M., Bourgeois, M., Viegas-Péquignot, E., Munnich, A., and Rötig, A., 1995, Mutation of a nuclear succinate dehydrogenase gene results in mitochondrial respiratory chain deficiency, *Nature Genet.* **11**:114–149.

Bodnar, A. G., Cooper, J. M., Leonard, J. V., and Schapira, A. H. V., 1995, Respiratory-deficient human fibroblasts exhibiting defective mitochondrial DNA replication, *Biochem. J.* **305**: 817–822.

Brown, W. T., 1992, Progeria: A human-disease model of accelerated ageing, *Am. J. Clin. Nutr.* **55**:1222S–1224S.

Brown, W. T., Ford, J. P., and Gershey, E. L., 1980, Variation of DNA repair capacity in progeria cells unrelated to growth conditions, *Biochem. Biophys. Res. Commun.* **97**:347–353.

Bu, X., and Rotter, J. I., 1991, X chromosome-linked and mitochondrial gene control of Leber hereditary optic neuropathy: Evidence from segregation analysis for dependence on X chromosome inactivation, *Proc. Natl. Acad. Sci. USA* **88**:8198–8202.

Bu, X., and Rotter, J. I., 1992, Leber hereditary optic neuropathy: Estimation of number of embryonic precursor cells and disease threshold in heterozygous affected females at the X-linked locus, *Clin. Genet.* **42**:143–148.

Campos, Y., Martin, M. A., Rubio, J. C., Ricard, C., Cabello, A., and Arenas, J., 1996, Multiple deletions of mitochondrial DNA in muscle from a patient with benign progressive external ophthalmoplegia, *J. Inherit. Metab. Dis.* **19**:366–367.

Carvalho, M. R. S., Müller, B., Rützer, E., Berninger, T., Kommerell, G., Blankenagel, A., Savontaus, M., Meitinger, T., and Lorenz, B., 1992, Leber's hereditary optic neuroretinopathy and the X-chromosomal susceptibility factor:no linkage to DXS7, *Hum. Hered.* **42**:316–320.

Casademont, J., Barrientos, A., Cardellach, F., Rötig, A., Grau, J. M., Montoya, J., Beltrán, B., Cervantes, F., Rozman, C., Estivill, X., Urbano-Márquez, A., and Nunes, V., 1994, Multiple deletions of mtDNA in two brothers with sideroblastic anemia and mitochondrial myopathy and in their asymptomatic mother, *Hum. Mol. Genet.* **3:**1945–1949.

Chalmers, R. M., Davis, M. B., Sweeney, M. G., Wood, N. W., and Harding, A. E., 1996, Evidence against an X-linked visual loss susceptibility locus in Leber hereditary optic neuropathy, *Am. J. Hum. Genet.* **59:**103–108.

Chen, J., and Denton, M. J., 1991, X-chromosomal gene in Leber hereditary optic neuroretinopathy, *Am. J. Hum. Genet.* **48:**692–693.

Chen, J., Cox, I., and Denton, M. J., 1989, Preliminary exclusion of an X-linked gene in Leber optic atrophy by linkage analysis, *Hum. Genet.* **82:**203–207.

Chomyn, A., Meola, G., Bresolin, N., Lai, S. T., Scarlato, G., and Attardi, G., 1991, *In vitro* genetic transfer of protein synthesis and respiration defects to mitochondrial DNA-less cells with myopathy patient mitochondria, *Mol. Cell. Biol.* **11:**2236–2244.

Chomyn, A., Lai, S. T., Shakeley, R., Bresolin, N., Scarlato, G., and Attardi, G., 1994, Platelet-mediated transformation of mtDNA-less human cells: Analysis of phenotyic variability among clones from normal individuals—and complementation behaviour of the tRNALys mutation causing myoclonic epilepsy and ragged red fibers, *Am. J. Hum. Genet.* **54:**966–974.

Ciataloni, E., Ricci, E., Servidei, S., Shansken, S., Silvestri, G., Manfredi, G., Schon, E. A., and DiMauro, S., 1991, Widespread tissue distribution of tRNA Leu (UUR) mutation in the mitochondrial DNA of a patient with MELAS syndrome, *Neurology* **41:**1663–1665.

Clark, M. A., and Weiss, A. S., 1995, Hutchinson–Gilford progeria types defined by differential binding of lectin DSA, *Biochim. Biophys. Acta* **1270:**142–148.

Clayton, D. A., 1992, Transcription and replication of animal mtDNAs, *Int. Rev. Cytol.* **141:**217–232.

Cock, H. R., Cooper, J. M., and Schapira, A. H. V., 1995a, Nuclear complementation in Leber's hereditary optic neuropathy, *Neurology* **45:**A294.

Cock, H. R., Cooper, J. M., and Schapira, A. H. V., 1995b, The 14,484 ND6 mtDNA mutation in Leber hereditary optic neuropathy does not affect fibroblast complex I activity, *Am. J. Hum. Genet.* **57:**1501–1502.

Cormier, V., Rötig, A., Tardieu, M., Colonna, M., Saudubray, J.-M., and Munnich, A., 1991, Autosomal dominant deletions of the mitochondrial genome in a case of progressive encepha-lomyopathy, *Am. J. Hum. Genet.* **48:**643–648.

Cornelissen, J. C., Wanders, R. J. A., Bolhuis, P. A., Bleeker-Wagermakers, E. M., Oostra, R., and Wijburg, F. A., 1993, Respiratory chain function in Leber's hereditary optic neuropathy: Lack of correlation with clinical disease, *J. Inherit. Metab. Dis.* **16:**531–533.

Corral-Debrinski, M., Horton, T., Lott, M. T., Shoffner, J. M., Beal, M. F., and Wallace, D. C., 1992, Mitochondrial DNA deletions in human brain: Regional variability and increase with ad-vanced age, *Nature Genet.* **2:**324–329.

Cortelli, P., Montagna, P., Avoni, P., Sangiorgi, S., Bresolin, N., Moggio, M., Zaniol, P., Mantovani, V., Barboni, P., Barbiroli, B., and Lugaresi, E., 1991, Leber's hereditary optic neuropathy: Genetic, biochemical and phosphorous magnetic spectroscopy study in an Italian family, *Neurology* **41:**1211–1215.

Cortopassi, G. A., and Arnheim, N., 1990, Detection of a specific mitochondrial DNA deletion in tissues of older humans, *Nucleic Acids Res.* **18:**6927–6933.

Cortopassi, G. A., Shibata, D., Soong, N.-W., and Arnheim, N., 1992, A pattern of accumulation of a somatic deletion of mitochondrial DNA in aging human tissues, *Proc. Natl. Acad. Sci. USA* **89:**7370–7374.

DeBusk, F. L., 1972, The Hutchinson–Gilford progeria syndrome, *J. Pediatr.* **80:**697–724.

Desjardins, P., Frost, E., and Morais, R., 1985, Ethidium bromide-induced loss of mitochondrial DNA from primary chicken embryo fibroblasts, *Mol. Cell. Biol.* **5:**1163–1169.

Dunbar, D. R., Moonie, P. A., Jacobs, H. T., and Holt, I. J., 1995, Different cellular backgrounds confer a marked advantage to either mutant or wild-type mitochondrial genomes, *Proc. Natl. Acad. Sci. USA* **92**:6562–6566.

Epstein, J., Williams, J. R., and Little, J. B., 1973, Deficient DNA repair in human progeroid cells, *Proc. Natl. Acad. Sci. USA* **70**:977–981.

Figarella-Branger, D., Pellissier, J. F., Scheiner, C., Wernert, F., and Desneulle, C., 1992, Defects of mitochondrial respiratory chain complexes in three pediatric cases with hypotonia and cardiac involvement, *J. Neurol. Sci.* **108**:105–113.

Gale, R. E., Wheadon, H., Boulos, P., amd Linch, D. C., 1994, Tissue specificity of X-chromosome inactivation patterns, *Blood* **83**:2899–2905.

Gilford, H., 1904, Ateleiosis and progeria: Continuous youth and premature age, *Br. Med. J.* **2**:914.

Giro, M., and Davidson, J. M., 1993, Familial co-segregation of the elastin phenotype in skin fibroblasts from Hutchinson–Gilford progeria, *Mech. Ageing Dev.* **70**:163–136.

Goldstein, S., Ballantyne, S. R., Robson, A. L., and Moerman, E. J., 1982, Energy metabolism in cultured human fibroblasts during aging *in vitro*, *J. Cell. Physiol.* **112**:419–424.

Greider, C. W., 1990, Telomeres, telomerase and senescence, *BioEssays* **12**:363–369.

Harding, A. E., Sweeney, M. G., Govan, G. G., and Riordan-Eva, P., 1995, Pedigree analysis in Leber hereditary optic neuropathy families with a pathogenic mtDNA mutation, *Am. J. Hum. Genet.* **57**:77–86.

Hayakawa, M., Hattori, K., Sugiyama, S., and Ozawa, T., 1992, Age-associated oxygen damage and mutations in mitochondrial DNA in human hearts, *Biochem. Biophys. Res. Commun.* **189**: 979–985.

Hayashi, J., Ohto, S., Kikuchi, A., Takemitsu, M., Goto, Y., and Nonaka, I., 1991, Introduction of disease-related mitochondrial DNA deletions into HeLa cells lacking mitochondrial DNA results in mitochondrial dysfunction, *Proc. Natl. Acad. Sci. USA* **88**:10614–10618.

Hayashi, J., Ohta, S., Takai, D., Miyabayashi, S., Sakuta, R., Goto, Y., and Nonaka, I., 1993, Accumulation of mtDNA with a mutation at position 3271 in tRNA Leu(UUR) gene intro-duced from a MELAS patient to HeLa cells lacking mtDNA results in progressive inhibition of mitochondrial respiratory function, *Biochem. Biophys. Res. Commun.* **197**:1049–1055.

Hayashi, J., Ohta, S., Kagawa, Y., Kondo, H., Kaneda, H., Yonekawa, H., Takai, D., and Miyaba-yashi, S., 1994, Nuclear but not mitochondrial genome involvement in human age-related mitochondrial dysfunction, *J. Biol. Chem.* **269**:6878–6883.

Hely, M. A., Morris, J. G. L., Reid, W. G. J., O'Sullivan, D. J., Williamson, P. M., Rail, D., Broe, G. A., and Margrie, S., 1994, The Sydney multicentre study of Parkinson's disease: A ran-domised, prospective five year study comparing low dose bromocriptine with low dose levodopa-carbidopa, *J. Neurol. Neurosurg. Psychiatry* **57**:903–910.

Hirano, M., Silvestri, G., Blake, D. M., Lombes, A., Minetti, C., Bonilla, E., Hays, A. P., Lovelace, R. E., Butler, I., Bertorini, T. E., Threlkeld, A. B., Mitsumoto, H., Salberg, L. M., Rowland, L. P., and DiMauro, S., 1994, Mitochondrial neurogastrointestinal encephalomyopathy (MNGIE): Clinical, biochemical, and genetic features of an autosomal recessive mitochondrial disorder, *Neurology* **44**:721–727.

Hofhaus, G., Johns, D. R., Hurko, O., Attardi, G., and Chomyn, A., 1996, Respiration and growth defects in transmitochondrial cell lines carrying the 11,778 mutation associated with Leber's hereditary optic neuropathy, *J. Biol. Chem.* **271**:13155–13161.

Holt, I. J., Miller, D. H., and Harding, A. E., 1989, Genetic heterogeneity and mitochondrial DNA heteroplasmy in Leber's hereditary optic neuropathy, *J. Med. Genet.* **26**:739–743.

Howell, N., Bindoff, L. A., McCullough, D. A., Kubacka, I., Mackey, D., Taylor, L., Turnbull, D. M., and Poulton, J., 1991, Leber hereditary optic neuropathy: Identification of the same mitochondrial ND1 Mutation in six pedigrees, *Am. J. Hum. Genet.* **49**:939–950.

Howell, N., Xu, M., Halvorson, S., Bodis-Wollner, I., and Sherman, J., 1994, A heteroplasmic LHON family: Tissue distribution and transmission of the 11,778 mutation, *Am. J. Hum. Genet.* **55**:203–206.

Hsieh, R., Hou, J., Hsu, H., and Wei, Y., 1994, Age-dependent respiratory function decline and DNA deletions in human muscle mitochondria, *Biochem. Mol. Int.* **32**:1009–1022.

Huoponen, K., Vilkki, J., Aula, P., Nikoskelainen, E. K., and Savontaus, M., 1991, A new mtDNA mutation associated with Leber hereditary optic neuroretinopathy, *Am. J. Hum. Genet.* **48:** 1147–1153.

Hutchinson, J., 1886, Congenital absence of hair and mammary glands with atrophic condition of the skin and its appendages in a boy whose mother had been almost totally bald from alopecia areata from the age of six, *Med. Chirurg. Trans.* **69**:36.

Imai, Y., and Moriwaki, D., 1936, A probable case of cytoplasmic inheritance in man: A critique of Leber's disease, *J. Genet. Hum.* **3**:163–167.

Johns, D. R., Neufeld, M. J., and Park, R. D., 1992, An ND6 mitochondrial DNA mutation associated with Leber hereditary optic neuropathy, *Biochem. Biophys. Res. Commun.* **187:** 1551–1557.

Johns, D. R., Smith, K. H., Miller, N. R., Sulewski, M. E., and Bias, W. B., 1994, Identical twins who are discordant for Leber's hereditary optic neuropathy, *Arch. Ophthalmol.* **111**:1491–1494.

Jun, A. S., Trounce, I. A., Brown, M. D., Shoffner, J. M., and Wallace, D. C., 1996, Use of transmitochondrial cybrids to assign a complex I defect to the mitochondrial DNA-encoded NADH dehydrogenase subunit 6 gene mutation at nucleotide pair 14459 that causes Leber hereditary optic neuropathy and dystonia, *Mol. Cell. Biol.* **16**:771–777.

Juvonen, V., Vilkki, J., Aula, P., Nikoskelainen, E. K., and Savontaus, M., 1993, Re-evaluation of the linkage of an optic atrophy susceptibility gene to X-chromosomal markers in Finish families with Leber hereditary optic neuroretinopathy (LHON), *Am. J. Hum. Genet.* **53**:289–292.

Kaukonen, J. A., Amati, P., Suomalainen, A., Rötig, A., Piscaglia, M.-G., Salvi, F., Weissenbach, J., Fratta, G., Comi, G., Peltonen, L., and Zeviani, M., 1996, An autosomal locus predisposing to multiple deletions of mtDNA on chromosome 3p, *Am. J. Hum. Genet.* **58**:763–769.

Kawai, H., Akaike, M., Yokoi, K., Nishida, Y., Kunishige, M., Mine, H., and Saito, S., 1995, Mitochondrial encephalomyopathy with autosomal dominant inheritance: A clinical and genetic entity of mitochondrial diseases, *Muscle Nerve* **18**:753–760.

Khalifa, M. M., 1989, Hutchinson-Gilford progeria syndrome: Report of a Libyan family and evidence of autosomal recessive inheritance, *Clin. Genet.* **35**:125–132.

King, M. P., and Attardi, G., 1989, Human cells lacking mtDNA: Repopulation with exogenous mitochondria by complementation, *Science* **246**:500–503.

Klopstock, T., Naumann, M., Schalke, B., Bischof, F., Seibel, P., Kottlors, M., Eckert, P., Reiners, K., Toyka, K. V., and Reichmann, H., 1994, Multiple symmetric lipomatosis: Abnormalities in complex IV and multiple deletions in mitochondrial DNA, *Neurology* **44**:862–866.

Larsson, N., Andersen, A., Holme, E., Oldfors, A., and Wahlstrvm, M. D., 1991, Leber's hereditary optic neuropathy and complex I deficiency in muscle, *Ann. Neurol.* **30**:704–708.

Lyon, M. F., 1996, Pinpointing the centre, *Nature* **379**:116–117.

Maaswinkel-Mooi, P. D., Van den Bogert, C., Scholte, H. R., Onkenhout, W., Brederoo, P., and Poorthuis, B. J. H. M., 1996, Depletion of mitochondrial DNA in the liver of a patient with lactic academia and hypoketotic hypoglycemia, *J. Pediatr.* **128**:679–683.

Mackey, D., 1993, Blindness in offspring of women blinded by Leber's hereditary optic neuropathy, *Lancet* **341**:1020–1021.

Macmillan, C. J., and Shoubridge, E. A., 1996, Mitochondrial DNA depletion: Prevalence in a pediatric population referred for neurologic evaluation, *Pediatr. Neurol.* **14**:203–210.

Majander, A., Huoponen, K., Savontaus, M., Nikoskelainen, E. K., and Wikstrvm, M., 1991, Electron transfer properties of NADH:ubiquinone reductase in the ND1/3,460 and the ND4/11,778 mutations of Leber hereditary optic neuroretinopathy (LHON), *FEBS Lett.* **292:** 289–292.

Mariotti, C., Tiranti, V., Carrara, F., Dallapiccola, B., DiDonato, S., and Zeviani, M., 1994, Defective respiratory capacity and mitochondrial protein synthesis in transformant cybrids harbouring the tRNALeu(UUR) mutation associated with maternally inherited myopathy and cardiomyopathy, *J. Clin. Invest.* **93:**1102–1107.

Mariotti, C., Uziel, G., Carrara, F., Mora, M., Prelle, A., Tiranti, V., DiDonato, S., and Zeviani, M., 1995, Early-onset encephalomyopathy associated with tissue-specific mitochondrial DNA depletion: A morphological, biochemical and molecular–genetic study, *J. Neurol.* **242:**547–556.

Mazziotta, R. M., Ricci, E., Bertini, E., Vici, C. D., Servidei, S., Burlina, A. B., Sabetta, G., Bartuli, A., Manfredi, G., Silvestri, G., Moraes, C. T., and DiMauro, S., 1992, Fatal infantile liver failure associated with mitochondrial DNA depletion, *J. Pediatr.* **121:**896–901.

Melegh, B., Bock, I., Gáti, I., and Méhes, K., 1996, Multiple mitochondrial deletions and persistent hyperthermia in a patient with Brachmann–de Lange phenotype, *Am. J. Med. Genet.* **65:** 82–88.

Mills, R. G., and Weiss, A. S., 1990, Does progeria provide the best model of accelerated ageing? *Gerontology* **36:**84–98.

Miranda, A. F., Ishii, S., DiMauro, S., and Shay, J. W., 1989, Cytochrome *c* oxidase deficiency in Leigh's syndrome: Genetic evidence for a nuclear DNA encoded mutation, *Neurology* **39:** 697–702.

Mita, S., Rizzuto, R., Moraes, C. T., Shanske, S., Arnaudo, E., Fabrizi, G. M., Koga, Y., DiMauro, S., and Schon, E., 1990, Recombination via flanking direct repeats is a major cause of large-scale deletions of human mitochondrial DNA, *Nucleic Acids Res.* **18:**561–567.

Moraes, C. T., Shanske, S., Tritschler, H.-J., Aprille, J. R., Andreetta, F., Bonilla, E., Schon, E. A., and DiMauro, S., 1991, mtDNA depletion with variable tissue expression: A novel genetic abnormality in mitochondrial diseases, *Am. J. Hum. Genet.* **48:**492–501.

Morris, A. A. M., Taanman, J.-W., Schapira, A. H. V., Cooper, J. M., Lake, B. D., Malone, M., Clayton, P. T., and Leonard, J. V., 1996, Liver failure due to mtDNA depletion: Three new pedigrees, (abstract), *J. Inherit. Metab. Dis.* **19**(Suppl. 1):73.

Morten, K. J., Freeman-Emmerson, C., and Poulton, J., 1996, Functional mtDNA replication defect in a fibroblast line from a patient with mtDNA depletion, *J. Inherit. Metab. Dis.* **19:**123–126.

Nikoskelainen, E. K., Hoyt, W. F., and Nummelin, K. U., 1983, Ophthalmoscopic findings in Leber's hereditary optic neuropathy II. The fundus findings in the affected family members, *Arch. Ophthalmol.* **101:**1059–1068.

Ohno, K., Tanaka, M., Sahashi, K., Ibi, T., Sato, W., Yamamoto, T., Takahashi, A., and Ozawa, T., 1991, Mitochondrial DNA deletions in inherited recurrent myoglobinuria, *Ann. Neurol.* **29:** 364–369.

Oldfors, A., Larsson, N.-G., Lindberg, C., and Holme, E., 1993, Mitochondrial DNA deletions in inclusion body myositis, *Brain* **116:**325–336.

Oshima, J., Brown, W. T., and Martin, G. M., 1996, No detectable mutations at Werner helicase locus in progeria, *Lancet* **348:**1106–XXX.

Otsuka, M., Niijima, K., Mizuno, Y., Yoshida, M., Kagawa, Y., and Ohta, S., 1990, Marked decrease of mitochondrial DNA with multiple deletions in a patient with familial mitochondrial myopathy, *Biochem. Biophys. Res. Commun.* **167:**680–685.

Ozawa, T., 1995, Mitochondrial DNA mutations associated with ageing and degenerative disease, *Exp. Gerontol.* **30:**269–290.

Ozawa, T., Katsumata, K., Hayakawa, M., Yoneda, M., Tanaka, M., and Sugiyama, S., 1995, Mitochondrial DNA mutations and survival rate, *Lancet* **345:**189–XXX.

Paquis-Flucklinger, V., Pellissier, J. F., Camboulives, J., Chabrol, B., Saunihres, A., Montfort, M. F., Giudicelli, H., and Desnuelle, C., 1995, Early-onset fatal encephalomyopathy associated with severe mtDNA depletion, *Eur. J. Pediatr.* **154**:557–562.

Parrot, F., Pedespan, L., Moretto, B., Coquet, M., Letellier, T., Negrier, M. L., and Mazat, J. P., 1996, mtDNA depletion in infantile encephalopathy associated with carbohydrate intolerance and fumaricaciduria (abstract), *J. Inherit. Metab. Dis.* **19**(Suppl. 1):73.

Pegoraro, E., Schimke, R. N., Garcia, C., Stern, H., Cadaldini, M., Angelini, C., Barbosa, E., Carroll, J., Marks, W. A., Neville, H. E., Marks, H., Appleton, S., Toriello, H., Wessel, H. B., Donnelly, J., Johnson, P. C., Taber, J. W., Weiss, L., and Hoffman, E. P., 1995, Genetic and biochemical normalization in female carriers of Duchenne muscular dystrophy: Evidence for failure of dystrophin production in dystrophin competent myonuclei, *Neurology* **45**:677–690.

Pegoraro, E., Carelli, V., Zeviani, M., Cortelli, P., Montagna, P., Barboni, P., Angelini, C., and Hoffman, E. P., 1996, X-inactivation patterns in female Leber's hereditary optic neuropathy patients do not support a strong X-linked determinant, *Am. J. Med. Genet.* **61**:356–362.

Pitkänen, S., Merante, F., McLeod, D. R., Applegarth, D., Tong, T., and Robinson, B. H., 1996, Familial cardiomyopathy with cataracts and lactic acidosis: A defect in complex I (NADH-dehydrogenase) of the mitochondria respiratory chain, *Pediatr. Res.* **39**:513–521.

Polymeropoulos, M. H., Swift, R. G., and Swift, M., 1994, Linkage of the gene for Wolfram syndrome to markers on the short arm of chromosome 4, *Nature Genet.* **8**:95–97.

Poulton, J., Morten, K., Freeman-Emmerson, C., Potter, C., Sewry, C., Dubowitz, V., Kidd, H., Stephenson, J., Whitehouse, W., Hansen, F. J., Paris, M., and Brown, G., 1994, Deficiency of the human mitochondrial transcription factor h-mtTFA in infantile mitochondrial myopathy is associated with mtDNA depletion, *Hum. Mol. Genet.* **3**:1763–1769.

Poulton, J., Sewry, C., Potter, C. G., Bourgeron, T., Chretien, D., Wijburg, F. A., Morten, K. J., and Brown, G., 1995, Variation in mitochondrial DNA levels in muscle from normal controls. Is depletion of mtDNA in patients with mitochondrial myopathy a distinct clinical syndrome? *J. Inherit. Metab. Dis.* **18**:4–20.

Puck, J. M., Nussbaum, R. L., and Conley, M. E., 1987, Carrier detection in X-linked severe combined immunodeficiency based on pattern of X-chromosome inactivation, *J. Clin. Invest.* **79**:1395–1400.

Riordan-Eva, P., and Harding, A. E., 1995, Leber's hereditary optic neuropathy: The clinical relevance of different mitochondrial DNA mutations, *J. Med. Genet.* **32**:81–87.

Riordan-Eva, P., Sanders, M., Govan, G. G., Sweeney, M. G., Da Costa, J., and Harding, A. E., 1995, The clinical features of Leber's hereditary optic neuropathy defined by the presence of a pathogenic mitochondrial DNA mutation, *Brain* **118**:319–337.

Saito, H., and Moses, R. E., 1991, Immortalization of Werner syndrome and progeria fibroblasts, *Exp. Cell Res.* **192**:373–379.

Schapira, A. H. V., 1994, Respiratory chain abnormalities in human disease, in *Mitochondria: DNA, Proteins and Disease* (V. Darley-Usmar and A. H. V. Schapira, eds.), pp. 241–278, Portland Press, London.

Schon, E. A., Rizzuto, R., Moraes, C. T., Nakase, H., Zeviani, M., and DiMauro, S., 1989, A direct repeat is a hot spot for large-scale deletions of human mitochondrial DNA, *Science* **244**:346–349.

Seedorf, T., 1985, The inheritance of Leber's disease, *Acta Ophthalmol.* **63**:135–145.

Servidei, S., Zeviani, M., Manfredi, G., Ricci, E., Silvestri, G., Bertini, E., Gellera, C., DiMauro, S., DiDonato, S., and Tonali, P., 1991, Dominantly inherited mitochondrial myopathy with multiple deletions of mitochondrial DNA: Clinical, morphologic, and biochemical studies, *Neurology* **41**:1053–1059.

Shadel, G. S., and Clayton, D. A., 1993, Mitochondrial transcription initiation—Variation and conservation, *J. Biol. Chem.* **268**:16083–16086.

Shoffner, J. M., Brown, M. D., Stugard, C., Jun, A. S., Pollock, S., Haas, R. H., Kaufman, A., Koontz, D., Kim, Y., Graham, J. R., Smith, E., Dixon, J., and Wallace, D. C., 1995, Leber's hereditary optic neuropathy plus dystonia is caused by a mitochondrial DNA point mutation, *Ann. Neurol.* **38**:163–169.

Smith, K. H., Johns, D. R., Heher, K. L., and Miller, N. R., 1993, Heteroplasmy in Leber's hereditary optic neuropathy, *Arch. Ophthalmol.* **111**:1486–1490.

Smith, P. R., Cooper, J. M., Govan, G. G., Harding, A. E., and Schapira, A. H. V., 1994, Platelet mitochondrial function in Leber's hereditary optic neuropathy, *J. Neurol. Sci.* **122**:80–83.

Smith, P. R., Cooper, J. M., Govan, G. G., Riordan-Eva, P., Harding, A. E., and Schapira, A. H. V., 1995, Antibodies to human optic nerve in Leber's hereditary optic neuropathy, *J. Neurol. Sci.* **130**:134–138.

Suomalainen, A., Majander, A., Haltia, M., Somer, H., Lvnnqvist, J., Savontaus, M.-L, and Peltonen, L., 1992a, Multiple deletions of mitochondrial DNA in several tissues of a patient with severe retarded depression and familial progressive external ophthalmoplegia, *J. Clin. Invest.* **90**:61–66.

Suomalainen, A., Paetau, A., Leinonen, H., Majander, A., Peltonen, L., and Somer, H., 1992b, Inherited idiopathic dilated cardiomyopathy with multiple deletions of mitochondrial DNA, *Lancet* **340**:1319–1320.

Suomalainen, A., Kaukonen, J., Amati, P., Timonen, R., Haltia, M., Weissenbach, J., Zeviani, M., Somer, H., and Peltonen, L., 1995, An autosomal locus predisposing to deletions of mitochondrial DNA, *Nature Genet.* **9**:146–151.

Sweeney, K. J., and Weiss, A. S., 1992, Hyaluronic acid in progeria and the aged phenotype? *Gerontology* **38**:139–152.

Sweeney, M. G., Davis, M. B., Lashwood, A., Brockington, M., Toscano, A., and Harding, A. E., 1992, Evidence against an X-linked locus close to DXS7 determining visual loss susceptibility in British and Italian families with Leber hereditary optic neuropathy, *Am. J. Hum. Genet.* **51**:741–748.

Taanman, J.-W., Turina, P., and Capaldi, R. A., 1994, Regulation of cytochrome *c* oxidase by interaction with ATP at two binding sites, one on subunit VIa, *Biochemistry* **33**:11833–11841.

Telerman-Toppet, N., Biarent, D., Bouton, J.-M., De Meirleir, L., Elmer, C., Noel, S., Vamos, E., and DiMauro, S., 1992, Fatal cytochrome *c* oxidase-deficient myopathy in infancy associated with mtDNA depletion. Differential involvement of skeletal muscle and cultured fibroblasts, *J. Inherit. Metab. Dis.* **15**:323–326.

Tiranti, V., Munaro, M., Sandon', D., Lamantea, E., Rimoldi, M., DiDonato, S., Bisson, R., and Zeviani, M., 1995, Nuclear DNA origin of cytochrome *c* oxidase deficiency in Leigh's syndrome: Genetic evidence based on patients'-derived rho^0 transformants, *Hum. Mol. Genet.* **4**:2017–2023.

Tritschler, H.-J., Andreetta, F., Moraes, C. T., Bonilla, E., Arnaudo, E., Danon, M. J., Glass, S., Zelaya, B. M., Vamos, E., Telerman-Toppet, N., Shanske, S., Kadenbach, B., DiMauro, S., and Schon, E. A., 1992, Mitochondrial myopathy of childhood associated with depletion of mitochondrial DNA, *Neurology* **42**:209–217.

Uncini, A., Servidei, S., Silvestri, G., Manfredi, G., Sabatelli, M., DiMuzio, A., Ricci, E., Mirabella, M., DiMauro, S., and Tonali, P., 1994, Ophthalmoplegia, demyelinating neuropathy, leukoencephalopathy, myopathy, and gastrointestinal dysfunction with multiple deletions of mitochondrial DNA: A mitochondrial multisystem disorder in search of a name, *Muscle Nerve* **17**:667–674.

Van Senus, A. H. C., 1963, Leber's disease in the Netherlands, *Doc. Ophthalmol.* **17**:1–162.

Vergani, L., Martinuzzi, A., Carelli, V., Cortelli, P., Montagna, P., Schievano, G., Carrozzo, R., Angelini, C., and Lugaresi, E., 1995, MtDNA mutations associated with Leber's hereditary optic neuropathy: Studies on cytoplasmic hybrid (cybrid) cells, *Biochem. Biophys. Res. Commun.* **210**:880–888.

Vilkki, J., Ott, J., Savontaus, M., Aula, P., and Nikoskelainen, E. K., 1991, Optic atrophy in Leber hereditary optic neuroretinopathy is probably determined by an X-chromosomal gene closely linked to DXS7, *Am. J. Hum. Genet.* **48:**486–491.

Villee, D. B., Nichols, G., and Talbot, N. B., 1969, Metabolic studies in two boys with classical progeria, *Pediatrics* **43:**207–216.

Wallace, D. C., 1970, A new manifestation of Leber's disease and a new explanation for the agency responsible for its unusual pattern of inheritance, *Brain* **93:**121–132.

Wallace, D. C., 1987, Maternal genes: Mitochondrial diseases, *Birth Defects* **23:**137–190.

Wallace, D. C., 1992, Mitochondrial genetics: A paradigm for ageing and degenerative diseases?, *Science* **256:**628–632.

Wallace, D. C., Singh, G., Lott, M. T., Hodge, J. A., Schurr, T., Lezza, A. M. S., Elsas II, L. J., and Nikoskelainen, E. K., 1988, Mitochondrial DNA mutation associated with Leber's hereditary optic neuropathy, *Science* **242:**1427–1430.

Wang, S., Nishigori, C., Yagi, T., and Takebe, H., 1991, Reduced DNA repair in progeria cells and effects of γ-ray irradiation on UV-induced unscheduled DNA synthesis in normal and progeria cells, *Mutat. Res.* **256:**59–66.

Yu, C., Oshima, J., Fu, Y., Wijsman, E. M., Hisama, F., Alisch, R., Matthews, S., Nakura, J., Miki, T., Ouais, S., Martin, G. M., Mulligan, J., and Schellenberg, G. D., 1996, Positional cloning of the Werner's syndrome gene, *Science* **272:**258–262.

Yuzaki, M., Ohkoshi, N., Kanazawa, I., Kagawa, Y., and Ohta, S., 1989, Multiple deletions in mitochondrial DNA at direct repeats of non-D-loop region in cases of familial mitochondrial myopathy, *Biochem. Biophys. Res. Commun.* **164:**1352–1357.

Zeviani, M., 1992, Nucleus-driven mutations of human mitochondrial DNA, *J. Inherit. Metab. Dis.* **15:**456–471.

Zeviani, M., Servidei, S., Gellera, C., Bertini, E., DiMauro, S., and DiDonato, S., 1989, An autosomal dominant disorder with multiple deletions of mitochondrial DNA starting at the D-loop region, *Nature* **339:**309–311.

Zeviani, M., Bresolin, N., Gellera, C., Bordoni, A., Pannacci, M., Amati, P., Moggio, M., Servidei, S., Scarlato, G., and DiDomanto, S., 1990, Nucleus-driven multiple large-scale deletions of the human mitochondrial genome: A new autosomal dominant disease, *Am. J. Hum. Genet.* **47:**907–914.

Zeviani, M., Amati, P., Comi, G., Fratta, G., Mariotti, C., and Tiranti, V., 1995, Searching for genes affecting the structural integrity of the mitochondrial genome, *Biochim. Biophys. Acta* **1271:** 153–158.

Zhang, Y., Marcillat, O., Giulivi, C., Ernster, L., and Davies, J. A., 1990, The oxidative inactivation of mitochondrial electron transport chain components and ATPase, *J. Biol. Chem.* **265:**16330–16336.

Zhu, D., Economou, E. P., Antonarakis, S. E., and Maumenee, I. H., 1992, Mitochondrial DNA mutation and heteroplasmy in type I Leber hereditary optic neuropathy, *Am. J. Med. Genet.* **42:**173–179.

CHAPTER 26

Strategy toward Gene Therapy of Mitochondrial DNA Disorders

Peter Seibel, Martina Schliebs, and Adrian Flierl

1. INTRODUCTION

The major function of mitochondria is to generate energy. Mitochondria fulfill this main task by synthesizing ATP from the energy released during the oxidation of substrates and cofactors, a process known as oxidative phosphorylation. Essential for this process is the respiratory chain, a four-enzyme system composed of multiprotein complexes located at the inner mitochondrial membrane. The complexes are responsible for the redox processes that are accompanied by electron transfer and proton pumping activities leading to a membrane potential across the inner mitochondrial membrane. The entire oxidative phosphorylation system is completed by a fifth complex, the ATP synthase, which converts the membrane potential into ATP by condensing ADP and phosphate. The genes required for this energy-generating system are distributed throughout two genomes—the nuclear and the mitochondrial genome. While mitochondria contain the only source of endogenous extra-chromosomal DNA, its genes contain the information for only 13 polypeptides of the oxidative phosphorylation system (OXPHOS) system as well as a set of 2

Peter Seibel, Martina Schliebs, and Adrian Flierl • Wissenschaftliche Nachwuchs-gruppe, Theodor-Boveri-Institut, 97074 Würzburg, Germany.

Frontiers of Cellular Bioenergetics, edited by Papa *et al* Kluwer Academic/Plenum Publishers, New York, 1999.

rRNAs and 22 tRNAs essential for their expression. The human mtDNA has evolved to show remarkable economy of organization, containing only a short section of noncoding DNA that is known as the D-loop. However, this region does contain sequences important for replication and transcription of the genome.

In most cases, primary genetic lesions linked to mitochondrial encephalo-myopathies have been identified to occur as a defect of the mitochondrial DNA (mtDNA). The clinical spectrum of these diseases is broad, since patient's symptoms can vary from fatal lactic acidosis in infancy to a dementing illness in adulthood (Munnich *et al.*, 1992). Contributing to the variety of clinical symptoms are mainly two factors: the kind of the mutation that occurred as well as its proportion within a cell and distribution throughout different tissues. Since a mitochondrion can contain up to ten copies of its genome and a single cell can contain hundreds of mitochondria, the intracellular number of mtDNA copies greatly varies. A mixture of genetically different genomes is known as heteroplasmy. When a heteroplasmic cell divides, mitochondria harboring mutated genomes are randomly partitioned into the daughter cells. As a result, over many cell divisions, the proportion of mutated genomes can shift either toward pure mutant or toward pure wild-type mtDNA, a process described as mitotic segregation, which can lead in turn again to homoplasmic populations of the mitochondrial DNA (Wallace, 1986). For deleterious mtDNA mutations, as the proportion of mutated mtDNA increases, the mitochondrial ATP-generating capacity declines. Ultimately, as the energy output becomes insufficient to sustain cells vital functions, the functioning of cells and tissues declines and concomitantly the patient's symptoms appear (Wallace, 1992a,b). Depending on their energy demand, different tissues exhibit different bioenergetic thresholds before clinical symptoms become manifest. Therefore, as the percentage of mutated mtDNA focally fluctuates in different tissues, the nature and severity of clinical symptoms can vary within a wide range, with the brain first being affected, since it relies most on oxidative energy supply, subsequently followed by other organ systems, as for instance heart, skeletal muscle, kidney, and the endocrine system (Wallace, 1992a).

While there is strong evidence that a portion of these metabolic disorders is caused by nuclear gene defects, a distinct number of diseases has been linked to mtDNA variations. Wallace and co-workers (1988) were the first to describe the abnormality of the mitochondrial genome that is linked to a maternally inherited disease, known as Leber's hereditary optic neuropathy (LHON). Subsequently, many other mtDNA variations have been reported that are associated with either sporadic or maternally inherited diseases. Genetically, mtDNA variations associated with OXPHOS diseases can be divided into two major subgroups: mutations in genes for proteins and mutations that affect the rate of mitochondrial protein synthesis, for instance, mutation affecting genes for mitochondrial rRNAs and mitochondrial tRNAs.

To date, one of best understood diseases that has been linked to a mitochondrial DNA variation is the myoclonic epilepsy and ragged-red fiber (MERRF) syndrome. Genetically, a single base substitution in the mitochondrial tRNA^{Lys} at position 8344 has been linked to the disease. When the proportion of mutant DNA is high enough, this mutation is associated with the clinically distinct phenotype of MERRF disease, while patients harboring a smaller portion of the mutated DNA present in a milder form, consistent with neurosensory hearing loss and mitochondrial myopathy and ragged-red fibers (Wallace, 1995). At the molecular level, the A-to-G transition mutation causes an alteration in the T-ψ-C loop of the tRNA. Biochemically, the single base replacement leads to a greatly impaired mitochondrial protein synthesis rate in Epstein–Barr virus (EBV) transformed B cells and primary fibroblasts of the patient (Seibel *et al.*, 1991). Further elucidation of this phenomena has been carried out by transferring mitochondria harboring the mutation into a mtDNA-less cell line. Subsequently, Enriquez and co-workers (1995) found that this gene defect leads to a 50–60% reduction in aminoacylation capacity per cell. Although the molecular mechanism is not yet fully understood, it is not surprising, however, that the severe protein synthesis impairment in 8344 mutant cells is linked to a premature termination of translation, near or at each lysine, due to a limited supply of tRNA^{Lys}.

As a main consequence, the rate of synthesis of functional oxidative phosphorylational subunits encoded for by mitochondrial DNA and synthesized in mitochondria is decreased. Thus, they form the rate-limiting components in the assembly of functional enzyme complexes. Hence, OXPHOS activity diminishes, leading to a severe decrease in oxidative energy supply and to a cellular energy crisis.

2. TREATMENT OF OXPHOS DEFECTS

2.1. Current Approaches

Despite the advances that have been made in the diagnosis and characterization of the OXPHOS defects and exploring their pathogenic nature, there is no satisfactory and reliable treatment for the vast majority of the patients. There have been reports of objective improvements in the symptoms of patients with common oxidative phosphorylation defects, as judged by clinical examinations and measurement of biochemical parameters, after treatment with cofactors and oxidizable substrates, activation of pyruvate dehydrogenase, or prevention of damage to mitochondrial membranes by reactive oxygen species. However, it is commonly seen that symptoms of the patients continue to progress, often leading to severe disability and death. While the defects biochemically involve the final reaction centers of the aerobic energy metabolism, the lack of treatment regimens is not surprising, since bypassing this bottleneck by giving alternative energy sources is impossible. Given the limited improvement of patients with mtDNA defects observed after such biochemical treatment, a somatic gene therapy approach as a more permanent treatment for these diseases is highly desirable.

2.2. Somatic Gene Therapy Approaches for mtDNA Diseases

Gene therapy strategies on mitochondrial DNA diseases theoretically can be divided into two categories, namely, nuclear cytoplasmically acting vectors complementing mtDNA defects and mitochondrial acting vectors modulating mtDNA defects (Chrzanowska-Lightowlers *et al.*, 1995). Each gene therapy approach relies strictly on an information transfer into the innermost compartment of mitochondria, where mtDNA is present and biochemical defects occur. While both approaches aim for similar goals, they follow completely different routes, and thus display unique difficulties that need to be overcome for a successful therapy.

2.2.1. Nuclear Cytoplasmically Acting Vectors Complementing mtDNA Defects

While no viral shuttle systems for the nucleic acid transfer directly into mitochondria have been described to date, a variety of viral transfer vectors is known to integrate their own genetic material into cellular systems. In some cases, such as retroviruses, they readily insert their own genes into the human genome. While this natural procedure aims to insert therapeutic genes where they are wanted, misdirected gene integrations as well as functions associated with viral multiplication must be deleted from the shuttle system so that the viral vectors themselves cannot promote diseases. Therefore, the therapeutic gene must be inserted into the viral DNA, so that it is transported across the cell membrane with the entire virus. While the therapeutic gene uses the same shell as the viral DNA to pass into the cell, the DNA payload needs to be modified to match cytoplasmic and mitochondrial needs. Based on the vector system available, gene transfer of a corrected mitochondrial gene into the nucleus, its expression, translation, and finally translocation of the gene product into the mitochondrial matrix seems feasible. Indeed, most of the mitochondrial proteins are nuclear encoded, cytoplasmically synthesized, and translocated into the mitochondrial compartment. The import and submitochondrial localization of many of these proteins have been shown to be mediated by their amino-terminal presequences, which contain targeting and localization signals. The import of precursor proteins involves specific binding to receptor molecules present in the outer mitochondrial membrane (Pfanner *et al.*, 1991), transloca-tion through one or both mitochondrial membranes by a process requiring energy at the step of inner membrane passage, and proteolytic cleavage of the NH_2-terminal leader sequence releasing the mature protein (Glick *et al.*, 1992; Glick and Schatz, 1991; Pfanner and Neupert, 1990; Neupert *et al.*, 1990), unless the leader sequence motif itself is part of the mature protein. While utilizing the common protein import pathway by the corrected artificial fusion proteins seems to be an obvious straightforward approach that is currently pursued by Jacobs and colleagues (Jacobs, 1995), a whole set of difficulties is revealed on a close-up. First, since there is a switch in codon usage from cytoplasm to mitochondria, when expressed in the nucleus and translated on cytoplasmic polysomes, the introduced mitochondrial gene has to be altered to match the cytoplasmic codon usage. Second, the translation product needs to be directed into the mitochondrial matrix or inner membrane. Thus, the altered gene has to be fused to a mitochondrial targeting sequence, thereby acknowl-edging the cleavage motif for the mitochondrial preprotein-processing pep-tidases to ensure a proper proteolytic cleavage. Third, the introduced gene needs to match the expression levels of its mitochondrial counterparts. Thus, a viral promoter to regulate the gene expression is not an optimal choice. Fourth and most important, the overall strategy is limited to DNA variations occurring in mitochondrial protein genes, since only the translated transcripts carry the targeting information capable of directing the protein into its destination com-partment. The major portion of mtDNA diseases, however, is linked to mtDNA rearrangements and single base substitutions in tRNA genes. Hence, the appli-cation of the outlined strategy is not applicable to the vast majority of mtDNA variations.

While there are major complications to realizing this strategy, the poten-tial value for the treatment of mtDNA defects has been demonstrated by Nagley *et al.* (1988) in their investigations on respiratory mutants of yeast. They have been able to demonstrate that an mtDNA defect of the endogenous ATPase-8 gene could be rescued by introducing a vector expressing the appro-priate gene so that a functional ATPase complex could be reestablished. How-ever, it has to be stressed in this context that the yeast mutant may not necessarily provide a good model for predicting the success of the outlined strategy. This is due to the fact that the defect in yeast was the result of the total absence of the protein (knock out). Thus, competition between the mutant and the complementing intact protein played no role for assembly of the functional complex, but could become a contributing factor when applying the approach onto mammalian cells.

2.2.2. Mitochondrially Acting Vectors that Modulate mtDNA Defects

An alternative approach for silencing mtDNA defects could be carried out by importing nucleic acids into mitochondria, thus acting directly at the place where DNA defects occur. The first step for any such approach is the targeting and uptake of nucleic acids across the mitochondrial membrane system. While the outer membrane of mitochondria is freely permeable to molecules of molecular weights below 5000 Da, large biomolecules, such as proteins, require a unique recognition and transportation system. While uptake and utilization of cytoplasmic tRNAs are observed in plants, protozoa, and yeast (Mahapatra *et al.*, 1994; Bordonne *et al.*, 1987; Small *et al.*, 1992; Tarassov *et al.*, 1995a,b), all mammalian mitochondrial tRNAs are supplied by their own genome (Anderson *et al.*, 1981). However, there have been reports of nucleic acids also entering mammalian mitochondria, involving the nuclear-encoded RNA component of the RNA processing enzyme RNAse mitochondrial ribonucleoprotein (MRP) (Schmitt and Clayton, 1993; Clayton, 1991; Hsieh *et al.*, 1990; Chang and Clayton, 1987, 1989; Li *et al.*, 1994). Using an artificial construct, Vestweber and Schatz (1989) have been able to demonstrate that mitochondrial ribonucleoproteins required for RNA processing could indeed enter the matrix via the protein import route. In expanding their experiments, we have been able to develop a nucleic acid shuttle system that takes advantage of the highly efficient and specific protein import machinery of mitochondria (Seibel *et al.*, 1995; Seibel and Seibel, 1994, 1995a,b). DNA coupled covalently to a mitochondrial targeting peptide can mimic cytoplasmically expressed mitochondrial preproteins, so that these chimeric molecules are recognized by the import receptors on the outer mitochondrial surface and finally are translocated into the mitochondrial matrix via the protein import pathway. The tested chimeric molecule was composed of the 32 amino-terminal residues of rat ornithine transcarbamylase (Takiguchi *et al.*, 1987), a mitochondrial matrix enzyme of the urea cycle, previously characterized as imported into mitochondria without an energy requirement on inner membrane passage (Pak and Weiner, 1990). Its carboxy-terminal end was extended by an extra cysteine, necessary for the coupling of peptide and DNA. The covalent linkage between both moieties was mediated by a heterobifunctional cross-linker that reacted with the carboxy-terminal cysteine of the peptide and an artificial amino-group of a synthetic oligonucleotide right in the center of its palindromic sequence. Thus, a loop–stem-like structure (hairpin loop) of the oligonucleotide can be enforced, which upon coupling with the targeting peptide forms an overall linear chimerical molecule (see Fig. 1), capable of directing attached passenger molecules into the matrix of mitochondria. Utilizing the described system, we have been able to show that passenger molecules of up to 3000 base pair (bp) in size are highly efficiently translocated into the matrix of mitochondria (P. Seibel *et al.*, unpublished data).

amino terminus   carboxy terminus   3'-end   5'-end

cross
linker

S

FIGURE 1. Model of the shuttle vector system to traffic nucleic acids into mitochondria.

In contrast to all nuclear cytoplasmically acting vectors, this shuttle system has the advantage of not being limited to mutations occurring in genes for proteins but theoretically being able to overcome any mutation of the mitochondrial genome. Based on the developed nucleic acid transfer system, three different somatic gene therapy approaches can be outlined.

2.2.3. Antisense RNA–DNA Approach to Modulating Mitochondrial Gene Defects

Based on the observation that mutated (deleted) mitochondrial genomes have a replicative advantage over wild-type genomes (Moraes *et al.*, 1992; Yoneda *et al.*, 1992), antisense DNA molecules could in principle be designed that interfere with mtDNA replication such that mutant genome replication is impaired in contrast to wild-type mtDNA replication. Considering the unique way of mtDNA replication, antisense DNA oligonucleotides could be designed that have a high affinity for the mutated mitochondrial genome during the single-stranded replication phase of the mtDNA, so that these oligonucleotides will selectively inhibit subsequent duplication of the mutated genomes, while wild-type DNA replication should be largely unaffected. Once specific targeting of the mutated genomes is achieved, the proportion of wild-type mtDNA relative to the mutated genomes could be promoted. This approach is of utmost interest for diseases associated with heteroplasmic mtDNA deletions, such as adult onset diabetes mellitus, chronic progressive external ophthalmoplegia (CPEO) or Kearns–Sayre's syndrome, since antisense oligodeoxynucleotides can be designed that are highly specific for the fusion breakpoint of the deleted mitochondrial genomes. Hence, by eliminating the replicative advantage of mutated genomes (Yoneda *et al.*, 1992), a segregation of heteroplasmic mtDNA populations toward homoplasmic wild-type mtDNA populations could be induced. While the antisense DNA molecules will most likely not lead to a complete elimination of the defective genome, this therapy approach has to be constantly repeated, if any treatment to be more than a transient phenomenon. Since the mutant mtDNA seems to be functionally recessive, an alternative approach could be to silence transcription of defective genes so that the wild-type gene is favorably processed. Again, this approach would rely on repetitive treatments, since a constant switch of the phenotype is desired.

2.2.4. Gene Addition Approach to Modulating Mitochondrial Gene Defects

Using the translocation vector for a gene addition approach would consist of introducing a construct that is composed of the corrected gene and a mitochondrial promoter to induce subsequent competitive transcription in *trans*. Ideally, the construct would have to be introduced as an artificial plasmid containing the origins of mtDNA replication, thus ensuring coordinated replication together with its endogenous counterparts. Hence, the autonomously replicating plasmid would not lead to a transient phenomena, but could lead in turn to a permanent cure of the disease. As for the modulation of gene defects by the antisense RNA–DNA approach, this approach will most likely not lead to a complete correction of all mitochondria. However, since the corrected gene will be constantly expressed in competing amounts, a permanent switch of phenotype could be achievable, eliminating the disadvantage of repetitive treatments.

2.2.5. Modulation of Gene Defects by a Gene Replacement Approach

Once transfer of the corrected gene into mitochondria has been achieved, a *cis*acting complementation could be considered as well. For *cis* complementation, the introduced DNA would be required to recombine with the mutated DNA, resulting in a restored wild-type mtDNA allele. In fact, recombination events have been reported recently to occur in mammalian mitochondria by Schon and King (1995), although to a very limited extent. Thus, a *cis*-acting complementation would require a highly specific and efficient transfection rate so that the chances of recombination of introduced genes can be enhanced. As for the gene addition approach, repetitive treatments may not be required.

3. OUTLOOK

All strategies described here will most likely not lead to a 100% correction of all defective mitochondria. This in turn is not required, because it has been demonstrated that clinical symptoms only become manifest when the OXPHOS capacity falls below a tissue-specific energetic threshold (Wallace, 1992a). Thus, overcoming the energetic barrier by applying the outlined strategies alone or together might be sufficient to switch phenotype expression. Although mtDNA variations display a heterogeneous group of diseases, gene therapy approaches can be outlined involving either the nuclear or the mitochondrial genome with major differences in the subsequent strategies. Primarily, the sublocalization of the DNA into a specialized compartment of the cytoplasm has been one of the major hurdles in the treatment of mtDNA defects. However, by using the novel mitochondrial shuttle vector system for nucleic acids (Seibel *et al.*, 1995; Seibel and Seibel, 1994, 1995a,b), this can in principle now be overcome.

ACKNOWLEDGMENTS. This work in part was supported by the Deutsche Forschungsgemeinschaft (Se 780/1-1 and Se 780/1-2).

4. REFERENCES

Anderson, S., Bankier, A. T., Barrell, B. G., de Bruijn, M. H., Coulson, A. R., Drouin, J., Eperon, I. C., Nierlich, D. P., Roe, B. A., Sanger, F., Schreier, P. H., Smith, A. J., Staden, R., and Young, I. G., 1981, Sequence and organization of the human mitochondrial genome, *Nature* **290:** 457–465.

Bordonne, R., Bandlow, W., Dirheimer, G., and Martin, R. P., 1987, A single base change in the extra-arm of yeast mitochondrial tyrosine tRNA affects its conformational stability and impairs aminoacylation, *Mol. Gen. Genet.* **206:**498–504.

Chang, D. D., and Clayton, D. A., 1987, A mammalian mitochondrial RNA processing activity contains nucleus-encoded RNA, *Science* **235:**1178–1184.

Chang, D. D., and Clayton, D. A., 1989, Mouse RNAase MRP RNA is encoded by a nuclear gene and contains a decamer sequence complementary to a conserved region of mitochondrial RNA substrate, *Cell* **56:**131–139.

Chrzanowska-Lightowlers, Z. M., Lightowlers, R. N., and Turnbull, D. M., 1995, Gene therapy for mitochondrial DNA defects: Is it possible?, *Gene Ther.* **2:**311–316.

Clayton, D. A., 1991, Nuclear gadgets in mitochondrial DNA replication and transcription, *Trends Biochem. Sci.* **16:**107–111.

Enriquez, J. A., Chomyn, A., and Attardi, G., 1995, MtDNA mutation in MERRF syndrome causes defective aminoacylation of tRNA(Lys) and premature translation termination, *Nat. Genet.* **10:**47–55.

Glick, B., and Schatz, G., 1991, Import of proteins into mitochondria, *Annu. Rev. Genet.* **25:**21–44.

Glick, B. S., Beasley, E. M., and Schatz, G., 1992, Protein sorting in mitochondria, *Trends Biochem. Sci.* **17:**453–459.

Hsieh, C. L., Donlon, T. A., Darras, B. T., Chang, D. D., Topper, J. N., Clayton, D. A., and Francke, U., 1990, The gene for the RNA component of the mitochondrial RNA-processing endoribonuclease is located on human chromosome 9p and on mouse chromosome 4, *Genomics* **6:**540–544.

Jacobs, H. T., 1995, Allotropic expression of mitochondrial genes in the nucleus: Towards a gene therapy strategy for mitochondrial diseases, EUROMIT III, Abstractbook, p. 9.

Li, K., Smagula, C. S., Parsons, W. J., Richardson, J. A., Gonzalez, M., Hagler, H. K., and Williams, R. S., 1994, Subcellular partitioning of MRP RNA assessed by ultrastructural and biochemical analysis, *J. Cell Biol.* **124:**871–882.

Mahapatra, S., Ghosh, T., and Adhya, S., 1994, Import of small RNAs into leishmania mitochondria *in vitro*, *Nucleic Acids Res.* **22:**3381–3386.

Moraes, C. T., Ricci, E., Petruzzella, V., Shanske, S., DiMauro, S., Schon, E. A., and Bonilla, E., 1992, Molecular analysis of the muscle pathology associated with mitochondrial DNA deletions, *Nat. Genet.* **1:**359–367.

Munnich, A., Rustin, P., Rotig, A., Chretien, D., Bonnefont, J. P., Nuttin, C., Cormier, V., Vassault, A., Parvy, P., and Bardet, J., 1992, Clinical aspects of mitochondrial disorders, *J. Inherit. Metab. Dis.* **15**:448–455.

Nagley, P., Farrell, L. B., Gearing, D. P., Nero, D., Meltzer, S., and Devenish, R. J., 1988, Assembly of functional proton-translocating ATPase complex in yeast mitochondria with cyto-plasmically synthesized subunit 8, a polypeptide normally encoded within the organelle, *Proc. Natl. Acad. Sci. USA* **85**:2091–2095.

Neupert, W., Hartl, F. U., Craig, E. A., and Pfanner, N., 1990, How do polypeptides cross the mitochondrial membranes? *Cell* **63**:447–450.

Pak, Y. K., and Weiner, H., 1990, Import of chemically synthesized signal peptides into rat liver mitochondria, *J. Biol. Chem.* **265**:14298–14307.

Pfanner, N., and Neupert, W., 1990, The mitochondrial protein import apparatus, *Annu. Rev. Biochem.* **59**:331–353.

Pfanner, N., Söllner, T., and Neupert, W., 1991, Mitochondrial import receptors for precursor proteins, *Trends Biochem. Sci.* **16**:63–67.

Schmitt, M. E., and Clayton, D. A., 1993, Nuclear RNase MRP is required for correct processing of pre-5.8S rRNA in *Saccharomyces cerevisiae*, *Mol. Cell Biol.* **13**:7935–7941.

Schon, E. A., and King, M. P., 1995, Physical communication between mammalian mitochondria: A genetic approach, EUROMIT III, Abstractbook, p. 7.

Seibel, P., and Seibel, A., 1994, Chimäres Peptid-Nukleinsäure-Fragment, Verfahren zu seiner Herstellung und Verfahren zur zielgerichteten Nukleinsäureeinbringung in Zellorganellen und Zellen, German Patent P 44 21 079.

Seibel, P., and Seibel, A., 1995a, Replikatives und transkriptionsaktives Peptid-Nukleinsäure-plasmid, sowie sein Verwendung zur Einbringung in Zellen und Zellorganellen, German Patent P 195 208 15.

Seibel, P., and Seibel, A., 1995b, Chimäres Peptid-Nukleinsäure-Fragment, Verfahren zu seiner Herstellung, sowie seine Verwendung zur zielgerichteten Nukleinsäureeinbringung in Zellorganellen und Zellen, International Patent PCT DE 95/00775.

Seibel, P., Degoul, F., Bonne, G., Romero, N., Francois, D., Paturneau Jouas, M., Ziegler, F., Eymard, B., Fardeau, M., Marsac, C., and Kadenbach, B., 1991, Genetic biochemical and pathophysiological characterization of a familial mitochondrial encephalomyopathy (MERRF), *J. Neurol. Sci.* **105**:217–224.

Seibel, P., Trappe, J., Villani, G., Klopstock, T., Papa, S., and Reichmann, H., 1995, Transfection of mitochondria: Strategy towards a gene therapy of mitochondrial DNA diseases, *Nucleic Acids Res.* **23**:10–17.

Small, I., Marechal Drouard, L., Masson, J., Pelletier, G., Cosset, A., Weil, J.H., and Dietrich, A., 1992, *In vivo* import of a normal or mutagenized heterologous transfer RNA into the mito-chondria of transgenic plants: Towards novel ways of influencing mitochondrial gene expres-sion? *EMBO J.* **11**:1291–1296.

Takiguchi, M., Murakami, T., Miura, S., and Mori, M., 1987, Structure of the rat ornithine carbamoyltransferase gene, a large, X chromosome-linked gene with an atypical promoter, *Proc. Natl. Acad. Sci. USA* **84**:6136–6140.

Tarassov, I., Entelis, N., and Martin, R. P., 1995a, Mitochondrial import of a cytoplasmic lysine–tRNA in yeast is mediated by cooperation of cytoplasmic and mitochondrial lysyl–tRNA synthetases, *EMBO J.* **14**:3461–3471.

Tarassov, I., Entelis, N., and Martin, R. P., 1995b, An intact protein translocating machinery is required for mitochondrial import of a yeast cytoplasmic tRNA, *J. Mol. Biol.* **245**:315–323.

Vestweber, D., and Schatz, G., 1989, DNA-protein conjugates can enter mitochondria via the protein import pathway, *Nature* **338**:170–172.

Wallace, D. C., 1986, Mitotic segregation of mitochondrial DNAs in human cell hybrids and expression of chloramphenicol resistance, *Somat. Cell Mol. Genet.* **12**:41–49.

Wallace, D. C., 1992a, Diseases of the mitochondrial DNA, *Annu. Rev. Biochem.* **61:**1175–1212.

Wallace, D. C., 1992b, Mitochondrial genetics: A paradigm for aging and degenerative diseases? *Science* **256:**628–632.

Wallace, D. C., 1995, 1994 William Allan Award Address. Mitochondrial DNA variation in human evolution, degenerative disease, and aging, *Am. J. Hum. Genet.* **57:**201–223.

Wallace, D. C., Singh, G., Lott, M. T., Hodge, J. A., Schurr, T. G., Lezza, A. M., Elsas, L. J., and Nikoskelainen, E. K., 1988, Mitochondrial DNA mutation associated with Leber's hereditary optic neuropathy, *Science* **242:**1427–1430.

Yoneda, M., Chomyn, A., Martinuzzi, A., Hurko, O., and Attardi, G., 1992, Marked replicative advantage of human mtDNA carrying a point mutation that causes the MELAS encephalomyopathy, *Proc. Natl. Acad. Sci. USA* **89:**11164–11168.

CHAPTER 27

The F_0F_1-ATP Synthase in Cell Proliferation and Aging

Ferruccio Guerrieri

1. INTRODUCTION

In normal conditions, the cellular ATP is provided mainly by the mitochondrial oxidative phosphorylation (OXPHOS), which is an integrated process between energy production by the respiratory chain and energy utilization by the F_0F_1-ATP synthase complex of the inner mitochondrial membrane, with the remainder met by glycolysis. Oxidative phosphorylation capacity and the ATP needs vary from tissue to tissue. The brain is the organ with the highest demand for aerobic ATP, but heart muscle and kidney also have a high oxidative phosphorylation capacity (Wallace, 1992). The contribution of OXPHOS to the cellular energy demand changes, however, in the life span and in response to the physical activity (Papa, 1996).

In fetal tissues (Valcarce *et al.*, 1988; Izquierdo *et al.*, 1990) and under some pathophysiological conditions, for example, in rapidly growing tumors (Uriel, 1979), in regenerating liver (Uriel, 1979; Buckle *et al.*, 1986; Guerrieri *et al.*, 1994, 1995), during myocardial ischemia (Rouslin, 1983; Rouslin and Broge, 1989; Rouslin *et al.*, 1995), and in various tissues from aged mammals

Ferruccio Guerrieri • Institute of Medical Biochemistry and Chemistry, and Centre for the Study of Mitochondria and Energy Metabolism, University of Bari, 70124 Bari, Italy.

Frontiers of Cellular Bioenergetics, edited by Papa *et al.* Kluwer Academic/Plenum Publishers, New York, 1999.

(Nohl and Kramer, 1980; Hansford, 1983; Guerrieri *et al.*, 1992a), the contribution of mitochondrial oxidative phosphorylation to cellular energy decreases. Morphological (Müller-Höcker, 1992) and biochemical (Byrne *et al.*, 1991; Cooper *et al.*, 1992; Torii *et al.*, 1992; Boffoli *et al.*, 1994, 1996) studies show a progressive decline, with age, of the mitochondrial respiratory system in human tissues.

The mitochondrial enzyme system of oxidative phosphorylation consists of five oligomeric protein complexes: four (complex I, II, III, and IV) concern the respiratory chain and the fifth is the F_0F_1-ATP synthase complex, which utilizes the transmembrane proton gradient generated by the respiratory chain to form ATP from ADP and Pi.

This chapter reviews the changes in structure and function of mitochondrial F_0F_1-ATP synthase in heart and liver of aged rats (see also Guerrieri *et al.*, 1992a,b, 1996; Capozza *et al.*, 1994) and during cell proliferation using liver regeneration after partial hepatectomy as a model (see also Buckle *et al.*, 1986; Guerrieri *et al.*, 1994, 1995). In particular the possible causative effect of reactive oxygen species (ROS) is examined.

2. AGE-RELATED CHANGES OF MITOCHONDRIAL F_0F_1-ATP SYNTHASE

In rat heart and liver mitochondria, no age-dependent changes in the basal respiratory rate (absence of ADP) is observed (Capozza *et al.*, 1994; Guerrieri *et al.*, 1996). The ADP-stimulated respiratory rate (state 3) changes with the age, decreasing in the mitochondria of aged rats as compared to that of adult rats (Capozza *et al.*, 1994, Guerrieri *et al.*, 1996), as a consequence the respiratory control index (RCI) decreases in mitochondria of aged rats (Guerrieri *et al.*, 1996). Both the age-dependent decreases of the RCI and the respiratory rate in state 3 are more pronounced in heart than in liver mitochondria (Capozza *et al.*, 1994; Guerrieri *et al.*, 1996).

Direct measurements of the rate of ATP synthesis in succinate-supplemented isolated mitochondria, using a trapping system (presence of an excess of glucose plus hexokinase) for the ATP produced by the mitochondrial ATP synthase (Papa *et al.*, 1969), in mitochondria isolated from heart and liver of aged rats, show rates lower than in mitochondria isolated from young rats (Table I). No change in the ~P–O ratio is observed (Capozza *et al.*, 1994). The mitochondrial synthesis of ATP is a complex process that depends on (1) the activity of the respiratory chain, which generates the transmembrane proton gradient; (2) the integrity of the mitochondrial inner membrane toward proton impermeability; (3) the activity of carriers for Pi and adenine nucleotides; and (4) the activity of the F_0F_1-ATP synthase.

Table I
Age-Related Changes of Mitochondrial F$_0$F$_1$-ATP Synthase Activities in Rat Liver and Rat Heart

	Liver		Heart	
	3 months	24 months	3 months	24 months
ATP synthesis	268 ± 11	262 ± 6	440 ± 13	323 ± 7
ATP hydrolysis	1.57 ± 0.1	1.43 ± 0.14	2.10 ± 0.14	1.10 ± 0.1*
H$^+$ permeability	1.43 ± 0.13	1.66 ± 0.11	0.52 ± 0.04	1.19 ± 0.07*

[a]The data reported are the means of six experiments ± SEM. ATP synthesis (expressed as nmole ATP/min per mg protein) in mitochondria and ATP hydrolysis (expressed as μmole ATP/min per mg protein) in ESMP were measuered as reported by Guerrieri *et al.* (1996). H$^+$ permeability was determined by following the oligomycin-sensitivity anaerobic release of the transmembrane proton gradient setup by respiratory chain in succinate supplemented ESMP (see Guerrieri *et al.*, 1992a) and is expressed as reciprocal value of the $t_{1/2}$ of the process (sec^{-1}). The asterisks indicate significant differences in old/young rats at $p < 0.05$.

Age-dependent declines in the activities of complexes I, II, and IV of the respiratory chain (Ferrándiz *et al.*, 1994; Boffoli *et al.*, 1994, 1996; Paradies *et al.*, 1994), in the content of cytochrome *c* oxidase (Müller-Höcker, 1992) as well as in the activities of the adenine nucleotide carrier (Nohl and Kramer, 1980; Paradies *et al.*, 1994) and of the Pi carrier (Paradies *et al.*, 1992) have already been observed. These age-dependent changes have been attributed either to increased mutated mitochondrial DNA (mtDNA) (Chapters 28 and 31, this volume; Kadenbach *et al.*, 1995; Papa, 1996) or to altered composition of mitochondrial lipid membranes, in particular of the cardiolipin levels (Paradies *et al.*, 1992, 1994).

The possibility that age-dependent alterations of mitochondrial F$_0$F$_1$-ATP synthase could directly contribute to the age-dependent decline of synthesis of ATP has been verified by analysis of oligomycin-sensitive ATP hydrolase activity of this enzyme in inside-out submitochondrial particles (ESMP), which directly expose the catalytic F$_1$ sector of the complex to the suspension medium (Lee and Ernster, 1968). This activity of the F$_0$F$_1$-ATP synthase is age dependent, showing an increase in ESMP isolated from heart, liver, and brain of adult rat as compared to ESMP isolated from the same tissues of young rats and a decrease in ESMP isolated from senescent rats as compared to those isolated from adult rats (Fig. 1) (see also Guerrieri *et al.*, 1994). The decline is much more pronounced in ESMP isolated from brain and heart than from liver.

The anaerobic release of transmembrane proton gradient set up by the respiratory chain in these particles is 70–80% inhibited by the F$_0$ inhibitor oligomycin (Pansini *et al.*, 1978; Guerrieri *et al.*, 1992a), indicating that this process mainly represents the proton permeability through the F$_0$ sector. In

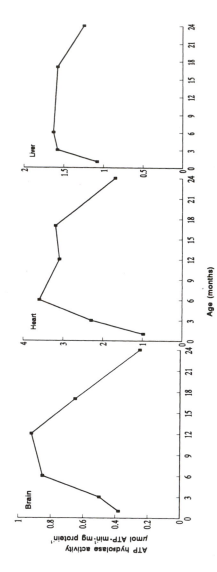

FIGURE 1. Age dependence of ATP hydrolase activity of ESMP from rat brain, rat heart, and rat liver. For experimental details, see Guerrieri et al. (1992a,b).

ESMP prepared from mitochondria of aged rats, the anaerobic release of the transmembrane proton gradient was faster than in ESMP from mitochondria of young animals (Table I), as it is in F_1-depleted submitochondrial particles (Pansini *et al.*, 1978; Guerrieri *et al.*, 1993).

Immunoblot analysis, using an antiserum against beef-heart-isolated F_1 (Guerrieri *et al.*, 1989), showed that the age-related changes in activities of ATP synthase were accompanied by parallel changes in the amount of immuno-detected β subunit of the complex (Fig. 2), which were more pronounced in heart than in liver mitochondria. These and other observations (Guerrieri *et al.*, 1992a, 1993, 1996; Capozza *et al.*, 1994) show that in the inner mitochondrial membrane of various tissues of aged rats the content of F_0F_1 complex is lower than in the inner mitochondrial membrane of adult rats. This causes a decrease in the capacity of the OXPHOS process in mitochondria of aged rats.

3. POSSIBLE INVOLVEMENT OF ROS IN AGE-RELATED ALTERATIONS OF MITOCHONDRIAL F₀F₁-ATP SYNTHASE

The free radical theory of aging (Harman, 1956) postulates that the age-linked cellular decline is caused by a progressive damage of cellular macro-molecules, such as DNA, lipids, and proteins, by accumulation of ROS, which are produced essentially at the level of the respiratory chain (Chance *et al.*, 1979; Papa, 1996). Since mitochondria are directly involved in the production of ROS, one should expect that the mitochondrial structures are easily altered if insufficient scavengers or antioxidant mechanisms are present.

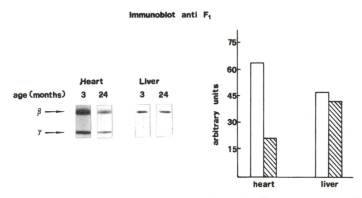

FIGURE 2. Immunoblots of heart or liver mitochondria isolated from young (3-month-old) or aged (24-month-old) rats. For isolation of mitochondria, preparation of antiserum against isolated bovine F_1, immunoblot and densitometric analysis (at 590 nm), see Guerrieri *et al.* (1992a,b). Semiquantitative analysis by densitometry is reported on the right panel. ☐, Mitochondria from 3-month-old rats; ▨, mitochondria from 24-month-old rats.

The most important natural antioxidant and scavenger is glutathione (GSH) (Viña, 1990; Richie, 1992), a tripeptide that is synthesized in cytosol and imported in mitochondria, where it represents the main antioxidant system (Viña, 1990). Age-dependent changes in the intramitochondrial content of GSH have been observed both in heart and liver (Capozza *et al.*, 1994). However, the age-dependent decrease of intramitochondrial GSH was much more significant in heart than in liver (Capozza *et al.*, 1994). *In vitro* exposure of mitochondria, isolated from heart and liver of young rats (3 month old rats), to ROS generated by ^{60}Co-irradiation (Zhang *et al.*, 1990; Capozza *et al.*, 1994; Guerrieri *et al.*, 1996) causes an increase of mitochondrial malondialdehyde (MDA) production (Table II), which is an index of increased lipid peroxidation by ROS. The increase of MDA production is higher in ^{60}Co-irradiated heart mitochondria than in ^{60}Co-irradiated liver mitochondria (Table II). This is accompanied by a decrease of mitochondrial content of GSH related to the decrease in the rate of ATP synthesis (Table II). These ^{60}Co-induced changes are much more pronounced in heart than in liver mitochondria (Table II).

It has been reported that oxidative stress of isolated rat liver mitochondria by ter-butyl-hydroperoxide causes a binding of GSH to free thiol groups of proteins, which is related to the loss of mitochondrial functions (Olafsdottir *et al.*, 1988). Similarly, oxidative stress induced by ^{60}Co-irradiation of isolated mitochondrial increases the level of GSH bound to thiol groups of intramitochondrial membrane proteins (Table II). Direct exposure of sonic submitochondrial particles (ESMP) to ^{60}Co-irradiation causes damage of the structure and function of mitochondrial F_0F_1-ATP synthase (Guerrieri *et al.*, 1993, 1996; Capozza *et al.*, 1994).

In conclusion, the *in vitro* experiments on exposure of isolated mitochondria to ^{60}Co-irradiations support the hypothesis that age-dependent changes

Table II
Effect of [60]Co Irradiation of Isolated Rat Liver and Rat Heart Mitochondria[a]

	Liver		Heart	
	Control	[60]Co (3000 rad)	Control	[60]Co (3000 rad)
MDA	1.10 ± 0.15	1.97 ± 0.20	1.58 ± 0.18	6.07 ± 0.31
GSH	5.67 ± 0.30	4.10 ± 0.35	4.15 ± 0.27	1.45 ± 0.21
ATP synthesis	268 ± 11	200 ± 23	440 ± 13	120 ± 18
GS-S protein	0.09 ± 0.01	0.31 ± 0.02	0.10 ± 0.02	0.39 ± 0.01

[a]For preparation of mitochondria, ^{60}Co irradiation, determination of rate of ATP synthesis (nmole ATP/min per mg protein), GSH level (nmole/mg protein), and content of GS-S protein (nmole GSH bound/mg protein), see Guerrieri *et al.* (1996). For determination of malonaldehyde (nmole/mg protein), see Slater and Sawyer (1971). The data reported are the means of six experiments ± SEM.

could be related to free radical damage of cellular structures and, in particular, of mitochondrial components. This can be favored by a progressive decrease of cellular (Viña, 1990) and mitochondrial GSH (Capozza *et al.*, 1994) observed during aging.

The increase of GSH bound to mitochondrial membrane proteins following the ^{60}Co-irradiation of isolated mitochondria seems to suggest the possibility that accumulation of ROS in mitochondria causes the formation of protein-S-radicals in membrane proteins directly exposed to the mitochondrial matrix (i.e., F$_1$ subunit of ATP synthase complex) forming denatured proteins (i.e., GS-S protein complexes), which can represent a better substrate for mitochondrial proteases (Goldberg, 1992). The difference between liver and heart mitochondria suggests that the rapidly replicating liver cell and organelles are better protected from ROS damage and aging process (Miquel and Fleming, 1986).

4. MITOCHONDRIAL ENERGY METABOLISM DURING LIVER REGENERATION

Liver regeneration can be easily experimentally induced by any treatment, surgical or chemical, that will remove or kill a large percentage of the parenchyma (Bucher and Malt, 1971). Loss of the parenchyma rapidly induces cell proliferation until the original total mass is restored (Michalopoulos, 1990). The process is characterized by two different phases: (1) a prereplicative phase, in which the cells of the remaining liver show drastic metabolic changes; and (2) a proliferative phase, characterized by an active DNA synthesis (Michalopoulos, 1990). Although the process has been known for a long time (Michalopoulos, 1990), the exact mechanism by which liver regeneration is initiated and modulated is still not very well known.

It has been suggested that one of the possible events that initially triggers liver regeneration is early changes in cellular energy metabolism (Ove *et al.*, 1967; Ngala-Kenda *et al.*, 1984; Skullman *et al.*, 1991). During liver regeneration the demand for cellular energy is increased to support the biosynthesis of cellular components (Lai *et al.*, 1992; Bláha *et al.*, 1992; Nagino *et al.*, 1989; Tsai *et al.*, 1992). Figure 3 shows that the time course of liver regeneration after partial hepatectomy (PH) is sigmoidal, with a lag in the early phase. The lag is age dependent, being longer in aged rats (Guerrieri *et al.*, 1994). Mitochondria isolated from the liver during this early phase of liver regeneration have an RCI that progressively lowers until the minimum is reached 15–24 hr after PH (Fig. 3) (see also Guerrieri *et al.*, 1994, 1995). After this phase, the recovery of the liver mass is accompanied by the recovery of the respiratory control index (Fig. 3) (see also Guerrieri *et al.*, 1994, 1995).

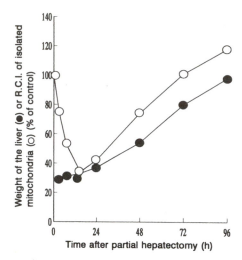

FIGURE 3. Time course of recovery of liver mass (●) and of respiratory control index (○) after partial hepatectomy. The mass of the liver is expressed as a percentage of the weight of the liver of sham-operated rats (12.3 ± 1 g). The respiratory control index (RCI) is the ratio between the rate of oxygen consumption of succinate-supplemented mitochondria in state 3 and the rate in state 4 and is expressed as percentage of RCI in the mitochondria from sham-operated rats (5.85 ± 0.21). For other experimental details, see Guerrieri *et al.* (1995).

Changes in RCI are an index of altered mitochondrial oxidative phosphorylation. During liver regeneration, direct measurements of the rate of ATP synthesis, in succinate-supplemented mitochondria, showed a pattern similar to that observed for RCI (Guerrieri *et al.*, 1995). A similar pattern was also observed for oligomycin-sensitive ATP hydrolase activity in ESMP (Buckle *et al.*, 1986; Guerrieri *et al.*, 1992b, 1994), suggesting that the changes in mitochondrial energy metabolism observed during liver regeneration can be due to alteration in the activity of F_0F_1-ATP synthase.

Immunoblot analyses using polyclonal antibodies against β-F_1 subunit (Fig. 4) (see also Guerrieri *et al.*, 1995) or F_0F_1-PVP protein (Guerrieri *et al.*, 1995) show a decrease of the mitochondrial content of these subunits in the early phase of liver regeneration (Fig. 4) (see also Guerrieri *et al.*, 1995). No decrease in the level of transcripts for β-F_1 subunits is observed in this phase (Fig. 4) (see also Guerrieri *et al.*, 1995). Twenty-four hours after PH, a progressive recovery of the intramitochondrial level of β-F_1 subunit starts. This recovery is parallel with an increase of the level of β-F_1 mRNA until, at 96 hr after PH, both intramitochondrial content of the protein and the levels of its transcripts become equal to the control (Fig. 4).

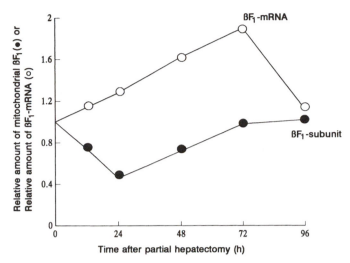

FIGURE 4. Time course of changes of immunodetected β-F₁ subunit in liver mitochondria (●) and the level of its mRNA (○), after partial hepatectomy. The mRNA was quantified by densitometric analysis of Northern blot hybridization of total in mRNA to a mRNA probe specific for β-F₁ subunit. For experimental details, see Guerrieri et al. (1995).

The fact that the decrease of the intramitochondrial content of β-F₁ subunit observed in the early phase of liver regeneration was not synchronous with a decrease in the cellular level of its mRNA suggests that during the early phase of liver regeneration an alteration of the turnover of β-F₁ subunit occurs, related to either a defect of the translation process or to an enhanced proteolytic breakdown.

5. POSSIBLE INVOLVEMENT OF ROS IN THE ALTERATION OF MITOCHONDRIAL F₀F₁-ATP SYNTHASE DURING THE EARLY PHASE OF LIVER REGENERATION

It has been reported that free radical damage of the cellular structure occurs during the early phase of liver regeneration (Tsai *et al.*, 1992; Steer, 1995; Rastogi *et al.*, 1995). Figure 5 shows that in the early phase of liver regeneration, when the decrease of activity and content of F₀F₁-ATP synthase occurs (Guerrieri *et al.*, 1995), a sharp increase of mitochondrial MDA production and a decrease of mitochondrial GSH (see also Vendemiale *et al.*, 1995) is observed. 18 hr after PH, both phenomena tend to return to control levels (Fig. 5). No significant change in the content of mitochondrial oxidized glutathione

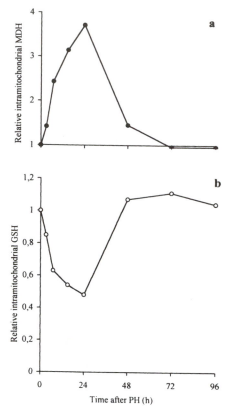

FIGURE 5. Time course of changes in MDA (●) and of GSH (○) contents in liver mitochondria after partial hepatectomy. MDA was detected spectrometrically, as reported by Vendemiale *et al.* (1996). GSH was measured enzymatically as reported by Vendemiale *et al.* (1995). The data reported in the figure are the ratio between the mitochondrial level of MDA or GSH at various times after partial hepatectomy and their levels in liver mitochondria from sham-operated rats that were 1.4 ± 0.15 nmole MDA/mg protein and 6.1 ± 0.3 nmole GSH/mg protein, respectively.

(GSSG) is observed during liver regeneration (Vendemiale *et al.*, 1995). Early liver regeneration is characterized also by a decrease of free thiol groups of membrane-bound proteins and an increase of glutathione bound by a disulfide bridge to membrane mitochondrial proteins (Fig. 6). Twenty-four hours after PH, both free proteic SH groups and glutathione bound to proteic thiol groups tended to approach progressively the values for control contents (Fig. 6).

The decrease of mitochondrial GSH observed during the early phase of liver regeneration appears to be linearly related to the decrease of immuno-

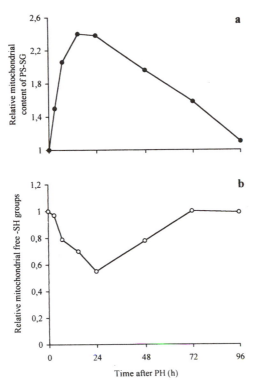

FIGURE 6. Time course of changes of GSH bound to mitochondrial proteins (●) and free-SH groups in liver mitochondria after partial hepatectomy (○). Free-SH groups were determined by the Elmann procedure. GSH bound to mitochondrial proteins was determined as reported by Guerrieri *et al.* (1996). The data reported on the figure are the ratio between the mitochondrial level of free thiol groups or of GSH bound to mitochondrial proteins at various times after partial hepatectomy and their levels in liver mitochondria isolated from sham-operated rats, that were 3.4 ± 0.2 nmole-SH/mg protein and 100 ± 12 pmole of Pro S-SG/mg protein, respectively.

detected catalytic β-F_1 subunits (Vendemiale *et al.*, 1995). These data indicate that the initial prereplicative phase of liver regeneration is characterized by an increase of mitochondrial ROS production that can be related to surgical stress and/or to the decreased blood flow in the liver during surgical partial hepatectomy. This ROS increase is synchronous with decrease of mitochondrial defense (decrease of GSH) and an increase of oxidized proteins. This can be related to the altered function of the F_0F_1-ATP synthase activity during this early phase of liver regeneration. In fact, the enzyme, whose F_1 moiety is largely exposed to the mitochondrial matrix, can be modified by ROS attack and the modified proteins can be actively degraded by the mitochondrial proteases (Goldberg, 1992).

In the replicative phase characterized by DNA synthesis (Steer, 1995), the ROS-modified mitochondrial proteins are progressively replaced by newly synthesized proteins, so that the normal mitochondrial function and the normal level of mitochondrial GSH, following the increase of the cytosolic GSH (Vendemiale $et\ al.$, 1995), are restored.

6. CONCLUDING REMARKS

Common features of various tissues from aged animals and of rapidly growing tissues (i.e., fetal tissues, regenerating liver, tumors) are the high glycolytic activity and the low mitochondrial oxidative phosphorylation. This is related to alterations of structure and function of mitochondrial F_0F_1-ATP synthase complex (Papa, 1996; Papa $et\ al.$, 1997). The F_0F_1-ATP synthase of inner mitochondrial membrane is made up of a peripheral catalytic sector (F_1 moiety), consisting of five subunits, and a membrane integral sector (F_0 moiety), consisting of ten subunits. Two F_0 subunits (ATPase 6 and A6L) are encoded by the mitochondrial genoma (Attardi and Schatz, 1988), with the others encoded by nuclear genes.

The biogenesis of the F_0F_1-ATP synthase depends on the concerted expression of nuclear and mitochondrial genes (Attardi and Schatz, 1988). Synthesis of the nuclear-encoded subunits of F_1 sector of the F_0F_1-ATP synthase appears to control the increased phosphorylation capacity of mitochondria after birth (Valcarce $et\ al.$, 1988; Izquierdo $et\ al.$, 1990). Therefore, altered turnover (i.e., synthesis and/or proteolysis) of F_0F_1-ATP synthase subunits could contribute to the changes in the oxidative phosphorylation capacity in tissues of aged animals and in cell proliferation.

The data reviewed in the present chapter show that mitochondria from senescent rats are characterized by a decrease in F_1 content in mitochondria from various tissues. The decrease of F_0F_1-ATP synthase subunits in mitochondria from aged animals appears to be associated with decrease of intramitochondrial GSH. Both phenomena—decrease of F_0F_1-ATP synthase subunits and decrease of GSH—seem to be related to oxidative damage of the mitochondrial proteins, which can result in an increase of proteolysis of damaged proteins (Goldberg, 1992; Stadtman, 1992).

The oxidative damage of F_0F_1-ATP synthase is much more pronounced in heart than in liver mitochondria from aged rats, suggesting that the liver mitochondria are better protected from ROS damage and the aging process (Miquel and Fleming, 1986). Oxidative damage of F_0F_1-ATP synthase can also be observed in the early phase of liver regeneration, characterized by retro-differentiation of hepatocytes, which change from an oxidative to a fermentative metabolism (Uriel, 1979). The question remains open whether the ROS damage of F$_0$F$_1$-ATP synthase observed in the early phase of liver regeneration is simply an accidental event bound to the surgical stress of hepatectomy or if it interferes or even plays a direct role in cellular signal transduction for cell proliferation (Ammendola *et al.*, 1995).

ACKNOWLEDGMENTS. I would like to thank Prof. Sergio Papa for helpful discussions and suggestions. The part of the reviewed work conducted in this laboratory was supported in part by M.U.R.S.T., Italy (40% and 60 %) and by C.N.R. and was run in collaboration with Drs. P. Cantatore, G. Capozza, I. Grattagliano, and G. Vendemiale. The help of Drs. E. Adorisio and G. Pellecchia for preparing some of the figures is acknowledged.

7. REFERENCES

Ammendola, R., Fiore, F., Esposito, F., Caserta, G., Mesuraca, M., Russo, T., and Cimino, F., 1995, Differentially expressed mRNAs as a consequence of oxidative stress in intact cells, *FEBS Lett.* **371:**209–213.

Attardi, G., and Schatz, G., 1988, Biogenesis of mitochondria, *Annu. Rev. Cell. Biol.* **4:**289–333.

Bláha, V., Simek, J., and Zadák, Z., 1992, Liver regeneration in partially hepatectomized rats infused with carnitine and lipids, *Exp. Toxic Pathol.* **44:**165–168.

Boffoli, D., Scacco, S. C., Vergari, R., Solarino, G., Santacroce, G., and Papa, S., 1994 Decline with age of the respiratory chain activity in human skeletal muscle, *Biochim. Biophys. Acta* **1226:**73–82.

Boffoli, D., Scacco, S. C., Vergari, R., Persio, M. T., Solarino, G., Laforgia, R., and Papa, S., 1996, Ageing is associated in females with a decline in the content and activity of the bc_1 complex in skeletal muscle mitochondria, *Biochim. Biophys. Acta* **1315:**66–72.

Bucher, N. R. L., and Malt, R. A., 1971, Regeneration of liver and kidney, in *Thirty Years of Liver Regeneration: A Distillate* (N. L. R. Bucker, ed.), pp. 15–26, Little, Brown and Co., Boston.

Buckle, M., Guerrieri, F., Pazienza, A., and Papa, S., 1986, Studies on polypeptide composition, hydrolytic activity and proton conduction of mitochondrial F$_0$F$_1$ H$^+$-ATPase in regenerating rat liver. *Eur. J. Biochem.* **155:**439–445.

Byrne, E., Dennet, X., and Trounce, I., 1991, Oxidative energy failure in fixed post-mitotic cells: A major factor in senescence, *Rev. Neurol.* **147:**6–7, 532–535.

Capozza, G., Guerrieri, F., Vendemiale, G., Altomare, E., and Papa, S., 1994, Age related changes of the mitochondrial energy metabolism in rat liver and heart, *Arch. Gerontol. Geriatr. Suppl.* **4:**31–38.

Chance, B., Sies, H., and Boveris, A., 1979, Hydroperoxide metabolism in mammalian organs, *Physiol. Rev.* **59:**527–605.

Cooper, J. M., Mann, V. M., and Schapira, A. V. H., 1992, Analyses of mitochondrial respiratory chain function and mitochondrial DNA deletion in human skeletal muscle: Effect of ageing, *J. Neurol. Sci.* **113:**91–98.

Ferrándiz, M.L., Martínez, M., De Juan, E., Díez, A., Bastos, G., and Miquel, J., 1994, Impairment of mitochondrial oxidative phosphorylation in the brain of aged mice, *Brain Res.* **644:** 335–338.

Goldberg, A. L., 1992, The mechanism and functions of ATP-dependent proteases in bacterial and animal cells, *Eur. J. Biochem.* **203:**9–23.

Guerrieri, F., Kopecky, J., and Zanotti, F., 1989, Functional and immunological characterization of mitochondrial F_0F_1-ATP synthase, in *Organelles in Eukaryotic Cells: Molecular Structure and Interactions* (J. M. Tager, A. Azzi, and S. Papa, eds.), pp. 197–208, Plenum, New York, London.

Guerrieri, F., Capozza, G., Kalous, M., Zanotti, F., Drahota, Z., and Papa, S., 1992a, Age-dependent changes in the mitochondrial F_0F_1-ATP synthase, *Arch. Geront. Geriatr.* **14:**299–308.

Guerrieri, F., Capozza, G., Kalous, M., and Papa, S., 1992b, Age-dependent changes in the mitochondrial F_0F_1-ATP synthase, *Ann. NY Acad. Sci.* **671:**395–402.

Guerrieri, F., Capozza, G., Fratello, A., Zanotti, F., and Papa, S., 1993, Functional and molecular changes in F_0F_1 ATP-synthase of cardiac muscle during aging, *Cardioscience* **4:**93–98.

Guerrieri, F., Kalous, M., Capozza, G., Muolo, L., Drahota, Z., and Papa, S., 1994, Age dependent changes in mitochondrial F_0F_1-ATP synthase in regenerating rat-liver, *Biochem. Molec. Biol. Int.* **33:**117–129.

Guerrieri, F., Muolo, L., Cocco, T., Capozza, G., Turturro, N., Cantatore, P., and Papa, S., 1995, Correlation between rat liver regeneration and mitochondrial energy metabolism, *Biochim. Biophys. Acta* **1272:**95–100.

Guerrieri, F., Vendemiale, G., Turturro, N., Fratello, A., Furio, A., Muolo, L., Grattagliano, I., and Papa, S., 1996, Alteration of mitochondrial F_0F_1-ATP synthase during aging, *Ann. NY Acad. Sci.* **786:**62–71.

Hansford, R. G., 1983, Bioenergetics in aging, *Biochim. Biophys. Acta* **726:**41–80.

Harman, D., 1956, Aging: A theory based on free radical and radiation chemistry, *J. Gerontol.* **11:**298–300.

Izquierdo, J. M., Luis, A. M., and Cuezva, J. M., 1990, Postnatal mitochondrial differentiation in rat-liver, *J. Biol. Chem.* **265:**9090–9097.

Kadenbach, B., Münscher, C., Frank, V., Müller-Höcker, J., and Napiwotzki, J., 1995, Human aging is associated with stochastic somatic mutations of mitochondrial DNA, *Mutat. Res.* **338:**161–172.

Lai, H. S., Chen, W. J., and Chen, K. M., 1992, Energy substrate for liver regeneration after partial hepatectomy in rats: Effects of glucose vs fat, *J. Parenter. Enter. Nutr.* **16:**152–156.

Lee, C. P., and Ernster, L., 1968, Studies of the energy transfer system of submitochondrial particles. Effects of oligomycin and aurovertin, *Eur. J. Biochem.* **3:**391–409.

Michalopoulos, G. K., 1990, Liver regeneration: Molecular mechanisms of growth control, *FASEB J.* **4:**176–187.

Miquel, J., and Fleming, J., 1986, Theoretical and experimental support for an oxygen radical mitochondrial injury. Hypothesis of cell aging, in *Free Radicals, Aging and Degenerative Diseases* (J. E. Johnson, R. Walford, D. Harman, and J. Miquel, eds.), pp. 51–74, Alan Liss, New York.

Müller-Höcker, J., 1992, Mitochondria and ageing, *Brain Pathol.* **2:**149–152.

Nagino, M., Tanaka, M., Nishikimi, M., Nimura, Y., Kubota, H., Kanai, M., Kato, T., and Ozawa, T., 1989, Stimulated rat liver mitochondrial biogenesis after partial hepatectomy, *Cancer Res.* **49:**4913–4918.

Ngala-Kenda, J. F., de Hamptinne, B., and Lambotte, L., 1984, Role of metabolic overload in the initiation of DNA synthesis following partial hepatectomy in the rat, *Eur. Surg. Res.* **16:** 294–302.

Nohl, H., and Kramer, R., 1980, Molecular basis of age-dependent changes in the activity of adenine nucleotide translocase, *Mech. Ageing Dev.* **14:**137–144.

Olafsdottir, K., Pascoe, G. A., and Reed, D. J., 1988, Mitochondrial glutathione status during Ca^{2+} ionophore-induced injury to isolated hepatocytes, *Arch. Biochem. Biophys.* **263**:226–235.

Ove, P., Takai, S., Umeda, T., and Lieberman, I., 1967, Adenosine triphosphate in liver after partial hepatectomy and acute stress, *J. Biol. Chem.* **242**:4963–4971.

Pansini, A., Guerrieri, F., and Papa, S., 1978, Control of proton conduction by the H$^+$-ATPase in the inner mitochondrial membrane, *Eur. J. Biochem.* **92**:545–551.

Papa, S., 1996, Mitochondrial oxidative phosphorylation changes in the life span. Molecular aspects and physiopathological implications, *Biochim. Biophys. Acta* **1276**:87–105.

Papa, S., Tager, J. M., Guerrieri, F., and Quagliariello, E., 1969, Effect of monovalent cations on oxidative phosphorylation in submitochondrial particles, *Biochim. Biophys. Acta* **172**: 194–186.

Papa, S., Guerrieri, F., Capuano, F., and Zanotti F., 1998, The mitochondrial ATP synthase in normal and neoplastic cell growth, in *Cell Growth and Oncogenesis* (P. Bannash, D. Kanduc, S. Papa, and J. M. Tager, eds.), pp. 31–46, Birkhäuser Verlag, Basel.

Paradies, G., Ruggiero, F. M., Gadaleta, M. N., and Quagliariello, E., 1992, The effect of aging and acetyl-L-carnitine on the activity of the phosphate carrier and on the phospholipid composition in rat heart mitochondria, *Biochim. Biophys. Acta* **1103**:324–326.

Paradies, G., Ruggiero, F. M., Petrosillo, G., Gadaleta, M. N., and Quagliariello, E., 1994, Effect of aging and acetyl-L-carnitine on the activity of cytochrome oxidase and adenine nucleotide translocase in rat heart mitochondria, *FEBS Lett.* **350**:213–215.

Rastogi, R., Saksena, S., Garg, N. K., and Dhawan, B. N., 1995, Effect of picroliv on antioxidant-system in liver of rats, after partial hepatectomy, *Phytother. Res.* **9**:364–367.

Richie, J. P., Jr., 1992, The role of glutathione in aging and cancer, *Exp. Gerontol.* **27**:615–626.

Rouslin, W., 1983, Protonic inhibition of the mitochondrial oligomycin-sensitive adenosine 5′-triphosphatase in ischemic and autolyzing cardiac muscle, *J. Biol. Chem.* **258**:9657–9661.

Rouslin, W., and Broge, C. W., 1989, Regulation of mitochondrial matrix pH and adenosine 5′-triphosphatase activity during ischemic in slow heart-rate hearts, *J. Biol. Chem.* **264**: 15224–15229.

Rouslin, W., Broge, C. W., Guerrieri, F., and Capozza, G., 1995, ATPase activity, IF$_1$ content and proton conductivity of ESMP from control and ischemic slow and fast heart-rate hearts, *J. Bioenerget. Biomembr.* **27**:459–466.

Skullman, S., Ihse, I., and Larsson, J., 1991, Availability of energy substrates during liver regeneration in malnourished rats, *Scand. J. Gastroenterol.* **26**:1152–1156.

Slater, T., and Sawyer, B., 1971, The stimulatory effect of carbon tetrachloride and other halogeno-alkanes on peroxidative reactions in rat liver fractions *in vitro*, *Biochem. J.* **123**:805–814.

Stadtman, E. R., 1992, Protein oxidation and aging, *Science* **257**:1220–1224.

Steer, C. J., 1995, Liver regeneration, *FASEB J.* **9**:1396–1400.

Torii, K., Sugiyama, S., Tagagi, K., Satake, T., and Ozawa, T., 1992, Age-related increase in respiratory muscle mitochondrial function in rats, *Am. J. Cell. Mol. Biol.* **6**:88–92.

Tsai, I. L., King, K. L., Chang, C. C., and Wei, Y., 1992, Changes of mitochondrial respiratory functions and superoxide dismutase activity during liver regeneration, *Biochem. Int.* **28**: 205–217.

Uriel, J., 1979, Retrodifferentation and the fetal patterns of gene expression in cancer, *Adv. Cancer Res.* **29**:127–174.

Valcarce, C., Navarete, R. M., Encabo, P., Loeches, E., Satrùstegui, J., and Cuezva, J. M., 1988, Postnatal development of rat liver mitochondrial function. The roles of protein synthesis and adenine nucleotides, *J. Biol. Chem.* **263**:7767–7775.

Vendemiale, G., Guerrieri, F., Grattagliano, I., Didonna, D., Muolo, L., and Altomare, E., 1995, Mitochondrial oxidative phosphorylation and intracellular glutathione compartmentation during rat liver regeneration, *Hepatology* **21**:1450–1454.

Vendemiale, G., Grattagliano, I., Altomare, E., Turturro, N., and Guerrieri, F., 1996, Effect of acetaminophen administration on hepatic glutathione compartmentation and mitochondrial energy metabolism in the rat, *Biochem. Pharmacol.* **52:**1147–1154.

Viña, J., ed., 1990, *Glutathione: Metabolism and Physiological Functions*, CRC Press, Boston.

Wallace, D. C., 1992, Diseases of the mitochondrial DNA, *Annu. Rev. Biochem.* **61:**1175–1212.

Zhang, Y., Marcillat, O., Gulivi, C., Ernster, L., and Davies, J. A., 1990, The oxidative inactivation of mitochondrial electron transport chain components and ATPase, *J. Biol. Chem.* **265:**16330–16336.

CHAPTER 28

Age-Linked Changes in the Genotype and Phenotype of Mitochondria

Maria N. Gadaleta, Bernhard Kadenbach,
Angela M. S. Lezza, Annette Reith, Palmiro Cantatore,
Domenico Boffoli, and Sergio Papa

1. INTRODUCTION

Photophosphorylation and oxidative phosphorylation (OXPHOS) are the main energy-generating systems of higher organisms. During the evolution of eukaryotic cells, the energy (ATP)-producing enzyme complexes related to the production of oxygen—photosynthetic phosphorylation in chloroplasts—as well as to the consumption of oxygen—OXPHOS in mitochondria—became encoded partly by nuclear DNA (nDNA) and partly by extranuclear DNA, the chloroplast DNA and the mitochondrial DNA (mtDNA), respectively. Mammalian mitochondrial DNA (mtDNA) is a molecule of about 16,000 nucleotide pairs (np). It codes for two rRNAs, 22 tRNAs, and 13 proteins, which represent exclusively subunits of the four proton-pumping enzyme complexes of OXPHOS:

Maria N. Gadaleta, Angela M. S. Lezza, and Palmiro Cantatore • Department of Biochemistry and Molecular Biology, University of Bari, 70125 Bari, Italy. **Bernhard Kadenbach and Annette Reith** • Fachbereich Chemie, Philipps-Universität, D-35032 Marburg, Germany. **Domenico Boffoli and Sergio Papa** • Institute of Medical Biochemistry and Chemistry, University of Bari, I-70124 Bari, Italy.

Frontiers of Cellular Bioenergetics, edited by Papa *et al.* Kluwer Academic/Plenum Publishers, New York, 1999.

seven subunits of NADH-dehydrogenase (complex I), one of cytochrome c reductase (complex III), three of cytochrome c oxidase (complex IV), and two of ATP synthase (complex V) (Wallace *et al.*, 1995). Since all eukaryotic cells synthesize ATP also by glycolysis, the enzyme complexes involved in OXPHOS, and thus the extrachromosomal genome (mtDNA), are not essential for the survival of individual cells. In fact, animal cells lacking mtDNA but still containing mitochondria (rho^0 cells) (King and Attardi, 1989) or containing pathogenetic mutations of mtDNA (Chomyn *et al.*, 1991) can be kept in culture indefinitely. The life of an animal and the proper function of its organs, however, are inevitably dependent on mitochondrial OXPHOS.

Aging is a complex phenomenon characterized by energy deficiencies that phenotypically appear in humans as muscular weakness, presbycardia, memory loss, presbycusis, and reduced ocular functions. It has been hypothesized that these deficits might be due to mitochondrial OXPHOS defects (Linnane *et al.*, 1989; Wallace, 1992a). In this chapter, after a brief introduction on the characteristics of mtDNA and of the mitochondrial genetic system, a review of the literature data from other laboratories as well as from our own and some new data on age-linked changes of the genotypic and phenotypic features of mitochondrial OXPHOS enzyme complexes will be presented.

2. THE MITOCHONDRIAL GENETIC SYSTEM

2.1. The mtDNA

Animal mtDNA is an extremely compact molecule. The genes are mostly present on one strand (the heavy strand, H), are close to each other, and contain only exons, with short overlappings. Only one noncoding region is present on mtDNA: the displacement loop (D-loop) region. The D-loop region is characterized by a short triple-stranded structure containing a small sequence of newly synthesized H-strand DNA, the 7S DNA, which displaces the parental H-strand. The D-loop is the major regulatory region of mtDNA and includes the origin of replication for the H-strand (Ori-H) and the transcription promoters for both H- and L-strands (HSP and LSP).

MtDNA replication occurs as a bidirectional, asynchronous process. It has two separate and distinct replication origins, Ori-H in the D-loop region and Ori-L localized 10 kilobases (kb) clockwise from Ori-H, in a small cluster of tRNAs. Replication starts at the unidirectional Ori-H utilizing a RNA primer (7S RNA) transcribed from the LSP and proceeds until the H-strand has been completed for two thirds of the replication. Initiation of the L-strand synthesis occurs after Ori-L is exposed on the displaced strand, leading to the synthesis of two identical double-stranded molecules (Attardi and Schatz, 1988).

The mtDNA transcription is complete and symmetric. Although both strands are transcribed, most of the processed transcripts derive from the H-strand; the L-strand, in fact, codifies only for one mRNA and eight tRNAs. The mitochondrial transcription mechanism includes two transcription units for the H-strand (one for the two rRNAs and two tRNAs and another for the entire strand and one unit for the L-strand. The transcription initiation sites are localized in the D-loop region, close to Ori-H. The mtRNAs are synthesized as high-molecular-weight precursors that are processed by a RNase-P-like endo-ribonuclease. The processing signals are presumably due to the cloverleaf structures of the tRNAs functioning as punctuation among the different genes. Mitochondrial mRNAs show some important characteristics: they lack non-coding sequences both at the 5' and 3' ends and in most cases they also lack a complete stop codon. The mRNAs, ending with U or UA, present a complete stop codon only after polyadenylation (Cantatore and Saccone, 1987). Replication and transcription of mtDNA are closely linked since the 7S RNA produced by the L-strand promoter functions as primer for the H-strand replication (Attardi and Schatz, 1988).

The mitochondrial transcripts are translated in the mitochondria by a complete protein synthesis apparatus made by mtDNA-encoded rRNAs and tRNAs and nuclear DNA-encoded ribosomal proteins. The mitochondrial mRNAs are translated by a genetic code slightly different from the "universal" one. This prevents their translation in the cytosol (Attardi and Schatz, 1988). Mitochondrial RNA and DNA polymerases as well as proteins involved in the regulation of mtDNA replication, transcription, and translation are coded for by the nuclear genome, synthesized in the cytoplasm, and transported into the mitochondria.

Mitochondrial gene expression can be regulated at the level of replication, transcription, processing, and translation. In the last few years, several factors controlling mtDNA replication have been identified. They include a human γ mtDNA polymerase (Gray and Wong, 1992), a single-stranded DNA-binding protein (SSB) (Tiranti et al., 1993), an ATP-dependent DNA-helicase (Hehman and Hauswirth, 1992), a protein that specifically binds to the termination sequence of the D-loop (TAS binding-protein) (Madsen et al., 1993), an RNA polymerase (Kelly et al., 1986), and a ribonucleoprotein (MRP-RNase), presumably involved in the processing of the RNA primers for replication (Clayton, 1991) (Table I). The human mtDNA polymerase gene has been cloned (Ropp and Copeland, 1996). As far as human mtRNA polymerase, the gene has been recently identified (Tiranti et al., 1997; Chapter 29, this volume).

Regulation at the level of mtDNA transcription may occur in several cases, for example, during development (Cantatore et al., 1986; Renis et al., 1989; Ostronoff et al., 1995) or under the control of hormones (Gadaleta et al., 1985; Mutvei et al., 1989; Wiesner et al., 1992; Demonacos et al., 1996). The

Table I
Proteins Involved in the Replication of Mammalian Mitochondrial DNA

Protein	Function
DNA polymerase γ	Catalyzes the mitochondrial DNA synthesis
RNA polymerase	Catalyzes the synthesis of RNA primers needed for the H-strand replication
mtTFA	Activates RNA synthesis and causes topological alterations of mitochondrial DNA
MRP-RNase	Processes the replication primers
DNA helicase	Catalyzes, using ATP, the progressive vectorial unwinding of duplex DNA
SSB	Binds single-stranded DNA, keeping it in this conformation during replication
TAS-binding protein	Binds double-stranded DNA, causing the termination of the D-loop expansion.

regulation may involve changes in the efficiency of promoters or in the stability of mtRNAs.

mtDNA-binding proteins that affect mtRNA synthesis have been characterized in mammals. One, mtTFA, was at first reported to stimulate H- and L-strand transcription by binding upstream the two H- and L-strand initiation sites (Parisi and Clayton, 1991), whereas the other, mTERF, causes the termination of the ribosomal transcription unit, binding to a region of the tRNA[Lev(UUR)] gene located immediately downstream the 3′ end of the 16S rRNA gene (Daga *et al.*, 1993). Further studies on the role of mtTFA showed that this protein belongs to a class of proteins called HMG (high-mobility group) that has the property to cause structural alterations of the bound DNA (Clayton, 1991). In agreement with this observation, *in vitro* studies showed that mtTFA was able to bend, wrap, and unwind the mtDNA (Fisher *et al.*, 1992). Moreover, studies by Ghivizzani *et al.* (1994) showed that this protein was also able to bind the entire regulatory region of human mtDNA, in a phased fashion without sequence specificity, suggesting the specific packaging of the mtDNA control region *in vivo* as primary a function of human mtTFA. Interestingly, the amount of mtTFA protein but not of the DNA polymerase has been reported to correlate with the number of copies of mtDNA per organelle (Davis *et al.*, 1996). Finally, down-regulation of mtTFA during spermatogenesis in humans has been reported, which could have implications for the understanding of maternal transmission of mtDNA (Larsson *et al.*, 1997).

Recently, another DNA-binding protein was described in *Xenopus laevis* mitochondria (Antoshechkin and Bogenhagen, 1995). This factor (mtTFB) seems similar to the yeast transcription-activating factor mtTFB. The factor does not recognize a specific sequence in mtDNA, but allows initiation of RNA synthesis at a much higher rate than mtTFA. It is conceivable that a homologue of mtTFB may exist in mammals. Cantatore *et al.* (1995) reported multiple protein contact sites in the rat mtDNA D-loop region; protein contact sites in the rat mtDNA ND2 region (NADH-DH subunit 2) and a profound helix distortion in the nearest Ori-L region were also reported by Cingolani *et al.* (1997). G. Gadaleta *et al.* (1996) purified from rat liver a protein that, on the basis of its partial sequence, has been identified as rat mtTFA (G. Gadaleta, personal communication); it contacts the Ori-L and the ATPase6-CoxIII regions of rat liver mtDNA (G. Gadaleta, personal communication). Furthermore, *in organello* footprinting experiments in rat liver showed the existence of a single-stranded DNA structure in correspondence of the main transition site from the RNA primer to the H-strand DNA (Cantatore *et al.*, 1995). This result suggests an alternative mechanism for the generation of the 3′ end of the RNA primer that instead of being formed by a RNA processing activity (Clayton, 1991) should be due to a RNA polymerase pause, which would generate the mtDNA open structure at conserved sequence block-I (CSB-I). Regulation at the level of mtDNA translation probably constitutes a final adjustment of mitochondrial gene expression. It likely operates at the level of mRNA utilization and might be involved in providing the proper stoichiometry of the respiratory complexes (Cantatore and Saccone, 1987; Attardi *et al.*, 1990).

Recently, two nuclear respiratory factors (NRF-1 and NRF-2) have been characterized that enhance not only the synthesis of the respiratory complexes subunits encoded by the nucleus, but also the synthesis of nuclear gene products such as mtTFA and MRP RNase, thus possibly regulating the maintenance of mtDNA and the coordinate expression of the two genetic systems (see Chapter 22, this volume). The NRF factors might receive signals from the cytoplasm on the energy state of the cell and then they might pass them to the nuclear transcription apparatus (Allen, 1993). The nucleus–cytoplasm interrelationship is an active field of research today.

2.2. Characteristics of the Mitochondrial Genetic System

The mitochondrial genetic system has peculiar features with respect to the nuclear genetic system (Table II) (DiMauro and Wallace, 1993). The mtDNA is maternally inherited as all the zygote mtDNA derives from the oocyte. However, a very small contribution from the father has been observed (Gyllensten *et al.*, 1991). The mtDNA is present in a large number of copies per mitochondrion (from 1 to 10). If this number is multiplied for the average number of mitochondria in a somatic cell, which is about 1000, it is clear that the information carried by the mtDNA is present in tens of thousands of copies per cell (polyploidy). The mtDNA molecules in one mitochondrion or in one cell

Table II
Characteristics of the Mitochondrial Genetic System

Maternal inheritance

Polyploidy (homoplasmy, heteroplasmy)

Mitotic segregation: it implies that the phenotypic expression of a mutation can change with time
   (during development or during life) and with space (between cells or tissues, mosaicism)

High mutation rate

Threshold effect: 10% wild-type mtDNA seems enough to ensure its function

---

can be all of the same type, wild type or mutant (homoplasmy), or can be partly mutant and partly wild type (heteroplasmy).

When a cell replicates, the mitochondria and the mitochondrial genomes are distributed randomly between the daughter cells (replicative, mitotic, or meiotic segregation); therefore, in the case of a cell containing heteroplasmic mutant molecules, the random distribution of such molecules to daughter cells will generate homoplasmic (all wild type or all mutant) or heteroplasmic (wild type plus mutant) mitochondrial genotypes in these cells. Therefore, the percentage of mutant mtDNA molecules per cell in the same individual may change both along time (during embryonal development or during the life span) or along space (between cells or tissues, mosaicism). From this point of view, mitochondrial genetics reminds population genetics.

The mtDNA replicates autonomously, independently of the nDNA. MtDNA synthesis does not appear to be restricted to any particular phase of the cell cycle (Clayton, 1991). MtDNA replicates also in terminally differentiated, postmitotic cells such as muscle cells and neurons (Wallace, 1992a).

The mtDNA has a high mutational rate. In fact, the mtDNA is "naked," that is, unlike nDNA, it is not protected by histonelike or nonhistone proteins. Therefore, it is exposed to the mutagenic effects of genotoxic compounds or to the oxygen radicals generated by the OXPHOS system; furthermore, the repair of the mtDNA seems to be less efficient than that of the nDNA (Wallace, 1992a; Richter, 1995).

MtDNA exhibits a threshold effect due to mtDNA polyploidy. Given the high copy number of mtDNA per cell, it is important to know which is the minimum percentage of mutant molecules that can lead to a dysfunction in the mitochondrial respiratory apparatus. This value is strictly related to the kind of tissue and to the age of individuals. A tissue highly dependent on the oxidative metabolism will have a lower bioenergetic threshold than other tissues (Wallace, 1992a). The percentage of mutant molecules able to induce a mutated phenotype depends also on the age of the individuals. In this case the effects of the mtDNA mutations will add to other types of damage already present at the mitochondrial and/or cellular level, such as peroxidation of membrane lipids, cross-linking between polypeptides of respiratory complexes, and/or cross-linking inside mtDNA molecules (Wallace *et al.*, 1995).

3. THE MITOCHONDRIAL OXPHOS SYSTEM AND AGING

In mitochondria, where 90% of the cellular oxygen is consumed, from 1 to 4% of this oxygen is converted into reactive oxygen species (ROS). The first species, produced by complexes I, II, and III of the respiratory chain is the O_2^- free radical anion. The superoxide anion, O_2^- is then converted to H_2O_2 by the mitochondrial and cytosolic superoxide dismutases. Since catalase, which cleaves H_2O_2 into H_2O and O_2, is confined to peroxisomes but is absent in the mitochondrial matrix (see, however, Radi *et al.*, 1991), the H_2O_2 not consumed there by glutathione peroxidase is converted by the Fenton reaction into the hydroxyl radical $OH^•$, which is the most aggressive ROS for DNA, proteins, and lipids. For this reason O_2^- and H_2O_2, produced in the mitochondrial matrix, are very deleterious for the mitochondria.

In his paper "The Biological Clock: the Mitochondria?" Harman (1972), the father of the free radical theory of aging, suggested that

> the maximal life span of a given mammalian species is largely an expression of genetic control over the rate of oxygen utilization. The latter determines the rate of accumulation of mitochondrial damage produced by free radical reactions, the rate increasing with the rate of oxygen consumption, which ultimately causes death. (p. 145)

The relationship between oxidative stress in mitochondria and animal longevity has been extensively studied. The production of H_2O_2 in the mitochondria has been demonstrated to be inversely proportional to the maximum life span potential (MLSP) of many animal species (Sohal *et al.*, 1990). Orr and Sohal (1994) showed that the *Drosophila melanogaster* double transgenic for the CuZn-dependent superoxide dismutase and catalase exhibited as much as one third extension of life span over the control, a longer mortality rate doubling time, a lower amount of protein oxidative damage, and a delayed loss in physical performance. Rats on caloric restricted diet are longer-lived and show much less oxidative damage than rats on a normal diet (Yu, 1995).

In the mitochondria, like in the cytosol, enzyme scavengers such as Mn-dependent superoxide dismutase, Se-dependent glutathione peroxidase, and nonenzymatic antioxidants such as vitamins C and E, glutathione, and ubiquinol-10 are present (Richter, 1995). It is generally accepted, however, that in mitochondria, with increasing age, a condition of oxidative stress occurs in which the defense capacities against ROS become insufficient (Shigenaga *et al.*, 1994).

Age-linked changes in the mitochondrial energetic metabolism have been studied in animals and humans by means of different parameters: activities of the complexes of the OXPHOS system, content of cytochromes, O_2 consumption in the presence of different substrates, substrate transport and membrane lipid composition (Boffoli *et al.*, 1994, 1996; Papa, 1996). Studies carried out in various laboratories have shown that in tissues of different mammals (Torii *et al.*, 1992; Petruzzella *et al.*, 1992; Guerrieri *et al.*, 1992; Bowling *et al.*, 1993) as well as in humans (for a review, see Papa, 1996) there is an age-linked decrease in the activities of the mitochondrial respiratory enzymes and F_0F_1 ATP synthase. The general decline of the OXPHOS capacity is associated with the appearance in senescent tissues of marked structural changes in mitochondria, like enlargement, matrix vacuolization, shortened cristae, and so on. Since only a part of these damaged mitochondria (light mitochondria as compared to normal heavy mitochondria) can be recovered in the isolation procedure, it is possible that differences in the respiratory and OXPHOS capacities of mitochondria isolated from old versus young animals are underestimated by selective loss of damaged organelles. This may be one reason for the apparent lack of age-associated biochemical changes in mitochondria reported by some authors (Hansford, 1983; Zucchini *et al.*, 1995).

The most extensively screened tissue for biochemical measurements is the human skeletal muscle because of the relative ease of obtaining biopsy samples from subjects exempt from clinical symptoms of mitochondrial diseases. The biochemical analysis of these samples has shown that aging is associated with a decline of the mitochondrial respiration (Trounce *et al.*, 1989; Cooper *et al.*, 1992; Hsie *et al.*, 1994; Lezza *et al.*, 1994; Boffoli *et al.*, 1994, 1996).

An epidemiological study of the activities of respiratory enzymes and ATP synthase in mitochondria isolated from biopsy samples of the proximal vastus lateralis muscle has been carried out by Papa and co-workers (Boffoli *et al.*, 1994, 1996; Papa, 1996) in "normal" humans (those exempt from overt clinical symptoms of neuromuscular disorders), ranging in age from a few years to about 90 years. The large number of screened subjects, about 200, has allowed the authors to examine the sex dependence of the age-linked pattern of enzymatic activities and cytochrome contents. Plots of the enzyme activities and cytochrome contents as a function of age show the measured values to be quite scattered for both sexes and at each age; see, for example, the activity of cytochrome *c* oxidase in mitochondria (Fig. 1). There is evidently a large variability of genotype and phenotype for OXPHOS enzymes in humans. This large variability can encompass cases of mitochondrial diseases that could have escaped standard clinical inspection. The possibility emerging from these observations is that individuals, with activities of OXPHOS enzymes that are borderline with respect to overt mitochondrial diseases (i.e., Lezza *et al.*, 1994), are relatively frequent and their cases should be seriously considered. These

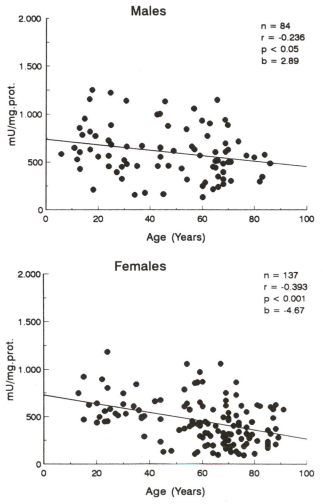

FIGURE 1. Statistical analysis of the activity of cytochrome c oxidase (complex IV) in human skeletal muscle mitochondria as a function of age in males and females.

situations, which might be without pathological consequences under normal conditions, could result in acute crisis of the energetic capacity of tissues and sudden organ failure when the individual attempts an intense physical effort or is put under stress.

The variability of mitochondrial respiratory capacity can also reflect differences in lifestyle, such as dietary habits, smoking, physical activity, and exposure to infective agents and to toxic substances. Statistical analysis of the data as a function of age shows, however, a linear negative correlation for the various activities of the OXPHOS enzymes, with a few exceptions discussed later (Boffoli *et al.*, 1994, 1996; Papa, 1996). Evidently, various genetic and acquired factors contribute to the overall decline of OXPHOS observed in the population. Barrientos *et al.* (1996), for example, have reported observations indicating that the tobacco consumption and sedentary habit can contribute to the decline of skeletal muscle respiratory activities in humans.

With age, a significant decrease of the mitochondrial respiratory rates with pyruvate plus malate, succinate and ascorbate plus TMPD occurs both in males and females (Table III); the decline with all three respiratory substrates appears to be more marked in males than in females. Aging both in males and females is associated with a decline in the activity of NADH–cytochrome *c* reductase (complex I plus III) and succinate–cytochrome *c* reductase (complex II plus III) (Table IV). In females the activity of quinol–cytochrome *c* reductase (complex III) exhibits a marked decrease with age that is not observed in males (Table IV). Thus, the decrease with age of the NADH–cytochrome *c* reductase and succinate–cytochrome *c* reductase, which is observed in males, is due to a decline in the activity of NADH–ubiquinone reductase (complex I) and succinate dehydrogenase (complex II), respectively. In females, too, the activity of complex I, which represents the rate-limiting step for NADH–cytochrome *c* reductase, apparently declines with age (Table IV). The activity of cytochrome *c* oxidase (complex IV), determined both by polarographic measurements of oxygen consumption (Table III) and spectrophotometric measurements of

Table III
Statistical Data of State III-Coupled Respiratory Activities in Human Skeletal Muscle Mitochondria from Subjects of Age Varying from a Few Years to around 90 Years[a]

	n	a	b	r	p
Pyruvate + malate					
Males	79	212	−1.42	−0.318	< 0.01
Females	121	162	−0.80	−0.208	< 0.05
Succinate					
Males	76	294	−2.30	−0.452	< 0.001
Females	123	264	−1.72	−0.345	< 0.001
Ascorbate + TMPD					
Males	81	1253	−6.94	−0.303	< 0.01
Females	121	805	−3.80	−0.265	< 0.001

[a]*n*, number of subjects; *a*, respiratory activities at the intercept with the ordinate of their linear plots as a function of age; *b*, slope, change per year expressed in ng atom O_2/min per mg protein; *r*, correlation coefficient; *p*, statistical significance.

Table IV

Statistical Data of Mitochondrial Enzymatic Activities and Cytochrome Contents in Human Skeletal Muscle Mitochondria from Subjects of Age Varying from a Few Years to around 90 Years[a]

	n	a	b	r	p
NADh-cyt.c reductase					
Males	70	228	−1.16	−0.355	<0.01
Females	132	223	−1.60	−0.397	<0.001
Succ.-cyt.c reductase					
Males	85	562	−2.03	−0.219	<0.05
Females	144	588	−3.40	−0.349	<0.001
UQH_2-cyt.c reductase					
Males	91	420	+0.85	+0.114	N.S.
Females	145	586	−2.99	−0.326	<0.001
Cytochrome b_{562}					
Males	71	247	−0.21	−0.044	N.S.
Females	97	269	−1.13	−0.333	<0.001
Cytochrome $c+c_1$					
Males	70	521	−1.65	−0.207	N.S.
Females	92	393	−0.53	−0.079	N.S.
Cytochrome c oxidase					
Males	84	739	−2.89	−0.236	<0.05
Females	137	729	−4.67	−0.393	<0.001
Hemes $a+a_3$					
Males	68	466	−1.80	−0.241	<0.05
Females	92	434	−2.14	−0.312	<0.01
ATP hydrolase					
Males	45	286	−1.14	−0.312	<0.05
Females	83	220	−0.54	−0.105	N.S.

[a]n, number of subjects; a, enzymatic activities at the intercept with the ordinate of their linear plots as a function of age; b, slope, change per year expressed in nmoles/min per mg protein or pmoles/mg protein; r, correlation coefficient; p, statistical significance.

ferrocytochrome c oxidation (Table IV), declines with age both in males and females. This is associated with an age-linked decline in the content of cytochrome aa_3 (Table IV).

Statistical analysis of the activity of the F_0F_1 ATP synthase (complex V, measured as the rate of ATP hydrolysis) shows a significant decrease in males that is not observed in females (Table IV). The decline in the activity of complex III observed in aging females is associated with a significant decrease in the content of cytochrome b (Table IV), while that of cytochromes c plus c_1 does not show a significant decrease (Table IV). In males, who do not show a decline with age of the activity of complex III, there is no change in the content of cytochrome b nor is there a significant decrease of cytochromes $c + c_1$. It is therefore evident that the age-linked decline of the activity of complex III in females is associated with the decrease in the content of cytochrome b.

Cytochrome b is the first largest protein of the three conserved subunits of bc_1 complexes, the other two being cytochrome c and the Rieske Fe-S proteins (Trumpower and Gennis, 1993). Cytochrome b is also the subunit of the complex where molecular coupling between electron flow and proton pumping takes place (Papa *et al.*, 1994). In the mammalian enzyme, which is made up of 11 subunits, cytochrome b is the only one encoded by the mitochondrial genome (Trumpower and Gennis, 1993; Wallace, 1992b). The structural gene coding for cytochrome b is located on the H-strand of mtDNA, near the D-loop control region (Wallace, 1992b).

It can be noted that the decrease in the rate of oxidation of respiratory substrates is more marked in males than in females. The age-linked decrease in the specific activities of the respiratory chain complexes is, on the contrary, more accentuated in females than in males. This would indicate that the rate-limiting step in the decrease of the rate of respiratory substrates oxidation is represented by the matrix dehydrogenases whose activity would decline more markedly in males than in females. The relative activities of the respiratory chain complexes are in the order complex IV > complex III > complex I. In particular, the finding that in females the decline with age of complex III does not intensify the decrease in the rate of oxidation of respiratory substrates as compared to males shows that the activity of this complex does not represent a rate-limiting step for mitochondrial respiration.

The decline of complex III observed in females with aging, together with the observation that the activity of this complex in young females is significantly higher than in young males, would indicate that the content of cytochrome b in mitochondria and of an active complex III is directly or indirectly controlled by female sex hormones. It could be mentioned with respect to this that an accumulation of deleted mtDNA in the human ovary during menopause has been observed (Kitagawa *et al.*, 1993). The more pronounced decrease of the respiratory activities and of the F_0F_1 synthase complex found in males, who have a shorter life expectancy than females, would be in favor of a critical role of the OXPHOS capacity in the age-linked decline of postmitotic tissues.

Oxygen superoxide, O_2^- is generated by complexes I, II, and III of the respiratory chain but not by cytochrome c oxidase, which seems, on the other hand, to display SOD activity itself (Papa, 1996). Thus, the tendency to produce O_2^-, and the other ROS deriving from it, might depend on the relative ratio between the activities of complex I and III on one hand and of complex IV on the other. Enhancement of this ratio would favor the production of O_2^-. Decrease of the ratio, as it occurs in females, would contribute to prevent the production of O_2^-.

4. mtDNA AND AGING

It is assumed that ROS may attack mtDNA earlier and more drastically than nDNA for the following reasons: (1) mitochondria are the main source of ROS in the cell; (2) mtDNA is not protected by histonelike proteins; and (3) mtDNA is relatively deficient in repair systems (Wallace *et al.*, 1995). Furthermore, mtDNA is more prone to mutations because: (1) γ mtDNA polymerase has a high insertion error rate of about 1/7000 bases (Kunkel and Loeb, 1981); (2) mtDNA contains an unusually high amount of direct repeats, supporting spontaneous deletions via a recombinational event or a slipped mispairing (Schon *et al.*, 1989; Shoffner *et al.*, 1989; Johns et al. 1989); and (3) the single-stranded D-loop replication mechanism of mtDNA could favor point mutations at "hot spots" in tRNA genes (Lauber *et al.*, 1991). Furthermore, since the genetic information of mtDNA is tightly packed, most mutational changes can have injurious effects.

In 1986, Miquel and Fleming, in their "oxygen radicals–mitochondrial injury" hypothesis of cell aging, suggested that alterations of mtDNA may accumulate with time, thus reducing the turnover of the polypeptides of respiratory complexes codified by mtDNA and compromising the mitochondrial energy production and the functional performance of cells and tissues. According to this hypothesis, tissues with high oxidative metabolism and/or terminally differentiated tissues such as cerebral tissue and skeletal and cardiac muscle should be particularly vulnerable. In fact, in the case of the sporadic appearance of mutant mtDNA molecules in a differentiated cell of such a tissue, for example, in a neuron, such molecules should replicate and accumulate in the same cell without being washed out or diluted as happens in cells of tissues with a high mitotic index (Wallace *et al.*, 1995).

An age-related higher level of oxidative damage to mtDNA than to nDNA has been demonstrated in different tissues after setting up quantitative methods for the measurement of OH^8dG, one of about 20 possible *in vitro* oxidized bases of DNA (Richter, 1995; Beckman and Ames, 1996). The content of OH^8dG is 16 times higher in mtDNA than in nDNA of aged rat liver and OH^8dG accumulates with age in mtDNA of rat heart and liver, mouse liver, and human heart and brain (Shigenaga *et al.*, 1994; Richter, 1995).

One of the first studies on the structural alterations of mtDNA in aging was an electron microscope analysis carried out in mouse liver showing that single-stranded loops and "knobs" indicative of deletions–insertions are five times more frequent in aged than in young mice (Pikò *et al.*, 1988). Later on, the discovery of the mitochondrial pathologies gave impetus to the study of structural mtDNA alterations in aging (Holt *et al.*, 1988). These are mainly encephalomyopathies, often associated with defects of the OXPHOS system, in which deletions, duplications, and point mutations of mtDNA were found (Wallace, 1992a; DiMauro and Hirano, 1995). It was hypothesized that aging might be the most common mitochondrial disease (Harding, 1992), and many studies were performed to search for mtDNA mutations in aging tissues. The 4977 bp deletion (ΔmtDNA^{4977}) or "common deletion," originally identified by Southern blot hybridization and characterized in patients with sporadic mitochondrial myopathies such as the Kearns–Sayre's syndrome (Zeviani *et al.*, 1988), has been particularly studied in aging human tissues. The presence of this mtDNA deletion in aging tissues was first ascertained by Cortopassi and Arnheim (1990) using the highly sensitive polymerase chain reaction (PCR) technique: the level of this deletion was, in fact, below the sensitivity of the Southern blot hybridization. Several deletions of different sizes have subsequently been reported in human aged tissues (Table V). Recently, a new qualitative PCR technique, the long PCR (Cheng *et al.*, 1994), which amplifies the whole mtDNA molecule by means of a single PCR reaction, has been used to evidentiate synchronously more deletions in a sample. A wide spectrum of mtDNA rearrangements has been found with this technique in each single sample of aged human skeletal muscles including both known deletions and not previously characterized deletions (Melov *et al.*, 1995; Gadaleta *et al.*, 1999). Most deletions appear in human tissues only in advanced age; others, like the common deletion, are already present at early ages and even in oocytes (Chen *et al.*, 1995a).

Extensive studies on mtDNA deletions have been performed also in other organisms, particularly in rodents (Table VI). About 50 different deletions have

Table V
Deletions of mtDNA in Tissues of Old Humans

Deletion (base pair)	Tissue (reference)[a]
4977 ("common")	Liver (a), muscle (a, b–h), brain (b, i–k), heart (b, i, l, m), lungs (d), testis (a), skin (n), oocytes (o)
3396	Muscle (p)
3610	Muscle (q)
6063	Liver (r), muscle (h)
7436	Heart (b, m, s–u), muscle (b, h), brain (b, k)
10422	Heart (m)
5756–8044[b]	Muscle (b, v)
8037–8048[b]	Heart (g), muscle (g), brain (g)

[a] (a) Lee, H.-C., *et al.*, 1994; (b) Zhang *et al.*, 1992; (c) Cooper *et al.*, (1992); (d) Simonetti *et al.*, 1992; (e) Lezza *et al.*, 1994; (f) Pallotti *et al.* 1996; (g) Baumer *et al.*, 1994; (h) Hsie *et al.*, 1994; (i) Cortopassi and Arnheim, 1990; (j) Soong *et al.*, 1992; (k) Corral-Debrinski *et al.*, 1992a; (l) Corral-Debrinski *et al.*, 1991; (m) Corral-Debrinski *et al.*, 1992b; (n) Pang *et al.*, 1992; (o) Chen *et al.*, 1995a; (p) Torii *et al.*, 1992; (q) Katayama *et al.*, 1991; (r) Yen *et al.*, 1992; (s) Hattori *et al.*, 1991; (t) Sugiyama *et al.*, 1991; (u) Hayakawa *et al.*, 1993; (v) Zhang *et al.*, 1997a.
[b] Size range of different deletions identified in the considered tissue.

Table VI
Deletions of mtDNA in Tissues of Old Animals

Deletion (base pairs)	Tissue
4834	Rat: Liver (Gadaleta *et al.*, 1992; Lezza *et al.*, 1993; Edris *et al.*, 1994; Filser *et al.*, 1997), heart (Gadaleta *et al.*, 1995), brain, skeletal muscle (Gadaleta *et al.*, 1995; Filser *et al.*, 1997), intestinal mucose, bone marrow, kidney, spleen, pancreas (Filser *et al.*, 1997)
4562	Skeletal muscle (Gadaleta *et al.*, 1995)
4949	Skeletal muscle (Gadaleta *et al.*, 1995)
6893	Brain (Gadaleta *et al.*, 1995)
9088	Liver, brain, heart (Gadaleta *et al.*, 1995)
4423–5240[a]	Brain (Gudikote and Van Tuyle, 1996)
6548–9977[a]	Brain (Van Tuyle *et al.*, 1996)
3867	Mouse: Skeletal muscle, brain, heart, kidney, liver (Chen *et al.*, 1993)
3276–4236[a]	Brain (Brossas *et al.*, 1994)
6553–7111[a]	Skeletal muscle (Chung *et al.*, 1994)
3821–5252[a]	Liver, brain, heart (Tanhauser and Laipis, 1995)
3822–7306[a]	Skeletal muscle, brain (Chung *et al.*, 1996)
5744–8027[a]	*M. mulatta*: skeletal muscle (C. M. Lee *et al.*, 1994)
807	*D. melanogaster*: whole organism (Chen *et al.*, 1993)
3231–3899[a]	*C. elegans*: whole organism (Melov *et al.*, 1994)

[a]Size range of different deletions identified in the considered species.

been characterized in rat (Gadaleta *et al.*, 1995; Van Tuyle *et al.*, 1996; Gudikote and Van Tuyle, 1996) and mouse mtDNAs (Brossas *et al.*, 1994; Tanhauser and Laipis, 1995; Chung *et al.*, 1996). It seems that under normal circumstances nearly the entire mitochondrial genome of these rodents is subject to large-scale rearrangements. In the old mouse and rat, about twofold more deletions in the major arc of mtDNA, which is the region enclosed by the Ori-H and the Ori-L, have been described than in the minor arc, which is the region enclosed by Ori-L and Ori-H and including HSP and LSP.

Direct repeats of different lengths have often been found at the breakpoints of deleted molecules: Direct repeats at the breakpoints seem to play an important role, although not essential, in all these rearrangements. Small (2–6 nucleotides) direct repeats have been found to flank precisely the breakpoints of a large number of deletions in rodents. Rearrangements resulting in the loss of the HSP and LSP have been detected in mouse, albeit, only in 1-day-old pup samples. The persistence in aging of molecules of mtDNA lacking the LSP would not be expected, because the LSP performs the dual roles of initiating L-strand transcription and priming mtDNA replication. It is possible that the molecules lacking LSP are the final product of sequential duplication and deletion events (Gudikote and Van Tuyle, 1996). At the moment, it is unclear which is the precise mechanism for the formation of these deletions, but it seems that only one mechanism (Shoffner *et al.*, 1989; Schon *et al.*, 1989; Baumer *et al.*, 1994) may not be able to explain all the reported rearrangements of mtDNA. Recently, a "replication jumping" model has been suggested as another possible mechanism of deletions formation, which takes into account the presence of oxidized bases on the mtDNA of aging individuals (Chung *et al.*, 1996).

Because of the polyploidy of mtDNA, quantification of deletions is a very important point. Some deletions have been quantified in different human tissues. In Table VII, most of the literature data about the levels of the ΔmtDNA^{4977} in humans have been summarized. The range of ages studied by different authors, the total number of examined subjects, and the range of the values observed are reported. The highest content of this deletion has not

Table VII
Age-Dependent Accumulation of ΔmtDNA^{4977} in Different Human Tissues

Tissue	Age range (years)	N	% ΔmtDNA^{4977a}
Skeletal muscle	0.5–84	9	0.0000094–0.1 (Simonetti *et al.*, 1992)
Skeletal muscle	21–78	3	0.00134–0.02318 (Cooper *et al.*, 1992)
Skeletal muscle	0–79	59	0.060018 (Lee *et al.*, 1994)
Skeletal muscle	17–89	13	0–0.35 (Lezza *et al.*, 1994)
Skeletal muscle	1–70	13	0–0.044 (Pallotti *et al.*, 1996)
Liver	20–79	39	0–0.000763–0.007631 (Lee *et al.*, 1994)
Testis	0–89	47	0–0.05289 (Lee *et al.*, 1994)
Putamen	39–82	6	0.001–0.23 (Soong *et al.*, 1992)[b]
Substantia nigra	39–82	6	0.003–0.046 (Soong *et al.*, 1992)
Cerebellar cortex	39–82	6	0.00011–0.0013 (Soong *et al.*, 1992)
Frontal cortex	39–82	6	0.0033–0.028 (Soong *et al.*, 1992)
Parietal cortex	39–82	6	0.0035–0.057 (Soong *et al.*, 1992)
Putamen	67–94	6	0.16–12 (Corral-Debrinski *et al.*, 1992a)
Heart	0–78	10	0–0.035 (Corral-Debrinski *et al.*, 1991)
Heart	30–81	10	0.0001–0.007 (Corral-Debrinski *et al.*, 1992b)
Oocytes	34–42	15	0–0.1 (Chen *et al.*, 1995a)
Optic nerve	1–95	15	0.000168–0.0133 (Soong *et al.*, 1996)
Retina	1–95	15	0.000004–0.001197 (Soong *et al.*, 1996)
Cerebellum	24–86	6	0.0012–0.0067 (Corral-Debrinski *et al.*, 1992a)
Frontal cortex	24–94	6	0.054–2.6 (Corral-Debrinski *et al.*, 1992a)
Parietal cortex	24–94	6	0.021–1.2 (Corral-Debrinski *et al.*, 1992a)
Occipital cortex	24–94	6	0.013–0.82 (Corral-Debrinski *et al.*, 1992a)
Temporal cortex	24–94	4	0.0067–3.4 (Corral-Debrinski *et al.*, 1992a)

[a]Percentage of mtDNA molecules bearing a specific deletion with respect to total mtDNA molecules.
[b]The results of eight other brain areas are not reported here.

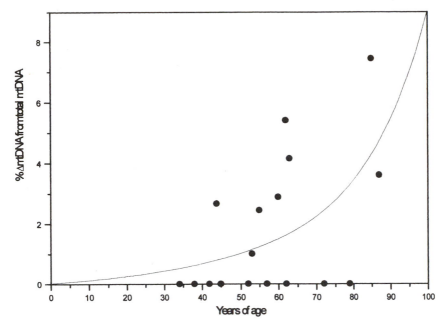

FIGURE 2. Increase of the "common deletion" of mtDNA in skeletal muscle with increasing age. Biopsy and autopsy samples of human skeletal muscles, taken between 12 and 48 hr after death, were obtained from individuals without indications for mitochondrial diseases. DNA was isolated according to Wallace *et al.* (1988). The amounts of wild-type mtDNA and mtDNA^{4977} were determined by PCR as described previously (Seibel *et al.*, 1991). The amount of mtDNA related to the absorbance at 260 nm of the isolated total DNA of different autopsy samples was constant, indicating no degradation of mtDNA in skeletal muscle within 48 hr after death of the individuals. In all points along the abscissa, the percentage of mtDNA^{4977} was below 0.01%.

always been found in the oldest subjects, although the maximum values, generally between 0.0013% and 0.82% of total ΔmtDNA molecules, have been reported after 70 years. Much higher values have been sporadically reported by some authors (Corral-Debrinski *et al.*, 1992a). In Fig. 2, the percentage of the common deletion in the skeletal muscle from 17 subjects of 34–79 years of age, determined in the laboratory of Kadenbach, is shown. The percentage of ΔmtDNA^{4977} was determined by a new quantitative method (Becker *et al.*, 1996). An exponential increase of the deletion with increased age is clearly present, in spite of a very large variation of individual data, and values higher than so far reported have been found. About 50% of the samples showed no detectable common deletion (less than 0.001%), including a 62- and a 79-year-old individual. Differences of one or two orders of magnitude are reported by different laboratories in the same tissue (Table VII); this might depend on the fact that each laboratory set up its own strategy of detection and quantification of deletions (Zhang *et al.*, 1996; Gadaleta *et al.*, 1999). Furthermore, the highest contents of this deletion have not always been found in the most metabolically active tissues: A very low level of the common deletion recently has been reported in the free radical-rich retinal environment without showing any age-dependence by Soong *et al.* (1996) (Table VII).

The other mtDNA deletion quantified in human tissues is the 7436 bp deletion (ΔmtDNA^{7436}) (Sugiyama *et al.*, 1991; Corral-Debrinski *et al.*, 1992b). The levels of this deletion were more or less the same as reported for the ΔmtDNA^{4977}. A positive correlation between the age-dependent increase of the ΔmtDNA^{7436} content and that of the mtDNA OH^{8}dG has been reported in human heart by Hayakawa *et al.* (1993) and between the ΔmtDNA^{4977} and the mtDNA OH^{8}dG contents in the frontal cortex of human brain (Lezza *et al.*, submitted).

Some point mutations have also been reported and quantified in human skeletal muscle. Contrasting data have been reported about their increase with age (Table VIII). However, also in the case of point mutations the highest reported value is 2.4% (Münscher *et al.*, 1993).

Qualitative and quantitative comparisons of deletions have been carried out in the same mouse and rat tissues with not always consistent results (Tanhauser and Laipis, 1995; Zhang *et al.*, 1997b; Filser *et al.*, 1997) (Fig. 3). A

Table VIII
Age-Dependent Accumulation of Point Mutations
in mtDNA of Human Skeletal Muscle

Nucleotide (pathology)[a]	Age range (years)	N	Percentage[b]
8344 (MERRF)	74–89	2	2–2.4[c] (Münscher *et al.*, 1993)
8344 (MERRF)	16–87	12	0–2.0[c] (Kadenbach *et al.*, 1995)
3243 (MELAS)	80 min–87	11	0–0.1[c] (Zhang *et al.*, 1993)
3243 (MELAS)	16–87	12	0–0.9[c] (Kadenbach *et al.*, 1995)
3243 (MELAS)	1–70	13	0–0.6[d] (Pallotti *et al.*, 1996)
8993 (NARP)	1–70	13	0–0.04[d] (Pallotti *et al.*, 1996)

[a]The reported mitochondrial pathologies in which point mutations studied in aging have been found to be pathogenetic are in parentheses. MERRF, myoclonic epilepsy with ragged-red fibers (RRF); MELAS, mitochondrial encephalomyopathy, lactic acidosis, and strokelike episodes; NARP, neuropathy, ataxia, and retinitis pigmentosa.
[b]Percentage of mtDNA molecules bearing a special point mutation with respect to total mtDNA molecules.
[c]Age-dependent accumulation.
[d]Causal, not age-dependent accumulation.

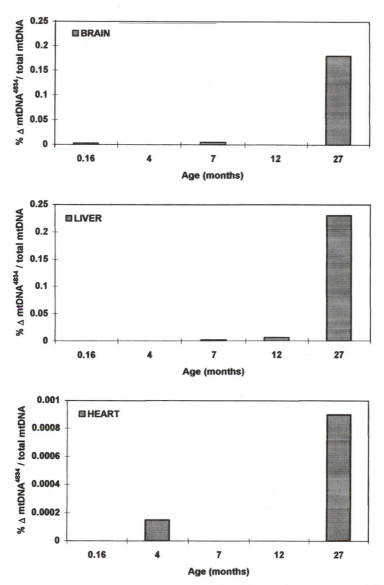

FIGURE 3. Age-dependent increase of the 4834 bp deletion of rat mtDNA in different tissues. Values are expressed as the percentage of the ratio of deleted molecules to total mtDNA molecules.

comparison of prevalence of age-associated mtDNA deletions between human and rat recently has been reported demonstrating different results depending on the considered tissues (Zhang *et al.*, 1997b). Interestingly, mice on caloric restricted diet have far fewer deletions than their control counterparts (Melov *et al.*, 1997).

The low percentage of deletions and point mutations so far found in aged tissues mean that only about a few hundred of the 2,000 to 10,000 molecules of mtDNA in a cell are mutated in a specific tissue of an old individual versus the 27–95% reported in mitochondrial pathologies (Moraes *et al.*, 1989). This extremely low number of mutant molecules does not support the hypothesis that mtDNA mutations and especially deletions are the cause of the reduced activity of the OXPHOS system and of the energy failure of tissues with age. However, in aging, contrary to what generally occurs in mitochondrial pathologies (Brown and Wallace, 1994) where a single mutation is observed in all affected cells and tissues of one patient, several different species of mtDNA carrying deletions, point mutations, and probably partial duplications accumulate in the cell. Moreover, in aging, the distribution of the mitochondrial mutant molecules in each cell or tissue seems to be mosaiclike (Linnane *et al.*, 1989). In fact, a great difference has been found in the percentage of $\Delta mtDNA^{4977}$ molecules between the cerebellum and the substantia nigra or the putamen of the same subject (Table VII), and up to three different types of cytochrome *c* oxidase-deficient muscle fibers have been identified as coexisting in a single skeletal muscle in a study on extraocular muscles from elderly individuals (Müller-Höcker *et al.*, 1993). These fibers were, respectively, characterized by: (1) the presence of mtDNA harboring the common deletion; (2) the total lack of mtDNA; and (3) the presence of mtDNA not carrying the 4977 bp deletion, but suggesting the presence of other mtDNA mutations than the common deletion or nDNA mutations. Recently, a gross mosaic pattern of mtDNA deletions in the skeletal muscle tissue of a human subject has been reported by Zhang *et al.* (1997a). Such a gross mosaic pattern might be explained by the localized clonal expansion of each single mtDNA molecule in which a deletion appeared. In fact, in autosomal dominant progressive external ophthalmoplegia (adPEO), characterized by multiple mtDNA deletions, the clonal expansion of different mtDNA deleted molecules has been reported: The deletions varied between individual muscle fibers and also along the different sections of the same fiber (Moslemi *et al.*, 1996). It is possible that similar clonal expansions of mtDNA mutations take place also in aging tissues.

The percentages of the mutated mtDNA molecules so far reported in aging refer to tissue mean values. More refined techniques are needed to quantify mutations of mtDNA at a single cell or mitochondrion level. In the terminally differentiated cells, like neurons, the accumulation of a high level of mutated mtDNA molecules could lead to death. The death of some neurons due to oxidative stress has been hypothesized in some degenerative, nonmitochon-drial, pathologies like Alzheimer's disease and Huntington's disease (Flint Beal, 1992). In Huntington's disease patients the load of ΔmtDNA[4977] has been reported to be lower than in age-matched controls by Chen et al. (1995b) (Table IX). It has been suggested by these authors that "at-risk" neurons, more sensitive to oxidative stress, might have accumulated so much damage that they die, clearing the way to glial cells. The glial cells should be less prone to accumulate deletions; fewer deletions, in fact, have been found in the white matter than in the gray matter of old human brains (Soong et al., 1992). A lower content of the ΔmtDNA[4977], in the presence of a higher mtDNA OH[8]dG content, has been found also by some of us in different brain areas of Alz-heimer's disease patients with respect to control subjects (Lezza et al., submitted). However, also in the case of the above-reported degenerative pathologies as well as in aging (Table VII) and in Parkinson's disease (Table X), the data of different laboratories are not in agreement.

Table IX
Percentage of ΔmtDNA[4977] in Huntington's and Alzheimer's Diseases

Pathology	Tissue	Age range (years)	N	%mtDNA[4977]
Huntington	Cerebral cortex	36–39	3	0–0.0007 (Chen et al., 1995b)
Control	Cerebral cortex	27–42	3	0.0006–0.0049 (Chen et al., 1995b)
Huntington	Frontal cortex	24–71	22	0.12–2.2 (Horton et al., 1995)
Control	Frontal cortex	24–77	25	0.12[a] (Horton et al., 1995)
Alzheimer	Cerebellum	72–86	6	0.0047–0.084 (Corral-Debrinski et al., 1994)
Alzheimer	Parietal cortex	59–88	11	0.08–1.8 (Corral-Debrinski et al., 1994)
Alzheimer	Frontal cortex	59–93	20	0.007–4.9 (Corral-Debrinski et al., 1994)
Control	Frontal cortex	51–90	13	0.019–1.2 (Corral-Debrinksi et al., 1994)
Alzheimer	Gyrus frontalis	65–102	25	0.01–2.92 (Cavelier et al., 1995)
Control	Gyrus frontalis	53–94	6	0.003–0.22 (Cavelier et al., 1995)
Alzheimer	Nucleus caudatus	62–102	28	0.04–2.94 (Cavelier et al., 1995)
Control	Nucleus caudatus	53–94	6	0.01–2.04 (Cavelier et al., 1995)
Alzheimer	Cerebellum	69	1	0.0016 (Lezza et al., submitted)
Control	Cerebellum	81	1	0.0008 (Lezza et al., submitted)
Alzheimer	Parietal cortex	51–79	6	0.003–0.323 (Lezza et al., submitted)
Control	Parietal cortex	71–86	4	0.182–0.5 (Lezza et al., submitted)
Alzheimer	Frontal cortex	69–78	4	0.019–0.250 (Lezza et al., submitted)
Control	Frontal cortex	63–81	5	0.177–2.2 (Lezza et al., submitted)

[a]Average value reported by the authors.

Table X

Percentage of ΔmtDNA^{4977} in Parkinson's Disease

Pathology	Tissue	Age range (years)	N	%mtDNA^{4977}
Parkinson	Striatum	73	1	5.0 (Ozawa *et al.*, 1990)
Control	Striatum	38	1	0.3 (Ozawa *et al.*, 1990)
Parkinson	Substantia nigra	71–84	4	3.5–4.7 (Mann *et al.*, 1992)
Control	Substantia nigra	72–86	4	3.5–4.6 (Mann *et al.*, 1992)
Parkinson	Frontal cortex	72	1	0.11 (Di Donato *et al.*, 1993)
Control	Frontal cortex	≈70	1	0.07 (Di Donato *et al.*, 1993)
Parkinson	Skeletal muscle	42–72	13	0.02[a] (Di Donato *et al.*, 1993)
Control	Skeletal muscle	56.3[a]	7	0.04[a] (Di Donato *et al.*, 1993)

[a]Average value reported by the authors.

An increased percentage of ΔmtDNA^{4977} molecules has also been found in the sun-exposed skin, in the hypoxemic and atherosclerotic heart, and in the skeletal muscle of oculopharyngeal muscular dystrophy (OPMD) and of chronic fatigue syndrome (CFS) patients (Table XI). Also in these cases, the different values found do not exceed 12.5%. However, new mtDNA deletions appear in such pathological or stress conditions. The above-reported data about the increase of mtDNA deletions in aging and in some nonmitochondrial pathologies allow the following conclusions: (1) an increase of mtDNA deletions is not exclusively found in aging or in pathologies where an oxidative stress is present, although the oxidative stress seems to be heavily implicated (Shigenaga *et al.*, 1994; Filser *et al.*, 1997); (2) in aging, as in other reported conditions, the summing up of the percentages of all kinds of deleted molecules also does not permit one to reach the values ranging between 27 and 90% and more, which are responsible in mitochondrial pathologies for the pathological phenotype (Wallace, 1992a); and (3) deletions of mtDNA have to be considered a marker of mitochondrial suffering, the consequence of a more general alteration of mitochondrial and/or cellular metabolism (Gadaleta *et al.*, 1990a; Wallace, 1992a; Tanhauser and Laipis, 1995). Therefore, mtDNA deletions cannot be the cause of tissues aging, although their accumulation, together with the accumulation of other molecular damages, oxidative or not, may render the tissues of an old individual less able to face stressful conditions (Martin *et al.*, 1996).

5. mtDNA EXPRESSION AND AGING

Expression of mtDNA in aging could play a role that so far might have been underestimated (Harding, 1992). Expression of mtDNA in aging has been

Table XI
Percentages of mtDNA Deleted Molecules in Different Pathological Conditions

Deletion (base pair)	Tissue	Condition	N (age, years)	Percentage[a]
4977 bp	Skin	Not sun-exposed	1 (86)	0.4 (Pang et al., 1994)
4977 bp	Skin	Sun-exposed	1 (86)	12.5 (Pang et al., 1994)
4977 bp	Skin	Precancerous	1 (86)	1.6 (Pang et al., 1994)
4977 bp	Skin	Cancerous	1 (86)	3.1 (Pang et al., 1994)
4977 bp	Skeletal muscle	OPMD[b]	2 (61; 63)	4.5; 2.7 (Lezza et al., 1997)
4977 bp	Skeletal muscle	Control	2 (58; 65)	0.07; 0.06 (Lezza et al., 1997)
7436 bp	Skeletal muscle	OPMD	2 (61; 63)	0.002; 0.00007 (Lezza et al., 1997)
7436 bp	Skeletal muscle	Control	2 (58; 65)	0; 0 (Lezza et al., 1997)
4977 bp	Skeletal muscle	CFS[c]	2 (24; 35)	0.01; 0.26 (Gadaleta et al., unpublished)
4977 bp	Skeletal muscle	Control	1 (34)	0.00005 (Gadaleta et al., unpublished)
4977 bp	Heart	Hypoxemic	7 (48–63)	0.02–0.85 (Corral-Debrinski et al., 1991)
4977 bp	Heart	Control	10 (0–78)	0–0.0035 (Corral-Debrinski et al., 1991)
4977 bp	Heart	Atherosclerotic	5 (42–69)	0.008–0.22 (Corral-Debrinski et al., 1992b)
4977 bp	Heart	Control	10 (30–81)	0.0001–0.007 (Corral-Debrinski et al., 1992b)
7436 bp	Heart	Atherosclerotic	2 (49; 56)	0.25; 0.15 (Corral-Debrinski et al., 1992b)
10422 bp	Heart	Atherosclerotic	2 (49; 56)	0.15; 0.1 (Corral-Debrinski et al., 1992b)

[a]Percentage of mtDNA molecules bearing a specific deletion with respect to total mtDNA molecules.
[b]OPMD, Oculopharyngeal muscular dystrophy.
[c]CFS, chronic fatigue syndrome.

extensively studied in the laboratory of several of the authors and in few other laboratories. Gadaleta *et al.* (1990a) demonstrated that in the rat the steady-state levels of two mitochondrial transcripts, each representing one of the two H-strand transcription units of mtDNA, namely, the 12S ribosomal RNA (12S rRNA) and the messenger RNA coding for subunit I of the cytochrome *c* oxidase (CoI mRNA), underwent an age-dependent decrease in heart and maximally in brain. No appreciable age-dependent change was observed in liver, thus suggesting a tissue specificity in the age-related reduction of the level of transcripts. The administration of acetyl-L-carnitine (ALCAR) at the dosage of 300 mg/kg body weight to aged rats brought back the mitochondrial transcripts levels to values similar to those of the adult animals 1 hr after the treatment and to values higher than the adult ones 3 hr after the treatment, maintaining such high levels even after 24 hr (Gadaleta *et al.*, 1990b, 1994). The reduced transcription of mtDNA in rat brain and heart did not depend on a reduced availability of mtDNA molecules as template, since the same authors (Gadaleta *et al.*, 1992) reported that the mtDNA amount corresponding to a set quantity of nuclear DNA in aged rat tissues was the same if not even higher than that present in adult rat. This result was confirmed by Pikò (1992). Using an *in vitro* system with free (nonsynaptic) mitochondria from cerebral hemispheres it was shown that the decreased level of transcripts was partially due to the reduced rate of RNA synthesis (Fernandez-Silva *et al.*, 1991). The *in vivo* ALCAR treatment was able to restore RNA synthesis in this system (Gadaleta *et al.*, 1994). The results obtained with ALCAR suggested that the reduction of transcription in aged rat brain and heart could not be due to irreversible damage of the mitochondrial transcriptional apparatus, i.e., to a high level of mutated mtDNA molecules (Gadaleta *et al.*, 1990a). Gadaleta *et al.* (1992) were able to demonstrate mtDNA molecules carrying deletions of different sizes in the liver, heart, and brain of aged rat, among which the 4834 bp deletion was very common (Gadaleta *et al.*, 1992, 1995). The number of deleted molecular species as well as the percentage of the 4834 bp mtDNA deletion increased with age. The maximum value, 0.25%, of ΔmtDNA[4834] was reached in the liver and the brain of old rats, whereas it was much lower in the heart (Fig. 3). The value of 0.25%, however, was too low in absolute and furthermore it was the same in liver and brain and higher than in heart; therefore, it was not consistent with the age-related, tissue-specific reduction of the steady-state levels of mitochondrial transcripts previously reported in these tissues (Gadaleta *et al.*, 1990a). The content of 4834-bp-deleted mtDNA molecules in various old rat tissues has also been evaluated by Filser *et al.* (1997), confirming, with a different method, the results reported in Fig. 3.

MtDNA is characterized by the existence of a triplex strand structure in the D-loop due to the newly synthesized 7S DNA in a certain number of molecules. Since several authors suggested a link between the content of the triplex strand structure and the developmental (Callen *et al.*, 1983) and the metabolic (Annex and Williams, 1990) cellular conditions, the content of mtDNA molecules that had lost the triplex strand structure in the cerebral hemispheres of aged rat as well as the content of the 7S DNA in the heart and in the brain of young and aged rats were measured. These studies showed that both heart and cerebral hemispheres of old rats had a reduced content of 7S mtDNA with respect to younger animals. Such a reduction was much higher in brain than in heart, which is reminiscent of the behavior of the steady-state levels of mitochondrial transcripts in aged rat tissues (Petruzzella *et al.*, 1995). Since, as it has been pointed out before, the D-loop-containing region harbors the promoters of transcription, it was suggested that conformational changes of the mtDNA, induced by the third strand, might affect the binding of *trans*-acting factors required for an accurate and efficient transcription. Furthermore, transcription may be directly affected by a change in the negative superhelix density resulting from the presence of the third strand (Buzan and Low, 1988; Barat-Gueride *et al.*, 1989). The decreased proportion of triplex strand containing molecules might explain both the reduced efficiency of mtDNA transcription in aged rat and the effect of ALCAR. In this case, ALCAR might act by delivering acetyl-CoA equivalents to the tricarboxylic acids cycle or facilitating the mitochondrial β-oxidation of fatty acids, thereby increasing the production of ATP (Siliprandi *et al.*, 1965): This could provide energy to aged cells for mtDNA supercoiling and therefore for an efficient mtDNA transcription.

Reduced steady-state levels of mtDNA transcripts, at least in part tissue specific, have been reported also in aging *Drosophila melanogaster* (Calleja *et al.*, 1993), confirming the age-related change of expression at the transcriptional level. Mitochondrial protein synthesis was also reduced in aged rat (Gadaleta *et al.*, 1995) and mouse brain (Takai *et al.*, 1995).

The lower efficiency of mtDNA transcription and translation in aged rats might be a consequence of altered metabolic conditions in the aged cell. ALCAR might function by removing such alterations. In this respect it is useful to remember that ALCAR, 3 hr after administration and at the doses reported above, can restore the respiratory deficit of state IV mitochondria isolated from aged rat brain, suggesting a protective action of this molecule toward the mitochondrial membrane fluidity and/or permeability (Petruzzella *et al.*, 1992). An alteration of the cholesterol–phospholipid ratio as well as a reduced cardiolipin content, corrected by the administration of ALCAR, were reported in the heart mitochondria of aged rats by Paradies *et al.* (1992). Furthermore, a reduced transport in aged rat heart mitochondria of adenine nucleotides, pyruvate, phosphate, and acylcarnitines, restored by *in vivo* ALCAR administration, was found by Paradies *et al.* (1992, 1994, 1995). The decreased content of cardiolipin might be responsible for the reduced activities of the adenine nucleotide carrier and of the cytochrome *c* oxidase of aged rat heart (Paradies *et al.*, 1994, 1997), as well as of the ATP synthase of aged rat heart and brain (Guerrieri *et al.*, 1992).

An altered composition of the mitochondrial membranes and a modification of the Ca^{2+} ion availability might reduce the ATP production required for mtDNA transcription and translation. A regulatory role for transcription has been ascribed to the ATP availability (Enriquez *et al.*, 1996), and a dependence of mitochondrial protein synthesis on Ca^{2+} ion and ATP has been recently confirmed (Joyal *et al.*, 1995). Furthermore, mitochondrial protein synthesis as well as mtDNA transcription are processes dependent on proteins and protein factors synthesized out of the mitochondrion: The import of such proteins, including ribosomal proteins, requires ATP and the right fluidity of the mitochondrial membranes (Attardi and Schatz, 1988). Recently, the redox potential of the cell also has been suggested to play a role in the "cross-talk" between nDNA and mtDNA: Redox sensors and regulators of the redox answer have been postulated (Allen, 1993, 1995; Wallace *et al.*, 1995). DNA-binding proteins have been suggested as examples of such regulators: By being reversibly phosphorylated (Allen, 1993) and/or acetylated or in some way modified, they should regulate the gene expression. In this context, important results have been reported by Pisano *et al.* (1996) on ALCAR. These authors demonstrated that high levels of carnitine and ALCAR were present in the nucleus. Furthermore, they demonstrated that in some cell cultures ALCAR might help to modulate, by means of the acetylation of histone H_4, the expression of genes already programmed for that specific cell. As reported in Section 2.1, two protein factors, NRF-1 and NRF-2, can recognize sequences present in several genes coding for important mitochondrial proteins (Scarpulla *et al.*, 1996, and Chapter 22, this volume). Therefore, it might be possible to hypothesize that the ALCAR, through histone acetylation, might modulate the activity of NRF-1 and NRF-2, which in turn might promote mitochondrial biogenesis and/or differentiation. Furthermore, the ALCAR, still through the histone acetylation, might activate the expression of the cardiolipin synthase gene, and thus might explain the effect of the ALCAR on the lipid composition of the mitochondrial membranes and eventually on the transport of various metabolites. Cardiolipin synthase has already been shown to be regulated by thyroid hormones action (Hostetler, 1991) and an additional effect of thyroid hormones and ALCAR on the steady-state level of CoI mRNA in hypothyroid rat has been reported by Gadaleta *et al.* (1990c).

6. CONCLUSIONS

It appears that the aging process represents an event that depends on a large number of genetic, environmental, and possibly nutritional factors, including a history of individual diseases. Genotypic and phenotypic alterations of mitochondria seem to be involved in this process. However, many aspects of their genetics and the cross-talk between the mitochondrial and nuclear genomes in the eukaryotic cell should be clarified before a conclusion on the role of genotypic and phenotypic mitochondrial defects in the aging process might be reached.

ACKNOWLEDGMENTS. This work was supported by funds from P.F. "Invecchiamento" Code No. 971725, from CNR Grant No. 97.04123.CT04 to M.N.G., from CNR Strategic project Grant No. 95.04679,ST75 on collection of Health Data to S.P., and from MURST Italy, COFIN '98 "Regolazione della biogenesi dei mitochondri velle 'invecchiamento' e in condizioni di stress ossidativo" to M.N.G. and COFIN '98 S.P. The authors are grateful to Ms. R. Longo for word processing and Mr. F. Fracasso for technical assistance.

7. REFERENCES

Allen, J. F., 1993, Control of gene expression by redox potential and the requirement for chloroplast and mitochondrial genomes, *J. Theor. Biol.* **165**:609–631.

Allen, C. A., Håkansson G., and Allen, J. F., 1995, Redox conditions specify the proteins synthesised by isolated chloroplasts and mitochondria, *Redox Rep.* **1**:119–123.

Annex, B. H., and Williams, R. S., 1990, Mitochondrial DNA structure and expression in specialized subtypes of mammalian striated muscle, *Mol. Cell. Biol.* **10**:5671–5678.

Antoshechkin, I., and Bogenhagen, D. F., 1995, Distinct roles for two purified factors in transcription of Xenopus mitochondrial DNA, *Mol. Cell. Biol.* **15**:7032–7042.

Attardi, G., and Schatz, G., 1988, Biogenesis of mitochondria, *Annu. Rev. Cell. Biol.* **4**:289–333.

Attardi, G., Chomyn, A., King, M. P., Kruse, B., Loguercio Polosa, P., and Narasimhan Murdter, N., 1990, Biogenesis and assembly of the mitochondrial respiratory chain: Structural, genetic and pathological aspects, *Biochem. Soc. Trans.* **18**:509–513.

Barat-Gueride, M., Dufresne, C., and Rickwood, D., 1989, Effect of DNA conformation on the transcription of mitochondrial DNA, *Eur. J. Biochem.* **183**:297–302.

Barrientos, A., Casademont, J., Rötig, A., Miró, O., Urbano-Márquez, A., Rustin, P., and Cardellach, F., 1996, Absence of relationship between the level of electron transport chain activities and aging in human skeletal muscle, *Biochem. Biophys. Res. Commun.* **229**:536–539.

Baumer, A., Zhang, C., Linnane, A. W., and Nagley, P., 1994, Age-related human mtDNA deletions: A heterogeneous set of deletions arising at a single pair of directly repeated sequences, *Am. J. Hum. Genet.* **54**:618–630.

Becker, A., Reith, A., Napiwotzk, J., and Kadenbach, B., 1996, A quantitative method of determining initial amounts of DNA by polymerase chain reaction cycle titration using digital imaging and a novel DNA stain, *Anal. Biochem.* **237**:204–207.

Beckman, K. B., and Ames, B. N., 1996, Detection and quantification of oxidative adducts of mitochondrial DNA, *Methods Enzym.* **264**:442–453.

Boffoli, D., Scacco, S. C., Vergari, R., Solarino, G., Santacroce, G., and Papa, S., 1994, Decline with age of the respiratory chain activity in human skeletal muscle, *Biochim. Biophys. Acta* **1226**:73–82.

Boffoli, D., Scacco, S. C., Vergari, R., Persio, M. T., Solarino, G., Laforgia, R., and Papa, S., 1996, Ageing is associated in females with a decline in the content and activity on the b-c1 complex in skeletal muscle mitochondria, *Biochim. Biophys. Acta* **1315:**66–72.

Bowling, A. C., Mutisya, E. M., Walker, L. C., Price, D. L., Cork, L. C., and Flint Beal, M., 1993, Age-dependent impairment of mitochondrial function in primate brain, *J. Neurochem.* **60:** 1964–1967.

Brossas J.-Y., Barreau, E., Courtois, Y., and Tréton, J., 1994, Multiple deletions in mitochondrial DNA are present in senescent mouse brain, *Biochem. Biophys. Res. Commun.* **202:**654–659.

Brown, M. D., and Wallace, D. C., 1994, Molecular basis of mitochondrial disease, *J. Bioenerg. Biomembr.* **26:**273–289.

Buzan, J. M., and Low, R. L., 1988, Preference of human mitochondrial RNA polymerase for superhelical templates with mitochondrial promoters, *Biochem. Biophys. Res. Commun.* **152:** 22–29.

Calleja, M., Peña, P., Ugalde, C., Ferreiro, C., Marco, R., and Garesse, R., 1993, Mitochondrial DNA remains intact during *Drosophila* aging, but the levels of mitochondrial transcripts are significantly reduced, *J. Biol. Chem.* **268:**18891–18897.

Callen, J. C., Tourte, M., Dennebouy, N., and Monoulou, J. C., 1983, Changes in D-loop frequency and superhelicity among the mitochondrial DNA molecules in relation to organelle biogenesis in oocytes of *Xenopus laevis*, *Exp. Cell Res.* **143:**115–125.

Cantatore, P., and Saccone, C., 1987, Organization, structure, and evolution of mammalian mitochondrial genes, *Int. Rev. Cytol.* **108:**149–208.

Cantatore, P., Loguercio Polosa, P., Fracasso, F., Flagella, Z., and Gadaleta, M. N., 1986, Quantitation of mitochondrial RNA species during rat liver development: The concentration of cytochrome oxidase subunit I (CoI) mRNA increases at birth, *Cell Different.* **19:**125–132.

Cantatore, P., Daddabbo, L., Fracasso, F., and Gadaleta, M. N., 1995, Identification by *in organello* footprinting of protein contact sites and of single-stranded DNA sequences in the regulatory region of rat mitochondrial DNA, *J. Biol. Chem.* **270:**25020–25027.

Cavelier, L., Jazin, E. E., Eriksson, I., Prince, J., Båve, U., Oreland, L., and Gyllensten, U., 1995, Decreased cytochrome *c* oxidase activity and lack of age-related accumulation of mitochondrial DNA deletions in the brains of schizophrenics, *Genomics* **29:**217–224.

Chen, X., Simonetti, S., DiMauro, S., and Schon, E. A., 1993, Accumulation of mitochondrial DNA deletions in organisms with various lifespans, *Bull. Mol. Biol. Med.* **18:**57–66.

Chen, X., Prosser, R., Simonetti, S., Sadlock, J., Jagiello, G., and Schon, E. A., 1995a, Rearranged mitochondrial genomes are present in human oocytes, *Am. J. Hum. Genet.* **57:**239–247.

Chen, X., Bonilla, E., Sciacco, M., and Schon, E. A., 1995b, Paucity of deleted mitochondrial DNAs in brain regions of Huntington's disease patients, *Biochim. Biophys. Acta* **1271:**229–233.

Cheng, S., Higuchi, R., and Stoneking, M., 1994, Complete mitochondrial genome amplification, *Nature Genet.* **7:**350–351.

Chomyn, A., Meola, G., Bresolin, N., Lai, S. T., Scarlato, G., and Attardi, G., 1991, *In vitro* genetic transfer of protein synthesis and respiration defects to mitochondrial DNA-less cells with myopathy-patient mitochondria, *Mol. Cell. Biol.* **11:**2235–2244.

Chung, S. S., Windruch, R., Schwarze, S. R., McKenzie, D. I., and Aiken, J. M., 1994, Multiple age-associated mitochondrial DNA deletions in skeletal muscle of mice, *Aging Clin. Exp. Res.* **6:**193–200.

Chung, S. S., Eimon, P. M., Windruch, R., and Aiken, J. M., 1996, Analysis of age-associated mitochondrial DNA deletion breakpoint regions from mice suggests a novel model of deletion formation, *Age* **19:**117–128.

Cingolani, G., Capaccio, L., D'Elia, D., and Gadaleta, G., 1997, *In organello* footprinting analysis of rat mitochondrial DNA:protein interaction upstream of the Ori-L, *Biochem. Biophys. Res. Commun.* **231:**856–860.

Clayton, D. A., 1991, Replication and transcription of vertebrate mitochondrial DNA, *Annu. Rev. Cell. Biol.* **2:**453–478.

Cooper, J. M., Mann, V. M., and Schapira, A. H. V., 1992, Analyses of mitochondrial respiratory chain function and mitochondrial DNA deletion in human skeletal muscle: Effect of ageing, *J. Neurol. Sci.* **113:**91–98.

Corral-Debrinski, M., Stepien, G., Shoffner, J. M., Lott, M. T., Kanter, K., and Wallace, D. C., 1991, Hypoxemia is associated with mitochondrial DNA damage and gene induction. Implications for cardiac disease, *J. Am. Med. Assoc.* **266:**1812–1816.

Corral-Debrinski, M., Horton, T., Lott, M. T., Shoffner, J. M., Flint Beal, M., and Wallace, D. C., 1992a, Mitochondrial DNA deletions in human brain: Regional variability and increase with advanced age, *Nature Genet.* **2:**324–329.

Corral-Debrinski, M., Shoffner, J. M., Lott, M. T., and Wallace, D. C., 1992b, Association of mitochondrial DNA damage with aging and coronary atherosclerotic heart disease, *Mutat. Res.* **275:**169–180.

Corral-Debrinski, M., Horton, T., Lott, M. T., Shoffner, J. M., McKee, A. C., Flint Beal, M., Graham, B. H., and Wallace, D. C., 1994, Marked changes in mitochondrial DNA deletion levels in Alzheimer brains, *Genomics* **23:**471–476.

Cortopassi, G. A., and Arnheim, N., 1990, Detection of a specific mitochondrial DNA deletion in tissues of older humans, *Nucleic Acids Res.* **18:**6927–6933.

Daga, A., Micol, V., Hess, D., Aebersold, R., and Attardi, G., 1993, Molecular characterization of the transcription termination factor from human mitochondria, *J. Biol. Chem.* **268:**8123–8130.

Davis, A. F., Ropp, P. A., Clayton, D. A., and Copeland, W. C., 1996, Mitochondrial DNA polymerase γ is expressed and translated in the absence of mitochondrial DNA maintenance and replication, *Nucleic Acids Res.* **24:**2753–2759.

Demonacos, C. V., Karayanni, N., Hatzoglou, E., Tsiriyiotis, C., Spandidos, D. A., and Sekeris, C. E., 1996, Mitochondrial genes as sites of primary action of steroid hormones, *Steroids* **61:** 226–232.

Di Donato, S., Zeviani, M., Giovannini, P., Savarese, N., Rimoldi, M., Mariottini, C., Girotti, F., and Caraceni, T., 1993, Respiratory chain and mitochondrial DNA in muscle and brain in Parkinson's disease patients, *Neurology* **43:**2262–2268.

DiMauro, S., and Hirano, M., 1995, Diseases due to mutations of mitochondrial DNA: Problems in pathogenesis, *Bull. Mol. Biol. Med.* **20:**169–175.

DiMauro, S., and Wallace, D. C., eds., 1993, *Mitochondrial DNA in Human Pathology*, Raven Press, New York.

Edris, W., Burgett, S., Stine, O. C., and Filburn, C. R., 1994, Detection and quantitation by competitive PCR of an age-associated increase in a 4.8-kb deletion in rat mitochondrial DNA, *Mutat. Res.* **316:**69–78.

Enriquez, J. A., Fernandez-Silva, P., Perez-Martos, A., Lopez-Perez, M. J., and Montoya, J., 1996, The synthesis of mRNA in isolated mitochondria can be maintained for several hours and is inhibited by high levels of ATP, *Eur. J. Biochem.* **237:**601–610.

Fernandez-Silva, P., Petruzzella, V., Fracasso, F., Gadaleta, M. N., and Cantatore, P., 1991, Reduced synthesis of mtRNA in isolated mitochondria of senescent rat brain, *Biochem. Biophys. Res. Commun.* **176:**645–653.

Filser, N., Margue, C., and Richter, C., 1997, Quantification of wild-type mitochondrial DNA and its 4.8-kb deletion in rat organs, *Biochem. Biophys. Res. Commun.* **233:**102–107.

Fisher, R. P., Lisowsky, T., Parisi, M. A., and Clayton, D. A., 1992, DNA wrapping and bending by a mitochondrial high mobility group-like transcriptional activator protein, *J. Biol. Chem.* **276:**3358–3367.

Flint Beal, M., 1992, Does impairment of energy metabolism result in excitotoxic neuronal death in neurodegenerative illnesses? *Ann. Neurol.* **31:**119–130.

Gadaleta, G., D'Elia, D., Capaccio, L., Saccone, C., and Pepe, G., 1996, Isolation of a 25-kDa protein binding to a curved DNA upstream the origin of the L strand replication in the rat mitochondrial genome, *J. Biol. Chem.* **271**:13537–13541.

Gadaleta, M. N., Loguercio Polosa, P., Lezza, A., Fracasso, F., and Cantatore, P., 1985, Mitochondrial transcription in rat liver under different physiological conditions, in *Achievements and Perspectives of Mitochondrial Research* (E. Quagliariello, E. C. Slater, F. Palmieri, C. Saccone, and A. M. Kroon, eds.), pp. 417–425, Elsevier, Amsterdam.

Gadaleta, M. N., Petruzzella, V., Renis, M., Fracasso, F., and Cantatore, P., 1990a, Reduced transcription of mitochondrial DNA in the senescent rat. Tissue dependence and effect of acetyl-L-carnitine, *Eur. J. Biochem.* **187**:501–506.

Gadaleta, M. N., Petruzzella, V., Renis, M., Fracasso, F., and Cantatore, P., 1990b, Reduced mitochondrial DNA transcription in two brain regions of senescent rat: Effect of acetyl-L-carnitine, in *Structure, Function and Biogenesis of Energy Transfer Systems* (E. Quagliariello *et al.*, eds.), pp. 135–138, Elsevier, Amsterdam.

Gadaleta, M. N., Petruzzella, V., Fracasso, F., Fernandez-Silva, P., and Cantatore, P., 1990c, Acetyl-L-carnitine increases cytochrome oxidase subunit I mRNA content in hypothyroid rat liver, *FEBS Lett.* **277**:191–193.

Gadaleta, M. N., Rainaldi, G., Lezza, A. M. S., Milella, F., Fracasso, F., and Cantatore, P., 1992, Mitochondrial DNA copy number and mitochondrial DNA deletion in adult and senescent rat, *Mutat. Res.* **275**:181–193.

Gadaleta, M. N., Petruzzella, V., Daddabbo, L., Olivieri, C., Fracasso, F., Loguercio Polosa, P., and Cantatore, P., 1994, Mitochondrial DNA transcription and translation in aged rat: Effect of acetyl-L-carnitine, *Ann. NY Acad. Sci.* **717**:150–160.

Gadaleta, M. N., Rainaldi, G., Lezza, A. M. S., Marangi, L. C., Milella, F., Daddabbo, L., Fracasso, F., Loguercio Polosa, P., and Cantatore, P., 1995, Structure and expression of mitochondrial DNA in aging rat: DNA deletions and protein synthesis, in *Progress in Cell Research*, vol. 5 (F. Palmieri *et al.*, eds.), pp. 231–235, Elsevier, Amsterdam.

Gadaleta, M. N., Lezza, A. M. S., and Cantatore, P., 1999, Mitochondrial DNA deletions, in *Methods in Aging Research* (B. Yu, ed.), pp. 475–511, CRC Press, Boca Raton, FL.

Ghivizzani, S. C., Madsen, C. S., Nelen, M. R., Ammini, C. V., and Hauswirth W. W., 1994, *In organello* footprint analysis of human mitochondrial DNA: Human mitochondrial transcription factor A interactions at the origin of replication, *Mol. Cell. Biol.* **14**:7717–7730.

Gray, H., and Wong, T. W., 1992, Purification and identification of subunit structure of human mitochondrial DNA polymerase, *J. Biol. Chem.* **267**:5835–5841.

Gudikote, J. P., and Van Tuyle, G. C., 1996, Rearrangements in the shorter arc of rat mitochondrial DNA involving the region of the heavy and light strand promoters, *Mutat. Res.* **356**:275–286.

Guerrieri, F., Capozza, G., Kalous, M., Zanotti, F., Drahota, Z., and Papa, S., 1992, Age-dependent changes in the mitochondrial F_0F_1 ATP synthase, *Arch. Gerontol. Geriatr.* **14**:299–308.

Gyllensten, U., Wharton, D., Josefsson, A., and Wilson, A. C., 1991, Paternal inheritance of mitochondrial DNA in mice, *Nature* **352**:255–257.

Hansford, R. G., 1983, Bioenergetics in aging, *Biochim. Biophys. Acta* **726**:41–80.

Harding, A. E., 1992, Growing old: The most common mitochondrial disease of all? *Nature Genet.* **2**:251–252.

Harman, D., 1972, The biological clock: The mitochondria? *J. Am. Geriat. Soc.* **20**:145–147.

Hattori, K., Tanaka, M., Sugiyama, S., Obayashi, T., Ito, T., Satake, T., Hanaki, T., Asai, J., Nagano, M., and Ozawa, T., 1991, Age-dependent increase in deleted mitochondrial DNA in the human heart: Possible contributory factor to presbycardia, *Am. Heart J.* **121**:1735–1742.

Hayakawa, M., Sugiyama, S., Hattori, K., Takasawa, M., and Ozawa, T., 1993, Age-associated damage in mitochondrial DNA in human hearts, *Mol. Cell. Biochem.* **119**:95–103.

Hehman, G. L., and Hauswirth, W. W., 1992, DNA helicase from mammalian mitochondria, *Proc. Natl. Acad. Sci. USA* **89**:8562–8566.

Holt, I. J., Harding, A. E., and Morgan-Hughs, J. A., 1988, Deletions of muscle mitochondrial DNA in patients with mitochondrial myopathies, *Nature* **331**:717–719.

Horton, T. M., Graham, B. H., Corral-Debrinski, M., Shoffner, J. M., Kaufman, A. E., Flint Beal, M., and Wallace, D. C., 1995, Marked increase in mitochondrial DNA deletion levels in the cerebral cortex of Huntington's disease patients, *Neurology* **45**:1879–1883.

Hostetler, K. Y., 1991, Effect of thyroxine on the activity of mitochondrial cardiolipin synthase in rat liver, *Biochim. Biophys. Acta* **1086**:139–140.

Hsie, R.-H., Hou, J.-H., Hsu, H.-S., and Wei, Y.-H., 1994, Age-dependent respiratory function decline and DNA deletions in human muscle mitochondria, *Biochem. Mol. Biol. Int.* **32**:1009–1022.

Johns, D. R., Rutledge S. L., Stune, O. C., and Hurko, O., 1989, Directly repeated sequences associated with pathogenic mitochondrial DNA deletions, *Proc. Natl. Acad. Sci. USA* **86**:8059–8062.

Joyal, J. L., Hagen, T., and Aprille, J. R., 1995, Intramitochondrial protein synthesis is regulated by matrix adenine nucleotide content and requires calcium, *Arch. Biochem. Biophys.* **319**:322–330.

Kadenbach, B., Münscher, C., Frank, V., Müller-Höcker, J., and Napiwotzki, J., 1995, Human aging is associated with stochastic somatic mutations of mitochondrial DNA, *Mutat. Res.* **338**:161–172.

Katayama, M., Tanaka, M., Yamamoto, H., Ohbayashi, T., Nimura, Y., and Ozawa, T., 1991, Deleted mitochondrial DNA in the skeletal muscle of aged individuals, *Biochem. Int.* **25**:47–56.

Kelly, J. L., Greenleaf, A. L., and Lehman, I. R., 1986, Isolation of the nuclear gene encoding a subunit of the yeast mitochondrial RNA polymerase, *J. Biol. Chem.* **261**:10348–10351.

King, M. P., and Attardi, G., 1989, Human cells lacking mtDNA: Repopulation with exogenous mitochondria by complementation, *Science* **246**:500–503.

Kitagawa, T., Suganuma, N., Nawa, A., Kikkawa, F., Tanaka, M., Ozawa, T., and Tomoda, Y., 1993, Rapid accumulation of deleted mitochondrial deoxyribonucleic acid in postmenopausal ovaries, *Biol. Reprod.* **49**:730–736.

Kunkel, T. A., and Loeb, L. A., 1981, Fidelity of mammalian DNA polymerases, *Science* **213**:765–768.

Larsson, N.-G., Oldfors, A., Garman, J. D., Barsh, G. S., and Clayton, D. A., 1997, Down-regulation of mitochondrial transcription factor A during spermatogenesis in humans, *Hum. Mol. Genet.* **6**:185–191.

Lauber, F., Marsac, C., Kadenbach, B., and Seibel, P., 1991, Mutations in mitochondrial tRNA genes: A frequent cause of neuromuscular disease, *Nucleic Acids Res.* **19**:1393–1397.

Lee, C. M., Eimon, P., Weindruch, R., and Aiken, J. M., 1994, Direct repeat sequences are not required at the breakpoints of age-associated mitochondrial DNA deletions in rhesus monkeys, *Mech. Ageing Dev.* **75**:69–79.

Lee, H.-C., Pang, C.-Y., Hsu, H.-S., and Wei, Y.-H., 1994, Differential accumulations of 4,977 bp deletion in mitochondrial DNA of various tissues in human ageing, *Biochim. Biophys. Acta* **1226**:37–43.

Lezza, A. M. S., Rainaldi, G., Cantatore, P., and Gadaleta, M. N., 1993, Quantitative determination of a 4.8 Kb deletion in mtDNA of aging rat liver, *Bull. Mol. Biol. Med.* **18**:67–80.

Lezza, A. M. S., Boffoli, D., Scacco, S., Cantatore, P., and Gadaleta, M. N., 1994, Correlation between mitochondrial DNA 4977-bp deletion and respiratory chain enzyme activities in aging human skeletal muscles, *Biochem. Biophys. Res. Commun.* **205**:772–779.

Lezza, A. M. S., Cormio, A., Gerardi, P., Silvestri, G., Servidei, S., Serlenga, L., Cantatore, P., and Gadaleta, M. N., 1997, Mitochondrial DNA deletions in oculopharyngeal muscular dystrophy, *FEBS Lett.* **418**:167–170.

Linnane, A. W., Marzuki, S., Ozawa, T., and Tanaka, M., 1989, Mitochondrial DNA mutations as an important contributor to ageing and degenerative diseases, *Lancet* **1**:642–645.

Madsen, C. S., Ghivizzani, C., and Hauswirth, W. W., 1993, Protein binding to a single termination-associated sequence in the mitochondrial DNA D-loop region, *Mol. Cell. Biol.* **13**:2162–2171.

Mann, V. M., Cooper, J. M., and Schapira, A. H. V., 1992, Quantitation of a mitochondrial DNA deletion in Parkinson's disease, *FEBS Lett.* **299**:218–222.

Martin, G. M., Austad, S. N., and Johnson, T. E., 1996, Genetic analysis of ageing: role of oxidative damage and environmental stress, *Nature Genet.* **13**:25–34.

Melov, S., Hertz, G. Z., Stormo, G. D., and Johnson, T. E., 1994, Detection of deletions in the mitochondrial genome of *Caenorhabditis elegans*, *Nucleic Acids Res.* **22**:1075–1078.

Melov, S., Shoffner, J. M., Kaufman, A., and Wallace, D. C., 1995, Marked increase in the number and variety of mitochondrial DNA rearrangements in aging human skeletal muscle, *Nucleic Acids. Res.* **23**:4122–4126.

Melov, S., Hinerfeld, D., Esposito, L., and Wallace, D. C., 1997, Multi-organ characterization of mitochondrial genomic rearrangements in ad libitum and caloric restricted mice show striking somatic mitochondrial DNA rearrangements with age, *Nucleic Acids Res.* **25**:974–982.

Miquel, J., and Fleming, J. E., 1986, Theoretical and experimental support for an "oxygen radical-mitochondrial injury hypothesis of cell aging," in *Free Radicals, Aging and Degenerative Disease* (J. E. Johnson, Jr., R. Walford, D. Harman, and J. Miquel, eds.), pp. 51–74, Liss, New York.

Moraes, C. T., DiMauro, S., Zeviani, M., Lombes, A., Shanske, S., Miranda, A. F., Nakase, H., Bonilla, E., Werneck, L. C., Servidei, S., Nonaka, I., Koga, Y., Spiro, A. J., Brownell, K. W., Schmidt, B., Schotland, D. L., Zupanc, M., DeVivo, D. C., Schon, E. A., and Rowland, L. P., 1989, Mitochondrial DNA deletions in progressive external ophthalmoplegia and Kearns–Sayre syndrome, *N. Engl. J. Med.* **320**:1293–1299.

Moslemi, A.-R., Melberg, A., Holme, E., and Oldfors, A., 1996, Clonal expansion of mitochondrial DNA with multiple deletions in autosomal dominant progressive external ophthalmoplegia, *Ann. Neurol.* **40**:707–713.

Müller-Höcker, J., Seibel, P., Schneiderbanger, K., and Kadenbach, B., 1993, Different *in situ* hybridization patterns of mitochondrial DNA in cytochrome *c* oxidase-deficient extraocular muscle fibers in the elderly, *Virchows Archiv A Pathol. Anat.* **422**:7–15.

Münscher, C., Lieger, T., Müller-Höcker, J., and Kadenbach, B., 1993, The point mutation of mitochondrial DNA characteristic for MERRF disease is found also in healthy people of different ages, *FEBS Lett.* **317**:27–30.

Mutvei, A., Kuzela, S., and Nelson, B. D., 1989, Control of mitochondrial transcription by thyroid hormone, *Eur. J. Biochem.* **180**:235–240.

Orr, W. C., and Sohal, R. S., 1994, Extension of life-span by overexpression of superoxide dismutase and catalase in *Drosophila melanogaster*, *Science* **263**:1128–1130.

Ostronoff, L. K., Izquierdo, J. M., and Cuezva, J. M., 1995, Mt-mRNA stability regulates the expression of the mitochondrial genome during liver development, *Biochem. Biophys. Res. Commun.* **217**:1094–1098.

Ozawa, T., Tanaka, M., Ikebe, S., Ohno, K., Kondo, T., and Mizuno, Y., 1990, Quantitative determination of deleted mitochondrial DNA relative to normal DNA in parkinsonian striatum by a kinetic PCR analysis, *Biochem. Biophys. Res. Commun.* **172**:483–489.

Pallotti, F., Chen, X., Bonilla, E., and Schon, E. A., 1996, Evidence that specific mtDNA point mutations may not accumulate in skeletal muscle during normal human aging, *Am. J. Hum. Genet.* **59**:591–602.

Pang, C.-Y., Lee, H.-C., Yang, J.-H., and Wei, Y.-H., 1994, Human skin mitochondrial DNA deletions associated with light exposure, *Arch. Biochem. Biophys.* **312**:534–538.

Papa, S., 1996, Mitochondrial oxidative phosphorylation changes in the life span. Molecular aspects and physiopathological implications, *Biochim. Biophys. Acta* **1276**:87–105.

Papa, S., Lorusso, M., and Capitanio, N., 1994, Mechanistic and phenomenological features of proton pumps in the respiratory chain of mitochondria, *J. Bioenerg. Biomembr.* **26**:609–618.

Paradies, G., Ruggiero, F. M., Gadaleta, M. N., and Quagliariello, E., 1992, The effect of aging and acetyl-L-carnitine on the activity of the phosphate carrier and on the phospholipid composition in rat heart mitochondria, *Biochim. Biophys. Acta* **1103**:324–326.

Paradies, G., Ruggiero, F. M., Petrosillo, G., Gadaleta, M. N., and Quagliariello, E., 1994, Effect of aging and acetyl-L-carnitine on the activity of cytochrome oxidase and adenine nucleotide translocase in rat heart mitochondria, *FEBS Lett.* **350**:213–215.

Paradies, G., Ruggiero, F. M., Petrosillo, G., Gadaleta, M. N., and Quagliariello, E., 1995, Carnitine acylcarnitine translocase activity in cardiac mitochondria from aged rats: The effect of acetyl-L-carnitine, *Mech. Ageing Dev.* **84**:103–112.

Paradies, G., Ruggiero, F. M., Petrosillo, G., and Quagliariello, E., 1997, Age-dependent decline in the cytochrome *c* oxidase activity in rat heart mitochondria: Role of cardiolipin, *FEBS Lett.* **406**:136–138.

Parisi, M. A., and Clayton, D. A., 1991, Similarity of human mitochondrial transcription factor 1 to high mobility group proteins, *Science* **252**:965–969.

Petruzzella, V., Baggetto, L. G., Penin, F., Cafagna, F., Ruggiero, F. M., Cantatore, P., and Gadaleta, M. N., 1992, *In vivo* effect of acetyl-L-carnitine on succinate oxidation, adenine nucleotide pool and lipid composition of synaptic and non-synaptic mitochondria from cerebral hemispheres of senescent rats, *Arch. Gerontol. Geriatr.* **14**:131–144.

Petruzzella, V., Fracasso, F., Gadaleta, M. N., and Cantatore, P., 1995, Decrease of D-loop frequency in heart and cerebral hemispheres mitochondrial DNA of aged rat, *Mol. Chem. Neuropathol.* **24**:193–202.

Pikò, L., 1992, Accumulation of mtDNA defects and changes in mtDNA content in mouse and rat tissues with aging, *Ann. NY Acad. Sci.* **663**:450–452.

Pikò, L., Houghman, A. J., and Bulpitt, K. J., 1988, Studies of sequence heterogeneity of mitochondrial DNA from rat and mouse tissues: Evidence for an increased frequency of deletions/additions with aging, *Mech. Ageing Dev.* **43**:279–293.

Pisano, C., Camerini, B., Castorina, M., Morabito, E., Carbonetti, A., Pucci, A., and Calvani, M., 1996, Acetyl-L-carnitine modulates histone H4 acetylation: Effects on gene expression and differentiation, in *24th Meeting of the Federation of European Biochemical Society*, FEBS '96, Barcelona, 7–12 July 1996. Abstract p. 155.

Radi, R., Turrens, J. F., Chang, L. Y., Bush, K. M., Crapo, J. D., and Freeman, B. A., 1991, Detection of catalase in rat heart mitochondria, *J. Biol. Chem.* **266**:22028–22034.

Renis, M., Cantatore, P., Loguercio Polosa, P., Fracasso, F., and Gadaleta, M. N., 1989, Content of mitochondrial DNA and of three mitochondrial RNAs in developing and adult rat cerebellum, *J. Neurochem.* **52**:750–754.

Richter, C., 1995, Oxidative damage to mitochondrial DNA and its relationship to aging, in *Molecular Aspects of Aging* (K. Esser and G. M. Martin, eds.), pp. 100–108, John Wiley & Sons, Chichester, England.

Ropp, P. A., and Copeland, W. C., 1996, Cloning and characterization of the human mitochondrial DNA polymerase, DNA polymerase γ, *Genomics* **36**:449–458.

Scarpulla, R. C., 1996, Nuclear respiratory factors and the pathways of nuclear-mitochondrial interaction, *Trends Cardiovasc. Med.* **6**:39–45.

Schon, E. A., Rizzuto, R., Moraes, C. T., Nakase, H., Zeviani, M., and DiMauro, S., 1989, A direct repeat is a hot spot for large-scale deletion of human mitochondrial DNA, *Science* **244**:346–349.

Seibel, P., Mell, O., Hannemann, A., Müller-Höcker, J., and Kadenbach, B., 1991, A method for quantitative analysis of deleted mitochondrial DNA by PCR in small tissue samples, *Methods Mol. Cell. Biol.* **2**:147–153.

Shigenaga, M. K., Hagen, T. M., and Ames, B. N., 1994, Oxidative damage and mitochondrial decay in aging, *Proc. Natl. Acad. Sci. USA* **91:**10771–10778.

Shoffner, J. M., Lott, M. T., Voljavec, A. S., Soueidan, S. A., Costigan, D. A., and Wallace, D. C., 1989, Spontaneous Kearns-Sayre/chronic external ophthalmoplegia plus syndrome associated with a mitochondrial DNA deletion: A slip replication model and metabolic therapy, *Proc. Natl. Acad. Sci. USA* **86:**7952–7956.

Siliprandi, N., Siliprandi, D., and Ciman, M., 1965, Stimulation of oxidation of mitochondrial fatty acids and of acetate by acetylcarnitine, *Biochem. J.* **96:**777–780.

Simonetti, S., Chen, X., Di Mauro, S., and Schon, E. A., 1992, Accumulation of deletions in human mitochondrial DNA during normal aging: Analysis by quantitative PCR, *Biochim. Biophys. Acta* **1180:**113–122.

Sohal, R. S., Svensson, I., and Brunk, U. T., 1990, Hydrogen peroxide production by liver mitochondria in different species, *Mech. Ageing Dev.* **53:**209–215.

Soong, N. W., Hinton, D. R., Cortopassi, G., and Arnheim, N., 1992, Mosaicism for a specific somatic mitochondrial DNA mutation in adult human brain, *Nature Genet.* **2:**318–323.

Soong, N. W., Dang, M. H., Hinton, D. R., and Arnheim, N., 1996, Mitochondrial DNA deletions are rare in the free radical-rich retinal environment, *Neurol. Aging* **17:**827–831.

Sugiyama, S., Hattori, K., Hayakawa, M., and Ozawa, T., 1991, Quantitative analysis of age-associated accumulation of mitochondrial DNA with deletion in human hearts, *Biochem. Biophys. Res. Commun.* **180:**894–899.

Takai, D., Inoue, K., Shisa, H., Kagawa, Y., and Hayashi, J.-I., 1995, Age-associated changes of mitochondrial translation and respiratory function in mouse brain, *Biochem. Biophys. Res. Commun.* **217:**668–674.

Tanhauser, S. M., and Laipis, P. J., 1995, Multiple deletions are detectable in mitochondrial DNA of aging mice, *J. Biol. Chem.* **270:**24769–24775.

Tiranti, V., Rocchi, M., Di Donato, S., and Zeviani, M., 1993, Cloning of human and rat cDNAs encoding the mitochondria single stranded DNA-binding protein (SSB), *Gene* **126:**219–225.

Tiranti, V., Savoia, A., Forti, F., D'Apolito, M. F., Centra, M., Rocchi, M., and Zeviani, M., 1997, Identification of the gene encoding the human mitochondrial RNA polymerase (H-mtRPOL) by cyberscreening of the expressed sequence tags data base, *Hum. Mol. Genet.* **6:**615–625.

Torii, K., Sugiyama, S., Takagi, K., Satake, T., and Ozawa, T., 1992, Age-related decrease in respiratory muscle mitochondrial function in rats, *Am. J. Respir. Cell Mol. Biol.* **6:**88–92.

Trounce, I., Byrne, E., and Marzuki, S., 1989, Decline in skeletal muscle mitochondrial respiratory chain function: Possible factor in aging, *Lancet* **1:**637–639.

Trumpower, B. L., and Gennis, R. B., 1993, Energy transduction by cytochrome complexes in mitochondrial and bacterial respiration: The enzymology of coupling electron transfer reactions to transmembrane proton translocation, *Annu. Rev. Biochem.* **63:**675–716.

Van Tuyle, G. C., Gudikote, J. P., Hurt, V. R., Miller, B. B., and Moore, C. A., 1996, Multiple, large deletions in rat mitochondrial DNA: Evidence for a major hot spot, *Mutat. Res.* **349:**95–107.

Wallace, D. C., 1992a, Mitochondrial genetics: A paradigm for aging and degenerative diseases? *Science* **126:**628–632.

Wallace, D. C., 1992b, Diseases of the mitochondrial DNA, *Annu. Rev. Biochem.* **61:**1175–1212.

Wallace, D. C., Zheng, X., Lott, M. T., Shoffner, J. M., Hodge, J. A., Kelley, R. L., Epstein, C. M., and Hopkins, L. C., 1988, Familial mitochondrial encephalomyopathy (MERRF): Genetic, pathophysiological and biochemical characterization of a mitochondrial DNA disease, *Cell* **55:**601–610.

Wallace, D. C., Bohr, V. A., Cortopassi, G., Kadenbach, B., Linn, S., Linnane, A. W., Richter, C., and Shay, J. W., 1995, The role of bioenergetics and mitochondrial DNA mutations in aging and age-related diseases, in *Molecular Aspects of Aging* (K. Esser and G. M. Martin, eds.), pp. 199–225, John Wiley & Sons, Chichester, England.

Wiesner, R. J., Kurowski, T. T., and Zak, R., 1992, Regulation by thyroid hormone of nuclear and mitochondrial genes encoding subunits of cytochrome *c* oxidase in rat liver and skeletal muscle, *Mol. Endocrinol.* **6:**1458–1467.

Yen, T.-C., Pang, C.-Y., Hsieh, R.-H., Su, C.-H., King, K.-L., and Wei, Y.-H., 1992, Age-dependent 6kb deletion in human liver mitochondrial DNA, *Biochem. Int.* **26:**457–468.

Yu, B. P., 1995, Putative interventions, in *Aging* (E. J. Masoro, ed.), pp. 613–631, *Handbook of Physiology*, Oxford University Press.

Zeviani, M., Moraes, C. T., DiMauro, S., Nakase, H., Bonilla, E., Schon, E. A., and Rowland, L. P., 1988, Deletions of mitochondrial DNA in Kearns–Sayre syndrome, *Neurology* **38:**1339–1345.

Zhang, C., Baumer, A., Maxwell, R. J., Linnane, A. W., and Nagley, P., 1992, Multiple mitochondrial DNA deletions in an elderly human individual, *FEBS Lett.* **297:**34–38.

Zhang, C., Linnane, A. W., and Nagley, P., 1993, Occurrence of a particular base substitution (3243 A to G) in mitochondrial DNA of tissues of ageing humans, *Biochem. Biophys. Res. Commun.* **195:**1104–1110.

Zhang, C., Peters, L. E., Linnane, A. W., and Nagley, P., 1996, Comparison of different quantitative PCR procedures in the analysis of the 4977-bp deletion in human mitochondrial DNA, *Biochem. Biophys. Res. Commun.* **223:**450–455.

Zhang, C., Liu, V. W. S., and Nagley, P., 1997a, Gross mosaic pattern of mitochondrial DNA deletions in skeletal muscle tissues of an individual adult human subject, *Biochem. Biophys. Res. Commun.* **233:**56–60.

Zhang, C., Bills, M., Quigley, A., Maxwell, R. J., Linnane, A. W., and Nagley, P., 1997b, Varied prevalence of age-associated mitochondrial DNA deletions in different species and tissues: A comparison between human and rat, *Biochem. Biophys. Res. Commun.* **230:**630–635.

Zucchini, C., Pugnaloni, A., Pallotti, F., Solmi, R., Crimi, M., Castaldini, C., Biagini, G., and Lenaz, G., 1995, Human skeletal muscle mitochondria in aging: Lack of detectable morphological and enzymic defects, *Biochem. Mol. Biol. Int.* **37:**607–616.

CHAPTER 29

Nucleus-Driven Lesions of mtDNA and Disorders of Nucleus-Encoded Energy Genes

Massimo Zeviani, Vittoria Petruzzella, Monica Munaro, and Francesca Forti

1. INTRODUCTION

Oxidative phosphorylation (OXPHOS) is a genetically unique metabolic pathway, as it is the product of the complementation of two distinct genetic systems, the nuclear and the mitochondrial genomes (Attardi and Schatz, 1988). Nuclear genes encode most of the protein components of the respiratory chain complexes, the factors controlling the cytoplasm–mitochondrial protein trafficking, as well as those carrying out and controlling the maintenance, propagation, and transcription of mtDNA. The latter provides the genetic information for only 13 proteins that are essential components of four of the five respiratory chain complexes and for the RNA components of the mito-chondrial autochthonous translational apparatus (Anderson *et al.*, 1981). The

Massimo Zeviani and Vittoria Petruzzella • Unit of Molecular Medicine, Children's Hospital "Bambino Gesù," Rome 00165, Italy. **Monica Munaro and Francesca Forti** • Division of Biochemistry and Genetics, National Neurological Institute "C. Besta," Milan 20133, Italy.

Frontiers of Cellular Bioenergetics, edited by Papa *et al.* Kluwer Academic/Plenum Publishers, New York, 1999.

dual genetic control of OXPHOS accounts for the existence of three groups of human OXPHOS-related disorders: (1) sporadic or maternally inherited disorders due to mtDNA mutations; (2) biochemically or genetically defined disorders due to defects of OXPHOS-related nuclear genes; and (3) nucleus-driven mutations of mitochondrial DNA (mtDNA).

The most relevant contribution to the understanding of mitochondrial disorders has come from the discovery of an impressive and ever-expanding number of disease-associated mutations affecting genes that either control the synthesis of or code for the 13 respiratory subunits specified by the mitochondrial genome (Zeviani *et al.*, 1997). On the other hand, defects of nuclear genes involved in the synthesis and function of OXPHOS enzymes are believed to be responsible for familial cases compatible with autosomal recessive inheritance and/or cases characterized by severe enzyme defects not associated with any known mutations of mtDNA. However, with the exception of a single report of a defect of complex II due to a point mutation in a nuclear succinate dehydrogenase gene (Bourgeron *et al.*, 1995), the attribution of these disorders to nuclear gene defects remains speculative and the current classification is based on biochemical findings only. Finally, the third group of disorders is characterized by the presence of mtDNA abnormalities associated with disease and transmitted as mendelian traits. Mendelian inheritance indicates the presence of transmissible mutations in nuclear genes that can damage the structural integrity of the mtDNA molecule, or affect the mtDNA copy number.

The following discussion will concern the last two groups of disorders, i.e. those that are caused by defects in nuclear genes, either directly, or via the induction of mtDNA mutations (nucleus-driven mutations of mtDNA).

2. NUCLEUS-DRIVEN MUTATIONS OF mtDNA

Mendelian traits can be associated with either multiple large-scale rearrangements of mtDNA (qualitative abnormalities) or with alterations of the mtDNA copy number (quantitative abnormalities).

2.1. Qualitative Abnormalities: Multiple Familial mtDNA Deletions

Autosomal dominant progressive external ophthalmoplegia (adPEO) is a neuromuscular disorder due to the accumulation of multiple deletions of mtDNA (ΔmtDNA) in stable tissues. Mendelian transmission indicates that the primary genetic defect resides in a nuclear gene. On the other hand, the presence of molecular lesions affecting the mtDNA molecule suggests that the defective gene is somehow involved in the control of the biogenesis or in the maintenance of the structural integrity of mtDNA (Zeviani *et al.*, 1989, 1990; Servidei *et al.*, 1991).

2.1.1. adPEO: Clinical Features

Invariant clinical features of symptomatic individuals are adult-onset PEO and proximal muscle weakness and wasting. Tremor, ataxia, extrapyramidal signs, hearing loss, and chronic sensorimotor peripheral neuropathy are reported in some families. Elevated levels of serum lactate at rest are detected in severely affected patients. Symptoms seem to progress with the age of the patients. Examination of muscle biopsies shows the presence of ragged-red fibers (RRF) due to the subsarcolemmal accumulation of abnormal mitochondria. In addition, the histochemical reaction to cytochrome *c* oxidase (COX, complex IV) is decreased or absent in scattered fibers, and neurogenic changes are also evident.

Biochemically, the activities of both respiratory complexes I and IV in muscle homogenates can range from normal to about 50% of the normal mean. Presymptomatic patients appear normal at the clinical examination, but usually have laboratory, electrophysiological, morphological, and biochemical features of a subclinical mitochondrial encephalomyopathy. Families with a similar phenotype have been reported by others (Suomalainen *et al.*, 1992; Moslemi *et al.*, 1996).

2.1.2. Molecular Features

Southern blot analysis of mtDNA from symptomatic and presymptomatic patients show the presence of "pleioplasmic" rearrangements, i.e., the coexistence of multiple forms of ΔmtDNAs in addition to a major 16.5 kb species corresponding to the wild-type molecule. An example of these results is shown in Fig. 1. Both Southern blot and polymerase chain reaction (PCR)-based

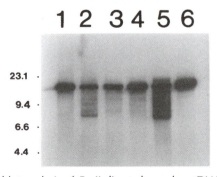

FIGURE 1. Southern blot analysis of *Pvu*II-digested muscle mtDNA of an adPEO family (Zeviani *et al.*, 1989). Patients (lanes 2–5) show multiple mtDNA species, while normal individuals (lanes 1, 6) contain only homoplasmic wild-type mtDNA.

strategies have been used to map the deletions throughout the circular mtDNA chromosomes, and to sequence the abnormal ΔmtDNAs around the deletion breakpoints (Zeviani *et al.*, 1990). These studies yielded the following results.

1. The mtDNA regions corresponding to probes spanning the major region from the origin of H-strand replication (O_H) to the origin of L-strand replication (O_L) are preferentially contained within deleted areas.
2. There are several mutational "hot spots" scattered throughout a wide area of mtDNA.
3. The D-loop, a crucial region containing both the O_H and the transcriptional promoters, is conserved in the mutant species.
4. The molecular rearrangements occur across flanking direct repeats of variable length.

2.1.3. adPEO: Tissue Distribution

In adPEO, multiple ΔmtDNAs are present in stable, postmitotic, highly specialized tissues including the skeletal muscle, myocardium, and brain (Suomalainen *et al.*, 1992). By contrast, the viscera (kidney, liver) contain hardly any deleted genomes. Moreover, cultured fibroblasts, peripheral blood cells, and cultured myoblasts from our patients consistently failed to show any deleted mtDNA species (Servidei *et al.*, 1991), and ΔmtDNAs were absent in cultured clones of both postmitotic muscle cell lines (myotubes) and *in vitro* innervated muscle cells.

2.1.4. adPEO: Pathogenetic Considerations

An etiologic relation between the mtDNA pleioplasmy observed in adPEO and the clinical phenotype is suggested by the fact that all of the affected and none of the unaffected tested individuals had deletions. Moreover, the PEO phenotype is commonly associated with mtDNA deletions (Moraes *et al.*, 1989; Holt *et al.*, 1989). Genetic analysis of our pedigrees indicates that the mitochondrial genomes were damaged by a dominantly inherited mutation of a still unknown nuclear gene. Although the molecular mechanism leading to deletions is still obscure, at least three hypotheses can be offered.

First, the mutated gene product can act by facilitating and amplifying an intrinsic propensity of mtDNA to undergo rearrangements. In experiments based on PCR amplification of mtDNA from sections of single muscle fibers, Moslemi *et al.* (1996) showed that within a single COX-deficient muscle fiber segment only one single deletion could be detected. However, different deletions were identified in different segments. These results indicate clonal expansion of a single ΔmtDNA in each COX-deficient muscle fiber segment. A two-hit mechanism can therefore be proposed consisting of the combination of a nuclear factor, which somehow predisposes to mtDNA deletions, with the subsequent clonal expansion of each ΔmtDNA in muscle and other stable tissues. According to this model, the number of somatic mutational events leading to mtDNA deletions should be relatively low. A combination of low mutational rate and active selection against deletion-containing cells could explain the absence of ΔmtDNAs in *in vitro* systems such as fibroblast and muscle cell cultures, or in cells with a rapid turnover such as lymphocytes. The functional status of any given tissue will ultimately depend on the accumulation of deleted mtDNA species. The mechanism leading to such accumulation is presently unknown. However, it seems reasonable that it can occur independently of the first hit because of the primary nuclear mutation and it can be influenced by local conditions within individual cells or areas in a tissue. This could explain the regional variation of detectable deletions in muscle fibers and brain, as well as the variability of neurological symptoms in different individuals. Similar to single sporadic mtDNA deletions, multiple familial deletions occur across flanking direct repeats. In contrast with sporadic cases, however, in familial cases the repeats are frequently nonperfect, often very short (triplets or quadruplets), sometimes slipped, and occasionally interrupted or inverted. Furthermore, deletions take place in hot-spot regions that have never been described in single, sporadic deletions (Mita *et al.*, 1990). These observations suggest that the molecular mechanism by which multiple mtDNA deletions are produced is probably identical or very similar to that causing single deletions, but both the frequency of and the propensity for deletional events are very greatly potentiated in the familial cases.

Second, the mutation could affect a gene encoding a hypothetical factor the function of which is to select and eliminate abnormal mtDNAs. This possibility is supported by the observation that, with the exceptions mentioned above, in normal mammalian tissues mtDNA heterogeneity is rare and usually confined to small-length variations in the D-loop region (Greenberg *et al.*, 1983; Ashley *et al.*, 1981), suggesting that extensive intraindividual mtDNA heterogeneity may not be tolerated and that heteroplasmic genomes may be rapidly sorted out by either cellular or molecular selection (Upholt and Dawid, 1977; Takahata and Muruyama, 1981; Chapman *et al.*, 1982).

A third hypothesis is that accumulation of multiple ΔmtDNAs occurs as a result of malfunctioning of the scavenging systems that protect the mitochondrial environment from the action of oxygen free radicals. It has been suggested that the formation of free radicals results in mtDNA mutations in normal subjects. Oxygen free radicals, particularly the hydroxyl radical, are generated as by-products of the oxidative metabolism, and being compounds of high chemical reactivity, they cause peroxidation of membrane lipids, accumulation of age-related pigments, and DNA fragmentation (reviewed by Harman, 1994). A defective OXPHOS pathway might in turn favor the accumulation of free radicals by hampering the reduction of oxygen to metabolic water. If free radicals accounted for the accumulation of multiple mtDNA deletions in familial PEO, the defective nuclear protein that predisposes mtDNA to deletions could be a protein normally eliminating the free radicals or protecting mtDNA from them or even a protein that when defective could create them.

2.1.5. Identification of adPEO Disease Loci

Irrespective of the pathogenetic mechanism(s) leading to deletions, the fundamental issue for adPEO is the identification of the responsible gene. The availability of a number of adPEO families has made it possible to develop a strategy based on positional cloning that aims toward the identification of the disease locus (or loci). We have been collecting data on several Italian families who fulfilled the following criteria: typical clinical features; unequivocally demonstrated autosomal dominant pattern of inheritance; presence in the tested affected members of multiple mtDNA deletions in muscle; and absence of the lesions in non-affected family members.

Given the small size of the chromosome pool available for the initial primary linkage analysis, significant results for each marker could be obtained only by pooling the LOD scores calculated in each family. This procedure was based on the arbitrary assumption that the trait was homogeneous in the tested families. However, genetic heterogeneity could not be excluded a priori. Indeed, both our initial exclusion data and a comparison with results of linkage analysis performed on a large Finnish family by Suomalainen and co-workers strongly indicated that the disease is in fact heterogeneous and that more than one locus is associated with adPEO, possibly on the basis of geographic or ethnic segregation. Moreover, evidence has been gained that incomplete penetrance, and variability of the age of onset could further complicate the interpretation of the results. Later, the size of the data for some of our families were substantially expanded by collecting data on new individuals, and a second linkage screening was undertaken to identify the locus specific for each family.

A primary linkage analysis carried out on a single large and clinically well-characterized Finnish family allowed Suomalainen and co-workers (1995) to identify a first adPEO locus on chromosome 10q23.3-24.3. This locus was excluded in two Italian families, indicating that the disease was genetically heterogeneous. A second adPEO locus was later found in three Italian families on chromosome 3p14.1-21.2 (Kaukonen *et al.*, 1996). Evidence indicating the existence of at least one more locus was provided in the same study by exclusion data of additional adPEO families. The presence of genetic heterogeneity is not unexpected, considering the extraordinary complexity of the nuclear control of mitochondrial biogenesis. The physical cartography of the two identified adPEO loci, the genetic characterization of new families, and the search for genes candidates by position and function are complementary strategies aiming to elucidate the genetic basis of this disease.

2.1.6. Clinical and Genetic Variants

Childhood-onset autosomal-recessive (ar)PEO complicated by mild facial and proximal limb weakness and severe cardiomyopathy associated with multiple ΔmtDNAs in muscle has been reported in six patients from two unrelated Arabian families (Bohlega *et al.*, 1996). The association of a nuclear gene defect with multiple ΔmtDNAs has been proposed also in Wolfram's syndrome, another autosomal recessive trait (Barrientos *et al.*, 1996), and in a few other conditions (see Zeviani *et al.*, 1997).

2.2. Quantitative Abnormalities: mtDNA Depletion

Depletion of mtDNA is a tissue-specific disorder whose clinical manifestations fall into three groups: (1) a fatal infantile congenital myopathy with or without a DeToni–Fanconi renal syndrome; (2) a fatal infantile epatopathy leading to rapidly progressive liver failure; and (3) a late infantile or childhood myopathy, with onset after 1 year of age, characterized by a progressive myopathy causing respiratory failure and death by 3 years of age. The presence of affected siblings born from healthy parents suggested an autosomal recessive mode of inheritance, possibly affecting a nuclear gene involved in the control of the mtDNA copy number. Southern blot analysis is diagnostic, demonstrating the severe reduction of mtDNA in affected tissues (up to 98% in the most severe forms) (Moraes *et al.*, 1991; Tritschler *et al.*, 1992; Mariotti *et al.*, 1995; Bakker *et al.*, 1996). An example of muscle-specific mtDNA depletion in a baby boy studied by us (Mariotti *et al.*, 1995) is shown in Fig. 2.

2.2.1. Etiological Considerations

The etiology of mtDNA depletion in humans remains obscure. However, the involvement of siblings of both sexes and the lack of involvement of their parents suggest that mtDNA depletion is transmitted as an autosomal recessive trait. The nonmaternal mode of inheritance suggests a mutation in a nuclear DNA as the primary cause of the mtDNA depletion.

Bodnar *et al.* (1993) observed the repopulation to normal mtDNA levels of transmitochondrial cybrids obtained by fusing cytoplasts derived from mtDNA-depleted fibroblasts with an mtDNA-less human tumor cell line. This result

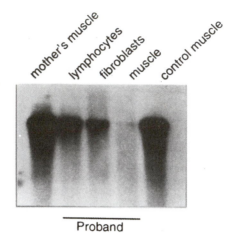

Proband

FIGURE 2. Southern blot analysis of *Pvu*II-digested mtDNA from several tissues of a baby boy with severe depletion of muscle mtDNA (Mariotti *et al.*, 1995).

suggests the nuclear origin of the defect in their case. However, the case of Bodnar and co-workers (1995) is the only example reported in the literature of mtDNA depletion found in cultured fibroblasts; in all the remaining cases the amount of mtDNA in cells characterized by a rapid turnover was normal. For instance, in one early-onset myopathic case studied by us (Mariotti *et al.*, 1995), fibroblasts, myoblasts, and lymphocytes showed normal amounts of mtDNA, in spite of a reduction to 2–10% of the normal content of mito-chondrial genomes in mature skeletal muscle (Fig. 2). Likewise, myoblast-derived myotubes were unaffected, even after prolonged innervation. In addi-tion, a complete and rapid recovery of mtDNA content was observed after mtDNA depletion was produced in the patient's myoblasts by exposure to ethidium bromide. These considerations and the observation that mtDNA depletion can be associated with a wide spectrum of clinical presentations suggest that this condition is probably heterogeneous.

Genes controlling the mtDNA copy number, e.g., those involved in mtDNA replication (mtTFA, mtDNA polymerase, and so on) or in hier-archically higher control mechanisms [nuclear respiratory factors (NRF-1, NRF-2)] are attractive candidates for the genes responsible for this condition. Reduction of the mtTFA gene product has been observed in one case of mtDNA depletion. However, expression of the gene is not affected, and the de-crease of immunodetectable protein is probably a secondary effect due to absence of substrate (i.e., mtDNA) (Larsson *et al.*, 1994).

2.3. Nucleus-Driven mtDNA Mutations: Search for Candidate Genes

The characterization of the human mitochondrial proteome* is a matter of convergent interest between basic scientists involved in the elucidation of the fundamental mechanisms of nucleomitochondrial intergenomic signaling and clinically oriented researchers interested in mitochondrial disorders. Several protein products involved in mtDNA replication, transcription, and translation have been identified and partially purified in vertebrates. However, the cDNAs coding for only a few proteins involved in mtDNA housekeeping functions have been cloned and characterized in humans or mammals. These include the cDNAs specifying the transcription factor mtTFA (Parisi and Clayton, 1991), mitochondrial single-stranded DNA binding protein (mtSSB) (Tiranti et al., 1993), and a new mitochondrial endonuclease, endonuclease G, similar to the yeast Nuc1 endonuclease (Coté and Ruiz-Carrillo, 1993). Very recently, the cDNAs encoding two additional important factors, the mitochondrial transcription terminator (mTERF) (Fernandez-Silva et al., 1997) and the γ-DNA polymerase (Ropp and Copeland, 1996), have been found in humans.

The molecular dissection of the mammalian mitochondrial proteome is hampered by a number of problems. First, many housekeeping mitochondrial functions are still unknown in higher eukaryotes. Second, even when the function is well established, purification of the corresponding polypeptide is hampered by protein lability, low amount, or difficulty in separating the activities that belong to mitochondria from analogous activities present in the nucleus or in other compartments of the cellular cytoplasm. These problems can make it technically very difficult, or even impossible, to apply cloning strategies that rely upon the availability of sequence information on the corresponding protein, such as hybridization of cDNA libraries or PCR amplification with degenerate oligonucleotides.

2.3.1. dbEST Cyberscreening: Cloning the Human mtRNA Polymerase

In contrast to the scarcity of information on human genes related to mitochondrial biogenesis, a remarkable wealth of information on such genes has been accumulated over the last two decades in lower eukaryotes, particularly in the yeast Saccharomyces cerevisiae. We have exploited this knowledge, together with the spectacular expansion of sequence information on expressed human genes provided by the expressed sequence tags database (dbEST), to work out a cloning strategy for important mitochondrial genes by means of dbEST "cyberscreening." Human dbEST as well as dbESTs for other metazoan species are reachable through numerous web sites. Likewise, as the

---

*Mitochondrial proteome is the collection of human proteins related to mitochondria.

nucleotide sequence of the entire yeast genome has recently been completed, all the expressed genes of *S. cerevisiae* are now available in public databases. We have been exploiting a computer-based cyberscreening of dbEST using yeast proteins as a "cyberprobe"*: The yeast protein sequences are "probed" against the entire database to identify homologous sequences by means of suitable softwares such as the Basic Local Alignment Search Tool (BLAST). This approach has recently allowed us to clone the full-length cDNA encoding the human ortholog gene to the yeast mitochondrial RNA polymerase. The Yeast Protein Database (YPD) worldwide web site (http//:www.ypd.org/proteome.html) was used to retrieve the predicted protein sequence deduced from the RP04 open reading frame (ORF) (yeast mitochondrial RNA polymerase gene). The TBLASTN option (protein sequence vs. six-frame nucleotide translated database sequences) of BLAST was used for a first dbEST searching at the TIGEMnet worldwide web site (http//:www.tigem.it).

We first identified two EST sequences that were used to start a "cDNA walking" based on both traditional screening of cDNA libraries and further screening of dbEST. The cDNA is 3831 bp in length and contains a potentially translatable continuous ORF of 3690 nucleotides. The predicted protein, consisting of 1230 amino acid residues, was identified as the bona fide precursor of the human mtRPOL, the central component of the mitochondrial transcriptional apparatus. The human protein sequence shares significant identity and similarity with the yeast mtRPOL, especially in the carboxy-terminal half of the sequences. The amino-terminus of the human protein sequence has the typical features of a mitochondrial leader peptide, suggesting that the protein is directed to and imported into mitochondria.

We were able to demonstrate that this indeed is the case. We allowed COS-7 cells to express a chimeric recombinant protein consisting of the first 330 amino acid residues of the human-mtRPOL precursor fused to an immunogenic "tag" at the C-terminus. By means of mitochondrial import experiments *in vivo* and immunofluorescence *in situ*, we showed that energy-dependent mitochondrial targeting of the fusion protein occurred and that cleavage of the leader peptide took place (Tiranti *et al.*, 1997).

One distinct advantage of this method is that it is based on protein cyberprobes instead of nucleotide probes used in traditional screening of cDNA libraries. Since protein sequences are less divergent than the corresponding nucleotide sequences, the chance to find significant homology is substantially increased even between evolutionarily distant organisms. In particular, many proteins of mitochondria are likely to be conserved throughout evolution, given the essential role of the mitochondrial energy metabolism in all the eukaryotic lineages. Other advantages of the cyberscreening approach are the rapidity and reliability of the results, due to the availability of powerful computational methods.

---

*Cyberprobe is a nucleotide (or amino acid) sequence serving as a "probe."

The rapid expansion of EST databases and other genetic tools linked to the human genome project are expected to increase the power and effectiveness of cloning strategies, such as that outlined here, to accelerate the molecular dissection of the human mitochondrial proteome. This is an important goal for both basic and clinical research.

3. NUCLEAR GENES AND OXPHOS DISORDERS: CELLULAR MODELS

The observation of familial cases with mendelian inheritance and severe isolated defects of the respiratory chain complexes, not associated with mtDNA lesions, suggests the presence of mutations of nuclear genes as a source of respiratory chain deficiencies. However, no mutations in nuclear genes related to the respiratory chain have been reported to date. The only exception was a point mutation in the nuclear-encoded flavoprotein subunit gene of succinate dehydrogenase that was recently found in two siblings affected with Leigh's syndrome associated with deficiency of complex II (Bourgeron et al., 1995). Therefore, for those mitochondrial disorders that still lack a molecular genetic definition, the classification is based on biochemical criteria only (DiMauro and Moraes, 1993).

The clinical presentation of defects of the respiratory chain is heterogenous, with the time of onset ranging from the neonatal period to adult life. The clinical features include fatal infantile multisystem syndromes, encephalomyopathies, or isolated myopathies sometimes associated with cardiopathies. In pediatric patients the most frequent clinical features are severe psychomotor delay, generalized hypotonia, lactic acidosis, and signs of cardiorespiratory failure. Patients with later onset usually show signs of myopathy associated with variable involvement of the CNS (ataxia, hearing loss, seizures, polyneuropathy, pigmentary retinopathy, and, rarely, movement disorders). Other patients complain only of muscle weakness and/or wasting with exercise intolerance.

Fatal infantile myopathy is characterized by severe muscle weakness, lactic acidosis, and ventilatory insufficiency causing death before 1 year of age. In some cases, the disease is associated with DeToni–Fanconi's syndrome. This form must be differentiated from a reversible (benign) form, initially indistinguishable from the former and characterized by a progressive spontaneous improvement with remission by age 3 years (Zeviani et al., 1987; Tritschler et al., 1991). The biochemical defect is most often a severe reduction of cytochrome c oxidase (COX, complex IV) activity in muscle. Defects of complex I [NADH-coenzyme Q reductase (NADH-CoQ RD)], or a combination of complex I and IV have been reported, but more rarely.

The most common disorder of the respiratory chain in infancy and childhood is Leigh's syndrome (LS) or subacute necrotizing encephalomyopathy. LS is characterized by the predominant involvement of CNS. Biochemically, a generalized defect of respiratory complex IV is found in the majority of patients. However, deficiencies of pyruvate dehydrogenase complex or respiratory complex I, or the presence of mtDNA point mutations have also been reported (Rahman *et al.*, 1996; Morris *et al.*, 1996). Thus, the nuclear versus mitochondrial origin of the primary defect in LS still must be established. We have recently worked out two strategies aiming to demonstrate which gene complement is responsible for the defect in COX-defective (COX$^{(-)}$) LS.

3.1. "Customized" Cybrids Using Patient-Derived rho^0 Cells

The first strategy (Fig. 3) is based on the fusion of nuclear DNA-less cytoplasts derived from normal fibroblasts with mtDNA-less transformant fibroblasts derived from the proband (Tiranti *et al.*, 1995). We first obtained an SV40-transformed, neomycin-resistant (Neo^r) derivative cell line from fibroblasts expressing a severe COX deficiency. The isolated defect of COX, found in the progenitor cell line, was maintained in the SV40 transformant Neo^r cell line as well. We then obtained a permanent rho^0 cell line by pharmacological treatment of our patient-derived SV40 transformant cell line. This cell line was fused with cytoplasts derived from a normal control, and the resulting cybrid line was tested for COX activity both histochemically and biochemically. Figure 4 shows the results of this strategy. The reexpressed OXPHOS defect was specific, since complex I showed normal activities, similar to what was found in the patient's fibroblasts and in the transformant derivative that served as the nuclear parent of the COX$^{(-)}$ cybrid line.

The COX defect detected in a cell homogenate was confirmed *in situ* by COX-specific histochemistry. The biochemical phenotype segregated entirely with the patient's nuclear genome, because the defect of COX was absent in "traditional" transmitochondrial cybrids (King and Attardi, 1989), obtained by complementing the patient's mitochondrial genome with the nuclear genome of a rho^0 cell line derived from COX$^{(+)}$ 143B.TK$^-$ cells. Therefore, using the rho^0 transformant cells and an OXPHOS-competent mitochondrial donor, we created a "customized" transmitochondrial cybrid system, in which we reproduced an isolated defect of COX. This study provides a positive, direct demonstration of the nuclear origin of an OXPHOS defect associated with LS.

3.2. Patient/143B.rho^0 Hybrids

The second strategy (Munaro *et al.*, 1997) is based on the ability of a COX$^{(-)}$, mtDNA-less 143B-derived rho^0 cell line to complement the COX$^{(-)}$ phenotype of transformant fibroblasts derived from LS patients (Fig. 5). We

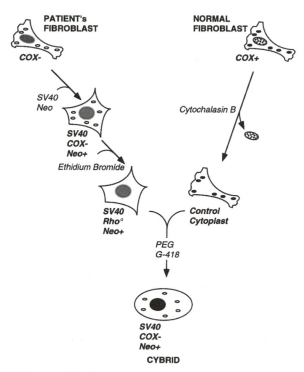

FIGURE 3. Cybrids derived by fusing control cytoplasts with patient-derived rho⁰ transformant cells. Experimental outline: Patient's COX$^{(-)}$ fibroblasts are transfected with a plasmid containing the SV40 genome and the neomycin-resistance (Neo) gene. Transformants are treated with ethidium bromide to produce rho⁰ derivatives. The latter are fused by polyethylene glycol (PEG) with control cytoplasts under continuous selection with G-418.

used seven cell lines derived from as many COX-defective LS patients. Each of the remaining seven patient-derived parental cell lines were fused with a COX$^{(-)}$, mtDNA-less rho⁰ clone derived from 143B.TK tumor cells. The patients' cell lines were transformed with an SV40^{ori-} construct, as above, and all were made puromycin-resistant (Puro^r), while the 143B.rho⁰ cells were made neomycin-resistant (Neo^r), by transfection of suitable recombinant vectors expressing the corresponding resistance-conferring gene.

The selection of the heterokaryon hybrids was obtained by prolonged treatment with both G-418 and puromycin: Only hybrids derived from the fusion of a Neo^r parental cell with a Puro^r parental cell could survive this treatment, while both unfused parental cells and homokaryon hybrids were eliminated by exposure to the toxic effect of the one antibiotic for which they

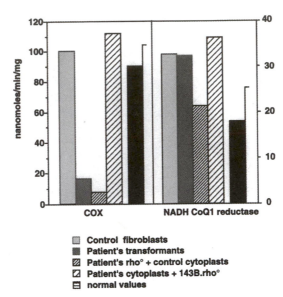

FIGURE 4. Biochemical results of the experiment outlined in Fig. 3. The data demonstrate the reexpression of the COX$^{(-)}$ phenotype in "customized" cybrids containing the patient's nuclear gene complement versus the mitochondrial gene complement of a normal control. Cybrids containing the nuclear gene complement of 143B tumor cells versus the mitochondrial gene complement of the patient are COX$^{(+)}$. See Tiranti *et al.* (1995).

lacked resistance. After selection, heterokaryons were demonstrated by DNA genotyping of the highly polymorphic locus D11S533. Hybrids were then tested for the restoration of COX competency by COX-specific histochemistry *in situ* and by a COX-specific biochemical assay on cell homogenates. Hybrids were consistently COX$^{(+)}$, indicating that (1) the 143B.rho^0 nuclear genome was able to complement the OXPHOS defect, and (2) the nuclear gene mutation behaved as a recessive trait in all cases. An identical result was obtained in a hybrid generated by fusing a COX$^{(-)}$, Puro^r cell line with a COX$^{(+)}$, Neo^r normal cell line, again indicating the recessive behavior of the COX defect (Munaro *et al.*, 1997). In this system, correction of the COX$^{(-)}$ phenotype could only have been provided by the nucleus of our rho^0 cells, since these cells completely lack mtDNA and are themselves OXPHOS incompetent.

3.3. COX$^{(-)}$ LS: How Many Genes?

In order to gain further insight into the genetic complexity of COX$^{(-)}$ LS, we have developed a third strategy (Munaro *et al.*, 1997), aiming to demonstrate whether individual cases of COX$^{(-)}$ LS are due to a defect in the same

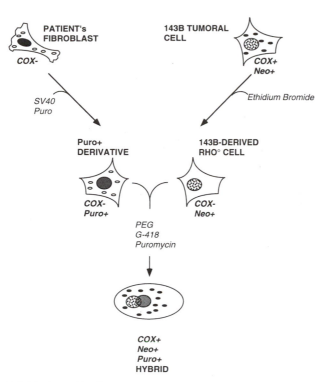

FIGURE 5. Hybrids derived by fusing 143B-derived, mtDNA-less (rho⁰) cells with patients-derived transformant cells. Experimental outline: Patient's COX$^{(-)}$ fibroblasts are transfected with a plasmid containing the SV40 genome and the puromycin-resistance (Puro) gene. The 143B rho⁰ cells are transfected with a plasmid containing the neomycin-resistance (Neo) gene. After fusion by polyethylene glycol (PEG), the heterodikaryon hybrids are selected in a medium containing both G-418 and puromycin.

gene or in different genes (Fig. 6). We created hybrids obtained by fusing a [Neo^r, SV40^{ori-}, COX$^{(-)}$] cell line of a patient with each of seven different [Puro^r, SV40^{ori-}, COX$^{(-)}$] cell lines from other patients. After selection by a double-antibiotic system, the presence of different cell complementation classes was evaluated in the heterokaryon hybrids by COX-specific histochemical and biochemical assays. Interestingly, in contrast to the results in patient–rho⁰ hybrids, histochemical and biochemical analyses showed that our patient–patient hybrids maintained the COX$^{(-)}$ phenotype of the parental cell lines, indicating that no complementation of the COX defect occurred in any case (Fig. 7).

These data demonstrate that our eight LS cases belonged to a single cell complementation class, suggesting that a major disease locus can account for

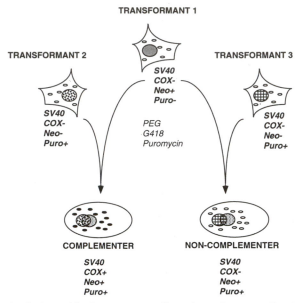

FIGURE 6. Hybrids derived by fusing patients-derived transformant cells with each other. Experimental outline: Patient's COX$^{(-)}$ fibroblasts are transfected with a plasmid containing the SV40 genome and either the puromycin-resistance (Puro) gene or the neomycin-resistance (Neo) gene. After fusion by polyethylene glycol (PEG), the heterodikaryon hybrids are selected in a medium containing both G-418 and puromycin.

several, perhaps most, of the cases of infantile COX$^{(-)}$ LS. However, the existence of other disease loci associated with either LS or other less well-defined generalized defects of COX is by no means excluded, and is indeed suggested by other reports (Brown and Brown, 1996).

The methods described here, based on the screening of somatic cell hybrids for COX complementation, provides a powerful means to characterize a genetically homogeneous population of patients, in spite of relatively heterogeneous clinical presentations. We think that this is an obligatory step toward the identification of the responsible gene by means of linkage analysis in selected families or other approaches based on positional cloning.

4. CONCLUSIONS

The identification of numerous mutations of human mtDNA associated with disease has revolutionized our concepts and approaches to mitochondrial disorders. On the other hand, the establishment of a rational diagnostic strategy

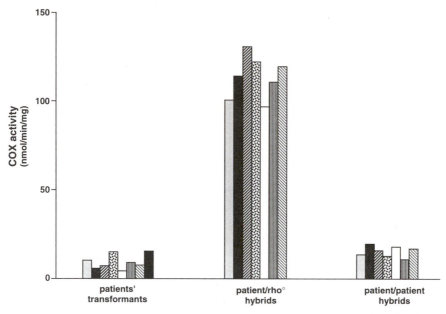

FIGURE 7. Biochemical results of the experiment outlined in Figs. 5 and 6. The data demonstrate the maintenance of the COX$^{(-)}$ phenotype in patient–patient hybrids. Hybrids obtained by fusing the nuclear gene complement of patients with that of rho^0 143B tumor cells are COX$^{(+)}$. See Munaro *et al.* (1997).

and pathogenetic interpretation of mtDNA abnormalities is based on knowledge of the principles dictating mtDNA genetics, expression, and control. A better understanding of mitochondrial biogenesis is also important for a genetic approach to therapy.

Conversely, mtDNA mutations can be considered as "experiments of nature" that permit one to gain insight into some fundamental aspects of mtDNA biology. For instance, the different effects of several tRNA–gene mutations on mitochondrial protein translation can assist in identifying the functional domains of mitochondrial tRNAs. Likewise, mtDNA deletions and mtDNA depletion have been shown to produce opposite effects on the expression and stability of mtTFA, revealing the existence of control loops connecting mtDNA to the principal modulator of its own expression and perpetuation.

Some mtDNA abnormalities are associated with mendelian disorders. These disorders are proposed to be due to defects of nuclear genes functionally related to mtDNA maintenance or protection. The identification of the genes responsible for these nucleus-driven mtDNA rearrangements will help us understand the molecular basis of nucleus–mitochondria communication.

A better understanding of the fundamental mechanisms of mitochondrial biogenesis therefore is essential to elucidate the pathogenesis of mitochondrial disorders and to explain numerous puzzling questions in this expanding area of human pathology. These questions include the influence of nuclear genes in the clinical expression of mtDNA mutations and the importance of mitochondrial abnormalities in the pathogenesis of diseases or conditions of great social and emotional impact, such as inherited deafness, diabetes mellitus, some neuro-degenerative disorders, the idiopathic cardiomyopathies, and the global progressive failure of organs and apparatuses linked to the aging process.

Studies on the tissue distribution of mutations and on the factors and conditions that can drive or influence their accumulation or instead, their complementation by normal genomes are essential steps toward the possibility of a rational prognostic evaluation of mitochondrial disorders and the provision of reliable genetic counseling and prenatal diagnosis. Finally, the molecular dissection of the human mitochondrial proteome will help identify the nuclear genes responsible for the many mitochondrial phenotypes that are not caused by mtDNA mutations. The availability of genes controlling mitochondrial DNA replication, gene expression, and possibly repair will allow the creation of recombinant animal or cellular models of mitochondrial disorders to test rational therapeutic strategies.

ACKNOWLEDGMENTS. We thank the I.M.A.G.E. Consortium and TIGEM for providing us with some of the genetic tools described in this chapter. We are indebted to Ms. B. Geehan for revising the manuscript. Partially supported by EU Human Capital and Mobility network grant to M.Z. on "Mitochondrial Biogenesis in Development and Disease," and Telethon-Italy (grant No. 767 to M.Z.).

5. REFERENCES

Anderson, S., Bankier, A. T., Barrel, B. G., de Bruijin, M. H. L., Coulson, A. R., Drouin, J., Eperon, I. C., Nierlich, D. P., Roe, B. A., Sanger, F., Schreier, P. H., Smith, A. J. H., Staden, R., and Young, I. G., 1981, Sequence and organization of the human mitochondrial genome, *Nature* **290:**457–465.

Ashley, M. V., Laipis P. J., and Hauswirth, W. W., 1981, Rapid segregation of heteroplasmic bovine mitochondria, *Nucleic Acids Res.* **17:**7325–7331.

Attardi, G., and Schatz, G., 1988, Biogenesis of mitochondria, *Annu. Rev. Cell Biol.* **4:**289–333.

Bakker, H. D., Scholte, H. R., Dingemans, K. P., Spelbrink, J. N., Wijkbur, F. A., and Van den Bogert, C., 1996, Depletion of mitochondrial deoxyribonucleic acid in a family with fatal neonatal liver disease, *J. Pediatr.* **128:**683–687.

Barrientos, A., Casademont, J., Saiz, F., Cardellach, F., Volpini, V., Solans, A., Tolosa, E., Urbano-Marquez, A., Estivill, X., and Nunes, V., 1996, Autosomal recessive Wolfram syndrome associated with an 8.5 kb mtDNA single deletion, *Am. J. Hum. Genet.* **58:**963–970.

Bodnar, A. G., Cooper, J. M., Holt, I. J., Leonard, J. V., and Schapira, A. H. V., 1993, Nuclear complementation restores mtDNA levels in cultured cells from a patient with mtDNA depletion, *Am. J. Hum. Genet.* **53:**663–669.

Bodnar, A. G., Cooper, J. M., Leonard, J. V., and Schapira, A. H. V., 1995, Respiratory deficient human fibroblasts exhibiting defective mitochondrial DNA replication, *Biochem. J.* **305:** 817–822.

Bohlega, S., Tanji, K., Santorelli, F. M., Hirano, M., al-Jishi, A., and Di Mauro, S., 1996, Multiple mitochondrial DNA deletions associated with autosomal recessive ophthalmoplegia and severe cardiomyopathy, *Neurology* **46:**1329–1334.

Bourgeron, T., Rustin, P., Chretien, D., Birch-Machin, M., Bourgeois, M., Viegas-Péquinot, E., Munnich, A. and Rötig, A., 1995, Mutation of a nuclear succinate dehydrogenase gene results in mitochondrial respiratory chain deficiency, *Nature Genet.* **11:**144–149.

Brown, R. M., and Brown, G. K., 1996, Complementation analysis of systemic cytochrome oxidase deficiency presenting as Leigh syndrome, *J. Inherit. Dis.* **19:**752–760.

Chapman, R. W., Stephens, J. C., Lansmann, R. A., and Avise, J. C., 1982, Models of mitochondrial DNA transmission genetics and evolution in higher eukaryotes. *Genet. Res.* **40:**41–57.

Coté, J., and Ruiz-Carrillo, A., 1993, Primers for mitochondrial DNA replication generated by endonuclease G, *Science* **261:**765–769.

DiMauro, S., and Moraes, C. T., 1993, Mitochondrial encephalomyopathies, *Arch. Neurol.* **50:**1197–1208.

Fernandez-Silva, P., Martinez-Azorin, F., Micol, V., and Attardi, G., 1997, The human mitochondrial transcription factor (mTERF) is a multizipper protein but binds to DNA as a monomer, with evidence pointing to intramolecular leucine zipper interactions, *EMBO J.* **16:**1066–1079.

Greenberg, B. D., Newbold, J. E., and Sugino, A., 1983, Intraspecific nucleotide sequence variability surrounding the origin of replication in human mitochondrial DNA, *Gene* **21:**33–49.

Harman, D., 1994, Free-radical theory of aging. Increasing the functional life span, *Ann. NY Acad. Sci.* **717:**1–15.

Holt, I. J., Harding, A. E., Cooper, J. M., Schapira, A. H. V., Toscano, A., Clark, J. B. and Morgan-Hughes, J. A., 1989, Mitochondrial myopathies: Clinical and biochemical features of 30 patients with major deletions of muscle mitochondrial DNA, *Ann. Neurol.* **26:**699–708.

Kaukonen, J. A., Amati, P., Suomalainen, A., Rötig, A., Piscaglia, M. G., Salvi, F., Weissenbach, J., Fratta, G., Comi, G., Peltonen, L., and Zeviani, M., 1996, An autosomal locus predisposing to multiple deletions of mtDNA on chromosome 3p, *Am. J. Hum. Genet.* **58:**763–769.

King, M., and Attardi, G., 1989, Human cells lacking mitochondrial DNA: Repopulation with exogenous mitochondria by complementation, *Science* **246:**500–503.

Larsson N. G., Oldfors, A., Holme, E., and Clayton, D. A., 1994, Low levels of mitochondrial transcription factor A in mitochondrial DNA depletion, *Biochem. Biophys. Res. Commun.* **200:**1374–1381.

Mariotti, C., Uziel, G., Carrara, F., Mora, M., Prelle, A., Tiranti, V., DiDonato, S., and Zeviani, M., 1995, Early-onset encephalomyopathy associated with tissue-specific mitochondrial DNA depletion: A morphological, biochemical and molecular-genetic study, *J. Neurol.* **242:**547–556.

Mita, S., Rizzuto, R., Moraes, C. T., Shanske, S., Arnaudo, E., Fabrizi, G. M., Koga, Y., DiMauro, S., and Schon, E. A., 1990, Recombination via flanking direct repeats is a major cause of large-scale deletions of human mitochondrial DNA, *Nucleic Acids Res.* **18:**561–567.

Moraes, C. T., DiMauro, S., Zeviani, M., Lombes, A., Shanske, S., Miranda, A. F., Nakase, H, Bonilla, E., Werneck, L. C., Servidei, S., Nonaka, I., Koga, Y., Spiro, A. J., Bromwell, K. W., Schmidt, B., Schotland, D. L., Zupanc, M., DeVivo, D. C., Schon, E. A., and Rowland, L. P., 1989, Mitochondrial DNA deletions in progressive external ophthalmoplegia and Kearns–Sayre syndrome, *N. Engl. J. Med.* **320:**1293–1299.

Moraes, C. T., Shanske, S., Tritschler, H.-J., Aprille, J. R., Andreetta, F., Bonilla, E., Schon, E. A., and DiMauro, S., 1991, mtDNA depletion with variable tissue expression: A noval genetic abnormality in mitochondrial diseases, *Am. J. Hum. Genet.* **48:**492–501.

Morris, A. A. M., Leonard, J. V., Brown, G. K., Bidouki, S. K., Bindoff, L. A., Woodward, C. E., Harding, A. E., Lake, B. D., Harding, B. N., Farrell, M. A., Bell, J. E., Mirakhur, M., and Turnbull, D. M., 1996, Deficiency of respiratory chain complex I is a common cause of Leigh disease, *Ann. Neurol.* **40:**25–30.

Moslemi, A.-R., Melberg, A., Holme, E., and Oldfors, A., 1996, Clonal expansion of mitochondrial DNA with multiple deletions in autosomal dominant progressive external ophthalmoplegia, *Ann. Neurol.* **40:**707–713.

Munaro, M., Tiranti, V., Sandonà, D., Lamantea, E., Uziel, G., Bisson, R., and Zeviani, M., 1997, A single cell complementation class is common to several cases of cytochrome *c* oxidase defective Leigh's syndrome, *Hum. Mol. Genet.* **6:**221–228.

Parisi, A. M., and Clayton, D. A., 1991, Similarity of human mitochondrial transcription factor 1 to high mobility group proteins, *Science* **252:**965–969.

Rahman, S., Blok, R. B., Dahl, H.-H. M., Danks, D. M., Kirby, D. M., Chow, C. W., Christodoulou, J., and Thorburn, D. R., 1996, Leigh syndrome: Clinical features and DNA abnormalities, *Ann. Neurol.* **39:**343–351.

Ropp, P. A., and Copeland, W. C., 1996, Mitochondrial DNA polymerase gamma is expressed and translated in the absence of mitochondrial DNA maintenance and replication, *Nucleic Acids Res.* **24:**2753–2759.

Servidei, S., Zeviani, M., Manfredi, G., Ricci, E., Silvestri, G., Bertini, E., Gellera, C., DiMauro, S., DiDonato, S., and Tonali, P., 1991, Dominantly inherited mitochondrial myopathy with multiple deletions of mitochondrial DNA: Clinical, morphologic, and biochemical studies, *Neurology* **41:**1053–1059.

Suomalainen, A., Majander, A., Haltia, M., Somer, H., Lonnqvist, J., Savontaus, M. L., and Peltonen, L., 1992, Multiple deletions of mitochondrial DNA in several tissues of a patient with severe retarded depression and familial progressive external ophthalmoplegia, *J. Clin. Invest.* **90:**61–66.

Suomalainen, A., Kaukonen, J. A., Amati, P., Timonen, R., Haltia, M., Weissenbach, J., Zeviani, M., Somer, H., and Peltonen, L., 1995, An autosomal locus predisposing to deletions of mitochondrial DNA, *Nature Genet.* **9:**146–151.

Takahata, N., and Maruyama, T., 1981, A mathematical model of extranuclear genes and genetic variability maintained in finite population, *Genet. Res.* **37:**291–302.

Tiranti, V., Rocchi, M., DiDonato, S., and Zeviani, M., 1993, Cloning of human and rat cDNAs encoding the mitochondrial single-stranded DNA-binding protein (SSB), *Gene* **126:**219–225.

Tiranti, V., Munaro, M., Sandonà, D., Lamantea, E., Rimoldi, M., DiDonato, S., Bisson, R., and Zeviani, M., 1995, Nuclear DNA origin of cytochrome *c* oxidase deficiency in Leigh syndrome: Genetic evidence based on patient's derived rho⁰ transformants, *Hum. Mol. Genet.* **4:**2017–2023.

Tiranti, V., Savoia, A., Forti, F., D'Apolito, M.-F., Centra, M., Rocchi, M., and Zeviani, M., 1997, Identification of the gene encoding the human mitochondrial RNA polymerase (h-mtRPOL) by cyberscreening of the expressed sequence tags database, *Hum. Mol. Genet.* **6:**615–625.

Tritschler, H.-J., Bonilla, E., Lombes, A., Andreetta, F., Servidei, S., Schneyder, B., Miranda, A. F., Schon, E. A., Kadenbach, B., and DiMauro, S., 1991, Differential diagnosis of fatal and benign cytochrome *c* oxidase-deficient myopathies of infancy: An immunohistochemical approach, *Neurology* **41:**300–305.

Tritschler, H.-J., Andreetta, F., Moraes, C. T., Bonilla, E., Arnaudo, E., Danon, M. J., Glass, S., Zelaya, B. M., Vamos, E., Telerman-Toppet, N., Shanske, S., Kadenbach, B., DiMauro, S., and Schon, E. A., 1992, Mitochondrial myopathy of childhood associated with depletion of mitochondrial DNA, *Neurology* **42:**209–217.

Upholt, W. B., and Dawid, I. B., 1977, Mapping of mitochondrial DNA of individual sheep and goats: Rapid evolution in the D-loop region. *Cell* **11:**571–583.

Zeviani, M., Peterson, P., Servidei, S., Bonilla, E., and DiMauro, S., 1987, Benign reversible muscle cytochrome c oxidase deficiency: A second case, *Neurology* **37:**64–67.

Zeviani, M., Servidei, S., Gellera, C., Bertini, E., DiMauro, S., and DiDonato, S., 1989, An autosomal dominant disorder with multiple deletions of mitochondrial DNA starting at the D-loop region, *Nature* **339:**309–311.

Zeviani, M., Bresolin, N., Gellera, C., Bordoni, A., Pannacci, M., Amati, P., Moggio, M., Servidei, S., Scarlato, G., and DiDonato, S., 1990, Nucleus-driven multiple large-scale deletions of the human mitochondrial genome: A new autosomal dominant disease, *Am. J. Hum. Genet.* **47:** 904–914.

Zeviani, M., Fernandez, P., and Tiranti, V., 1997, Disorders of mitochondria and related metabolism, *Curr. Opin. Neurol.* **10:**160–167.

CHAPTER 30

Aging and Degenerative Diseases
A Mitochondrial Paradigm

Douglas C. Wallace

1. INTRODUCTION

Over the past 9 years, a large number of mitochondrial DNA (mtDNA) mutations have been linked to human degenerative diseases. Clinical manifestations that have been identified in mtDNA disease patients include vision loss, deafness, dementia, movement disorders, seizures, strokes, myopathy, cardiomyopathy, endocrine disorders, and renal failure. One of the striking features of many mtDNA diseases is that they have a delayed onset and progressive course and that many of the symptoms seen in mitochondrial disease patients are also common features of the aging process. This has led to the hypothesis that many of the age-related degenerative diseases of humans as well as aging itself are related to defects in the mitochondrial bioenergetic pathway: oxidative phosphorylation (OXPHOS) (Wallace, 1992a, 1992b, 1994, 1995, 1997).

Douglas C. Wallace • Center for Molecular Medicine, Emory University School of Medicine, Atlanta, Georgia 30322.

Frontiers of Cellular Bioenergetics, edited by Papa *et al.* Kluwei Academic/Plenum Publishers, New York, 1999.

2. MITOCHONDRIAL BIOENERGETICS

Mitochondrial OXPHOS generates energy by oxidizing NADH and FADH$_2$ via the electron transport chain (ETC) (complexes I to IV) and using the energy released to synthesize ATP via complex V. In the ETC, electrons flow from complex I (NADH–ubiquinone oxidoreductase) or Complex II (succinate–ubiquinone oxidoreductase), to ubiquinone [coenzyme Q (CoQ)], to complex III (ubiquinol–cytochrome c oxidoreductase), to cytochrome c, to complex IV (cytochrome c oxidase), and then to oxygen. The energy released is used to pump protons out of the mitochondrion, creating an electrochemical gradient ($\Delta\mu^{+/+}$). The energy stored in this electrochemical gradient is utilized by complex V (ATP synthase) to condense ADP and Pi to create ATP. The ATP is then exchanged for cytosolic ADP by the adenine nucleotide translocator (ANT) (Shoffner and Wallace, 1995; Wallace, 1992a,b; Wallace et al., 1996a).

Because the oxygen consumption of the ETC is coupled to ATP synthesis through the ATP synthase, the oxygen consumption rate is regulated by the mitochondrial matrix concentration of ADP. In the presence of oxidizable substrates, the oxygen consumption rate in the absence of ADP is slow (state IV respiration) but increases when ADP is added (state III respiration). The ratio of state III to state IV respiration is called the respiratory control ratio (RCR) and, together with the ratio of ADP phosphorylated to O$_2$ consumed (P/O ratio), is indicative of the efficiency of OXPHOS. Uncouplers such as 2,4-dinitrophenol (DNP) bypass the ATP synthase and permit the ETC to run at its maximum rate (Wallace, 1997).

Superoxide anion is a toxic by-product of OXPHOS, generated by the transfer of electrons from the ETC directly to O$_2$ to give superoxide anion (O$_2^-$). Between 1 and 4% of the oxygen consumed by the mammalian respiratory chain is converted to oxygen radicals. The primary electron donors are reduced NADH dehydrogenase (complex I) and half-reduced CoQ (ubisemiquinone) (Bandy and Davison, 1990; Boveris and Turrens, 1980; Ksenzenko et al. 1983; Turrens and Boveris, 1980; Turrens et al., 1985; Wallace, 1997). Superoxide anion can be converted to H$_2$O$_2$ by the mitochondrial manganese superoxide dismutase (MnSOD) or cytosolic Cu/ZnSOD, and H$_2$O$_2$ can be reduced to water by glutathione peroxidase (GPx) or catalase. However, H$_2$O$_2$, in the presence of transition metals, can also be converted to the highly reactive hydroxyl radical (OH$^\bullet$) by the Fenton reaction (Bandy and Davison, 1990; Goldhaber and Weiss, 1992; Wallace, 1997).

Superoxide anion production and H$_2$O$_2$ release are highest during state IV respiration and lowest during state III respiration (Boveris, 1984; Boveris et al., 1972; Cadenas and Boveris, 1980; Chance et al., 1979). Moreover, inhibition of the electron transport chain by the complex III inhibitor antimycin A stimulates oxygen radical production (Boveris et al., 1972; Cadenas and Boveris, 1980;

Ksenzenko *et al.*, 1983). Thus, defects in the OXPHOS enzyme complexes should increase oxygen radical production and toxicity. This has been confirmed by demonstrating that patients with OXPHOS defects have elevated MnSOD in skeletal muscle (Ohkoshi *et al.*, 1995) and patients with partial complex I defects have elevated oxygen radical production and MnSOD mRNA (Pitkanen and Robinson, 1996).

3. mtDNA GENETICS AND DISEASE

The mitochondrial OXPHOS complexes are composed of multiple polypeptides, mostly encoded by the nuclear DNA (nDNA), but a total of 13 are encoded by the 16,569 (np) circular mtDNA. The mtDNA also codes for the 12S and 16S rRNAs and 22 tRNAs for mitochondrial protein synthesis. The 13 mtDNA polypeptides include seven subunits (MTND1, 2, 3, 4, 4L, 5, 6) of complex I, one subunit (MTCYB) of complex III, three subunits (MTCOI, II, III) of complex IV, and two subunits (MTATP6 and 8) of complex V. The mtDNA also contains a 1121-np "control region" containing the H- and L-strand promoters and the H-strand origin of replication (Wallace *et al.*, 1996b).

Each human cell contains hundreds of mitochondria and thousands of mtDNAs. The mtDNAs are maternally inherited and accumulate base substitution and rearrangement mutations much faster than nDNA. When a new mtDNA mutation arises in a cell, a mixed intracellular population of molecules (heteroplasmy) is generated. As heteroplasmic cells replicate, the mutant and normal molecules are randomly distributed into the daughter cells so that the proportion of mutant mtDNAs can drift by the process of replicative segregation. In individuals bearing mutant mitochondria, the cellular energy output reflects the severity of the mtDNA mutation and the proportion of mutant mtDNAs. As the severity of the mtDNA bioenergetic defect increases, the energetic capacity of the cells declines until there is insufficient energy for normal tissue and organ function, resulting in symptoms. Each tissue appears to have a different "bioenergetic threshold" with the central nervous system (CNS) being most sensitive to OXPHOS defects, followed by heart, skeletal muscle, the endocrine system, and kidney (Shoffner and Wallace, 1995; Wallace, 1992b; Wallace *et al.*, 1996a).

The high mtDNA mutation rate generates a high degree of mtDNA sequence variation in both the female germline and the somatic tissues. The germline variation arises as random mutations in individual mtDNAs, which then undergo replicative segregation. If the mutant mtDNAs become enriched in the germ line, then they can affect metabolism. Deleterious mutations result in disease and are rapidly lost. Neutral mutations can segregate to homoplasmy and through drift can become established in the population at polymorphic frequencies. Since the mtDNA is uniparentally inherited, independent mtDNAs cannot become mixed, and hence do not recombine. Consequently, all genetic variation of the mtDNA has accumulated by sequential mutations along radiating maternal lineages (Wallace, 1994).

4. mtDNA VARIATION IN HUMAN POPULATIONS

A large number of mtDNA sequence polymorphisms have accumulated as women migrated out of Africa and into the various continents over the past 150,000 years. Since many of these "naturally" occurring variants alter tRNA, rRNA, and protein gene sequences, it has been essential to identify the naturally occurring variants before it was possible to recognize pathogenic mutations. Moreover, we have found that certain mtDNA lineages increase the penetrance of certain pathogenic mtDNA mutations, and thus are essential in expressing the disease phenotype. Consequently, a detailed knowledge of normal population variation is essential for understanding mtDNA diseases (Wallace, 1994, 1995).

Continent-specific mtDNA variation has been characterized by screening for restriction site polymorphisms (RFLPs), by sequencing the control region, and by sequencing selected mtDNAs. The mtDNA is amplified in nine overlapping polymerase chain reaction (PCR) fragments, and each fragment is analyzed with 14 restriction endonucleases (HpaI, AvaII, BamHI, DdeI, HaeII, HaeIII, HhaI, HinfI, HpaI, HpaII/MspI, MboI, RsaI, and TaqI). The identified polymorphic sites are combined to define an individual haplotype. Haplotypes are then compared by phylogenetic analysis that clusters related haplotypes into continent-specific branches called haplogroups (Ballinger *et al.*, 1992a; Torroni *et al.*, 1992).

4.1. African mtDNA Variation

African mtDNA variation has been examined in 214 individuals including 101 Senegalese, 39 East and West African pygmies, and 74 Khoisan-speaking !Kung and Khwe (Chen *et al.*, 1995, in preparation). These studies revealed that about 75% of all sub-Saharan African mtDNAs belong to a macrohaplogroup, designated "L," and defined by an African-specific HpaI site at np 3592 (C to T at np 3594) and a DdeI site at np 10394 (A to G at np 10398). Macrohaplogroup L is subdivided into haplogroups L1 and L2, L1 defined by a HinfI site at np 10806 and L2 by a HinfI site gain at np 16389. The L1 lineage encompasses mtDNAs with two-length polymorphisms: a 9-np deletion between the COII and tRNA^{Lys} genes (np 8272–8289) and a 10- to 12-bp insertion between the tRNA^{Tyr} and COI genes (nps 5895–5899) (Chen *et al.*, 1995). A distinctive sublineage of L1, designated "α," is specific for the !Kung Bushman (Chen *et al.*, 1995). About 25% of sub-Saharan African mtDNAs lack the HpaI np 3592 site and are "non-L." One non-L haplogroup, which lacks the DdeI site at np 10394, appears to be the progenitor of roughly half of the European, Asian, and Native American mtDNAs. Generally, control region sequence variation is consistent with the RFLP data (Chen *et al.*, 1995), and the overall sequence divergence of the African mtDNAs is 0.292%, giving an age for the origin of African mtDNAs of between 101,000 and 131,000 years before present (YBP) (Torroni *et al.*, 1994d).

4.2. European mtDNA Variation

Analyses of 175 European mtDNAs by RFLP analysis has revealed that about 99% can be subsumed into ten haplogroups (H, I, J, K, M, T, U, V, W, and X), each defined by specific markers. The entire European mtDNA tree is subdivided by the presence or absence of the DdeI at site 10394, with I, J, K, and M having the site, while H, T, U, V, W, and X do not. Haplogroup H it is the most common haplogroup, accounting for about 40% of mtDNAs and is defined by the absence of the DdeI 10394 site and also by the presence of an AluI site at np 7025. Haplotype J encompasses about 9–10% of European mtDNAs and is defined by a BstNI site at np 13704 and a HinfI site at np 16065. Haplogroup I is defined by five distinctive restriction site polymorphisms and has the novel characteristic of being predisposed to a somatic duplication of 270 np of the control region between two runs of C at np 567 and np 302. Again, control region variants are consistent with the RFLP phylogenic delineations. The overall sequence diversity of European mtDNAs is about 0.113%, giving an age of about 39,000–51,000 YBP (Torroni *et al.*, 1994b, 1996).

4.3. Asian mtDNA Variation

The mtDNA variation of Asians has been analyzed in 153 Central and Southeastern Asians, 54 Tibetans, and 411 Siberians from 10 aboriginal populations. All Asian mtDNAs can be divided into two major groups: those that harbor the DdeI site at np 10394 and an adjacent AluI site at np 10397 (C to T at np 10400) and those that lack both sites. Within these macrohaplogroups are multiple distinctive haplogroups. The 10394–10397 minus macrohaplogroup harbors haplotypes A defined by a HaeIII site gain at np 663 (A to G at 663), B defined by an independent 9-np deletion between COII/tRNA^{Lys}, and F defined by a combined HpaI/HancII site loss at np 1206 (G to A at np 12406). The 10394–10397 positive macrohaplogroup includes haplotypes C, defined by a combined HincII site loss at np 13259 and an AluI site gain at np 13262 (A to G at np 13262), D by an AluI site loss at np 5176 (C to A at np 5178), and G by the a combined HaeIII and HhaI site gain at np 48301–4831. These lineages are also delineated by control region variants. Interestingly, as you move from Southeast to Northeast Asia, haplogroup F declines in frequency, while haplogroups A, C, D, and G increase, until in Siberia the later four haplogroups predominant. Furthermore, haplogroup B is dispersed along the Pacific Coast (Hertzberg *et al.*, 1989). The overall sequence divergence of the major 10394–10397 positive macrohaplogroup, which is unique to Asia, is 0.161%, which puts the age of the Asian population at 56,000–73,000 YBP (Ballinger *et al.*, 1992a; Schurr *et al.*, 1998; Starikovskaya *et al.*, 1998; Torroni *et al.*, 1993b, 1994c,d).

4.4. Native American mtDNAs

We have analyzed 743 Native American mtDNAs by RFLP and control region analysis. The vast majority fall into four major haplogroups derived from the Asian haplogroups A, B, C, and D. Asians and Native Americans share only the nodal or founding haplotype of each haplogroup. The mtDNA data suggest that the Americas were populated in four migrations. An initial migration from Siberia across the Bering land bridge bringing haplogroups A, C, and D occurred about 26,000–34,000 YBP (0.075%). A second coastal migration bringing B occurred about 12,000–15,000 YBP (0.034%). These two migrations created the "Paleo-Indians." A third migration, which crossed the Bering land bridge bringing a new infusion of haplogroup A and a modified A (RsaI site loss at np 16329), created the Na Dene and occurred around 7200–9000 YBP (0.021%). A final migration, which again brought haplogroups A and D from Siberia, resulted in the Eskimos and Aleuts (Schurr *et al.*, 1990, 1998; Starikovskaya *et al.*, 1998; Torroni *et al.*, 1992, 1993a, 1994a,d; Torroni and Wallace, 1995; Wallace and Torroni, 1992).

5. mtDNA MUTATIONS AND DEGENERATIVE DISEASE

This knowledge of naturally occurring mtDNA variation has permitted us to gain new insights into pathogenic mtDNA variation. Pathogenic mutations in the mtDNA fall into three broad categories: missense mutations, protein synthesis mutations, and rearrangements (Wallace, 1994, 1995).

5.1. Diseases Resulting from Missense Mutations

Three major clinical phenotypes have been associated with missense mutations: Leber's hereditary optic neuropathy (LHON), dystonia, and Leigh's syndrome, together with NARP (Shoffner and Wallace, 1995; Wallace, 1995; Wallace *et al.*, 1996a,b).

5.1.1. Leber's Hereditary Optic Neuropathy

The first LHON mutation was described in 1988 (Wallace *et al.*, 1988a). This is a G to A transition at np 11778 in the MTND4 gene (MTND4*LHON11778A), which converts the highly conserved arginine at codon 340 to a histidine (R340H). The 11778 mutation accounts for approximately 50% of all European LHON cases and >90% of all the remaining cases in the world. This LHON mutation is heteroplasmic in about 14% of cases and shows a spontaneous recovery rate of about 4% (Brown and Wallace, 1994; Wallace, 1994, 1995; Wallace *et al.*, 1996a). In Europeans, two other complex I mutations also play an important role in the etiology of LHON. One is a G to A transition at np 3460 in the MTND1 gene (MTND1*LHON3460A) (Howell *et al.*, 1991; Huoponen *et al.*, 1991). This mutation accounts for about 13% of European cases, is occasionally heteroplasmic, and has a visual recovery rate of about 22% (Brown *et al.*, 1994; Wallace, 1994; Wallace *et al.*, 1996a). The other is a T to C transition at np 14484 in the MTND6 gene (MTND6*LHON14484C) (Johns *et al.*, 1992). This is the mildest of the three complex I LHON mutations, accounts for about 15% of cases, is rarely heteroplasmic, and has a 37% visual recovery rate (Cooper 1992; Wallace, 1994; Wallace *et al.*, 1996a).

Analysis of the background mtDNA haplotypes of 47 European patients revealed that the 11778 and 3460 mutations can occur on any of the Europeans haplogroups, indicating that new LHON mutations are continually arising. By contrast, the 14484 mutation was found to cluster predominantly on one mtDNA haplogroup J (Brown *et al.*, 1995). Further studies revealed that 85% of all patients with the 14484 mutation were associated with haplogroup J, and 25–30% of 11778 patient mtDNAs were also haplogroup J. Moreover, we have found only one MTND6*LHON14484C case associated with African haplogroup L (Torroni *et al.*, 1996). Since haplogroup J is present in less than 10% of the European population, this association is highly statistically significant (Brown *et al.*, 1997). This result indicates that LHON mutations that arise on haplogroup J have a greater probability of being expressed as blindness than those that occur on other haplogroups. Further, the milder the LHON mutation, the greater the effect of haplogroup J on expressivity. Thus, haplogroup J must be carrying sequence variants that augment the LHON mutations.

Sequencing multiple haplogroup J mtDNAs has revealed three potentially important variants: MTND1*LHON4216C, which changes the nonconserved Y304H; MTND5*LHON3708A, which changes a weakly conserved A458T; and MTCYB*LHON15257A, which converts a very highly conserved N171R. Consequently, the MTCYB*LHON15257A mutation has been hypothesized to be a fourth pathogenic LHON mutation (Brown *et al.*, 1992b; Johns and Neufeld, 1991). Some LHON pedigrees lack any of the primary mutations (11778, 3460, and 14484), yet still harbor mtDNAs with haplogroup J. Sequencing the mtDNAs from two of these pedigrees revealed that both had the 4216+13708+15257 mutations. In addition, one had a heteroplasmic mutation at 5244 (MTND2*LHON5244A), which changed the highly conserved G259S (Brown *et al.*, 1992a), and this pedigree plus another also harbored a novel MTND4L mutation at np 10663, which changed a poorly conserved valine to an alanine (Brown *et al.*, 1995). These results suggest that any or all of these variants may play varying roles in causing LHON.

5.1.2. Dystonia

The first mtDNA dystonia mutation was found in a large Hispanic pedigree in which some maternal relatives presented with LHON, while others developed a childhood onset (mean age of 4 years) progressive rigidity (generalized dystonia) associated with bilateral basal ganglia degeneration (striatal necrosis). Of the 42 maternal relatives in the pedigree, 19% had LHON, 31% dystonia, and 2% had both (Novotny *et al.*, 1986). Sequencing the mtDNA from a dystonia patient revealed 40 nucleotide differences relative to the "Cambridge" sequence, several of which could have been the pathogenic mutation. Concurrent studies of Native American mtDNA variation revealed that this LHON–dystonia pedigree harbored a Native American haplogroup D mtDNA. By comparing the patient's variants to those of other haplogroup D mtDNAs, we were able to rule out all but one of the missense mutations. This was a G to A transition at np 14459 in MTND6, changing a conserved alanine at codon 72 to a valine (A72V). The mutation was heteroplasmic and segregating in the pedigree (Jun *et al.*, 1994). A subsequent survey of dystonia and LHON patients revealed two additional pedigrees with this mutation, one haplogroup L African-American family with LHON and a single haplogroup I European child with generalized dystonia. A muscle biopsy of one of the African-American LHON patients revealed a marked complex I defect (Shoffner *et al.*, 1995b).

OXPHOS enzyme analyses of lymphoblastoid cell lines from these three MTND6*LDYT14459A families revealed a consistent reduction in complex-I-specific activity of about 55%. For all three cases, this defect transferred along with the mtDNA to our rho⁰ lymphoblastoid cell line, WAL2A-rho⁰, in *trans*-mitochondrial cybrids (Jun *et al.*, 1996; Trounce *et al.*, 1996). Kinetic studies of the mutant complex I revealed that for the NADH substrate, the K_m was the same but the V_{max} was reduced about 30–50%. Moreover, titration with the CoQ analogue substrate (2,3-dimethyl-5-decyl 6-methylbenzoquinone, DB) revealed that the patient's enzyme reached a maximum activity at 5 nM, after which it declined, while the control enzyme activity increased to a plateau between 5 and 10 nM. These data indicate that the primary defect for the 14459 mutation is at the CoQ binding site of complex I (Jun *et al.*, 1996).

5.1.3. Leigh's Disease and NARP

A np 8993 T to G mutation in the MTATP6 gene (MTATP6*NARP8993G) converts highly conserved L156R and causes a range of neurodegenerative manifestations including neurogenic muscle weakness, ataxia, and retinitus pigmentosa (NARP) as well as Leigh's syndrome (Holt *et al.*, 1990). This mutation is invariably heteroplasmic, and its segregation results in phenotypes ranging from mild retinitus pigmentosa to more severe macular degeneration, mental retardation, and olivopontocerebellor atrophy to lethal childhood Leigh's disease associated with classical symmetrical basal ganglia lesions with vascular proliferation (subacute necrotizing encephalopathy) (Ortiz *et al.*, 1993; Shoffner *et al.*, 1992; Starikovskaya *et al.*, 1998).

Respiration studies of lymphoblastoid cell lines harboring the MTATP6*NARP8993G mutation revealed that uncoupled respiration through complexes I, III, and IV was normal. However, coupled respiration in the presence of ADP and ADP/O ratio were reduced 30–40%, resulting in an overall reduction in ATP synthesis efficiency of 70% in homoplasmic cell lines. This defect was transferred from heteroplasmic patient cell lines to our rho^0 osteosarcoma 143B recipient cell line (143BTK$^-$rho^0) by cybrid fusion. Cybrids whose mtDNAs had a G at np 8993 had the defect, while those with a T were normal (Trounce *et al.*, 1994). The leucine at codon 156 in MTATP6 is positioned halfway through the membrane, adjacent to a glutamate in the nuclear-encoded ATP9 subunit. The negative charge of this glutamate is essential for proton transport through the membrane. Conversion of the L at 156 to R neutralizes the negative charge of the adjacent glutamate and blocks the proton channel (Tatuch and Robinson, 1993).

5.2. Diseases Resulting from Protein Synthesis Mutations

The two best-characterized pathogenic tRNA mutations are the MERRF (myoclonic epilepsy and ragged-red fibers) and the MELAS (mitochondrial encephalomyopathy, lactic acidosis, and stroke) mutations. The MERRF mutation was first identified in a large Caucasian pedigree (Wallace *et al.*, 1988b). The heteroplasmic MERRF mutation changed np 8344 in the tRNA^{Lys} from an A to G (MTTK*MERRF8344G) (Shoffner *et al.*, 1990).

This MERRF family was then used to determine the relationship between the genotype (percentage heteroplasmy), the biochemical defect, and the phenotype (Fig. 1). Clinical manifestations of maternal relatives ranged from

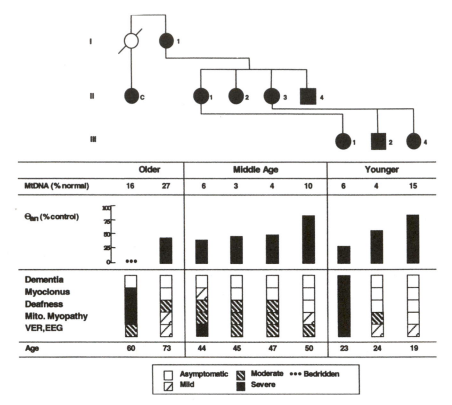

FIGURE 1. The interrelationship between phenotype, bioenergetic defect, genotype, and age in a MTTK*MERRF8344G pedigree. The upper panel shows the maternal pedigree with affected individuals represented by filled symbols. Below each individual's pedigree symbol is his or her mtDNA genotype, muscle bioenergetic capacity, clinical manifestations, and age in years. The subjects are broadly grouped by age (older, middle age, and younger), the exact ages at examination provided on the bottom line. The mtDNA genotype [mtDNA (%) normal)] is the percentage of normal mtDNAs in the individual's skeletal muscle. The muscle bioenergetic capacity [Θ_{an} (% control)] is an indication of the muscle mitochondrial energetic capacity as determined by exercise stress test, while monitoring VO_2 consumed and VCO_2 exhaled. The clinical manifestations that are common in these patients are listed and the severity of symptoms for each manifestation are given by the shading of the appropriate box. Mito, mitochondrial; VER, visual evoked response abnormalities; EEG, electroencephalograph abnormalities. From Wallace et al. (1994), with permission.

essentially normal, through mitochondrial myopathy with RRFs and abnormal mitochondria with or without sensory neural deafness, to severe myoclonic jerking, mitochondrial myopathy, deafness, dementia, renal dysfunction, and cardiomyopathy. The genotype was found to be related to the phenotype by two variables: the individual's percentage mutant mtDNAs and age. For example, in this pedigree, the 19-year-old with 15% normal and 85% mutant mtDNAs had normal muscle energetic capacity (anaerobic threshold, Θ_{an}) and was phenotypically normal, while her comparably aged cousin with 6% normal mtDNAs had greatly reduced bioenergetic capacity and severe clinical symptoms. Similarly, the 73-year-old with 27% normal mtDNAs had a partial deficiency in muscle energetics and moderate disease, while her 60-year-old maternal relative with 16% normal mtDNAs had substantially reduced muscle energetics and very severe disease. Thus, for each age group there was a correlation between the percentage of residual normal mtDNAs, the tissue bioenergetic capacity, and the clinical phenotype. However, comparison of individuals with comparable genotype, 15–16% normal mtDNAs, of different ages (19 versus 60), revealed a marked age-related decline in bioenergetics and a concomitant increase in disease severity. This demonstrates an important principle: All mtDNA protein synthesis diseases have a delayed onset and progress with age in both the severity of the biochemical defect and the symptoms (Shoffner and Wallace, 1995; Wallace *et al.*, 1994, 1996a; Wallace 1995).

The MELAS syndrome has also been found to be caused by a tRNA mutation, an A to G at np 3243 in tRNA$^{Leu(UUR)}$ (MTTL1*MELAS3243G) (Goto *et al.*, 1990; Kobayashi *et al.*, 1990). When the percentage mutant mtDNAs is high, the individuals have MELAS, but when the percentage is low (1 to 30% mutant), then they can manifest diabetes mellitus and deafness (van de Ouweland *et al.*, 1992, 1994). About 1.4% of all diabetes mellitus cases harbor the np 3243 mutation (Gerbitz *et al.*, 1995; Wallace *et al.*, 1996a).

One of the most severe tRNA$^{Leu(UUR)}$ mutations is due to the deletion of one of three Ts that make up the anticodon loop of (MTTL1*PEM3271ΔT). In this case, the proband developed childhood hearing loss and seizures that progressed to mitochondrial myopathy, retinitus pigmentosa, glaucoma, dementia, and severe cerebral calcification (Fahr's disease) and death at age 28 of tonic–clonic seizures, renal failure, and sepsis. Surprisingly, while the proband was about 70% mutant in muscle, his mother had 100% normal mtDNAs (Shoffner *et al.*, 1995a).

In addition to severe heteroplasmic protein synthesis mutations, we have also identified mild, homoplasmic mutations that are associated with late-onset degenerative disease, some of which occur on specific mtDNA haplogroups. One of these is a tRNA^{Glu} mutation at np 4336 (MTTQ*ADPD4336C), which is associated with about 5% of Alzheimer's disease (AD) patients but only 0.4–0.7% controls (Shoffner *et al.*, 1993). These frequencies have been confirmed in an independent study (Hutchin and Cortopassi, 1995). This A to G transition is always associated with a specific sublineage of haplogroup H, and sequence diversity in this lineage suggests the mutation arose about 15,000–20,000 years ago. Additional mutations identified in AD patients' mtDNAs include MTND1*ADPD3397G, which changes a highly conserved methionine to a valine and has arisen twice, once with MTTQ*ADPD4336G and once without, and an insertion of Cs in the 12S rRNA between np 956 and 965, which also occurred on the MTTQ*ADPD4336C lineage (Brown et al., 1996; Shoffner et al., 1993).

5.3. Diseases Resulting from Rearrangement Mutations

Studies of the ocular myopathies [Kearns–Sayre's syndrome (KKS) and chronic progressive external ophthalmoplegia (CPEO)] and Pearson's marrow–pancreas syndrome patients have frequently revealed a "common 5-kb" deletion. This heteroplasmic deletion of 4977 nps occurs between two 13 np direct repeats at np 8470–8482 and np 13447–13459 and may arise by slip mispairing replication (Holt et al., 1988; Schon et al., 1989; Shoffner et al., 1989). To date, over 100 different rearrangements have been reported in the ocular myopathy and Pearson's syndrome patients (Wallace et al., 1996b).

In addition to these "spontaneous" systemic rearrangements, a three-generation African-American pedigree has been reported with a maternally transmitted rearrangement associated with sensory neural hearing loss and adult-onset diabetes mellitus. This pedigree provided the first proof that mtDNA mutations could cause diabetes. Molecular analysis of this family revealed that the maternal relatives harbored a trimolecular heteroplasmy involving normal molecules, molecules with a 10.4-kb deletion, and others with a 6.1-kb duplication. The breakpoints occurred between np 4389 in the tRNA^{Gln} gene and np 14822 in the MTCYB gene and involved a 10-bp direct repeat. Patients along the maternal lineage had higher levels of the duplication in blood, but increased deletion in muscle. Moreover, lymphoblastoid cell lines from patients tended to lose the deleted and normal molecules and drift toward primarily duplicated mtDNAs. These cells had a defect in mitochondrial protein synthesis that was proportional to the percentage mutant molecules (Ballinger et al., 1992b, 1994).

5.4. Induction of OXPHOS Gene Expression in Mutant Tissues

Many patients with severe mtDNA protein synthesis defects, either due to tRNA point mutations or rearrangements, show a selective proliferation of mutant mitochondria in muscle associated with mitochondrial myopathy including RRF and abnormal mitochondria. This has been shown to be associated with a coordinate induction of both nDNA and mtDNA OXPHOS gene expression (Heddi et al., 1993, 1994).

5.5. Somatic mtDNA Mutations and the Age-Related Decline of OXPHOS

The delayed onset and progressive course mtDNA protein synthesis diseases suggest that aging must have a significant effect on OXPHOS. This has been supported by the demonstration of the age-related decline in OXPHOS enzyme levels in a number of human tissues (Boffoli *et al.*, 1994; Bowling *et al.*, 1993; Cooper *et al.*, 1992; Trounce *et al.*, 1989; Yen *et al.*, 1989).

This decline of OXPHOS in postmitotic tissues has been correlated with the progressive accumulation of somatic mtDNA rearrangement mutations including the common 5-kb deletion (Corral-Debrinski *et al.*, 1992a,b; Corto-passi and Arnheim, 1990; Cortopassi *et al.*, 1992; Linnane *et al.*, 1990; Soong *et al.*, 1992; Zhang *et al.*, 1992) and base substitution mutations including the np 8344 (MERRF), np 3243 (MELAS), and np 10006 RNA^{Gly} (Munscher *et al.*, 1993a,b; Zhang *et al.*, 1993) mutations. The heterogeneity of the somatic mtDNA rearrangements has been demonstrated using long extension-PCR (LX-PCR) to amplify full length mtDNAs (Cheng *et al.*, 1994). Muscle from individuals younger than 40 harbored primarily full-length molecules, while muscle from individuals between 50 and 80 harbored a wide range of mtDNA rearrangements. The significance of this observation was confirmed by Southern blot analysis of undigested mtDNA that revealed a significant proportion of the mtDNAs of older individuals migrated in the same region of the gel as rearranged mtDNAs (Melov *et al.*, 1995).

The relative levels of mutant mtDNAs has been determined by quantitating the common 5-kb deletion by the dilution-PCR procedure. In normal hearts, the 5-kb deletion was low prior to age 40, but increased progressively after age 40 (Corral-Debrinski *et al.*, 1991, 1992b). Similarly, in brain, the 5-kb deletion remained low in cerebellum, but increased exponentially in the cerebral cortex after age 75. Moreover, it increased dramatically in the basal ganglion, reaching between 10 and 12% by age 80 (Corral-Debrinski *et al.*, 1992a). Thus, we do accumulate somatic mtDNA mutations in postmitotic tissues as we age.

One likely cause of these somatic mutations is oxygen radical damage. In normal brain the levels of the DNA oxidation product 8-hydroxy-2-deoxy-guanosine (8-OhdG) were found to increase with age and were higher in the mtDNA then nDNA (Wallace, 1997). In patients with chronic ischemic heart disease associated with bursts of oxygen radicals within the mitochondria (Das *et al.*, 1989), the mtDNA deletion level was increased between 8- and 2200-fold (Corral-Debrinski *et al.*, 1991, 1992b). Hence, mtDNA mutations do accumulate with age, possibly due to mutations caused by oxygen radical damage.

5.6. Mitochondrial Defects in Common Degenerative Diseases

A variety of neurodegenerative diseases have been associated with defects in mitochondrial OXPHOS. Dystonia, involving generalized or localized rigidity frequently associated with basal ganglia degeneration, has been associated with complex I defects (Benecke *et al.*, 1992) as well as a complex III defect (Nigro *et al.*, 1990).

Parkinson's disease (PD), involving bradykinesia, rigidity, and tremor associated with the death of the dopaminergic neurons of the substantia nigra, can be induced in animals and humans by activated 1-methyl-4-phenyl-1,2,3,6-tetrahydropyridine (MPTP) or 1-methyl-4-phenyl-pyridinium (MPP^+). MPP^+ is selectively transported into dopaminergic neurons and their mitochondria where it inhibits complex I and stimulates the generation of oxygen radicals (Cleeter *et al.*, 1992; Langston *et al.*, 1983; Wallace *et al.*, 1997). PD patients have also been found to harbor complex I and possibly III defects in the substantia nigra (Schapira *et al.*, 1990; Wallace *et al.*, 1997), skeletal muscle (Bindoff *et al.*, 1989; Wallace *et al.*, 1992; Wallace, 1997) and blood platelets (Boveris, 1984; Parker *et al.*, 1989; Wallace *et al.*, 1997). The substantia nigra of PD patients also show an induction of MnSOD and a reduction in reduced glutathione (Jenner *et al.*, 1992).

Huntington's disease (HD) involves movement disorders, cognitive decline, and psychiatric problems and is caused by a chromosome 4, intragenic, trinucleotide repeat expansion (The Huntington's Disease Collaborative Research Group, 1993). However, it also has been associated with a decline in energy metabolism in the basal ganglia and cortex (Brennan *et al.*, 1985; Wallace, 1997; Wallace *et al.*, 1997), a complex I defect in platelets (Parker *et al.*, 1990a), and can be induced in rodents and primates by the complex II inhibitors malonate and 3-nitropropionic acid (3-NP) (Beal, 1994, 1995; Greene and Greenamyre, 1996).

Alzheimer's disease (AD) involves progressive dementia associated with neurofibrillary tangles and senile plaques. It has been associated with reduced pyruvate dehydrogenase and respiration (Wallace, 1997; Wallace *et al.*, 1997) and complex IV defects in patient platelets (Parker *et al.*, 1990b). Moreover, AD brains show an increase in 8-OHdG levels in DNA relative to controls (Mecocci *et al.*, 1994).

Support for OXPHOS decline as a component of these late-onset neurodegenerative diseases has come from analysis of the somatic mtDNA deletion levels in brains of HD and AD patients. Quantitation of the common 5-kb deletion in HD brains revealed cortical deletion levels 5- to 11-fold higher in patients than age-matched controls (Horton *et al.*, 1995). Similarly, in AD patient brains, the cortical levels of the 5-kb deletion were 10- to 15-fold higher than controls for patients younger than 75 years and declined to one fifth that of controls, after age 75, possibly because of loss of the neurons containing deletions (Corral-Debrinski *et al.*, 1994).

5.7. Mitochondrial Defects in Degenerative Diseases and Aging

These observations suggest a hypothesis that relates degenerative diseases and aging to defects in mitochondrial OXPHOS (Fig. 2). Each individual is envisioned as born with an OXPHOS genotype involving both nuclear and mtDNA genes. If the individual's genotype is good, then he starts with a high bioenergetic capacity; but if his genotype includes a deleterious mitochondrial

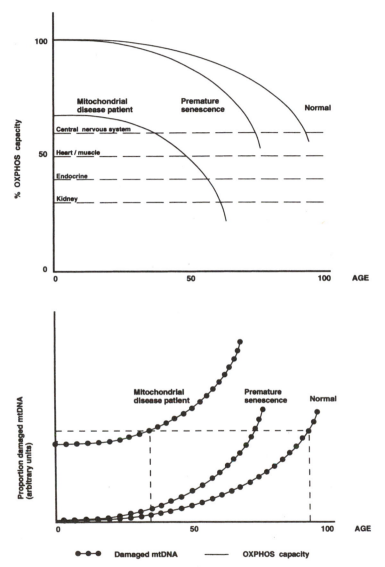

FIGURE 2. Hypothetical paradigm relating mitochondrial mutations, both inherited and somatically acquired, to the age-related decline in mitochondrial OXPHOS and the onset and progression of degenerative disease and senescence. The dashed horizontal lines in both panels represent tissue-specific bioenergetic expression thresholds. The top panel shows the expected age-related decline in OXPHOS for an individual born with a normal mito-chondrial genotype, a mutant mitochondrial genotype, and an increased mtDNA somatic mutation rate. The bottom panel gives the relative proportion of defective mtDNAs for each class of individual that accumulated during their lifetime. From Wallace *et al.* (1996a), with permission.

mutation, then he starts with a lower bioenergetic capacity. As the individual ages, he accumulates somatic mtDNA mutations that erode his bioenergetic capacity. Eventually, his mitochondrial capacity falls below tissue bioenergetic thresholds, resulting in clinical symptoms. The lower the initial OXPHOS capacity, the earlier the symptoms appear. Moreover, if an individual has an elevated somatic mtDNA mutation rate, possibly due to increased oxygen radical production, then his tissues will fall below their minimum bioenergetic thresholds prematurely, possibly resulting in AD, PD, and HD.

ACKNOWLEDGMENTS. This work was supported by NIH grants NS21328, HL45572, and AG13154.

6. REFERENCES

Ballinger, S. W., Schurr, T. G., Torroni, A., Gan, Y. Y., Hodge, J. A., Hassan, K., Chen, K. H., and Wallace, D. C., 1992a, Southeast Asian mitochondrial DNA analysis reveals genetic continuity of ancient mongoloid migrations, *Genetics* **130**:139–152.

Ballinger, S. W., Shoffner, J. M., Hedaya, E. V., Trounce, I., Polak, M. A., Koontz, D. A., and Wallace, D. C., 1992b, Maternally transmitted diabetes and deafness associated with a 10.4 kb mitochondrial DNA deletion, *Nature Genet.* **1**:11–15.

Ballinger, S. W., Shoffner, J. M., Gebhart, S., Koontz, D. A., and Wallace, D. C., 1994, Mitochondrial diabetes revisited, *Nature Genet.* **7**(4):458–459.

Bandy, B., and Davison, A. J., 1990, Mitochondrial mutations may increase oxidative stress: Implications for carcinogenesis and aging? *Free Radic. Biol. Med.* **8**(6):523–539.

Beal, M. F., 1994, Neurochemistry and toxin models in Huntington's disease, *Curr. Opin. Neurol.* **7**(6):542-547.

Beal, M. F., 1995, Aging, energy, and oxidative stress in neurodegenerative diseases, *Ann. Neurol.* **38**(3):357–366.

Benecke, R., Strumper, P., and Weiss, H., 1992, Electron transfer complex I defect in idiopathic dystonia, *Ann. Neurol.* **32**:683–686.

Benecke, R., Strumper, P., and Weiss, H., 1993, Electron transfer complexes I and IV of platelets are abnormal in Parkinson's disease but normal in Parkinson-plus syndromes, *Brain* **116**: 1451–1463.

Bindoff, L. A., Birch-Machin, M., Cartlidge, N. E. F., Parker, W. D., Jr., and Turnbull, D. M., 1989, Mitochondrial function in Parkinson's disease, *Lancet* **1**:49.

Boffoli, D., Scacco, S. C., Vergari, R., Solarino, G., Santacroce, G., and Papa, S., 1994, Decline with age of the respiratory chain activity in human skeletal muscle, *Biochim. Biophys. Acta* **1226**:73–82.

Boveris, A., 1984, Determination of the production of superoxide radicals and hydrogen peroxide in mitochondria, *Methods Enzymol.* **105**:429–435.

Boveris, A., and Turrens, J. F., 1980, Production of superoxide anion by the NADH-dehydrogenase of mammalian mitochondria, in *Chemical and Biochemical Aspects of Superoxide and Superoxide Dismutase. Developments in Biochemistry* (J. V. Bannister and H. A. O. Hill, eds.), pp. 84–91, Elsevier-North Holland, New York.

Boveris, A., Oshino, N., and Chance, B., 1972, The cellular production of hydrogen peroxide, *Biochem. J.* **128**(3):617–630.

Bowling, A. C., Mutisya, E. M., Walker, L. C., Price, D. L., Cork, L. C., and Beal, M. F., 1993, Age-dependent impairment of mitochondrial function in primate brain, *J. Neurochem.* **60:**1964–1967.

Brennan, W. A., Jr., Bird, E. D., and Aprille, J. R., 1985, Regional mitochondrial respiratory activity in Huntington's disease brain, *J. Neurochem.* **44:**1948–1950.

Brown, M. D., and Wallace, D. C., 1994, Spectrum of mitochondrial DNA mutations in Leber's hereditary optic neuropathy, *Clin. Neurosci.* **2**(3–4):138–145.

Brown, M. D., Voljavec, A. S., Lott, M. T., MacDonald, I., and Wallace, D. C., 1992a, Leber's hereditary optic neuropathy: A model for mitochondrial neurodegenerative diseases, *FASEB J.* **6:**2791–2799.

Brown, M. D., Voljavec, A. S., Lott, M. T., Torroni, A., Yang, C.-C., and Wallace, D. C., 1992b, Mitochondrial DNA complex I and III mutations associated with Leber's hereditary optic neuropathy, *Genetics* **130:**163–173.

Brown, M. D., Torroni, A., Reckord, C. L., and Wallace, D. C., 1995, Phylogenetic analysis of Leber's hereditary optic neuropathy mitochondrial DNA's indicates multiple independent occurrences of the common mutations, *Hum. Mutat.* **6:**311–325.

Brown, M. D., Shoffner, J. M., Kim, Y. L., Jun, A. S., Graham, B. H., Cabell, M. F., Gurley, D. S., and Wallace, D. C., 1996, Mitochondrial DNA sequence analysis of four Alzheimer's and Parkinson's disease patients, *Am. J. Hum. Genet.* **61**(3):283–289.

Brown, M. D., Sun, F., and Wallace, D. C., 1997, Clustering of Caucasian Leber's Hereditary Optic Neuropathy patients containing the 11778 or 14484 mutations on a mitochondrial DNA lineage, *Am. J. Hum. Genet.*

Cadenas, E., and Boveris, A., 1980, Enhancement of hydrogen peroxide formation by protophores and ionophores in antimycin-supplemented mitochondria, *Biochem. J.* **188**(1):31–37.

Chance, B., Sies, H., and Boveris, A., 1979, Hydroperoxide metabolism in mammalian organs, *Physiol. Rev.* **59**(3):527–605.

Chen, Y. S., Torroni, A., Excoffier, L., Santachiara-Benerecetti, A. S., and Wallace, D. C., 1995, Analysis of mtDNA variation in African populations reveals the most ancient of all human continent-specific haplogroups, *Am. J. Hum. Genet.* **57**(1):133–149.

Chen, Y.-S., Schurr, T. G., Olckers, A., Kogelnik, A., Huoponen, K., and Wallace, D. C., Mitochondrial DNA variation in the South African !Kung and Khwe and their genetic relationships to other African populations, *Am. J. Human Genet.* (in press).

Cheng, S., Higuchi, R., and Stoneking, M., 1994, Complete mitochondrial genome amplification, *Nature Genet.* **7**(3):350–351.

Cleeter, M. W., Cooper, J. M., and Schapira, A. H., 1992, Irreversible inhibition of mitochondrial complex I by 1-methyl-4-phenylpyridinium: Evidence for free radical involvement, *J. Neurochem.* **58**(2):786–789.

Cooper, J. M., Mann, V. M., and Schapira, A. H. V., 1992, Analyses of mitochondrial respiratory chain function and mitochondrial DNA deletion in human skeletal muscle: effect of ageing, *J. Neurol. Sci.* **113:**91–98.

Corral-Debrinski, M., Stepien, G., Shoffner, J. M., Lott, M. T., Kanter, K., and Wallace, D. C., 1991, Hypoxemia is associated with mitochondrial DNA damage and gene induction, *J. Am. Med. Assoc.* **266:**1812–1816.

Corral-Debrinski, M., Horton, T., Lott, M. T., Shoffner, J. M., Beal, M. F., and Wallace, D. C., 1992a, Mitochondrial DNA deletions in human brain: Regional variability and increase with advanced age, *Nature Genet.* **2:**324–329.

Corral-Debrinski, M., Shoffner, J. M., Lott, M. T., and Wallace, D. C., 1992b, Association of mitochondrial DNA damage with aging and coronary atherosclerotic heart disease, *Mutat. Res.* **275:**169–180.

Corral-Debrinski, M., Horton, T., Lott, M. T., Shoffner, J. M., McKee, A. C., Beal, M. F., Graham, B. H., and Wallace, D. C., 1994, Marked changes in mitochondrial DNA deletion levels in Alzheimer brains, *Genomics* **23**(2):471–476.

Cortopassi, G. A., and Arnheim, N., 1990, Detection of a specific mitochondrial DNA deletion in tissues of older humans, *Nucleic Acids Res.* **18**:6927–6933.

Cortopassi, G. A., Shibata, D., Soong, N. W., and Arnheim, N., 1992, A pattern of accumulation of a somatic deletion of mitochondrial DNA in aging human tissues, *Proc. Natl. Acad. Sci. USA* **89**:7370–7374.

Das, D. K., George, A., Liu, X. K., and Rao, P. S., 1989, Detection of hydroxyl radical in the mitochondria of ischemic-reperfused myocardium by trapping with salicylate, *Biochem. Biophys. Res. Commun.* **165**(3):1004–1009.

Gerbitz, K. D., van den Ouweland, J. M., Maassen, J. A., and Jaksch, M., 1995, Mitochondrial diabetes mellitus: A review, *Biochim. Biophys. Acta* **1271**(1):253–260.

Goldhaber, J. I., and Weiss, J. N., 1992, Oxygen free radicals and cardiac reperfusion abnormalities, *Hypertension* **20**(1):118–127.

Goto, Y., Nonaka, I., and Horai, S., 1990, A mutation in the tRNALeu(UUR) gene associated with the MELAS subgroup of mitochondrial encephalomyopathies, *Nature* **348**:651–653.

Greene, J. G., and Greenamyre, J. T., 1996. Bioenergetics and glutamate excitotoxicity, *Prog. Neurobiol.* **48**(6):613–634.

Heddi, A., Lestienne, P., Wallace, D. C., and Stepien, G., 1993, Mitochondrial DNA expression in mitochondrial myopathies and coordinated expression of nuclear genes involved in ATP, *J. Biol. Chem.* **268**:12156–12163.

Heddi, A., Lestienne, P., Wallace, D. C., and Stepien, G., 1994, Steady-state levels of mitochondrial and nuclear oxidative phosphorylation transcripts in Kearns–Sayre syndrome, *Biochim. Biophys. Acta* **1226**(2):206–212.

Hertzberg, M., Mickleson, K. N. O., Serjeantson, S. W., Prior, J. F., and Trent, R. J., 1989, An Asian specific 9-bp deletion of mitochondrial DNA is frequently found in Polynesians, *Am. J. Hum. Genet.* **44**:504–510.

Holt, I. J., Harding, A. E., and Morgan-Hughes, J. A., 1988, Deletions of muscle mitochondrial DNA in patients with mitochondrial myopathies, *Nature* **331**:717–719.

Holt, I. J., Harding, A. E., Petty, R. K., and Morgan-Hughes, J. A., 1990, A new mitochondrial disease associated with mitochondrial DNA heteroplasmy, *Am. J. Hum. Genet.* **46**:428–433.

Horton, T. M., Graham, B. H., Corral-Debrinski, M., Shoffner, J. M., Kaufman, A. E., Beal, B. F., and Wallace, D. C., 1995, Marked increase in mitochondrial DNA deletion levels in the cerebral cortex of Huntington's disease patients, *Neurology* **45**(10):1879–1883.

Howell, N., Bindoff, L. A., McCullough, D. A., Kubacka, I., Poulton, J., Mackey, D., Taylor, L., and Turnbull, D. M., 1991, Leber hereditary optic neuropathy: Identification of the same mitochondrial ND1 mutation in six pedigrees, *Am. J. Hum. Genet.* **49**:939–950.

Huoponen, K., Vilkki, J., Aula, P., Nikoskelainen, E. K., and Savontaus, M. L., 1991, A new mtDNA mutation associated with Leber hereditary optic neuroretinopathy, *Am. J. Hum. Genet.* **48**:1147–1153.

Hutchin, T., and Cortopassi, G., 1995, A mitochondrial DNA clone is associated with increased risk for Alzheimer disease, *Proc. Natl. Acad. Sci. USA* **92**(15):6892–6895.

Jenner, P., Dexter, D. T., Sian, J., Schapira, A. H., and Marsden, C. D., 1992, Oxidative stress as a cause of nigral cell death in Parkinson's disease and incidental Lewy body disease, *Ann. Neurol.* **32**(Suppl):S82–87.

Johns, D. R., and Neufeld, M. J., 1991, Cytochrome b mutations in Leber hereditary optic neuropathy, *Biochem. Biophys. Res. Commun.* **181**:1358–1364.

Johns, D. R., Neufeld, M. J., and Park, R. D., 1992, An ND-6 mitochondrial DNA mutation associated with Leber hereditary optic neuropathy, *Biochem. Biophys. Res. Commun.* **187**(3):1551–1557.

Jun, A. S., Brown, M. D., and Wallace, D. C., 1994, A mitochondrial DNA mutation at np 14459 of the ND6 gene associated with maternally inherited Leber's hereditary optic neuropathy and dystonia, *Proc. Natl. Acad. Sci. USA* **91**(13):6206–6210.

Jun, A. S., Trounce, I. A., Brown, M. D., Shoffner, J. M., and Wallace, D. C., 1996, Use of transmitochondrial cybrids to assign a complex I defect to the mitochondrial DNA-encoded NADH dehydrogenase subunit 6 gene mutation at nucleotide pair 14459 that causes Leber hereditary optic neuropathy and dystonia, *Mol. Cell. Biol.* (in press).

Kobayashi, Y., Momoi, M. Y., Tominaga, K., Momoi, T., Nihei, K., Yanagisawa, M., Kagawa, Y., and Ohta, S., 1990, A point mutation in the mitochondrial tRNALeu(UUR) gene in MELAS (mitochondrial myopathy, encephalopathy, lactic acidosis and stroke-like episodes), *Biochem. Biophys. Res. Commun.* **173:**816–822.

Ksenzenko, M., Konstantinov, A. A., Khomutov, G. B., Tikhonov, A. N., and Ruuge, E. K., 1983, Effect of electron transfer inhibitors on superoxide generation in the cytochrome bc1 site of the mitochondrial respiratory chain, *FEBS Lett.* **155**(1):19–24.

Langston, J. W., Ballard, P., Tetrud, J. W., and Irwin, I., 1983, Chronic parkinsonism in humans due to a product of meperidine-analog synthesis, *Science* **219:**979–980.

Linnane, A. W., Baumer, A., Maxwell, R. J., Preston, H., Zhang, C., and Marzuki, S., 1990, Mitochondrial gene mutation: The aging process and degenerative diseases, *Biochem. Int.* **22:** 1067–1076.

Mecocci, P., MacGarvey, U., Kaufman, A. E., Koontz, D., Shoffner, J. M., Wallace, D. C., and Beal, M. F., 1993, Oxidative damage to mitochondrial DNA shows marked age-dependent increases in human brain, *Ann. Neurol.* **34**(4):609–616.

Mecocci, P., MacGarvey, U., and Beal, M. F., 1994, Oxidative damage to mitochondrial DNA is increased in Alzheimer's disease, *Ann. Neurol.* **36**(5):747–751.

Melov, S., Shoffner, J. M., Kaufman, A., and Wallace, D. C., 1995, Marked increase in the number and variety of mitochondrial DNA rearrangements in aging human skeletal muscle, *Nucleic Acids Res.* **23**(20):4122–4126.

Munscher, C., Muller-Hocker, J., and Kadenbach, B., 1993a, Human aging is associated with various point mutations in tRNA genes of mitochondrial DNA, *Biol. Chem. Hoppe Seyler* **374:**19959–1104.

Munscher, C., Rieger, T., Muller-Hocker, J., and Kadenbach, B., 1993b, The point mutation of mitochondrial DNA characteristic for MERRF disease is found also in healthy people of different ages, *FEBS Lett.* **317:**27–30.

Nigro, M. A., Martens, M. E., Awerbuch, G. I., Peterson, P. L., and Lee, C. P., 1990, Partial cytochrome *b* deficiency and generalized dystonia, *Pediatr. Neurol.* **6**(6):407–410.

Novotny, E. J., Singh, G., Wallace, D. C., Dorfman, L. J., Louis, A., Sogg, R. L., and Steinman, L., 1986, Leber's disease and dystonia: A mitochondrial disease, *Neurology* **36:**1053–1060.

Ohkoshi, N., Mizusawa, H., Shiraiwa, N., Shoji, S., Harada, K., and Yoshizawa, K., 1995, Superoxide dismutases of muscle in mitochondrial encephalomyopathies, *Muscle Nerve* **18** (11):1265–1271.

Ortiz, R. G., Newman, N. J., Shoffner, J. M., Kaufman, A. E., Koontz, D. A., and Wallace, D. C., 1993, Variable retinal and neurological manifestations in patients harboring the mitochondrial DNA 8993 mutation. *Arch. Ophthalmol.* **111:**1525–1530.

Parker, W. D., Jr., Boyson, S. J., and Parks, J. K., 1989, Abnormalities of the electron transport chain in idiopathic Parkinson's disease, *Ann. Neurol.* **26:**719–723.

Parker, W. D., Jr., Boyson, S. J., Luder, A. S., and Parks, J. K., 1990a, Evidence for a defect in NADH: ubiquinone oxidoreductase (complex I) in Huntington's disease, *Neurology* **40:**1231–1234.

Parker, W. D., Jr., Filley, C. M., and Parks, J. K., 1990b, Cytochrome oxidase deficiency in Alzheimer's disease, *Neurology* **40:**1302–1303.

Pitkanen, S., and Robinson, B. H., 1996, Mitochondrial complex I deficiency leads to increased production of superoxide radicals and induction of superoxide dismutase, *J. Clin. Invest.* **98**(2):345–351.

Schapira, A. H., Cooper, J. M., Dexter, D., Clark, J. B., Jenner, P., and Marsden, C. D., 1990, Mitochondrial complex I deficiency in Parkinson's disease, *J. Neurochem.* **54:**823–827.

Schon, E. A., Rizzuto, R., Moraes, C. T., Nakase, H., Zeviani, M., and DiMauro, S., 1989, A direct
    repeat is a hot spot for large-scale deletion of human mitochondrial DNA, *Science* **244**:346–349.
Schurr, T. G., Ballinger, S. W., Gan, Y. Y., Hodge, J. A., Merriwether, D. A., Lawrence, D. N.,
    Knowler, W. C., Weiss, K. M., and Wallace, D. C., 1990, Amerindian mitochondrial DNAs
    have rare Asian mutations at high frequencies, suggesting they derived from four primary
    maternal lineages, *Am. J. Hum. Genet.* **46**:613–623.
Schurr, T. G., Sukernik, R. I., Starikovskaya, Y. B., and Wallace, D. C., 1998, Mitochondrial DNA
    variation in Koryaks and Itel'men: Population replacement in the Okhotsk Sea-Bering Sea
    region during the Neolithic, *Am. J. Phys. Anthropol.* (in press).
Shoffner, J. M., and Wallace, D. C., 1995, Oxidative phosphorylation diseases, in *The Metabolic
    and Molecular Basis of Inherited Disease* (C. R. Scriver, A. L. Beaudet, W. S. Sly, and D.
    Valle, eds.), pp. 1535–1609, McGraw-Hill, New York.
Shoffner, J. M., Lott, M. T., Voljavec, A. S., Soueidan, S. A., Costigan, D. A., and Wallace, D. C.,
    1989, Spontaneous Kearns–Sayre/chronic external ophthalmoplegia plus syndrome associ-
    ated with a mitochondrial DNA deletion: A slip-replication model and metabolic therapy,
    Proc. Natl. Acad. Sci. USA **86**:7952–7956.
Shoffner, J. M., Lott, M. T., Lezza, A. M., Seibel, P., Ballinger, S. W., and Wallace, D. C., 1990,
    Myoclonic epilepsy and ragged-red fiber disease (MERRF) is associated with a mitochondrial
    DNA tRNALys mutation, *Cell* **61**:931–937.
Shoffner, J. M., Fernhoff, M. D., Krawiecki, N. S., Caplan, D. B., Holt, P. J., Koontz, D. A., Takei,
    Y., Newman, N. J., Ortiz, R. G., Polak, M., Ballinger, S. W., Lott, M. T., and Wallace, D. C.,
    1992, Subacute necrotizing encephalopathy: Oxidative phosphorylation defects and the ATP-
    ase 6 point mutation, *Neurology* **42**:2168–2174.
Shoffner, J. M., Brown, M. D., Torroni, A., Lott, M. T., Cabell, M. R., Mirra, S. S., Beal, M. F.,
    Yang, C., Gearing, M., Salvo, R., Watts, R. L., Juncos, J. L., Hansen, L. A., Crain, B. J., Fayad,
    M., Reckord, C. L., and Wallace, D. C., 1993, Mitochondrial DNA variants observed in
    Alzheimer disease and Parkinson disease patients, *Genomics* **17**:171–184.
Shoffner, J. M., Bialer, M. G., Pavlakis, S. G., Lott, M. T., Kaufman, A., Dixon, J., Teichberg, S.,
    and Wallace, D. C., 1995a, Mitochondrial encephalomyopathy associated with a single
    nucleotide pair deletion in the mitochondrial tRNALeu(UUR) gene, *Neurology* **45**(2):
    286–292.
Shoffner, J. M., Brown, M. D., Stugard, C., Jun, A. S., Pollok, S., Haas, R. H., Kaufman, A.,
    Koontz, D., Kim, Y., Graham, J., Smith, E., Dixon, J., and Wallace, D. C., 1995b, Leber's
    hereditary optic neuropathy plus dystonia is caused by a mitochondrial DNA point mutation
    in a complex I subunit, *Ann. Neurol.* **38**(2):163–169.
Simonetti, S., Chen, X., DiMauro, S., and Schon, E. A., 1992, Accumulation of deletions in human
    mitochondrial DNA during normal aging: Analysis by quantitative PCR, *Biochim. Biophys.
    Acta* **1180**:113–122.
Soong, N. W., Hinton, D. R., Cortopassi, G., and Arnheim, N., 1992, Mosaicism for a specific
    somatic mitochondrial DNA mutation in adult human brain, *Nature Genet.* **2**:318–323.
Starikovskaya, Y. B., Sukernik, R. I., Schurr, T. G., Kogelnik, A. M., and Wallace, D. C., 1998,
    MtDNA diversity in Chukchi and Siberian Eskimos: Implications for the genetic history of
    ancient Beringia and peopling of the New World, *Am. J. Hum. Genet.* **63**:1473–1491.
Tatuch, Y., Christodoulou, J., Feigenbaum, A., Clarke, J. T. R., Wherret, J., Smith, C., Rudd, N.,
    Petrova-Benedict, R., and Robinson, B. H., 1992, Heteroplasmic mtDNA mutation (T-G) at
    8993 can cause Leigh disease when the percentage of abnormal mtDNA is high, *Am. J. Hum.
    Genet.* **50**:852–858.
Tatuch, Y., and Robinson, B. H., 1993, The mitochondrial DNA mutation at 8993 associated with
    NARP slows the rate of ATP synthesis in isolated lymphoblast mitochondria, *Biochem. Bio-
    phys. Res. Commun.* **192**:124–128.

The Huntington's Disease Collaborative Research Group, 1993, A novel gene containing a trinucleotide repeat that is expanded and unstable on Huntington's disease chromosomes, *Cell* **72**(6):971–983.

Torroni, A., and Wallace, D. C., 1995, MtDNA haplogroups in Native Americans [see also comment, *Am. J. Hum. Genet.* **56**:1236–1238, 1995]; *Am. J. Hum. Genet.* **56**(5):1234–1236.

Torroni, A., Schurr, T. G., Yang, C.-C., Szathmary, E. J., Williams, R. C., Schanfield, M. S., Troup, G. A., Knowler, W. C., Lawrence, D. N., and Weiss, K. M., 1992, Native American mitochondrial DNA analysis indicates that the Amerind and the Nadene populations were founded by two independent migrations, *Genetics* **130**:153–162.

Torroni, A., Schurr, T. G., Cabell, M. F., Brown, M. D., Neel, J. V., Larsen, M., Smith, D. G., Vullo, C. M., and Wallace, D. C., 1993a, Asian affinities and continental radiation of the four founding Native American mtDNAs, *Am. J. Hum. Genet.* **53**:563–590.

Torroni, A., Sukernik, R. I., Schurr, T. G., Starikovskaya, Y. B., Cabell, M. F., Crawford, M. H., Comuzzie, A. G., and Wallace, D. C., 1993b, MtDNA variation of aboriginal Siberians reveals distinct genetic affinities with Native Americans, *Am. J. Hum. Genet.* **53**:591–608.

Torroni, A., Chen, Y., Semino, O., Santachiara-Beneceretti, A. S., Scott, C. R., Lott, M. T., Winter, M., and Wallace, D. C., 1994a, MtDNA and Y-chromosome polymorphisms in four native American populations from southern Mexico, *Am. J. Hum. Genet.* **54**(2):303–318.

Torroni, A., Lott, M. T., Cabell, M. F., Chen, Y., Laverge, L., and Wallace, D. C., 1994b, MtDNA and the origin of Caucasians. Identification of ancient Caucasian-specific haplogroups, one of which is prone to a recurrent somatic duplication in the D-loop region, *Am. J. Hum. Genet.* **55**(4):760–776.

Torroni, A., Miller, J. A., Moore, L. G., Zamudio, S., Zhuang, J., Droma, R., and Wallace, D. C., 1994c, Mitochondrial DNA analysis in Tibet. Implications for the origin of the Tibetan population and its adaptation to high altitude, *Am. J. Phys. Anthropol.* **93**(2):189–199.

Torroni, A., Neel, J. V., Barrantes, R., Schurr, T. G., and Wallace, D. C., 1994d, A mitochondrial DNA "clock" for the Amerinds and its implication for timing their entry into North America, *Proc. Natl. Acad. Sci. USA* **91**(3):1158–1162.

Torroni, A., Carelli, V., Petrozzi, M., Terracina, M., Barboni, P., Malpassi, P., Wallace, D. C., and Scozzari, R., 1996, Detection of the mtDNA 14484 mutation on an African-specific haplotype: Implications about its role in causing Leber hereditary optic neuropathy, *Am. J. Hum. Genet.* **59**(1):248–252.

Torroni, A., Huoponen, K., Francalacci, P., Petrozzi, M., Morelli, L., Scozzari, R., Obinu, D., Savontaus, M.-L., and Wallace, D. C., 1996, Classification of European mtDNAs from an analysis of three European populations, *Genetics* **144**:1835–1850.

Trounce, I., Byrne, E., and Marzuki, S., 1989, Decline in skeletal muscle mitochondrial respiratory chain function: Possible factor in ageing, *Lancet* **1**:637–639.

Trounce, I., Neill, S., and Wallace, D. C., 1994, Cytoplasmic transfer of the mtDNA nt 8993 TG (ATP6) point mutation associated with Leigh syndrome into mtDNA-less cells demonstrates cosegregation with a decrease in state III respiration and ADP/O ratio, *Proc. Natl. Acad. Sci. USA* **91**(18):8334–8338.

Trounce, I. A., Kim, Y. L., Jun, A. S., and Wallace, D. C., 1996, Assessment of mitochondrial oxidative phosphorylation in patient muscle biopsies, lymphoblasts, and transmitochondrial cell lines, *Methods Enzymol.* **264**:484–509.

Turrens, J. F., and Boveris, A., 1980, Generation of superoxide anion by the NADH dehydrogenase of bovine heart mitochondria, *Biochem. J.* **191**(2):421–427.

Turrens, J. F., Alexandre, A., and Lehninger, A. L., 1985, Ubisemiquinone is the electron donor for superoxide formation by complex III of heart mitochondria, *Arch Biochem Biophys.* **237**:408–414.

van den Ouweland, J. M., Lemkes, H. H. P., Ruitenbeek, W., Sandkjujl, L. A., deVijldcr, M. F.,

Struyvenberg, P. A. A., van de Kamp, J. J. P., and Maassen, J. A., 1992, Mutation in mitochondrial tRNALeu(UUR) gene in a large pedigree with maternally transmitted type II diabetes mellitus and deafness, *Nature Genet.* **1**:368–371.

van den Ouweland, J. M., Lemkes, H. H., Trembath, R. C., Ross, R., Velho, G., Cohen, D., Froguel, P., and Maassen, J. A., 1994, Maternally inherited diabetes and deafness is a distinct subtype of diabetes and associates with a single point mutation in the mitochondrial tRNALeu(UUR) gene, *Diabetes* **43**(6):746–751.

Wallace, D. C., 1992a, Diseases of the mitochondrial DNA, *Annu. Rev. Biochem.* **61**:1175–1212.

Wallace, D. C., 1992b, Mitochondrial genetics: A paradigm for aging and degenerative diseases? *Science* **256**:628–632.

Wallace, D. C., 1994, Mitochondrial DNA sequence variation in human evolution and disease, *Proc. Natl. Acad. Sci. USA* **91**(19):8739–8746.

Wallace, D. C., 1995, 1994 William Allan Award Address. Mitochondrial DNA variation in human evolution, degenerative disease, and aging, *Am. J. Hum. Genet.* **57**(2):201–223.

Wallace, D. C., 1997, Mitochondrial DNA mutations and bioenergetic defects in aging and degenerative diseases, in *The Molecular and Genetic Basis of Neurological Disease* (R. N. Rosenberg, S. B. Prusiner, S. DiMauro, and R. L. Barchi, eds.), pp. 237–267, Butterworth and Heinemann, Boston.

Wallace, D. C., and Torroni, A., 1992, American Indian prehistory as written in the mitochondrial DNA: A review, *Hum. Biol.* **64**:403–416.

Wallace, D. C., Singh, G., Lott, M. T., Hodge, J. A., Schurr, T. G., Lezza, A. M., Elsas, L. J., and Nikoskelainen, E. K., 1988a, Mitochondrial DNA mutation associated with Leber's hereditary optic neuropathy, *Science* **242**:1427–1430.

Wallace, D. C., Zheng, X., Lott, M. T., Shoffner, J. M., Hodge, J. A., Kelley, R. I., Epstein, C. M., and Hopkins, L. C., 1988b, Familial mitochondrial encephalomyopathy (MERRF): Genetic, pathophysiological, and biochemical characterization of a mitochondrial DNA disease, *Cell* **55**:601–610.

Wallace, D. C., Shoffner, J. M., Watts, R. L., Juncos, J. L., and Torroni, A., 1992, Mitochondrial oxidative phosphorylation defects in Parkinson's disease, *Ann. Neurol.* **32**:113–114.

Wallace, D. C., Lott, M. T., Shoffner, J. M., and Ballinger, S., 1994, Mitochondrial DNA mutations in epilepsy and neurological disease, *Epilepsia* **35**(Suppl.1):S43–S50.

Wallace, D. C., Brown, M. D., and Lott, M. T., 1996a, Mitochondrial Genetics, in *Emery and Rimoin's Principles and Practice of Medical Genetics* (D. L. Rimoin, J. M. Connor, R. E. Pyeritz, and A. E. H. Emery, eds.), pp. 277–332, Churchill Livingstone, London.

Wallace, D. C., Lott, M. T., and Brown, M. D., 1996b, Report of the committee on human mitochondrial DNA, in *Human Gene Mapping 1995, a Compendium* (A. J. Cuticchia, M. A. Chipperfield, and P. A. Foster, eds.), pp. 1280–1331, The Johns Hopkins University Press, Baltimore.

Wallace, D. C., Lott, M. T., and Brown, M. D., 1997, Mitochondrial defects in neurodegenerative diseases and aging, in M. F. Beal, N. Howell, and I. Bodis-Wollner (eds.), *Mitochondria and Free Radicals in Neurodegenerative Diseases*, pp. 283–307, Wiley-Liss, New York.

Yen, T. C., Chen, Y. S., King, K. L., Yeh, S. H., and Wei, Y. H., 1989, Liver mitochondrial respiratory functions decline with age, *Biochem. Biophys. Res. Commun.* **165**:944–1003.

Zhang, C., Baumer, A., Maxwell, R. J., Linnane, A. W., and Nagley, P., 1992, Multiple mitochondrial DNA deletions in an elderly human individual, *FEBS Lett.* **297**:4–8.

Zhang, C., Linnane, A. W., and Nagley, P., 1993, Occurrence of a particular base substitution (3243 A to G) in mitochondrial DNA of tissues of ageing humans, *Biochem. Biophys. Res. Commun.* **195**:1104–1109.

CHAPTER 31

Perspectives on the Permeability Transition Pore, a Mitochondrial Channel Involved in Cell Death

Paolo Bernardi

1. INTRODUCTION

The permeability transition (PT) is a familiar event for the mitochondrial scientist studying energy conservation or ion transport in isolated mitochondria. As a result of Ca^{2+} accumulation, or simply as a consequence of *in vitro* aging, mitochondria tend to undergo an increase of basal respiration not coupled to ATP synthesis or to ion accumulation. This phenomenon is linked to a generalized increase of permeability of the inner membrane, which leads to increased H^+ cycling, membrane depolarization, substrate depletion, and equilibration of species with molecular weight ≈ 1500 Da or lower. In turn, this causes swelling of the matrix space, particularly when the osmotic support of the incubation medium is provided by sugars, which is the standard condition in studies on isolated mitochondria. The end result of the PT is a swollen, uncoupled mitochondrion unable to synthesize ATP (see Zoratti and Szabo',

Paolo Bernardi • CNR Unit for the Study of Biomembranes, Department of Biomedical Sciences, University of Padova Medical School, I-35121 Padova, Italy.

Frontiers of Cellular Bioenergetics, edited by Papa *et al.* Kluwer Academic/Plenum Publishers, New York, 1999.

1995, for a thorough review). Despite early results indicating that swelling in saline media could be reversed by ATP (e.g., Azzone and Azzi, 1965), it was generally felt that the PT was due to "deterioration" of mitochondria following an *in vitro* artifactual process. And because of the lack of selectivity of the permeability pathway(s), the PT was also widely considered as an unspecific form of membrane damage. This attitude is rapidly changing because of a series of discoveries that have modified our perspectives on mitochondrial ion channels in general and on the PT pore in particular. This chapter provides a synthetic account of how the field developed and presents recent results suggesting that the PT plays a key role in a variety of forms of cell death.

1.1. Chemiosmosis and Mitochondrial Channels

Throughout the 1960s and 1970s, the chemiosmotic hypothesis of energy coupling in mitochondria and chloroplasts has been subjected to rigorous scrutiny by a variety of experimental approaches in many laboratories. In retrospect, there is little doubt that the basic postulates of chemiosmosis were indeed used "... as the target for critical experiments designed to show that the chemiosmotic hypothesis may be untenable," as Mitchell (1966, p. 451) had forecast over 30 years ago.

In summarizing his fourth basic postulate of chemiosmotic energy conservation, Mitchell (1966) stated that the ATPase systems of mitochondria and chloroplasts, their membrane-located oxidoreduction chain systems coupling reversibly the translocation of protons across the membrane to the flow of reducing equivalents, and the exchange-diffusion carrier systems for anions and cations "... are located in a specialized coupling membrane which has a low permeability to protons and to anions and cations generally" (p. 452). This basic concept has implied that the mitochondrial inner membrane did not possess channels for cations, which would (1) dissipate the membrane potential because of electrophoretic cation uptake along the electrochemical gradient, and (2) lead to osmotic burst of the mitochondrion. The existence of electroneutral H^+ cation exchangers (third postulate) may allow osmotic regulation and prevent swelling but not depolarization, which would be sustained by cation cycling. For these reasons, the existence of mitochondrial ion channels has been widely perceived as not compatible with chemiosmosis, and the acceptance of chemiosmosis as the basic mechanism of energy coupling has paradoxically slowed down research on mitochondrial channels in general and on the PT in particular. Indeed, not only was the latter rightly perceived as catastrophic for energy conservation; its lack of selectivity and seemingly intractable complexity did not make it a popular subject in bioenergetics. Today, we know that mitochondria possess a variety of ion channels and that their tight control allows ion transport and volume homeostasis without disruption of the membrane potential. Among the mitochondrial channels, the PT pore is emerging as a key player in a variety of forms of cell death.

1.2. The Permeability Transition: Lipid or Protein?

The PT is greatly facilitated by the accumulation of Ca^{2+} in the matrix. Although it is clear today that Ca^{2+} is not an absolute requirement and that part of the apparent Ca^{2+} dependence is in fact due to the accompanying H^+ extrusion (Bernardi et al., 1992, 1993; Petronilli et al., 1993a), one of the most puzzling features of the PT remained its facilitation by a variety of compounds and conditions (so-called inducers) that have no common structural or functional features. In their thorough review of the field, Gunter and Pfeiffer (1990) listed 43 different classes of inducers, ranging from metabolites like phospho-enol-pyruvate and oxaloacetate, to sulfhydryl reagents like N-ethylmaleimide and mersalyl, to organic hydroperoxides like tert-butylhydroperoxide, to oxidants like diamide and redox-cycling quinones, to fatty acids. Likewise, the PT could also be inhibited by a variety of compounds like adenine nucleotides (ADP being more potent than ATP), reduced pyridine nucleotides, H^+, serum albumin, Me^{2+} ions other than Ca^{2+}, and local anesthetics (Gunter and Pfeiffer, 1990; Zoratti and Szabo', 1995).

In the attempt to provide a unifying hypothesis of the PT and to account for the variety of its modulators, Pfeiffer and co-workers proposed the "membrane" theory of the PT, which was extremely popular in the early 1980s. In synthesis, the defect was localized to the membrane itself, which would undergo major changes of permeability as a result of the accumulation of acyllyso-phospholipids following activation of Ca^{2+}-dependent phospholipase A_2 (Pfeiffer et al., 1979; Beatrice et al., 1980). In this model, pore reversibility could be explained by reacylation upon removal of Ca^{2+} (Pfeiffer et al., 1979). A variety of pore inducers and inhibitors could be accommodated within this theory, which also accounted for the lack of selectivity of the permeability pathway. An alternative theory, already proposed in the early 1970s (Massari and Azzone, 1972), considered the PT as linked to reversible opening of a proteinaceous pore, regulated by a variety of effectors at discrete sites. The pore theory was fully developed in the late 1970s, mainly by Hunter and Haworth (1979a,b), but did not gain much consensus. In fact, the 1978 Nobel prize award to Peter Mitchell might have played an indirect role in the early dismissal of the pore theory of the PT.

In 1987, Fournier et al. (1987) introduced the immunosuppressant cyclosporin A (CsA) in mitochondrial studies and clearly showed that in the presence of this drug mitochondria accumulated large Ca^{2+} loads that would otherwise collapse the membrane potential and cause Ca^{2+} efflux because of a "membrane damage." Inhibition of the mitochondrial PT by CsA was subsequently demonstrated by Crompton et al. (1988) and Broekemeier et al. (1989). Since CsA inhibited the PT but not phospholipase A_2 (Broekemeier et al., 1989), the pore hypothesis of the PT, favored by Crompton and co-workers throughout the 1980s (Crompton et al., 1987, 1988; Crompton and Costi, 1988), gained new ground.

In the same year, another conceptual barrier was broken with the demonstration that inner mitochondrial membranes possess ion channels that can be studied by electrophysiological techniques (Sorgato et al., 1987). The subsequent discovery of the mitochondrial megachannel (Petronilli et al., 1989; Kinnally et al., 1989), a high-conductance inner membrane channel, and the demonstration of its inhibition by CsA (Szabo' and Zoratti, 1991) took the pore hypothesis of the PT full circle. This was considerably strengthened by a joint effort between Zoratti's and Bernardi's group, demonstrating that most known effectors of the PT were also affecting the megachannel in the same way (Bernardi et al., 1992; Szabo' et al., 1992; Szabo' and Zoratti, 1992). A number of major issues remained open, however. How can a channel be modulated by over 40 unrelated compounds? What is the molecular nature of the pore? What is the physiological meaning, if any, of a high-conductance channel in the coupling membrane? Does a PT ever occur in vivo? Why should such a dangerous system have been conserved in evolution? Some answers to these intriguing questions are now beginning to emerge.

2. MECHANISTIC ASPECTS OF PORE REGULATION

2.1. Matrix pH

Since the early work of Hunter and Haworth (1979a), it had been appreciated that the mitochondrial PT is inhibited at acidic pH values. Our initial effort was to determine whether the inhibitory site was on the cytosolic or on the matrix side of the inner membrane. Our finding that inhibition is exerted from the *matrix* side (Bernardi et al., 1992) through one or more critical histidyl residue(s) (Nicolli et al., 1993) had two important implications.

1.   In energized mitochondria, Ca^{2+} uptake is an electrophoretic process accompanied by membrane depolarization, H^+ ejection, and, in the absence of weak acids, by the buildup of a large inside-alkaline proton gradient that can be as high as 100 mV (see, e.g., Bernardi and Pietrobon, 1982). In other words, as a result of Ca^{2+} accumulation, a potent pore inhibitor (H^+) is removed while an activator (Ca^{2+}) is added. Since matrix pH is a complex function of H^+ pumping and of secondary transport of cations (including Ca^{2+}) and anions (including Pi), the finding that the pore open–closed transitions were modulated by matrix pH represented a first indication that the pore could be controlled by the proton electrochemical gradient.

2.   It was already known that addition of uncouplers to Ca^{2+}-loaded mitochondria can induce pore opening if ruthenium red (the inhibitor of the Ca^{2+} uniporter) is added before uncoupler. Since pore opening was not observed in the absence of ruthenium red, this finding was considered as evidence

that matrix Ca^{2+} retention was required (Igbavboa and Pfeiffer, 1988). Indeed, in the absence of ruthenium red addition of uncoupler is followed by Ca^{2+} efflux. We reasoned that this Ca^{2+} current must be charge compensated, and that due to the high H^+ conductance created by the uncoupler, charge compensation is mostly provided by H^+ influx, resulting in matrix acidification and pore inhibition. Under these conditions, any effect(s) of membrane depolarization on the pore would be offset by matrix acidification. This consideration paved the way to the experimental demonstration that the pore is modulated by the membrane potential (Bernardi, 1992; Petronilli et al., 1993a).

2.2. Membrane Potential

It is now widely accepted that the PT pore behaves like a voltage-dependent channel, which favors the closed conformation at high membrane potentials and the open conformation after depolarization (see Bernardi et al., 1994, for a review). The voltage dependence appears to be an intrinsic property of the pore. This has been described both in isolated mitochondria, by modulating the membrane potential with uncouplers (Bernardi, 1992) or valinomycin (Scorrano et al., 1997), and in patch-clamp experiments of single channels, by modulating the applied voltage (Szabo' and Zoratti, 1993). Based on these findings we postulated the existence of a sensor that decodes the voltage changes into variations of the probability of pore opening (Petronilli et al., 1993b). Irrespective of the molecular structure of this hypothetical sensor, it appears that many effectors are able to modify the threshold voltage (the "gating potential") at which opening occurs. Many pore inducers (for example, oxidative agents) shift the apparent gating potential to more negative (physiological) values, thereby favoring pore opening, while many pore inhibitors (like reducing agents) have the opposite effect and favor its closure. Thus, pore opening can be obtained either by depolarization or by changing the threshold potential at which opening occurs. The recent proposal that the pore also senses changes of the surface potential represents a substantial improvement in our understanding of how pore responses can be modulated by a variety of heterogeneous agents (Broekemeier and Pfeiffer, 1995). For example, amphipathic anions (like fatty acids) behave as pore activators with an effect that cannot be explained by depolarization (Broekemeier and Pfeiffer, 1995), while polycations (like spermine) (Lapidus and Sokolove, 1992, 1993), positively charged peptides (Rigobello et al., 1995), and amphipathic cations (like sphingosine and trifluoroperazine) (Broekemeier and Pfeiffer, 1995) inhibit pore opening, the latter acting independently of inhibition of phospholipase A_2. All these data are consistent with the view that a more positive surface potential favors pore closure, while a more negative surface potential favors its opening (Bernardi et al., 1994; Broekemeier and Pfeiffer, 1995).

2.3. Divalent Cations

As mentioned before, the PT is greatly favored by accumulation of Ca^{2+} ions in the matrix, while it is counteracted by other Me^{2+} ions like Mg^{2+}, Sr^{2+}, and Mn^{2+}. The Ca^{2+} requirement is not as strict as has been previously assumed. In general, pore opening *in vitro* can be easily achieved at micromolar Ca^{2+} concentrations, and the Ca^{2+} requirement therefore should not be intended in the sense of a massive overload. The general effects of Me^{2+} ions on the pore easily can be rationalized with the existence of two sites (Bernardi *et al.*, 1993): (1) an external site—occupation of this site (apparent $I_{50} = 0.2$ mM) by an Me^{2+} ion (including Ca^{2+} itself) decreases the probability of pore opening; and (2) an internal site—occupation of this site by Ca^{2+} increases the probability of pore opening, while other Me^{2+} (Sr^{2+}, Mn^{2+}) are inhibitory and apparently compete with Ca^{2+}; the apparent I_{50} of this site is harder to estimate but could be in the micromolar range. The well-known inducing effects of Pi have been partly explained by its buffering effect on matrix pH and partly by its ability to decrease the intramitochondrial free (Mg^{2+}) (Petronilli *et al.*, 1993a).

2.4. Oxidative Stress

Oxidative stress long has been known to favor the PT (Zoratti and Szabo', 1995). Pore opening is favored by oxidants of both pyridine nucleotides (like acetoacetate and oxaloacetate), glutathione (like *tert*-butylhydroperoxide), and of dithiols (like diamide), as well as by dithiol cross-linkers (like phenylarsine oxide and arsenite). Since mitochondrial levels of pyridine nucleotides and glutathione are connected through energy-linked transhydrogenase and glutathione reductase, it was debated whether NADH, NADPH, or GSH levels were the relevant factor in pore modulation (Zoratti and Szabo', 1995). After defining the importance of dithiol–disulfide interconversion for pore opening (Petronilli *et al.*, 1994a), we have recently readdressed this long-standing issue by independently modulating the levels of reduced pyridine nucleotides and glutathione in both energized (Costantini *et al.*, 1996) and deenergized mitochondria (Chernyak and Bernardi, 1996). Our results indicate that two sites can be experimentally distinguished, and that both contribute to pore modulation by oxidating and reducing agents.

1. A first site (which we dubbed the "S-site") coincides with the aforementioned oxidation–reduction-sensitive dithiol. Cross-linking of the S-site by arsenite or phenylarsine oxide or its oxidation by oxidized glutathione favors the PT under conditions where the pyridine nucleotides pool is demonstrably kept in the fully reduced state (Costantini *et al.*, 1996). Dithiothreitol can fully revert the effects of cross-linking or oxidation at this site, which is blocked by low concentrations (10–20 μM) of *N*-ethylmaleimide (Petronilli *et al.*, 1994a),

monobromobimane (Costantini *et al.*, 1995a), or benzoquinone (Palmeira and Wallace, 1997). Oxidized glutathione appears to be the immediate oxidant of the S-site, and many pore agonists (like organic hydroperoxides) affect the pore through changes in the levels of reduced glutathione rather than by direct oxidation (Chernyak and Bernardi, 1996).

2. A second site (which we dubbed the "P-site") is in apparent oxidation–reduction equilibrium with the pool of pyridine nucleotides. Pyridine nucleotides oxidation favors the PT under conditions where the glutathione pool is kept in the fully reduced state or when the dithiol is reacted with arsenite. In contrast to the S-site, this site cannot be blocked by monobromobimane or dithiothreitol, while it is sensitive to N-ethylmaleimide in the same concentration range as the S-site (10–20 μM). Oxidation of pyridine nucleotides does not affect the chemical reactivity of the S-site toward monobromobimane, suggesting that the P- and S-sites do not share common oxidation-reduction intermediates (Costantini *et al.*, 1996; Chernyak and Bernardi, 1996).

2.5. CsA and Cyclophilin

Most biological effects of CsA are mediated by its binding to a family of intracellular receptors, the cyclophilins, which all possess peptidyl-prolyl-*cis-trans*-isomerase activity (reviewed by Galat and Metcalfe, 1995). The enzymatic activity is inhibited as a result of CsA binding, but this effect is not related to immunosuppression. Rather, the latter is due to the Ca^{2+}–calmodulin-dependent binding of the CsA–cyclophilin complex to calcineurin, a cellular phosphatase, which in turn blocks transcription of nuclear genes that depend on nuclear translocation of nuclear factor of activated T cells (Emmel *et al.*, 1989).

A rat mitochondrial cyclophilin with a unique amino-terminus has been isolated (Connern and Halestrap, 1992) and recently shown to be the rat homologue of human cyclophilin D (Woodfield *et al.*, 1997), but a major question remained open. Was CsA directly inhibiting the pore or was cyclophilin D modulating pore activity in a CsA-sensitive manner? Initial reports by Crompton and co-workers of a 10-kDa membrane receptor for CsA (Andreeva and Crompton, 1994) have not been substantiated (Andreeva *et al.*, 1995). It appears that this protein was a degradation product of cyclophilin D or of contaminating cyclophilin B from the endoplasmic reticulum (Nicolli *et al.*, 1996). Furthermore, affinity methods have failed to reveal the existence of mitochondrial integral membrane proteins specifically binding to CsA (Nicolli *et al.*, 1996). The most plausible hypothesis is that the pore is modulated by cyclophilin binding from the matrix side of the inner membrane and that CsA and H^+ may be affecting the pore indirectly, by determining unbinding of cyclophilin (Connern and Halestrap, 1996; Nicolli *et al.*, 1996). Whether or not this model will prove to be correct, it appears that calcineurin inhibition is not

involved in pore modulation, since N-methylVal-4-cyclosporin, a CsA derivative that binds cyclophilin but not calcineurin (Schreier et al., 1993), is as
effective as CsA itself at pore inhibition (Nicolli et al., 1996).

3. MOLECULAR NATURE OF THE PORE: AN OPEN QUESTION

Among the effectors of the PT pore, it is notable to find two inhibitors of
the adenine nucleotide translocase: atractylate and bongkrekate (Zoratti and
Szabo', 1995). The two inhibitors affect the pore in opposite directions: atractylate favoring and bongkrekate inhibiting the transition. These findings indicate
that if the effects of the inhibitors are mediated by the translocase, these must
occur through a conformational effect. The adenine nucleotide carrier can
adopt different conformations during the translocation cycle. In the presence of
atractylate the carrier is locked in the so called "c" conformation, while in the
presence of bongkrekate the carrier adopts the "m" conformation (Schultheiss
and Klingenberg, 1984). Interestingly, ADP favors orientation of the translocase in the "m" conformation (Scherer and Klingenberg, 1974) and, like
bongkrekate, inhibits the pore.

Because of the effects of bongkrekate and atractylate, the suggestion has
been made that the pore might in fact be formed by the translocase itself
(Halestrap and Davidson, 1990). This interpretation appears rather simplistic,
however. In the first place, one wonders why most pore inducers and inhibitors,
including CsA, do not affect the activity and the conformation of the translocase at all; furthermore, the translocase is a very abundant protein, and its
"c" to "m" transition may profoundly affect general membrane properties like
the surface potential. It remains thus possible that the carrier conformation
might affect the pore indirectly, possibly through a modulation of the putative
voltage sensor (Bernardi et al., 1994).

In 1994, Skulachev and co-workers reported that treatment of the translocase with mersalyl caused the appearance of unselective, high-conductance
channels after incorporation of the modified protein in black lipid membranes
(Tikhonova et al., 1994). However, pore opening in mitochondria does not
require mersalyl or other sulfhydryl reagents, although the latter can induce the
PT in the presence of Ca^{2+} (Zoratti and Szabo', 1995). Furthermore, modification of sulfhydryl groups can induce a channel-type behavior in many mitochondrial carriers (Dierks et al., 1990a,b; Herick and Krämer, 1995), which
makes it harder to correlate the experiments of Tikhonova et al. (1994) on the
translocase with the properties of the pore in mitochondria.

In the same context, an interesting contribution recently came from the
Klingenberg laboratory with the demonstration that the unmodified translocase
reconstituted in giant liposomes exhibits a striking high-conductance channel

activity that is stimulated by Ca^{2+} and displays a marked voltage dependence (Brustovetski and Klingenberg, 1996). The reconstituted channel exhibits prominent gating effects, with abrupt closures at high-membrane potentials of either sign, which is consistent with what is observed in patch–clamp experiments of mitoplasts (Szabo' and Zoratti, 1993), and with the reported voltage dependence of the pore in intact mitochondria (Petronilli *et al.*, 1993b,1994a).

A third relevant observation came from the Brdiczka laboratory, with the finding that complexes prepared by low-detergent extraction of mitochondria and enriched in hexokinase, porin, and the adenine nucleotide translocase exhibit Ca^{2+}-dependent and CsA-sensitive high-conductance channel activity when reconstituted in planar lipid bilayers (Beutner *et al.*, 1996). This preparation also catalyzed Ca^{2+}-dependent and CsA-sensitive ATP and malate diffusion after incorporation in proteoliposomes, although the most active fractions were not enriched in translocase or porin but rather in a 67-kDa-species, which may be composed of heterodimers of these proteins (Beutner *et al.*, 1996).

Finally, it must be mentioned that inner membrane preparations from translocase-deficient *Saccaromyces cerevisiae* mitochondria exhibited high-conductance channels with the same properties as those from wild-type cells (Lohret *et al.*, 1996). Whether these channels may be identified with the PT pore remains questionable, however, since a PT comparable to that occurring in mammalian mitochondria has not been observed in yeast mitochondria (Jung and Pfeiffer, 1997).

In summary, the molecular nature of the PT pore remains undefined at present. It appears conceivable that the pore is a supramolecular structure involving components from both the outer and inner membranes and from the intermembrane and matrix spaces. It appears possible that the adenine nucleotide translocase is part of the complex or that it substantially affects its function, but a precise definition of the molecular nature of the pore remains matter for future work.

4. THE PERMEABILITY TRANSITION PORE: A ROLE IN CALCIUM HOMEOSTASIS?

Pore opening in the high-conductance mode appears to play a role in the activation of the cell suicide program known as apoptosis, or programmed cell death. This represents a potential physiological role for the pore, which will be discussed later. An open question is whether the pore has any physiological role in cells not committed to die. An attractive hypothesis is that the PT pore may play a role in mitochondrial Ca^{2+} homeostasis under physiological conditions.

Mitochondria participate in intracellular Ca^{2+} homeostasis and they can rapidly accumulate Ca^{2+} after cell stimulation by a variety of signals increasing cytosolic Ca^{2+} in the physiological range (Rizzuto *et al.*, 1992, 1993, 1994; Jouaville *et al.*, 1995). This is made possible by the existence of a set of sophisticated Ca^{2+} transport systems:

1. The Ca^{2+} uniporter. This channel allows Ca^{2+} transport along its electrochemical gradient, thus driving Ca^{2+} accumulation at physiological membrane potentials; it has a high V_{max} (more than 1200 nmole Ca^{2+}/mg protein per min), and allows fast removal of extramitochondrial Ca^{2+} when its concentration rises above about 0.5–1.0 μM.
2. The rapid uptake mode (RaM), a fast system for sequestration of submicromolar Ca^{2+} pulses that is sensitive to their frequency and may be different from the uniporter (Sparagna *et al.*, 1995).
3. The Na^{+}–Ca^{2+} exchanger. This antiporter exchanges Na^{+} for Ca^{2+} ions and catalyzes Ca^{2+} efflux with a V_{max} of 3–18 nmole Ca^{2+}/mg protein per min, depending on the tissue.
4. The H^{+}–Ca^{2+} exchanger. This antiporter exchanges two or more H^{+} ions per Ca^{2+} ion and may have an active (energy-requiring) component; in energized mitochondria it catalyzes Ca^{2+} efflux, with a V_{max} lower than 2 nmole Ca^{2+}/mg protein per min (see Gunter *et al.*, 1994 for review).

The combined rate of the efflux pathways is adequate to compensate the rate of Ca^{2+} uptake *in vitro* when extramitochondrial free $[Ca^{2+}]$ is in the submicromolar range. However, a kinetic imbalance becomes apparent when the steady-state Ca^{2+} concentration is modified. Whereas elevation of extramitochondrial $[Ca^{2+}]$ is followed by fast Ca^{2+} uptake on the Ca^{2+} uniporter with reestablishment of the steady state within seconds, a decrease of external $[Ca^{2+}]$ requires several minutes for reestablishment of the steady state, because the Ca^{2+} efflux pathways are so slow (e.g., Nicholls, 1978). This kinetic imbalance would pose the problem of progressive mitochondrial Ca^{2+} overload *in vivo*. This does not occur under physiological circumstances, because the Ca^{2+} efflux pathways operate at a slow rate (Rizzuto *et al.*, 1992, 1993).

We have proposed that fast mitochondrial Ca^{2+} release *in vivo* can be achieved through transient opening of the PT pore operating as an inducible mitochondrial Ca^{2+} release channel [see Bernardi and Petronilli (1996) for a thorough discussion of this hypothesis, and Ichas *et al.* (1994) for conditions under which the pore appears to reversibly release Ca^{2+}]. The key features of this system can be summarized as follows:

1.   The pore is switched to the open conformation (by Ca^{2+} uptake and/or by other cellular or mitochondrial signals), followed by depolarization and Ca^{2+} release. Depolarization is an essential feature of the release process, because it allows Ca^{2+} efflux down its concentration gradient, while the large

size of the pore would serve the dual purpose of allowing both for Ca^{2+} efflux and for charge compensation (by monovalent cations and protons), thus ensuring Ca^{2+} release at zero potential (i.e., even for vanishingly low Ca^{2+} concentration gradients).

2. Ca^{2+} efflux is followed by pore closure and repolarization once intramitochondrial $[Ca^{2+}]$ has returned below threshold levels and/or the stimulatory signals have been removed or inhibitory ones have been generated. Closure also would be favored by the increased concentrations of ADP and Mg^{2+} secondary to ATP hydrolysis during the phase of pore opening, and/or by inhibitory cellular or mitochondrial signals.

3. No futile Ca^{2+} cycling occurs during the phase of Ca^{2+} release, because the proton electrochemical gradient is collapsed as long as the pore remains in the open state, and therefore there is no driving force for Ca^{2+} uptake until the pore switches back to the closed state and the inner membrane repolarizes.

4. Despite the apparent lack of selectivity displayed by the PT pore in isolated mitochondria, the system operates as a *selective* Ca^{2+} release channel because no K^+ or Na^+ gradients are maintained across the inner membrane. This would explain why no evolutionary pressure has been posed on the development of cation selectivity (Bernardi and Petronilli, 1996). Alternatively, the PT pore might be more selective *in vivo* because of the effects of regulatory proteins (cyclophilin D or other unknown regulators) (see, e.g., Broekemeier and Pfeiffer, 1995) or of other unidentified diffusible factor(s) lost during isolation of mitochondria.

If the pore is involved in Ca^{2+} homeostasis as a Ca^{2+} release channel, an appealing link can be identified with a variety of pathological conditions characterized by Ca^{2+} overload. In this scenario, pore opening might contribute to cellular Ca^{2+} overload because of both mitochondrial release and ATP depletion, causing in turn impaired Ca^{2+} clearance by both the endoplasmic reticulum and the plasma membrane.

5. MITOCHONDRIA IN CELL DEATH

Opening of the PT pore *in vivo* would undoubtedly represent a bioenergetic catastrophe. Not only would mitochondria cease to make ATP, but due to the short circuiting of the proton-motive force, mitochondria would consume ATP at the maximal attainable rates and would contribute to ATP depletion and onset of cell death (see Di Lisa and Bernardi, 1998, for a recent review of the role of mitochondria in cell death). It is not surprising that initial evidence for an involvement of the PT pore in cell death came from established models like ischemia–reperfusion or treatment with toxicants (e.g., organic hydro-peroxides or redox cycling quinones).

5.1. Necrosis

Early suggestions for an involvement of mitochondria in the initial phases of cell death induced by a variety of toxicants came from studies of Ca^{2+} homeostasis (Shen and Jennings, 1972; Lemasters et al., 1987; Farber, 1990; Nicotera and Orrenius, 1992). Following Lehninger's suggestion that oxidation of mitochondrial pyridine nucleotides could activate a selective Ca^{2+} release pathway (Lehninger et al., 1978), a number of studies have investigated the nature of mitochondrial Ca^{2+} release induced by oxidants like redox-cycling quinones, organic hydroperoxides, and heavy metals of toxicological interest (Lötscher et al., 1979, 1980; Bellomo et al., 1982, 1984; G. A. Moore et al., 1987; M. Moore et al., 1985; Rizzuto et al., 1987; Carbonera et al., 1988; Chavez et al., 1991; Hermes Lima et al., 1991, 1992; Zazueta et al., 1994; Costantini et al., 1995b). While Lehninger's work provided a testable bio-chemical mechanism linking mitochondrial Ca^{2+} release to metabolism through NAD(H) levels, further studies were undoubtedly stimulated by the prevailing view that mitochondria represented high-capacity Ca^{2+} storage organelles, an idea that did not stand the test of time. For instance, in resting cardiac myocytes, mitochondrial $[Ca^{2+}]$ was found to be lower than cytosolic $[Ca^{2+}]$ (e.g., Miyata et al., 1991).

It has been proposed (1) that Ca^{2+} release induced by oxidants occurs through a selective pathway activated by ADP-ribosylation of a membrane protein following oxidation of pyridine nucleotides and their hydrolysis by a matrix glycohydrolase; (2) that a high-membrane potential (measured in mito-chondria treated with oxidants in the presence of EGTA) would be maintained despite Ca^{2+} efflux; and (3) that "excessive" Ca^{2+} cycling would then cause mitochondrial dysfunction through an undefined mechanism (e.g., Richter et al., 1992, and references therein). We note that (1) the contention that oxidation of pyridine nucleotides and their hydrolysis precedes Ca^{2+} efflux is much weakened by the findings that all mitochondrial glycohydrolase activity is located in the intermembrane space (Scott Boyer et al., 1993), and that in many cases pyridine nucleotide oxidation is the consequence rather than the cause of pore opening (Costantini et al., 1996; Chernyak and Bernardi, 1996); (2) that the argument that mitochondria are not depolarized during the phase of Ca^{2+} efflux is untenable if the membrane potential measurements are performed in the presence of EGTA, since EGTA reseals the pore and allows repolarization (Crompton et al., 1987), and therefore does not allow conclusions regarding the mechanism(s) of Ca^{2+} efflux in its absence; and (3) that Ca^{2+} cycling as such is not harmful to mitochondria (Bernardi and Pietrobon, 1982). Today, most investigators agree that oxidative stress causes mitochondrial dysfunction through opening of the PT pore, which is followed by deenergization and by release of Ca^{2+} (Crompton *et al.*, 1987; Broekemeier *et al.*, 1992; Imberti *et al.*, 1993; Scott Boyer *et al.*, 1993; Weis *et al.*, 1994; Savage and Reed, 1994a,b; Hoek *et al.*, 1995; Bernardi and Petronilli, 1996).

An involvement of the PT pore in ischemic cell death had been suggested in early studies by Crompton and co-workers (Duchen *et al.*, 1993). The hypothesis that a similar mechanism could be underlying cell death in a variety of models and the availability of CsA as a selective inhibitor have recently led to an exponential increase of the studies exploring the potential role of mitochondria in general and of the pore in particular in a variety of models of disease (Lemasters *et al.*, 1987; Pastorino *et al.*, 1993, 1994, 1995; Griffiths and Halestrap, 1993; Imberti *et al.*, 1993; Zoeteweij *et al.*, 1993; Chacon *et al.*, 1994; Fuji *et al.*, 1994; Botla *et al.*, 1995; Di Lisa *et al.*, 1995; Nieminen *et al.*, 1995; Ankarkrona *et al.*, 1995; Saxena *et al.*, 1995; Zahrebelski *et al.*, 1995; Aguilar *et al.*, 1996; Kristian and Siesjo, 1996; Schinder *et al.*, 1996). Although it must be stressed that evidence of pore involvement remains indirect, taken together these studies strongly suggest that pore opening may be a critical event in cell death. Part of the results can be explained by the ATP depletion following pore opening, that is, by a purely bioenergetic mechanism aggravating or even causing cellular Ca^{2+} overload (e.g., Imberti *et al.*, 1993; Nieminen *et al.*, 1995). In other models, however, pore opening could be dissociated from ATP depletion and CsA prevented cell death irrespective of the level of cellular ATP or of mitochondrial (de)energization (Pastorino *et al.*, 1993, 1995). It has been speculated that the ATP-independent alteration precipitating cell death might involve interactions of mitochondria with the cytoskeleton (Pastorino *et al.*, 1993). In keeping with this, the anticancer drugs taxol and colchicine, both inhibitors of microtubule assembly, are able to prevent closure of the PT in permeabilized Ehrlich ascites tumor cells (Evtodienko *et al.*, 1996). A further consequence of pore opening that appears to play a key role in induction of cell death independently of ATP depletion has been recently described by Kroemer and co-workers, and will be treated in detail in Section 5.2 (Susin *et al.*, 1996).

5.2. Apoptosis

Apoptosis, or programmed cell death, is a highly conserved and regulated process whereby cells "decide to die" following an initial, appropriate stimulus (see Sluyser, 1996, for a comprehensive coverage of this subject). After the induction phase, a variety of pro- and antiapoptotic intracellular factors compete to either "approve" or override the death stimulus. If the former signals prevail, the cell enters the effector phase, which is then followed by the final degradation phase where the biochemical and morphological hallmarks of apoptosis become apparent (see Kroemer *et al.*, 1997, for a recent review). The bioenergetics of apoptosis has been the subject of recent interest, mainly (but not exclusively) because of the development and the application of very sensitive techniques allowing monitoring of the mitochondrial membrane potential *in situ* at the level of isolated cells with fluorescent probes (Farkas *et al.*, 1989; Petit *et al.*, 1990, 1995; Glab *et al.*, 1993; Cossarizza *et al.*, 1993, 1994, 1995; Loew *et al.*, 1993, 1994; Chacon *et al.*, 1994; Newmeyer *et al.*, 1994; Vayssière *et al.*, 1994; Di Lisa *et al.*, 1995; Zamzami *et al.*, 1995a,b, 1996a,b; Susin *et al.*, 1996; Marchetti *et al.*, 1997).

In studies aimed at determining whether changes in mitochondrial membrane potential are part of the apoptotic process, seemingly conflicting results have been obtained by different research groups. In one study on rat thymocytes treated *in vitro* with dexamethazone, mitochondrial depolarization *in situ* was detected well after the nuclear signs of apoptosis had appeared, suggesting that integrity of mitochondria was required for the apoptotic process to proceed (Cossarizza *et al.*, 1994). On the other hand, early changes of mitochondrial function (as gauged by depolarization) and structure (as studied by morphological alterations) have been shown to precede rather than follow nuclear degradation in spontaneous as well as in dexamethazone-induced apoptosis (Vayssière *et al.*, 1994; Petit *et al.*, 1995; Zamzami *et al.*, 1995a,b, 1996a,b; Susin *et al.*, 1996; Marchetti *et al.*, 1997). While the reason for this discrepancy remains unclear, the latter observations are of particular relevance in the context of this review.

After *in vivo* dexamethazone treatment, a subpopulation of cells with a lower mitochondrial membrane potential could be sorted (Zamzami *et al.*, 1995a), which would then undergo apoptosis in culture (Zamzami *et al.*, 1995b). Addition of CsA prevented depolarization and apoptosis, suggesting a key role for the PT in the apoptotic cascade (Kroemer *et al.*, 1995). A very strong correlation has been subsequently established between onset of the PT and appearance of the nuclear signs of apoptosis in a reconstituted system where nuclei were incubated with isolated mitochondria. With no exception, nuclear degradation was observed when a PT occurred, while nuclei remained intact when mitochondria did not undergo a PT irrespective of the methods used to induce or inhibit the pore (Zamzami *et al.*, 1996a,b). This correlation has now been extended to pore inducers acting at the S-site (see Section 2.4). In perfect match with pore modulation at this site (Petronilli *et al.*, 1994; Costantini *et al.*, 1995a, 1996; Chernyak and Bernardi, 1996) apoptosis induced by dithiol oxidants (like diamide) could be blocked by monofunctional thiol reagents (Marchetti *et al.*, 1997). These observations strengthened the earlier general theory of the mitochondrial control of apoptosis and the proposal that the PT is a key event in the effector phase of programmed cell death (Kroemer *et al.*, 1995, 1997).

The potential biochemical link between opening of the PT pore and nuclear degradation has been provided by the Kroemer group. In a recent report, they have shown that mitochondria that have undergone the PT release a factor able to induce the signs of apoptosis in isolated nuclei. This factor has been purified to homogeneity and shown to correspond to a 50-kDa protein [apoptosis-inducing factor (AIF)], associated with markers of the mitochondrial outer membrane (Susin *et al.*, 1996). AIF is a protease devoid of endonuclease activity, it appears to act through proteolytic activation of a nuclear endonuclease, and it is inhibited by *N*-benzyloxycarbonyl-Val-Ala-Asp-fluoromethylketone (an inhibitor of interleukin-converting enzyme-like proteases) but not by specific inhibitors of known Ca^{2+}, serine, or cysteine proteases including interleukin-converting enzyme itself and CPP32/Yama (Susin *et al.*, 1996), all of which are involved in the apoptotic cascade (see Martin and Green, 1995, for a review on protease activation in apoptosis). Intriguingly, it appears that the protein product of the *bcl-2* proto-oncogene, which largely localizes to the outer mitochondrial membrane in human lymphoma cells (Riparbelli *et al.*, 1995), prevents release of AIF when induced by some but not all PT-inducing stimuli, suggesting a mechanism for the antiapoptotic effect of *bcl-2* itself (Susin *et al.*, 1996).

A second intriguing finding linking mitochondria to programmed cell death is the specific requirement for dATP and cytochrome *c* in induction of the apoptotic program in a reconstituted system (Liu *et al.*, 1996). Although in this case a direct link with the PT has not been made, we note that mitochondrial swelling following pore opening in salt-containing media causes release of several matrix proteins (Igbavboa *et al.*, 1989) and of cytochrome *c* (Petronilli *et al.*, 1994b), which makes the pore a very attractive candidate for the release of cytochrome *c in vivo*.

6. CONCLUSIONS AND PERSPECTIVES

This review has tried to document how the PT has progressed from the status of *in vitro* artifact to that of a regulated channel and to illustrate how this phenomenon is emerging today as a promising candidate for the effector phase of programmed cell death. Pore involvement in accidental cell death as well may be a further indication that necrosis and apoptosis are less far apart than previously thought. In general, however, I think that research on the PT pore has substantially contributed to attract novel interest in mitochondria from immunologists, oncologists and physiopathologists. Whether or not our current speculations on pore (dys)function will turn out to be correct, it appears that the active role of mitochondria in the decision-making processes of cellular life and death can no longer be questioned.

ACKNOWLEDGMENT. Research in our laboratory is supported by the Italian Ministry for University and Scientific Research, the National Research Council, Telethon-Italy (Grant 847), the Harvard Foundation, and NATO's Scientific Affairs Division (Grant CRG 960668).

7. REFERENCES

Aguilar, H. I., Botla, R., Arora, A. S., Bronk, S. F., and Gores, G. J., 1996, Induction of the mitochondrial permeability transition by protease activity in rats: A mechanism of hepatocyte necrosis, *Gastroenterology* **110:**558–566.

Andreeva, L., and Crompton, M., 1994, An ADP-sensitive cyclosporin-A-binding protein in rat liver mitochondria, *Eur. J. Biochem.* **221:**261–268.

Andreeva, L., Tanveer, A., and Crompton, M., 1995, Evidence for the involvement of a membrane-associated cyclosporin-A-binding protein in the Ca^{2+}-activated inner membrane pore of heart mitochondria, *Eur. J. Biochem.* **230:**1125–1132.

Ankarcrona, M., Dypbukt, J. M., Bonfoco, E., Zhivotovsky, B., Orrenius, S., Lipton, S. A., and Nicotera, P., 1995, Glutamate-induced neuronal death: A succession of necrosis or apoptosis depending on mitochondrial function, *Neuron* **15:**961–973.

Azzone, G. F., and Azzi, A., 1965, Volume changes in liver mitochondria, *Proc. Natl. Acad. Sci. USA* **53:**1084–1089.

Beatrice, M. C., Palmer, J. W., and Pfeiffer, D. R., 1980, The relationship between mitochondrial membrane permeability, membrane potential, and the retention of Ca^{2+} by mitochondria, *J. Biol. Chem.* **255:**8663–8671.

Bellomo, G., Jewell, S. A., and Orrenius, S., 1982, The metabolism of menadione impairs the ability of rat liver mitochondria to take up and retain calcium, *J. Biol. Chem.* **257:**11558–11562.

Bellomo, G., Martino, A., Richelmi, P., Moore, G. A., Jewell, S. A., and Orrenius, S., 1984, Pyridine-nucleotide oxidation, Ca^{2+} cycling and membrane damage during tert-butyl hydroperoxide metabolism by rat-liver mitochondria, *Eur. J. Biochem.* **140:**1–6.

Bernardi, P., 1992, Modulation of the mitochondrial cyclosporin A-sensitive permeability transition pore by the proton electrochemical gradient. Evidence that the pore can be opened by membrane depolarization, *J. Biol. Chem.* **267:**8834–8839.

Bernardi, P., and Petronilli, V., 1996, The permeability transition pore as a mitochondrial Ca^{2+} release channel: A critical appraisal, *J. Bioenerg. Biomembr.* **28:**129–136.

Bernardi, P., and Pietrobon, D., 1982, On the nature of Pi-induced, Mg^{2+}-prevented Ca^{2+} release in rat liver mitochondria, *FEBS Lett.* **139:**9–12.

Bernardi, P., Vassanelli, S., Veronese, P., Colonna, R., Szabo', I., and Zoratti, M., 1992, Modulation of the mitochondrial permeability transition pore. Effect of protons and divalent cations, *J. Biol. Chem.* **267:**2934–2939.

Bernardi, P., Veronese, P., and Petronilli, V., 1993, Modulation of the mitochondrial cyclosporin A-sensitive permeability transition pore. I. Evidence for two separate Me^{2+} binding sites with opposing effects on the pore open probability, *J. Biol. Chem.* **268:**1005–1010.

Bernardi, P., Broekemeier, K. M., and Pfeiffer, D. R., 1994, Recent progress on regulation of the mitochondrial permeability transition pore: A cyclosporin-sensitive pore in the inner mitochondrial membrane, *J. Bioenerg. Biomembr.* **26:**509–517.

Beutner, G., Rück, A., Riede, B., Welte, W., and Brdiczka, D., 1996, Complexes between kinases, mitochondrial porin and adenylate translocator in rat brain resemble the permeability transition pore, *FEBS Lett.* **396:**189–195.

Botla, R., Spivey, J. R., Aguilar, H., Bronk, S. F., and Gores, G. J., 1995, Ursodeoxycholate (UDCA) inhibits the mitochondrial membrane permeability transition induced by gly-cochenodeoxycholate: A mechanism of UDCA cytoprotection, *J. Pharmacol. Exp. Ther.* **272**: 930–938.

Broekemeier, K. M., and Pfeiffer, D. R., 1995, Inhibition of the mitochondrial permeability transition by cyclosporin A during long time frame experiments: Relationship between pore opening and the activity of mitochondrial phospholipases, *Biochemistry* **34**:16440–16449.

Broekemeier, K. M., Dempsey, M. E., and Pfeiffer, D. R., 1989, Cyclosporin A is a potent inhibitor of the inner membrane permeability transition in liver mitochondria, *J. Biol. Chem.* **264**: 7826–7830.

Broekemeier, K. M., Carpenter Deyo, L., Reed, D. J., and Pfeiffer, D. R., 1992, Cyclosporin A protects hepatocytes subjected to high Ca^{2+} and oxidative stress, *FEBS Lett.* **304**:192–194.

Brustovetsky, N., and Klingenberg, M., 1996, Mitochondrial ADP/ATP carrier can be reversibly converted into a large channel by Ca^{2+}, *Biochemistry* **35**:8483–8488.

Carbonera, D., Angrilli, A., and Azzone, G. F., 1988, Mechanism of nitrofurantoin toxicity and oxidative stress in mitochondria, *Biochim. Biophys. Acta* **936**:139–147.

Chacon, E., Reece, J. M., Nieminen, A. L., Zahrebelski, G., Herman, B., and Lemasters, J. J., 1994, Distribution of electrical potential, pH, free Ca^{2+}, and volume inside cultured adult rabbit cardiac myocytes during chemical hypoxia: A multiparameter digitized confocal microscopic study, *Biophys. J.* **66**:942–952.

Chavez, E., Zazueta, C., Osornio, A., Holguin, J. A., and Miranda, M. E., 1991, Protective behavior of captopril on Hg^{++}-induced toxicity on kidney mitochondria. *In vivo* and *in vitro* experiments, *J. Pharmacol. Exp. Ther.* **256**:385–390.

Chernyak, B. V., and Bernardi, P., 1996, The mitochondrial permeability transition pore is modulated by oxidative agents through both pyridine nucleotides and glutathione at two separate sites, *Eur. J. Biochem.* **238**:623–630.

Connern, C. P., and Halestrap, A. P., 1992, Purification and N-terminal sequencing of peptidyl-prolyl *cis-trans*-isomerase from rat liver mitochondrial matrix reveals the existence of a distinct cyclophilin, *Biochem. J.* **284**:381–385.

Connern, C. P., and Halestrap, A. P., 1996, Chaotropic agents and increased matrix volume enhance binding of cyclophilin D to the inner mitochondrial membrane and sensitize the mitochondrial permeability transition to [Ca^{2+}], *Biochemistry* **35**:8172–8180.

Cossarizza, A., Baccarani Contri, M., Kalashnikova, G., and Franceschi, C., 1993, A new method for the cytofluorimetric analysis of mitochondrial membrane potential using the J-aggregate forming lipophilic cation 5,5',6,6'-tetrachloro-1,1',3,3'-tetraethylbenzimidazol carbocyanine iodide (JC-1), *Biochem. Biophys. Res. Commun.* **197**:40–45.

Cossarizza, A., Kalashnikova, G., Grassilli, E., Chiappelli, F., Salvioli, S., Capri, M., Barbieri, D., Troiano, L., Monti, D., and Franceschi, C., 1994, Mitochondrial modifications during rat thymocyte apoptosis: A study at the single cell level, *Exp. Cell Res.* **214**:323–330.

Cossarizza, A., Franceschi, C., Monti, D., Salvioli, S., Bellesia, E., Rivabene, R., Biondo, L., Rainaldi, G., Tinari, A., and Malorni, W., 1995, Protective effect of N-acetylcysteine in tumor necrosis factor-alpha-induced apoptosis in U937 cells: The role of mitochondria, *Exp. Cell Res.* **220**:232–240.

Costantini, P., Chernyak, B. V., Petronilli, V., and Bernardi, P., 1995a, Selective inhibition of the mitochondrial permeability transition pore at the oxidation-reduction sensitive dithiol by monobromobimane, *FEBS Lett.* **362**:239–242.

Costantini, P., Petronilli, V., Colonna, R., and Bernardi, P., 1995b, On the effects of paraquat on isolated mitochondria. Evidence that paraquat causes opening of the cyclosporin A-sensitive permeability transition pore synergistically with nitric oxide, *Toxicology* **99**:77–88.

Costantini, P., Chernyak, B. V., Petronilli, V., and Bernardi, P., 1996, Modulation of the mitochondrial permeability transition pore by pyridine nucleotides and dithiol oxidation at two separate sites, *J. Biol. Chem.* **271:**6746–6751.

Crompton, M., and Costi, A., 1988, Kinetic evidence for a heart mitochondrial pore activated by Ca^{2+}, inorganic phosphate and oxidative stress. A potential mechanism for mitochondrial dysfunction during cellular Ca^{2+} overload, *Eur. J. Biochem.* **178:**489–501.

Crompton, M., Costi, A., and Hayat, L., 1987, Evidence for the presence of a reversible Ca^{2+}-dependent pore activated by oxidative stress in heart mitochondria, *Biochem. J.* **245:**915–918.

Crompton, M., Ellinger, H., and Costi, A., 1988, Inhibition by cyclosporin A of a Ca^{2+}-dependent pore in heart mitochondria activated by inorganic phosphate and oxidative stress, *Biochem. J.* **255:**357–360.

Dierks, T., Salentin, A., Heberger, C., and Krämer, R., 1990a, The mitochondrial aspartate/glutamate and ADP/ATP carrier switch from obligate counterexchange to unidirectional transport after modification by SH-reagents, *Biochim. Biophys. Acta* **1028:**268–280.

Dierks, T., Salentin, A., and Krämer, R., 1990b, Pore-like and carrier-like properties of the mitochondrial aspartate/glutamate carrier after modification by SH-reagents: Evidence for a preformed channel as a structural requirement of carrier-mediated transport, *Biochim. Biophys. Acta* **1028:**281–288.

Di Lisa, F., and Bernardi, P., 1998, Mitochondrial function as a determinant of recovery or death in cell response to injury, *Mol. Cell Biochem.* **184:**379–391.

Di Lisa, F., Blank, P. S., Colonna, R., Gambassi, G., Silverman, H. S., Stern, M. D., and Hansford, R. G., 1995, Mitochondrial membrane potential in single living adult rat cardiac myocytes exposed to anoxia or metabolic inhibition, *J. Physiol.* **486:**1–13.

Duchen, M. R., McGuinness, O., Brown, L. A., and Crompton, M., 1993, On the involvement of a cyclosporin A sensitive mitochondrial pore in myocardial reperfusion injury, *Cardiovasc. Res.* **27:**1790–1794.

Emmel, E. A., Verweij, C. L., Durand, D. B., Higgins, K. M., Lacy, E., and Crabtree, G. R., 1989, Cyclosporin A specifically inhibits function of nuclear proteins involved in T cell activation, *Science* **246:**1617–1620.

Evtodienko, Yu. V., Teplova, V. V., Sidash, S. S., Ichas, F., and Mazat, J.-P., 1996, Microtubule-active drugs suppress the closure of the permeability transition pore in tumour mitochondria, *FEBS Lett.* **393:**86–88.

Farber, J. L., 1990, The role of calcium ions in toxic cell injury, *Environ. Health Perspect.* **84:**107–111.

Farkas, D. L., Wei, M. D., Febbroriello, P., Carson, J. H., and Loew, L. M., 1989, Simultaneous imaging of cell and mitochondrial membrane potentials, *Biophys. J.* **56:**1053–1069.

Fournier, N., Ducet, G., and Crevat, A., 1987, Action of cyclosporine on mitochondrial calcium fluxes, *J. Bioenerg. Biomembr.* **19:**297–303.

Fuji, Y., Johnson, M. E., and Gores, G. J., 1994, Mitochondrial dysfunction during anoxia/reoxygenation injury of liver sinusoidal endothelial cells, *Hepatology* **20:**177–185.

Galat, A., and Metcalfe, S. M., 1995, Peptidylproline *cis/trans* isomerases, *Prog. Biophys. Molec. Biol.* **63:**67–118.

Glab, N., Petit, P. X., and Slonimski, P. P., 1993, Mitochondrial dysfunction in yeast expressing the male sterility T-urf13 gene from maize: analysis at the population and individual cell level, *Mol. Gen. Genet.* **236:**299–308.

Griffiths, E. J., and Halestrap, A. P., 1993, Protection by cyclosporin A of ischemia/reperfusion-induced damage in isolated rat hearts, *J. Mol. Cell Cardiol.* **25:**1461–1469.

Gunter, T. E., and Pfeiffer, D. R., 1990, Mechanisms by which mitochondria transport calcium, *Am. J. Physiol.* **258:**C755–786.

Gunter, T. E., Gunter, K. K., Sheu, S. S., and Gavin, C. E., 1994, Mitochondrial calcium transport: physiological and pathological relevance, *Am. J. Physiol.* **267:**C313–339.

Halestrap, A. P., and Davidson, A. M., 1990, Inhibition of Ca^{2+}-induced large-amplitude swelling of liver and heart mitochondria by cyclosporin is probably caused by the inhibitor binding to mitochondrial-matrix peptidyl-prolyl *cis-trans* isomerase and preventing it interacting with the adenine nucleotide translocase, *Biochem. J.* **268:**153–160.

Herick, K., and Krämer, R., 1995, Kinetic and energetic characterization of solute flux through the reconstituted aspartate/glutamate carrier from beef heart mitochondria after modification with mercurials, *Biochim. Biophys. Acta* **1238:**63–71.

Hermes Lima, M., Valle, V. G., Vercesi, A. E., and Bechara, E. J., 1991, Damage to rat liver mitochondria promoted by delta-aminolevulinic acid-generated reactive oxygen species: Connections with acute intermittent porphyria and lead-poisoning, *Biochim. Biophys. Acta* **1056:**57–63.

Hermes Lima, M., Castilho, R. F., Valle, V. G., Bechara, E. J., and Vercesi, A. E., 1992, Calcium-dependent mitochondrial oxidative damage promoted by 5-aminolevulinic acid, *Biochim. Biophys. Acta* **1180:**201–206.

Hoek, J. B., Farber, J. L., Thomas, A. P., and Wang, X., 1995, Calcium ion-dependent signalling and mitochondrial dysfunction: Mitochondrial calcium uptake during hormonal stimulation in intact liver cells and its implication for the mitochondrial permeability transition, *Biochim. Biophys. Acta* **1271:**93–102.

Hunter, D. R., and Haworth, R. A., 1979a, The Ca^{2+}-induced membrane transition in mitochondria. I. The protective mechanisms, *Arch. Biochem. Biophys.* **195:**453–459.

Hunter, D. R., and Haworth, R. A., 1979b, The Ca^{2+}-induced membrane transition in mitochondria. III. Transitional Ca^{2+} release, *Arch. Biochem. Biophys.* **195:**468–477.

Ichas, F., Jouaville, L. S., Sidash, S. S., Mazat, J.-P., and Holmuhamedov, E. L., 1994, Mitochondrial calcium spiking: A transduction mechanism based on calcium-induced permeability transition involved in cell calcium signalling, *FEBS Lett.* **348:**211–215.

Igbavboa, U., and Pfeiffer, D. R., 1988, EGTA inhibits reverse uniport-dependent Ca^{2+} release from uncoupled mitochondria. Possible regulation of the Ca^{2+} uniporter by a Ca^{2+} binding site on the cytoplasmic side of the inner membrane, *J. Biol. Chem.* **263:**1405–1412.

Igbavboa, U., Zwizinski, C. W., and Pfeiffer, D. R., 1989, Release of mitochondrial matrix proteins through a Ca^{2+}-requiring, cyclosporin-sensitive pathway, *Biochem. Biophys. Res. Commun.* **161:**619–625.

Imberti, R., Nieminen, A. L., Herman, B., and Lemasters, J. J., 1993, Mitochondrial and glycolytic dysfunction in lethal injury to hepatocytes by t-butylhydroperoxide: Protection by fructose, cyclosporin A and trifluoperazine, *J. Pharmacol. Exp. Ther.* **265:**392–400.

Jouaville, L. S., Ichas, F., Holmuhamedov, E. L., Camacho, P., and Lechleiter, J. D., 1995, Synchronization of calcium waves by mitochondrial substrates in *Xenopus laevis* oocytes, *Nature* **377:**438–441.

Jung, D. W., and Pfeiffer, D. R., 1997, Properties of cyclosporin-insensitive permeability transition pore in yeast mitochondria, *J. Biol. Chem.* **272:**21104–21112.

Kinnally, K. W., Campo, M. L., and Tedeschi, H., 1989, Mitochondrial channel activity studied by patch-clamping mitoplasts, *J. Bioenerg. Biomembr.* **21:**497–506.

Kristian, T., and Siesjo, B. K., 1996, Calcium-related damage in ischemia, *Life Sci.* **59:**357–367.

Kroemer, G., Petit, P., Zamzami, N., Vayssière, J. L., and Mignotte, B., 1995, The biochemistry of programmed cell death, *FASEB J.* **9:**1277–1287.

Kroemer, G., Zamzami, N., and Susin, S. A., 1997, Mitochondrial control of apoptosis, *Immunol. Today*, in press.

Lapidus, R. G., and Sokolove, P. M., 1992, Inhibition by spermine of the inner membrane permeability transition of isolated rat heart mitochondria, *FEBS Lett.* **313:**314–318.

Lapidus, R. G., and Sokolove, P. M., 1993, Spermine inhibition of the permeability transition of isolated rat liver mitochondria: An investigation of mechanism, *Arch. Biochem. Biophys.* **306:** 246–253.

Lehninger, A. L., Vercesi, A., and Bababunmi, E. A., 1978, Regulation of Ca^{2+} release from mitochondria by the oxidation-reduction state of pyridine nucleotides, *Proc. Natl. Acad. Sci. USA* **75:**1690–1694.

Lemasters, J. J., Di Giuseppi, J., Nieminen, A. L., and Herman, B., 1987, Blebbing, free Ca^{2+} and mitochondrial membrane potential preceding cell death in hepatocytes, *Nature* **325:**78–81.

Liu, X., Kim, C. N., Yang, J., Jemmerson, R., and Wang, X., 1996, Induction of apoptotic program in cell-free extracts: Requirement for dATP and cytochrome *c*, *Cell* **86:**147–157.

Loew, L. M., Tuft, R. A., Carrington, W., and Fay, F. S., 1993, Imaging in five dimensions: Time-dependent membrane potentials in individual mitochondria, *Biophys. J.* **65:**2396–2407.

Loew, L. M., Carrington, W., Tuft, R. A., and Fay, F. S., 1994, Physiological cytosolic Ca^{2+} transients evoke concurrent mitochondrial depolarizations, *Proc. Natl. Acad. Sci. USA* **91:** 12579–12583.

Lohret, T. A., Murphy, R. C., Drgon, T., and Kinnally, K. W., 1996, Activity of the mitochondrial multiple conductance channel is independent of the adenine nucleotide translocator, *J. Biol. Chem.* **271:**4846–4849.

Lötscher, H. R., Winterhalter, K. H., Carafoli, E., and Richter, C., 1979, Hydroperoxides can modulate the redox state of pyridine nucleotides and the calcium balance in rat liver mitochondria, *Proc. Natl. Acad. Sci. USA* **76:**4340–4344.

Lötscher, H. R., Winterhalter, K. H., Carafoli, E., and Richter, C., 1980, Hydroperoxide-induced loss of pyridine nucleotides and release of calcium from rat liver mitochondria, *J. Biol. Chem.* **255:**9325–9330.

Marchetti, P., Decaudin, D., Macho, A., Zamzami, N., Hirsch, T., Susin, S. A., and Kroemer, G., 1997, Redox regulation of apoptosis: Impact of thiol redoxidation on mitochondrial function, *Eur. J. Immunol.*, in press.

Martin, S. J., and Green, D. R., 1995, Protease activation during apoptosis: Death by a thousand cuts? *Cell* **82:**349–352.

Massari, S., and Azzone, G. F., 1972, The equivalent pore radius of intact and damaged mitochondria and the mechanism of active shrinkage, *Biochim. Biophys. Acta* **283:**23–29.

Mitchell, P., 1966, Chemiosmotic coupling in oxidative and photosynthetic phosphorylation, *Biol. Rev. Camb. Philos. Soc.* **41:**445–502.

Miyata, H., Silverman, H. S., Sollot, S. J., Lakatta, E. G., Stern, M. D., and Hansford, R. G., 1991, Measurement of mitochondrial free Ca^{2+} concentration in living single rat cardiac myocytes, *Am. J. Physiol.* **261:**H1123–H1134.

Moore, G. A., Rossi, L., Nicotera, P., Orrenius, S., and O'Brien, P. J., 1987, Quinone toxicity in hepatocytes: Studies on mitochondrial Ca^{2+} release induced by benzoquinone derivatives, *Arch. Biochem. Biophys.* **259:**283–295.

Moore, M., Thor, H., Moore, G., Nelson, S., Moldeus, P., and Orrenius, S., 1985, The toxicity of acetaminophen and *N*-acetyl-*p*-benzoquinone imine in isolated hepatocytes is associated with thiol depletion and increased cytosolic Ca^{2+}, *J. Biol. Chem.* **260:**13035–13040.

Newmeyer, D. D., Farschon, D. M., and Reed, J. C., 1994, Cell-free apoptosis in *Xenopus* egg extracts: Inhibition by Bcl-2 and requirement for an organelle fraction enriched in mitochondria, *Cell* **79:**353–364.

Nicholls, D. G., 1978, The regulation of extramitochondrial free calcium ion concentration by rat liver mitochondria, *Biochem. J.* **176:**463–474.

Nicolli, A., Petronilli, V., and Bernardi, P., 1993, Modulation of the mitochondrial cyclosporin A-sensitive permeability transition pore by matrix pH. Evidence that the pore open-closed probability is regulated by reversible histidine protonation, *Biochemistry* **32:**4461–4465.

Nicolli, A., Basso, E., Petronilli, V., Wenger, R. M., and Bernardi, P., 1996, Interactions of cyclophilin with the mitochondrial inner membrane and regulation of the permeability transition pore, a cyclosporin A-sensitive channel, *J. Biol. Chem.* **271:**2185–2192.

Nicotera, P., and Orrenius, S., 1992, Ca^{2+} and cell death, *Ann. NY Acad. Sci.* **648**:17–27.

Nieminen, A. L., Saylor, A. K., Tesfai, S. A., Herman, B., and Lemasters, J. J., 1995, Contribution of the mitochondrial permeability transition to lethal injury after exposure of hepatocytes to *t*-butylhydroperoxide, *Biochem. J.* **307**:99–106.

Palmeira, C. M., and Wallace, K. B., 1997, Benzoquinone inhibits the voltage-dependent induction of the mitochondrial permeability transition caused by redox-cycling naphtoquinones, *Toxicol. Appl. Pharmacol.* **143**: in press.

Pastorino, J. G., Snyder, J. W., Serroni, A., Hoek, J. B., and Farber, J. L., 1993, Cyclosporin and carnitine prevent the anoxic death of cultured hepatocytes by inhibiting the mitochondrial permeability transition, *J. Biol. Chem.* **268**:13791–13798.

Pastorino, J. G., Simbula, G., Gilfor, E., Hoek, J. B., and Farber, J. L., 1994, Protoporphyrin IX, an endogenous ligand of the peripheral benzodiazepine receptor, potentiates induction of the mitochondrial permeability transition and the killing of cultured hepatocytes by rotenone, *J. Biol. Chem.* **269**:31041–31046.

Pastorino, J. G., Snyder, J. W., Hoek, J. B., and Farber, J. L., 1995, Ca^{2+} depletion prevents anoxic death of hepatocytes by inhibiting mitochondrial permeability transition, *Am. J. Physiol.* **268**: C676–685.

Petit, P. X., O'Connor, J. E., Grunwald, D., and Brown, S. C., 1990, Analysis of the membrane potential of rat- and mouse-liver mitochondria by flow cytometry and possible applications, *Eur. J. Biochem.* **194**:389–397.

Petit, P. X., Lecoeur, H., Zorn, E., Dauguet, C., Mignotte, B., and Gougeon, M. L., 1995, Alterations in mitochondrial structure and function are early events of dexamethasone-induced thymocyte apoptosis, *J. Cell Biol.* **130**:157–167.

Petronilli, V., Szabo', I., and Zoratti, M., 1989, The inner mitochondrial membrane contains ion-conducting channels similar to those found in bacteria, *FEBS Lett.* **259**:137–143.

Petronilli, V., Cola, C., and Bernardi, P., 1993a, Modulation of the mitochondrial cyclosporin A-sensitive permeability transition pore. II. The minimal requirements for pore induction underscore a key role for transmembrane electrical potential, matrix pH, and matrix Ca^{2+}, *J. Biol. Chem.* **268**:1011–1016.

Petronilli, V., Cola, C., Massari, S., Colonna, R., and Bernardi, P., 1993b, Physiological effectors modify voltage sensing by the cyclosporin A-sensitive permeability transition pore of mitochondria, *J. Biol. Chem.* **268**:21939–21945.

Petronilli, V., Costantini, P., Scorrano, L., Colonna, R., Passamonti, S., and Bernardi, P., 1994a, The voltage sensor of the mitochondrial permeability transition pore is tuned by the oxidation-reduction state of vicinal thiols. Increase of the gating potential by oxidants and its reversal by reducing agents, *J. Biol. Chem.* **269**:16638–16642.

Petronilli, V., Nicolli, A., Costantini, P., Colonna, R., and Bernardi, P., 1994b, Regulation of the permeability transition pore, a voltage-dependent mitochondrial channel inhibited by cyclosporin A, *Biochim. Biophys. Acta* **1187**:255–259.

Pfeiffer, D. R., Schmid, P. C., Beatrice, M. C., and Schmid, H. H., 1979, Intramitochondrial phospholipase activity and the effects of Ca^{2+} plus *N*-ethylmaleimide on mitochondrial function, *J. Biol. Chem.* **254**:11485–11494.

Richter, C., Schlegel, J., and Schweizer, M., 1992, Prooxidant-induced Ca^{2+} release from liver mitochondria. Specific versus nonspecific pathways, *Ann. NY Acad. Sci.* **663**:262–268.

Rigobello, M. P., Barzon, E., Marin, O., and Bindoli, A., 1995, Effect of polycation peptides on mitochondrial permeability transition, *Biochem. Biophys. Res. Commun.* **217**:144–149.

Riparbelli, M. G., Callaini, G., Tripodi, S. A., Cintorino, M., Tosi, P., and Dallai, R., 1995, Localization of the Bcl-2 protein to the outer mitochondrial membrane by electron microscopy, *Exp. Cell Res.* **221**:363–369.

Rizzuto, R., Pitton, G., and Azzone, G. F., 1987, Effect of Ca^{2+}, peroxides, SH reagents, phosphate and aging on the permeability of mitochondrial membranes, *Eur. J. Biochem.* **162**:239–249.

Rizzuto, R., Simpson, A. W., Brini, M., and Pozzan, T., 1992, Rapid changes of mitochondrial Ca^{2+} revealed by specifically targeted recombinant, *Nature* **358**:325–327.

Rizzuto, R., Brini, M., Murgia, M., and Pozzan, T., 1993, Microdomains with high Ca^{2+} close to IP_3-sensitive channels that are sensed by neighboring mitochondria, *Science* **262**:744–747.

Rizzuto, R., Bastianutto, C., Brini, M., Murgia, M., and Pozzan, T., 1994, Mitochondrial Ca^{2+} homeostasis in intact cells, *J. Cell Biol.* **126**:1183–1194.

Savage, M. K., and Reed, D. J., 1994a, Oxidation of pyridine nucleotides and depletion of ATP and ADP during calcium- and inorganic phosphate-induced mitochondrial permeability transition, *Biochem. Biophys. Res. Commun.* **200**:1615–1620.

Savage, M. K., and Reed, D. J., 1994b, Release of mitochondrial glutathione and calcium by a cyclosporin A-sensitive mechanism occurs without large amplitude swelling, *Arch. Biochem. Biophys.* **315**:142–152.

Saxena, K., Henry, T. R., Solem, L. E., and Wallace, K. B., 1995, Enhanced induction of the mitochondrial permeability transition following acute menadione administration, *Arch. Biochem. Biophys.* **317**:79–84.

Scherer, B., and Klingenberg, M., 1974, Demonstration of the relationship between the adenine nucleotide carrier and the structural changes of mitochondria as induced by adenosine 5'-diphosphate, *Biochemistry* **13**:161–170.

Schinder, A. F., Olson, E. C., Spitzer, N. C., and Montal, M., 1996, Mitochondrial dysfunction is a primary event in glutamate neurotoxicity, *J. Neurosci.* **16**:6125–6133.

Schreier, M. H., Baumann, G., and Zenke, G., 1993, Inhibition of T cell signaling pathways by immunophilin drug complexes: Are side effects inherent to immunosuppressive properties? *Transpl. Proc.* **25**:502–507.

Schultheiss, H. P., and Klingenberg, M., 1984, Immunochemical characterization of the adenine nucleotide translocator. Organ specificity and conformation specificity, *Eur. J. Biochem.* **143**:599–605.

Scorrano, L., Petronilli, V., and Bernardi, P., 1997, On the voltage dependence of the mitochondrial permeability transition pore. A critical appraisal, *J. Biol. Chem.* **272**: in press.

Scott Boyer, S., Moore, G. A., and Moldeus, P., 1993, Submitochondrial localization of the NAD+ glycohydrolase. Implications for the role of pyridine nucleotide hydrolysis in mitochondrial calcium fluxes, *J. Biol. Chem.* **268**:4016–4020.

Shen, A. C., and Jennings, R. B., 1972, Myocardial calcium and magnesium in acute ischemic injury, *Am. J. Pathol.* **67**:417–421.

Sluyser, M. (ed.), 1996, *Apoptosis in Normal Development and Cancer*, Taylor & Francis Ltd, London.

Sorgato, M. C., Keller, B. U., and Stühmer, W., 1987, Patch-clamping of the inner mitochondrial membrane reveals a voltage-dependent ion channel, *Nature* **330**:498–500.

Sparagna, G. C., Gunter, K. K., Sheu, S. S., and Gunter, T. E., 1995, Mitochondrial calcium uptake from physiological-type pulses of calcium. A description of the rapid uptake mode, *J. Biol. Chem.* **270**:27510–27515.

Susin, S. A., Zamzami, N., Castedo, M., Hirsch, T., Marchetti, P., Macho, A., Daugas, E., Geuskens, M., and Kroemer, G., 1996, Bcl-2 inhibits the mitochondrial release of an apoptogenic protease, *J. Exp. Med.* **184**:1331–1341.

Szabo', I., and Zoratti, M., 1991, The giant channel of the inner mitochondrial membrane is inhibited by cyclosporin A, *J. Biol. Chem.* **266**:3376–3379.

Szabo', I., and Zoratti, M., 1992, The mitochondrial megachannel is the permeability transition pore, *J. Bioenerg. Biomembr.* **24**:111–117.

Szabo', I., and Zoratti, M., 1993, The mitochondrial permeability transition pore may comprise VDAC molecules. I. Binary structure and voltage dependence of the pore, *FEBS Lett.* **330**:201–205.

Szabo', I., Bernardi, P., and Zoratti, M., 1992, Modulation of the mitochondrial megachannel by divalent cations and protons, *J. Biol. Chem.* **267**:2940–2946.

Tikhonova, I. M., Andreyev, A. Y., Antonenko, Yu. N., Kaulen, A. D., Komrakov, A. Y., and Skulachev, V. P., 1994, Ion permeability induced in artificial membranes by the ATP/ADP antiporter, *FEBS Lett.* **337**:231–234.

Vayssière, J. L., Petit, P. X., Risler, Y., and Mignotte, B., 1994, Commitment to apoptosis is associated with changes in mitochondrial biogenesis and activity in cell lines conditionally immortalized with simian virus 40, *Proc. Natl. Acad. Sci. USA* **91**:11752–11756.

Weis, M., Kass, G. E., and Orrenius, S., 1994, Further characterization of the events involved in mitochondrial Ca^{2+} release and pore formation by prooxidants, *Biochem. Pharmacol.* **47**: 2147–2156.

Woodfield, K.-Y., Price, N. T., and Halestrap, A. P., 1997, cDNA cloning of rat mitochondrial cyclophilin, *Biochim. Biophys. Acta*, in press.

Zahrebelski, G., Nieminen, A. L., al Ghoul, K., Qian, T., Herman, B., and Lemasters, J. J., 1995, Progression of subcellular changes during chemical hypoxia to cultured rat hepatocytes: A laser scanning confocal microscopic study, *Hepatology* **21**:1361–1372.

Zamzami, N., Marchetti, P., Castedo, M., Decaudin, D., Macho, A., Hirsch, T., Susin, S. A., Petit, P. X., Mignotte, B., and Kroemer, G., 1995a, Sequential reduction of mitochondrial transmembrane potential and generation of reactive oxygen species in early programmed cell death, *J. Exp. Med.* **182**:367–377.

Zamzami, N., Marchetti, P., Castedo, M., Zanin, C., Vayssière, J. L., Petit, P. X., and Kroemer, G., 1995b, Reduction in mitochondrial potential constitutes an early irreversible step of programmed lymphocyte death *in vivo*, *J. Exp. Med.* **181**:1661–1672.

Zamzami, N., Marchetti, P., Castedo, M., Hirsch, T., Susin, S. A., Masse, B., and Kroemer, G., 1996a, Inhibitors of permeability transition interfere with the disruption of the mitochondrial transmembrane potential during apoptosis, *FEBS Lett.* **384**:53–57.

Zamzami, N., Susin, S. A., Marchetti, P., Hirsch, T., Gomez Monterrey, I., Castedo, M., and Kroemer, G., 1996b, Mitochondrial control of nuclear apoptosis, *J. Exp. Med.* **183**:1533–1544.

Zazueta, C., Reyes Vivas, H., Bravo, C., Pichardo, J., Corona, N., and Chavez, E., 1994, Triphenyltin as inductor of mitochondrial membrane permeability transition, *J. Bioenerg. Biomembr.* **26**:457–462.

Zoeteweij, J. P., van de Water, B., de Bont, H. J., Mulder, G. J., and Nagelkerke, J. F., 1993, Calcium-induced cytotoxicity in hepatocytes after exposure to extracellular ATP is dependent on inorganic phosphate. Effects on mitochondrial calcium, *J. Biol.Chem.* **268**:3384–3388.

Zoratti, M. and Szabo', I., 1995, The mitochondrial permeability transition, *Biochim. Biophys. Acta* **1241**:139–176.

Author Index

Subject Index